16.6.94   TL

£65·00

D1639368

30150   011881866

Plant Breeding

PLANT BREEDING SERIES

Edited by Professor P.D.S. Caligari

Head of the Department of Agricultural Botany
University of Reading, UK

Modern plant breeders need to have a wide grasp of scientific knowledge but in a form which is relevant to their activities and needs. Although there are numerous published books which cover areas relevant to plant breeding, very few are written for the student who is learning the subject of plant breeding or the breeder who is trying to improve a particular crop species. The *Chapman & Hall Plant Breeding Series* will provide an up-to-date, comprehensive set of books covering this diverse but important area. The books in the series are written in an easily accessible form, specifically with plant breeders and students of plant breeding in mind.

Forthcoming titles

Statistical Methods for Plant Variety Evaluation
Edited by R. Kempton and P. Fox

Selection Methods in Plant Breeding
I. Bos and P.D.S. Caligari

Seed Science and Technology for Plant Breeding and Genetic Conservation
E. Roberts and R. Ellis

Cellular and Molecular Approaches to Crop Improvement
W. Powell

Plant Breeding

Principles and prospects

EDITED BY

M.D. Hayward

AFRC IGER
Welsh Plant Breeding Station,
Aberystwyth, UK

N.O. Bosemark

Hilleshög AB,
Landskrona, Sweden

I. Romagosa

UdL-IRTA,
Lleida, Spain

Coordinating editor

M. Cerezo, CIHEAM,
Zaragoza, Spain

CHAPMAN & HALL

London · Glasgow · New York · Tokyo · Melbourne · Madras

Published by Chapman & Hall, 2–6 Boundary Row, London SE1 8HN

Chapman & Hall, 2–6 Boundary Row, London SE1 8HN, UK

Blackie Academic & Professional, Wester Cleddens Road, Bishopbriggs, Glasgow G64 2NZ, UK

Chapman & Hall Inc., 29 West 35th Street, New York NY10001, USA

Chapman & Hall Japan, Thomson Publishing Japan, Hirawacho Nemoto Building, 6F, 1–7–11 Hirakawa-cho, Chiyoda-ku, Tokyo 102, Japan

Chapman & Hall Australia, Thomas Nelson Australia, 102 Dodds Street, South Melbourne, Victoria 3205, Australia

Chapman & Hall India, R. Seshadri, 32 Second Main Road, CIT East, Madras 600 035, India

First edition 1993

© 1993 Chapman & Hall

Typeset by Photoprint, Torquay
Printed in Great Britain at the University Press, Cambridge

ISBN 0 412 43390 7

A catalogue record for this book is available from the British Library

Library of Congress Cataloging-in-Publication data

Plant breeding : principles and prospects / edited by M. Hayward,
   N. Bosemark, I. Romagosa ; coordinating editor, M. Cerezo.
      p.   cm. -- (Plant breeding ; v. 1)
   Includes bibliographical references and index.
   ISBN 0–412–43390–7
   1. Plant breeding.   2. Plant genetics.   I. Hayward, M. D.
II. Bosemark, N.   III. Romagosa, I.   IV. Series.
SB123.P547   1993
631.5'23--dc20   92–42654
                                                              CIP

⊚ Printed on permanent acid-free text paper, manufactured in accordance with the proposed ANSI/NISO Z 39.48–199X and ANSI Z 39.48–1984

Contents

Contributors

E. Acevedo
CIMMYT
Apartado 6–641
C.P. 06600
México 6 D.F.
MEXICO

P. Arus
IRTA
Centre d'Investigació Agrària de Cabrils
Carretera de Cabrils s/n
08348 Cabrils, Barcelona
SPAIN

R.B. Austin
15 Wingate Way
Trumpington
Cambridge CB2 2HD
UNITED KINGDOM

J. Azcon-Bieto
Universidad de Barcelona
Facultad de Biología
Depto. de Biología Vegetal
Avda. Diagonal 645
08028 Barcelona
SPAIN

C.H. Bornman
Hilleshög AB
PO Box 302
S–26123 Landskrona
SWEDEN

N.O. Bosemark
Hilleshög AB
PO Box 302
S–26123 Landskrona
SWEDEN

E.L. Breese
AFRC IGER
Welsh Plant Breeding Station
Plas Gogerddan, Near Aberystwyth
SY23 3EB Dyfed
UNITED KINGDOM

A. Caballero
Universidad de Barcelona
Facultad de Biología
Depto. de Fisiología Vegetal
Avda. Diagonal 645
08028 Barcelona
SPAIN

J.M. Clarke
Agriculture Canada
Research Station
PO Box 1030
Swift Current
Saskatchewan S9H 3X2
CANADA

J.I. Cubero
Universidad de Cordoba
ETS. Ing. Agrónomos
Depto. de Genética
Alameda del Obispo
Apartado 3048
14080 Córdoba
SPAIN

G.E. van Dijk
Centre for Plant Breeding and
Reproduction Research (CPRO–DLO)
Droevendaalsesteeg 1
P.O. Box 16
6700 AA Wageningen
THE NETHERLANDS

B. Donini
ENEA
C.R.E. Casaccia
Via Anguillarese 301
000160 Santa Maria di Galeria, Roma
ITALY

P.R. Ellis
Institute of Horticultural Research
Wellesbourne
Warwick CV35 9EF
UNITED KINGDOM

J.T. Esquinas-Alcázar
FAO
Commission on Plant Genetic Resources
Via delle Terme di Caracalla
00100 Roma
ITALY

E. Fereres
Universidad de Córdoba
ETS Ing. Agrónomos
Depto. de Agronomía
Apartado 3048
14080 Córdoba
SPAIN

B. Foroughi-Wehr
Biologische Bundesanstalt
Institut für Resistenzgenetik
D–8059 Grünbach
GERMANY

P.N. Fox
CIMMYT
Apartado 6–641
C.P. 06600
México 6 D.F.
MEXICO

M.D. Hayward
AFRC IGER
Welsh Plant Breeding Station
Plas Gogerddan, Near Aberystwyth
SY23 3EB Dyfed
UNITED KINGDOM

L. Herrera-Estrella
CINVESTAV – IPN
Unidad Irapuato
Apdo. 629
36500 Irapuato, Guanajuato
MEXICO

M.J. Kearsey
The University of Birmingham
Department of Genetics
PO Box 363
Birmingham, B15 2TT
UNITED KINGDOM

J.M. Lasa
CSIC – EEAD
Departamento de Genética y Producción
    Vegetal
Apartado 202
50080 Zaragoza
SPAIN

E. Lozoya-Gloria
CINVESTAV – IPN
Unidad Irapuato
Apdo. 629
36500 Irapuato, Guanajuato
MEXICO

T.N. McCaig
Agriculture Canada
Research Station
PO Box 1030
Swift Current
Saskatchawan S9H 3X2
CANADA

A. Micke
Joint FAO/IAEA Division
Plant Breeding and Genetic Section
Wagramerstrasse 5
PO Box 100
A–1400 Vienna
AUSTRIA

J. Moreno-González
Investigaciones Agrarias
Mabegondo
Apartado 10
15080 La Coruña
SPAIN

M. Motto
Istituto Sperimentale per la Cerealicoltura
Sezione di Bergamo
Via Stezzano 24
POB 164
24100 Bergamo
ITALY

L. Munck
The Royal Veterinary and Agricultural
  University
Department of Dairy and Food Science
Thorvaldsensvej 40
DK–1871 Frederiksberg C
DENMARK

L.R. Mytton
AFRC IGER
Welsh Plant Breeding Station
Plas Gogerddan, Near Aberystwyth
SY23 3EB Dyfed
UNITED KINGDOM

D. de Nettancourt
CEC
Directorate-General for Science, Research
  and Development
Biotechnology Division
Rue de la Loi 200
1040 Brussels
BELGIUM

A.P.M. den Nijs
Centre for Plant Breeding and
Reproduction Research (CPRO–DLO)
Droevendaalsesteeg 1
PO Box 16
6700 AA Wageningen
THE NETHERLANDS

R.E. Niks
Agricultural University
Department of Plant Breeding
PO Box 386
6700 AJ Wageningen
THE NETHERLANDS

The late E. Ottaviano
Universita degli Studi di Milano
Dipt. di Genetica e di Biologia dei
  Microrganismi
Via Celoria 26
20133 Milano
ITALY

J.E. Parlevliet
Agricultural University
Department of Plant Breeding
PO Box 386
6700 AJ Wageningen
THE NETHERLANDS

G. Pelletier
INRA
Laboratoire de Biologie Cellulaire
Route de Saint-Cyr
78026 Versailles Cedex
FRANCE

M. Pérez de la Vega
Universidad de León
Departamento de Ecología, Genética y
  Microbiología
Area de Genética
Campus Universitario de Vegazana
24071 León
SPAIN

B. Pickersgill
University of Reading
Dept. of Agricultural Botany
Whiteknights
PO Box 221
Reading, RG6 2AS
UNITED KINGDOM

I. Potrykus
ETH
Institut für Pflanzenwissenschaften
ETH–Zentrum
Universitätstrasse 2
8092 Zürich
SWITZERLAND

H. Rees
School of Agricultural Sciences
University College of Wales
Aberystwyth
Dyfed SY23 3DD
UNITED KINGDOM

I. Romagosa
UdL – IRTA
Departamento de Producción Vegetal
Avenida Rovira Roure 177
25006 Lleida
SPAIN

F. Salamini
Max Planck Institut für Züchtungsforschung
D–5000 Köln 30
GERMANY

E. Sánchez-Monge
E.T.S. Ing. Agrónomos
Depto. Genética Agraria
Ciudad Universitaria
28040 Madrid
SPAIN

M. Sari-Gorla
Universita degli Studi di Milano
Dipt. di Genetica e di Biologia dei
  Microrganismi
Via Celoria 26
20133 Milano
ITALY

L. Skøt
AFRC IGER
Welsh Plant Breeding Station
Plas Gogerddan, Near Aberystwyth
SY23 3EB Dyfed
UNITED KINGDOM

H. Thomas
AFRC IGER
Welsh Plant Breeding Station
Plas Gogerddan, Near Aberystwyth
SY23 3EB Dyfed
UNITED KINGDOM

G. Wenzel
Biologische Bundesanstalt
Institut für Resistenzgenetik
D–8059 Grünbach
GERMANY

B. Xoconostle-Cázares
CINVESTAV – IPN
Unidad Irapuato
Apdo. 629
36500 Irapuato, Guanajuato
MEXICO

Series foreword

Our requirement for plant breeders to be successful has never been greater. However one views the forecasted numbers for future population growth we will need, in the immediate future, to be feeding, clothing and housing many more people than we do, inadequately, at present. Plant breeding represents the most valuable strategy in increasing our productivity in a way that is sustainable and environmentally sensitive.

Plant breeding can rightly be considered as one of the oldest multidisciplinary subjects that is known to humans. It was practised by people who first started to carry out a settled form of agriculture. The art, as it must have been at that stage, was applied without any formal underlying framework, but achieved dramatic results, as witnessed by the forms of cultivated plants we have today. We are now learning how to apply successfully the results of yet imperfect scientific knowledge. This knowledge is, however, rapidly developing, particularly in areas of tissue culture, biotechnology and molecular biology. Plant breeding's inherent multifaceted nature means that alongside obvious subject areas like genetics we also need to consider areas such as: statistics, physiology, plant pathology, entomology, biochemistry, weed science, quality, seed characteristics, reproductive biology, trial design, selection and computing.

It therefore seems apparent that modern plant breeders need to have a grasp of a wide range of scientific knowledge and expertise if they are successfully to exploit the techniques, protocols and strategies which are open to them. This book is the first in a series which aims to provide plant breeders with up-to-date, comprehensive information an an accessible and integrated form.

The present book is aimed not only at the student who is learning the subject of plant breeding, but also at the breeder who is trying to improve our crop species. It has been written by a carefully selected group who were asked to present their thoughts at an advanced course on plant breeding staged at the International Centre for Advanced Mediterranean Studies, Zaragoza, Spain. The book thus represents an integrated set of topics covering the main aspects of plant breeding as formulated by leading international experts.

This volume gives an excellent overall presentation of the subject and, at the same time, provides a basis on which further books in the series can build to cover specific aspects and elements relevant to plant breeders.

Peter D.S. Caligari
Editor, *Plant Breeding Series*

Foreword

The International Centre for Advanced Mediterranean Agronomic Studies (CIHEAM), established in 1962, is an intergovernmental organization of fourteen countries; Albania, Algeria, Egypt, France, Greece, Italy, Lebanon, Malta, Morocco, Portugal, Spain, Tunisia, Turkey and Yugoslavia.

Four institutes (Bari, Italy; Chania, Greece; Montpellier, France and Zaragoza, Spain) provide post-graduate education at the Master of Science level. CIHEAM promotes research networks on Mediterranean agricultural priorities, supports the organization of specialized education in member countries, holds seminars and workshops bringing together technologists and scientists involved in Mediterranean agriculture and regularly produces diverse publications including the series 'Options Médirranéennes'. Through these activities, CIHEAM promotes North/South dialogue and international co-operation for agricultural development in the Mediterranean region.

The Mediterranean Agronomic Institute of Zaragoza (IAMZ) has been organizing, since 1982, an advanced course on Plant Breeding. The objective of this course is to provide the integrated knowledge and training needed to initiate any plant breeding programme, designed for countries with different degrees of development. Its main asset has, undoubtedly, been its faculty. Given the international nature of our Institution, it has been possible to select, as lecturers, a significant number of scientists with well recognized expertise in their subject areas from many private companies and public research institutions. Finally, the diversity of participants of many nationalities, often active plant breeders, has allowed a continuous interaction with the lecturers, enriching the development of the course. This course, particularly through the interaction of its objectives, faculty and participants, has been the origin of this book.

We would like to express our gratitude to everyone who has, directly or indirectly, contributed towards this endeavour. Particularly to the contributors of the chapters, mostly lecturers of the course, and to Drs Hayward, Bosemark and Romagosa, the scientific editors, who have made this book possible not only through their scientific experience but also through their enthusiasm and perseverance.

We would also like to express our satisfaction with the efficient collaboration initiated with Chapman & Hall whom we thank for their interest in and attention to this project.

This project, as well as having intrinsic technical interest, is considered by CIHEAM as a significant and successful example of international collaboration. We hope that this book will encourage the co-operative attitude amongst its readers throughout the world.

Miguel Valls
Director
Mediterranean Agronomic Institute of Zaragoza,
Spain

Preface

There are over a quarter of a million plant species of which approximately 5000 are cultivated and only a hundred can be considered as important crops. The great majority of cultivated plants are the result of the early domestication and selection of food plants carried out in prehistoric times. Thus 'Plant Breeding' is often defined as plant evolution directed by man. As a commercial activity, plant breeding has only been in existence for the last two or three centuries. In fact with the significant exception of hybrid varieties, the basic breeding methods have their origin in empirical, often unconscious, pre-Mendelian systems. By the end of the 19th Century, artificial crossing, bulk and pedigree methods and alternative progeny testing schemes were already used in both private and public breeding programmes.

In the present Century plant breeding has rapidly co-evolved with the rediscovery of Mendel's Laws and the development of the science of genetics. Genetics has provided the breeder with a better knowledge of the processes involved in generating variability, and the means by which such variation may be manipulated in a breeding programme. Genetic advances thus allowed breeders to improve upon the empirical techniques that they had traditionally used. At the same time plant breeding has answered many fundamental genetic questions and posed new ones.

Plant breeding is nowadays at a crossroads. In the past the genetic contribution to yield gains has often been of secondary importance compared with the improvements through fertilizers, irrigation and the extensive use of agrochemicals. Nevertheless in the last 50 years it has been estimated that half the yield gains in most major crops is attributable to the introduction of new cultivars as a result of plant breeding. However, while considerable genetic improvements have been attained for crops grown under optimum conditions, at least in absolute terms, response to selection has been limited for crops grown in marginal situations. At the same time, the need to preserve the environment for future generations, together with new economic constraints, is leading to a more comprehensive and integrated agriculture for which yield alone will no longer be its sole objective.

Plant breeding has become a very wide discipline and is undergoing a rapid evolution, particularly due to the advancements of the novel recombinant DNA technologies. Thus, the aim of this book is to present an up to date account of the philosophy and principles underlying current and future plant breeding. To do so, it is structured in a series of Parts to bring together many of the long standing principles and to consider them in the light of the current and developing technologies. At the same time these new technologies are presented in the context in which their integration into breeding is foreseen.

The foundation of plant breeding is the manipulation of existing or newly generated genetic variation, the causal bases of which are numerous and diverse.

In natural systems they include hybridization, introgression, nuclear and cytoplasmic mutations, variation in chromosome number and/or form and their combination and recombination. Similarly in breeding programmes variability is generated by artificial crossings, backcrosses, mutagenesis, somatic *in vitro* methods, chromosome manipulation or genetic engineering, etc. The effective use of such variability in applied breeding requires a thorough knowledge of the genetic architecture of the target species. Genetic progress from selection can only be maximized if full knowledge of the genetic architecture of the raw material is known. Thus the origin and development of variation, which includes the new gene technologies, are presented in Parts One to Three whereas the assessment and characterization of this variation is considered in Part Four.

In order properly to design a breeding programme of any crop it is crucial to know not only the nature of its reproductive system but the alternative ways in which it may be manipulated. Sexual systems, such as self- and cross-fertility and male sterility, together with the possible exploitation of facultative and obligate apomixis and *in vitro* techniques, such as micropropagation, andro- and gynogenesis, are reviewed in Part Five. Selection strategies and choice of breeding method, the use of molecular markers in selection, the possibilities offered by gametophytic versus sporophytic techniques and *in vitro* selection are discussed in Part Six.

It has been estimated that 60–80% of the seasonal variation in crop productivity is due to weather fluctuations. In fact, in order to close the gap between actual and potential yields, it is also essential for any plant breeder to know all limiting factors, that is the biotic and abiotic stresses, their frequency and intensity. In order to design an appropriate breeding programme for stress-prone environments, the implications of genotype × environment interactions and the utilization of statistical models for assessing genotypic adaptation have to be considered. Knowledge of the limitations of yield-based selection and the effectiveness of the exploitation of the physiology of the crop and its environmental responses is of major importance. In order to reduce the dependence of agriculture on inputs of undesirable agrochemicals, etc. the role of resistance breeding will increase. All these aspects of adaptation are considered in Part Seven.

Although details of the breeding of specific crops have been deliberately omitted, the principles underlying the breeding of the specific traits, such as photosynthetic and respiratory efficiency, root systems, symbiotic nitrogen fixation and the utilization of renewable plant resources are considered in Part Eight. These have been identified as exemplifying some of the likely areas in which further breeding effort will be rewarding. The achievement of these goals will however require a continual review of breeding methods as proposed in the final chapter.

The stimulus for this book came from the organizers of the course on plant breeding at the International Centre for Advanced Mediterranean Agronomic Studies, Zaragoza, Spain and we most gratefully acknowledge their support in this venture. We must also thank the contributors of the various chapters, many of whom are involved in the presentation of the course. It is with great sadness that we record the untimely death of Professor E. Ottaviano during the preparation of his joint chapter with Professor M. Sari-Gorla. His contribution to the study of gametophytic selection has been a landmark in the development of this science.

The major part of the co-ordination of the manuscripts and their preparation for publication has been carried out by Manuela Cerezo of the CIHEAM, Zaragoza. We owe her a great debt of gratitude for her devotion to this most onerous task.

Michael D. Hayward   **N.Olof Bosemark**   **Ignacio Romagosa**

Part One

Introduction

Introduction

E. Sánchez-Monge

PLANT BREEDING

Plant breeding is essentially an election made by man of the best plants within a variable population as a potential cultivar. In other words plant breeding is a 'selection' made possible by the existence of 'variability'. The main objective, in the past, has been to increase yield by improvement of the potential productivity followed by the successful expression of that potential.

The potential productivity of a plant has traditionally been increased by modifying its morphological characteristics such as the number of kernels per ear in a cereal or the weight of individual seeds within the pod of a pulse, and also by modifying physiological traits such as harvest index, the utilization of nutrients, or tolerance to stress. With changing demands on agriculture, objectives are now being altered to take account of these and new requirements. Quality and nutritive value are now of increasing importance, particularly in association with improved efficiency of production.

Modern agriculture is highly mechanized and for this reason some breeding programmes include objectives to make the crop more amenable to mechanical handling. For example the development of monogerm beets for mechanical sowing thus eliminating the need for thinning, or the introduction of 'jointless' tomatoes for mechanical harvesting. Similarly the

Plant Breeding: Principles and prospects. Edited by M.D. Hayward, N.O. Bosemark and I. Romagosa. Published in 1993 by Chapman & Hall, London. ISBN 0 412 43390 7.

use of some agrochemicals is often coupled with the need for specific crop characteristics e.g. herbicide resistance.

The extension of a crop to an area to which it was not previously adapted, clearly necessitates the production of new cultivars. The introduction of stone fruits to warmer regions was only made possible by selective breeding, eliminating the cold requirement for fruiting. An understanding of the problems of adaptation are a prerequisite to the objective of 'yield stability' for varieties to be grown in areas of variable agroclimatology. It is extremely difficult to introduce wide scale testing into the early phases of a breeding programme.

The success of a breeding programme in meeting the various objectives briefly outlined above is dependent upon two main factors. First having available the necessary variation and second, being able to manipulate it to produce a stable new cultivar. In the past the variation exploited in most breeding programmes derived from naturally occurring variants and the wild relatives of our main crop species. The distribution of variability among such species was found by Vavilov to follow a well defined pattern.

THE VAVILOV CONCEPT

Nikolai I. Vavilov (1887–1942), the Russian botanist and plant breeder, demonstrated the existence of 'centres of origin' of cultivated plants (more correctly named today as 'centres of diversity'), in which can be found the highest level of genetic variability of a species. This variability, which arises in nature by mutation, spontaneous hybridization, introgression and changes in chromosome form and number, provides the means by which adaptation to heterogeneous environments can occur. It allows the breeder to identify sources of variation for specific characteristics. The extension of this principle to related species was formulated by Vavilov in his 'law of homologous series of variation'. This law allows the prediction of the appearance of a given type of mutation in a plant species when such a type has been found in another species phylogenetically related to the first. As a result of his studies Vavilov defined plant breeding as 'plant evolution directed by man'. He recognized that in a breeding programme, by growing variable populations in conditions favouring the expression of the characters, selection may be facilitated. Also, by creating variability in a parallel way to nature the breeder exploits genetical methods. Plant breeding can thus also be defined as 'applied plant genetics'.

THE EVOLUTION OF PLANT BREEDING

Plant breeding started with sedentary agriculture and the domestication of the first agricultural plants, the cereals, which were chosen by early man. This led to the rapid elimination of undesirable characters such as seed-shattering and dormancy. It seems natural that the need to collect fruits, seeds and roots for nutrition brought with it the knowledge to identify plants, aspects of their biology and their potential as food, poison or other uses. The need for pollination for fruit setting in dates was recognized some nine centuries before Christ, as can be deduced from the observation of an Asyrian bas-relief showing the pollination of female date palms with male pollen.

Until the first decenium of this century plant breeding was in the hands of 'experts' who in many cases were unable to make progress due to a lack of knowledge of the fundamental plant processes such as the reproductive system. In spite of this, however, some important goals were reached before the rediscovery of

Mendel's laws and the consecration of genetics as a science. Some examples are worth mentioning.

In France, in the 17th century, several varieties of 'heading lettuce' were developed, some of which are still in cultivation.

A family of French seed growers, the Vilmorins, established in 1727 the first company devoted to plant breeding and the production of new varieties. A member of the family, Louis de Vilmorin, was the first, some years later, to use the progeny test, evaluating a selection by the study of its descendants.

The sexuality of plants was described by Caesalpinus in 1583 and in 1696 Camerarius published an essay entitled 'De sexo plantarum', but it was Kolreuter, a German botanist who was the first to exploit this knowledge in the production of the first artificial plant hybrids in *Nicotiana*. A well-known example of premendelian plant breeding is that of sugar beet. In 1747 Margraaf discovered that the roots of the forage beet contained about 6% of sucrose. By the beginning of the 19th century, Achard, by means of mass selection, increased this to 11%. Application of the Vilmorin pedigree method had further increased this to 16% by 1811.

The use of artificial crosses in premendelian breeding can be exemplified by the case of *Fragaria* × *ananassa* developed in the botanical garden of Paris by Duchesne, in the 17th century by crossing *F. chiloense* with *F. virginiana*. In England at about the same time new varieties of fruits, wheat and peas were being obtained by artificial hybridization.

PLANT BREEDING POST-MENDEL

After the rediscovery of the work of Mendel in 1900 it was some six years before Bateson, who coined the name 'genetics' for the new science, realized that this new discipline could give a scientific basis and new openings to plant breeding methods. Modern plant breeding is, as has already been stated, applied genetics, but its scientific basis is broader and uses, as conceptual and technical tools, cytology, systematics, physiology, pathology, entomology, chemistry and statistics and has also developed its own technology. There have been many advances in these various areas which have made a substantial contribution to plant breeding. These include the demonstration of the auto- and allopolyploid nature of many crop plants and their origin. This opened the way to the production of artificial polyploid forms and their direct or indirect exploitation. The discovery of the possibility of increasing the frequency of mutations by means of radiation or chemicals was hailed as a means of increasing the variability available to the breeder. There have been many other developments which have all contributed to the progress and success of plant breeding too numerous to exemplify in detail but include such areas as quantitative inheritance and genotype × environment interaction, resistance breeding, and the conservation and characterization of genetic resources.

Genetics has given to breeding a better knowledge of the processes involved in the mechanisms of variability and the necessary information for regulating and increasing such variability. More recent genetical advances than those outlined above have allowed breeders to design new methods as is well exemplified by the developments in such areas as RFLP mapping and marker-assisted selection, gene cloning and genetic transformation. These new enabling technologies, all adding to the overall science of plant breeding, are considered in some detail in this book. They hold out great promise for the future. The ceilings for yield that some authorities claim have been reached will be surpassed by the combined application of the classical and the new methods available to the breeder. Where yield as such is not the prime criterion of a breeding programme they will allow the production of more efficient cultivars which can maintain production without the requirement for expensive inputs. At the same time breeding can and will make a positive contribution to halting the decline in the agricultural environment.

Part Two

Genetic Systems and Population Structure

1

Genetic systems, recombination and variability

H. Rees

1.1 INTRODUCTION

Information from many sources, from archaeologists, cytologists and geneticists, taxonomists and others provides us with detailed knowledge of the origin and ancestry of many of our domesticated plants and animals. We have, as well, information about the ways and means by which these organisms evolved in the face of pressures imposed upon them by a variety of factors relating to their cultivation and utilization. One can, on the one hand, identify particular changes in morphology and physiology whereby they became adapted to particular conditions of growth and management. On the other hand, one can determine also the causal basis of such variation and by what means it was made available, that is, exposed to the action of selection. In plants the causes are many and varied. They include nuclear and cytoplasmic mutation, their combination and recombination, modifications in breeding behaviour, structural and numerical changes in chromosomes. They embrace, in short, the same spectrum of changes by which 'natural' species evolve and adapt to various environments. We come to appreciate, above all, the astonishing range of variation available for exploitation by the breeder and, equally astonishing, the rapidity with which it may be

Plant Breeding: Principles and prospects. Edited by M.D. Hayward, N.O. Bosemark and I. Romagosa. Published in 1993 by Chapman & Hall, London. ISBN 0 412 43390 7.

exploited. Within a mere 10 000 years during which man has engaged in agriculture a plethora of new forms, indeed new species, have been developed and established.

In the main it is the plant breeder's deliberate and considered selection towards achieving a particular objective that determines the direction in which the cultivated forms evolve. It is not, however, the only kind of selection that is effective. Weather, the soil in which the plants grow and, above all, the system of husbandry to which they are subjected, impose their own constraints and requirements for change in plant form and functions. The effects of grazing upon forage grasses and clovers is a good example. These fortuitous consequences of selection may on occasion of course be translated to good fortune. By judicious collection one might acquire material from natural pastures well nigh ready-made for sale and distribution: more often, perhaps, valuable material for incorporation into a breeding programme.

1.2 THE SOURCES OF VARIATION

Whatever the nature of the selection pressure, its effectiveness will clearly depend upon the nature and the extent of the variation in plant form and function to which it is directed. We are now in a position to explain how such variation is generated and what factors influence its exploitation by the breeder. We do so, of course, in genetic terms. Before considering them it is both salutary and rewarding to take account, albeit briefly, of the work and thoughts of plant breeders of the 18th and, particularly, the 19th century. They were active at a time before the word genetics was coined, unaware of the principles established by Mendel, indeed ignorant of the mechanism of gamete formation and of fertilization. Yet their achievements were remarkable and it would be a mistake to suppose that they operated in a spirit of 'suck it and see'. Far from it. Many of them grasped with remarkable probity certain essential and important principles which contributed not only to their own success but, without doubt, influenced profoundly those who later were to lay the foundations of modern genetics and its application to the manipulation of genetic systems in plant breeding. An excellent account of these early breeders is given by Roberts (1929). I shall deal with but one of them.

Patrick Shirreff from the lowlands of Scotland, was a breeder of wheat and oats. An account of his work appeared in 1873 and is summarized by Roberts (1929). He was a successful breeder and, with it, a shrewd observer, a skilful experimenter, a thinker of considerable substance and sound judgement. The conclusions which he drew from the results of his experiments and observations no doubt contributed to his success as plant breeder. They also make good sense a hundred years later. In the first place he appreciated that the only variation to respond effectively to selection was variation of a heritable nature. That 'plants can be altered by skilful treatment' he dismisses with conviction. In 1993 such a statement is trite enough. Not so in Shirreff's time. His experiments were begun in 1819 when many subscribed to Lamarckian views or, at least, hesitated to refute them. Given this sound foundation Shirreff, again on the strength of his own experimental results and observations, went on to identify the sources of such variation. They were 'natural sports' and the 'diversity effected by crossing'. In effect he recognizes and distinguishes between two very different and exclusive sources of heritable change. 'Natural sports' we now recognize as consequences of gene and chromosome mutations, the 'diversity effected

by crossing' as consequences of recombination at meiosis. They are two of the ingredients of what many years later Darlington (1939) came to describe as the genetic system, that system which regulates the variability of populations. The third ingredient is of course the nature of the breeding system. Shirreff's emphasis upon crossing as distinct 'from merely selecting from the crops on the farm' shows that he was well aware at least of one aspect of its relevance.

C.D. Darlington's genius was to recognize that these three ingredients which make up the genetic system are, in nature, themselves subject to heritable change. He points out that whereas such changes do not affect the fitness of the individuals in which they arise their consequences to the progenies most certainly do. The consequences of change in each ingredient of the genetic system when considered in isolation are readily stated. Very simply and crudely an increase in the mutation rate leads to a greater variety of genes and of chromosome complements. An increase in recombination releases more rapidly the potential variability of heterozygotes to the gametes and progenies. A decrease has the reverse effect and in the extreme leads to failure of chromosome pairing and to infertility. As for the breeding system a change from sexual to asexual reproduction will 'fix' whatever variation there is in heterozygous or in homozygous form. A change from outbreeding to inbreeding generates a variety of disparate homozygous populations. The reverse, a change from inbreeding to outbreeding, gives rise to genotypes with a high potential variation composed of heterozygous gene combinations. To consider change in any one constituent of the genetic system in isolation is, however, unrealistic. They are 'bound up together in each group of organisms' (Darlington, 1939). For example, the homozygous populations resulting from inbreeding are disparate partly because of the dissipation of the potential variation of heterozygotes by recombination. And it goes without saying that there would be nothing to recombine without the mutations that constitute the raw material of genetic diversity.

The references above to the variability of

heterozygotes and homozygotes to potential variability and to the release of variability by recombination are, of necessity, brief and blunt. They are considered in detail by Hayward and Breese (Chapter 2).

1.3 THE MANIPULATION OF GENETIC SYSTEMS

In nature, changes in the constitutent elements of genetic systems are matters of chance. To the plant breeder they are not. They may be imposed and the genetic systems thereby manipulated. For the early breeders, Shirreff, Knight, Naudin and the Vilmorins, the manipulation impinged mainly upon the breeding system with the emphasis on hybridization. For them, therefore, the variability from which the varieties were constructed derived from two main sources. The first was that obtainable from farm crops at home and abroad. The second was that generated from crosses. To this day the vast majority of new varieties are bred from selection practised upon variability deriving from the very same two sources. One might have expected the conscious manipulation of other elements of the genetic system to have contributed more to the practice of plant breeding than has been the case to date. The problems, the possibilities and the prospects for such manipulations are considered with respect to recombination.

1.4 THE CONTROL OF RECOMBINATION

That the recombination of genes, be they genes of major effect or components of polygenic complexes, is of paramount importance in plant breeding is unquestionable. Indeed to a very large degree the breeder's aim is to construct by selection among hybrids and their progenies those combinations of genes most suited to his purposes. The speed and the efficiency with which they are built up will depend on the number of genes involved and, in particular, whether or not the genes are linked. If, as is likely, there is linkage, recombination will depend upon the frequency with which chiasmata form between them. The greater the likelihood of the formation of such chiasmata the more effective and expeditious the selection.

(a) A correlated response

Because the breeder is selecting for recombinants one might expect that he would, unconsciously, be selecting plant genotypes with high chiasma frequencies, these being the most likely to yield most rapidly the recombinants sought. The frequency and, as well, the distribution of chiasmata, are of course under genotypic control (Rees, 1961). For this reason one would expect also that the progenies of these plants, which ultimately constitute the bred varieties, would have higher chiasma frequencies than the original populations from which they derive. There is good evidence to this effect: (1) highly specialized, short-lived populations of *Lolium perenne* and *Festuca pratensis* have higher chiasma frequencies than their more perennial ancestors (Rees and Dale, 1974); (2) intensively bred strains of *Lolium perenne* such as Irish rye grass have higher chiasma frequencies than strains such as Kent rye grass, which represent virtually the products of a natural pasture (Rees and Dale, 1974); (3) disruptive selection, for early and late flowering, among outbreeding populations of *Brassica campestris* is accompanied by increases in the chiasma frequencies of the selection lines (Harinarayana and Murty, 1971).

While a consideration of the breeding and domestication of mammals is well outside the compass of this chapter, it is of interest to draw attention to a survey by Burt and Bell (1987) of chiasma frequencies among mammalian species. What stands out is that the intensively bred, domesticated species such as the dog, the cat and the chicken have chiasma frequencies, which are very high indeed, in comparison with those of wild, undomesticated species.

This information signifies that breeding in itself may be instrumental in effecting change in genetic systems, in this case the recombination component. It is an unpremeditated and unconscious response to selection pressures directed toward the achievement of quite unrelated objectives. It is, at the same time, compelling confirmation of the effectiveness of an increase in recombination to generate the variability upon which effective selection depends. This being so, why not as a deliberate policy, incorporate into breeding stocks genes with the capacity to boost recombination at meiosis, either locally or generally within the complement? There are difficulties. Before considering them it is appropriate to give details of an exceptional case where recombination has been controlled ingeniously and effectively to achieve a notable success in plant breeding.

(b) Riley's Compair

'Chinese Spring', a variety of bread wheat, the hexaploid *Triticum aestivum* ($2n = 2x = 42$), is susceptible to yellow rust caused by the fungus *Puccinia striiformis*. A wild diploid relative *Aegilops comosa* ($2n = 2x = 14$) carries a gene which confers resistance to the disease. By repeated backcrossing of the hybrid 'Chinese Spring' × *Ae. comosa*, using 'Chinese Spring' as the recurrent parent, a line resistant to yellow rust was isolated which contained one *Ae. comosa* chromosome in addition to the full 'Chinese Spring' complement of 42 chromosomes. This additional chromosome was designated *M2*, *M* because *Ae. comosa* carries the *M* genome and 2 because it compensates for homoeologous group 2 chromosomes of 'Chinese Spring'. While *M2* carried the gene for rust resistance it also carried many others with undesired effects. The transfer of the gene for rust resistance, along with the very minimum of unwanted *Ae. comosa* genes to wheat, by recombination to a wheat chromosome of homoeologous group 2, could not be achieved, however, because the 'Chinese Spring' complement carries the *Ph* locus, which suppresses pairing and chiasma formation at first metaphase of meiosis between homoeologous, as distinct from strictly homologous chromosomes (Riley and Chapman, 1958; Sears and Okamoto, 1958). This particular obstacle was overcome by crossing the *M2* addition line to *Aegilops speltoides*, a diploid with a complement of 14 chromosomes which, in turn, suppresses the

activity of the *Ph* locus. This 29 chromosome hybrid was backcrossed repeatedly to 'Chinese Spring'. A plant with 21 bivalents at meiosis was isolated from among the progeny which was resistant to yellow rust. This plant was heterozygous for the dominant gene, *Yr8*, from *Ae. comosa* which conferred the resistance to rust. Selfing produced resistant homozygotes which constituted the variety 'Compair' (Riley *et al.*, 1968).

This achievement, ingenious certainly, is also instructive. It shows how demanding the condition for controlling recombination may be. First, one required a gene, or possibly genes, to facilitate effective pairing between homoeologous chromosomes. Having acquired the desired recombinant, this gene (from *Ae. speltoides*) had to be removed. Its retention, at least in an active state, would have rendered 'Compair' largely infertile by virtue of multivalent formation due to homoeologous pairing at meiosis. Apart from the demanding requirements for genes with the appropriate qualities there is also the problem of transferring them across species barriers. This was feasible in the creation of 'Compair'. It is by no means always so.

1.5 POSSIBILITIES AND PROSPECTS

As mentioned earlier many aspects of meiosis are under genotypic control. Unfortunately, much of the variation, in chiasma formation for example, is under polygenic control (e.g. Rees, 1961) and consequently more difficult to manipulate than would be the case were the control exercised by major genes. Given that suitable major genes were available there are tasks to which future endeavours might well be directed with profit.

1.5.1 Allopolyploids

Many of our most valuable crop plants are allopolyploids. They are endowed with the advantages of hybridity without the handicap of infertility. In the bread wheat both properties are largely dependent on the activity of one locus, the *Ph* locus on chromosome *5B* (Riley and Chapman, 1958; Sears and Okamoto, 1958). As was mentioned earlier, its immediate effect is to suppress pairing at first metaphase of meiosis between homoeologous as distinct from homologous chromosomes. There are grounds for believing that similar diploidizing genes operate in most if not all allopolyploids. Indeed the failure by breeders to date to synthesize useful allopolyploids, and there have been numerous attempts, is almost certainly due to the failure to acquire effective diploidizing genes. Triticale (Larter, 1976) and certain *Lolium* hybrids (Breese *et al.*, 1981) are exceptions to the general rule, but even these have not the stability and regularity of chromosome behaviour at meiosis that is characteristic of many 'natural' allopolyploids.

Apart from the *Ph* locus in wheat there is evidence for comparable loci in *Avena* (e.g. Rajhathy and Thomas, 1972) and *Festuca* polyploids (Jahuar, 1975). Surprisingly, diploidizing elements have also been found in supernumerary *B* chromosomes; in *Lolium* (Evans and Macefield, 1973) and in *Aegilops* species (Dover and Riley, 1972). For example, the introduction of *B* chromosomes into the tetraploid hybrid *Lolium temulentum* × *L. perenne* transforms meiosis from that typical of an autotetraploid with multivalents to that characteristic of an allopolyploid with bivalents composed of strictly homologous pairs of chromosomes. With *Ph* and the diploidizing elements in *B* chromosomes the prospects of transfering them in an effective state to alien species, and to do so without contamination by undesired genetic material, have, up to now, been very poor at best. As will be mentioned later the prospects in the future may be very much better. The creation of fertile polyploids from interspecific and even intergeneric hybrids remains a challenge well worth the continued attention of breeders.

1.5.2 Diploids

Apart from polyploids one should perhaps consider also the possibility of utilizing genes to

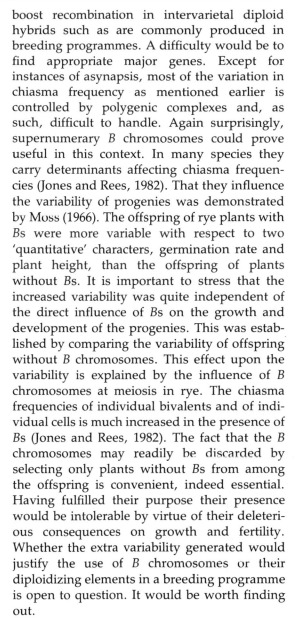

boost recombination in intervarietal diploid hybrids such as are commonly produced in breeding programmes. A difficulty would be to find appropriate major genes. Except for instances of asynapsis, most of the variation in chiasma frequency as mentioned earlier is controlled by polygenic complexes and, as such, difficult to handle. Again surprisingly, supernumerary *B* chromosomes could prove useful in this context. In many species they carry determinants affecting chiasma frequencies (Jones and Rees, 1982). That they influence the variability of progenies was demonstrated by Moss (1966). The offspring of rye plants with Bs were more variable with respect to two 'quantitative' characters, germination rate and plant height, than the offspring of plants without Bs. It is important to stress that the increased variability was quite independent of the direct influence of Bs on the growth and development of the progenies. This was established by comparing the variability of offspring without *B* chromosomes. This effect upon the variability is explained by the influence of *B* chromosomes at meiosis in rye. The chiasma frequencies of individual bivalents and of individual cells is much increased in the presence of Bs (Jones and Rees, 1982). The fact that the *B* chromosomes may readily be discarded by selecting only plants without Bs from among the offspring is convenient, indeed essential. Having fulfilled their purpose their presence would be intolerable by virtue of their deleterious consequences on growth and fertility. Whether the extra variability generated would justify the use of *B* chromosomes or their diploidizing elements in a breeding programme is open to question. It would be worth finding out.

1.6 THE NEW DIMENSION

During recent years a battery of new techniques have become available to the plant breeder. The following are the most important: (1) transformation following the isolation and cloning of genes; (2) the use of transposons; (3) clonal propagation of somatic cells and tissues followed by regeneration to produce mature plants; (4) regeneration from pollen; (5) protoplast fusion and subsequent regeneration; (6) cybrid formation, involving fusion between nuclei and alien cytoplasm.

There is no doubt that these methods will be deployed to advantage and, in some instances, with dramatic results. Transformation, for example, offers opportunities for incorporating useful genes into alien species without accompanying unwanted genetic material. If clones were to become available of the *Ph* locus or of the diploidizing elements from *B* chromosomes, they might well prove invaluable for controlling recombination and constructing new allopolyploids following transformation. This is not the place, however, for dealing with details of particular procedures and projects. It is on the other hand worthwhile considering in very general terms how these new methods relate to the genetic systems of crop plants and facilitate their manipulation.

The six items in the list above would, on the face of it, appear to fall outside the compass of the three variables of genetic systems with which we are familiar, namely mutation, recombination and the breeding system. This impression is more apparent than real. In fact all six of these new techniques may readily be accommodated under one or the other of the three headings which define the components of genetic systems. Transposons provide new opportunities for inducing *mutation*. For transformation the generation and incorporation into chromosomes of alien genes are achieved by variations on the theme of *recombination*. Protoplast fusion and cybrid formation, regeneration from pollen and from clones of somatic cells and tissues are exotic, unusual modes of hybridization and reproduction, unorthodox *breeding systems*. Some of the six techniques fall under more than one heading. The somaclonal variation, associated with the regenerated products of cell and tissue cultures, represents an additional source of mutation. The transfer of alien genes by vectors or other means in the course of transformation represents a bizarre form of hybridization, albeit of a grossly asymmetrical kind.

While such a classification serves little

purpose in itself, the principle on which it is based is important. It tells us that none of the new techniques generates variability of a fundamentally different kind from that which derives from familiar and traditional breeding practices, nor is this variability the product of fundamentally different genetic mechanisms. The new techniques, even so, provide opportunities for manipulating more effectively different components of genetic systems. Put in another way, they supplement in useful fashion the traditional procedures. Not withstanding enthusiastic claims they do not supplant them. The old and the new will be deployed together for the foreseeable future.

REFERENCES

Breese, E.L., Lewis, E.J. and Evans, G.M. (1981), Interspecies hybrids and polyploidy. *Philos. Trans. R. Soc. Lond. B*, **292**, 487–497.

Burt, A. and Bell, G. (1987), Mammalian chiasma frequencies as a test of two theories of recombination. *Nature*, **326**, 803–805.

Darlington, C.D. (1939), *Evolution of Genetic Systems*, Cambridge University Press, Cambridge.

Dover, G.A. and Riley, R. (1972), The prevention of pairing of homoeologous chromosomes of wheat by a genetic activity of supernumerary chromosomes of *Aegilops*. *Nature*, **240**, 159–161.

Evans, G.M. and Macefield, A.J. (1973), The effect of *B* chromosomes on homoeologous pairing in species hybrids. I. *Lolium temulentum* × *Lolium perenne*. *Chromosoma*, **41**, 63–73.

Harinarayana, G. and Murty, B.R. (1971), Cytological regulation of recombination in *Pennisetum* and *Brassica*. *Cytologia*, **36**, 435–448.

Jahuar, P.P. (1975), Genetic control of diploid like meiosis in hexaploid tall fescue. *Nature*, **254**, 595–597.

Jones, R.N. and Rees, H. (1982), *B Chromosomes*, Academic Press, New York.

Larter, E.N. (1976), Triticale. In *Evolution of Crop Plants*, Simmonds, N.W. (ed.), Longman, London, pp. 117–120.

Moss, J.P. (1966), The adaptive significance of *B* chromosomes in rye. In *Chromosomes Today*, Vol. I, Darlington, C.D. and Lewis, D. (eds), Oliver and Boyd, London, pp. 15–23.

Rajhathy, T. and Thomas, H. (1972), Genetic control of chromosome pairing in hexaploid oats. *Nature New Biology*, **239**, 217–219.

Rees, H. (1961), Genotypic control of chromosome form and behaviour. *Bot. Rev.*, **27**, 288–318.

Rees, H. and Dale, P.J. (1974), Chiasmata and variability in *Lolium* and *Festuca* populations. *Chromosoma*, **47**, 335–351.

Riley, R. and Chapman, V. (1958), Genetical control of cytologically diploid behaviour of hexaploid wheat. *Nature*, **182**, 713–715.

Riley, R., Chapman, V. and Johnson, R. (1968), The incorporation of alien disease resistance in wheat by genetic interference with the regulation of meiotic chromosome synapsis. *Genet. Res. Camb.*, **12**, 199–219.

Roberts, H.F. (1929), *Plant Hybridisation Before Mendel*, Princeton University Press, Princeton.

Sears, E.R. and Okamoto, M. (1958), Intergenomic pairing and fertility in hybrids and amphidiploids in the Triticinae. *Res. Bull. Missouri Agric. Expt. Sta.*, **337**, 20.

2

Population structure and variability[*]

M.D. Hayward and E.L. Breese

2.1 INTRODUCTION

The genetic structure of a population, defined as a community of individuals which share a common gene pool, determines its capacity to be improved or otherwise changed by selection. It will itself have evolved through the action of past selective forces on the genes controlling variability. It follows that an understanding of this process is of fundamental importance in deciding plant breeding options and selection strategies. This is manifestly true for more conventional plant breeding methods involving crossing and selection: it is no less true for the successful integration and expression of genes introduced by the increasingly sophisticated methods offered by the developing technologies described in some of the later chapters.

The essential features are first the type of variability and its underlying genetic control and second the mode of selection. These will be considered in turn bearing in mind that they are interdependent in their effects on population structure. Since the vast majority of crop plants are diploid, or amphidiploid displaying disomic inheritance patterns, the theme is disomic segregation and the manner in which repro-

ductive mode, breeding system and chromosomal organization are deployed to govern the maintenance and release of variability.

2.2 VARIABILITY AND ITS CONTROL

2.2.1 Genic control of variability

The important variation within and between natural and artificial populations is for the most part continuous, differences in expression being of degree rather than absolute. Such quantitative (or metrical) variation is displayed by virtually all characters of an organism affecting growth, development and reproduction, including chromosome form and function (as discussed in Chapter 1). Copious experiments have now established that characters displaying quantitative variation are mediated by the joint action of a number of supplementary genes each having a small effect in relation to the total variation (genetic and environmental) and have been designated as polygenic systems by Mather (1941).

Since the effects of individual genes cannot be identified, except in special conditions, polygenic variation cannot be handled by classical Mendelian techniques and, consequently, biometrical methods involving statistical measures (means, variances and covariances) have been developed as discussed in Chapter 11. These techniques have demonstrated that polygenes (as defined) exhibit the same properties in terms of dominance, interaction (epistasis) and linkage as classical Mendelian genes. Typ-

[*] This chapter is dedicated to the late professors Sir Kenneth Mather and John Jinks who individually and jointly did so much to establish the foundations on which this topic is based.

Plant Breeding: Principles and prospects. Edited by M.D. Hayward, N.O. Bosemark and I. Romagosa. Published in 1993 by Chapman & Hall, London. ISBN 0 412 43390 7.

ically, their segregation patterns indicate that they are located on the chromosomes although some quantitative differences may be traceable to non-nuclear elements in the cytoplasm.

Major gene differences, which give discrete and easily identifiable phenotypic classes, have provided the foundation for the classical study of gene segregation, gene action and linkage. These gene differences do not necessarily have useful or important phenotypic effects in terms of adaptation or plant breeding but they nevertheless continue to provide markers for the study of genome organization and transmission. As will be shown in later chapters these markers, which are rapidly being extended by the application of biochemical and molecular techniques, are becoming of increasing importance in the development of breeding strategies.

2.2.2 Gene action, characters and fitness

In so far as individual genic differences between genotypes affect growth and development, their effects on the phenotype can vary from the gross (including developmental failure and lethals) to those so slight that they can only be detected under strictly controlled conditions, as in the case of polygenes. A great deal of biochemical research is now directed towards elucidating how genes and their products act and interact at the cell, tissue and organ level. From the genetical viewpoint these studies have served to show that there is rarely a simple relationship between the gene and the character it affects; the complexity of the relationships depending in the first instance on the number of developmental stages between initial action and final expression. During these stages the effect of the genes primary action on the final character will be subject to modification by the action of other genes and environmental agencies. A single gene may also have more than one primary action, while the same primary action may be shared by a number of genes. It follows that the inheritance of polygenic characters can ultimately only be considered by reference to the polygenic system controlling

it's expression. It also follows that all characters of an organism will to a greater or lesser extent have a developmental (pleiotropic) connection.

Quantitative characters (or traits) may be defined and measured morphologically, physiologically (as responses to light, temperature, water, etc.) or agronomically (yield, etc.). Attempts to simplify genetic analysis are often made through the 'somatic analysis' of particular characters into component subcharacters; thus grain yield in cereals may be considered as the product of the number of ears, the number of grains per ear and the average weight per grain. However the subcharacters are generally also polygenic and show varying degrees of correlation with other traits; in a breeding programme therefore they have to form part of a multitrait approach to selection (see Chapters 18 and 28). Such correlations may imply the sharing of a common precursor by the different biosynthetic pathways, pleiotropic gene action or, as will be considered later, linkage of the genes concerned.

The concept of the relationship of genes and characters within the context of the organism as a whole has been developed by Mather and Jinks (1982). Characters and subcharacters are obviously interrelated features of the total phenotype and ultimately cannot be considered in isolation from each other. Thus in plant breeding, harvestable yield may be targeted as a primary character but finally it has to be adjudged in relation to quality, resistance to disease, tolerance to pests and environmental stresses, etc., which jointly establish the economic value of the end product. Yield *per se* then becomes a component of what Mather and Jinks (1982) term a 'supercharacter' which gives a measure of the overall merit, and has to be handled as such during a breeding programme. This may be achieved by a weighted discriminant function determined by statistical analysis or, perhaps as occurs more often, by empirical intuitive assessment. Clearly the way a supercharacter is defined, the nature of the component characters and the weight given to each, will depend on the crop and the improvement sought. For instance, in grain crops fertility will be an integral subcharacter in maximizing yield: in root and forage crops fertility and seed yield become secondary characters weighted in relation to the required price of seed.

Under natural selection the supercharacter must equate with fitness of the phenotype in a particular environment; where fitness is defined in the Darwinian sense of the capacity to survive and leave offspring. Although we may be able to infer on *a priori* grounds that a particular feature of the environment may give prominence to a specific component character, e.g. winter hardiness or drought tolerance, the relationship between supercharacter and component characters can only be deduced by assessing the effects on overall fitness of changes in individual characters. This requires an understanding of the genetic control of the various characters, their genetic architecture, and their interrelationships; thus ultimately the genetic structure of the population as a whole.

Consideration of population structure must ultimately centre on genic variation but, since natural selection operates at the level of the phenotype, prime attention must be given to the special properties of polygenic systems where the value of each constituent gene (regardless of its mode of action) is conditional on all other genes in the genome and on the environment.

2.2.3 Properties of polygenic systems: gene combinations and states of genetic variability

In outbreeding populations, a single gene with two allelomorphs, *A*:*a*, will give rise to three genotypes, *AA*, *Aa* and *aa*. Under completely random mating, where each individual has an equal opportunity to contribute to the next generation, the frequency of the three genotypes will be according to the Hardy–Weinberg equilibrium formula:

$$AA + Aa + aa$$
$$p^2 + 2pq + q^2$$

where p is the frequency of *A* and q the frequency of *a*, and $p + q = 1$. When the gene frequencies are equal (as in F_2 populations), the proportion of the genotypes are 1/4:1/2:1/4. Assuming no dominance, two states of variabi-

lity are evident. The two homozygous extremes represent 'free variability' which can be seized upon by selection and fixed. The intermediate heterozygote has the potential to generate fresh variability through segregation. This 'hidden' or 'potential' heterozygotic variability is of course regenerated in outbreeding populations through the crossing of the two homozygotes, and at equilibrium (i.e. random mating and no selection) the rate of crossing and segregation will be balanced to maintain the proportion of free and potential variability at 1/2:1/2. The structure of the population is thus maintained as a dynamic flow of crossing and segregation according to the Hardy–Weinberg formula.

If we consider more than one gene affecting the same quantitative character, a third state of variability emerges. Consider first a pair of alleles at each of two independent loci, *A*-*a* and *B*-*b*, each having a similar effect on the character, with equal gene frequencies and no dominance. If *A* and *B* represent the alleles adding positive increments and *a* and *b* the alleles with negative increments, there are nine genotypes in a random mating population which fall into five classes as depicted below:

Genotypes:	*AABB* :	2*AABb* :	4*AaBb* :	2*Aabb* :	*aabb*
	2*AaBB* :	*AAbb* :	2*aaBb*		
		aaBB			
Phenotypic value:	++++	+++−	++−−	+−−−	−−−−
Frequency:	1	4	6	4	1

This is, of course, the simplest form of a polygenic system and as we would expect, intermediate expressions are the commoner, giving rise to the typical normal curve of continous variation. The completely free variability stemming from the extreme reinforcing homozygotes, *AABB* and *aabb*, now only forms one-eighth of the total variation. Most importantly, it will be observed that the intermediate phenotype is formed not only by the balancing + and − alleles of the double heterozygote *AaBb*, but also by the balancing effects of ++ and −− alleles of the opposition homozygotes *AAbb* and *aaBB*. These phenotypically similar, but contrasting genotypes, clearly contain latent variability, termed 'homozygotic potential variability', which however will not be expressed in the free state until they are crossed to form the heterozygote *AaBb* and which in turn is allowed to segregate in the subsequent (F$_2$) generation. It thus takes two generations to convert homozygotic potential variability into its free state and available for the action of selection.

In a randomly mating population all three states of variability subsist in dynamic equilibrium, each state being constantly replenished through crossing, segregation and recombination. Following Mather (1973), this flow of variability can be graphically portrayed as in Fig. 2.1 for the two locus situation. It will be

seen that only a small proportion of the variability, one-eighth, is available for selection in the fully free state, the remaining seven-eighths is hidden in the heterozygotic or homozygotic potential states. It will also be noted that the homozygotic potential variability must pass through the heterozygotic state (following crossing) before it is converted by segregation into the free state.

With more than two loci affecting the same character the situation obviously becomes more complex. Thus as the numbers (*n*) of genes in the system increases, the proportions of the variability which is freely expressed decreases. It can be shown that the ratio of free:heterozygotic potential:homozygotic potential is (1:*n*:*n*−1) (Mather, 1973). For instance, with ten polygenes the proportion of visible (free) variability is only 1/20th of the total, 95% being hidden in the heterozygotic or homozygotic potential states; and the chance of obtaining the extreme positive homozygote is $(1/4)^{10}$ which is of the order of one in a million. Although estimating the numbers of polygenes in a system is not easy using the classical methods of selection experiments (Mather and Jinks, 1982), the recent developments in the analysis of quantitative trait loci (QTLs) using saturated linkage maps of molecular markers will alleviate this problem. Already estimates of the numbers of QTLs controlling some yield and quality characters in tomato (Tanskley *et al.*, 1989; Paterson *et al.*, 1991) indicate that the number of loci is at least 11 for one of the characters, mass per fruit. The relationship of QTLs to polygenes, in the classical concept of the latter which we are considering here, will only be clarified with more detailed analysis of the proposed QTL at the molecular level. It may well be that QTLs equate to the 'effective factors' of Mather and Jinks (1982) and represent groups of tightly linked polygenes.

So far we have assumed dominance and non-allelic interaction to be absent. If we now take the two allelic pairs illustrated above and assume full dominance in the positive direction, i.e. phenotypes *AA* = *Aa* = *BB* = *Bb*; then the nine genotypes in a randomly mating population will now form three phenotypic classes with the following frequencies:

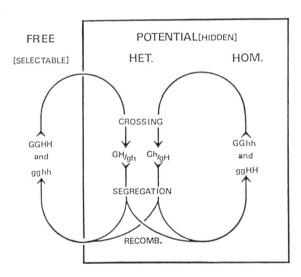

Fig. 2.1. The states of variability. Free variability is expressed as phenotypic differences among the genetic classes, heterozygotic potential variability is concealed by the balancing effects of alleles and is therefore a property of heterozygotes, and homozygotic potential is concealed by the balancing effects of non-allelic genes. Free variability is available to the action of selection, but both types of potential variability (enclosed in the box) are incapable of being acted on by selection. The arrows show the directions of flow of the variability, the rates of flow being governed by crossing, segregation and recombination. Reproduced by kind permission of Dr R.R. Mather and Chapman & Hall.

Genotypes:	AABB+4AaBb+ 2AaBB+AABb	2Aabb+2aaBb+ AAbb+aaBB	aabb
Phenotype value:	++++	++--	----
Frequency:	9	6	1

Compared with the no-dominance situation two main differences emerge.

First, since the contrasting alleles at a locus no longer balance each other, the heterozygotes now add to the visible rather than the cryptic variability. In fact half the visible variability is now in the heterozygous potential state which cannot be fixed by selection. This has the effect of reducing the rate of response to phenotypic selection at least in the same direction as the dominance effect. This of course, follows from the fact that fixable homozygotes cannot be distinguished from the heterozygotes except by a breeding test.

Second, class frequencies are no longer symmetrically arranged around the central phenotype but rather give a skewed distribution of phenotypes with the preponderance having a greater expression of the character. Consequently, the mean of the population is shifted in a positive direction. In polygenic systems, the dominance can equally well be preponderantly in the reverse direction giving a negatively skewed distribution; or it may be ambidirectional, with the heterozygotes at different loci having a balancing effect on character expression resulting in an increased number of intermediate phenotypes (positive kurtosis).

The effects of non-allelic interaction (epistasis) will not be considered in detail here. It is sufficient to say that by and large they can reinforce the effects of dominance in affecting the distribution of variability and mean expression of the population. Clearly, the genetic architecture of a character in terms of dominance and interaction can have a profound effect on the genetic structure of a population. As such, both these effects not only affect the way a population responds to selection but are themselves a target of selection, as we shall see later.

The above treatment is abbrieviated and the conclusions largely intuitive. A more detailed treatment and algebraic proof is offered in Chapter 11, together with methods for estimating the fixable and unfixable components of variation and so predicting the response to selection as outlined in Chapter 18.

2.2.4 Regulating recombination: linkage and breeding systems

The above serves to illustrate that in outbreeding populations, polygenic systems are capable of storing vast amounts of cryptic variability through the numbers and action of the constituent genes. This is only gradually made available for exploitation by selection through the processes of crossing, segregation and recombination. As is made abundantly clear in Chapter 1, and is obvious from Fig. 2.1, the key to the flow of variability from the cryptic to the free state is the rate of recombination, and this in turn depends on the linkage of genes on the chromosomes and on the breeding system.

(a) Linkage

Consider two linked genes with a recombination value of r, then segregation in the second generation depends on the initial cross as shown below:

Initial cross	Free	Homozygous potential
(i) AABB × aabb	1 (1−r)	2r
(ii) AAbb × aaBB	2r	2(1−r)

Thus if genes are linked in repulsion (balancing = ii), the flow of variability from the homozygous potential to the free state depends on how tightly the genes are linked, being at its maximum when $r = 0.5$ and recombination is free, and diminishing as r becomes smaller. Clearly with more than two loci closely linked on the same chromosome, the flow of variability would be severely impeded.

(b) The breeding system

The amount of crossing which actually takes place, which is determined by the breeding system, will influence the flow of variability (Fig. 2.1). If all crossing is prevented by enforced selfing (i.e. complete inbreeding), then through segregation the proportion of heterozygotes will be halved in each generation, and after 8 or 9 generations, will be virtually nil. If no selection is applied the population contains the same amount of variability as an outbreeding population, but much of this will be frozen in the homozygotic potential state (as balanced +− homozygotes) which is unavailable for selection.

2.2.5 Genetic adjustment of the regulators: with special reference to the breeding system

Since the breeding system and chromosomal linkage are integral elements of the machinery governing the flow of genetic variability, we can expect that they are themselves under genetic control and will be adjusted by natural selection during population differentiation. The control of recombination and how it may be artificially manipulated has been discussed in depth in Chapter 1, whilst the genetic adjustment of the breeding system will now be considered.

As has already been indicated, the population structure of crop species and the ensuing states of variability, which predispose response to selection and systems of variety construction, are conditioned to a large extent by the mode of reproduction. Sexual reproduction is the commonest form among crops but asexual, by vegetative or allied means, is of importance in many species and is likely to become of greater significance as systems of mass propagation and manipulation of apomixis is extended to more crops. Sexual reproduction is the basic mechanism by which variability may be made available for selection, whether natural or artificial. It is not surprising therefore that the sexual mode, whether it be inbreeding, outbreeding or a mixture of the two, has itself become subject to genetic control.

Inbreeding, is generally achieved by mechanisms which ensure that self-pollination takes place. These may range from cleistogamy (the shedding of pollen and fertilization within closed florets), as in many of the major cereal crops such as wheat and barley, through to chasmogamy where the flowers are open when anther dehiscence occurs and the stigma is receptive as in the tomato. Inbreeding mechanisms are generally under precise genetic control but can be influenced by both the background genotype and environmental factors. Needless to say these slight variations may allow a breakdown in the predominance of self-fertilization. In barley for example Allard and his co-workers have shown that an outcrossing rate of about 1% occurs allowing novel recombinants to appear (Allard and Hansche, 1965) with further potential for selection and evolutionary change.

Numerous mechanisms exist to promote outcrossing and with it the potential of the heterozygous state. These mechanisms may range from the relatively simple monoecious state (often coupled with protandry and protogyny) which, whilst encouraging outcrossing, do not prevent selfing occurring if self-pollen should alight on the stigma, through to the complex, but very precise genetically controlled incompatibility systems (see Chapter 13). Whatever mechanism is applicable the genetic systems determining their expression are open to manipulation which may change the state of variability of the species concerned. This is well exemplified by the tomato whose wild progenitors are outbreeding. In cultivation, however, it is inbreeding with a consequent loss of variability as revealed by recent analyses of molecular polymorphisms (Miller and Tanskley, 1990). Clearly the reproductive system is a major determinant in the maintenance and flow of variability and hence the manner in which a species responds to both natural and artificial selection.

2.2.6 Ploidy

Compared with the haploid condition of primitive organisms, the diploid state and disomic segregation has allowed the evolution of a sophisticated genetic system for the expression, storage and ordered retrieval of genetic information as discussed above. Higher levels of polyploidy are common in plant taxa (Stebbins, 1957) but these are rarely found in conjunction

with the potentially associated levels of polysomic segregation. They are usually allopolyploids, derivatives of wide crosses (race or
species hybrids, see Chapter 6). In almost all
natural cases, progressive polyploidy (chromosome doubling) is accompanied by the genetic
enforcement of preferential pairing between the
homologous chromosomes of each contributing
genome to form amphdiploids as discussed in
Chapter 1. The resultant forms are permanent
hybrids stabilized over sexual reproduction.

Autopolyploids with functional polysomic
segregation are rare and generally confined to
long-lived species with a wide faculty for
asexual reproduction. Obviously polysomy has
not featured as an evolutionary pathway. The
reason is clear. Tetrasomic inheritance (let alone
that deriving from higher somic levels) involves
complex segregation patterns which vastly
retard the recovery and fixation of desirable
gene combinations during sexual reproduction.
Polysomy may excel in the storage of genetic
information but is relatively inefficient at
orderly retrieval and exploitation compared
with disomy. In plant breeding, induced autopolyploidy has thus chiefly been exploited more
for the physiological (*gigas*) effects of chromosome doubling than for continued genetic
manipulation (Breese *et al.*, 1981).

2.3 SELECTION: MODE AND ACTION

Under natural circumstances selection is concerned with increasing the fitness of the species
whilst, in plant breeding, artificial selection is
usually directed towards a particular goal. The
response to selection in both situations is clearly
dependent on the nature and extent of the
variability present and the form of selection
pressure applied. The quantitative genetic
theory underlying selection is considered in
detail in Chapter 11 so we will restrict the
discussion here to a consideration of the differing forms of selection.

In plant breeding, the aim will be to change
existing populations, varieties, genetic stocks,
etc., to achieve specific objectives. Selection will
therefore be 'directional' towards maximal

expression of the targeted character(s), or an
optimal expression within the constraints of the
management system envisaged. This is the kind
of selection with which we are most familiar as
plant breeders; it is best understood and most
easily accommodated in terms of assessing the
potentials and predicting advances.

Under natural conditions there will also be
directional selection in adapting to new environmental conditions, whether in pioneering
new areas or in accommodating progressive
change in the original habitat. Mather (1953,
1973) recognizes that under natural conditions
this phase can normally be succeeded by two
different types of selection pressures, 'stabilizing' and 'disruptive'. Each of these will have
quite different effects on the genetic structure of
the population and even on the genetic architecture of the characters.

2.3.1 Directional selection

Under artificial selection the achievement of the
targetted expression requires the reassociation
of genes affecting a number of connected
characters in producing a fully balanced phenotype. Once achieved, these 'co-adapted gene
complexes' are protected against further change
by genetic linkage, and the tighter the linkage
the greater the resistance to change, i.e. the
greater the genetic inertia of the population.
Further resistance to change would derive from
an alteration in the breeding system from
outbreeding to inbreeding, such that a fully
inbred (selfed) homozygous population would
be inacapable of further change (until again
crossed); the whole gene complex would in
effect be a single linkage group.

An important distinction must be made
between the action of selection on the predominantly homozygous gene combinations of
inbreeders and the heterozygous combinations
of outbreeders (Mather, 1973). In the former
case the correct 'homozygous balance' between
the genes affecting both the selected and
correlated characters has to be established
'internally' within the chromosomes since each
of the paired chromosomes are exact replicas:
whereas in the latter case, 'heterozygous

balance' is 'relational' between the dissimilar paired chromosomes. The differential consequences are fundamental. In heterozygous populations genes showing dominance in the appropriate direction will be favoured over other alleles (Fisher, 1930); similarly genes at different loci exhibiting positive interactions are also likely to be seized on by selection. We might expect therefore, that characters subjected to prolonged directional selection will exhibit high directional dominance and/or genic interaction. Thus Breese and Mather (1960) were able to demonstrate high directional dominance and reinforcing duplicate type epistasis for the 'fitness' character viability, in their study of polygenic activity in chromosome III of *Drosophila*. Similarly Breese (1960) found that there was high directional dominance for increased yield of seed-producing tillers in perennial ryegrass.

2.3.2 Stabilizing selection

For characters which directly affect the fitness of the organism in the Darwinian sense (e.g. viability and fertility in sexually reproducing species), we may expect that selection will continually be directional towards maximum expression. For other characters, when directional selection has produced the optimal phenotype for any particular habitat, selection will operate to perpetuate this phenotype, at least as long as the habitat itself remains stable. Extreme expression, whether positive or negative, will then be at a disadvantage and selection will thus be for the population mean, i.e. selection will be 'stabilizing' rather than 'directional'. This, of course presupposes that the 'optimal' expression does not correspond with the 'maximal' expression, and for many characters there is good evidence that this is the case e.g. flowering time in many crop species.

Characters which have been subjected to stabilizing selection show quite a different genetic architecture to those which have been subjected to directional selection. Here dominance will be low or absent and if present, may be ambidirectional in nature, whereas epistasis will not generally occur. In natural populations

of *Lolium perenne* for example, vegetative features such as rate of tillering and leaf size, which could be presumed to have been under stabilizing selection, displayed largely additive gene action (Hayward and Breese, 1968) compared with the high directional dominance found for the fitness character, inflorescence production (Breese, 1960).

2.3.3 Disruptive selection

The aim in cultivation is to obtain homogeneous cropping areas and standardized managements of the crops. Natural habitats are rarely, if ever, likely to be so uniform and may encompass a number of distinct 'ecological niches' each selecting different phenotypic optima in form or function. Differences may be spatial (microniches), temporal (e.g. seasonal or long-term cycles), or functional. The subject is complex, often involving several related processes such as density-dependent and/or frequency-dependent interactions between genes or genotypes, inevitably the process favours diversity rather than uniformity within the populations.

The consequences of such 'disruptive selection' on population structure depends basically on two factors (Mather, 1973): (1) whether the different optimal phenotypes are independent of, or dependent on, each other for maintenance or functioning; (2) the rate at which genes are exchanged between the differentially selected genotypes.

The most obvious co-existing differences (polymorphisms) in most animal species and dioecious plants are male and female forms. Occasionally the switch mechanism which determines development into one or other form is environmentally triggered as in some tropical grasses. Most often it is controlled by a gene or supergene (the chromosomal $X:Y$ system as in *Humulus*). Both male and female forms are of course, completely interdependent in reproduction and so gene exchange is 100% enforced. They thus share a common genetic background which has to ensure that both channels of development (male or female) are equally adapted to the prevailing conditions. Thoday and his colleagues (Thoday, 1972), working with *Drosophila*, have demonstrated experimentally that such polymorphisms sharing a common genotype can be built up by continued disruptive selection with 50% gene flow between groups, i.e. cross-mating the selections. Ultimately a 'supergene' switch mechanism is brought into being through the close linkage of genes governing development into one or other channels of expression.

The precise genetically controlled polymorphisms considered above are built up in response to selection for consistent interdependent functions particularly in governing the reproductive processes. Apart from sexual dimorphism in sexual species, another example is polystylic forms in *Primula*. Disruptive selection resulting from less exact forms of interdependence results in a more diffuse system of gene polymorphisms. Thus in species where outbreeding is controlled by incompatibility genes, the rarer the allele at a locus the greater its chance of securing a compatible mating and *vice versa* for the more common alleles. Such frequency-dependent selection is capable of building up a large number of incompatibility alleles in a population.

A most important form of frequency-dependent polymorphisms in wild populations relates to disease and pest resistance genes, in that a relatively low proportion of resistant plants curtails the spread of disease and so offers protection to the population at large. Challenge by different pathogen races thus builds up a large number of alleles which buffer the population against epidemic attack. This is one of the few areas in which genetic heterogeneity has been successfully used to give 'ecological protection' to otherwise uniform crop plants (see Chapter 25).

Another form of disruptive selection occurs when a population seeks to exploit small differences in time or space (microniches) within the habitat. This involves density and frequency dependent interactions and can result in stable mixtures of co-adapted genotypes having different and complementary requirements for resources, such as nutrients and light and thus escaping competitive elimination. Such ecological combining ability of mutually interdependent genotypes has been

shown to maintain high levels of variability in wild populations of inbreeding plants (Allard and Adams, 1969), and in land races of inbred cereals (Simmonds, 1979). Inbreeding can, of course, maintain the co-adapted genotypes as homozygotes over seed generations as is well exemplified by the studies of Allard on the composite cross series of populations of barley (Allard, 1988). The faculty to develop co-adapted associations is also possible for outbreeding species provided they are perennial and long-lived or are propagated asexually as with perennial pasture species (Clements *et al.*, 1983): the co-adapted genotypes, however, cannot be preserved as such over sexual generations but add to the genetic heterogeneity of the seed population, and thus increase the population's capacity to respond to further selective change in heterogeneous environments (Breese, 1983). A further dimension of co-adaptation occurs at the interspecific level in mixed plant communities, e.g. grass and legume associations. The very considerable problems of breeding compatible grass and clover components of pastures has been considered by Hill and Michaelson-Yeates (1987).

If spatial and temporal differences are on a sufficient scale to fragment the habitat into independent niches, then integrated disruptive selection becomes two-way directional selection for independent phenotypic optima. Differentiation can occur over very short distances and despite considerable gene flow from the adjacent sites if the selection pressures are high enough (Bradshaw, 1984). If continued long enough, genetic barriers to crossing may arise and the new populations will become genetically isolated and may even become distinct species.

2.3.4 Selection and gene action: genetic architecture

It is clear from the above discussion that differing selective forces will result in different types of gene action. Thus directional selection for a character will lead to the establishment of dominance and epistasis, whereas stabilizing selection will promote largely additive variation. Disruptive selection, in that it is directional in nature may again lead to dominance but, more importantly, establishes polymorphisms or even integrated supergenes. It follows that the genetic anatomy of a character is in itself diagnostic of its previous selection history. In turn it determines the method of exploitation in a breeding programme.

From the above it is clear that the pattern of genetic variation and reproductive system are intimately interrelated and both are the resultant of past forms of selection. Together they provide a dynamic mechanism for continuing adaptation irrespective of whether it be to the natural environment or to a plant breeding objective.

2.4 ADAPTATION AND POPULATION STRUCTURE

2.4.1 Integregated genetic systems

Adaptation in its strict Darwinian concept of 'fitness' is not always synonymous with the usage of the term in a plant breeding context where it more often than not implies ability of a crop to produce in an agricultural environment. The latter may involve the processes of establishment, tolerance to pest and diseases, and so on, with the product being only perhaps indirectly related to the capacity for reproduction. However, in many crops adaptation follows the 'fitness' pattern in that fecundity as revealed by seed yield may be the prime component of agricultural yield. Adaptation in its plant breeding context will be considered in detail in Chapter 22. Here we will only briefly examine its role in relation to the integrated adjustment of genetic systems during population differentiation and evolution. Particular reference is made to reproductive systems and gene action; adjustments in chromosomal form and behaviour (linkage) have been considered in Chapter 1.

Inbreeding is generally associated with annual or short-lived life-forms. It has evolved from outbreeding in response to the environmental conditions to which such species are typically subjected. Often inbreeders are found at the extremes of the species distribution, in marginal habitats and particularly in stress environments (Stebbins, 1957). This is well exemplified by many of the cereal crop species whose progenitors were adapted to the Mediterranean climate where typically a period of winter/spring growth is followed by a reproductive phase before the onset of the extremes of summer drought. Summer growing annuals likewise survive the extremes of winter cold through seed. Under these conditions immediate fitness of the population is most easily secured by inbreeding which guarantees a high proportion of adapted genotypes in successive generations. The capacity for evolutionary response to changing conditions depends on the degree of occasional outcrossing.

By contrast, adapted genotypes of long-lived and perennial species survive vegetatively (many can be vegetatively propagated). Consequently they can tolerate high variability during sexual regeneration to provide long-term flexibility. Such species are thus generally outbreeders. Certainly outbreeding is also of widespread occurrence among short-lived species, but in these instances chromosomal restrictions on recombination may preserve valuable gene combinations over generations (see Chapter 1).

During the development of many crops, both unconscious and directed selection has led to a change from outbreeding to inbreeding and, with it, consequences for the states of variability to be found within populations or cultivars of the species concerned. As we have seen, the breeding system, in conjunction with selection forces, largely determines the genetic architecture of the characters. Thus, dominance is likely to be a pronounced feature of outbreeding rather than inbreeding species and generally this is found to be the case. Nevertheless, for many characters dominance is still exhibited by inbreeding species to a greater or lesser extent, and this is now generally accepted as evidence of ancestral outbreeding progenitors.

Obviously, therefore, balanced inbreeders may be artificially developed from outbreeding populations, as the theory discussed earlier indicates, and is an approach now being proposed for some crops, e.g. forage grasses and brassicae (see Hayward, 1990). This can, however, only be achieved at a cost. Whereas relationally balanced heterozygous gene combinations with directional dominance and epistasis will give a high proportion of adapted genotypes, only one or a few homozygous combinations will give the same internal balance (this is reflected in the low success rate of obtaining androgenetic haploids from outbreeding species compared with inbreeders). In passing it should be noted that many of the less vigorous inbred lines produced will be unbalanced in different ways; crossing them will restore relational balance and so produce heterosis. Obviously an abrupt transition from outbreeding to close inbreeding would involve the production of a preponderance of unbalanced genotypes and a high cost in the selective elimination of so many unfit phenotypes, as breeders of selfed crops are well aware. It is perhaps therefore no surprise to find that the occurrence of close inbreeders in the wild tend to be limited to those species which have very high reproductive rates.

A further dimension to adaptation is provided by the mechanisms of 'developmental plasticity' where the capacity to switch from one growth form to another is environmentally triggered. Perhaps the best examples are those aquatic plants which have different forms and anatomy of submerged and aerial leaves. This type of facility has rarely been considered as a plant breeding target; normally the requirements are for a uniform end product. Nevertheless, in pasture plants a switch from early-season erect, large-leaved forms suitable for conservation to more prostrate, smaller-leaved forms for later season grazing has been sought with some success in hybrids between Italian and perennial ryegrasses (Breese *et al.*, 1981).

Whereas plasticity in expression can be advantageous in many growth characters, in traits more closely associated with fitness, particularly reproductive, the requirement is for stability of expression. For such characters the successful organism will have to be genotypically buffered against environmental changes likely to be encountered during its lifetime. Obviously perennial species will encounter seasonal changes which are avoided by annuals and ephemerals and thus have different requirements for winter-hardiness, etc. The types of selective forces and the outcome in terms of population structure will thus be a resultant of the life cycles and life styles of the organisms concerned.

2.4.2 Population structure and scope for genetic manipulation

Knowledge of the genetic structure of populations can provide valuable guidelines for the formulation of many aspects of breeding strategies. For example, the acquisition of genetic resources or the synthesis of new cultivars requires information on the organization of genes and their potential for evolving novel genotypes as required, whether this is through the procedures of recombination and segregation, or by the application of some of the biotechnological methods outlined in later chapters.

The study of plant population structure has progressed rapidly with the development of appropriate biometrical methods for the genetic assessment of quantitative characters and biochemical/molecular techniques for the determination of polymorphisms at the single gene, restriction fragment or nucleotide sequence level. Some aspects of the quantitative and biochemical characterization of populations are presented in Chapters 11 and 12 respectively. With the notable exception of the extensive work of Allard and his colleagues in California, particularly on barley but also including a wide diversity of other crop and wild species (Allard, 1988), and our own more restricted programme on *Lolium* (Hayward, 1985), few detailed studies of population genetic structure have combined these two approaches. However, by drawing upon these, together with information from other species, patterns of population genetic organization are beginning to emerge which can be clearly related to variability in the reproductive system and accord with the principles already considered.

A most significant feature to emerge from both the studies, when considering continuous variation, is the high level of additive gene action found for many quantitative differences especially in vegetative traits. Dominance effects are most apparent in component characters of the reproductive system which directly bear on evolutionary fitness. As discussed, dominance is, after all, a mechanism to maintain a high frequency of a desirable genotype over sexual generations. Whilst these effects have been determined at the gross 'effective factor' level, and do not necessarily imply the action of individual QTLs or polygenes, they undoubtedly reflect the expression of many differing gene combinations. As we have already seen, similar phenotypes may arise from different genotypes and the varied combinations which evolution has established offer both free and potential variability for further selection.

The capacity for even a few individuals of an outbreeding species to contain such a wealth of variability in the heterozygous condition is dramatically revealed by the selection experiments of Cooper (1959) on *Lolium* for timing of inflorescence emergence. From a starting population of only four plants, each of the modal phenotype, he was able to produce divergent selection lines which extended well beyond the range of the original populations in as little as four generations. In inbreeders the same principles are revealed by the continuing improvements which are being made in such crops as barley in Europe, where, by using single biparental matings followed by pedigree selection from the F_2, yield gains are still being achieved from what many breeders consider to be a very restricted gene pool.

2.5 IMPLICATIONS FOR BREEDING STRATEGIES AND TACTICS

An oft quoted and somewhat cynical concept of the scope of many traditional breeding programmes is that they are still largely a matter of crossing the best with the best and hoping for the best!

This may have been a fair summary of early endeavours but largely ignores the marketing experience which is now required to define the aims, the agronomic expertise to measure and assess what is 'best', and the physiological and biochemical research to translate broad aims into plant characters and processes which more tangibly define plant breeding objectives. More within the scope of this book; while there will always be statistical elements in prediction, the greater understanding of genetic systems, and the application of ever growing batteries of techniques and technologies for their manipulation, can certainly increase confidence in the outcome of any programme.

In the past the constraints set on breeding programmes by previous selection histories have been appreciated if not fully understood. Thus artificial selection methods largely matched the breeding system of the species under improvement. With selfers the ultimate aim has been to develop true breeding lines by some form of pedigree or related selection method. For outbreeders various systems of selection have been practised largely differing as to whether selection was phenotypic, thus depending on exploiting additive variation, or whether some form of progeny test is employed to establish 'prepotence' (see Chapter 18). We have seen that the genetic architecture of a character, in terms of the dominance and interaction properties of the underlying polygenic system, is the resultant of the breeding system and past selection: which will in turn influence the choice of breeding method for further manipulation. The phenomenon of 'hybrid vigour' has long been appreciated and exploited in both plant and animal breeding; while the associated 'inbreeding depression' and its attendant problems have been recognized since the dawn of recorded history, as exemplified by the restrictions embodied in religious or social codes on the mating of close relatives in human populations. Yet it has only recently been realized that both phenomena can be explained entirely in terms of gene action (dominance with or without genic interaction) and the way the genes are combined in the parents. It is now clear that it is possible to recombine the genes so that the hybrid vigour stemming from heterozygous (relational) balance can be matched or exceeded by inbred lines with the appropriate homozygous (internal) balance; as has been demonstrated by Jinks and his coworkers (Jinks, 1981). Such procedures may however take a considerable length of time in terms of practical plant breeding programmes.

The ease with which recombination can be achieved depends upon the complexity of the character and the linkage relationships of the constituent genes (see Chapter 1). In the short term it may be more efficient to capitalize on dominance (where present) by the use of hybrid lines than to attempt reassociating the genes in a balanced homozygous inbred, especially if the crop has a recent history of outbreeding. This immediately raises the question as to the extent to which the breeding system can itself be manipulated, and to what end (Hayward, 1990).

The manipulation of the breeding system is discussed in several chapters and may range from the induction of male sterility to the breakdown of apomixis. An undercurrent throughout is the need to establish uniform populations for the imperatives of managing and marketing of new varieties. Natural populations and genetic resources on the other hand, must retain an optimum of heterogeneity if they are to meet the evolutionary demands of changing circumstances.

Shifts in the economic and physical climates are inherent in agricultural and horticultural systems. An adequate response by the breeder requires appropriate genetic resources together with the knowledge of how to manipulate them for the production of new superior cultivars.

New forms of genetic engineering provide exciting possibilities for future crop improvements. Genes may be identified, excised, cloned and moved over greater distances than has been possible by conventional techniques. Yet, their effective deployment in obtaining an integrated phenotype will still require a knowledge of gene action, genetic architecture and population structure. They must be seen as supplementary and not supplanting existing manipulative skills and techniques.

REFERENCES

Allard, R.W. (1988), Genetic changes associated with the evolution of adaptedness in cultivated plants and their wild progenitors. *J. Hered.*, **79**, 225–238.

Allard, R.W. and Adams, J. (1969), Population studies in predominantly self-pollinating species. XIII. Intergenotypic competition and population structure in barley and wheat. *Am. Nat.*, **103**, 621–645.

Allard, R.W. and Hansche, P.E. (1965), Population and biometrical genetics in plant breeding. In *Genetics Today*, Vol. 3, Proc. XIth Int. Cong. Genetics, The Hague, The Netherlands, 1963, Geerts, S.J. (ed.), Pergamon Press, Oxford, pp. 665–668.

Bradshaw, A.D. (1984), The importance of evolutionary ideas in ecology and *vice versa*. In *Evolutionary Ecology*, Shorrocks, B. (ed.), Blackwell Sci. Publ., Oxford.

Breese, E.L. (1960), The genetic assessment of breeding material. In *Proc. VIIIth Int. Grassland Congr.*, pp. 45–49.

Breese, E.L. (1983), Exploitation of genetic resources through breeding: *Lolium* species. In *Proc. Symp. on Genetic Resources of Forage Plants*, Bray, R.A. and McIvor, J.G. (eds), CSIRO, Melbourne, pp. 275–288.

Breese, E.L., Lewis, E.J. and Evans, G.M. (1981), Interspecies hybrids and polyploidy. *Philos. Trans. R. Soc. Lond. B*, **292**, 487–497.

Breese, E.L. and Mather, K. (1960), The organization of polygenic activity within a chromosome in *Drosophila*. II. Viability. *Heredity*, **14**, 375–399.

Clements, R.J., Hayward, M.D. and Byth, D. (1983), Genetic adaptation in pasture plants. In *Proc. Symp. on Genetic Resources of Forage Plants*, Bray, R.A. and McIvor, J.G. (eds), CSIRO, Melbourne, pp. 101–115.

Cooper, J.P. (1959), Selection and population structure in *Lolium*. III. Selection for date of ear emergence. *Heredity*, **13**, 461–479.

Fisher, R.A. (1930), *The Genetical Theory of Natural Selection*, Clarendon Press, Oxford.

Hayward, M.D. (1985), Adaptation, Differentiation and Population Structure in *Lolium perenne*. In *Genetic Differentiation and Dispersal in Plants*, NATO ASI Series G, Vol. 5, Jacquard, P. and Heims, G. (eds), Springer-Verlag, Berlin, pp. 83–93.

Hayward, M.D. (1990), Genetic strategy and future prospects for breeding cross-pollinated species. *Norwegian Agricultural Research*, Supplement 9, 77–84.

Hayward, M.D. and Breese, E.L. (1968), Genetic organisation of natural populations of *Lolium perenne* L. III. Productivity. *Heredity*, **23**, 357–368.

Hill, J. and Michaelson-Yeates, T.P.T. (1987), Effects of competition upon the productivity of white clover/perennial ryegrass mixtures. Genetic Analysis. *Plant Breeding*, **99**, 239–250.

Jinks, J.L. (1981), The genetic framework of plant breeding. *Philos. Trans. R. Soc. Lond. B*, **292**, 407–419.

Mather, K. (1941), Variation and selection of polygenic characters. *J. Genet.*, **41**, 159–193.

Mather, K. (1953), The genetical structure of populations. *Symp. Soc. Exp. Biol.*, **7**, 66–95.

Mather, K. (1973), *Genetical Structure of Populations*, Chapman and Hall, London.

Mather, K. and Jinks, J.L. (1982), *Biometrical Genetics*, 3rd ed, Chapman and Hall, London.

Miller, J.C. and Tanskley, S.D. (1990), RFLP analysis of phylogenetic relationships and genetic variation in the genus *Lycopersicon*. *Theor. Appl. Genet.*, **80**, 437–448.

Paterson, A.H., Damon, S., Hewitt, J.D., Zamir, D., Rabinowitch, H.D., Lincoln, S.E., Lander, E.S. and Tanskley, S.D. (1991), Mendelian factors underlying quantitative traits in tomato: comparison across species, generations and environments. *Genetics*, **127**, 181–197.

Simmonds, N.W. (1979), *Principles of Crop Improvement*, Longman, London.

Stebbins, G.L. (1957), *Variation and Evolution in Plants*, Columbia University Press, New York.

Tanskley, S.D., Young, N.D., Patterson, A.H. and Bonierbale, M.W. (1989), RFLP mapping in plant breeding: new tools for an old science. *Biotechnology*, **7**, 257–264.

Thoday, J.M. (1972), Disruptive selection. *Proc. R. Soc. Lond. B*, **182**, 109–143.

Part Three

Sources of Variation

3

Plant genetic resources

J.T. Esquinas-Alcázar

3.1 INTRODUCTION

Plant genetic resources represent both the basis for agricultural development and a reservoir of genetic adaptability that acts as a buffer against environmental change. The erosion of these resources threatens world food security. The need to conserve and utilize plant genetic resources as a safeguard against an unpredictable future is well recognized. The prospect of dwindling plant genetic diversity, coupled with increased demands on these resources, has propelled them into the centre of global discussions on the environment and sustainable development.

From an agricultural utilitarian standpoint, plant genetic resources can be considered limited and perishable natural resources. They provide the raw material (genes) which, when used and combined in the right way, produce new and better plant varieties, and are an irreplaceable source of such characteristics as resistance . to disease, local adaptation, and productivity. Plant genetic resources are now, and will continue to be in the future, of inestimable value independently of whether scientists use them by means of conventional plant breeding or modern genetic engineering. These genes are dispersed throughout local cultivars and natural plant populations that have been selected over thousands of years by farmers and nature for their characteristics of adaptation, resistance and/or productivity.

Plant Breeding: Principles and prospects. Edited by M.D. Hayward, N.O. Bosemark and I. Romagosa. Published in 1993 by Chapman & Hall, London. ISBN 0 412 43390 7.

In recent years the appearance of new technologies, the replacement of local varieties by imported ones, the settlement of new lands, changes in cultivation methods, etc., have been causing rapid and extreme plant genetic erosion. This affects both cultivated species and many wild ones that have a direct, indirect or potential agricultural use. The erosion of these resources could lead to the extinction of valuable material which has not yet been exploited. The road to a continuous increase in the production and quality of food lies through the protection and efficient utilization of plant genetic resources; this requires their conservation, evaluation, documentation and exchange.

3.2 PLANT GENETIC RESOURCES: THEIR IMPORTANCE AND INCREASING EROSION

The age of the earth is estimated at ca 5000 million years, and the emergence of the first traces of life on our planet goes back more than 3000 million years. The appearance of *Homo sapiens*, is a relatively recent event that took place less than a million years ago. Agriculture is a more modern phenomenon, barely 10 000 years old, which arose when humans began to cultivate wild plants having food value. This set off an evolutionary process which has created countless cultivars adapted to their local conditions, which today constitute an incalculable reserve of genetic material. Until this latest stage had been reached, the evolutionary processes on earth were only controlled by natural selection which screened the existing genetic variability produced by mutation, migration and recombination. The emergence of agriculture signalled the domestication of those species of major interest to humanity; natural selection now operated together with 'artificial' selection. As a consequence, the further evolution of these species was directed by and for man.

Although most probably, agriculture began independently in several parts of the world, the process is best known in the Near East and Central America. The first domesticated plants were cereals, legumes and other species used for their fruits or roots. These and other crops were extended and adapted up to their present ecological limits, at first being transported through migration movements, and later through the trade routes, often over long distances.

As the first crops extended into new regions, they encountered wide differences in climate, soil and other environmental factors. Natural geographical barriers frequently separated and isolated agricultural populations. The genetic variants that appeared within certain populations evolved freely and independently; some of them grew like weeds among or around the crops, leading to the creation of even more variable populations capable of tolerating extreme conditions of cold, drought, pests and diseases. Thousands of years of selection performed by farmers and nature, have produced local varieties and genotypes adapted to the different places and to cultivation practices which were determined by the climate and other environmental factors. Today, the spectrum of valuable variation is enormous and this visible variation conceals an even greater genetic diversity. To intervarietal variation must be added a wide intravarietal genetic diversity, which is the cause of the well-known morphological heterogeneity of primitive landraces. This heterogeneity reflecting local adaptation also exists for other characteristics which are not so easily observable, like disease resistance, oil and protein content, amino acid composition, etc.

Until relatively recently, as reflected above, a constant increase of diversity was favoured; however, in recent years, many factors have contributed to a drastic reversal of this tendency. Industrial development and the consequent migration of agricultural manpower to industry, combined with the increasingly marked separation between the areas of production and consumption, tends to eliminate self-sufficient agricultural production units. This adds a new dimension to the transport and marketing of agricultural products, encouraging the homogenization and standardization of cultivated varieties. Furthermore, the growing mechanization of agricultural activities and operations requires varieties having uniform characteristics in their cultivation needs, harvest periods, etc. The mechanization of post-harvest processes is also based on machines designed for universal plant and fruit models.

Following market demand, plant breeders in commercial seed firms as well as national and international institutes have joined their efforts to provide uniform, and generally more productive, new varieties to replace a vast assortment of heterogeneous, primitive landraces more adapted to the needs of earlier times. This phenomenon is occurring or has occurred in the developing or already developed countries of both the East and the West.

However, we must not forget that the heterogenous varieties of the past are the plant breeder's raw material. With them he begins his job of creating new varieties through patient and careful selection of plants that are carriers of the desired characteristics. Afterwards, through a long process of crosses and selections among the progenies, all these characteristics are combined in a uniform commercial variety. That variety, especially in autogamous plants and in vegetatively propagated species, is reproduced generation after generation and its evolution is practically nil. It can be said that it remains fixed in a mould determined by the plant breeder.

Plant improvement based on controlled crosses, and not on the simple selection of genotypes from cultivated fields, began in the eighteenth and nineteenth centuries in Europe. At the beginning of the twentieth century, many of the cultivated areas of industrialized Europe and North America were planted with varieties that had been obtained or selected by professional plant breeders. However, until the 1940s, this process had scarcely affected the regions located in warmer zones containing the greatest genetic diversity. Around 1950 intense and widespread agricultural development, largely financed by international assistance programmes, began to reduce the areas devoted to primitive local varieties, and the need to conserve the remaining genetic variability began to be recognized. That need became more obvious when, in the 1960s, millions of hectares in Asia and the Near East (where the centres of diversity of many major crops are found) were planted with commercial semi-dwarf wheat varieties, and at the same time new rice varieties were being introduced in the plains of Southeast Asia and modern cultivation methods were extending into South America and Africa.

No one can deny, however, that much of today's growing and underfed world population is dependent on the introduction of improved, high-yielding varieties and that this is a key element in the fight against hunger. In this context, the 'green revolution' permitted an enormous increase in productivity for the most important crops during the 1960s and 1970s (Table 3.1). This increase has made it possible to

Table 3.1. Changes in average worldwide yields of seven important crops (kg/ha) (from FAO Production Yearbooks)

Crop	1961–65	1969–71	1974–76	1982
Wheat	1 209	1 540	1 684	2 009
Barley	1 466	1 875	1 946	2 068
Rice	2 038	2 331	2 471	2 871
Maize	2 170	2 472	2 722	3 465
Soya	1 144	1 487	1 538	1 772
Beans	916	961	1 070	1 142
Potatoes	11 939	13 855	13 895	14 421

boost food production in the world, but at a very high cost, which may rise even higher in the future, since there is a greater energy and technological dependence on costly inputs such as fertilizers, pesticides, irrigation and the like. Moreover, many heterogeneous local varieties have been lost forever.

Primitive landraces are often able to withstand conditions that would seriously damage many modern varieties, thereby conferring greater productive stability. Their greatest value for humanity, now and in the future, lies fundamentally in the genes they contain that are not only the source of features like disease resistance, nutritional quality and adaptability to adverse environmental conditions, but also those which, although unrecognized at present, may some day be considered invaluable. Up to now the primitive varieties and related wild populations have been a fruitful, sometimes the sole, source of genes for pest and disease resistance, adaptations to difficult environments, and other agricultural traits like the dwarf type in rice, wheat and other grains, that have contributed to the green revolution in many parts of the world. With the substitution and consequent loss of a primitive variety, the genetic diversity contained in it is eliminated. To prevent such losses, samples of the replaced landraces should be adequately conserved for possible future use.

From a more theoretical standpoint, the importance of maintaining this genetic diversity is based on the relationships of the binomial variation-selection. In fact, variation is the basis of all selection. To select is to choose an alternative, and this is only possible when there are several options, in other words when variation exists. In the same way all genetic selection requires the existence of genetic variation. The greater the genetic variation existing in a population, the greater the margin of action for selection, whether natural (driving motor of evolution) or man-made (driving motor for agricultural improvement). Fisher (1930) was the first to quantify and demonstrate these principles mathematically. Ayala (1963, 1968) proved experimentally the positive correlation between the quantity of genetic variation and the rate at which evolutionary changes occur. In the same way a dependency can be established between the effectiveness of the genetic improvement by breeding of a population and the genetic variation for the trait intended to be improved in that population.

Many experimental examples of the successful use of natural variability of populations were already available in the 1950s and 1960s (Brewbaker, 1964; Falconer, 1964; Lerner, 1968 and others). Sometimes the selection has been made in two opposite directions (divergent selection) (see Chapter 2) from the same population. More recent studies have confirmed that in maize and other crops extraordinary genetic gains for certain traits can be obtained by selection in limited populations (Chapter 2). It could be argued that these results seem to contradict the need to preserve genetic resources, but it should be noted that they have been obtained from cultivars, heterogenous for specific quantitative traits, in highly allogamic species. Furthermore, the gains observed vary when the material is grown in different agroecological conditions where local varieties often have comparative advantages.

The tendency to eliminate the genetic diversity contained in primitive landraces of plants jeopardizes the possible development of future varieties adapted to tomorrow's unforeseeable needs. Some examples involving problems that can endanger the national economies of the affected countries may better illustrate the importance of the preceding remarks.

One of the most tragic cases on record is the potato famine of the mid-nineteenth century, when more than two million Irish people starved to death as a consequence of a massive attack of blight (*Phytophthora infestans*) that destroyed the potato crops. The potato had been the main staple of the Irish diet during the preceding centuries. The underlying cause of the catastrophe was the narrow genetic base of the potato plants in that country. They all originated from a small quantity of uniform material brought from Latin America in the sixteenth century (Hawkes, 1979).

Tarwi (*Lupinus mutabilis*) is one of the Andean crops which has formed the staple diet of the farmers of Inca origin for thousands of years. Because of its high protein content, tarwi is indispensable to the local diet. The local varieties cultivated for family home consumption have been selected by farmers over many generations for the quantity and quality of their proteins (as much as 40%). Although of less interest to the farmer, tarwi also has a high fat content (as much as 26%). There is however, a

negative correlation between productivity and oil content in tarwi seeds. In 1977, in an effort to industrialize the crop, an experimental factory for the extraction of tarwi oil was established about 80 km south of Lima. The commercial production of new varieties of this crop, offering better characteristics for oil extraction was encouraged, and the farmers consequently replaced their seeds, which were usually highly varied and rich in protein, with the new, uniform oil-rich/protein-poor selections. The experiment failed and the factory was closed in 1979; the farmers found themselves with no alternative but to use the new varieties as food. In many cases, two years were enough to lose permanently some of the germplasm which had been selected for human consumption over thousands of years. Fortunately, some of the potential loss was avoided because samples of local varieties had previously been collected and kept viable through storage (M. Tapia, Cuzco, personal communication).

Other examples are surfacing with increasing frequency throughout the world, and in many cases germplasm has been lost (Harlan, 1975). This increase is a consequence of the dominance in national and international markets of a small group of uniform varieties for each species which have dramatically narrowed the genetic base of the leading crops.

3.3 DISTRIBUTION OF PLANT GENETIC RESOURCES

The genetic variability of cultivated plants is not randomly distributed throughout the world. In the 1920s Vavilov (1926, 1951) first identified areas with similar physiographic characteristics where there was maximum variability for the major cultivated species. Studies conducted subsequently have not introduced substantial modifications in the distribution maps he proposed. Zhukovsky (1965), a close associate of Vavilov, identified 12 megagene centres of crop-plant diversity (Fig. 3.1) and a number of microgene centres of wild growing species related to our crop plants, where the cultigens first originated, were also demarcated. Zeven and Zhukovsky (1975) have further elaborated

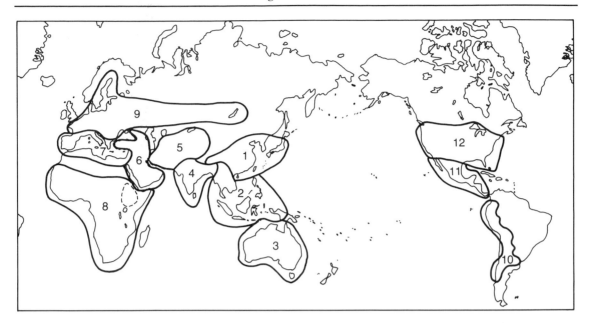

Fig. 3.1. The twelve megacentres of cultivated plants (Zeven and Zhukovsky, 1975).

on this in the *Dictionary of Cultivated Plants and Their Centres of Diversity*, listing species for different megagene centres and the range and extent of distribution of its diversity. Table 3.2 lists the main cultivated plants for which much diversity appears in these 12 regions.

3.4 CLASSIFICATION OF PLANT GENETIC RESOURCES

The plant genetic resources used or potentially usable by man can be grouped in cultivated and wild species.

3.4.1 Cultivated species

1. Commercial varieties: these are the standardized and commercialized varieties or cultivars which in general have been obtained by professional plant breeders. Many of them are characterized by high productivity when subjected to intensive cultivation systems requiring heavy investments (of fertilizers, irrigation, pesticides, etc.) and most by a uniformity which gives them high genetic vulnerability.

2. Landraces or traditional local varieties: these are primitive varieties or cultivars which have evolved over centuries or even millennia, and have been influenced decisively by migrations and both natural and artificial selection. There is a large diversity between and within these varieties which, being adapted to survive in unfavourable conditions, have low but stable levels of production and are therefore characteristic of subsistence agriculture.

3. Breeding lines: this is the material obtained by plant breeders as intermediate products. These lines usually have a narrow genetic base because, in general, they have originated from a small number of varieties or populations.

4. Special genetic stocks: this category includes other genetic combinations such as the genic, chromosomal and genomic mutants, naturally or artificially produced, and in most cases conserved in plant breeders' collections. This material can be of value in itself or as an instrument in the hands of the plant breeder. An example of the latter case can be seen in marker genes and trisomic or monosomic individuals.

Among cultivated species the 'commercial varieties' category, including obsolete ones, has been, and continues to be, used to obtain new varieties. It constitutes a valuable genetic patrimony, but one that is rather limited due to uniformity and the close interrelationship among them. Generally commercial varieties are well represented in existing collections.

In recent years, the greatest attention has been given to 'traditional local varieties' for two reasons: the abundance of potentially useful genetic variation they contain in already co-adapted gene complexes, and the speed with which they are disappearing when replaced by commercial varieties. With a few exceptions, it is an understatement to say that these varieties are not adequately represented in existing collections today. This is because: (a) in many collections, more importance is given to pure lines and selected material; (b) many of the

Table 3.2.   Cultivated plants and their regions of diversity. Based on Zeven and Zhukovsky (1975) and Zeven and de Wet (1982)

1   Chinese–Japanese Region
   – Prosomillet, Fox tail millet, Naked oat
   – Soybean, Adzuki bean
   – Leafy mustard
   – Orange/*Citrus*, Peach, Apricot, Litchi
   – Bamboo, Ramie, Tung oil tree, Tea

2   Indochinese–Indonesian Region
   – Rice
   – Rice bean, Winged bean
   – Cucurbits/Ash gourd
   – Mango, Banana, Rambutan, Durian, Bread fruit, *Citrus*/Lime, Grapefruit
   – Bamboos, Nutmeg, Clove, Sago-palm, Ginger, Taros and Yams, Betel nut, Coconut

3   Australian Region
   – *Eucalyptus, Acacia, Macadamia* nut

4   Hindustani Region
   – Rice, Little millet
   – Black gram, Green gram, Moth bean, Rice bean, *Dolichos* bean, Pigeonpea, Cowpea, Chickpea, Horse gram, Jute
   – Eggplant, Okra, Cucumber, Leafy mustard, Rat's tail radish, Taros and Yams
   – *Citrus*, Banana, Mango, Sunnhemp, Tree cotton
   – Sesame, Ginger, Turmeric, Cardamom, Arecanut, Sugarcane, Black pepper, Indigo

5   Central Asian Region
   – Wheat (Bread/Club/Shot), Rye
   – *Allium*/Onion, Garlic, Spinach, Peas, Beetroot, Faba bean
   – Lentil, Chickpea
   – Apricot, Plum, Pear, Apple, Walnut, Almond, Pistachio, Melon, Grape, Carrot, Radish
   – Hemp/*Cannabis*, Sesame, Flax, Safflower

6   Near Eastern Region
   – Wheat (Einkorn, Durum, Poulard, Bread), Barley, Rye/*Secale*
   – Faba bean, Chickpea, French bean, Lentil, Pea
   – *Brassica oleracea, Allium*, Melon, Grape, Plum, Pear, Apple, Apricot, Pistachio, Fig, Pomegranate, Almond
   – Safflower, Sesame, Flax
   – Lupins, Medics

Table 3.2 continued

7 Mediterranean Region
   – Wheat (Durum, Turgidum), Oats
   – *Brassica oleracea*, Lettuce, Beetroot, Colza
   – Faba bean, Radish
   – Olive, *Trifolium*/Berseem, Lupins, *Crocus*, Grape, Fennel, Cumin, Celery, Linseed

8 African Region
   – Wheat, (Durum, Emmer, Poulard, Bread)
   – African rice, Sorghum, Pearl millet, Finger millet, Teff
   – Cowpea, Bottle gourd, Okra, Yams, Cucumber
   – Castor bean, Sesame, Niger, Oil palm, Safflower, Flax
   – Cotton, Kenaf, Coffee
   – Kola, Bambara groundnut, Date palm, Ensete, Melons

9 European–Siberian Region
   – Peach, Pear, Plum, Apricot, Apple, Almond, Walnut, Pistachio, Cherry
   – Cannabis, Mustard (black), Chicory, Hops, Lettuce

10 South American Region
   – Potato, Sweet potato, *Xanthosoma*
   – Lima bean, Amaranth, *Chenopodium*, *Cucurbita*, Tomato, Tobacco, Lupin
   – Papaya, Pineapple
   – Groundnut, Sea island cotton
   – Cassava, Cacao, Rubber tree, Passion fruit

11 Central American and Mexican Region
   – Maize, French bean, Potato, *Cucurbita*, Pepper/Chilli, Amaranth, *Chenopodium*, Tobacco, Sisal hemp, Upland cotton

12 North American Region
   – Jerusalem artichoke, Sunflower, Plum, Raspberry, Strawberry

populations collected in the field have been subject to selection before being stored, thereby decreasing their genetic variability; (c) most collections have been maintained traditionally through periodic multiplications in small adjoining fields with a consequent genetic erosion due to hybridization, natural selection and the genetic drift characteristic of small populations.

3.4.2 Wild species

(a) For direct use

These are wild species which man uses but does not plant or cultivate. Genetic erosion in this category does not occur by chance, but select-ively, against the most valuable material. In fact within the natural populations of these species, man often selects and consumes the plants possessing the most desired characteristics. Such consumption frequently involves the destruction of either the seed or the plants before seeds have been produced, setting off a negative selection that ends with the elimination of those characters in a few generations. For example, a typical case in this category can be found in the many species of *Durio* spp. which grow in Malaysia and Indonesia and reach a height of 20 metres; the local people pull down the most productive trees in order to collect their fruits. Also, in the case of timber species, the individuals having the best habit and quality are the first to be cut. Another example is offered by sexually reproducing forage species existing in natural pastures subject to intensive exploitation; livestock eliminate the most appetizing plant specimens before seeds are produced. In these three examples drastic selection operates against plants with the most desirable characteristics.

(b) For indirect use

These are wild species related to cultivated species possessing beneficial characters that can be transferred to their cultivated relatives. In species reproducing by seeds, this transfer is carried out mainly through sexual crossings. When the affinity between the natural and the cultivated species is not enough to allow crossing, somatic hybridization or genetic engineering may be resorted to. An outstanding example of this category is genus *Lycopersicon* which comprises wild species related to the tomato. The following species, which generally cross successfully with the cultivated tomato (*L. esculentum*), have been used as donors of genes: for fungus resistance (*L. hirsutum, L. pimpinellifolium* and *L. peruvianum*); virus resistance (*L. chilense* and *L. peruvianum*); nematode resistance (*L. peruvianum*); insect resistance (*L. hirsutum*); quality (*L. chmielewskii*); adaptation to adverse environments (*L. cheesmanii*), etc. (Esquinas-Alcázar, 1981a). Other noteworthy examples are provided by wild species related to potato, rice, sunflower, sugarcane, etc. which have successfully solved various problems of these crops.

In vegetatively reproduced species, the wild relatives of cultivated plants, which are generally hardier, can sometimes be used as rootstock for grafting. Through this system, the crop can be extended to marginal areas having adverse soil and climatic conditions; areas that would not be tolerated by the cultivated species. For example, *Poncirus trifoliata*, used as a rootstock for several cultivated citrus plants, provides cold resistance. By this method it is also possible to prevent certain infectious root diseases, to which some cultivated plants are vulnerable whereas some of their related wild species are resistant. This is the case with *Vitis labrusca* and some other American wild vines which are used as rootstocks for the cultivated vine (*V. vinifera*) to combat *Philoxera vastatrix*.

(c) Potentially utilizable

These species which are not used today have characteristics or composition which make their use in the future probable. This is the case for some quick-growing wild species for which the present energy crisis has opened up wide prospects of utilization as profitable agro-energy producers. This category also includes many wild species for which analysis in pharmaceutical laboratories has revealed contents of certain medicinal substances which are higher than in species traditionally used to obtain these products.

Little attention has been given to the preservation of directly, indirectly or potentially useful 'wild species' since it was believed that they were less threatened than cultivated species. Unfortunately, in many cases, this is no longer true, and factors, such as settlement of new lands, intensive deforestation, desertification, the use of weed-killers, the increase of harmful substances in the atmosphere, etc., are drastically changing the ecological balances in some areas thereby endangering the genetic diversity and even the very existence of many useful or utilizable wild species.

3.5 CONSERVATION OF PLANT GENETIC RESOURCES

The conservation of genetic resources goes far beyond the preservation of a species. The objective must be to conserve sufficient diversity within each species to ensure that its genetic potential will be fully available in the future. Resistance to the 'Grassy Stunt' rice virus, for example, has been found in only one population of *Oryza nivara*.

Plant genetic resources can be conserved *in situ* as well as *ex situ*; and the two systems should be considered complementary, not antagonistic. *In situ* conservation consists in the legal protection of the area and habitat in which the species grows. This is the preferred technique for wild plants. The great advantage is that the evolutionary dynamics of the species are maintained. The principal drawback is the cost, and the social and political difficulties which occasionally arise. This system can be considered cost-effective however, if the important point is to conserve all species, rather than just a few, within a specific area. *Ex situ* conservation implies the collection of representative samples of the genetic variability of a population or cultivar, and their maintenance in germplasm banks or botanical gardens as seeds, shoots, *in vitro* culture, plants, etc. *Ex situ* conservation is mainly used for cultivated plants multiplied by seed. The major advantage is the control of the material in a small space under intensive care. Another advantage is that the material is easily accessible to plant breeders. The major drawback is that the germplasm ceases to evolve, and the natural processes of selection and continuous adaptation to local habitat are halted. Further drawbacks are genetic drift (random loss of diversity due to the fact that the samples collected and multiplied are necessarily very small) and selection pressure (the material is usually multiplied in phytoecogeographical areas different from those where it was collected). Both phenomena, genetic drift and selection pressure, produce a cumulative genetic erosion which may sometimes exceed the genetic erosion actually taking place in the field.

In the current context of plant breeding, emphasis is placed on the *ex situ* conservation, therefore, although it is extremely important, *in situ* conservation and management will not be discussed further here. Activities related to *ex situ* conservation of plant resources include collection, maintenance, multiplication, evaluation and exchange.

3.5.1 Collection

For most species the material to be collected consists of seeds, although in some cases it may be bulbs, tubers, cuttings, whole plants, pollen grains or even *in vitro* tissues depending on the characteristics of the species and the manner in which the material is to be conserved. The collecting team should have adequate know-ledge of botany, ecology, population genetics, plant breeding and plant pathology. It is also important for collectors to be familiar with the species to be collected and to have a good knowledge of the country or region where the expedition is conducted, including socioecological and cultural aspects of the farming societies. Good knowledge of the plant habit and breeding system will allow a better genetic interpretation of the diversity observed. Tactics and procedures for collecting have been elaborately dealt with by Frankel and Bennet (1970), Hawkes (1980), Chang (1985) and Arora (1981, 1991).

Sampling is a most important aspect of the collection of material, since a sample must be representative of the population's genetic variability. The main decisions to be taken are: (a) the number of seeds to be collected from each plant; (b) the number and distribution of the plants to be collected in each site; (c) the number and distribution of the sites within a given area where collecting will be carried out.

The answers are not always the same and will depend on the specific circumstances of each case. The principles used in selecting the sites, plants and sample numbers in relation to genetic diversity have been discussed by Marshall and Brown (1975), Hawkes (1980), Chang (1985) and Arora (1991).

A breeder will usually look for 'useful' agronomic characteristics (selective sampling), whereas the population geneticist may try to collect randomly (random sampling). It should be noted here that the concept of 'usefulness' is a relative one, that may vary according to the objectives and information available to the collectors. The examples that follow are very illustrative.

One local variety of wheat from Turkey, collected by J.R. Harlan in 1948, was ignored for many years because of its many negative agricultural characteristics. It has only recently been discovered that the variety carries genes for resistance to *Puccinia striiformis*, to 35 strains of *Tilletia caries* and *T. foetida*, and to 10 varieties of *T. controversa*, and is also tolerant to certain species of *Urcocystis*, *Fusarium* and *Typhula*. It was therefore used as a source of resistance to a whole array of diseases (Myer, 1983).

Wild relatives of our present crop plants, although agronomically undesirable, may also have acquired many desirable stress-resistant characteristics as a result of their long exposure to nature's pressures. Early US collection activities for wild forms of *Beta* were conducted by George H. Coons (USDA-ARS) in 1925 and 1935. These collections were mainly wild forms of the section *Beta*. When *Rhizomania*, a devastating root disease, became important in California in the 1980s, E.D. Whitney found that several accessions of the Coons' collection were highly resistant to this disease. Interestingly, these same accessions have subsequently been found to exhibit *Erwinia* root rot resistance, sugar beet root maggot tolerance, and moderate leaf spot resistance (Doney and Whitney, 1990).

It is obvious from these two examples that the value of both cultivated and wild germplasm is not always immediately apparent. Other illustrative examples are provided by Esquinas-Alcázar (1987).

It must be kept in mind that the ultimate objectives are to collect the maximum genetic diversity and, if possible, to obtain samples that maintain the allelic frequencies of the collected populations or varieties. However, these objectives may be difficult to achieve in practice since we are necessarily collecting a limited number of individuals and seeds per population and, consequently, genetic drift may be unavoidable.

In taking samples, a number of field 'passport' data including climatic and soil characteristics, type of vegetation and diseases present in the collection site must also be noted. The data will accompany the samples in the future, and can be very useful to plant breeders. The information provided by farmers and field workers living in collecting areas can also be of unique value (see Nabhan *et al.*, 1989 and Berg *et al.*, 1991). When varieties disappear, the knowledge about those varieties also disappears. Thus genetic erosion is usually associated with a corresponding erosion of knowledge.

In vitro collecting techniques (i.e. of zygotic embryos) are being developed for some crops. This method has obvious advantages for transportation from remote areas in the case of big-seeded species or for those with recalcitrant seeds (i.e. coconut). The same technology can facilitate the safe movement of germplasm between genebanks. For more information see Whithers (1991a).

3.5.2 Maintenance

The main task of a germplasm bank is to conserve germplasm in a state in which it can be indefinitely propagated. Although scientists specialized in the field have no doubts concerning the importance of conserving populations or varieties indefinitely, independently of their current value, there is some controversy in the case of breeding lines (F_1, F_2, products of backcrossing, etc.) obtained as intermediate products of breeding programmes. Such material may be valuable, although they have a narrow genetic base because they often derive from very few varieties or populations. The enormous number of lines or samples of this material that sometimes exist raises space problems. The present tendency is for breeding lines to be maintained in the working collection of the plant breeder, reserving long-term storage for the original populations or varieties which, it is assumed, contain all genes introduced in the breeding lines.

In general, the term 'base collection' is applied to collections stored under long-term conditions, whereas the term 'active collection' is used for collections stored under medium-term conditions and 'working collection' refers to breeders' collections usually stored under short-term conditions. For safety reasons, duplicates of the base collections should be conserved in other germplasm banks.

Maintenance of the material and the collection and collation of the data on the accessions, which is necessary for their effective utilization, become increasingly demanding operations. Using the available genetic resources of a crop species and its wild relatives can be a daunting prospect when trying to make use of collections of many thousands of accessions. Frankel and Brown (1984) and Brown (1989) have developed the concept of a 'core collection'. A core collection represents, with a minimum of repetitiveness, the genetic diversity of crop species and its wild relatives. The objective in developing such a collection is to describe the genetic diversity of a crop species and its wild relatives and to identify a subset of accessions which contains, or is believed to contain, most of the known genetic diversity. This subset of the whole collection provides potential users with a large amount of the available genetic variation of a crop genepool in a workable number of accessions. It can therefore be used by plant breeders seeking new characters which require screening techniques not possible with large collections. Because each of the accessions in a core collection is, to some extent, representative of a number of accessions (from a particular area of the world or with some shared characters), the core can also be used as a point of entry to the available collections of a crop. It is important to emphasize that the core collection should not replace existing collections.

The core collection concept has aroused considerable worldwide interest and debate within the plant genetic resources community. It has been welcomed as a way of making existing collections more accessible through the development of a small group of accessions that would be the focus of evaluation and use and provide an entry point to the large collections which it aims to represent. However, the concern that still remains is that the available knowledge of genetic diversity in any crop is insufficient to enable a meaningful core to be developed and that the most useful characters often occur at such a low frequency that they would be omitted from any small core collection. Additional information on the core collection concept can be found in Hodgkin, 1991.

The conservation requirements and treatment to be applied for storage vary considerably depending upon whether seed or vegetatively-propagated species are involved.

(a) Seed propagated species

It has been established by Harrington (1963, 1972) that a predictable increase in longevity in most seeds of cultivated plants can be induced by decreasing their storage temperature and humidity. Roberts (1973) coined the names 'orthodox' for seeds having this type of behaviour together with tolerating a sizeable decrease in moisture content and 'recalcitrant' for those unable to tolerate such treatment because reduction of internal moisture provokes irreversible chemical processes that kill the seeds (Chin and Roberts, 1980).

Safe long-term storage of 'orthodox' seeds requires careful control of the environment in which the seeds are kept. Seed moisture content (MC) is the most important factor affecting seed storage life. In general, when the seed MC decreases from 12 to 5%, seed storage life is increased (Zhang and Tao, 1988). In order to maintain the MC at constant levels during storage, seeds should be kept in hermetically sealed and moisture-proof containers after having been properly dried (Tao, 1985; Cromarty *et al.*, 1990). According to Roberts and Ellis (1977) high quality 'orthodox' seeds stored under these conditions in rooms at about $-18°C$ can maintain good viability for a century or longer. For medium-term storage the International Board for Plant Genetic Resources (IBPGR) advises storage of seeds at 5% MC at a temperature below 15°C.

The period of viability of 'recalcitrant' seeds varies between two weeks and several months. Among plants producing this type of seed, there are some of major economic value, i.e. cocoa, coconut, rubber, etc. Species such as coffee and oil palm are on the borderline and can be considered as 'sub-orthodox' with a period of germinability of no more than two years. It is therefore important to find alternative techniques that can lengthen their conservation period.

Other conservation techniques which are still at an experimental stage are: (a) maintenance of seeds at ultralow temperatures (cryopreservation), e.g. in liquid nitrogen ($-196°C$) (Sakai and Noshiro, 1975); (b) conservation of pollen grains; (c) storage of *in vitro* cultures derived from zygotic embryos or vegetative tissues, which can be combined with cryopreservation for long-term storage; (d) other methods described below can also be applied to seed propagated species.

A general problem in seed conservation is the minimum size of samples to be conserved. This depends on such factors as allelic frequencies and the degree of heterozygosity of the population, which are greatly influenced by the breeding system. In polymorphic populations, the probability of including all the alleles present in the population increases with the size of the sample. Normally, available space and the possibilities of managing the samples, rather than other theoretical considerations, impose the main size limitations.

(b) Vegetatively propagated species

The conservation of vegetatively propagated species, such as fruit trees, artichokes, strawberries, potato, cassava, *Musa*, etc. deserves separate mention. A characteristic of these cultivated species is that for centuries, nature and the farmer have selected not only genes, but also the most 'suitable' allelic combinations in very heterozygous genotypes which are maintained through the vegetative propagation. It is these genotypes, and not only the genes constituting them, that concerns us, and therefore it is not enough to keep the seeds which, when coming from heterozygous families, would produce high segregation with the appearance of new genotypes and the consequent loss of the initial genotype. Some of the solutions adopted to solve this problem are:

1. Maintenance of vegetatively growing collections in the field as in the case of an arboretum, a field genebank, a botanical garden or a natural reserve.
2. Maintenance of cuttings, bulbs or tubers under controlled humidity and temperature conditions. However, this method is only practical for the short and medium-term or used in conjunction with a field genebank.
3. Use of *in vitro* tissue culture techniques. These techniques, although in their early stages for most species, are developing rapidly. *In vitro* storage involves the establishment of cultures, preferably of the type that carries a minimal risk of somaclonal variation and their transfer to slow growth conditions or cryopreservation. *In vitro* conservation using meristems and shoot tip cultures under minimum media, low

temperatures and low light intensity provides adequate slow growth methods for medium-term germplasm conservation. These techniques are already available for many species. Such methods are being used systematically in the programme for conservation and exchange of potato and sweet potato germplasm in the Centro Internacional de la Papa (CIP) in Peru. A similar programme for cassava germplasm is operational in the Centro Internacional de Agricultura Tropical (CIAT) in Colombia and in the International Institute for Tropical Agriculture (IITA) in Nigeria. Cryopreservation in liquid nitrogen for long-term storage is at a more experimental stage. However, the technique has great potential and, according to Chandel and Pandey (1991) more than 50 species have been successfully freeze-preserved and more than half of them can be cryostored in the form of cell cultures. For more information on *in vitro* techniques and cryopreservation, see Withers (1991b,c).

3.5.3 Rejuvenation and multiplication

The loss of the germination capacity of stored seeds necessitates their periodic rejuvenation. The frequency with which this must be done obviously varies according to species and in some species, such as lettuce and onion, it can also depend considerably on the variety involved. In order to determine the maximum rejuvenation periods, germination tests are indispensable. As the seed ages, before losing germination capacity, mutations increase and if rejuvenation of the material does not take place soon enough, the genetic structure of the population can vary. Seed multiplication may also be necessary in order to meet the demand by plant breeders for samples. Techniques for multiplying or rejuvenating populations, in order to avoid genetic contamination, must take into account the reproductive characteristics of each species, particularly the outcrossing rate. The multiplication site should have ecological characteristics similar to those where the material was collected, in order to prevent selection that can change the allelic frequencies, even eliminating those alleles most sensitive to certain soil-climatic factors. The choice of site is particularly important for species with requirements for specific photoperiod, heat integral or vernalization. Decisions on the number of seeds needed and the choice of the appropriate technique and site for multiplication or rejuvenation should be taken in each case, after appropriate tests, in order to reduce the loss of genetic variability to a minimum. Finally, it is extremely important to take advantage of the process of rejuvenation or multiplication of the material in order to eliminate viral or other infectious diseases. *In vitro* (i.e. meristem-tip culture) techniques can be very useful for this purpose.

3.5.4 Evaluation

The preceding discussions concerned collection methods and maintenance of plant genetic resources, whereby their loss is avoided. However, to be able to use plant genetic resources with maximum efficacy, stored genetic variability must be known; in other words it must be evaluated. The evaluation of a population starts at the moment of collection and never ends. Whereas collection and conservation time is limited because of genetic erosion, the time for evaluation is much more flexible. It can be accomplished more or less rapidly, depending on needs and/or available resources. It can deal with one or several possible aspects, i.e. agronomic, pathological, morphological, biochemical, cytological, etc., and be carried out at various stages. Nevertheless, the sooner and more completely it can be done, the more useful it will be and the sooner the initial investment will pay off. All available data, from those compiled by the collector to those emerging from the latest evaluation, can be used as traits to describe a population. All the data can also help toward detecting duplications and differences among the conserved samples.

The term 'descriptor' is used increasingly often in referring to each of those characters considered important and/or useful in the description of a population. Descriptors differ according to species as to whether they have been selected by plant breeders, botanists, geneticists or experts in other disciplines. Plant breeders tend to choose descriptors of agronomic importance, useful for plant improvements which generally have polygenic regulation; botanists choose morphological characters independent of their genetic regulation; whereas geneticists try to choose qualitative monogenic traits with discriminatory power. Naturally the degree of usefulness of any descriptor depends on the objectives. Today, there is a tendency toward accepting compromise solutions through the selection of a minimum number of universally accepted descriptors that can facilitate the exchange of information and material. The recent development of techniques such as electrophoresis enables us to know the alloenzymatic composition of individuals (zymotypes) and genetic distances between populations (Esquinas-Alcázar, 1981b, 1986) and allows us, for uniform populations and varieties to determine their 'fingerprints' (see Chapter 12 and Brown, 1978).

The term 'pre-breeding' is often used to designate the phase between evaluation and breeding in the strict sense. Many programmes which aim to facilitate the utilization of plant germplasm include the process of pre-breeding, also called 'development breeding' or 'germplasm enhancement'. Though the end products of pre-breeding are usually deficient in certain desirable characters, they are attractive to plant breeders due to their greater potential for direct utilization in a breeding programme when compared to the original sources.

3.5.5 Documentation

A good documentation system is the key to the effective utilization of the material deposited in a germplasm bank. The use of a central computer to store the data obtained during the collection and evaluation of populations enables plant breeders, having access by terminal to the computer network, to obtain almost immediately all the information available on each of the samples existing in a large number of germplasm banks. For several years, constructing one ideal and universally accepted documentation system, that would allow a rapid and efficient exchange of information within a worldwide germplasm bank network, was considered necessary. However, the different realities and needs of each germplasm centre, combined with the fact that documentation systems are soon obsolete, forced a change of attitude: the fact that a diversity of systems is necessary and unavoidable became accepted. On the other hand, rapid technological development in this field now allows the exchange of data between systems which previously were incompatible. At present there are many genetic resources documentation systems in the world, and sometimes the same centre even has a different system for different crops. None of this makes impossible the exchange of data between centres provided there are agreements for the utilization of the same descriptors.

3.5.6 Exchange

The purpose of the entire process described above is the utilization of appropriately documented samples and related technologies by plant breeders. The exchange of the existing material documented with available information and of the technology needed for its efficient utilization should be the ultimate goal of germplasm banks and related data banks. Such exchanges, however, cannot be achieved without the consent and/or agreement of the parties involved and often require international co-operation and agreements (section 3.6). The exchange of material also requires adequate inspection and testing services as well as quarantine facilities that can reduce to a minimum the risks of spreading pests and diseases (Hewitt and Chiarappa, 1977 and the FAO/IBPGR series for Technical Guidelines for the Safe Movement of Germplasm). *In vitro* techniques for transfer of germplasm can also reduce these risks and are widely used for some crops (i.e. potato and sweet potato in CIP and cassava in CIAT and IITA).

3.6 INTERNATIONAL CO-OPERATION AND AGREEMENTS

Owners and users of germplasm are not necessarily the same. In addition, most of the existing genetic diversity is to be found in the tropical and subtropical areas where a large number of developing countries are located, whereas most modern technologies and economic capacity to utilize germplasm are to be found in the industrialized countries of the North (Fowler and Mooney, 1990). No country or region can be self-sufficient in its needs for genetic diversity; according to recent studies, the average crop genetic resources dependency (crop production relying on exotic germplasm) among regions of the world is more than 50%, and for some regions it may go up to 100% for the most important crops (Kloppenburg, 1988). Local information and traditional technologies developed by farmers and farming communities for their specific varieties and species are also of utmost importance for a better understanding and wise utilization of this germplasm (Berg *et al.*, 1991 and Nabhan, 1989). This is therefore an area in which all countries are at the same time donors and recipients and where international co-operation is of vital importance.

Most of the activities described in the previous sections of this chapter can only be properly carried out through co-operation among countries and regions. The collection of germplasm, as well as its rejuvenation, characterization and evaluation in appropriate agroecological conditions can only be done through international co-operation and adequate agreements among the countries involved. Exchanges of germplasm and related information and technology should ideally involve all countries of the world.

Since the 1950s a number of agencies and organizations, especially the Food and Agriculture Organization (FAO), of the United Nations, have promoted and supported this kind of co-operation. In order to overcome a number of technical problems, FAO hosted in 1967, 1973 and 1981 three International Technical Conferences; each Conference led to the publication of a book summarizing the state of knowledge and technical advances (Frankel and Bennet, 1970; Frankel and Hawkes, 1975; Holden and Williams, 1984).

The non-governmental International Agricultural Research Centres (IARCs) established under the aegis of the Consultative Group of International Agriculture Research (CGIAR) in the 1970s have also promoted and facilitated international technical co-operation for the genetic resources of crops under their mandate, i.e. IRRI for rice, CIAT for *Phaseolus*, cassava and fodder plants for acid soils, CIP for potatoes and sweet potatoes, CIMMYT for wheat and maize, ICRISAT for sorghum, millet, *Cajanus*, chickpeas, and groundnuts, ICARDA for barley, chickpea, lentil, faba beans and some forages, and IITA for cowpea, African rice, maize and some root crops. One of these research centres, the International Board for Plant Genetic Resources (IBPGR), established in FAO in 1974, was given a specific mandate to promote international co-operation on the conservation of plant genetic resources. Currently, IBPGR is being transformed into the International Plant Genetic Resources Institute (IPGRI), a fully independent institute that will work in close co-operation with FAO and other IARCs.

However, questions regarding international co-operation are not only technical. The last few years have seen a growing acknowledgement of the greatly increased value of plant germplasm, due to the fact that rapid genetic erosion has shown that germplasm is not an unlimited or replenishable resource. This has already resulted in a number of formal or practical restrictions on the availability of germplasm. In addition, questions such as the safety of the material, the ownership of collections and the development of national laws restricting the exportation of certain species or protecting intellectual property rights for new varieties, are the subject of continuing debate. Since the relative value of plant genetic resources will continue to grow rapidly, it has become clear that plant germplasm needs to be protected for the use of future generations, and its availability for scientific purposes ensured through equitable international agreements and regulations that guarantee governmental commitments.

The Commission on Plant Genetic Resources was established in FAO in 1983 to provide the intergovernmental forum that was needed to negotiate, develop and monitor this kind of international agreements and regulations at a global level. To this date, 114 countries are members of the Commission. International agreements on PGR of agricultural interest include:

1. 'The International Undertaking on Plant Genetic Resources', approved in FAO in 1983 is a non-binding formal agreement based on the principle that PGR are a heritage of mankind and should be available without restriction for plant breeding and scientific purposes. It also defines the responsibility of countries and provides the framework for international co-operation in this field. Within the context of the Undertaking, FAO and IBPGR are developing an international network of base collections in genebanks under the auspices of FAO that guarantee unrestricted access to the germplasm therein contained. Up to date, 107 countries have adhered to the principles and articles of the Undertaking. Annexes to the undertaking, unanimously approved by all member countries of FAO, recognize the rights of both donors of germplasm and technology to be compensated for their contribution. They define Farmers' Rights as 'rights arising from past, present and future contributions of farmers in conserving, improving and making available plant genetic resources'.

2. An International 'Code of Conduct for Collecting and Transfer of PGR' has also been developed and negotiated through the Commission on PGR. The Code aims to promote rational collection and safe transfer of germplasm and sets out procedures which collectors and governments should follow to request or to issue licenses for collecting missions (FAO, 1991a).

In order to promote and monitor systematic co-operation and co-ordination of PGR activities at a global level, as envisaged in the international agreements mentioned above, the Commission on PGR has developed a compre-

hensive Global System for the Conservation
and Utilization of PGR (FAO, 1991b). Currently, 132 countries are part of the Global
System. Within that context global information
and early warning mechanisms, as well as
networks for *in situ* conservation areas and *ex
situ* collections and periodical reports on the
State of the World's PGR are being developed
by FAO in close co-operation with IBPGR.

3.7 PROSPECTS FOR THE FUTURE

The prospects for the future are somewhat
contradictory. On the one hand, the risk and
the rate of genetic erosion are at catastrophic
proportions at intraspecific, interspecific and
even ecosystem levels. This is happening at a
time when genetic diversity is becoming essential as the basic material needed to move
towards sustainable agricultural development
and face expected climatic change and environmental stress. On the other hand, the technology available to conserve and utilize
germplasm is more powerful and promising
than ever before. The public is becoming aware
of the value of PGR and international efforts to
deal with these problems are steadily increasing. They are, however, by any standards,
technically, economically, and politically insufficient.

3.7.1 Technical aspects

In the realm of technology, the rapid development of powerful new biotechnologies has
contributed to the increase in the value of PGR,
especially for wild species. Molecular genetics,
tissue cultures, and cryopreservation techniques have greatly expanded the technological
basis for the conservation and utilization of
germplasm. Genetic engineering techniques are
overcoming the genetic barriers between different species making it possible to transfer genes
between unrelated species, genera and even
kingdoms, increasing the economic interest of
preserving and utilizing wild species as potential donors of useful agricultural traits for crops
(Chapter 10; Hobbelink, 1991; Sasson and
Costarini, 1989; FAO, 1991c).

Biotechnology also has applications in the
conservation and use of PGR through *in vitro*
techniques. Mass clonal propagation *in vitro* has
become a useful commercial technique for the
reproduction and distribution of bred material.
In vitro conservation is particularly valuable for
crops that produce recalcitrant seeds and those
that are clonally propagated. *In vitro* conservation can be considered as a package of
technologies commencing with collecting and
leading through disease indexing, disease
eradication and storage, to distribution. *In vitro*
collection is useful in overcoming problems
related to safety and transportation of germplasm gathered in the field. Long-term preservation of meristems, shoot tips, embryos (both
zygotic and somatic), embryonic meristems,
protoplast and cell suspension culture can be
successfully achieved by freezing them in liquid
nitrogen ($-196°C$) and plantlets could be established at ease. Cryopreservation of seeds,
pollen, excised embryos/embryonic meristems
and buds has also proved feasible and practical
in many cases. A new approach to cryopreservation called 'vitrification' (rapid cooling) offers
the potential to overcome problems experienced with highly organized material such as
shoot-tips. Another exciting new departure is
the involvement of artificial seed technology in
conservation work (Redenbaugh, 1990;
Withers, 1991b). The new molecular technologies (Chapter 10) promise to revolutionize some
areas of genetic resource management. Molecular markers can be of immense value in
detecting genetic variants, as well as genetic
similarities and distances and in identifying
duplicates in repositories (Chandel and
Pandey, 1991). The extraction of DNA segments
and its storage in DNA banks (or libraries) is
another promising and complementary technology for the conservation of genetic resources.

There is a need for an integrated strategy for
conservation of plant germplasm at several
levels: (a) genepools; (b) individuals; (c) tissues
and organs; (d) sequences of DNA, using in
each case appropriate technologies. In each
species and region such strategy would provide
different levels of conservation for different
needs and capacities (Swaminathan, 1983).

The development and application of user-friendly computer software to plant genetic resources has dramatically increased the capacity for storing, processing and exchanging data and information in these fields with a number of important new databases and information systems being established.

3.7.2 Economic aspects

Economic resources for *ex situ* and *in situ* conservation are well below adequate levels. In the future, the costs of *ex situ* conservation may be reduced by taking advantage of favourable environmental conditions such as natural caves in permafrost zones (i.e. Norway is establishing in Svalbard an International Seedbank under permanent natural freezing conditions) and high-altitude deserts, with their cold temperatures and for the latter with extremely low levels of humidity. These are common in some developing countries. The economic problems are especially serious in the case of wild plants that need to be conserved *in situ*, generally in developing countries. The lack of economic resources in these countries is not only an obstacle to the protection of the wild species but also a major cause of genetic erosion, as people search for fuelwood or convert virgin areas into farmland. The protection of these areas for *in situ* conservation of important species benefits all mankind, and therefore, all countries should participate in efforts to safeguard them. Recognizing that many countries rich in genetic resources are developing ones that cannot afford to bear the burden of conservation, the Resolution on Farmers' Rights (an annex to the International Undertaking on PGR considered earlier) has opened important economic perspectives. It has been agreed by the 161 member countries of FAO that these rights will be implemented through a "sustainable and substantial" International Fund "which will support plant genetic conservation and utilization programmes, particularly, but not exclusively in developing countries" (FAO, 1991d).

The cost of the necessary actions to conserve *ex situ* and *in situ* plant genetic diversity is very high, but the cost of not taking action would be much higher. It should be noted that our present economy is giving to PGR the treatment of a 'rent' that is permanently renewed, whereas we are starting to realize that it should be treated as limited capital that needs to be constantly replaced if it is going to be conserved and its depletion avoided. The 'value of use' of PGR has already been recognized, but the world economic system has not yet assigned it a 'value of exchange'.

3.7.3 Future prospects

The recognition of plant genetic resources as the common concern of mankind has been a major step forward in the status of germplasm preservation efforts. However, the channels of international co-operation which guarantee the safety of and access to these resources are still young and will need to be developed. The International Undertaking and the FAO Intergovernmental Commission are new and important efforts in the area of co-ordination and international guarantees. However, some countries have still not joined these common efforts. The network of germplasm banks under the auspices and/or jurisdiction of FAO (which endeavours to maintain copies of the most important crop species and to assure free access to germplasm regardless of political upheavals) is still embryonic and its legal structure is not adequately defined.

It is very important that national legislation takes a role in deterring genetic erosion, protecting indigenous germplasm and promoting access to genetic resources. Political and economic support for such legislation can be stimulated if the public is well informed about the importance of genetic diversity and the dangers involved in its depletion, and is encouraged to act to prevent genetic erosion. In any event, it should not be forgotten that plant genetic erosion is but one consequence of man's abusive exploitation of the planet's natural resources, which has broken the balance of many agro-ecological systems and brought on an increasing degradation of the biosphere. Nevertheless, safeguarding genetic resources by protecting them *ex situ* or *in situ* is crucial for sustainable agricultural development, food security and environmental stability. The fundamental problem remains man's lack of solidarity with the rest of nature, and any lasting solution will have to involve establishing a new relationship with our small planet, in full understanding and recognition of its limitations and fragility. If humanity is to have a future, it is imperative that children learn this in the primary schools and that adults make it part of their life.

REFERENCES

Arora, R.K. (1981), Plant genetic resources exploration and collection: planning and logistics. In *Plant Exploration and Collection*, Mehra, K.L., Arora, R.K. and Wadhi, S.R. (eds), NBPGR Sci. Monogr. 3, NBPGR, New Delhi, pp. 46–54.

Arora, R.K. (1991), Plant diversity in the Indian Gene Centre. In *Plant Genetic. Conservation and Management*, Paroda, R.S. and Arora, R.K. (eds), IBPGR/Regional Office for South and Southeast Asia, New Delhi, pp. 25–54.

Ayala, F.J. (1963), Relative fitness of populations of *Drosophila serrata* and *Drosophila birchii*. *Genetics*, 51, 527–544.

Ayala, F.J. (1968), Genotype, environment and population numbers. *Science*, 162, 1453–1459.

Berg, T., Bjørnslad, A., Fowler, C. and Skrøpa, T. (1991) *Technology Options and the Gene Struggle*. A Report to the Norwegian Council for Science and Humanities (NAVF), Oslo.

Brewbaker, J.L. (1964), *Agriculture Genetics*, Prentice-Hall, Englewood Cliffs, New Jersey.

Brown, A.H.D. (1978), Isozymes, Plant Population Genetic Structure and Genetics. *Theor. Appl. Genet.*, 52, 145–157.

Brown, A.H.D. (1989), The case for core collections. In *The Use of Plant Genetic Resources*, Brown, A.H.D., Marshall, D.R., Frankel, O.H. and Williams, J.T. (eds), Cambridge University Press, Cambridge, pp. 136–156.

Chandel, K.P.S. and Pandey, R. (1991), Plant genetic resources conservation: recent approaches. In *Plant Genetic Resources. Conservation and Management*, Paroda, R.S. and Arora, R.K. (eds), IBPGR/Regional Office for South and Southeast Asia, New Delhi, pp. 247–272.

Chang, T.T. (1985), Germplasm enhancement and utilisation. *Iowa State Journal of Research*, 54, 349–364.

Chin, H.F. and Roberts, E.H. (1980), *Recalcitrant Crop Seeds*, Tropical Press SDN, BHD, Malaysia.

Cromarty, A.S., Ellis, R.H. and Roberts, E.H. (1990), The design of seed storage facilities for genetic conservation. *Handbook for Genebanks*, No. 1, IBPGR, Rome.

Doney, D.L. and Whitney, E.D. (1990), Genetic enhancement in *Beta* for disease resistance using wild relatives. A strong case for the value of genetic conservation. *Economy Botany*, 44(4), 445–451.

Esquinas-Alcazar, J.T. (1981a), *Genetic Resources of Tomatoes and Wild Relatives*, IBPGR-FAO, Rome.

Esquinas-Alcazar, J.T. (1981b), Alloenzyme variation and relationships among Spanish land-races of *Cucumis melo* L. *Kulturpflanze*, XXIX, 337–352.

Esquinas-Alcazar, J.T. (1986), Melon's botanical varieties: alloenzyme distribution and taxonomic controversies. In *Proceedings of Symposium on Methods of Biochemical Evaluation of Germplasm Collections*, Plant Breeding and Acclimatization Institute, Radzikow, pp. 76-96.

Esquinas-Alcazar, J.T. (1987), Plant Genetic Resources: A Base for Food Security. *Review Ceres*, 118(29, 4), 39–45.

Falconer, D.S. (1964), *Introduction to Quantitative Genetics*, Oliver and Boyd, London.

FAO (1983), *International Undertaking on Plant Genetic Resource*, Doc. W/U 0672, FAO, Rome.

FAO (1991a), *Draft International Code of Conduct for Plant Germplasm Collecting and Transfer*, Doc. CPGR/91/10, FAO, Rome.

FAO (1991b), *FAO Global System for the Conservation and Utilization of Plant Genetic Resources – Progress Report and Matters for Decision by the Conference*, Doc. 91/24, FAO, Rome.

FAO (1991c), *Biotechnology and Plant Genetic Resources and Elements of a Code of Conduct for Biotechnology*, Doc. CPGR/91/12, FAO, Rome.

FAO (1991d), *Conference 26th Session. Report*, FAO, Rome.

Fisher, R.A. (1930), *The Genetical Theory of Natural Selection*, Clarendon, Oxford.

Fowler, C. and Mooney, P. (1990), *Shattering. Food, Politics and the Loss of Genetic Diversity*, The University of Arizona Press, Tucson.

Frankel, O.H. and Bennet, E. (1970), *Genetic Resources in Plants – Their Exploration and Conservation*, IBP Handbook, No. 11, Blackwell Scientific Publication, Oxford.

Frankel, O.H. and Brown, A.H.D. (1984), Current Plant Genetic Resources. A Critical Appraisal. In *Genetics: New Frontiers*, Proc. XV International Congress of Genetics, Vol. IV, Chopra, V.L., Joshi, B.C., Sharma, R.P. and Bansal, H.C. (eds),

Oxford and IBH Publishing Co., New Delhi, pp. 3–13.

Frankel, O.H. and Hawkes, J.G. (1975), *Crops Genetic Resources for Today and Tomorrow*, Cambridge University Press, Cambridge.

Harlan, J.R. (1975), Our vanished genetic resources. *Science*, **188**, 618–621.

Harrington, J.F. (1963), Practical advice and instructions on seed storage. *Proc. Int. Seed Testing Assoc.*, **28**, 989–994.

Harrington, J.F. (1972), Seed storage and longevity. In *Seed Biology*, Vol. 3, Kozlowski, T.T. (ed.), Academic Press, New York, pp. 145–245.

Hawkes, J.G. (1979), Genetic poverty of the potato in Europe. In *Proc. Conf. Broadening the Genetic Base of Crops*, Zaven, A.C. and Van Harten, A.M., PUDOC, Wageningen, pp. 19–28.

Hawkes, J.G. (1980), *Crop Genetic Resources. A Field Collection Manual*, IBPGR, Rome.

Hewitt, W.B. and Chiarappa, L. (1977), *Plant Health and Quarantine in International Transfer of Genetic Resources*, CRC Press, Cleveland, Ohio.

Hobbelink, H. (1991), *Biotechnology and the Future of World Agriculture*, Zed Books Ltd, London and New Jersey.

Hodgkin, T. (1991), The Core Collection Concept. In *Crop networks – searching for new concepts for collaborative genetic resources management*, EUCARPIA Symposium, Wageningen, 1990, IBPGR, Rome, pp 43–48.

Holden, J.H.W. and Williams, J.T. (eds) (1984), *Crop Genetic Resources: Conservation and Evaluation*, George Allen and Unwin, London.

Kloppenburg, J.R. (ed.) (1988), *Seeds and Sovereignty. The Use and Control of Plant Genetic Resources*, Duke University Press, Durham, London.

Lerner, I.M. (1968), *Heredity, Evolution and Society*, Freeman, San Francisco.

Marshall, D.R. and Brown, A.H.D. (1975), Optimum sampling strategies in genetic conservation. In *Genetic Resources for Today and Tomorrow*, Frankel, O.H. and Hawkes, J.G. (eds), Cambridge University Press, Cambridge, pp. 53–80.

Myers, N. (1983), *A Wealth of Wild Species*, Westriew Press, Colorado.

Nabhan, G.P. (1989) *Enduring Seeds: Native Plant Agriculture and Wild Plant Conservation*. North Point Press, San Francisco, CA.

Redenbaugh, K. (1990), Applications of artificial seed to tropical crops. *HortScience*, **25**, 251–255.

Roberts, E.H. (1973), Predicting the storage life of seed. *Seed Sci. Technol.*, **1**, 499–514.

Roberts, E.H. and Ellis, R.H. (1977), Prediction of seed longevity at sub-zero temperatures and genetic resources conservation. *Nature*, **268**, 431–432.

Sakai, A. and Noshiro, M. (1975), Some factors contributing to the survival of crop seeds cooled to the temperature of liquid nitrogen. In *Crop Genetic Resources for Today and Tomorrow*, Frankel, O.H. and Hawkes, J.G. (eds), Cambridge University Press, Cambridge, pp. 317–326.

Sasson, A. and Costarini, V. (eds) (1989), *Plant Biotechnologies for Developing Countries*, Technical Center for Agriculture and Rural Cooperation (CTA) and Food and Agriculture Organization of the United Nations (FAO), Wageningen.

Swaminathan, M.S. (1983), Genetic conservation: microbes to man, *Presidential Address at the V International Congress of Genetics*, New Delhi.

Tao, K.L. (1985), Standars for genebanks. *Plant Genetic Resources Newsl.*, **62**, 36–41.

Vavilov, N.I. (1926), Centers of Origin of Cultivated Plants. *Trudi po Prikl. Bot.*, **16**(2), 139–245.

Vavilov, N.I. (1951), Phytogeographic basis of plant breeding. The origin, variation, inmunity and breeding of cultivated plants. *Chronica Bot.*, **13**, 1–366.

Withers, L.A. (1991a), *In vitro* collecting: concept and background. In *In vitro Collection Techniques in the Conservation of Plant Genetic Resources*, IBPGR-CATIE Training Manual, IBPGR, Rome, CATIE, Turrialba, in press.

Withers, L.A. (1991b), Biotechnology and plant genetic resources conservation. In *Plant Genetic Resources Conservation and Management*, Paroda, R.S. and Arora, R.K. (eds), IBPGR/Regional Office for South and Southeast Asia, New Delhi, pp. 273–297.

Withers, L.A. (1991c), Maintenance of plant tissue cultures. In *Maintenance of Microorganisms*, Kirsop, B.E. and Doyle, A. (eds), Academic Press, London, in press.

Zeven, A.C. and de Wet, J.M.J. (1982), *Dictionary of Cultivated Plants and Their Regions of Diversity*, PUDOC, Wageningen.

Zeven, A.C. and Zhukovsky, P.M. (1975), *Dictionary of Cultivated Plants and Their Centres of Diversity*, PUDOC, Wageningen.

Zhang, X.Y. and Tao, K.L. (1981), Silical gel seed drying for germplasm conservation. Practical guidelines. *PGR Newsletter*, 75/76, 1–25.

Zhukovsky, P.M. (1965), Genetic and botanical irregularities in the evolution of cultivated plants. *Genetika*, **1**, 41–49 (in Russian, English summary).

4

Induced mutations

A. Micke and B. Donini

4.1 IS THERE A NEED FOR SUPPLEMENTING EXISTING GERMPLASM BY MUTAGENESIS?

Any attempt at plant improvement requires genetic variability. Ancient 'breeders' resorted to the genetic variants provided by nature and selected among them those that more suited their desire in terms of fruit or grain size, taste, ease of harvest, suitability for storage, yield, etc. Resistance to biotic attacks and tolerance to abiotic stresses were also selected. This is how domestication began more than 10000 years ago, departing from the rigid requirement of fitness dominating natural evolution, and adapting certain suitable plant species more and more to the increasingly artificial conditions of man-made habitats (FAO/IAEA, 1988).

In this process of domestication, man relied on spontaneously occurring variants coming from mutations ('sports'). Outcrossing, leading to the combination and recombination of traits, has certainly aided crop plant evolution. Selection favoured desired and eliminated undesired variants and led, in general, to a reduction of genetic variation. Conscious hybridization by man was performed rather late and only since the end of the 19th century have we understood the principles of hybridization, segregation and recombination. Varieties cultivated by farmers today comprise numerous desirable genes accumulated over centuries and may have reached an optimum plateau of a combination of good

Plant Breeding: Principles and prospects. Edited by M.D. Hayward, N.O. Bosemark and I. Romagosa. Published in 1993 by Chapman & Hall, London. ISBN 0 412 43390 7.

traits for a particular location and purpose. Cross-breeding among well adapted varieties is practised and selection of rare recombinants still gives some improvement. However, substantial progress in plant breeding during the second half of the 20th century came from the courageous crossing of farmers' varieties with exotic germplasm, where the term exotic may refer to varieties of exotic countries as well as to local landraces and weedy relatives of the crop species in question.

Although geneticists estimate the number of genes possessed by a higher plant as being in the order of 100 000, it is assumed that breeders are actually dealing only with a few hundred genes. As a consequence, when two established cultivars are crossed, the recombination that can be expected will be restricted to a relatively small fraction of the whole plant genome. Mutations, however, whether spontaneous or induced, may affect any of the 100 000 or more genes of a nuclear genome, and in addition also the ones located in cytoplasmic organelles. Mutagenesis is a unique force in creating variation, as it may alter even those genes that are common to all the varieties of a species and therefore do not show any segregation after a cross within that species ('house-keeping genes'). Therefore, when the environment (in a broad sense) imposes new fitness criteria upon a plant population, genetic variability from mutations is fundamental for the ability of the plant population to respond to the challenge (see Chapters 1 and 2). Without the capacity for undergoing mutation, the enforced selection might eventually lead to the extinction of a species. Application of mutagens can increase the mutation frequency, thus allowing the breeder to make use of mutations within the boundaries of his limited nursery and in the time span of a few plant generations.

It is important to understand the unique potential offered by induced mutations but also to realize the limitations of this tool. The first limitation is imposed by the pre-existing genome: genes that do not exist can neither be mutated nor eliminated. Another principal limitation of this tool comes from the fact that the action of present day mutagens cannot be directed to a specific gene. Mutagens affect the molecular structure of the DNA, but many induced changes in the DNA are repaired before they become manifest as mutations (gene mutations, translocations and other chromosomal aberrations) and are expressed phenotypically. Obviously, most alterations induced in a genome, having a combination of good traits, will cause disturbances in the highly sophisticated, balanced and synergistic system. However, favourable induced changes can be selected and also be incorporated, through crossing and selection, in a new improved genotype. In spite of the limitations of mutagenesis, one should conclude that future plant breeders cannot abstain from using induced mutations to supplement the existing genetic resources.

4.2 HOW TO INDUCE MUTATIONS

After Mendel showed that the elements of heredity remain intact when passing from generation to generation, researchers found it challenging to try to eliminate or modify these stable genes. Ionizing radiations, discovered around the turn of the 20th century were considered to be powerful enough, but it still took until 1926/27 to establish convincing evidence that changes caused by ionizing radiation lead to heritable alterations, 'mutations', in plants (Stadler, 1930). However, due to disappointment about the wide range of mostly useless genetic variation obtained from radi-ation treatments, researchers looked for more specific interactions between genes and mutagenic agents. Mildly effective chemical mutagens ('radiomimetic substances') were tried first, but were followed by more powerful ones ('super-mutagens') developed under the auspices of chemical/biological warfare. However, these chemical mutagens were also not able to fulfil the unrealistic hope for 'directed' mutagenesis (Röbbelen, 1959; FAO/IAEA, 1965).

4.2.1 The choice of the mutagen and dose

Today the plant breeder has at hand a number of effective physical and chemical mutagens. The *Manual on Mutation Breeding* (FAO/IAEA, 1977) contains a detailed discussion of aspects relevant for the choice of a mutagen. When properly applied, all of them will induce genetic variation. The choice of the mutagen is not really related to its effectiveness in terms of frequency of desired mutations but to the kind of material to be treated and to the availability of a mutagen. It is known that the frequency and the spectrum of mutations differ somewhat depending on the mutagen used and the dose applied. The physical mutagens, X-rays and gamma-rays (sparsely ionizing radiations), are widely used. They have the advantage of good penetration and precise dosimetry. Ultraviolet light (UV) has low penetration power and can be effectively used with materials such as pollen or *in vitro* cultured cells in a thin layer. Densely ionizing radiation, such as thermal and fast neutrons, cause more chromosomal aberrations. Chemical mutagens are known to produce a higher rate of gene mutations which is generally preferred. However, chemical mutagens present particular problems such as uncertain penetration to the relevant target cells, poor reproducibility, persistence of the mutagen or its metabolites in the treated material and finally the risk of safe handling. Tables 4.1 and 4.2 give examples of mutagens and doses used by plant breeders.

Table 4.1.   Examples of mutagens and doses used for mutation breeding in seed propagated crop plants

Species	Material	Recommended treatment	
		Mutagen	Dose or concentration
Arachis hypogaea peanut	dry seed	gamma-rays	20–30 krad
Capsicum annuum pepper	gametes	gamma-rays	750 rad
	dry seed	fast neutrons	2.4 krad
	dry seed	gamma-rays	14–22 krad
Cicer arietinum chickpea	dry seed	X-rays	10–16 krad
Glycine max soybean	dry seed	gamma-rays	10–20 krad
Phaseolus vulgaris french bean	dry seed	gamma-rays	8–14 krad
Lycopersicon esculentum tomato	dry seed	EMS	0.8% 24 h, 24°C
Lupinus albus lupine	dry seed	gamma-rays	16–28 krad
Pisum sativum pea	dry seed	DES	2‰, 15 h, 20°C
		X-rays	10 krad
	pollen	gamma-rays	600 rad
Avena sativa oat	dry seed	X-rays	12–24 krad
Hordeum vulgare barley	dry seed	X-rays	10–22 krad
	pollen	gamma-rays	0.6–2.0 krad
Oryza sativa rice	dry seed	X-rays	14–28 krad
Triticum aestivum bread wheat	dry seed	gamma-rays	10–25 krad
Triticum durum durum wheat	dry seed	gamma-rays	10–25 krad
		fast neutrons	600–800 rad
		EMS	3.76%
	pollen	gamma-rays	0.75–3.00 krad
Zea mays corn	dry seed	gamma-rays	14–28 krad

4.2.2 Starting material

A solid knowledge of the genetics of relevant traits as well as familiarity with the available germplasm would be helpful in the choice of the genotype and setting the optimal strategy of mutation breeding. Usually the advice given is to take the best varieties, outstanding in their agronomical fitness but requiring the least improvement. In some instances F_1 seeds are used; the advantage of this choice resides in the fact that two different genomes are exposed to a mutagen at the same time, which may increase gene recombination and reveals a wider spectrum of induced mutations.

In seed propagated plants, the physical or chemical treatment of seed is the most convenient and practical (Fig. 4.1). However, seed

Table 4.2. Examples of mutagens and doses used for mutation breeding in vegetatively propagated crop plants

Species	Material, explant	Recommended treatment	
		Mutagen	Dose or concentration
Solanum tuberosum potato	tubers	EMS	100–500 ppm 4 h, 25°C
	leaf rachis and leaf petioles (*in vitro*)	X-rays	1.5–2.0 krad
	leaflet-blade (*in vitro*)	X-rays	2.25–2.75 krad
	shoot tip (*in vitro*)	gamma-rays	2.5–3.5 krad
Chrysanthemum	pedicel segments (*in vitro*)	X-rays	800 rad
Begonia	leaf blades (*in vitro*)	X-rays	1.5–2.0 krad
Dianthus caryophyllus carnation	nodal stems	X-rays	1.5–2.0 krad
Corylus avellana hazelnut	dormant buds	gamma-rays	7.0–8.0 krad
Malus pumila apple	dormant graftwood	gamma-rays	6.0–7.0 krad
	bud dormant shoots	gamma-rays	2.5–5.0 krad
Olea europea olive tree	bud dormant shoots	gamma-rays	3.0–4.0 krad
Prunus cerasus sour cherry	dormant buds	gamma-rays	2.0–5.0 krad
Prunus avium sweet cherry	dormant buds	gamma-rays	3.5–4.5 krad
	pollen mother cells	X-rays	0.8 krad
Citrus sinensis orange	ovular callus (*in vitro*)	gamma-rays	8.0–16.0 krad
Vitis vinifera grape	dormant buds	gamma-rays or X-rays	2.5–3.5 krad
	dormant buds	EMS	0.15–0.20% 15 h, 20°C
Musa spp. banana	shoot tips (*in vitro*)	gamma-rays	1.0–2.5 krad
Saccharum sp. sugarcane	single-budded sets	gamma-rays	3.5–7.0 krad
Cynodon dactylon turf grass	dormant rhizomes	gamma-rays	7.0–9.0 krad
Ipomoea batatas sweet potato	cuttings	gamma-rays	20 krad
		EI	0.5%, 3 h
	shoot tips (*in vitro*)	gamma-rays	3.0–5.0 krad

is a multicellular organized structure, in which a mutation, induced in a single cell, will give rise to a chimeric plant (M_1). Competition between meristematic cells occurs and persist-ence of mutated cells in the meristem will depend on the inhibition caused by the mutagen treatment. A true gene mutation, which will be in its heterozygous state and does not affect cell primary functions such as DNA replication and cell growth, would not be at a disadvantage, while gross chromosomal changes, e.g. translocations, deletions, inversions, may strongly affect cell competition and cause 'diplontic selection'. Problems arising from such 'diplontic selection' are relatively more acute in diploids than in polyploids, since the latter, due to the presence of multiple copies at the genic as well as chromosomal level tend to tolerate chromosomal aberrations. A mutation, to have a chance for transmission to the subsequent generation, must be induced in a cell and cell lineage able to compete successfully during the ontogenesis of the plant and to participate in the formation of the sporogenic tissue from which the gametes originate.

However, there is further risk of elimination of mutations during the haplophase (e.g. pollen tube competition) and of course in the M_2 generation, starting with the zygote and pro-embryo development, where mutated genes could already be homozygous. Mutagen treatment of gametes (male or female, or both) or of zygotes, prior to the first zygotic division, leads to non-chimeric plants heterozygous for the induced mutation (Fig. 4.1).

In vegetatively propagated plants, which are usually highly heterozygous, crossing or selfing would cause wide segregation with recombinants suffering from the loss of outstanding characteristics of the original variety. For these crops the choice of mutation breeding can be viewed as the method to induce changes in one or few traits, but otherwise retaining the valuable cultivar unchanged. The material for treatment should come from a disease-free plant. The system of mutation breeding of vegetatively propagated plants is outlined in Fig. 4.2.

Clonal propagation, maintaining genetic uniformity of a variety, uses vegetative propagules (like budwoods, scions, cuttings, tubers, bulbs), or various explants of plant organs (shoot tips, meristems, epidermis, ovaries, nucellar tissue, etc.), cultured *in vitro* (Chapter 18 and Fig. 4.2). Those organs consist of many cells and after mutagenic treatment will give rise to a chimeric structure consisting initially of genetically different cells in the form of a sector within a histogenic layer (mericlinal chimera). A mericlinal chimera occurring in the M_1V_1 generation may convert into a periclinal chimera with a uniformly mutated cell layer. To favour this development and eventually obtain non-chimeric (homohistont) mutant plants, a number of established techniques can be applied, e.g. in woody perennials pruning or cutting back the primary shoot and using the axillary buds in propagation, or repeated budding of those buds coming from pre-formed primordia and located generally in the basal and middle part of the M_1V_1 shoot (Lapins, 1983). Another approach is to treat shoot meristems *in vitro*: uniformly mutated plants will be obtained relatively fast through propagation of axillary buds.

The adventitious bud technique, as advocated by Broertjes and van Harten (1988), assumes that adventitious buds develop from a single cell; therefore, an adventitious bud, developing on a mutagenized leaf, stem or root cutting, is expected to give rise to a non-chimeric homohistont mutant. Unfortunately the technique is not applicable for all plant species. *In vitro* cell and tissue cultures, particularly callus and single cell culture, are also means to induce mutations. The bottleneck in this approach is the regeneration into a whole mutated plant, as well as the fact that this genetic variation can be very complex and may require recombination to be useful for plant breeding (see Chapter 18 and Buiatti, 1989).

4.2.3 Types of induced changes

Directed induction of mutations has been an objective since artificial mutagenesis became possible. However, on the basis of present knowledge, one can only imagine a very limited control over the mutation spectrum. Some specificity of action is shown by different mutagens in the ratio of 'point mutations' to chromosomal aberrations. A tendency to affect particular heterochromatic regions is typical of certain chemicals. For example, certain *erectoides* and *eceriferum* loci in barley mutate preferentially according to the different ion densities of radiations used (Hagberg, 1960; Lundqvist *et al.*, 1968). The available results suggest that in

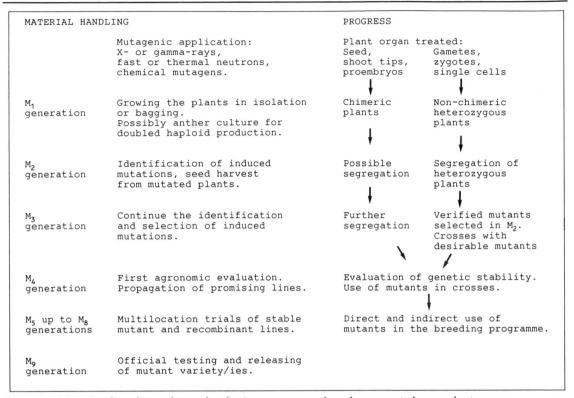

Fig. 4.1. Mutation breeding scheme for the improvement of seed propagated crop plants.

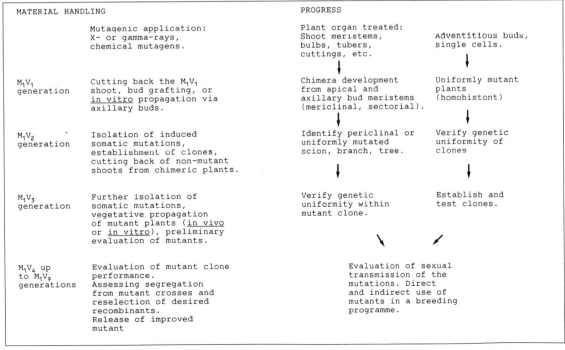

Fig. 4.2. Mutation breeding scheme for the improvement of vegetatively propagated plants (*in vivo* and *in vitro* methods).

higher plants different loci will respond differently to mutagens. However, the evidence for mutagenic specificity is very limited.

Other kinds of genetic variation can be obtained by using induced structural changes, especially translocations. The distribution of breaks and the frequency of reciprocal translocation of chromosome segments in barley was found to be random (Hagberg, 1960). However, by crossing different lines with translocations, a chromosome segment can be duplicated almost at will. This possibility of increasing gene dosage may be important for characters of agronomic importance conditioned by many genes. The induction of reciprocal translocations can be used for increasing chiasma frequencies in chromosome segments in which chiasmata are rare, and in heterozygous plants it ought to produce recombinant types not otherwise obtainable. In vegetatively propagated plants the useful types of induced genetic changes include deletions and duplications because in a heterozygous genotype they can have a strong phenotypic effect (FAO/IAEA, 1982).

Mutagen application also appears promising for purposes other than just creating genetic variation e.g. producing haploids and aneuploids, inducing sexuality in apomictic species, reducing self- or cross-incompatibility; genetically diploidizing polyploids through induction of mutations in homologous chromosomes. Also activation of transposons could be one of the effects of mutagen application.

4.3 HOW TO SELECT MUTATIONS

The recovery of desired mutations at a favourable cost/benefit ratio is important and depends to a large extend on the handling of the treated material and the selection methods applied. One has to take into account the breeding objectives, when deciding whether from M_1 plants all the seeds should be harvested, a single seed per plant, or one or more reproductive organs per M_1 plant (spike, capsule, pod, fruit, etc.). Consequently, the M_2 generation can be grown as a cost-saving bulk or be

planted in progenies (plant-to-row, spike-to-row, fruit-to-row). Progenies, of course, allow a better recognition of mutant segregation, but visibly different mutants may also be easily detected in a bulk population (Fig. 4.1). In the case of diploid hermaphrodite species self-pollination of M_1 plants will cause segregation leading to homozygous mutants in the M_2 generation. However, M_1 plants show some degree of sterility. This increases the potential outcrossing even in species classified as obligate self-pollinaters. Therefore, particular measures such as spatial isolation or bagging are necessary in the M_1 generation. For diploid species, M_2 progenies of 20 plants should be sufficient to obtain a certain number of segregants homozygous for a mutant trait. However, for polyploids, and in general for quantitatively inherited characters, it has been found that it is advantageous to allow more recombination to take place by delaying selection to the M_3 or M_4 generation. If the 'doubled-haploid system' can be efficiently used (Maluszynski, 1989), selection will be much facilitated, particularly for quantitatively inherited traits due to the absence of the heterozygous fraction of the population.

In diclinous species like maize, the chimeric structure of the M_1 plants does not allow self-pollination within a mutated sector and therefore no segregation of homozygous mutants in M_2. A practical procedure here would be an interpollination of M_1 plants, which would bring all induced mutations into heterozygous M_2 plants (Singleton, 1969). An alternative would be the mutagen treatment of pollen (Neuffer and Chang, 1989) and pollination of non-treated mother plants, which would lead to non-chimeric M_1 plants heterozygous for induced mutations. These plants should then be self-pollinated as usual for obtaining homozygous mutants. In cross-pollinated species that are self-incompatible mutation breeding should involve some form of recurrent selection.

Vegetatively propagated plants are very different from seed propagated plants with regard to selection of induced mutations. In the latter only mutations occurring in the sporogenous tissue and passing through the meiosis are transmitted to the progeny whereas in the former a mutation induced in a 'somatic cell' capable of forming a 'somatic descendant' can be manifest in a new clonally propagated organism. Mutant selection usually starts in the M_1V_2 generation. Each selected variant has to be further propagated to give in the M_1V_3, a basis for observation to confirm the mutant characteristics. In the M_1V_4 and subsequent generations one has to assess the stability and uniformity of the clone. Only uniform and stable clones should be planted for agronomic evaluation (Fig. 4.2).

4.4 HOW TO MAKE USE OF INDUCED MUTATIONS

The induction of mutations as such only leads to variation, not necessarily to any improvement. There is no guarantee that a 'mutant plant' carries only one genetic change. On the contrary, with the effective mutagens available today and the many thousand genes exposed to the mutagen, it is rather likely that there will be several induced mutations. One or two may be noticeable, but most are not. Therefore, where possible, it is advisable to 'backcross' a selected mutant to its original parent and to reselect the 'purified' desired mutant.

This procedure at the same time offers valuable information about the inheritance of the mutant character and is of additional assistance in eliminating contamination (admixtures or outcrosses). For full exploitation of the potential of mutagenesis, transfer of mutated genes into different genetic backgrounds should be seriously considered. A whole series of successful cultivars, often exhibiting surprising features, have been created by crossbreeding with mutants (FAO/IAEA, 1976; Donini *et al.*, 1984; Micke *et al.*, 1985, 1990).

In vegetatively propagated plants mutants are used primarily as improved clonal cultivars. The utilization of mutants as parents in crosses requires that the mutations involve the LII histogenic layer, which is responsible for the formation of sporogenous tissues. The crossing of the mutants would eventually lead to recombination of mutated genes but at the same time

may destroy the original valuable genotype which the breeder wanted to maintain. This has a number of important implications: first, the mutation dose should not be too high in order to avoid excessive occurrence of multiple mutations. Second, one should aim at a number of similar mutations to choose from.

Chimeric mutants may not be stable and thus unable to transmit the mutated characters to their progenies; therefore solid (non-chimeric) mutant clones with genetically uniform histogenic layers are very desirable (section 4.2.2.) Mutant clones produced from adventitious buds are most likely non-chimeric. However, mutagenic treatment can also be used to rearrange tissues of a chimera, e.g. to bring a mutated LI layer into the LII layer and vice versa, eventually leading to a non-chimeric (homohistont) shoot.

4.5 ECONOMICALLY VALUABLE RESULTS OF MUTAGENESIS

Up to 1960, there were few reports of improved cultivars developed by mutagenesis and those few cases were more or less by-products of radiobiology and fundamental mutagenesis research. An international symposium convened by FAO and IAEA in 1964 in Rome reviewed the results and prospects of practical application of mutagenesis for developing better crop cultivars (FAO/IAEA, 1965). The meeting and its published proceedings apparently stimulated the interest of plant breeders and five years later at another symposium, 77 'mutant cultivars' were reported. At an FAO/IAEA Symposium convened in 1990, records were presented of 1363 cultivars, which derived valuable traits from mutagenesis (Micke *et al.*, 1990; FAO/IAEA, 1991). Table 4.3 shows the countries where these cultivars were released and the different groups of plants. More than 90% of these cultivars are based on mutants induced by X- or gamma-rays. However, this should not be interpreted as if chemical mutagens were not useful. It is probably rather a reflection of the higher number of experiments performed with X- or gamma-rays.

So far, useful induced mutations reported

Table 4.3. Number of released varieties developed through induced mutations in different countries (from Micke *et al.*, 1990)

Countries	Seed propagated plants (A)			Vegetatively propagated plants (B)				Seed + Vegetative plants (A+B)
	Cereals (1)	Others (2)	Total (1+2)	Ornamentals (3)	Fruits (4)	Others (5)	Total (3+4+5)	
Algeria		1	1					1
Argentina	1	2	3		1		1	4
Australia	2	4	6					6
Austria	13	1	14		1		1	15
Bangladesh	2	5	7					7
Belgium		2	2	18		1	19	21
Brazil	29	1	30					30
Bulgaria	12	7	19		1		1	20
Burkina Faso	1		1					1
Burma	2	2	4					4
Cameroon	3		3					3
Canada	2	5	7	4	8		12	19
Chile	1		1					1
China	206	43	249	7	4	4	15	264
Costa Rica		1	1					1
Côte d'Ivoire	16		16					16
Czechoslovakia	31	2	33	1			1	34
Denmark	5		5					5
Egypt		1	1					1
Finland	10		10					10
France	5	2	7	14	7		21	28
Germany	27	8	35	50	1		51	86
Greece	1		1					1
Guyana	26		26					26
Haiti	1		1					1
Hungary	4		4	1			1	5
India	35	65	100	84		12	96	196
Indonesia	3	3	6					6
Italy	11	8	19		2	2	4	23
Japan	40	25	65	18		4	22	87
Kenya		2	2					2
Korea, Rep. of	2	3	5					5
Netherlands		2	2	169			169	171
Norway	2		2					2
Pakistan	4	7	11					11
Philippines	3		3					3
Poland		13	13					13
Portugal	1		1					1
Senegal	2		2					2
Sri Lanka	1		1					1
Switzerland	1		1					1
Thailand	3	1	4	3	1		4	8
Togo	1		1					1
United Kingdom	17		17	1			1	18
USA	30	9	39	27	1	8	36	75
USSR	37	33	70	18	8		26	96
Vietnam	6	3	9					9
Yugoslavia		1	1					1

and incorporated into commercial cultivars are mainly of the easily recognizable types, like those causing changes in plant architecture, flowering time, flower shape or colour, fruit size or colour, resistance to pathogens and pests (Donini *et al.*, 1984). However, many valuable mutants are known, where detection of the genetic change required chemical analysis. Outstanding examples are changes in oil content of sunflower and soybean, altered fatty acid composition in linseed, soybean and rapeseed (Röbbelen, 1990), as well as changes in protein content, amino acid composition and starch quality in the endosperm of cereals (rice, barley, maize, wheat). In maize and rice, inbred lines possessing better combining ability for hybrid vigour were produced, other induced mutants proved useful by causing male sterility or by restoring fertility, thus facilitating hybrid seed production (Konzak, 1984; Micke *et al.*, 1985; Broertjes and van Harten, 1988; Mutation Breeding Newsletter, FAO/IAEA, 1991). Among the mutant cultivars in seed propagated plants many are reported with higher yields, a result one would not have expected under the assumption that induced mutations affect only monogenically controlled traits.

The most successful cultivars (e.g. of rice, cotton, wheat) in China based on mutants were grown on several hundred thousand hectares. In Czechoslovakia nearly all malting barley is produced by induced mutant derived cultivars and because of their high yield and excellent malting quality these are also used as crossing parents all over Europe. Mutation breeding of cotton, mungbean and chickpea has been very successful in Pakistan.

In Italy, most of the durum wheat varieties used for making pasta were developed by crossbreeding with lodging resistant mutants. In California, over 80% of the 160 000 ha of rice were planted with cultivars carrying valuable traits from mutagenesis (FAO/IAEA, 1991).

For vegetatively propagated crop plants also a number of successful varieties were produced particularly in fruit tree species, such as apple, cherry, olive, orange, apricot. Examples are compact tree shape obtained in Canada, UK, Poland, USSR, Israel, Italy, Austria; earlier flowering, self-compatibility or seedlessness in cherry, citrus, almond, grape (Canada, USA, Israel, China, Italy); fruit shape, colour and maturity time in apple, peach, grapefruit, or blackberry (France, Canada, USA, Italy, Israel, Germany); resistance to diseases in sugarcane (smut disease), peppermint (*Verticillium* sp.), apple (powdery mildew), pear (black spot), grape (*Plasmopora viticola*) reported from India, USA, Canada, Japan, Portugal (Spiegel-Roy, 1990; FAO/IAEA, 1991).

A large number of mutants in ornamental plants were commercialized in The Netherlands, India, China, Thailand, Germany and other countries. They possess mutations affecting the size and the architecture of plants, organs, leaves and flowers, their colour, flowering period, temperature requirements etc. (Broertjes and van Harten, 1988). The number of such cultivars, particularly in The Netherlands, cannot be traced any more.

A major contribution of mutagenesis to plant breeding progress, which cannot be overestimated, is its use for the advancement of genetics. Also future applications of gene engineering technology will rely heavily upon mutagenesis as a tool to localize useful genes, to have effective markers and to provide the desired genetic diversity in molecular gene banks. By the same token genetic engineering procedures may themselves be mutagenic in the case of transposons and T-DNA mediated inversions.

It can be expected that modern mutation breeding technologies in combination with refined selection methods applicable to large populations will continue to be of great potential for the improvement of both seed and vegetatively propagated crop plants.

REFERENCES

Broertjes, C. and van Harten, A.M. (1988), *Applied Mutation Breeding for Vegetatively Propagated Crops*, Elsevier Science Publishers, Amsterdam.

Buiatti, M. (1989), Use of cell and tissue cultures for mutation breeding. In *Science for Plant Breeding*, Proc. 12th Congress of Eucarpia, Parey, Berlin, pp. 179–200.

Donini, B., Kawai, T. and Micke, A. (1984), Spectrum

of mutant characters utilized in developing improved cultivars. In *Selection in Mutation Breeding*, IAEA, Vienna, pp. 7–31.

FAO/IAEA (1965), *The Use of Induced Mutations in Plant Breeding*, Report of FAO/IAEA Technical Meeting, Rome, 1964, Pergamon Press, Oxford.

FAO/IAEA (1976), *Induced Mutations in Cross Breeding*, IAEA, Vienna.

FAO/IAEA (1977), *Manual on Mutation Breeding*, 2nd edn., IAEA, Vienna.

FAO/IAEA (1982), *Induced Mutations in Vegetatively Propagated Plants II*, IAEA, Vienna.

FAO/IAEA (1988), *Plant Domestication by Induced Mutations*, IAEA, Vienna.

FAO/IAEA (1991), *Plant Mutation Breeding for Crop Improvement*, Proc. of a Symposium, Vienna, 1990, IAEA, Vienna.

Hagberg, A. (1960), Cytogenetic analysis in induced mutations. *Gen. Agr.*, 12, 319–336.

Konzak, C.F. (1984), Role of induced mutations. In *Crop Breeding, a Contemporary Basis*, Vose, P.B. and Blixt, S.G. (eds), Pergamon Press, Oxford, pp. 216–292.

Lapins, K.O. (1983), Mutation breeding. In *Methods in Fruit Breeding*, Moore, J.N. and Janick, J. (eds), Purdue University Press, West Lafayette, pp. 74–99.

Lundqvist, U., Wettstein-Knowles, P. and von Wettstein, D. (1968), Induction of 'eceriferum' mutants in barley by ionizing radiations and chemical mutagens. *Hereditas*, 59, 473–504.

Maluszynski, M. (1989), *Current Options for Cereal Improvement: Doubled Haploids, Mutants and Heterosis*, Kluwer, Dordrecht.

Micke, A., Donini, B. and Maluszynski, M. (1990), Induced mutations for crop improvement. *Mutation Breeding Review*, No. 7, IAEA, Vienna, pp. 1–41.

Micke, A., Maluszynski, M. and Donini, B. (1985), Plant cultivars derived from mutation induction or the use of induced mutants in cross breeding. *Mutation Breeding Review*, No. 3, IAEA, Vienna, pp. 1–92.

Neuffer, M.G. and Chang, M.T. (1989), Induced mutations in biological and agronomic research. In *Science for Plant Breeding*, Proc. 12th Congress of Eucarpia, Parey, Berlin, pp. 165–178.

Röbbelen, G. (1959), 15 Jahre Mutationsauslösung durch Chemikalien. *Züchter*, 29, 92–95.

Röbbelen, G. (1990), Mutation breeding for quality improvement a case study for oilseed crops. *Mutation Breeding Review*, No. 6, IAEA, Vienna, pp. 1–43.

Singleton, W.R. (1969), Mutation induced by treating maize seeds with thermal neutrons. In *Induced Mutations in Plants*, Proc. IAEA/FAO Symposium, Pullman, IAEA STI, Pub. 231, pp. 479–483.

Spiegel-Roy, P. (1990), Economic and agricultural impact of mutation breeding in fruit trees. *Mutation Breeding Review*, No. 5, IAEA, Vienna, pp. 1–25.

Stadler, L.J. (1930), Some genetic effects of X-rays in plants. *J. Hered.*, 21, 3–19.

5

Interspecific hybridization by sexual means

B. Pickersgill

5.1 INTRODUCTION

The importance of interspecific hybridization in crop improvement varies greatly from one crop to another (Stalker, 1980; Prescott-Allen and Prescott-Allen, 1988). In tomato, virtually all cultivars grown on a commercial scale anywhere in the world carry resistance to *Fusarium* wilt introduced from one wild species (Goodman *et al.*, 1987), while six others are sources of characters ranging from salt tolerance to insect resistance. In the faba bean, on the other hand, improvement has come about entirely through crossing and selection within *Vicia faba*. It has not yet been possible to introduce genes from any other species into this crop. Whether fertile interspecific hybrids can or cannot be produced easily will influence the attitude of breeders of a crop to the use of alien germplasm. Harlan and de Wet (1971) introduced a concept of gene pools which has proved useful to both plant breeders and those working with crop genetic resources. Species and accessions in the primary gene pool of the crop cross easily and their hybrids are fertile. Species in the secondary gene pool cross with the crop with difficulty and their hybrids are partially sterile. Species in the tertiary gene pool can be crossed with the crop only by use of special techniques such as

Plant Breeding: Principles and prospects. Edited by M.D. Hayward, N.O. Bosemark and I. Romagosa. Published in 1993 by Chapman & Hall, London. ISBN 0 412 43390 7.

embryo rescue, and their hybrids are partially or completely sterile.

Breeders face further difficulties once an interspecific hybrid has been obtained, because genetic recombination in the F_1 and subsequent backcross generations can seldom be controlled precisely. Segments of chromosomes, or even whole chromosomes, not just a single gene, are usually introduced into the genotype of the crop. Breaking linkages with unwanted wild-type genes and restoring the genotype associated with commercial quality may take a very long time. In tomato, 12 years were needed to separate nematode resistance from undesirable fruit characters introduced by hybridization with a wild species. When an alien gene has been successfully transferred, further time-consuming crossing and selection are needed to produce, evaluate and multiply a new commercial cultivar. Improvement through interspecific hybridization therefore often takes longer than intraspecific breeding programmes. In practice, pre-breeding (also known as germplasm enhancement) may be separated from cultivar production and release. Pre-breeding covers the stages from initial production of an F_1 hybrid to release of lines with all the attributes of the crop plus the desired new character and is often carried out in public sector institutes. Commercial companies then take the enhanced germplasm lines released from pre-breeding programmes and develop them into new cultivars.

Alternatives to interspecific hybridization are

mutation breeding, and production of trans-genic plants (Chapters 9 and 10). Both have the advantage that changes are induced or incor-porated directly into the genotype of the crop, without problems from associated wild-type characters. However, neither technique is yet routine for all crop plants. Conventional inter-specific hybridization via the sexual process is therefore likely to remain useful to plant breeders.

5.2 BARRIERS TO INTERSPECIFIC HYBRIDIZATION

The techniques needed to produce an inter-specific hybrid will depend very much on the nature of the barriers involved in each cross. There is no point in applying chemicals to enhance pollen tube growth if the pollen tubes never penetrate the alien stigma. Fruits from interspecific pollinations may lack viable seeds because fertilization has not taken place, or because the seeds abort after fertilization. Ovule culture may enable embryos to be res-cued from inviable seeds, but only if fertiliza-tion has first produced an embryo to be rescued.

5.2.1 Barriers in stigma and style

Interspecific fertilizations frequently fail because the pollen tubes are incompatible with the pistil and stop growing before they reach the ovules. There are two different views on the relationship between intra- and interspecific incompatibilities, put forward by Pandey and Hogenboom respectively. Pandey (1981) believes that both phenomena are pleiotropic effects of the *S* gene (or supergene), which controls incompatibility. Hogenboom (1984), on the other hand, considers that alien pollen is rejected when pollen and pistil of different species do not complement one another prop-erly. Since this is quite different from the active rejection of self pollen, he argues that a differ-ent term, incongruity, should be used. Incon-gruities may involve a series of barriers (to

germination of pollen on the stigma, pene-tration of pollen tubes into the transmitting tract, passage of pollen tubes down the style, and so on) and the genetic control is thus likely to be complex, unlike the simple control via the *S* gene postulated by Pandey.

(a) Structure and function of the stigma

Before pollen grains can germinate on a stigma, they must adhere and take up water (Heslop-Harrison, 1987). Wet stigmas, such as those of Solanaceae, are covered with a fluid secretion in which compatible pollen grains readily hydrate. Dry stigmas, such as those of the Cruciferae, have no fluid secretion present at pollination, so compatible pollen grains apparently hydrate by withdrawing water from the stigma through discontinuities in the cuticle (Knox *et al.*, 1986). Some stigmas do not fit precisely into the categories of wet or dry. In legumes whose flowers require 'tripping' before they will set fruit, such as *Phaseolus coccineus* and most cultivars of *Vicia faba*, a stigmatic secretion accumulates under the cuticle and is released only when the cuticle is ruptured, for example by rubbing against the body of a visiting insect. These flowers have to be visited by an insect before they will set fruit, even though the plants may be self-compatible.

Proteins carried in the walls of pollen grains are released when the grains hydrate, and may be involved in recognition reactions with pro-teins on the stigma (Chapter 13). If the pollen is recognized as compatible, germination pro-ceeds and the pollen tubes penetrate the stigma. Pollen which is not compatible may not germinate, or may fail to penetrate the stigma. Once inside the stigma, pollen tubes grow towards, then along, the transmitting tract. The conformation of cells in the transmitting tract may orientate the pollen tubes toward the ovules (Heslop-Harrison and Heslop-Harrison, 1987). Maize pollen tubes are correctly orien-tated in maize stigmas, but the narrower pollen tubes of sorghum or pearl millet often enter the main axis of the maize stigma orientated towards the stigma tip and continue to grow in that direction instead of towards the ovary.

(b) Techniques for overcoming barriers presented by the stigma

Effective pollination

In experimental crosses, pollen is transferred by the experimenter, not by the normal pollinating agent. Pollen must, first, be transferred to the correct place. If the stigma is large, as in *Brassica*, this is not a problem. However, some species have small stigmas with an even smaller receptive area which may be difficult to pollinate accurately. Second, pollen should be transferred at the correct time. The period during which a stigma is receptive varies from a few hours (in mango) to about 10 days (in grape) (Sedgley and Griffin, 1989). The onset of receptivity, as well as the extent of the receptive area, may be indicated by the appearance of esterase activity (Knox *et al.*, 1986). In chickpea (Turano *et al.*, 1983), stigmas from green buds display no esterase activity. Activity appears in the white bud stage (when pollen is shed), but is confined to a few papillae at the distal end of the stigma. By the time the flowers are fully open, esterase activity has declined. Cross-pollinations at the white bud stage are thus more effective than pollinations of fully open flowers. Third, the pollen must hydrate properly. When the cuticle has to be ruptured to release the stigmatic exudate, it may be necessary to rub the stigma before or while the pollen is applied. Humid conditions often improve pollen germination, so it may be helpful to spray the stigma with water or with a pollen culture solution, and/or to protect the pollinated stigma by a gelatine capsule or other enclosure.

Mentor pollination

Mentor pollen is pollen which is fully compatible with the intended seed parent, but which has usually been killed or rendered inviable. When mixed with pollen from the intended male parent, the mentor pollen facilitates germination and penetration by pollen tubes of the alien species, possibly by releasing proteins involved in the recognition reaction (Stettler and Ager, 1984). Mentor pollen techniques have been used successfully in *Populus*, which has a large stigma. When the stigma is small, as in chickpea, mentor pollen will compete for the limited space with alien pollen grains. Mentor pollen leachate, produced by soaking mentor pollen so that pollen proteins diffuse into the leachate, may have the same effect as mentor pollen without occupying valuable space on small stigmas.

Other techniques for overcoming barriers on the stigma

Experimental self-pollination of *Brassica* buds, before proteins implicated in the incompatibility reaction have appeared on the stigma, can overcome self-incompatibility (Sastri, 1984). However, it is not yet known whether the protein(s) involved in interspecific inhibitions are the same as those involved in self-incompatibility and hence whether bud pollination can also overcome interspecific incompatibilities. If the stigma is treated with proteases, no pollen grains adhere or germinate, apparently because proteins vital to the compatible interaction have been removed (Shivanna, 1982; Heslop-Harrison, 1987). On the other hand, pollinating at 40°C has overcome both self- and interspecific incompatibility in *Trifolium*, presumably by inactivating proteins concerned with incompatibility while leaving proteins involved in other essential reactions able to function. Other interspecific crosses succeed better at low temperatures, so optimal conditions may have to be determined for each combination. Some stigmatic barriers, such as those associated with incorrect orientation of alien pollen tubes, can be overcome only by eliminating the stigma and style and pollinating the ovary or ovules directly.

(c) Structure and function of the style

In some families, such as the Solanaceae, the style is solid, with a central transmitting tract of loosely packed cells, each surrounded by an intercellular secretion, which contains proteins, carbohydrates and other compounds. The pollen tubes grow intercellularly through this secretion to the ovary. In hollow styles, like those of *Lilium* and *Rhododendron*, a continuous canal lined with secretory cells runs from the stigma surface to the ovary. Pollen tubes again grow through the secretion to the ovary (Knox *et al.*, 1986). In the legumes, the stigma is solid but the style often develops a canal within which secretion accumulates and through which pollen tubes grow.

Pollen tubes can grow for some distance using reserves stored in the pollen grain, but usually need an additional supply of nutrients, presumably from the style, to complete their journey to the ovules (Knox *et al.*, 1986). Maize pollen contains sufficient reserves to produce a pollen tube about 2 cm long, but the maize silk is several dm long. Sorghum pollen tubes grow sufficiently far in maize silks to indicate that they must be able to use some nutrients provided by the maize. On the other hand, pearl millet pollen tubes stop growing in maize silks once they have used up their own reserves (Heslop-Harrison and Heslop-Harrison, 1987). It is not yet possible to predict which crosses will fail because alien pollen tubes cannot obtain the nutrients they require from the style through which they are growing. However, this seems to explain the stylar inhibition of pollen tube growth observed in many incompatible crosses in the Solanaceae.

Successful fruit set may depend on ample pollination of the stigma (Sedgley and Griffin, 1989). In apple, about 50 pollen grains per flower are required for fruit set although there are many fewer than 50 ovules to be fertilized. The surplus pollen tubes may elevate to threshold levels the substance(s) involved in the signalling between pollen tubes and ovules that seems to occur in a pollinated pistil. In an interspecific cross, germination and growth of pollen may be too limited for signal and response to be effective, so the cross fails.

The time available for pollen tubes to reach the ovules may be limited. Alien pollen tubes often grow slowly in a style to which they are not adapted, so may not reach and fertilize the ovules before the style senesces. Likewise, the ovules may remain viable for a limited time, and may senesce before the alien pollen tubes reach them (Sedgely and Griffin, 1989). In cherry, for example, ovules start to lose viability 3–10 days after anthesis. Since the micropyle of most ovules admits only one pollen tube, once an ovule has been penetrated, it can usually be fertilized only by the sperm contained in that pollen tube.

Unilateral incompatibility/incongruity

If an interspecific cross succeeds only in one direction, while the reciprocal cross fails, the two species show unilateral incompatibility (or unilateral incongruity) (UI). UI is widespread in the Solanaceae, where it often accompanies self-incompatibility (Pandey, 1981). Pollen tubes of the self-compatible species are usually inhibited by the style of the self-incompatible species, whereas pollen tubes of the self-incompatible species grow successfully down the style of the self-compatible species. For example, the cultivated tomato is self-compatible. Tomato pollen tubes are unable to grow down the styles of self-incompatible *Lycopersicon peruvianum*, but tomato styles do not inhibit pollen tubes of *L. peruvianum* (Gradziel and Robinson, 1991).

(d) Techniques for overcoming stylar barriers

Reciprocal crosses

Interspecific crosses should always be attempted in both directions, since this will demonstrate whether the barriers are unilateral or not. Often a breeder is not concerned which parental species contributes the cytoplasm to an interspecific F_1 hybrid, but if it is important that the cross be made in the direction blocked by stylar barriers, then other ways of overcoming these barriers will be needed.

Intraspecific variation in interspecific crossability

Occasionally intraspecific variation in crossability is under simple genetic control. In wheat, cultivars with recessive alleles of the two crossability genes *Kr-1* and *Kr-2* cross readily with rye, whereas in cultivars with dominant alleles, rye pollen tubes are inhibited at the base of the wheat style and in the ovary (Sastri, 1984). More often, the genetic basis of differences in interspecific crossability is unknown. Wild accessions of *Phaseolus vulgaris* and *P. coccineus* will cross spontaneously, whereas embryo rescue is required to secure hybrids from crosses between domesticated accessions. The inheritance of this trait has not been studied. Unfortunately, it is not usually possible to predict which accessions are most likely to be successful parents in an interspecific breeding programme. It is therefore advisable to test for crossability as diverse a range of parental accessions as time, labour and materials permit (Williams, 1987).

Mentor pollination
Mentor pollen has occasionally been reported to overcome inhibition of pollen tubes in the style as well as on the stigma. In crosses of *Cucumis sativus* with *C. metuliferus* or *C. africanus*, inviable mentor pollen used alone produces fruit but no embryos, whereas inviable mentor pollen used with viable alien pollen produces some globular stage embryos. Mentor pollen may act by causing the style to mobilize nutrients for pollen tube growth, or by prolonging the viability of ovules, perhaps by raising the number of pollen tubes in the style above the critical level. Whatever the mechanism, the significant factor is probably the presence of compatible mentor pollen tubes in the style. This necessitates the use of inviable rather than killed mentor pollen (Stettler and Ager, 1984).

Use of plant growth regulators
Growth regulators, sprayed on or near flowers or applied to pedicel or ovary at or after pollination, may improve the set of fruits from interspecific pollination (Sastri, 1984). Gibberellins and auxins usually inhibit pollen germination and pollen tube growth, but occasionally are stimulatory (Pharis and King, 1985). Gibberellins (contained in or manufactured by the pollen tube) seem to be involved in the signalling system between pollen tubes and ovules that 'conditions' the ovules for fertilization (Jensen *et al.*, 1983). Alien pollen tubes, which often grow slowly or cease growth altogether in a style of a different species, may be inadequate sources of gibberellin. An exogenous supply might therefore compensate for this deficiency and thereby improve the frequency of interspecific fertilization. In pear, exogenous gibberellin appears to mimic the effect of pollen tubes growing down the style by extending the longevity of the ovules. In tomato, gibberellin may substitute for growing pollen tubes and stimulate auxin production in the ovary, which in turn delays abscission of the pollinated flower. Exogenous auxins are also often applied directly in order to delay abscission. The role, if any, of cytokinins in pre-fertilization events is obscure, and they are used less often than gibberellins or auxins in attempts to overcome barriers in stigma or style.

By-passing barriers in the style
Immature styles may not have developed the ability to inhibit incompatible pollen tubes which is characteristic of the mature organ. Bud pollination may therefore succeed where a cross with fully open flowers fails. Pollen may have to be applied in a medium favouring germination to compensate for deficiencies in the immature stigma. This technique increased the number of tomato pollen tubes which could grow past the upper part of the style of *Lycopersicon peruvianum*, where unilateral incompatibility is usually expressed. However, the tomato pollen tubes were then inhibited at the base of the style or in the ovary (Gradziel and Robinson, 1991). This illustrates that overcoming one barrier to interspecific hybridization may simply expose other, later-acting, barriers and supports Hogenboom's views on interspecific incompatibility.

A more drastic way of overcoming barriers in the style is to amputate the style and pollinate the cut stump (Shivanna, 1982). Germination is often poor, probably because the cut style cannot supply the substances present in or on the stigma. Pollen tubes also may not orientate correctly, because the physical cues provided by the stigma or the chemical cues emanating from transmitting tissue or 'conditioned' ovules may be lacking. However, this technique has succeeded in both *Lathyrus*, which has a hollow style, and *Prunus*, which has a solid style. If the style is relatively robust, and the experimenter manually dexterous, it may be possible to graft a compatibly-pollinated style and stigma on to an alien style cut below the zone in which incompatible pollen tubes would be inhibited (Knox *et al.*, 1986). The pollen tubes may then be able to reach the ovules. This technique has been successful in *Datura* and *Oenothera*, but in *Prunus* pollen tubes would not grow through the graft union.

Finally, it may be possible to by-pass stigma and style completely and apply pollen directly to the ovules, either by injecting a suspension of pollen in a nutrient solution into the ovary or by pollinating the surface of the ovules (Shivanna, 1982). Prior experiments are necessary to determine whether the number of fertilizations is greater when ovules are excised and isolated, or when the ovules remain attached to the placenta, or when the placenta remains attached to pedicel and/or calyx. *In vitro* fertilizations must be carried out under aseptic conditions, so succeed only when embryos can be grown in culture from fertilization to maturity. Interspecific and intergeneric fertilizations have been produced *in vitro* in the Caryophyllaceae, Primulaceae and Solanaceae, but the hybrid embryos usually abort at the globular stage (Zenkteler and Slusarkiewicz-Jarzina, 1986).

5.2.2 Fertilization

The number of ovules fertilized after interspecific pollination, *in vivo* or *in vitro*, is often very small. This may itself represent a barrier to interspecific hybridization, and the small number of hybrid embryos produced may in turn limit opportunities for study of post-fertilization barriers and application of techniques such as embryo rescue. It is therefore desirable to know more about the events associated with fertilization, in the hope that it may become possible to influence or control this process. However, fertilization in flowering plants occurs within the embryo sac, which is surrounded by the integuments of the ovule. Details of fertilization can therefore be observed only in cleared or sectioned ovules (Knox *et al.*, 1986).

(a) Structure of the mature embryo sac

Female meiosis in flowering plants produces a tetrad of megaspores, three of which usually degenerate, while the functional megaspore usually undergoes two mitotic divisions to produce an embryo sac containing eight haploid nuclei in seven cells (two synergids, one egg cell, three antipodal cells, and a large central cell containing the two polar nuclei). This is the *Polygonum*-type embryo sac, found in most crop species.

The two synergids, together with the egg cell, form the egg apparatus, which is located at the micropylar end of the embryo sac. Nutrient transfer into the embryo sac before fertilization may take place via the synergids. Synergids may also release compounds involved in the final attraction of pollen tubes to the ovule (Jensen *et al.*, 1983). A pollen tube grows through the micropyle, penetrates one synergid, and discharges its two sperm into that synergid. The cell wall of each synergid is incomplete at the point where the synergid, egg and central cell adjoin. The two sperm can thus move out of the synergid into which they were released. One sperm fuses with the egg, to give rise to the zygote, which will develop into the embryo. The other sperm fuses with the two polar nuclei, which have moved to lie near the egg apparatus, to form the triple fusion nucleus, which gives rise to the endosperm and so is known also as the primary endosperm nucleus.

The antipodal cells vary considerably, both in behaviour and in the length of time for which they persist, but often seem to be important in the early nutrition of the fertilized embryo sac. In the Gramineae, the three antipodals present at fertilization divide after fertilization to form a haploid tissue by means of which nutrients are transferred into the embryo sac. In the Solanaceae, the antipodals persist but do not divide. In the Leguminosae and Rosaceae, the antipodals degenerate before or shortly after fertilization, but the unfertilized embryo sac already contains a food reserve for early endosperm development because the central cell is filled with starch grains which disappear as the endosperm develops.

(b) Limits to fertilization in wide crosses

Critical study of ovules fixed at appropriate intervals after interspecific pollination has shown that fertilization takes place, although the products soon abort, in some surprisingly wide crosses (Table 5.1). Wide fertilizations

Table 5.1. Examples of wide crosses which have resulted in fertilization in some economically important genera

Family	Cross
Gramineae	*Hordeum vulgare* × *Alopecurus agrestis, Elymus repens, Lolium perenne, Pennisetum americanum*
	Oryza sativa × *Pennisetum americanum, Sorghum vulgare*
	Secale cereale × *Dactylis glomerata, Hordeum bulbosum, Zea mays*
	Triticum aestivum × *Alopecurus agrestis, Dactylis glomerata, Pennisetum americanum, Sorghum vulgare, Zea mays*
Leguminosae	*Pisum sativum* × *Vicia faba*
Solanaceae	*Nicotiana tabacum* × *Hyoscyamus albus, Lycopersicon esculentum, Petunia parodii*

may also be surprisingly frequent: up to 40% in crosses between wheat and maize (Laurie and Bennett, 1988a); up to 90% in crosses between wheat and sorghum (Laurie and Bennett, 1988b). On the other hand, there are some intrageneric crosses where pollen tubes reach the ovary but nevertheless fail to fertilize the ovules. It is not yet possible to predict when wide fertilizations will take place and when they will not.

Studies on wide hybridization are of theoretical interest in that they may permit more to be learned about the control of fertilization so that ultimately, it may be possible to manipulate when or whether fertilization takes place. They may also be of practical interest, even though

viable hybrid embryos may never be produced. In cereals, transposable elements might possibly transfer themselves from chromosomes of one species to those of another (for example, from maize to wheat) during the few division cycles in which both genomes are present together in a hybrid nucleus (Laurie and Bennett, 1988a). This might increase the possibilities for genetic manipulation of species which presently lack active transposable elements.

(c) Partial failure of fertilization

Double fertilization is usually required for normal seed development in flowering plants. In intraspecific crosses, one of the two fertilizations may occasionally fail, but in a few interspecific crosses single fertilizations are so frequent and so predictable that this can be exploited by plant breeders. For example, when *Solanum tuberosum*, which is tetraploid, is pollinated by its diploid relative *S. phureja*, both of the *phureja* sperm fuse with the polar nuclei of *S. tuberosum*. The primary endosperm nucleus which results is hexaploid (two genomes contributed by each *tuberosum* polar nucleus and one genome from each *phureja* sperm). This is the normal ploidy level for endosperm in a fertilized *tuberosum* ovule. Active division and development of the hybrid endosperm stimulates parthenogenetic development of the unfertilized *tuberosum* egg, giving rise eventually to a seedling with the reduced number of chromosomes (Williams *et al.*, 1987). This is now widely exploited as a source of 'haploid' (more strictly, polyhaploid) potatoes.

(d) Techniques for overcoming barriers to fertilization

Problems associated with fertilization are difficult to overcome because so little is known about its control or the extent to which it can be manipulated.

Application of exogenous plant growth regulators (PGRs), often as mixtures, has been reported to increase the numbers of hybrid embryos of sufficient size to be rescued by embryo culture (Sastri, 1984). However, few if any critical studies have been made of whether these PGRs actually affect the frequency of alien fertilizations, or whether they are simply improving post fertilization development of hybrid embryo and/or endosperm. Studies on this point would be of considerable interest. For example, when unfertilized ovules of cotton are cultured on a medium containing gibberellic acid (GA), one synergid degenerates (Jensen *et al.*, 1983). GA thus appears to mimic the effect of cotton pollen tubes growing down the style. Since the degenerated synergid may attract pollen tubes to the micropyles of receptive ovules, a shortage of endogenous GA might explain why pollen tubes may reach an alien ovary, but nevertheless fail to penetrate the ovules. This is obviously very speculative, but a possibly fruitful field for further research.

5.2.3 Barriers acting during seed development

Very often, seeds resulting from interspecific fertilization develop for a short time, then abort. In order to understand some of the possible reasons for this, and hence what techniques may be used to overcome post-fertilization barriers to interspecific hybridization, it is necessary first to understand the nature and function of tissues in a seed which is developing normally.

(a) Development of a viable seed

A seed contains tissues representing three different genotypes and two different ploidy levels. If both parents are diploids, the ploidy levels and genomic compositions of the various tissues are as shown in Table 5.2. These tissues each fulfil a different function and usually develop in a co-ordinated way. Normal, co-ordinated development may in turn depend on correct balance between the ploidies of the different tissues and/or between maternal and paternal genomes within these tissues (Williams 1987; Singh *et al.*, 1990).

Maternal tissues
For a short time after fertilization, the develop-

Table 5.2. Ploidy level and numbers of maternal and paternal genomes present in tissues of a seed produced by diploid parents

Tissue	Ploidy	Genomes present
Maternal tissues (integuments, nucellus)	$2x$	♀ ♀
Endosperm	$3x$	♀ ♀ ♂
Embryo	$2x$	♀ ♂

ing seed is not an active sink for nutrients. Early growth of endosperm and embryo uses reserves stored in the integuments, nucellus or embryo sac. The innermost cell layer of the integument may become specialized to form the endothelium, which seems to be involved in both digestion of food reserves stored in the integument and transport of digested reserves into the embryo sac. Organic and inorganic compounds from the maternal parent reach the developing seed by way of a vascular bundle in the funicle (the stalk which attaches the ovule to the placenta). This bundle ends in the integument. Compounds delivered in the xylem or phloem are either stored in the integument or conveyed across the integumentary cells into the embryo sac (Murray, 1988).

Endosperm
Nuclear divisions in the endosperm usually begin before any development of the zygote. If the endosperm nuclei divide but no cell walls are formed, as in pea or lupin, the endosperm is said to be free nuclear (also known as liquid or coenocytic). If nuclear division is accompanied by cell wall formation, as in cereals and other grasses, the endosperm is said to be cellular. Sometimes, as in the Solanaceae, the endosperm is free nuclear at first, but becomes cellular later. The endosperm may be short-lived, disappearing before the seed is mature (e.g. Leguminosae) or may form the main storage tissue in the seed (e.g. Gramineae). While present, it plays an important role in transferring nutrients from the tissues outside the embryo sac to the developing embryo (van Staden *et al.*, 1987; Williams, 1987).

Embryo

The first division of the zygote is transverse. The cell near the micropyle develops into a structure known as the suspensor. Suspensors vary greatly in form and size, but seem very important as a source of PGRs and as the pathway of nutrient transfer from the endosperm to the embryo proper (van Staden *et al.*, 1987; Williams, 1987; Murray, 1988). The other cell resulting from the initial division of the zygote develops into the embryo. The embryo progresses through a succession of stages known descriptively as the globular, heart, torpedo and cotyledonary stages. The heart stage marks the onset of differentiation and is also the stage at which the embryo becomes capable of independent, autotrophic, growth (Williams *et al.*, 1987).

(b) Abnormalities in the development of hybrid seeds

Embryo

Abnormalities in hybrid embryos are not usually the primary cause of abortion of hybrid seed. However, an apparent example occurs in the cross *Gossypium arboreum* × *G. hirsutum*, where the hybrid embryo increases in size but fails to differentiate, even when cultured *in vitro*. In some crosses, chromosomes from one parent are eliminated from the hybrid embryo early in development so that, although produced as a result of interspecific fertilization, the embryo is no longer a hybrid. Although now haploid, this embryo may nevertheless germinate successfully. For example, when *Hordeum vulgare* is crossed with *H. bulbosum*, the *bulbosum* chromosomes are lost from many of the hybrid embryos (Stalker, 1980). This has been exploited to produce haploid barley for use in breeding programmes. Haploid wheat has similarly been recovered from wheat × maize or wheat × pearl millet crosses, although not yet in sufficient frequency for the technique to be used as a routine (Laurie and Bennett, 1988a).

Endosperm

Failure of the endosperm to develop normally is the most common cause of abortion of hybrid seeds. The efficiency with which hybrid endosperms function and the stage at which they fail vary very considerably (Williams, 1987). These factors are crucial in determining whether the hybrid embryo reaches a sufficient size to be rescued successfully. For example, when faba bean ovules are fertilized by species of the *Vicia narbonensis* group, nuclei in the hybrid endosperm divide normally at first, but cease division about 3 days after pollination. The embryo dies a little later, when still too small for successful culture. On the other hand, in the cross between *Phaseolus vulgaris* and *P. acutifolius*, the hybrid endosperm survives long enough for the embryos to reach the heart stage. These seeds abort if left to develop *in vivo*, but many seedlings have been raised by embryo culture. An even closer approach to normal seed development occurs in triticale, the intergeneric hybrid between wheat and rye. Here, the endosperm develops sufficiently for the seeds to germinate without special rescue techniques, although grain shrivelling, caused by incomplete development of the endosperm, can reduce commercial yield.

There are no satisfactory general explanations of why hybrid endosperms abort in some crosses but function normally in others. Differences between the parental species in quantity and/or quality of DNA may be involved. Grain shrivelling in triticale has been associated with aberrant mitotic divisions in the endosperm, which may in turn be associated with presence of late-replicating heterochromatin at the ends of many of the rye chromosomes. Others have suggested that deficiencies, or alterations to the normal balance, of plant growth regulators may be implicated, though whether this is cause or effect of problems in the endosperm is seldom clear (van Staden *et al.*, 1987). For example, cytokinin is known to induce cell division and create active sinks. Seeds from the cross between *Phaseolus vulgaris* and *P. acutifolius*, which usually abort about 3 weeks after fertilization, contain much less cytokinin than normally developing seeds (Sastri, 1984). This could explain, at least partially, why the endosperm ceases to be able to support a potentially viable embryo in this cross. However, no one has explained why cytokinin content is low in these hybrid seeds.

Maternal tissues

Sometimes the most obvious abnormalities in seeds developing after interspecific hybridization occur in the maternal tissues (Williams, 1987), though this may be a consequence of less-visible abnormalities in one or both of the hybrid tissues. Abnormalities in the integument, particularly the innermost layer or endothelium, are very common in crosses in the Solanaceae. The endothelium proliferates into the embryo sac, destroying the endosperm and embryo, while starch, which normally is used to support embryo and endosperm development, persists in the integument. This behaviour has been observed in interspecific crosses between species at the same ploidy level, for example diploid *Lycopersicon esculentum* × diploid *L. peruvianum*, and also in crosses between species at different ploidy levels, for example hexaploid *Solanum demissum* × diploid *S. pinnatisectum*. It may also occur in intraspecific but interploidy crosses, for example in reciprocal crosses between synthetic autotetraploid *Capsicum chacoense* and its diploid progenitor. No simple theory of balance between ploidy levels of maternal tissues, endosperm and embryo can explain all these data.

In *Datura*, proliferation of the endothelium similar to that observed in interspecific, or intraspecific but interploidy, crosses can be induced by auxins. Auxins may be released by degenerating tissues. Abnormalities in maternal tissues may thus also trace back ultimately to inability of the hybrid endosperm to attract sufficient nutrients for its own development, leading to premature degeneration of the endosperm and proliferation of the endothelium.

(c) Techniques for overcoming post fertilization barriers

Removal of competing sinks

Many plants produce more flowers than can be matured into fruits. Surplus fertilized ovaries apparently do not abort at random, but in such a way as to conserve maternal resources, or quality of offspring produced, or both (Stephenson, 1981). Thus, if a plant is already carrying as many developing fruits as it can mature succesfully, any further fertilized ovaries will abort soon after fertilization. An interspecific cross therefore has a greater chance of success if the cross is made using the first flowers to open on the maternal parent, or if all immature fruits already set on the maternal parent are removed before the cross is made.

Fruits containing fewer seeds than normal and/or seeds which are developing slowly, are more likely to abort than fruits containing a full complement of vigorous seeds (Stephenson, 1981). An interspecific cross may thus abort if it has to compete with fruits set by later natural pollinations. It is therefore advisable to remove all other fruits from the vicinity of a fruit produced by wide crossing. Developing seeds are usually stronger sinks than vegetative meristems, so seeds are filled even if there is some vegetative die-back in consequence. However, seeds containing hybrid tissues which are developing unusually slowly may not be able to out-compete vegetative sinks. Pruning the maternal parent to remove all active growing points may therefore also help to channel resources to the feeble sinks represented by the hybrid seeds.

Reciprocal crosses

A cross which yields abortive seed in one direction may yield viable seed in the reciprocal direction. For example, when *Phaseolus coccineus* is pollinated by *P. vulgaris*, the embryos abort unless rescued by culture *in vitro*, but the reciprocal cross produces seed which may germinate normally. The genomic composition of the embryo is the same in the abortive and the viable cross, but the composition of the hybrid endosperm is different. The hybrid endosperm is also surrounded by different maternal tissues in the two crosses.

Results of this sort have led to much speculation about whether some balance between paternal and maternal genomes in the endosperm is necessary for normal development, or whether the balance must be between endo-

Table 5.3. Ploidy levels, endosperm balance numbers and results of interspecific crosses in *Solanum*. Endosperm balance numbers (EBN) of the haploid genomes of the parental species are assumed to be as follows: *S. acaule*, EBN = 0.5; *S. tuberosum* and *S. chacoense*, EBN = 1

| | | Endosperm | | |
| | | | | |
Cross	Ploidy	Genome ratio ♀ : ♂	EBN ratio ♀ : ♂	Seed development
2x *S. chacoense* × 2x *S. chacoense*	3x	2 : 1	2 : 1	Normal
4x *S. acaule* × 2x *S. chacoense*	5x	4 : 1	2 : 1	Normal
4x *S. acaule* × 4x *S. chacoense*	6x	4 : 2	2 : 2	Aborts
4x *S. tuberosum* × 4x *S. chacoense*	6x	4 : 2	4 : 2	Normal
4x *S. acaule* × 4x *S. tuberosum*	6x	4 : 2	2 : 2	Aborts
4x *S. acaule* × 2x *S. tuberosum*	5x	4 : 1	2 : 1	Normal
8x *S. acaule* × 4x *S. tuberosum*	10x	8 : 2	4 : 2	Normal

sperm and surrounding maternal tissues. Because there is no satisfactory general explanation of why endosperm breaks down in some interspecific crosses but not others, it is not usually possible to predict in advance whether reciprocal crossing will or will not overcome problems of endosperm abortion, nor to predict which direction of cross is the more likely to succeed. The practical plant breeder should consequently make his initial attempts at an interspecific cross in both directions.

Manipulations of ploidy level
Sometimes a cross between two species will produce inviable seeds but if the ploidy level of one of these species is changed, viable seeds result. For example, *Solanum acaule* is a tetraploid. When it is crossed with tetraploid *S. tuberosum* no viable seeds are produced. However, if a synthetic octaploid *S. acaule* is crossed with tetraploid *S. tuberosum*, viable seed results. Sufficient data are now available in the potato group for breeders to predict successfully when, and in which direction, ploidy levels may need to be changed to produce viable seeds from an interspecific cross Table 5.3).

Geneticists have not fully explained why these manipulations produce their effects. Once again, various theories have been put forward concerning the balance between gene dosages in embryo, endosperm and/or maternal tissues. The theory that fits the greatest number of experimental results assumes that normal endosperm development requires a balance between two maternally derived genomes and one paternally derived genome. If the ratios in the endosperm deviate significantly from this, in either direction, the seed aborts. Thus the diploid wild potato *Solanum chacoense* sets a full complement of viable seed after compatible intraspecific pollination. Endosperm of these seeds contains two maternal genomes, contributed by the two polar nuclei, and one paternal genome, contributed by the sperm. However, a cross between diploid *S. chacoense* and synthetic autotetraploid *S. chacoense* seldom or never produces viable seed. If the diploid is used as the seed parent, endosperm of the aborting seeds contains two maternal genomes (from the polar nuclei) and two paternal genomes (from the sperm of the autotetraploid). If the diploid is used as the pollen parent, endosperm of the aborting seeds contains four maternal genomes (from the polar nuclei of the autotetraploid) and only one paternal genome. Since both these ratios depart from normal, the seeds abort regardless of the direction in which the cross is made.

However, the theory of endosperm balance on its own does not explain why tetraploid *S. acaule* fails to produce viable seed when crossed with tetraploid *S. tuberosum*. There are many similar examples in other genera where, when species at the same ploidy level are crossed, the hybrid endosperm aborts early in seed development. It has therefore been suggested that the genome of each species may have its own characteristic 'strength' or 'value', independent of its level of ploidy. These different values are represented by different endosperm balance numbers (EBN). Thus, each haploid genome of *S. acaule* has been assigned an arbitrary EBN of 0.5, whereas each haploid genome of *S. tuberosum* has an EBN of 1.0. The endosperm in a cross between *S. acaule* and *S. tuberosum* therefore has a ratio of 2 maternal:2 paternal EBN (if *S. acaule* is the female parent) or 4 maternal:1 paternal EBN (if *S. tuberosum* is the female parent) and the seeds abort. However, in the cross between octaploid *S. acaule* and tetraploid *S. tuberosum* the normal balance of 2 maternal:1 paternal EBN is restored and the seeds are viable. Other examples are given in Table 5.3. Differences in EBN also seem to explain data on interspecific crossing in *Trifolium*, and in the wheat group (Williams, 1987; Singh *et al.*, 1990), even though no one has yet explained what an EBN actually represents.

An additional, and equally poorly understood, complication is that certain unspecified gene(s) concerned with early seed development seem to be expressed more actively when inherited from the paternal rather than the maternal parent. The most convincing data come from studies on maize endosperm, though a similar situation may occur in pea. The phenomenon has intriguing similarities with imprinting in mammalian embryos, whereby the expression of certain genes, or chromosome segments, involved in embryogenesis depends on whether they are inherited through the male or the female gamete. In flowering plants, with their double fertilization, imprinting might be expected in both endosperm and embryo. More work needs to be done to establish whether imprinting is widespread in crop plants, and if so, whether it can be combined with any of the balance theories to provide a retrospective explanation of results already obtained from interspecific crosses and a useful predictor of results from untried crosses.

Embryo rescue

Hybrid embryos may be excised from aborting seeds and cultured *in vitro* (Williams *et al.*, 1987). If the embryo has reached the heart or later stages, it can usually be cultured successfully, though much patient experimentation may be required to determine the optimum medium or, more usually, succession of media for continued growth and normal differentiation. Embryos which abort before the heart stage is reached are difficult to culture successfully. Such embryos are often fatally damaged by attempts to remove them from their surrounding tissues, possibly because of injury to the suspensor. Globular embryos are completely undifferentiated, whereas in heart stage embryos, the polarization which ultimately leads to differentiation of root and shoot is already established. Very little is known about what induces this polarization, though changes in osmotic potential, osmotic gradients or gradients in growth regulators within both ovule and embryo sac may be involved. It is sometimes difficult to induce proper polarization, and hence normal differentiation, of immature embryos *in vitro*. The heart stage also marks the period when the embryo switches from a heterotrophic phase, in which it is completely dependent on the endosperm for nutrients, to an autotrophic phase in which it is independent of the endosperm.

To reduce the risk of damage to very small embryos, a two-stage process of 'ovule culture' followed by embryo culture is often advocated (Collins *et al.*, 1984; Williams *et al.*, 1987). This has the additional advantage that the natural pathway of nutrient transfer within the ovule, and possibly also the normal osmotic environment, is maintained. The embryo is excised from the cultured ovule when it has reached a sufficiently advanced stage to be capable of surviving in culture on its own. This technique has been used successfully to rescue interspecific hybrids in *Gossypium* and *Allium*.

The endosperm contains a complex, and apparently genus- or species-specific, mixture of macro- and micronutrients, including various amino acids, and growth substances. Early attempts at embryo rescue, particularly rescue of very immature embryos, involved culture on complex media whose composition was designed to imitate that of the natural endosperm. This now seems unnecessary. Simpler media, including those used for somatic embryogenesis, may prove satisfactory also for rescue of sexual embryos. In general, solid or semi-solid media with low agar concentrations give better results than liquid media. Occasionally, as in cotton, a single medium can be used to support the embryo through all the stages of its development. More often, changing the composition of the medium as the embryo develops proves beneficial. The most important single factor seems to be the osmotic potential of the medium, but pH, and concentration and balance of the various plant growth regulators, may also be critical (Collins *et al.*, 1984).

Selection of an appropriate time and technique for embryo rescue requires some knowledge of when and why the embryo aborts *in vivo*. In *Solanum*, interspecifically fertilized ovules did not produce hybrids when cultured, though some embryos excised from these ovules before culture did yield hybrid plants. This might have been predicted, since in aborting hybrid seeds of Solanaceae, the maternal tissues in the ovule proliferate at the expense of the embryo and endosperm. Maintaining the fertilized ovule intact simply maintains whatever imbalances lead to abortion *in vivo*. The plant breeder needs in addition to know which genotypes of his intended parental species respond best to culture. Sufficient examples have been found of intraspecific variation in response to *in vitro* culture (Chapters 16 and 21) to suggest that effort invested in such initial screening is well worth while.

Use of plant growth regulators

Although exogenous PGRs are often applied to increase both quantity and quality of embryos from interspecific crosses, there are very few studies on precisely which processes in seed development are favourably affected by the added PGRs. Generalizations are also difficult because the effects of PGRs vary with concentration, with time of application, and with whether they are used singly or in combination. Gibberellins, auxins and cytokinins have all been extensively used, both *in vivo* and *in vitro*.

Gibberellic acids (GAs) are required by young embryos (Pharis and King, 1985). They stimulate cell division, and appear to attract assimilates to areas of high GA concentration. GAs may be made in the suspensor of young embryos (Williams, 1987), but the endosperm seems to be a major site of biosynthesis. GAs are then exported to both the embryo and the fruit, where they appear to control continued growth. Exogenous GAs may therefore promote retention, after interspecific pollination, of fruits which subsequently prove to be seedless. Exogenous GA may genuinely improve results from interspecific crosses by permitting hybrid embryos to develop *in vivo* to a stage at which they can be excised and cultured. Presumably the applied GA partially compensates for defects in development and nutrient transfer in the hybrid endosperm. Given this evidence of a requirement for, and manufacture of, gibberellins in young developing seeds, it is not surprising that GAs are usually necessary for *in vitro* culture of young embryos, though embryos of some species respond unfavourably. The balance between GA and other PGRs may be critical at these early stages. By the cotyledonary stage, addition of GA to the culture medium is more likely to inhibit embryo development.

Exogenous auxins, like gibberellins, may stimulate production of seedless, parthenocarpic fruit by substituting for the auxin normally produced by developing seeds. Auxins appear to be involved in DNA synthesis, and may be released from degenerating nuclei in dying or differentiating cells (Sheldrake, 1973). They are present in many parts of the developing seed, where they may be synthesized in the endosperm and possibly the suspensor. Auxins are important in the early development of the embryo, so application of exogenous auxin may stimulate the growth of hybrid embryos *in vivo* as well as *in vitro*. However, data from the Solanaceae suggest that excess auxin, whether endogenous or exogenous, early in seed development, may upset nutrient transport through integument, endosperm and embryo and the co-ordinated development of these tissues. Later in embryo development, auxin concentration, or the auxin:cytokinin balance, influences root formation.

Exogenous cytokinins may, like auxins and gibberellins, improve fruit set. Cytokinins are involved in cell division, and create active sinks (van Staden *et al.*, 1987). They may be deficient in inviable hybrid endosperms, but it is not yet certain whether they are synthesized in the endosperm or imported from the maternal plant. They accumulate in young embryos during the heart stage, and their presence in the culture medium usually enhances the survival of young embryos. However, concentration and balance with gibberellins and/or auxins are crucial for both early development and normal differentiation of the embryo. Although effects of cytokinins on embryos cultured *in vitro* have been quite extensively studied, much less is known about the movements and effects of externally applied cytokinins and whether they do actually reach their intended target site in a failing seed (van Staden *et al.*, 1987).

5.3 RETROSPECT AND PROSPECTS

The first recorded interspecific hybrid intentionally produced by man was Fairchild's mule pink (*Dianthus caryophyllus* x *D. barbatus*), synthesized in 1718. Since that time, much progress has been made in understanding and overcoming barriers to interspecific hybridization and extensive living collections of wild species in primary, secondary and to some extent tertiary gene pools of all the major and many of the minor crop plants have been assembled for use by breeders. Improvements in the sampling techniques used to collect genetic resources

from the wild mean that *ex situ* collections are now samples of allelic and not just taxonomic diversity. Accessions from most of these global collections are potentially available for breeding programmes anywhere in the world either as seeds or as *in vitro* cultures. Nevertheless, breeders have used these collections less than was originally hoped, partly because it is still difficult to locate and then to exploit potentially useful genes even when a collection has been assembled in a single place. Few breakthroughs in the laborious tasks of characterization and evaluation seem imminent, but techniques for identifying recombinants carrying desirable gene(s) from wild species have been greatly assisted by the use of molecular markers to construct saturated linkage maps. These, combined with accelerated breeding programmes, can do much to shorten the pre-breeding phase and thus overcome one widespread objection to the use of wild species to improve crops.

The most intractable barriers to interspecific gene transfer by sexual means remain those acting before fertilization, in the style, and after fertilization, in the endosperm. As understanding of the recognition processes involved in pollen germination, pollen tube growth and fertilization has improved, so have our abilities to manipulate these processes to secure hybrids which could not previously be produced. This co-operation between fundamental and applied research will undoubtedly continue, to the benefit of both. Similarly, advances in *in vitro* techniques and studies on how and why very different genomes do, or do not, coexist in the same nucleus, are helping us to understand, and then either to correct or to exploit, processes involved in post-fertilization breakdown. Some of the tools of molecular biology, such as monoclonal antibodies and *in situ* hybridization using various DNA probes, may soon make it possible to study the switching-on of various genes in the diverse tissues of the fertilized ovule, and the levels and movements of both exogenous and endogenous growth substances within the developing seed. This should enhance our abilities to overcome post-fertilization barriers to hybrid production.

It is sometimes suggested that, now that techniques in molecular biology and genetic engineering may be used to move genes not merely between species but even between kingdoms of living things, interspecific gene transfer by sexual means may be redundant. However, although genetic engineering has come a remarkably long way in a remarkably short time, it cannot yet be applied to all crops, nor yet to all characters. Moreover, much remains to be learned about the factors regulating the expression of alien genes in a new milieu. Most scientific revolutions improve rather than overturn old ways of doing things, and genetic engineering and interspecific sexual hybridization seem likely to continue to coexist as complementary ways of constructing the transgenic cultivars by means of which crop yields can be maintained, their resistance to pests and diseases improved, and their ecological adaptations adjusted to meet the challenges of agriculture on into the 21st century.

REFERENCES

Collins, G.B., Taylor, N.L. and de Verna, J.W. (1984), In vitro approaches to interspecific hybridization and chromosome manipulation in plants. In *Gene Manipulation in Plant Improvement*, 16th Stadler Genetics Symposium, Gustafsson, J.B. (ed.), Plenum Press, New York, pp. 323–383.

Goodman, R.M., Hauptli, H., Crossway, A. and Knauf, V.C. (1987), Gene transfer in crop improvement. *Science*, **236**, 48–54.

Gradziel, T.M. and Robinson, R.W. (1991), Overcoming unilateral breeding barriers between *Lycopersicon peruvianum* and cultivated tomato, *Lycopersicon esculentum*. *Euphytica*, **54**, 1–9.

Harlan, J.R and de Wet, J.M.J. (1971), Towards a rational classification of cultivated plants. *Taxon*, **20**, 509–517.

Heslop-Harrison, J. (1987), Pollen germination and pollen tube growth. In *Pollen: Cytology and Development*, Giles, K.L. and Prakash, J. (eds), Int. Rev. Cytol., **107**, 1–78.

Heslop-Harrison, J. and Heslop-Harrison, Y. (1987), Pollen-stigma interaction in the grasses. In *Grass Systematics and Evolution*, Soderstrom, T.R., Hilu, K.W., Campbell, C.S. and Barkworth, M.E. (eds), Smithsonian Institution Press, Washington DC, pp. 133–142.

Hogenboom, N.G. (1984), Incongruity: non-functioning of intercellular and intracellular partner relationships through non-matching information. In *Cellular Interactions*, Encyclopaedia of Plant Physiology, New Series, Vol. 17, Linskens, H.F. and Heslop-Harrison, J. (eds), Springer-Verlag, Berlin, pp. 609–623.

Jensen, W.A., Ashton, M.E. and Beasley, C.A. (1983), Pollen tube – embryo sac interaction in cotton. In *Pollen: Biology and Implications for Plant Breeding*, Mulcahy, D.L. and Ottaviano, E. (eds), Elsevier, New York, pp. 67–72.

Knox, R.B., Williams, E.G. and Dumas, C. (1986), Pollen, pistil and reproductive function in crop plants. In *Plant Breeding Reviews*, No. 4, Janick, J. (ed.), AVI Publishing Co., Westport, pp. 9–79.

Laurie, D.A. and Bennett, M.D. (1988a), The production of haploid wheat plants from wheat × maize crosses. *Theor. Appl. Genet.*, **76**, 393–397.

Laurie, D.A. and Bennett, M.D. (1988b), Cytological evidence for fertilization in hexaploid wheat × sorghum crosses. *Plant Breeding*, **100**, 73–82.

Murray, D.R. (1988), *Nutrition of the Angiosperm Embryo*, John Wiley, New York.

Pandey, K.K. (1981), Evolution of unilateral incompatibilities in flowering plants: further evidence in favour of twin specificities controlling intra- and interspecific incompatibility. *New Phytol.*, **89**, 705–728.

Pharis, R.P. and King, R.W. (1985), Gibberellins and reproductive development in seed plants. *Ann. Rev. Pl. Physiol.*, **36**, 517–568.

Prescott-Allen, R. and Prescott-Allen, C. (1988), *Genes from the Wild*, 2nd edn, Earthscan Publications, London.

Sastri, D.C. (1984), Incompatibility in angiosperms: significance in crop improvement. In *Advances in Applied Biology*, Coaker, T.H. (ed.), No. 10, Academic Press, London, pp. 71–111.

Sedgley, M. and Griffin, A.R. (1989), *Sexual Reproduction of Tree Crops*, Academic Press, London.

Sheldrake, A.R. (1973), The production of hormones in higher plants. *Biol. Rev.*, **48**, 509–559.

Shivanna, K.R. (1982), Pollen-pistil interaction and control of fertilization. In *Experimental Embryology of Vascular Plants*, Johri, B.M. (ed.), Springer-Verlag, Berlin, pp. 131–174.

Singh, A.K., Moss, J.P. and Smartt, J. (1990), Ploidy manipulations for interspecific gene transfer. *Adv. Agron.*, **43**, 199–240.

Stalker, H.T. (1980), Utilisation of wild species for crop improvement. *Adv. Agron.*, **33**, 111–147.

Stephenson, A.G. (1981), Flower and fruit abortion: proximate causes and ultimate functions. *Ann. Rev. Ecol. Systemat.*, **12**, 253–279.

Stettler, R.F. and Ager, A.A. (1984), Mentor effects in

pollen interactions. In *Cellular Interactions*, Encyclopaedia of Plant Physiology, New Series, Vol. 17, Linskens, H.F. and Heslop-Harrison, J. (eds), Springer-Verlag, Berlin, pp. 609–623.

Turano, M.J., Baird, L.M. and Webster, B.D. (1983), Characteristics of the stigma of chickpea. *Crop Sci.*, **23**, 1033–1036.

van Staden, J., Manning, J.C. and Dickens, C.W.S. (1987), A phylogenetic analysis of the role of plant hormones in the development and germination of legume seeds. In *Advances in Legume Systematics*, Part 3, Stirton, C.H. (ed.), Royal Botanic Gardens, Kew, pp. 387–442.

Williams, E.G. (1987), Interspecific hybridization in pasture legumes. In *Plant Breeding Reviews*, No. 5, Janick, J. (ed.), AVI Publishing Co., Westport, pp. 237–305.

Williams, E.G., Maheswaran, G. and Hutchinson, J.F. (1987), Embryo and ovule culture in crop improvement. In *Plant Breeding Reviews*, No. 5, Janick, J. (ed.), AVI Publishing Co., Westport, pp. 181–236.

Zenkteler, M. and Slusarkiewicz-Jarzina, A. (1986), Sexual reproduction in plants by applying the method of test-tube fertilization of ovules. In *Genetic Manipulation in Plant Breeding*, Horn, W., Jensen, C.J., Odenbach, W. and Schieder, O. (eds), Walter de Gruyter, Berlin, pp. 415–423

6

Chromosome manipulation and polyploidy

H. Thomas

6.1 INTRODUCTION

Overcoming the barriers associated with pollination and zygotic development to successfully produce interspecific hybrids between wild and cultivated species (Chapter 5), does not necessarily mean that the variation within the alien species can be used in breeding. The accessibility of alien variation to the breeder is dependent on the genetic relationship between the wild and cultivated species. Divergence between these forms, which impose restriction on chromosome pairing and hence recombination, limit the usefulness of hybrids between them in breeding programmes. The successful use of alien variation in breeding is related to the strength of internal barriers to the free flow of genes between the wild and cultivated species and the ingenuity of the plant breeder/cytogeneticist to overcome these barriers.

In diploid species the isolation imposed by domestication does not always result in differentiation and considerable variation in this respect is found between the cultivated form and its wild progenitor depending on the crop concerned. Within the plant kingdom polyploidy has played a significant role in speciation and a number of cultivated species are polyploids. The origin of the polyploid forms will often dictate whether variation within the

Plant Breeding: Principles and prospects. Edited by M.D. Hayward, N.O. Bosemark and I. Romagosa. Published in 1993 by Chapman & Hall, London. ISBN 0 412 43390 7.

diploid progenitors can be used for improving the polyploid species. The most significant factor is the control of chromosome pairing and recombination in polyploids and hybrid derivatives. As outlined in Chapter 1, success in transferring genes across ploidy levels often requires the manipulation of these control system to overcome restrictions on recombination in interspecific hybrids. Similarly the breeding strategies employed to improve polyploid crop species will also depend on whether their origin involved straightforward doubling of the chromosomes or a combination of interspecific hybridization and chromosome doubling.

6.2 GENE POOLS

Based on the genetic relationships between species, Harlan and de Wet (1971) developed the concept of gene pools of cultivated crops. In their classification the primary gene pool constitutes species which form fertile hybrids, with regular chromosome pairing and near normal gene segregation, making gene transfer easily attainable. In the secondary gene pool, species hybrids have a tendency to be sterile, chromosomes pair poorly and, sometimes, development is abnormal. Gene transfer in this group is possible and the variation can be used if the breeder is prepared to devote the required effort. At the level of the tertiary gene pool, access to alien variation is only possible by using special techniques to overcome the barriers isolating the wild species from the cultivated form. The classification of natural variation into distinct gene pools closely follows the concept of biological species.

6.3 DIPLOID SPECIES

Domestication has meant spatial isolation of the crop species from its progenitor, but this has not always been accompanied by the development of internal isolation barriers. The cultivated barley *Hordeum vulgare* crosses easily with its diploid progenitor *H. spontaneum*, forming a fertile hybrid and gene transfer is accomplished through backcrossing. There is little restriction of recombination in hybrids between tomato and related wild species and many genes conferring resistance to disease have been introduced into tomato from these sources by backcrossing.

In general, fertile and partly fertile F_1 hybrids are amenable to backcrossing, and it is an effective method of transferring genes from related species into cultivated forms. The genome of the recipient species is reconstituted through backcrossing and the goal is to transfer the smallest possible segment of the genome of the donor species to avoid the introduction of deleterious genes with the target gene. Up to six backcrosses are required to achieve this, but by using genetic markers to assist the selection of the genome of the recurrent species the number of generations of backcrossing required can be reduced (Chapter 19). Speeding up the process increases the efficiency of gene transfer and the use of interspecific hybridization at the primary gene pool level.

Desirable variation occurs in the secondary and tertiary gene pools, but hybrids between the cultivated species and selected donor species are sterile and have poor chromosome pairing. In hybrids between sugar beet, *Beta vulgaris*, and *B. procumbens*, sterility and poor chromosome pairing restricts the transfer of nematode resistance from the alien species into the recipient sugar beet. By producing a triploid hybrid between $4x$ *B. vulgaris* and *B. procumbens*, Savitsky (1975) was able to overcome these barriers to successful gene transfer. In the

triploid hybrids nine bivalents (9_{II}), arising from the associations between the *B. vulgaris* chromosomes, are formed and the nine *B. procumbens* chromosomes remain unpaired (9_I) with occasional cells having $8_{II} + 8_I + 1_{III}$. Such meiotic behaviour results in the production of some viable gametes, enabling the triploid to be backcrossed to diploid sugar beet. Of the 6750 backcross (BC) hybrid plants tested, four were found to be nematode resistant and were single chromosome addition lines with 19 chromosomes. The single chromosome of *B. procumbens*, carrying the gene for resistance, had been added to the sugar beet complement. Transmission of the alien chromosome averaged 12% over eight generations of backcrossing. Two diploid progeny resistant to the nematode were also isolated among the BC progeny. The resistant diploid progenies were the product of rare recombination. The meiotic behaviour of the triploid hybrid ensured the establishment of the single chromosome addition line which in turn provided the opportunity for recombination that was easily identified in the BC progeny. This example illustrates the effort required to achieve transfers within the secondary and tertiary gene pool in diploid species.

The successful transfer of the *B. procumbens* gene into sugar beet is a classical example of using an induced autopolyploid as a bridge to overcome the isolation of the two species. The bridge species is strategically used to optimize the chances of obtaining the rare recombinant. An autopolyploid bridge can also be used effectively to regulate recombination in diploid species where the restrictions imposed by sterility and poor chromosome pairing do not prevail. The triploid hybrid *Lolium multiflorum* ($4x$) × *Festuca pratensis* ($2x$) is sufficiently fertile to be used as the male parent in backcrosses to *L. multiflorum* ($2x$) and over 70% of the BC progeny were diploid and mainly resembled the ryegrass parent. Marker genes of *F. pratensis* were recovered in the diploid progeny, but at a lower frequency than expected. Although a high number of trivalents were formed in the triploid hybrid the functional gametes were mainly haploid since more than 70% of the BC progeny were diploid (H. Thomas and W.G. Morgan, in preparation). Alternate disjunction of the trivalents would produce the haploid gametes and their selective advantage over others with chromosome numbers in excess of the haploid number would ensure that the majority of the BC progenies were diploid. If the fescue chromosome occupied one of the terminal positions in chain trivalents or the stem in pan-handle configuration, only the recombined segment of the fescue chromosome would have a chance of being included in the functional haploid gametes. This reduces segregation and increases the chances of fixing recombinants in a predominatly *Lolium* genotype and thus reduces the number of backcrosses required. This scheme could be used wherever the triploid hybrid is fertile.

6.4 POLYPLOID SPECIES

Polyploidy has played a significant role in speciation within the plant kingdom and a number of our cultivated crops are polyploids. The breeding strategies employed in polyploid crops largely depend on their origin, the main criterion used by Stebbins (1950) in his classification of auto- and allopolyploids. Autopolyploidy originates from the doubling of the chromosome number of a diploid species or a hybrid between races of the same species, resulting in two pairs of homologous chromosomes. Alloploidy usually involves hybridization between diverged species followed by doubling of the chromosome number and thus combine two distinct genomes.

6.4.1 Autopolyploids

The consequence of chromosome doubling is that the four homologous chromosomes can form a quadrivalent, a trivalent and a univalent, or two bivalents at meiosis, which can lead to irregular meiosis in newly induced autopolyploids. However, many natural autopolyploids have evolved systems leading to regular meiosis. In *Lotus corniculatus* and *Phleum pratense*, chromosome pairing is mainly as bivalents but both species show polysomic inheritance, since the homologous chromosomes pair at random within the bivalents. Such a system ensures

regular chromosome disjunction and equal separation of the chromosomes at anaphase I and thus the absence of aneuploidy in the progeny. In *Dactylis glomerata* the number of quadrivalents formed at metaphase I is that expected of an autopolyploid, but a nearly 14:14 distribution to the anaphase poles is achieved (Jones, 1962). The regular alignment of the quadrivalents on the metaphase plate, which leads to the equal separation of the chromosomes at anaphase I, is rarely found in induced autopolyploids. The doubling of the chromosome number also leads to duplication of genetic loci. Genetic duplication and polysomic inheritance are factors that have to be taken into consideration in devising appropriate strategies for breeding autopolyploid crops.

(a) Unreduced gametes and the origin of autopolyploids

The production of functional gametes which have the somatic chromosome number, has been reported in a range of plant genera and their implication in the origin of polyploids the subject of much discussion. In a review of the origin of polyploidy, Harlan and de Wet (1975) proposed that almost all polyploids arise by way of unreduced gametes and that all other mechanisms are negligible. They claim that the most common event in this respect would be the fusion of $2n$ and n gametes to form a triploid which by backcrossing or selfing could yield a tetraploid. Spontaneous polyploids, mainly autopolyploids, appear repeatedly and their fate and chances of becoming established depends on their vigour and competitive ability. Normal meiotic behaviour in most species results in the formation of n gametes and the occurrence of unreduced gametes is generally rare. A survey of the literature (Harlan and de Wet, 1975) showed that the phenomenon was widely distributed throughout plant genera and there was evidence of genetic control.

Most known natural autopolyploid species are outbreeders and, as pointed out by Bingham (1980), such species show marked inbreeding depression unless heterozygosity is maintained. He proposed that the greater than predicted inbreeding depression in natural autopolyploids was due to loss of first order interactions from tri- and tetra-allelic loci. In alfalfa the performance of double cross (DC), single cross (SC) hybrids and parents (P) were always of the order DC > SC > P. Differences were significant and such consistent ranking is correlated with the theoretical increase in the frequency of tetra- and tri-allelic loci. Maximizing heterozygosity in autopolyploids should thus be an important aim in breeding.

(b) Unreduced gametes and maximizing heterozygosity

As indicated above, meiosis in most plant species is normal and haploid (n) gametes produced, but abnormalities can give rise to $2n$ gametes. The occurrence of $2n$ gametes is variable and is probably under genetic control. Unreduced gametes are common in the potato (Solanaceae) family and Peloquin _et al._ (1989) have made detailed studies of the occurrence and consequences of $2n$ gametes on the development of polyploidy in that family. These $2n$ gametes arise through First Division Restitution (FDR) or Second Division Restitution (SDR). FDR $2n$ gametes result from parallel spindle formation after normal first division in many cells. Cleavage furrows occur across the parallel spindles to form dyads and $2 \times 2n$ pollen. The consequence of this division is that most of the heterozygosity of the diploid hybrid is conserved in the $2n$ gametes (Fig. 6.1). SDR $2n$ gametes arise through another mechanism that affects meiosis. The first meiotic division is followed by cytokinesis but the second division is absent also resulting in a dyad with $2 \times 2n$ gametes. The consequence of FDR on the genetic composition of the $2n$ gametes is that they maintain most of the heterozygosity of the diploid hybrid, whereas this is greatly reduced in SDR $2n$ gametes. In the potato the performance of certain $4x$ progeny from $2n \times 2n$ FDR $2n$ gametes reflects this difference in heterozygosity because they exceeded the mid-parent yield and some of the best tetraploid cultivars, but the $4x$ progeny produced by SDR $2n$ gametes did not produce a similar heterotic effect. Bingham (1980) has reported similar effects of using FDR $2n$ gametes to construct

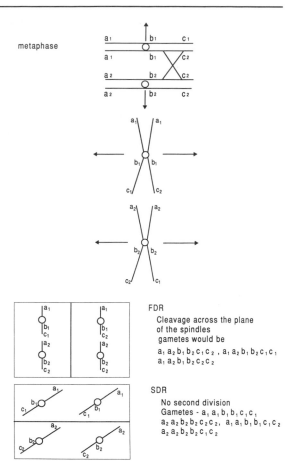

Fig. 6.1. Consequences of FDR and SDR in unreduced gametes.

highly heterozygous genotypes of alfalfa that consistently surpassed the parental and mid-parent values. Harnessing this heterotic effect would be more practical in potato since any elite lines produced from the fusion of two FDR $2n$ gametes could be multiplied asexually. In an autopolyploid crop like alfalfa, which is dependent on a seed crop for multiplication, the direct use of FDR to maximize heterozygosity would not be practical and other strategies would have to be followed such as hybrid seed production.

(c) Diploid–tetraploid–diploid cycles in gene transfer in autopolyploids

Meiotic irregularities that result in the reduction of autopolyploids ($4x$) to dihaploids ($2x$) have been reported in both potato and alfalfa. The reduction of an autopolyploid to a dihaploid means that the latter can be crossed to related diploid species and genes introgressed at the diploid level. The diploid genotype may then be repolyploidized using appropriate techniques. This ability to initiate such $2x–4x–2x$ cycles is an effective procedure for the introgression of genes from diploid species into tetraploid species wherever the set of mutant stocks affecting meiotic behaviour is available. In cocksfoot a slight modification of this procedure has been used to introgress genes from related diploid species into the autotetraploid *Dactylis glomerata*. Hybridizing diploid species with the tetraploid resulted in some $4x$ progeny and this was an effective means of utilizing variation from the diploid species in breeding cocksfoot (Thomas, 1981).

(d) Induced autopolyploids

Larger vegetative parts of the plant are features of autopolyploids that created the original stimulus for the research input into artificial polyploids. In general with these '*gigas* effects', which are mainly attributed to increased cell size, there is a reduction in fertility, a severe limitation in crops grown for grain, but less so in fodder crops which are cultivated mainly for vegetative growth. Levan (cited by Dewey, 1980) concluded that crops most amenable to improvement through chromosome doubling should: (a) have low chromosome number; (b) be harvested for vegetative growth; (c) be cross-pollinating and Dewey added; (d) have perennial habit; (e) have ability to reproduce vegetatively.

The ryegrasses meet all these five conditions and a number of tetraploid cultivars have been released in Europe. Comparison of the diploid and tetraploid cultivars showed that polyploids produced 18% higher yield of fresh weight but dry matter yields of both groups were identical (van Bogaert, 1975). The polyploids had superior disease resistance, palatability and digestibility, more structural carbohydrates and less crude fibre. Ryegrasses are cross-pollinating and perennial and their maintenance in a sward, post-seeding year, is by vegetative

reproduction. Although there is a reduction in fertility, seed yields are adequate for the development of cultivars. Red clover, *Trifolium pratense*, also satisfies most of the conditions for successful autopolyploid breeding but low seed yield has restricted their development as $4x$ cultivars.

Infertility is generally associated with irregular meiosis in artificial autopolyploids. A consequence of these irregularities is the failure of equal separation of the chromosomes at anaphase I and gametes with chromosomes deviating from the haploid number are formed. This in turn gives rise to aneuploidy in filial generations of the autopolyploids. In red clover, *T. pratense*, larger flower size means that bees can not pollinate the plants, reducing still further their seed production. Even perennial fodder crops grown for their vegetative growth are dependent on seed for initial establishment. Rye, *Secale cereale*, is the only grain-producing crop to be successfully developed as an autotetraploid. Triploid cultivars of sugar beet are prevalent in Europe but their superiority over diploids is not universal. Sterility of triploids has been exploited to produce high value crops such as seedless water melons.

The changes in the genetic system imposed by genome duplication would also have to be considered in devising strategies for breeding synthetic autotetraploids. As in newly produced polyploids the highest level of heterozygosity even on doubling the chromosome number of a diploid hybrid (a_ia_j) would be duplex ($a_ia_ia_ja_j$). Bingham (1980) proposed that a sufficient number of tetraploid lines should be synthesized from diverse diploid lines to insure 'allelic diversity in the breeding programme'. Single and double crosses of the diverse $4x$ lines would be required to produce cultivars with maximum hybrid vigour. In vegetatively reproducing crops heterozygosity of elite genotypes could be fixed instantly but in seed crops strategies based on maximizing heterozygosity within populations would have to be implemented. Aneuploidy in advanced generations would be another important factor to consider in devising breeding strategies for induced autopolyploids.

6.4.2 Allopolyploids

The origin of allopolyploidy is closely connected with interspecific hybridization as well as chromosome doubling. The mechanism of polyploidization is not always clear but Harlan and de Wet (1975) have made a strong case that unreduced gametes were the primary force behind the production of allopolyploids as well as autopolyploids. Whereas in the origin of autopolyploids the importance of the unreduced gametes would be in intraspecific hybrids, including those between different races of the same species, in allopolyploids it is concerned with hybrids between species. Studies of allopolyploids have revealed that the progenitor species could be closely related or show a high level of divergence. The net result of interspecific hybridization and polyploidization is that the genomes of distinct species are combined in the allopolyploid. Hybrids between diverged species are often sterile but polyploidization usually restores fertility and the resulting allopolyploid will breed true for a highly heterozygous genotype (Stebbins, 1950). Autopolyploids rely on the breeding system to generate heterozygosity but the internal hybridity of allopolyploids maintains heterozygosity even in inbreeding species like wheat, due to allelic differences at the loci of the constituent genomes.

Most natural allopolyploids have evolved genetic systems which ensure that chromosome associations are confined to homologous pairs of the same origin resulting in diploid-like bivalent pairing, and disomic inheritance. Such genetic control is best understood in hexaploid wheat *Triticum aestivum* ($2n = 6x = 42$) where the presence of a gene on chromosome *5B* controls diploid-like chromosome pairing. Deletion of chromosome *5B* results in associations between homoeologous, or corresponding chromosomes of the three constituent genomes, resulting in multivalents at metaphase I. The control of chromosome pairing maintains the integrity of the constituent genomes and a relaxation of the system results in random pairing between the homoeologues as in autopolyploids. This leads to the break down of favourable gene complexes and the internal hybridity associated with heterozygosity in the different genomes. Other natural allopolyploids such as oats, cotton and tall fescue have similar genetic control but these systems are not as well understood as in wheat.

(a) Induced allopolyploids

An understanding of the significance of interspecific hybridization in the evolution of allopolyploids, combined with the discovery that colchicine could be used to double the chromosome number of plants, led to an interest in the synthesis of allopolyploids by treating interspecific hybrids with colchicine. The attraction of this approach is to combine complementary characters of two species in a newly synthesized species. The success of induced allopolyploidy as a breeding technique has been limited since only triticale (wheat–rye amphiploid) and tetraploid hybrid ryegrass (*L. multiflorum* × *L. perenne*) have been released as cultivars and used in farming systems. The early effort to produce wheat–rye amphiploids was based on hexaploid wheat but, within the past 30 years attention has shifted to combining the complements of tetraploid *T. durum* and rye. Although hexaploid triticale is undoubtedly more successful than the octoploid form there are still problems of genetic instability usually associated with asynapsis. The hexaploid triticale has been hybridized with hexaploid wheat forming a hybrid in which the tetraploid component *AABB* is maintained together with the haploid complements of the *D* and *R* genomes (see review by Gupta and Priydarshan, 1982). Selections in which the third genome has a variable number of pairs of *D* and *R* chromosomes have been made. Some of the most successful cultivars available are derived from such breeding programmes. Triticale is a crop for marginal areas of wheat production and is not regarded as a replacement for wheat.

Tetraploid hybrids combining the genomes of Italian and perennial ryegrass are another example of a synthetic allopolyploid released as cultivars, which have had an impact on grassland farming in Europe. These amphiploids generally combine the complementary characters of the parental species and bridge the gap between them. The diploid F_1 hybrids between the species are fertile indicating the close relationship between the genomes of the two species. As a consequence, the chromosomes of the two genomes do not completely pair preferentially to form 14 bivalents at metaphase I, but form one to three quadrivalents. Breese *et al.* (1981) have demonstrated that a degree of preferential chromosome pairing superimposed on tetrasomic segregation would be effective in reducing segregation and maintain a high percentage of hybridity in the amphiploid through the generations of seed multiplication required prior to release of a cultivar. In contrast, efforts to produce diploid cultivars, combining species attributes from segregating populations of the interspecific hybrid, were not successful. Being outbreeders it was extremely difficult to fix such genotypes at the diploid level.

Although the Italian × perennial ryegrass allopolyploid has been successful, other amphiploids within the *Lolium/Festuca* complex have been disappointing. The main reason for the lack of success has been infertility and genetic instability in later generations of the amphiploids (Thomas and Humphreys, 1991). Similar reports of amphiploids within the perennial grasses of the Triticeae have been reported by Dewey (1980). Poor success in allopolyploid breeding is mainly due to the inability to stabilize meiotic behaviour to a level comparable to that found in natural allopolyploids. Irregular meiosis arises mainly as a result of homoeologous chromosome pairing leading to the formation of unbalanced gametes. The key to success in allopolyploid breeding is regular diploid-like chromosome pairing.

In a comprehensive review of polyploidy and plant breeding, Dewey (1980) proposed that tetraploid amphiploids combining the genomes of two diploid species had a greater chance of succeeding than higher levels of ploidy, and that hexaploid was the limit in terms of chromosome number as far as breeding allopolyploids was concerned. In perennial grasses of the Triticeae over 60% are tetraploid and Dewey suggested that this was a strong indicator of greater expectations when tetraploid allopolyploids are synthesized. Nevertheless, within the Triticeae and Festucae there is a small percentage of successful natural $8x$ and $10x$ allopolyploid species which would indicate that, if regular meiosis could be achieved, octoploids like *L. multiflorum* × *F. arundinacea* could be successfully developed into cultivars.

6.5 GENE TRANSFERS INVOLVING CHROMOSOMES OR CHROMOSOME SEGMENTS

Although allopolyploid breeding has only been partially successful in terms of the development of cultivars, they provide an effective bridge for controlled introgression of genes into cultivated species. The inclusion of only part of the genome rather than the whole has produced a more acceptable genotype. Controlled introgression aims to introduce the smallest possible segment of the alien genome into the cultivated species to avoid introducing any further genetic material that could be detrimental to the performance of the recipient. Where the possibilities of recombination are limited, the introduction of the whole chromosome can be an alternative method of using alien variation.

6.5.1 Chromosome addition lines

The progressive reduction in the amount of the alien genome retained when synthesized allopolyploids are backcrossed to the cultivated crop leads to a more agronomically acceptable progeny. This is well illustrated in Cauderon's (1977) description of using perennial intermediate wheat grass *Agropyron intermedium* ($2n = 42$) as a source of variation in wheat breeding. The three genomes of *A. intermedium* are not related to wheat. The amphiploid obtained by doubling the chromosome number of the F_1 hybrid was weak and only partially fertile. This was backcrossed twice to wheat and progeny selected for fertility and rust resistance. From this backcross a fairly stable octoploid was isolated, which was a partial amphiploid with 42 wheat and 14 *Agropyron* chromosomes. The partial amphiploid was fertile, resistant to rust and relatively stable, but its yield was only 50% of that of wheat. Further backcrosses produced single chromosome addition lines of *Agropyron* chromosomes to wheat, including the chromosomes on which the rust resistant genes are located. Reduction of the proportion of *Agropyron* genome present in the backcross progeny is correlated with a more agronomically acceptable wheat plant.

The majority of chromosome addition lines reported in the literature originate from backcrosses aimed at introducing a gene from an alien into a cultivated species. The absence of recombination and selection for the particular character invariably leads to the transfer of the whole chromosome. Some of the alien chromosome addition lines are agronomically comparable to the crop species but they are not sufficiently stable to be released as cultivars. Instability is correlated with the failure of the pair of alien chromosomes to synapse in a small proportion of pollen mother cells. The unpaired chromosomes usually fail to become incorporated into the second telophase nuclei and form micronuclei in the tetrads. This leads to the formation of haploid gametes which have a selective advantage over the normal $(n+1)$ gametes produced by the addition lines, and monosomic addition lines $(2n = 43)$ appear among the progeny. On selfing the $2n = 43$ plants produce up to 90% euploid progeny that do not express the transferred character, e.g. if it concerned resistance to rust there would be a high proportion of susceptible progeny in the population. The variation arising from the instability would be greater than the level tolerated for cultivar registration.

6.5.2 Chromosome substitution

Although alien chromosome addition lines are insufficiently stable to be developed as cultivars, they provide invaluable material for further chromosome manipulation techniques. An alternative to adding an alien chromosome would be to substitute a chromosome of the recipient species with an alien chromosome. Cultivars of wheat bred in Germany in the 1930s have been shown to be rye/wheat substi-

tution lines (Mettin *et al.*, 1973). A triticale line was used as a parent in the crossing programme and selection for disease resistance based on the presence of the rye chromosome, resulted in a series of cultivars which have since been shown to be spontaneous substitution lines.

The availability of a series of aneuploid lines in wheat makes it possible to produce specific substitution lines by following the crossing scheme in Fig. 6.2. The alien chromosome will only substitute for a wheat chromosome of the same homoeologous group. The rye chromosome compensates for the deleted wheat chromosome but substitutions involving non-homoeologous chromosomes show no evidence of compensation. If alien chromosome substitution lines are to be successful as cultivars, the alien chromosome has to be integrated into the genome of the recipient species, have no deleterious effect on development and the desired genes located on the chromosome must be expressed in the genotypic background of the recipient species. The spontaneous substitutions in the German wheat cultivars met all these criteria and became successful cultivars.

Intervarietal chromosome substitution lines, whereby a chromosome of one cultivar is replaced by one from another cultivar can be produced using the monosomic series $(2n = 6x\text{-}1 = 41)$ as outlined in Fig. 6.3. After six to eight backcrosses the genotypic background of variety 'A' is recovered except for the single chromosome of variety 'B'. Comparisons between variety 'A' and the intervarietal substitution line will provide a precise assay of the genes located on that chromosome (Law *et al.*, 1980). Wheat is the only species in which complete sets of aneuploids are available, but partial sets of monosomic lines have been isolated in other polyploid crop plants such as oats, cotton and tobacco.

6.5.3 Transferring part of the chromosome

Chromosome substitutions are generally more stable and successful than chromosome addition lines, but undesirable linkages limit their usefulness as breeding material. The ideal

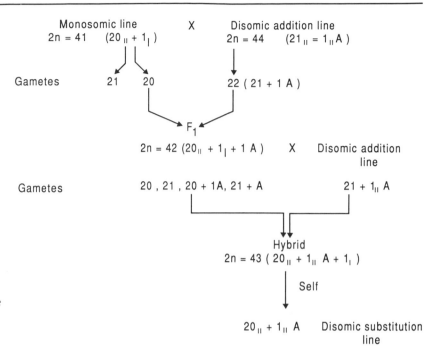

Monosomic line X Disomic addition line
$2n = 41$ $(20_{II} + 1_{I})$ $2n = 44$ $(21_{II} = 1_{II}A)$

Gametes 21 20 22 (21 + 1 A)

F_1
$2n = 42 (20_{II} + 1_{I} + 1 A)$ X Disomic addition line

Gametes 20 , 21 , 20 + 1A, 21 + A 21 + 1_{II} A

Hybrid
$2n = 43 (20_{II} + 1_{II} A + 1_{I})$

Self

$20_{II} + 1_{II}$ A Disomic substitution line

Fig. 6.2. Crossing scheme for producing alien chromosome substitution lines.

situation would be the introgression of the target gene and the exclusion of any extraneous genes carried by the chromosome. In secondary and tertiary gene pools, where genetic recombination is limited, this restriction has to be overcome to achieve transfers involving small segments of the alien chromosome.

6.5.4 Irradiation-induced transfers

The knowledge that irradiation can result in chromosome breakage and that subsequent reunion can give rise to reciprocal exchange of chromosome segments, has been used as a means of obtaining gene transfers. The transfer of leaf rust resistance from *Aegilops umbellulata* to bread wheat described by Sears (1956) was the first report of the successful use of irradiation to insert the critical segment of an alien chromosome into bread wheat. Chromosome addition lines provide the starting point for such procedures where dry seed or panicles at various stages of meiosis can be irradiated and the progeny screened for possible transfers. There are a number of reports of irradiation-induced transfers in the literature, most of which are in wheat, but others have been reported in tobacco and oats.

Irradiation-induced transfers are always based on reciprocal translocations between the alien chromosome and a chromosome of the recipient species due to chance chromosome breakage and reunion. Subsequent breeding and selection results in only the restructured chromosome bearing the transferred locus being retained, whereas the reciprocal restructured chromosome is eliminated. The consequence of this is that the pre-breeding line including the transferred gene has a segment of chromosome of the recipient species deleted and the segment of the alien chromosome as a duplication. The usefulness of such transfers hinges on the ability of the recipient species to tolerate such genetic deletion/duplication. Such anomalies can only be tolerated by polyploid species on account of their genetic duplication, whereas such a situation would be lethal in a diploid species.

A number of successful irradiation transfers have been used in breeding. There is firm evidence that the performance of genotypes

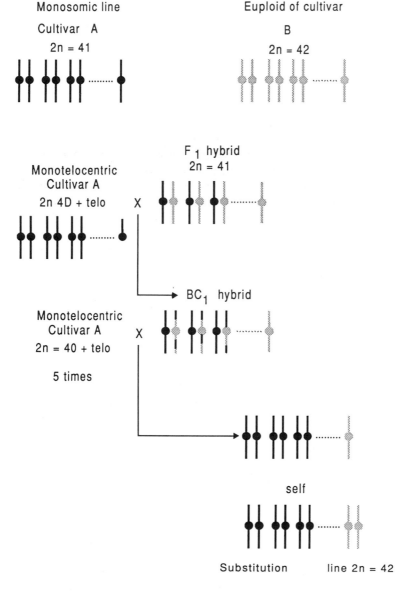

Monosomic line
Cultivar A
2n = 41

Euploid of cultivar
B
2n = 42

Monotelocentric
Cultivar A
2n 4D + telo

X

F₁ hybrid
2n = 41

BC₁ hybrid

Monotelocentric
Cultivar A
2n = 40 + telo

X

5 times

self

Substitution   line 2n = 42

Fig. 6.3. Method of producing intervarietal chromosome substitution lines.

including the transfer is influenced by genetic background. A transfer involving mildew resistance from *Avena barbata* into cultivated oat was backcrossed four times into a range of cultivars and the transmission of the translocation was normal in some genotypes but impaired in others. Yields of the different lines, when the advantage of mildew resistance was negated by spraying the original varieties with fungicide, also showed marked translocation/genotype interaction (Thomas *et al.*, 1980).

6.5.5 Genetically induced translocations

The genetic control of bivalent pairing in wheat is due to the *Ph* gene on chromosome *5B* that discriminates between homologues and homoeologues in the pairing process. The divergence between an alien chromosome and the corresponding chromosomes in wheat is often greater than among the three chromosome pairs of the homoeologous group and therefore the *Ph* gene would also prevent pairing between the alien chromosome and its homoeologues in wheat. This would account for the absence of recombinants in selfed progeny from alien chromosome addition lines. Deletion of the *5B* chromosome does result in associations between homoeologous chromosomes in wheat and alien chromosomes should also pair if they have not diverged beyond the homologous/homoeologous threshold over which the *5B* system operates (Riley and Law, 1965). Thus manipulating the *5B* system to relax the restriction on homoeologous chromosome pairing should increase the chances of recombining the chromosomes of wheat and the alien species.

The first successful transfer based on modifying the *5B* system was in the production of 'Compair' wheat which is fully described in Chapter 1. A similar approach has been applied to transfer mildew resistance from *A. barbata* into cultivated oat *A. sativa* by Thomas *et al.* (1980) using a genotype of *A. longiglumis* to induce homoeologous chromosome pairing. The addition line involving only a pair of telocentrics of the short arm of the *A. barbata* chromosome was crossed with the *A. sativa* x *A. longiglumis* 8x amphiploid and the hybrid backcrossed twice to *A. sativa*. Selection for mildew resistance resulted in a resistant derivative with 42 chromosomes. Previous attempts to isolate a recombinant from the progeny of the monosomic addition line (42 *A. sativa* + 1 *A. barbata* chromosome) had failed. The disadvantage of this procedure is that the *A. longiglumis* and *A. sativa* chromosomes also pair and this leads to the introduction of undesirable characters from the wild species which have to be eliminated by further backcrossing.

Induction of homoeologous chromosome pairing through the deletion of *5B* has been described by Sears (1977) as a method of transferring genes for leaf rust resistance from *Agropyron elongatum* into wheat. Using substitution lines of chromosomes *3Ag* and *7Ag* he established hybrids that were heterozygous for the *Ag* chromosomes, nullisomic for *5B* and trisomic for *5D*. The extra dose of *5D* compensates for the absence of *5B*. There was evidence of homoeologous chromosome pairing and a

range of transfers were identified which included variable lengths of *3Ag* and *7Ag*. The smaller the segment of the *Ag* chromosome that is transferred, the greater its utility in breeding since this reduces the chances of introducing deleterious genes with the rust resistance. In the case of the gene for rust resistance on *3Ag*, linkage with a gene for yellow endosperm has been a limitation in its use in breeding programmes. Repeating another cycle of induced homoeologous chromosome pairing using the isolated transfers could result in more acceptable transfers by reducing the length of the segment of the *Ag* chromosome through further recombination. A mutation at the *Ph* locus which increases homoeologous chromosome pairing reduces the effort required compared with establishing a nulli-*5B*-trisomic *5D* genotype.

In addition to providing more transfers, the ability to switch 'on or off' systems controlling genetic exchange in species hybrids improves the potential for use of such transfers compared with irradiation-induced translocations. In irradiation-induced translocations the resulting deletion/duplication of genetic material limits the usefulness of the technique, whereas in the induced recombinants the exchange is between genetically comparable chromosomes and the introduced segment compensates for the loss of the replaced chromosome of the recipient parent. Even in species that show marked chromosome divergence, there is strong evidence for the conservation of gene sequences that have persisted through the evolution of the species. Therefore, as long as the introduced segment is not linked to any deleterious genes and the size of the segment is not excessive, such transfers should stand a good chance of being of beneficial use in crop improvement.

6.5.6 Autopolyploids as bridge species in interspecific gene transfer

In hexaploid fescue species, *F. arundinacea* and *F. gigantea*, diploid-like meiosis is genetically controlled (Jahuar, 1975) but it is ineffective in the hemizygous state and, consequently, breaks down in F_1 hybrids between the hexaploid and diploid ryegrass. The main features of meiotic behaviour in the tetraploid hybrids are the prevalence of homoeologous chromosome pairing and hybrid sterility. Pentaploid hybrids between tetraploid ryegrass and hexaploid fescues have variable pollen fertility and certain genotypes have normal anther dehiscence that allows their use as pollen parents. Crossing the pentaploid hybrids as the male parent with diploid ryegrass produced BC_1 hybrids with 21 to 23 chromosomes with sufficient pollen fertility to be backcrossed again to diploid ryegrass. Over 80% of the BC_2 progeny are diploid, the majority of which are morphologically similar to ryegrass but fescue characters were expressed in some of them. Using alleles at the phosphoglucose isomerase (*PGI/2*) locus as genetic markers, Humphreys (1989) demonstrated that recombination between ryegrass and fescue chromosomes did occur since diploid BC_2 genotypes heterozygous for the fescue allele were recovered, albeit at a low frequency (2–4%). The fescue chromosomes were eliminated in two backcrosses to diploid ryegrass. The use of the pentaploid and BC_1 hybrid as the male parent in the crossing scheme is a distinct advantage since the capacity of the pollen/style interaction to act as a sieve in eliminating extraneous chromosomes is well documented (Khush, 1973). The low frequency of recombinants recovered in the BC_2 emphasizes the efficiency of the crossing scheme in eliminating the alien genome. The recovery of the diploid genome after only two backcrosses plus the evidence of recombination can be exploited to introgress genes from the hexaploid species into diploid ryegrass without recourse to a protracted backcrossing programme.

The level of recombination can be further manipulated by slight modification of the crossing scheme, such as interpollinating the pentaploid hybrids before using them as parents. The interpentaploid hybrid would have undergone three cycles of meiosis compared with a single cycle when using the original pentaploid and should have accumulated a higher frequency of recombinants. Interpollinating selected BC_1 hybrids expressing fescue characteristics should also increase the chances of obtaining suitable transfers. The transfer of characters that are polygenically controlled would benefit from such modifications but the original scheme would be more applicable to transfer single genes.

6.6 FUTURE PROSPECTS

Although a large proportion of cultivated crops are polyploid, the impact of polyploid induction as a plant breeding method over the past 50 years has not realized the breeders' expectations. The evolution of a genetic system in natural polyploid species to cope with the constraints imposed by genomic duplication is a key factor in their success. Failure to achieve a similar regularity of meiotic behaviour in induced polyploids is a limitation in their successful development as crops. In recent years knowledge has accumulated on the genetic control of chromosome pairing in polyploids and it has been extensively used to devise procedures to effect gene transfers within the tertiary gene pool. Studies of the synaptonemal complex and the spatial distribution of chromosomes within pollen mother cells in relation to initiation of pairing, by means of electron microscopy, are contributing to a greater understanding of the mechanism of chromosome pairing and its control, which should improve the chances of stabilizing meiotic behaviour in induced polyploids.

It is possible that genes controlling regular chromosome pairing in natural polyploids like the *Ph* locus in wheat will be isolated and cloned in the future. The introduction of such genes through transformation procedures into novel allopolyploids, would be invaluable in stabilizing agronomically promising allopolyploids, such as those combining whole genomes within the *Lolium/Festuca* complex. It is beyond the scope of this chapter to evaluate the procedures and difficulties involved in achieving such goals, but any advances made would generate renewed interest in establishing novel allopolyploid combinations as agricultural crops to meet future changes in climate and farming practices.

So far, the greatest benefit that has accrued from investigations of polyploidy is in their use in the introgression of genes across species barriers. This will remain the most important contribution since it does increase the portfolio of variation available to the breeder. The use of molecular markers to produce a more comprehensive map of the genomes of cultivated crops, especially the location of quantitative trait loci, which can then be used to improve the efficiency of selection procedures, would equally apply to introgressed genes or chromosome segments. Techniques of *in situ* hybridization can be used for the early recognition of introgressed segments of alien chromosomes in a crop species and assist selection at the seedling stage. Similarly, 'marked assisted selection' (Chapter 19) can be exploited in introgression procedures based on backcrossing, to ensure the accelerated reconstitution of the genome of the recurrent parent.

Although polyploidy, like induced mutations, has not lived up to the expectations of the breeder it will remain an important factor in breeding programmes because of its wide distribution among domesticated crops. The significance of polyploidy in plant breeding will depend on advances that will be made in understanding the factors controlling chromosome pairing and recombination. The approaches to gene transfer oulined in this chapter may be overtaken by the development of transformation procedures, in particular the transfer of single gene characters (Chapter 10), but they would undoubtedly still have a role in the transfer of quantitative traits.

REFERENCES

Bingham, E.T. (1980), Maximizing heterozygosity in autotetraploids. In *Polyploidy – Biological Relevance*, Lewis, W.H. (ed.), Plenum Press, New York, London, pp. 471–490.

Breese, E.L., Lewis, E.J. and Evans, G.M. (1981), Interspecific hybrids and polyploidy. *Philos. Trans. R. Soc. London B*, **292**, 487–497.

Cauderon, Y. (1977), Allopolyplody. In *Interspecific Hybridization in Plant Breeding. Proc. 8th Eucarpia Congress*, Sanchez-Monge, E. and Garcia-Olmedo, F. (eds), Madrid, pp. 131–143.

Dewey, D.R. (1980), Some applications and misappli-

cations of induced polyploidy to Plant Breeding. In *Polyploidy – Biological Relevance*, Lewis, W.H. (ed.), Plenum Press, New York, London, pp. 445–470.

Gupta, P.K. and Priydarshan, P.M. (1982), Triticale: present status and future prospects. *Adv. Genet.*, **21**, 255–345.

Harlan, J.R. and de Wet, J.M. (1971), Towards a rational classification of cultivated plants. *Taxon*, **20**, 509–517.

Harlan, J.R. and de Wet, J.M.J. (1975), On Ø. Winge and a prayer: The origins of polyploidy. *Bot. Rev.*, **41**, 361–390.

Humphreys, M.W. (1989), The controlled introgression of *Festuca arundinacea* genes into *Lolium multiflorum*. *Euphytica*, **42**, 105–116.

Jahuar, P.P. (1975), Genetic control of diploid like meiosis in hexaploid tall fescue. *Nature*, **254**, 595–597.

Jones, K. (1962), Chromosomal status, gene exchange, and evolution in *Dactylis*. 2. The chromosomal analysis of diploid, tetraploid, and hexaploid species and hybrids. *Genetica*, **32**, 272–295.

Khush, G.S. (1973), *Cytogenetics of Aneuploids*, Academic Press, New York, London.

Law, C.N., Snape, J.W. and Worland, A.J. (1980), Aneuploidy in wheat and its uses in genetic analysis. In *Wheat Breeding*, Lupton, F.G.H. (ed.), Chapman and Hall, London, New York, pp. 71–107.

Mettin, W.D., Bluthner, W.D. and Schlegel, C. (1973), Additional evidence on spontaneous 1B/R wheat/rye substitutions and translocations. In *Proc. 4th Int. Wheat Genet. Symp.*, Sears, E.R. and Sears, L.M.S. (eds), Missouri Agric. Expl. Stn, Columbia, pp. 179–184.

Peloquin, S.K., Yerk, G.L., Wesnev, J.E. and Dasmo, E. (1989), Potato breeding with haploid and 2*n* gametes. *Genome*, **31**, 1000–1004.

Riley, R. and Law, C.N. (1965), Genetic variation in chromosome pairing. *Adv. Genet.*, **13**, 57–114.

Savitsky, H. (1975), Hybridization between *Beta vulgaris* and *B. procumbens* and transmission of nematode (*Heterodera schachtii*) resistance to sugar beet. *Can. J. Genet. Cytol.*, **17**, 197–209.

Sears, E.R. (1956), The transfer of leaf-rust resistance from *Aegilops umbellulata* to wheat. *Brookhaven Symp. Biol.*, **9**, 1–22.

Sears, E.R. (1977), Analysis of wheat-*Agropyron* recombinant chromosomes. In *Interspecific Hybridisation in Plant Breeding. Proc. 8th Eucarpia*

Congress, Sanchez-Monge, E. and Garcia-Olmedo, F. (eds), Madrid, pp. 63–72.

Stebbins, G.L. (1950), *Variation and Evolution in Plants*, Columbia University Press, New York.

Thomas, H. (1981), Interspecific manipulation of chromosomes. *Philos. Trans. R. Soc. London B.*, **292**, 519–527.

Thomas, H. and Humphreys, M.O. (1991), Progress and potential of interspecific hybrids of *Lolium* and *Festuca*. *J. Agric. Sci. Camb.*, **117**, 1–8.

Thomas, H., Powell, W. and Aung, T. (1980), Interfering with regular meiotic behaviour in *Avena sativa* as a method of incorporating the gene for mildew resistance from *A. barbata*. *Euphytica*, **29**, 635–640.

van Bogaert, G. (1975), A comparison between colchicine induced tetraploid and diploid cultivars of *Lolium* species. In *Ploidy in Fodder Crops*, Neusch, B. (ed.), Eucarpia Report, Zurich.

7

Somatic hybridization

G. Pelletier

7.1 INTRODUCTION

Plant breeders have, for a long time, been tempted to cross very different species to create new living forms. Interspecific crosses have been extensively used and in some cases produced new species as, for example, triticale. To overcome sexual barriers, *in vitro* fertilization and *in vitro* embryo or ovary culture are technical refinements which have been used on a relatively limited scale, because each combination between two species brings specific problems. Protoplast fusion was proposed very early as a possible way to produce new hybrid forms (Kuster, 1909) by radically eliminating sexual barriers. Nevertheless it was known that interspecific incompatibilities occur not only before and during the fertilization process, but also later, when the embryo and the endosperm co-differentiate in the maternal tissues, and even later, when the young plantlet develops after germination.

Somatic hybridization needs the 'isolation' of intact protoplasts, the interparental 'fusion' of these naked cells, the 'sustained divisions' of fusion products before or after their 'selection', and the 'regeneration' of plants. In the fusion products a mixture of the different cell genomes (nucleus, plastome, chondriome) from the two parents is obtained and results in new 'nuclear' or 'cytoplasmic' genetic combinations in regenerated plants.

Plant Breeding: Principles and prospects. Edited by M.D. Hayward, N.O. Bosemark and I. Romagosa. Published in 1993 by Chapman & Hall, London. ISBN 0 412 43390 7.

7.2 ISOLATION OF PLANT PROTOPLASTS

The plant cell is turgid and the pressure exerted by the negative osmotic potential of the vacuole is equalized by the rigidity of the cell wall. When the cell wall is removed, the medium in which the protoplast is released must have a sufficiently high osmotic pressure to avoid water transfer into the vacuole which would result in the bursting of the protoplast. To obtain media of relatively high osmotic potential, substances like mannitol or sorbitol, at concentrations around 0.5 molar, are used. It is also possible to make use of mineral solutions composed of calcium or potassium chloride during the different phases of the protoplast isolation. High ionic concentrations must be used only during short periods of time as they drastically modify the cell metabolism and the plasmalemma structure. The effect of plasmolysis is to detach the protoplast from the cell wall and to give better access to the cellulolytic enzymes during the digestion process. The vacuole regresses, the cytoplasm loses a part of its water, and consequently the metabolism is altered. The optimal osmotic potential depends on the plant material used. Generally, with young tissues taken from plants regularly watered and protected from high light intensities, the absolute value of this potential is lower, which is a favourable condition for protoplast viability. The cell wall itself is composed of pure cellulose (20–30%), hemicelluloses and pectic compounds. The composition is not constant and varies with the growth conditions and the age of the cell. The chemical complexity of the cell wall explains the difficulties encountered with some plant materials. Commercial enzymatic preparations available (cellulase, hemicellulase, pectinase, 'Driselase', 'Pectolyase', 'Caylase') are crudely purified extracts of microorganisms, grown on different substrates (sawdust, molasses, cattle cakes). They also contain other enzymes like peroxidases and ribonuclease which could be deleterious at high concentrations.

One example of protoplast isolation is the enzymatic digestion of tobacco mesophyll cells. The best preparations are obtained with plants grown in shaded greenhouses at a temperature below 25°C and regularly watered. The upper third of the plant gives the most favourable leaves. These are disinfected by immersion in ethanol (70%) for a few seconds followed by five minutes in a solution of calcium hypochlorite (5%) and rinsed with sterile water. The lower epidermis is then removed aseptically with forceps and leaf fragments are placed on the maceration medium in petri dishes. The maceration medium contains 0.1% cellulase, 0.02% 'Macerozyme', and 0.05% 'Driselase' for an overnight digestion (approximately 16 h). The protoplasts are then separated from debris by filtration through a sieve (50–80 μm mesh) and rinsed by two successive centrifugations at 100 g with fresh medium without enzymes.

As mesophyll is not a convenient source of protoplasts in many species, other possibilities reside in the use of hypocotyls or roots. The advantage of these protoplasts is the absence of differentiated chloroplasts which renders them easily distinguishable from mesophyll protoplasts (Section 7.5). The protoplasts have a low density and they are collected by centrifugation either in a medium containing a high sucrose concentration (by pipetting the upper part of the medium in the centrifuge tube) or by sedimentation in a medium containing only mineral salts as osmoticum. Another source of protoplasts are cells grown *in vitro* in suspension cultures. This is particularly important in cereals where so far only protoplasts derived from embryogenic cell suspension cultures have given sustainable cell divisions and plant regeneration (Fujimura *et al.*, 1985).

7.3 PROTOPLAST FUSION

Plant protoplasts are uniformly negatively charged and spontaneous fusions are prevented by the resulting repulsion forces. Rapid modifications of the ionic environment of protoplasts were the first conditions in which aggregations followed by fusion were observed. The first somatic hybrid plants were produced through fusion mediated by iso-osmotic solutions of sodium nitrate (Carlson *et al.*, 1972). Aggregation, mediated by neutralization of plasmalemma negative charges by 50 mM Ca^{2+} mixed with 0.4 M mannitol in a solution of high pH (10.5), was the second fusion method applied to two tobacco mutants, resulting in the regeneration of the second known plant somatic hybrid (Melchers and Labib, 1974). This 'high Ca^{2+}', 'high pH' method emphasized the key role of calcium and pH in the process of membrane coalescence, a major step toward cytoplasmic fusion. The discovery of the powerful aggregation effect of polyethylene glycol (PEG) (Kao and Michayluck, 1974; Wallin *et al.*, 1974) provided the awaited reproducible tool.

A few minutes after being immersed in 20–40% (w/v) PEG (mol. wt. 1500–6000) virtually all protoplasts exhibit adhesion. However, actual fusion occurs upon dilution of the PEG with a 'high Ca^{2+}', 'high pH' elution medium. Variations of this method are now routinely used in the majority of laboratories. Depending on the manufacturer, PEG preparations are often toxic to plant cells. This problem can be overcome by deionization of the PEG solution or by using lower concentrations with the addition of a few percent of DMSO (dimethylsulphoxide).

More recently, electrofusion was developed by Zimmerman (1982). In this technique, protoplasts are first brought in close contact by a non-uniform alternating field between two electrodes, then fusion is initiated by a field pulse of high voltage (500–1000 V/cm) for a very short time (microseconds). This method has mostly been used with members of the Solanaceae and proved to be very efficient in some cases where the fusion rates were in excess of 50% (Bates and Hasenkampf, 1985). The fusion rate is affected by the conditions in which the experiment is performed (protoplast density, temper-

Table 7.1. Culture medium for tobacco mesophyll protoplasts from Bourgin *et al.* (1979)

	mg/l		mg/l
NH_4NO_3	825	Inositol	100
KNO_3	950	Calcium pantothenate	1
$CaCl_2, 2H_2O$	220	Biotin	0,01
$MgSO_4, 7H_2O$	185	Nicotinic acid	1
KH_2PO_4	85	Pyridoxine	1
$FeSO_4, 7H_2O$	27,85	Thiamine	1
Na_2EDTA	37,25	Naphthalene acetic acid	3
$ZnSO_4$	1	6-Benzylaminopurine	1
H_2BO_3	1	Sucrose	20 000
$MnSO_4, 4H_2O$	0,1	Mannitol	80 000
$CuSO_4, 5H_2O$	0,03	pH 5.5	
$AlCl_3$	0,03		
$NiCl_2, 6H_2O$	0,03		
KI	0,01		

ature, PEG concentration, electric field strength, and cell number) and also by the source of protoplasts used: mesophyll protoplasts are generally more suitable than callus or root protoplasts for fusion. The presence of a large vacuole or amyloplasts is detrimental for survival of the protoplast during the fusion process.

7.4 PROTOPLAST CULTURE AND PLANT REGENERATION

Protoplast culture media are not essentially different from tissue culture media used to obtain sustained cell divisions and callus formation from explants. The major difference is the osmotic potential needed at the beginning which is achieved by adding mannitol or sorbitol. Another distinctive feature is the necessity for relatively high doses of growth substances. In the case of tobacco, the best results are obtained with 3 mg/l of naphthalene acetic acid (15 μM) and 1 mg/l of benzyl aminopurine (5 μM) (Table 7.1). Protoplast division, as for other isolated cells, is only possible in a 'conditioned' medium where diffusible molecules released by each cell can reach a minimal concentration in the solution. The cell density (number of protoplasts per ml) must be adjusted precisely by counting the number of cells with a haemocytometer. With

tobacco protoplasts this density is around 60 000 protoplasts per ml, and it is directly and negatively correlated with the cell size. Practically for example, haploid tobacco protoplasts have a volume which is half that of diploid protoplasts and they are plated at a cell density which is twice the density of the latter. The plating efficiency, i.e. the percentage of dividing protoplasts, measures the quality of the growth conditions. Better results are generally obtained when protoplast suspensions are grown in the dark until the first division. This follows the cell wall regeneration and occurs after 2 to 6 days, according to the species used. The appearance of a bicellular structure is visible under an inverted microscope permitting the development of the culture to be observed directly in the petri dish if protoplasts are cultivated in a liquid medium. To study individual protoplasts it is possible to embed them in a solidified medium with agarose, although the first division is optically less visible in these conditions. The solidified medium has the advantage of giving better results for further divisions and growth of micro-colonies derived from protoplasts.

It is also possible to cultivate individual protoplasts in microdroplets or, after a few days (3 in the case of tobacco) to replace the culture medium by a low auxin medium and to lower the cell density. For tobacco cells derived from protoplasts, it is possible to decrease this cell density to 1 cell per ml which is a great advantage in further positive chemical selection of mutants, transformants, or somatic hybrids whose frequencies are generally low. In some cases it is necessary to use successive culture media which are added by dilution of the previous medium or by its complete replacement, such as when toxic compounds are produced by the cells as soon as the cell density is high. This is the case for *Brassica* protoplasts for which two or three successive media are necessary to obtain colonies capable of regeneration after transfer on to solid medium for caulogenesis.

Two morphogenetic pathways are followed by cells derived from protoplasts to give complete plants. In the case of mesophyll protoplasts the general rule is the neoformation of buds on colonies, a few weeks after protoplast isolation. These buds are rooted on a medium without growth substances or limited amounts of auxin. Nevertheless in species like alfalfa or eggplant, there is spontaneous embryogenesis. In protoplasts derived from embryogenic cell suspensions (which are generally used in the case of the Gramineae) the colonies obtained retain their embryogenic capacity and embryos are formed. The genetic conformity of plants obtained is generally good when all steps in the procedure from the protoplast to the plant are optimized so that the embryos or the meristems are formed in a few weeks (3–5 weeks). However, in non-optimal conditions a lot of genetic aberrations are seen to accumulate, the most frequent being changes in chromosome numbers. For example in the case of diploid potato protoplasts, the frequency of diploid regenerants is only about 10%, the remaining plants being mostly tetraploid.

7.5 SELECTION OF FUSION PRODUCTS

Assuming that two different plant cells have fused, the resulting cell, a heterokaryocyte consists of a mixture of cytoplasms in which is found a mixture of both types of chloroplasts, mitochondria and nuclei. An ideal fusion treatment should give a heterokaryocyte with only one nucleus from each parent. However, random protoplast aggregation often leads to multiple fusion products including various numbers of parental nuclei. Nuclear fusion occurs in only a fraction of heterokaryocytes. The proportion may be very small or surprisingly high depending on the species used as fusion partners. In the case of *Petunia parodii/P. parviflora* fusion the frequency is 10^{-4}, in *Nicotiana tabacum/N. tabacum* fusion 10^{-1}, in *Brassica campestris/B. oleracea* fusion 0.5.

To increase the chances of regenerating hybrid plants, different techniques of heterokaryon enrichment or methods of hybrid cell selection have been proposed. Gleba (1978) introduced the technique of mechanical isolation and cloning of single fusion products. Morphological differences, such as the presence or absence of differentiated chloroplasts in respectively mesophyll and callus or hypocotyl protoplasts, are sufficient to identify fusion products under the microscope. The isolation of such heterokaryons is possible, although tedious, with a micropipette. The use of fluorescent dyes (Galbraith and Galbraith, 1979) and of an automatic cell sorter (Redenbaugh *et al.*, 1982) generalizes this principle and allows the recovery of large numbers of heterokaryons marked by the fluorescein and rhodamine isothiocyanates used to stain each parent before the fusion. In favourable cases use was made of genetic or physiological complementation between the two partners. Physiological complementation was used by Power *et al.* (1977) and Schenck and Robbelen (1982) to select *Petunia* and *Brassica* hybrids respectively. In these cases hybrid cells regenerated to plants under specific conditions which were inefficient for parental cells. Genetic complementation uses different forms of auxotrophy: albino mutants or chlorophyll deficient mutants (Melchers and Labib, 1974), or nitrate reductase mutants deficient either for the apoenzyme or for the cofactor of this enzyme (Glimelius *et al.*, 1978).

Marker genes can be introduced into the parents of a fusion either by mutagenesis or, now, by gene transfer. Resistance to antibiotics (Hamill *et al.*, 1984), amino acid analogues (Harms *et al.*, 1981) and herbicides (Evola *et al.*,

1983) have been employed in some cases using induced or spontaneously occurring mutations. Masson *et al.* (1989) and Thomas *et al.* (1990) have utilized transformation of both fusion partners, one with kanamycin, the other with hygromycin resistance genes and selected hybrids by adding both antibiotics to the culture medium. A method adapted from animal cells to plant protoplasts allows the recovery of fusion products by complementation between protoplasts inactivated by different metabolic inhibitors. Additionally, combinations of nuclear destruction by heavy irradiation (X, τ) (Zelcer *et al.*, 1978) and cytoplasmic metabolic inhibitors can provide a means of selectively transferring an alien nucleus into a specific cytoplasm (Sidorov *et al.*, 1981).

The combination in the same genotype of an auxotrophic mutation (for example nitrate reductase deficiency or albinism) and a dominant trait such as those introduced by gene transfer (antibiotic or herbicide resistance) results in a so called 'universal hybridizer'. Hybrids between this genotype and any wild type genotype belonging to the same or another species can be directly selected after fusion by culture on a minimal medium supplemented with the antibiotic or the herbicide, where both parents are unable to grow normally. Such universal hybridizers have been successfully employed in *Nicotiana*, *Brassica* and *Lactuca*.

7.6 NUCLEAR HYBRIDS

Addition of nuclear genomes occurred in nature to create species like wheat, rapeseed or tobacco and these new species have been maintained by the strong selective pressure of agriculture. Somatic hybridization was first considered as a powerful tool to overcome sexual barriers and to create new species.

7.6.1 Symmetric hybrids

Interfamilial hybridization was tried with low success by Gleba and Sytnik (1984) and the regeneration of plants with a strict addition of both parental nuclear genomes was never observed. Similarly, the hybrids obtained between species belonging to different tribes were aneuploid and sterile. Moreover, the differentiation of buds from hybrid calli occurred only when chromosomes of one species were almost entirely eliminated. One exception to this rule, is the hybrid between *B. oleracea* and *Moricandia arvensis*, a species with an intermediate C_3–C_4 photosynthetic pathway (Toriyama *et al.*, 1987). A similar hybrid was recently obtained by sexual crossing and ovary culture (Takahata, 1990). Another example of a new species obtained by somatic hybridization is the pomato, resulting from cell fusion between potato and tomato (Melchers *et al.*, 1978). More recently another pomato hybrid was produced by fusing *Solanum tuberosum* and *Lycopersicon pimpinellifolium* mesophyll protoplasts (Okamura, 1988). These latter hybrids were fertile, unlike the first pomato plants which were obtained by fusing aneuploid potato callus protoplasts and tomato mesophyll protoplasts.

Numerous programmes of somatic hybridization have been performed and are still continuing with *Brassica* oil crops and wild or cultivated alien species. The three diploid *Brassica* species (*oleracea*, *nigra*, *campestris*) bearing respectively the C, B and A genomes are the natural progenitors of *B. carinata* (BC), *B. juncea* (AB) and *B. napus* (AC) (U, 1935). Resynthesized amphidiploids were produced by somatic hybridization with different cultivars (Schenck and Robbelen, 1982; Sundberg *et al.*, 1987; Pelletier *et al.*, unpublished). The trigenomic species (ABC) was obtained by fusion between *B. napus* and *B. nigra* (Sjödin and Glimelius, 1989a). Other combinations between *Eruca sativa* and *B. napus* (Fahlesson *et al.*, 1988) to transfer drought tolerance and insect resistance present in *Eruca*, or between *B. napus* and *Raphanus*, *Sinapis* or *Diplotaxis* species gave a majority of aneuploid plants and only some true hybrids.

The potato (*S. tuberosum*) offers several possibilities for the application of protoplast fusion techniques. Interspecific hybrids between this autotetraploid species and *S. brevidens*, a non-tuber-bearing diploid plant were obtained by Austin *et al.* (1985). The resulting tuber-bearing amphiploids (6*x*) were fertile and cross-compatible with *S. tuberosum*. The interest of these hybridizations is the presence in the wild species of genes conferring potato leaf roll virus resistance. Moreover, hybrid tubers have resistance to bacterial diseases (*Erwinia*) which was not predictable since *S. brevidens* does not form tubers.

A possibility offered by somatic hybridization to an autotetraploid species, according to a general schema proposed by Chase (1963) is the addition of two highly heterozygous diploid genomes which cannot be achieved by conventional breeding methods, where crossing and chromosome doubling by colchicine limits the level of heterozygosity that may be achived. Figure 7.1 shows that it is theoretically possible by using a combination of haploidization methods and somatic hybridization to reach the highest level of heterozygosity which appears to be, in the case of potato, a prerequisite for maximizing the plant vigour. A general method to obtain a great number of somatic hybrids between different selected diploid genotypes was proposed by Masson *et al.* (1989) and is based on the introduction of selectable markers by gene transfer.

7.6.2 Asymmetric hybrids

No barriers exist to the fusion of cells of different species, families, or even kingdoms. With the increasing number of somatic hybrids produced by plant protoplast fusion it became obvious that incompatibility reactions occurred at various levels after fusion. Hybrid combinations may undergo a few initial cell divisions, or grow successfully as undifferentiated calli but cannot regenerate plants or regenerate highly abnormal and sterile organisms. These cell fusion products cannot be incorporated into breeding schemes. Sometimes, when the final product of fusion between phylogenetically remote species is a plant with a regular vegetative morphology, it results from unpredictable but directed chromosome loss from one or both parents. Factors such as the suitability of *in vitro* culture conditions, sources of protoplast isolation and the number of subcultures may

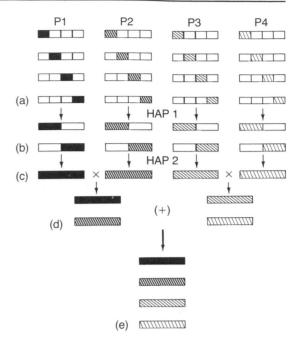

Fig. 7.1. Diagrammatic representation of a potato breeding scheme: (a) starting from four different populations (P1, P2, P3, P4) of tetraploid plants, a first haploidization step (HAP1) leads to selected diploid plants (b), and a second haploidization step (HAP2) to selected haploid (c) with accumulation of desirable traits in each case. After chromosome doubling, these four homozygous diploid clones are crossed, giving heterozygous diploids (d) the protoplasts of which are fused to obtain in one unique genome (e) all favourable characters (Chase, 1963; Wenzel, 1980).

influence this asymmetry. For example in *Datura innoxia/Atropa belladonna* somatic hybrid clones, whole plant structures were regenerated after more than one year of callus culture and contained no more than 2 to 6 *Atropa* chromosomes from the 72 present in the initial hybrid cell. If true and new hybrids between distant species are only possible in some individual combinations, the observation of spontaneous asymmetry suggested a search for methods for transferring only part of a plant genome to allow the synthesis of partial but fertile hybrid plants, more directly usable in plant breeding (Table 7.2).

In order to increase the number of highly asymmetric hybrids and to direct the process of

Table 7.2. Examples of somatic hybrids with potential agronomic interest

Combination	Resistance trait
Nicotiana tabacum (+) *N. repanda*	TMV
Solanum tuberosum (+) *S. brevidens*	PLRV
Solanum tuberosum (+)	
S. pimpinellifolium	*Fusarium*
Solanum melongena (+)	
S. sisymbrifolium	Nematodes
Lycopersicon esculentum (+)	
L. pennellii	Drought
Brassica napus (+) *B. nigra*	*Phoma*
Brassica napus (+) *Sinapis alba*	*Alternaria*
Brassica napus (+) *Eruca sativa*	Drought
Lactuca sativa (+) *L. perennis*	*Bremia*

chromosome elimination, several workers used irradiated donor protoplasts. The irradiation induces chromosome breakage into fragments which can be transferred into the recipient genome after protoplast fusion. By this method, Dudits *et al.* (1980) restored chlorophyll synthesis in a carrot albino mutant by fusion with irradiated parsley protoplasts. Somers *et al.* (1986) restored nitrate reductase activity in a NR$^-$ tobacco mutant by fusion with irradiated barley protoplasts. Such an approach was used to transfer the resistance to *Phoma lingam*, a fungus causing severe diseases in several cruciferous crops, to rapeseed. This fungus produces a toxin, sirodesmin PL, which inhibits the growth of susceptible cells *in vitro*. Complete resistance to the pathogen was known in *Brassica nigra* and the related species *B. carinata* and *B. juncea*. By fusing rapeseed protoplasts with irradiated *B. juncea* or *B. carinata* protoplasts, and selecting by adding sirodesmin in the culture medium, asymmetric hybrids were regenerated and proved to be fully resistant to pycnospore infections (Sjödin and Glimelius, 1989b).

7.7 CYTOPLASMIC HYBRIDS (CYBRIDS)

Plastids and mitochondria contain relatively small genomes compared to the nucleus, less

than one per thousand on a kilobase-pair basis. This DNA codes for approximately 10% of the polypeptides necessary to ensure organelle functioning, the remaining 90% being encoded by nuclear genes and imported through the membrane of the organelle. Genetic information is duplicated several times in each organelle and a plant cell may contain hundreds or thousands of cytoplasmic genomes which may therefore be considered as highly polyploid. At each cell generation, these genomes are transmitted to the progeny cells by a random distribution of organelles. Male and female gametes which equally participate to form the egg nucleus have an asymmetric role in cytoplasmic genome transmission, not only because of differences in cell size and number of organelles in the egg cell and the sperm cell, but also conceivably by some exclusion phenomena which would occur following fertilization. Maternal inheritance is the most frequent case for both organelles in the majority of crop species and thus the cytoplasmic genetic information which they transmit may, to all intents and purposes be considered a single entity in sexual reproduction.

Protoplast fusion appeared as a tool to systematically obtain a parasexual biparental inheritance of organelle genomes. From the fusion product in which nuclei, plastids, and mitochondria from both parents are present in a single cell, new associations between the nucleus of one species and organelles from a remote one could be produced. Moreover, interactions at the DNA level between organelle genomes would be possible allowing the creation of new genomes of interest in plant breeding.

7.7.1 Protoplast fusion and the exchange of cytoplasms

If the aim of the experiment is to combine the nucleus of one parent with the entire cytoplasm of the other, and if the frequency of nuclear hybridization is high, one can irradiate the cytoplasm donor by X- or gamma-rays before fusion (Zelcer *et al.*, 1978; Aviv and Galun, 1980). These treatments have no deleterious or mutagenic effects on organelle genomes recovered in the progeny of irradiated proto-plasts, probably because they are present at a high copy number in a plant cell. To retain only the cytoplasmic organelles of the irradiated parent, Medgyesy *et al.* (1980) proposed the use of the metabolic inhibition method first des-cribed in mammalian cell genetics (Wright, 1978) in order to eliminate the organelles of the other parent. Chemicals such as iodoacetate or iodoacetamide are used for this purpose and the method, combining irradiation of the cyto-plasm donor parent and metabolic inactivation of the nuclear recipient parent (the donor–recipient method) (Sidorov *et al.*, 1981), is now routinely used.

7.7.2 Protoplast fusion and the recombination of plastid genomes

The most frequent behaviour of plastid popula-tions mixed in a fusion product is a random segregation of parental types in the successive cell generations. In small colonies cells already exist with plastids exclusively from one or the other parent. Several reports have confirmed these findings in somatic hybrids (with hybrid nuclei) as well as in cytoplasmic hybrids (or cybrids) (Chen *et al.*, 1977; Bonnett and Glimelius, 1983).

Mixed populations of plastid genomes have been known for a long time in species with biparental plastid inheritance. In *Oenothera*, Chiu and Sears (1985) did not observe evidence of plastid recombination in a population of sexual hybrids. On the other hand, two reports from the same research group (Medgyesy *et al.*, 1985; Thanh and Medgyesy, 1989) have recently demonstrated the occurrence of plastid genome recombination in higher plants after somatic hybridization between two *Nicotiana* and between *Nicotiana* and *Solanum*. In both cases, they used plastid mutants, characterized by antibiotic or toxin resistance, and albinism (Fig. 7.2). The scarcity of plastid recombinants (only two cases) prevents any accurate estimation of the frequencies of these recombinations in heteroplastidic cells derived from fusion. It appears that the key to these successes was the

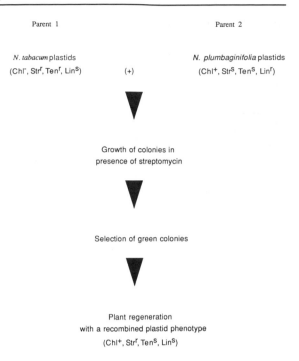

Fig. 7.2. Scheme employed to select recombinant plastid genomes in *Nicotiana* (Medgyesy *et al.*, 1985). The different plastid markers present in parental lines are: Chl^+/Chl^- = normal/albino; Str^r/Str^s = strepto-mycin resistance/susceptibility; Ten^r/Ten^s = tentoxin resistance/susceptibility; Lin^r/Lin^s = lincomycin resistance/susceptibility.

use of strong selection pressure, only possible when plastid mutants are available. Molecular analysis of the plastid genomes obtained in the *Nicotiana* recombinant revealed that it resulted from a high number of crossovers (Fejes *et al.*, 1990).

7.7.3 Protoplast fusion and the recombination of mitochondrial genomes

Cytoplasmic male sterility (CMS), a trait which is found in a wide range of species (Kaul, 1988), is the only genetic marker known to be encoded by the mitochondrial genome (Lonsdale, 1987) and used by cell biologists in protoplast fusion experiments to follow the fate of mitochondrial genomes. The lack of any selection pressure at the cell level is not an impediment to the recovery of mitochondrial recombinants. Interparental recombinations generally occur at a very high frequency, and were described very early in the story of plant somatic hybridization (Belliard *et al.*, 1978, 1979; Galun *et al.*, 1982). The molecular study of novel mitochondrial restriction fragments in *Petunia* (Rothenberg *et al.*, 1985) and in *Brassica* (Vedel *et al.*, 1986) somatic hybrids or cybrids, brought a definite proof of interparental DNA exchange. The result is a new rearranged mitochondrial genome with a unique combination of parental sequences in each regenerated cybrid.

These fusion experiments involved cytoplasmic male sterile plants. Phenotypically, CMS may be characterized by the stage at which disorders are visible during floral morphogenesis. In *Nicotiana*, male sterilities are induced by the cytoplasm of different *Nicotiana* species (Gerstel 1980), and in these alloplasmic combinations, stamens are transformed into feminized or petal-like structures or totally absent. Each cytoplasm is recognizable by a specific flower morphology. The interesting aspect of mitochondrial recombination, in fusions between normal and male sterile *Nicotiana*, is that it created new flower morphologies different from both parents (Belliard *et al.*, 1978, 1979; Pelletier, 1986; Kofer *et al.*, 1990). These studies could lead to the isolation of different mitochondrial fragments involved in flower modifications found in these systems (absence of stamens, pistilloidy, petaloidy, incision of petals, anther filament length) since each of these different features seems to behave as a single independent trait. Such recombinations between mitochondrial genomes have been found in a majority of species combinations, opening the possibility to truly exchange mitochondrial genes between species.

7.7.4 Application of somatic fusion to organelle manipulation in crop species

As illustrated in Fig. 7.3 there are several possibilities resulting from protoplast fusion.

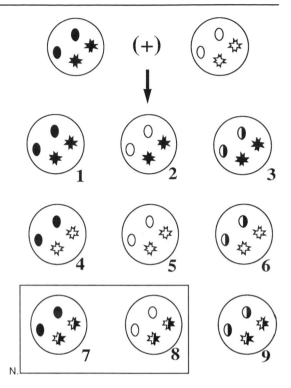

Fig. 7.3. The nine theoretical products of protoplast fusion between parents differing by plastid (● or ○) and mitochondrial (★ or ☆) genomes, considering three possibilities for each genome: exclusion of one or the other parental genome and interparental recombination (○ or ☆). Combinations No. 7 and No. 8 are the most frequently observed.

These possibilities derive from the application of simple rules. Plastid genomes mixed together in a single cell exceptionally undergo recombination and segregate in further cell generations, sometimes very quickly when one parental type is less competitive, or non-competitive compared to the other. In contrast, mitochondrial genomes very often undergo interparental recombination and after segregation, one recombined genome, among many others, is randomly retained in a regenerated cybrid plant or in its progeny. The most frequent cybrid constitutions are therefore those where mitochondrial genomes are recombinant ones associated with one or the other parental plastid genome (cases no. 7 and no. 8 in Fig. 7.3).

*(a) Transfer of male sterile cytoplasms
to different cultivars*

Protoplast fusion is a tool to transfer in one step a male sterile cytoplasm to different fertile varieties and to save several cycles of back crossing needed by conventional sexual methods. In rice, Yang *et al.* (1988, 1989), Akagi *et al.* (1989), and Kyozuka *et al.* (1989) succeeded in regeneration of cybrid plants possessing the nucleus of a fertile variety and mitochondria derived by recombination from both parents but retaining the CMS trait from the male sterile partner. These authors used the donor recipient method described by Sidorov *et al.* (1981). In carrot, Tanno-Suenaga *et al.* (1988) transferred the CMS trait (brown anther type) to a fertile cultivar by the donor recipient method. In *Brassica*, Barsby *et al.* (1987) transferred cytoplasmic male sterility of the 'Polima' type to different oilseed rape cultivars, and Primard *et al.* (1988) introduced an improved CMS cybrid cytoplasm previously obtained by somatic hybridization into winter oilseed rape.

In sugarbeet, the regeneration of entire plants from protoplasts (Krens *et al.*, 1990) opens up possibilities for the transfer of cytoplasms from different CMS sources to different nuclear backgrounds. The method would be of particular interest for this species, which is biennial, and where hybrid production world wide is based only on the Owen (1945) CMS source, which makes the crop potentially vulnerable to new races of pathogens. In tobacco, Kumashiro *et al.* (1988) succeeded in creating a new type of cytoplasmic male sterility by fusing *N. tabacum* protoplasts with irradiated *N. africana* protoplasts, a combination which is sexually incompatible, particularly when the latter species is used as the maternal parent. In citrus breeding, male sterility in cultivars with potential parthenocarpy would be of great economical interest. Vardi *et al.* (1989) succeeded in regenerating citrus cybrids with organelles (plastids and some mitochondrial sequences) from microcitrus. If male sterility is revealed among such plants, they would constitute a new tool for breeding seedless citrus varieties.

*(b) Improvement of cytoplasmic male
sterility systems in Cruciferae*

Brassica crops have great economic importance as vegetables (*B. oleracea, B. campestris*) or for oil production (*B. napus, B. juncea, B. campestris*). They represent the third most important oil source, after soybean and palm, for human consumption and industrial uses. F$_1$ hybrid varieties have been produced in these species using self-incompatibility systems but never on a large scale, and there was a real need for an efficient system, for control of pollination through male sterility. Cytoplasmic male sterility discovered in radish by Ogura (1968) was introduced into *B. oleracea* and *B. napus* by Bannerot *et al.* (1974, 1977) through interspecific sexual crosses, but the resulting plants, although perfectly male sterile, were practically unusable because of chlorophyll deficiency at low temperatures and flower malformations (inefficient nectar production and reduced female fertility).

Somatic fusion between *Brassica* protoplasts, bearing 'Ogura' radish organelles, and protoplasts with *Brassica* organelles was performed to exchange radish chloroplasts for *Brassica* chloroplasts, in order to recover normal chlorophyll synthesis (Pelletier *et al.*, 1983). Male sterile normal greening plants were screened among regenerated plants. They were shown to contain recombined mitochondrial genomes (Vedel *et al.*, 1986). Among these cybrids, some were selected by breeders because they were shown in field test with their progeny to be corrected for the other defects related to floral morphology. These plants produced enough nectar to be as attractive to bees as fertile plants and resulted in normal seed set. These characteristics were correlated with specific recombination events between the two parental mitochondrial genomes.

The results with *Brassica* cybrids confirm that from a plant breeding perspective, only a small part of a foreign genome has to be introgressed into the cultivated species, when it bears a trait of interest like CMS, and that entire organelle genomes bring together with this trait, other genes whose interaction with the nucleus of the cultivated species have a high probability of leading to agronomic defects. These results demonstrate the practical importance of mitochondrial recombination, which allows the elimination of these unfavourable traits. Such cytoplasms are now being used by breeders who may be able to produce F_1 hybrid varieties in *B. napus*, *oleracea* and *campestris*, with an 'improved' system of pollination control.

7.8 PROSPECTS

In the past two decades considerable progress has been made in protoplast technology: isolation, fusion, culture and plant regeneration in the major crop species including cereals (Vasil *et al.*, 1990). Somatic fusion allows sexually limited exchanges to be overcome by the production of asymmetric nuclear hybrids between distant species. This technique offers many possibilities if polyploidy or the absence of fertile sexual organs, which is often the case of vegetatively propagated plants, is an impediment to gene exchange. Moreover, protoplast fusion is an irreplaceable tool permitting the manipulation of organelle genomes and the possible creation of new combinations of nucleus and cytoplasm.

REFERENCES

Akagi, H., Sakamoto, M., Negishi, T. and Fujimura, T. (1989), Construction of rice cybrid plants. *Mol. Gen. Genet.*, **215**, 501–506.

Austin, S., Baer, M. and Helgeson, J.P. (1985), Transfer of resistance to potato leaf roll virus from *Solanum brevidens* into *Solanum tuberosum* by somatic fusion. *Plant Sci.*, **39**, 75–82.

Aviv, D. and Galun, E. (1980), Restoration of fertility in cytoplasmic male sterile (CMS) *Nicotiana sylvestris* by fusion with X-irradiated *N. tabacum* protoplasts. *Theor. Appl. Genet.*, **58**, 121–127.

Bannerot, H., Boulidard, L., Cauderon, Y. and Tempé, J. (1974), Transfer of cytoplasmic male sterility from *Raphanus sativus* to *Brassica oleracea*. *Proc. Eucarpia Meeting Cruciferae*, **25**, 52–54.

Bannerot, H., Boulidard, L. and Chupeau, Y. (1977), Unexpected difficulties met with the radish cytoplasm. *Cruciferae Newsl.*, **2**, 16.

Barsby, T.L., Yarrow, S.A., Kemble, R.J. and Grant, I. (1987), The transfer of cytoplasmic male sterility to winter type oilseed rape (*Brassica napus* L.) by protoplast fusion. *Plant Sci.*, **53**, 243–248.

Bates, G.W. and Hasenkampf, C.A. (1985), Culture of plant somatic hybrids following electrical fusion. *Theor. Appl. Genet.*, **70**, 227–233.

Belliard, G., Pelletier, G., Vedel, F. and Quetier, F. (1978), Morphological characteristics and chloroplast DNA distribution in different cytoplasmic parasexual hybrids of *Nicotiana tabacum*. *Mol. Gen. Genet.*, **165**, 231–237.

Belliard, G., Vedel, F. and Pelletier, G. (1979), Mitochondrial recombination in cytoplasmic hybrids of *Nicotiana tabacum* by protoplast fusion. *Nature*, **281**, 401–403.

Bonnett, H.T. and Glimelius, K. (1983), Somatic hybridization in *Nicotiana*: behaviour of organelles after fusion of protoplasts from male fertile and male sterile cultivars. *Theor. Appl. Genet.*, **65**, 213–217.

Bourgin, J.P., Chupeau, Y. and Missonier, C. (1979), Plant regeneration from mesophyll protoplasts of several *Nicotiana* species. *Physiol. Plant.*, **45**, 288–292.

Carlson, P.S., Smith, H.H. and Dearing, R.D. (1972), Parasexual plant hybridization. *Proc. Natl. Acad. Sci. USA*, **69**, 2292–2294.

Chase, S.S. (1963), Analytic breeding in *Solanum tuberosum* L. A scheme utilizing parthenotes and other diploid stocks. *Can. J. Bot.*, **5**, 359–363.

Chen, K., Wildman, S.G. and Smith, M.H. (1977), Chloroplast DNA distribution in parasexual hybrids as shown by polypeptide composition of fraction I protein. *Proc. Natl. Acad. Sci. USA*, **74**, 5109–5112.

Chiu, W.L. and Sears, B.B. (1985), Recombination between chloroplast DNAs does not occur in sexual crosses of *Oenothera*. *Mol. Gen. Genet.*, **198**, 525–528.

Dudits, D., Fejer, G., Hadlaczky, G., Lazar, G.B. and Horvath, G. (1980), Intergeneric gene transfer mediated by plant protoplast fusion. *Mol. Gen. Genet.*, **179**, 283–288.

Evola, S.V., Earle, E.D. and Chaleff, R.S. (1983), The use of genetic markers selected *in vitro* for the isolation and genetic verification of intraspecific somatic hybrids of *Nicotiana tabacum* L. *Mol. Gen. Genet.*, **189**, 441–446.

Fahleson, J., Rahlen, L. and Glimelius, K. (1988), Analysis of plants regenerated from protoplast fusions between *Brassica napus* and *Eruca sativa*. *Theor. Appl. Genet.*, **76**, 507–512.

Fejes, E., Engler, D. and Maliga, P. (1990), Extensive homologous chloroplast DNA recombination in the pt 14 *Nicotiana* somatic Thybrid. *Theor. Appl. Genet.*, **79**, 28–32.

Fujimura, T., Sakurai, M., Akagi, H., Negishi, T. and Hirose, A. (1985), Regeneration of rice plants from protoplasts. *Plant Tissue Culture Letter*, **2**, 74–75.

Galbraith, D.W. and Galbraith, J.E.C. (1979), A method for identification of fusion of plant protoplasts derived from tissue culture. *Z. Pflanzenphysiol.*, **93**, 149–158.

Galun, E., Arzee-Gonen, P., Fluhr, R., Edelman, M. and Aviv, D. (1982), Cytoplasmic hybridization in *Nicotiana*: mitochondrial DNA analysis of progenies resulting from fusion between protoplasts having different organelle constitutions. *Mol. Gen. Genet.*, **186**, 50–56.

Gerstel, D.U. (1980), Cytoplasmic male sterility in *Nicotiana* (a review). *North Carolina Technical Bulletin USA*, **263**, 1–31.

Gleba, Y.Y. (1978), Microdroplet culture: tobacco plants from single mesophyll protoplasts. *Naturwissenschaften*, **65** 158–159.

Gleba, Y.Y. and Sytnik, K.M. (1984), *Protoplast Fusion. Genetic Engineering in Higher Plants*, Monographs on Theroretical and Applied Genetics, No. 8, Springer Verlag, Berlin Heidelberg, New-York, Tokyo.

Glimelius, K., Eriksson, T., Grafe, R. and Müller, A. (1978), Somatic hybridization of nitrate reductase deficient mutants of *Nicotiana tabacum* by protoplast fusion. *Physiol. Plant.*, **44**, 273–277.

Hamill, J.D., Pental, D. and Cocking, E.C. (1984), The combination of a nitrate deficient nuclear genome with a streptomycin resistant chloroplast genome in *Nicotiana tabacum* by protoplast fusion. *J. Plant Physiol.*, **115**, 253–261.

Harms, C.T., Potrykus, I. and Widholm, J.M. (1981), Complementation and dominant expression of amino acid analogue resistance markers in somatic hybrid clones from *Daucus carota* after protoplast fusion. *Z. Pflanzenphysiol.*, **101**, 377–390.

Kao, K.N. and Michayluck, M.R. (1974), A method for high frequency intergeneric fusion of plant protoplasts. *Planta*, **115**, 355–367.

Kaul, M.L.M. (1988), *Male Sterility in Higher Plants*, Monographs on Theoretical and Applied Genetics, No. 10, Springer-Verlag, Berlin Heidelberg, New-York, London, Paris, Tokyo.

Köfer, W., Glimelius, K. and Bonnett, H.T. (1990), Modifications of floral development in tobacco induced by fusion of protoplasts of different male sterile cultivars. *Theor. Appl. Genet.*, **79**, 97–102.

Krens, F.A., Jamar, D., Rouwendal, G.J.A. and Hall, R.D. (1990), Transfer of cytoplasm from new *Beta* CMS sources to sugar beet by asymmetric fusion. 1 shoot regeneration from mesophyll protoplasts and characterization of regenerated plants. *Theor. Appl. Genet.*, **79**, 390–396.

Kumashiro, T., Asahi, T. and Komari, T. (1988), A new type of cytoplasmic male sterile tobacco obtained by fusion between *Nicotiana tabacum* and X irradiated *N. africana* protoplasts. *Plant Sci.*, **55**, 247–254.

Kuster, E. (1909), Uber die Verschmelzung nackter protoplasten. *Ber. Dtsch. Bot. Ges.*, **27**, 589–598.

Kyozuka, J., Kaneda, T, and Shimamoto, K. (1989), Production of cytoplasmic male sterile rice (*Oryza sativa* L.) by cell fusion. *Bio/Technology*, **7**, 1171–1174.

Lonsdale, D. (1987), Cytoplasmic male sterility: a molecular perspective. *Plant Physiol. Biochem.*, **25**, 265–272.

Masson, J., Lancelin, D., Bellini, C., Lecerf, M., Guerche, P. and Pelletier, G. (1989), Selection of somatic hybrids between diploid clones of potato (*Solanum tuberosum* L.) transformed by direct gene transfer. *Theor. Appl. Genet.*, **78**, 153–159.

Medgyesy, P., Fejes, E. and Maliga, P. (1985), Interspecific chloroplast recombination in a *Nicotiana* somatic hybrid. *Proc. Natl. Acad. Sci. USA*, **82**, 6960–6964.

Medgyesy, P., Menczel, L. and Maliga, P. (1980), The use of cytoplasmic streptomycin resistance: chloroplast transfer from *Nicotiana tabacum* into *Nicotiana sylvestris* and isolation of their somatic hybrids. *Mol. Gen. Genet.*, **179**, 693–698.

Melchers, G. and Labib, G. (1974), Somatic hybridization of plants by fusion of protoplasts. I. Selection of light resistant Thybrid of haploid light sensitive varieties of tobacco. *Mol. Gen. Genet.*, **135**, 271–294.

Melchers, G., Sacristan, M.D. and Holder, A.A. (1978), Somatic hybrid plants of potato and tomato regenerated from fused protoplasts. *Carlsberg Res. Commun.*, **43**, 203–218.

Ogura, H. (1968), Studies of the new male sterility in Japanese radish with special reference to the utilization of this sterility towards the practical raising of hybrid seeds. *Mem. Fac. Agric., Kayoshima Univ.*, **6**, 39–78.

Okamura, M. (1988), Regeneration and evolution of somatic hybrid plants between *Solanum tuberosum* and *Lycopersicon pimpinellifolium*. *Plant Cell Tissue Organ Cult.*, **12**, 213–214.

Owen, F.V. (1945), Cytoplasmically inherited male sterility in sugar beet. *J. Agric. Res.*, **71**, 423–440.

Pelletier, G.R. (1986), Plant organelle genetics through somatic hybridization. In *Oxford Surveys of Plant Molecular and Cell Biology*, Miflin, B. (ed.), Oxford University Press, Oxford, pp. 97–121.

Pelletier, G., Primard, C., Vedel F., Chetrit, P., Remy, R., Rousselle, P. and Renard, M. (1983), Intergeneric cytoplasmic hybridization in Cruciferae by protoplast fusion. *Mol. Gen. Genet.*, **191**, 244–250.

Power, J.B., Berry, S.F., Frearson, E.M. and Cocking, E.C. (1977), Selection procedures for the production of inter-species somatic hybrids of *Petunia hybrida* and *Petunia parodii*. I. Nutrient media and drug sensitivity complementation selection. *Plant Sci. Lett.*, **10**, 1–6.

Primard, C., Lépingle, A., Masson, J., Lancelin, D., Chèvre, A.M. and Pelletier, G, (1988), Transfer of male sterile cybrid cytoplasms into winter rape cultivar of *Brassica napus* by protoplast fusion. *Cruciferae Newsl.*, **13**, 78–79.

Redenbaugh, K., Ruzin, S., Batholomew, J. and Bassham, J.A. (1982), Characterization and separation of plant protoplasts via flow cytometry and cell sorting. *Z. Pflanzenphysiol.*, **107**, 65–80.

Rothenberg, M., Boeshore, M.L., Hanson, M.R. and Izhar, S. (1985), Intergenomic recombination of mitochondrial genomes in a somatic hybrid plant. *Curr. Genet.*, **9**, 615–618.

Schenck, H.R. and Röbbelen, G. (1982), Somatic hybrids by fusion of protoplasts from *Brassica oleracea* and *B. campestris. Z. Pflanzenzüchtg.*, **89**, 278–288.

Sidorov, V.A., Menczel, L., Nagy, F. and Maliga, P. (1981), Chloroplast transfer in *Nicotiana* based on metabolic complementation between irradiated and iodoacetate treated protoplasts. *Planta*, **152**, 341–345.

Sjödin, C. and Glimelius, K. (1989a), *Brassica naponigra*, a somatic Thybrid resistant to *Phoma lingam. Theor. Appl. Genet.*, **77**, 651–656.

Sjödin, C. and Glimelius, K. (1989b), Transfer of resistance against *Phoma lingam* to *Brassica napus* by asymmetric somatic hybridization combined with toxin selection. *Theor. Appl. Genet.*, **78**, 513–520.

Somers, D.A., Narayanan, K.R., Kleinhofs, A., Cooper-Blaud, S. and Cocking, E.C. (1986), Immunological evidence for transfer of the barley nitrate reductase structural gene to *Nicotiana tabacum* by protoplast fusion. *Mol. Gen. Genet.*, **204**, 296–301.

Sundberg, E., Landgren, M. and Glimelius, K. (1987), Fertility and chromosome stability in *Brassica napus* resynthesised by protoplast fusion. *Theor. Appl. Genet.*, **75**, 96–104.

Takahata, Y. (1990), Production of intergeneric hybrids between a C_3–C_4 intermediate species *Moricandia arvensis* and a C_3 species *Brassica oleracea* through ovary culture. *Euphytica*, **46**, 259–264.

Tanno-Suenaga, L., Ichikawa, H. and Imamura, J. (1988), Transfer of the CMS trait in *Daucus carota* L. by donor-recipient protoplast fusion. *Theor. Appl. Genet.*, **76**, 855–860.

Thanh, N.D. and Medgyesy, P. (1989), Limited chloroplast gene transfer via recombination overcomes plastome genome incompatibility between *Nicotiana tabacum* and *Solanum tuberosum. Plant Mol. Biol.*, **12**, 87–93.

Thomas, M.R., Johnson, L.B. and White, F.F. (1990), Selection of somatic hybrids of *Medicago* by using *Agrobacterium* transformed tissues. *Plant Sci.*, **69**(2), 189–198.

Toriyama, K., Hinata, K. and Kameya, T. (1987), Production of somatic hybrid plants 'Brassicomoricandia', through protoplast fusion between *Moricandia arvensis* and *Brassica oleracea. Plant Sci.*, **48**, 123–128.

U, N. (1935), Genome analysis in *Brassica* with special reference to the experimental formation of *B. napus* and peculiar mode of fertilization. *Jpn. J. Bot.*, **7**, 389–452.

Vardi, A., Arzee-Gonen, P., Frydman-Shani, A., Bleichman, S. and Galun, E. (1989), Protoplast fusion mediated transfer of organelles from microcitrus into citrus and regeneration of novel alloplasmic trees. *Theor. Appl. Genet.*, **78**, 741–747.

Vasil, V., Redway, F. and Vasil, I.K. (1990), Regeneration of plants from embryogenic suspension culture protoplasts of wheat (*Triticum estivum* L.). *Bio/Technology*, **8**, 429–434.

Vedel, F., Chetrit, P., Mathieu, C., Pelletier, G. and Primard, C. (1986), Several different mitochondrial DNA regions are involved in intergenomic recombination in *Brassica napus* cybrid plants. *Curr. Genet.*, **11**, 17–24.

Wallin, A., Glimelius, K. and Eriksson, T. (1974), The induction of aggregation and fusion of *Daucus carota* protoplasts by polyethylene glycol. *Z. Pflanzenphysiol.*, **74**, 64–80.

Wenzel, G. (1980), Protoplast techniques incorporated into applied breeding programs. In *Advances in Protoplast Research*, Ferenczy, L. and Farkas,

G.L. (eds), Pergamon Press, Oxford, pp. 327–340.

Wright, W.E. (1978), The isolation of heterokaryons and hybrids by a selective system using irreversible biochemical inhibitors. *Exp. Cell Res.*, **112**, 395–407.

Yang, Z.Q., Shikanai, T., Mori, K. and Yamada, Y. (1989), Plant regeneration from cytoplasmic hybrids of rice (*Oryza sativa* L.). *Theor. Appl. Genet.*, **77**, 305–310.

Yang, Z.Q., Shikanai, T. and Yamada, Y. (1988), Asymmetric hybridization between cytoplasmic male-sterile (CMS) and fertile rice (*Oryza sativa* L.) protoplasts. *Theor. Appl. Genet.*, **76**, 801–808.

Zelcer, A., Aviv, D. and Galun, E. (1978), Interspecific transfer of cytoplasmic male sterility by fusion between protoplasts of normal *Nicotiana sylvestris* and X-ray irradiated protoplasts of male sterile *N. tabacum*. *Z. Pflanzenphysiol.*, **90**, 397–407.

Zimmermann, V. (1982), Electric field mediated fusion and related electrical phenomena. *Biochim. Biophys. Acta*, **694**, 227–277.

8

Gene cloning and identification

B. Xoconostle-Cázares, E. Lozoya-Gloria and L. Herrera-Estrella

8.1 INTRODUCTION

Biological sciences in general and plant biology in particular underwent a revolutionary change with the development of recombinant DNA technology. This term can be applied to a variety of *in vivo* and *in vitro* techniques aimed at the production of new combinations of heritable material by the splicing of nucleic acid molecules *in vitro* (Old and Primrose, 1989). Gene manipulation started with the discovery of bacterial transformation, restriction and modification enzymes and techniques to monitor DNA cutting and joining reactions. In order to isolate specific genes from eukaryotic organisms using bacterial or yeast model systems, it is necessary to attach them to a genetic element which will allow the production of many copies of the isolated DNA fragment. Such elements are known as vectors or cloning vehicles. Plasmids and bacteriophages are so far the most versatile vectors available. Their maintenance in the cell does not necessarily require integration into the host genome and their DNA can be isolated independently of the host genome.

In its simplest form, molecular cloning requires the following steps: (a) vector DNA must be purified and cut open with restriction enzymes; (b) foreign DNA must be covalently joined to the vector; (c) recombinant molecules generated in this way must be introduced (transformed) to a bacterial host cell, where

Plant Breeding: Principles and prospects. Edited by M.D. Hayward, N.O. Bosemark and I. Romagosa. Published in 1993 by Chapman & Hall, London. ISBN 0 412 43390 7.

they can be amplified; (d) clones that carry the desired sequence must be selected.

The second fundamental process in the isolation and manipulation of genes is the information and strategy(s) that will allow the identification of the gene(s) of interest among all the sequences that compose the genome of a given organism. In this chapter the different types of cloning vectors and some of the essential steps for the construction of gene libraries, as well as some of the strategies that can be followed for the identification and isolation of genes will be described.

8.2 CONSTRUCTION OF GENE LIBRARIES

Nucleic acid isolation and analysis techniques are a fundamental requirement not only for the isolation of genes, but also to determine the physical structure and expression of genes. To carry out these types of studies it is necessary to isolate the gene of interest. The first step is the production of a genomic or cDNA library. In the following section we shall briefly describe how these libraries are constructed.

8.2.1 Genomic DNA libraries

A collection of recombinant molecules that contain DNA fragments that represent the entire genome of a given organism is called a genomic library or gene bank. Genomic libraries carry all DNA sequences present in the genome of the organism of interest: exons, introns and regulatory sequences and are useful to study DNA topology and the regulatory regions that govern the gene expression. In principle any sequence of interest should be present in a genomic library, however, in practice it has been found that certain eukaryotic sequences are difficult to clone and propagate in bacterial systems.

The number of independent recombinant clones that are required in a gene bank to have a given probability that the entire genome is represented can be theoretically calculated using Equation 8.1 (Dale, 1987):

$$N = \frac{\ln (1-p)}{\ln (1-t)\, n} \qquad (8.1)$$

where: N = number of independent recombinant clones required, p = probability that a particular sequence is represented (choose 0.95%), t = average length of the fragments to be cloned (kb), n = total amount of DNA per cell.

The number of clones that are required to have a complete gene library will depend on the size of the genome and the type of cloning vector. One of the crucial steps to successfully construct a gene library is the quality of the DNA to be used. In the following sections some of the important points for the isolation of plant nucleic acids will be discussed.

First, high molecular weight, purified DNA is required in order to generate genomic libraries. Details of gene cloning and analytical procedures have been described in detailed by Maniatis *et al.* (1989) and only a brief outline will be presented here.

(a) Preparation of total DNA from plant tissues

A critical factor in the isolation of plant DNA is the efficient disruption of the plant cell wall. Many techniques that have been described for breaking open cells also shear DNA, in this way, techniques must compromise between DNA integrity and yield. The isolation of high-molecular-weight DNA is only one part of the problem, since plant extracts often contain large amounts of polysaccharides, tannins and pigments, which are difficult to separate from DNA and inhibit the activity of most DNA modifying enzymes, leading to problems during genomic cloning procedures.

Plant cells are usually disrupted in aqueous solutions in the presence of chelating agents to inhibit nuclease action and detergents to solubilize membrane structures. After the dispersion of cell contents, proteins are denatured and precipitated from the extract using organic solvents. Some protocols incorporate protease treatments to help release DNA from chromatin. After deproteinization, DNA is still contaminated with RNA and carbohydrates. DNA can be cleaned in several different ways, depending to some extent both on the quantity of tissue extracted and the requirements of the experiment. If a large amount of pure DNA is required then crude DNA extracts can be cleaned by density centrifugation on a caesium chloride/ethidium bromide gradient (Draper and Scott, 1988). A technique described by Dellaporta *et al.* (1985), allows the preparation of DNA of acceptable quality, suitable for restriction and cloning. Dellaporta's method has been used to isolate total DNA from fresh materials of many plant species and is the most commonly used method, especially for small amounts of tissue.

Freeze-drying is a convenient way of storing plant material before DNA isolation, since this procedure prevents nucleolytic degradation and preserves DNA integrity. An extraction protocol widely used for freeze dried tissue is that of Saghai-Maroof *et al.* (1984).

(b) Isolation of nuclear DNA

For the construction of genomic libraries it is more convenient to use purified nuclear DNA rather than total DNA, since this avoids the cloning of organelle DNA, present in many copies and which may constitute up to 35% of total plant DNA. Nuclear DNA isolation techniques involve the disruption of fresh plant material in the presence of nuclear membrane stabilizing agents, filtration to remove non-disrupted tissue and isolation of nuclei by differential centrifugation, lysis, deproteinization and DNA purification on CsCl/EtBr gradients.

8.2.2 Construction of cDNA libraries

In order to construct a cDNA library, mature mRNA is isolated and a complementary copy of its sequence in the form of DNA is synthesized. This new molecule is called complementary DNA or cDNA, and can be cloned in a plasmid or phage vector. Therefore, a cDNA library is a collection of recombinant molecules that contain all the mRNAs present in an organism, tissue or cell line under defined growing conditions or developmental stages. A cDNA library will not represent all the genes of a given organism but only the genes that are functionally expressed in the cell type(s) from which the mRNA was isolated. This procedure avoids the cloning of 5' and 3' flanking regions and introns and produces molecules that can be used for direct expression of eukaryotic DNA in a bacterial system.

(a) Isolation of total plant RNA

The major problem in the isolation of RNA is the ubiquitous presence of ribonucleases that often lead to the partial degradation of RNA molecules during the purification protocol. To inactivate the endogenous ribonucleases, plant extracts are immediately treated with agents that have a potent ability to denature proteins. Many procedures have been published that use different protein denaturing agents, such as phenol, strong caotropic agents (such as guanidinium chloride) or protease treatments. During RNA isolation strict precautions to avoid ribonuclease activity must be taken and extreme cleanliness should be observed in the working area. A convenient procedure to isolate total plant RNA was developed by Logemann *et al.* (1987). Tissue is homogenized in the presence of the strong- protein-denaturing agent guanidine hydrochloride and 2–ß-mercaptoethanol. RNA is separated from proteins, DNA and polysaccharides by several ethanol precipitation steps, or by sedimentation through a discontinuous caesium chloride gradient.

(b) Isolation of poly(A)+ mRNA

Eukaryotic messenger RNA (mRNA) can be separated from other RNA species in a total RNA preparation by affinity chromatography based on the presence of a polyadenosine tail present at the 3' end of most eukaryote messenger RNAs. When a total RNA preparation is passed through an oligo(dT) cellulose or poly(U) sepharose column, poly(A)+ RNA will be retained, whereas tRNA, rRNA and other RNAs that lack a poly(A) tail will not bind to the column and can be displaced by washes. Later, the mRNA fraction can be recovered by washing with low salt buffers. A good guide to the purity of an mRNA preparation is to test its efficiency of translation in an *in vitro* cell free protein synthesizing system (Pelham and Jackson, 1976).

(c) cDNA synthesis

The cDNA first strand is synthesized by an RNA dependent DNA polymerase, called reverse transcriptase. This enzyme is used to make a DNA copy of poly(A)+ mRNA using as a primer oligo(dT). The product of the synthesis is a hybrid cDNA–mRNA molecule from which the RNA component is eliminated by chemical or enzymatic hydrolysis. Remnants of mRNA, random oligonucleotides or specifically designed oligonucleotides are usually used as primers for synthesis of the second strand, which is carried out by the Klenow fragment of *Escherichia coli* DNA polymerase I. Finally, the double stranded products of the reaction are ligated to a cloning vector by bacteriophage T4 DNA ligase.

8.3 CLONING VECTORS

Several vectors for the cloning of eukaryotic genomes have been developed, the choice, for DNA cloning, depends on several factors of

which the most important are: number of
recombinants required, the stability of the
cloned DNA in a particular cloning vector and
the procedure for screening the recombinant of
interest. The most commonly used vectors are
described below.

8.3.1 Plasmid vectors

Bacterial plasmids are double-stranded closed
circular DNA molecules ranging in size from 1
kb to more than 200 kb and from which the
most commonly used vectors for the *in vitro*
manipulation of eukaryotic DNA and the construction of cDNA libraries have been derived.
Plasmids replicate independently of the host
chromosome, since they have their own replication sequence called 'the origin of replication'.
In the laboratory, plasmid DNA can be introduced into bacteria by either transformation or
electroporation. To identify the bacterial cells
that have acquired the plasmid, genes that
confer resistance to antibiotics are present in
most plasmid cloning vectors. The most commonly used selectable markers are genes that
confer resistance to ampicillin, tetracycline,
chloramphenicol or kanamycin.

Plasmid vectors are generally small and
contain a number of useful restriction enzyme
cleavage sites that facilitate the cloning of DNA
fragments using a wide variety of combinations
of restriction enzymes. Almost all vectors now
contain a closely arranged series of synthetic
cloning sites, called polylinkers. During the
development of new, more versatile plasmid
cloning vectors, a series of ancillary sequences
have been incorporated that are used for a
variety of purposes, including visual identification of recombinant clones by histochemical
tests, generation of single-stranded DNA templates, DNA sequencing, transcription of foreign DNA sequences *in vitro* and the production
of large amounts of foreign proteins in bacterial
cells. For more information about the different
vectors that can be used for the construction of
cDNA libraries and manipulation of DNA
fragments see Maniatis *et al.* (1989).

8.3.2 Bacteriophage lambda-derived vectors

The *E. coli* bacteriophage λ is a linear double
stranded molecule of 50 kb in length which
terminates in single stranded sequences of 15
bases that are complementary to each other (cos
sites) and allow the phage to become circular.
Soon after entering a host bacterium, during the
early phase of infection, the linear genome of λ
is circularized. Two pathways can be followed
by the phage during this phase; either to enter
the lytic or the lysogenic growth cycle. In the
lytic cycle, phage DNA replicates, bacteriophage subunits are assembled and the cell
eventually lyses, releasing the new infectious
virus particles.

A wide variety of cloning vectors derived
from bacteriophage λ designed for the construction of cDNA and genomic libraries are now
available. The choice of which vector to use to
construct a genomic or cDNA library depends
on specific needs, which are: (a) the availability
of restriction sequences compatible with the
restriction enzyme(s) to be employed; (b) the
size of the foreign DNA fragments to be
inserted; (c) whether or not the vector is to be
used to express cloned DNA sequences in *E.
coli*.

Vectors that have a single target site for
insertion of foreign DNA are known as insertion vectors; those that have a pair of sites
flanking a segment of non-essential bacteriophage λ DNA that can be replaced by foreign
DNA are known as replacement vectors. The
EMBL and Charon vector series are typical
lambda replacement vectors with a cloning
capacity of 9-23 kb. These vectors have multiple
cloning sites flanking the non-essential λ region
(for a detailed description of these vectors see
Maniatis *et al.*, 1989). Replacement vectors with
properties additional to those present in the
classic EMBL and Charon vectors are constantly
being constructed. Examples of these new
vectors are λ 2001, λ DASH and λ FIX. All of
these vectors contain multiple cloning site
sequences. λ DASH and λ FIX harbour bacteriophage T7 and T3 promoters flanking the poly-
linker region and allow the production of RNA

probes from the ends of the insert DNA without subcloning (Karn *et al.*, 1984).

λ gt11 is the classical insertion vector that allow the expression of the inserted foreign DNA and is often used for immunological screening of cDNA libraries to identify the clones encoding the polypeptide of interest. New more versatile λ insertion vectors for the cloning and expression of cDNA fragments have recently been constructed (Short, 1988).

8.3.3 Cosmid vectors

Cosmid vectors are modified plasmids carrying the sites from the lambda bacteriophage required for DNA packaging (cos sites). Cosmids are useful for cloning because they combine the bacteriophage capacity to accept large fragments of DNA and the ability to replicate independently of the host chromosome as a plasmid. Segments of foreign DNA approximately 35–45 kb in length can be cloned and propagated in cosmid vectors. Cosmids have proven to be a useful tool to facilitate the selection of clones that cover large segments of a chromosome in techniques such as genomic walking and for the construction of physical genetic maps.

8.3.4 Yeast artificial chromosomes

A high capacity cloning system that is based on *in vitro* construction of linear DNA molecules that can be maintained as artificial chromosomes in yeast has recently been developed. Yeast artificial chromosomes (YACs) provide a means for cloning eukaryotic DNA fragments in the megabase range (Schlessinger, 1990).

These vectors contain the basic functional units of yeast chromosomes: centromeres, telomeres and Autonomous Replication Sequences (ARS), but also include the necessary sequences for replication in *E. coli*, cloning sites and selectable markers for the yeast host (Burke *et al.*, 1987). Individual YACs can contain inserts of up to 1 Mb or more. The most critical step in generating useful YAC libraries is the isolation of high molecular weight DNA to produce YACs with large inserts. To facilitate this, several techniques using DNA from cells lysed in agarose blocks, and cloning in the same agarose gel or beads have been developed. Prior to ligation to the YAC vector, large DNA fragments are generated by cleavage of genomic DNA with rare-cutting restriction enzymes such as those having 8 bp recognition sites. YAC vectors containing promoter sequences for T3 and T7 RNA polymerases, to permit the synthesis of RNA probes from the insert ends in order to facilitate chromosome walking, have been constructed (Marchuk and Collins, 1988). YAC vectors have become a useful tool for gene mapping and genome analysis, since they can carry large genomic inserts and most chromosomal sequences appear to be clonable in YACs. Using YAC overlapping clones, physical maps covering chromosomal regions of up to several megabases can be assembled and used for map-based cloning (see cloning strategies below) (Schlessinger, 1990).

8.4 THE POLYMERASE CHAIN REACTION

The polymerase chain reaction (PCR) is an *in vitro* method for producing large amounts of a specific DNA fragment of defined length and sequence from small amounts of a complex template (Saiki *et al.*, 1985). PCR is a powerful, extremely sensitive technique with applications in many fields such as molecular biology, diagnostics, population genetics and forensic analysis. Recombinant DNA techniques have revolutionized genetics by permitting the isolation and characterization of genes, allowing the detailed study of their function and expression during developmental processes, or as a response to environmental factors. Many of the cloning methods involved can be accelerated and sometimes even circumvented by using PCR, and novel applications of the technique now permit studies that were not possible before. In this section some of the most relevant applications of PCR to plant biology and plant breeding will be briefly reviewed, but reference is made to only a small fraction of the multiple uses of PCR. Further applications of this technique are presented in sections 12.1.3 and 19.7.1.

8.4.1 Principle of PCR

PCR is based on the enzymatic amplification of a DNA fragment by the use of two oligonucleotide primers that are complementary to the 5′ ends of both strands of the target sequence. These oligonucleotides are used as primers (PCR primers) to allow the template DNA to be copied by a DNA polymerase. To facilitate the annealing of the primers to the template it is necessary first to separate the strands of the substrate DNA by heating. The reaction temperature is then reduced to allow the pairing of the primer with the complementary target sequence and finally the polymerization reaction is carried out by a DNA polymerase in order to generate the complementary strand; this is known as a PCR cycle (Fig. 8.1a). Since the newly polymerized products started from each primer can serve as templates for the other primer, each cycle essentially doubles the amount of the DNA fragment produced in the previous cycle. Repeated cycles of heat denaturation of the template, annealing of the primers to their complementary sequences and extension of the annealed primers with a DNA polymerase result in the amplification of the segment defined by the 5′ ends of the PCR primers. This reaction results in the exponential accumulation of the specific target fragment, up to several millionfold in a few hours, which in practical terms means the production of microgram quantities of the target DNA from the initial picogram or lower levels present in the substrate. The method can be used with a complex template such as genomic DNA and can amplify a single-copy gene contained therein. It is also capable of amplifying a single molecule of target in a complex mixture of RNAs or DNAs. Under particular conditions fragments up to 10 kb long can be reproduced (Jeffreys *et al.*, 1988).

The use of the thermostable taq DNA polymerase isolated from the thermophilic bacterium *Thermus aquaticus* has greatly facilitated the application of PCR techniques. The heat stable taq polymerase avoids the need to add new polymerase after each cycle of heat denaturation of the template, as was done in the original method developed using the Klenow fragment of the *E. coli* polymerase I (Saiki *et al.*, 1988). This development has led to the automation of PCR by a variety of simple temperature-cycling devices, and consequently the use of PCR has expanded rapidly over the past three years. In addition, the specificity of the amplification reaction can be modified by adjusting to higher or lower temperatures for primer annealing and extension. The increased specificity improves the overall yield of amplified products by minimizing the competition by non-target fragments for enzyme and primers, whereas lowering the temperature leads to the amplification of families of related sequences.

Since the primers are physically incorporated into the amplified product and mismatches between the 5′ end of the primer and the template do not significantly affect the efficiency of the amplification, it is possible to alter the amplified sequence relative to the starting template. For instance, the addition of restriction enzyme sites to the 5′ end of each primer facilitates cloning of the final PCR. Modifications of the region around the initiation codon can improve the efficiency of translation of coding sequences in transgenic eukaryotic organisms. The incorporation of a T7 or similar promoter in the primer which is complementary to the 5′ of the sequence, allows synthesis of RNA copies of the PCR product to be generated without the need for cloning the amplified product in the bacterial system. The ability to add, delete or alter information from primers makes PCR ideal for introducing mutations at specific positions in a DNA fragment and facilitates the study of sequence-specific interactions of proteins and DNA, for example in transcription complexes (Higuchi *et al.*, 1988).

8.4.2 Strategies for the cloning of unknown sequences using PCR

Although it is necessary to have enough sequence information to design the PCR primers, the method can be used to amplify and

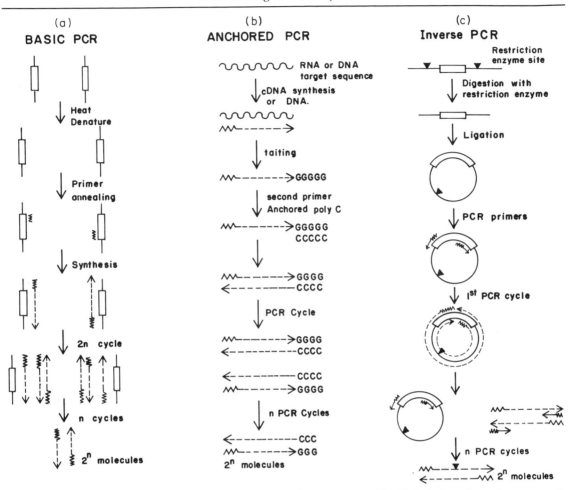

Fig. 8.1. PCR technology. (a) The basic polymerase chain reaction. The PCR is based on the enzymatic amplification of a DNA fragment that is flanked by two oligonucleotide primers that hybridize to opposite strands of the target sequence. Primers are oriented with their 3' ends pointing towards each other. Repeated cycles of heat denaturation of the template, annealing of the primers to their complementary sequence and extension of the annealed primers with a DNA polymerase results in the exponential amplification of the DNA segment defined by the 5' ends of the PCR primers. After n number of reaction cycles, 2^n molecules of the target sequence are produced. (b) Anchored PCR. In cases where only one sequence suitable as a primer-binding site is known, the DNA sequence to be amplified can be 'tailed' with poly(G) so that an anchored primer sequence including a poly(C) stretch can be used as the second primer. This technique has been used to amplify transcripts for which the amino acid sequence of one region of the protein is known or for those that share similar sequences with other genes. (c) Inverse PCR. PCR can be used to amplify regions of DNA of unknown sequence that flank known sequences. Primers complementary to the ends of the known sequences and whose 3' ends point away from each other are used. Circular molecules containing the target sequences are formed after digestion with restriction enzymes and ligation. PCR cycles amplify the sequences that flank the ends of the known sequence.

analyse unknown intervening sequences. For example, the use of evolutionary conserved portions of a gene for primer construction will allow the amplification of highly variable sequences that lie between the conserved templates for these primers. This approach has been successfully used to isolate plant genes encoding protein kinases by using information available on the conserved sequence of protein kinases from other organisms to design the PCR primers. Similarly, information on amino acid sequences that are conserved among different viruses, allows the design of degenerate primer mixtures which can be used to search for novel viruses that retain significant similarity at the amino acid but not at the nucleotide sequence level (Mack and Sninsky, 1988). RNA can also serve as a template for amplification after conversion to cDNA (Tood *et al.*, 1987). A mixture of degenerate primers derived from partial amino acid sequence data has been employed to amplify cDNA segments that can serve to clone directly the desired sequence or to produce probes for identifying clones or for hybridization studies (Lee *et al.*, 1988).

(a) Anchored PCR

In many cases when a protein is available in small quantities, information about its amino acid sequence can only be obtained for a single region of the protein, most often from the amino terminal part. In these cases, or in general when either the 5′ or the 3′ portion of the sequence of interest is not known, homopolymer tailing of the end of unknown sequence in the DNA or cDNA can effectively substitute for the unknown sequence information. The complementary sequence to the homopolymer tail or anchor sequence serves as a priming sequence for subsequent amplification in the so-called 'anchored PCR' technique as illustrated in Fig. 8.1b (Frohman *et al.*, 1988; Loh *et al.*, 1988).

(b) Inverse PCR

Although it may seem counterintuitive, sequences that are external to the primers can also be amplified by a modification of the PCR procedure (Triglia *et al.*, 1988). In this approach, the template is digested with a restriction enzyme that cuts outside the region to be amplified, the resulting restriction fragments are circularized by ligation, and amplified using

primer sequences whose ends point away from each other, that is, across the external segments that are now joined (Fig. 8.1c). This 'inverse PCR' or IPCR method should prove extremely useful for the cloning of sequences that are adjacent to transposon insertions that have produced mutations in genes of interest. It would require only the digestion of the genomic DNA of the mutant plant with a restriction enzyme that does not cut in the transposon sequence, ligation of the DNA, IPCR amplification and cloning of the amplified fragment, avoiding the cumbersome problem of constructing a genomic library for each of the mutants in which one is interested. It can also be used to study and clone the 5′ flanking transcriptional regulatory regions adjacent to the coding sequence of different members of gene families that encode highly conserved proteins. It could also be used for genetic mapping via chromosome walking or insertion inactivation.

8.4.3 Using PCR for sequencing

Several methods have been described that permit direct sequencing of amplified genomic DNA without cloning (Wong *et al.*, 1987; Stoflet *et al.*, 1988; Wrischnik *et al.*, 1988). These methods involve the use of a sequencing primer that lies within the amplified fragment. Although successful sequencing can be carried out on double-stranded PCR products, a modification of the PCR protocol in which one of the PCR primers is at a lower concentration than the other can generate a single-stranded template. By using primers based on cloning vector sequences flanking an insertion site, cDNA and short genomic clones can be rapidly analysed or sequenced (Friedman *et al.*, 1988; Saiki *et al.*, 1988).

8.4.4 Some considerations on PCR

The sensitivity of PCR permits relatively crude DNA samples to be used as templates for amplification. Potential inhibitors can be eliminated by simple procedures such as boiling and dilution. Furthermore, a single 5 μm thin section from a formalin-fixed paraffin embedded tissue block contains enough DNA, even after 40 years, to serve as a template for PCR and makes possible retrospective molecular analysis of archive material (Shibata *et al.*, 1988). DNA contained in hair, ancient or mummified specimens, museum or herbarium collections and single-celled organisms such as plankton, has also been amplified (Higuchi *et al.*, 1988; Paabo *et al.*, 1988; Steffan and Atlas, 1988). The sensitivity of the method requires great care to avoid amplifying DNA contaminants, such as amplified DNA from previous experiments. As many as 10^{12} molecules/reaction of a specific fragment can be produced, and these should be analysed in a different area of the laboratory from where new PCR experiments are carried out, since traces of a previous experiment may provide more template molecules than the 10^0–15^5 molecules that are normally present in a sample of eukaryotic DNA.

Several factors have been noted to affect the efficiency of the PCR reaction. A wide range of concentrations of nucleotide triphosphates have been used. A level of 200 μM works well under most conditions. The length of the primers is also important. Primers of 16 or more nucleotides work well provided that they contain at least 40% G/C, the longer the primer the better the specificity but also the higher the cost to make it. In general, sequences containing runs of four or more of a single nucleotide should be avoided because of the possibility of non-specific priming.

8.4.5 Using PCR for the construction of genetic maps

Genetic maps comprising closely-spaced phenotypic or molecular markers are useful for genome analysis. Markers that are linked to a trait of interest can be used for plant and animal breeding programmes, especially for trait introgression or for gene cloning (Landegren *et al.*, 1988; Tanksley *et al.*, 1989). Because in most organisms phenotypic markers are not sufficiently abundant to construct precise genetic maps, molecular markers (segments of DNA) are being used more and more frequently

(Chapter 19). The DNA markers most commonly used are restriction fragment length polymorphisms (RFLPs) (Botstein *et al.*, 1980). The detection of RFLPs requires the use of Southern blot hybridizations, that are laborious and technically difficult to set up. PCR has been used to avoid the need for hybridization, by using primers that correspond to the ends of the sequence of the RFLP probes (Vosberg, 1989; D'Ovidio *et al.*, 1990). Unfortunately the need of knowledge of the nucleotide sequence of the ends of the RFLP probes has made this approach not generally applicable. Recently a new PCR based technique that circumvents most of these problems has been developed. The Random Amplified Polymorphic DNA (RAPD) technique is based on the use of primers of arbitrary nucleotide sequence to amplify random segments of genomic DNA by means of PCR and thus to reveal polymorphisms (Welsh and McClellan, 1990; Williams *et al.*, 1990). Primers for RAPD analysis are 9–10 bases long. The polymorphisms observed using the RAPDs are believed to be due to single base changes that prevent pairing of the primer with target sequences, deletions of the priming site, insertions or deletions that modify the size of the amplified DNA or insertions that render priming sites too distant to support amplification. In contrast to RFLP markers that are in general 'co-dominant', the majority of RAPD markers behave as dominant Mendelian markers, since in most cases the amplified fragment is present in one individual but not in the other (Williams *et al.*, 1990). Co-dominant RAPD markers that produce different-sized DNA segments in different individuals are only rarely observed. Each primer may produce one or more detectable polymorphisms and in addition produce a banding pattern that can be used as a genetic fingerprint. One of the major advantages of this method is that the same set of oligonucleotides can be used for any species or organism, whereas the RFLP probes can only be used for a set of closely related species. The potential polymorphisms detectable by RAPD markers has also been estimated to be much higher than with RFLPs and allows detection of polymorphisms even within regions containing repeated sequences. A second major advantage is that after the 2–4 hour amplification, the polymorphisms can be directly observed by normal agarose gel electrophoresis, avoiding costly and time consuming hybridization techniques using radiolabelled probes.

8.5 GENE IDENTIFICATION

The molecular identification of a gene of interest between all the recombinants that are present in a cDNA or genomic library can be accomplished by different strategies. The best procedure to use will depend mainly upon the previous information available about the gene. When the gene product is known the isolation and cloning of the gene is relatively straightforward. However, there are other strategies available to identify and isolate genes for which no knowledge about the function of the gene is available. Below, different strategies for gene identification will be described. The references provided in this section do not by any means cover all the published works in which the different strategies have been used, and are meant to provide a guide to the reader to a few recent publications.

8.5.1 Identification by the gene product

Depending on the information available about the polypeptide encoded by a particular gene, different strategies can be used to identify the corresponding cDNA or genomic clones. For example, if the polypeptide has been purified partially or to homogeneity, specific monoclonal or polyclonal antibodies or the amino acid sequence can serve as the basis for gene identification.

(a) Detection of the product using antibodies

A very successful approach is the use of antibodies against the protein encoded by the gene of interest. Here it is first necessary to purify the protein and raise specific antibodies against it. The antibodies can later be used to select for a specific clone out of a cDNA library. The cDNA library should be constructed in either a plasmid or phage expression vector to have the corresponding gene product synthesized in *E. coli*. The plasmid or phage cDNA library is introduced in *E. coli* and plated in petri dishes containing solid media. The protein content of bacterial colonies or phage plaques is transferred to nylon or nitrocellulose membranes. These membranes are incubated with the specific antibody, and the plaque or colonies to which this antibody is bound are identified by means of an anti-antibody that is linked to an enzyme that produces a colour reaction such as alkaline phosphatase or peroxidase (Huynh *et al.*, 1985; Maniatis *et al.*, 1989). Using this approach cDNA clones for the 33 kDa protein of the oxygen-evolving complex (OEE 1) from tomato and *Arabidopsis* were identified (Ko *et al.*, 1990). With this strategy it was also possible to find new genes of *Lupinus polyphyllus* (Perrey *et al.*, 1990) even when the functions of the proteins were unknown.

(b) Using the amino acid sequence of the gene product to produce synthetic oligonucleotide probes

The amino acid sequence of a protein obtained after its purification or microsequencing directly from two dimensional gel electrophoresis can be used to design the nucleotide version of this sequence in the form of synthetic oligonucleotides. These oligonucleotides can be radioactively labelled and used as specific probes to find the corresponding gene within a cDNA-library (Fig. 8.2 and Maniatis *et al.*, 1989).

This strategy was used to identify cDNA clones coding for UDP-glucose pyrophosphorylase from potato tuber (Katsube *et al.*, 1990) and for the betaine-aldehyde dehydrogenase (*BADH*) gene from salt-stressed spinach leaves (Wesetilnyk and Hanson, 1990). Similar procedures were used to identify cDNA clones of glutamine synthase (*GS*) genes from barley (Freeman *et al.*, 1990) and a cDNA encoding the wheat CM16 protein (Gautier *et al.*, 1990). Oligonucleotides can also be designed to use the PCR technique, as described before in this

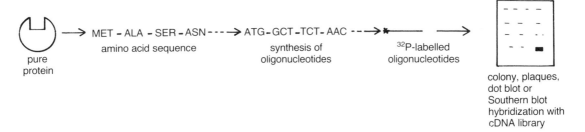

Fig. 8.2. Use of oligonucleotides based on the amino acid sequence of a protein to identify the corresponding gene. The amino acid sequence of a protein is obtained. Synthetic oligonucleotides are generated according to the amino acid sequence. Oligonucleotides are radioactively end-labelled and used as specific probes to screen a cDNA library using colony, plaques, dot blot or Southern blot hybridization.

chapter, to amplify a specific sequence out of an RNA population, genomic- or cDNA-libraries.

8.5.2 Identification of genes based on sequence characteristics

Even without any knowledge about the polypeptide encoded by the gene of interest in a given plant species, it is possible to identify genes using only their nucleotide sequence. This homology helps to find specific genes by hybridization either after elimination of common molecules by differential selection or by using probes with similar nucleotide sequences.

(a) Using heterologous probes to detect specific sequences

Another very common procedure to isolate genes is the use of heterologous probes, isolated and characterized in other organisms, to screen for similar genes in a genomic or cDNA library of the organism of interest. Proteins with the same functions of different organisms often have a similar primary structure. The amino acid sequence of these proteins probably arose from a common ancester and therefore they have related nucleotide sequences. Nucleotide sequences isolated from animal cells, *Drosophila* and yeast have been successfully used to isolate plant related genes. Using this strategy a gene for chlorophyll a/b-binding proteins of LHC-II

was identified in a cDNA library prepared from mRNA of Scots pine light-treated seedlings using a cDNA from pea (Jansson and Gustafsson, 1990).

A similar approach is the use of nucleotide sequences contained in databases such as the EMBL, GenBank and NBRF to search for conserved regions of genes or proteins from different organisms and to use this information to synthesize oligonucleotide probes (Devereux *et al.*, 1984). In this case the oligonucleotides should match preferentially with the most highly conserved region of the protein, such as the sequence of the active site of an enzyme. Codon usage should also be considered to cover all possibilities in order to increase the probability of specific hybridization (Murray *et al.*, 1989; Campbell and Gowri, 1990).

(b) Differential screening

The differential screening procedure consists in the elimination of mRNA transcripts (cDNAs) common to two populations of RNAs (cDNAs) obtained from different organs of the same plant or siblings grown under different environmental conditions or harvested at distinct developmental stages. With this procedure the aim is to select for a population of mRNAs (cDNAs) which are specific to a particular growing condition, developmental stage or tissue. The selection is based on the subtraction of RNA populations that are common to the two RNA samples by first preparing cDNA

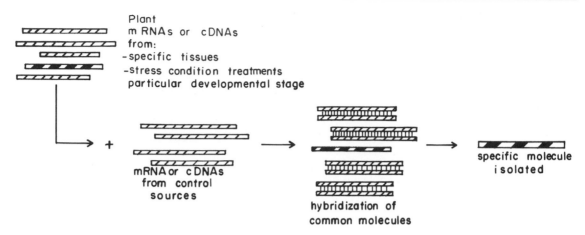

Fig. 8.3. Differential hybridization to isolate cell type or developmental specific genes. Specific tissues, stress conditions or particular developmental stages can induce some genes of interest. mRNA or cDNA libraries obtained from these sources can be used to identify those particular genes even without knowing the gene product. mRNAs and cDNAs obtained from induced and control sources will contain common molecules normally present in all conditions and tissues. Differential screening by hybridization between these samples allows the 'cleaning' of particular molecules (non-hybridized) from the common ones (hybridized). After several steps of differential screening it is possible to purify the particular molecule of interest.

copies of one of the populations and then allowing the hybridization of these cDNA molecules with the other RNA population. The cDNAs for which homologous sequences exist in this hybridization mixture will form double stranded molecules that can be separated from the RNA molecules that are present in only one of the two populations by chromatography on hydroxyapatite columns. The isolated specific mRNAs can be used as probes to screen a cDNA library or to construct a directed-cDNA library. A scheme of this strategy is shown in Fig. 8.3.

This type of approach was used to isolate a gene called *kin 1* from a cDNA library of *Arabidopsis thaliana* (Kurkela and Franck, 1990). A more complicated but complete differential screening was used to isolate desiccation-induced genes (Bartels *et al.*, 1990). This screening method has also been applied to the identification of tissue-specific genes. The identification of a new nodule-specific gene (*Nms-25*) of *Medicago sativa* (Kiss *et al.*, 1990) was achieved by this means. An interesting example was the differential screening of a cDNA library made of poly(A)+ mRNA from roots of peas

(Evans *et al.*, 1990) with transcripts from green and etiolated leaves and cotyledons.

8.5.3 Identification of genes based on the function of the gene product

There are several strategies that allow the identification of genes for which no antibodies or information about the amino acid or nucleotide sequence of the gene are available. One of these is based on knowledge of the enzymatic or structural function of the gene product. In this case the cDNA of interest is expressed in bacterial or yeast systems, where the enzymatic or structural activity of the gene product can be measured or observed as a distinguishable phenotype. This requires bacterial or yeast mutants altered with respect to the function of the gene of interest and, furthermore, that the phenotype of the mutants can be easily distinguished from that of the wild type. Identification of the target gene is carried out by complementation of the mutant bacteria or yeast. In this case a cDNA library must be constructed in a suitable expression vector that allows the production of the plant polypeptide in bacteria or yeast. Obviously the ideal system for application of this strategy would be the complementation of plant mutants. However, two major obstacles have hampered the use of plant systems to isolate genes by complementation, namely the lack of a transformation system with an efficiency high enough to allow the introduction of a whole genomic or cDNA library into plant cells and the practical problems in the handling of hundreds of thousands of transformants when direct selection cannot be used. The use of *E. coli* or yeast eliminates both these limitations.

A similar approach can be used to differentiate between very closely related genes such as those encoding isoenzymes. If an accurate assay for the isoenzyme is already established, the transformation of microorganisms with the putative, closely related cDNAs allows expression of the proteins. Specific enzymatic assays for the isoenzymes in an extract from the transformed yeast or bacteria can identify the respective cDNAs. This type of approach was used to establish the correlation between the genes and the two 4-coumarate: CoA ligase isoenzymes (4CL) of parsley (Lozoya *et al.*, 1988).

8.5.4 Molecular tagging of genes

Mobile genetic elements cause mutations when they transpose within or near the coding sequence of a gene. For the case in which the element has been characterized, they become a molecular tag that can be used to isolate the mutated gene by hybridization experiments, using the mobile element as a probe. Two types of mobile genetic elements have been used to isolate plant genes, namely transposons and the T-DNA of *Agrobacterium tumefaciens*.

(a) Tagging with Agrobacterium T-DNA

Agrobacterium tumefaciens has the capacity of transferring a defined segment (T-DNA) of its tumour inducing (Ti) plasmid to cells of plant species that are susceptible to infection by this bacterium. Several vectors for the transfer of foreign genes based on the Ti plasmid of *A. tumefaciens* have been developed. Upon insertion in the genome of plant cells the T-DNA can cause mutations. Genes mutated by the T-DNA can be identified and isolated using the T-DNA as a hybridization probe. This strategy is known as the T-DNA tagging procedure. To isolate a gene of interest a population of transgenic plants obtained by transformation with *A. tumefaciens* is screened for the plants showing the desired phenotype. After screening, the affected gene is isolated from a genomic library of the mutant plant using the T-DNA as a probe (Fig. 8.4). The isolated mutated gene can later be used to isolate the wild-type gene in a gene bank from a normal plant. In this case the probe will be the plant DNA found on the left or the right side of the T-DNA. Confirmation of the identity of the isolated gene can be carried out by complementation of mutant plants after transformation with the wild type gene.

Using this strategy the *GL-1* gene responsible for the normal trichome phenotype of *Arabidopsis thaliana* (Marks and Feldman, 1989) was identified and isolated. The identity of the gene (*GL-1*) was confirmed by complementation analysis (Herman and Marks, 1989). Using the same strategy a gene of *A. thaliana* involved in chloroplast development has been cloned. Transformation of a pale mutant with T-DNA vectors carrying the selected clones resulted in a normal green phenotype demonstrating positive complementation (Koncz *et al.*, 1990).

(b) Use of transposons for tagging genes

A strategy of gene isolation similar to T-DNA tagging using plant transposons has also been developed and very successfully applied for the isolation of maize genes. The insertion of a copy of transposable elements (e.g. *Ds*, *Ac* and others) may inactivate a gene producing a mutation (for a review of plant transposons see Döring and Starlinger, 1986; Gierl and Saedler, 1989). Phenotypic screening of mutants allows the function of the interrupted gene to be identified. Using the same strategy as that

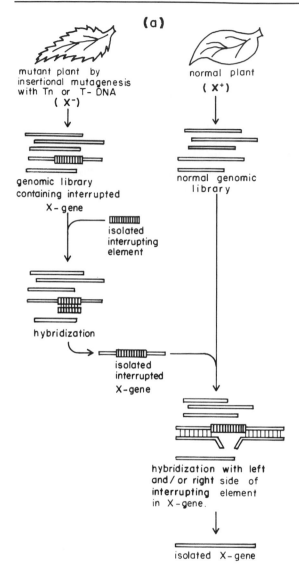

Fig. 8.4. Isolation of genes using transposon or T-DNA tagging. This strategy allows the isolation of genes for which no knowledge of the gene product is available and only the phenotype of mutants affected by the gene of interest is known. Mutants with a desired phenotype are isolated either by the insertion of transposons, in plants species where mobile elements are present, or by T-DNA insertion after transformation with *Agrobacterium*. Using the transposon or T-DNA as a specific probe to screen a genomic library from the mutant, it is possible to identify the affected gene by hybridization. The isolated interrupted gene is then used as a specific probe to identify the normal gene in a cDNA or genomic library made from the wild type plant.

described for T-DNA tagging, the mutated gene can be cloned using the Tn element as a hybridization probe. This technique is called 'transposon tagging' (Shaw, 1988). Among several other genes, this strategy was used to clone a gene producing the knotted phenotype in maize (Hake *et al.*, 1989).

8.5.5 Map-based cloning

A general strategy for the isolation of genes for which no biochemical or molecular information is available about the gene or the gene product has been recently developed. This strategy known as map-based cloning, relies on knowledge of the phenotype conferred by the gene of interest and its genetic map position. An outline of the scheme is presented in Fig. 8.5. The first step in this strategy is to identify tightly linked DNA sequences located at both sides of the target gene. These linked sequences are used as starting points for cloning of the chromosomal region where the target gene is located. RFLP or RAPD markers are the most useful flanking sequences for map-based cloning, since saturated maps of the target region using these markers can be relatively easily obtained. The second step is to isolate DNA clones covering the entire region between the two markers by a process known as chromosome walking. In this procedure, DNA sequences progressively further away from the RFLP or RAPD markers are identified by a series of overlapping cloning steps, based on DNA hybridization experiments. Then walking from one of the flanking sequences to the one on the other side will result in cloning all the DNA between these sequences including the target gene. Initially, chromosome walking was based on lambda or cosmid vectors capable of harbouring genomic inserts of up to 50 kb. Overlapping clones spanning over 200 kb could be isolated in this manner. Map-based cloning for the large genomes of most plant species, where the distances to be covered by chromosome walking are expected to be as great as 1000 kb, will be greatly facilitated when reliable YAC libraries with genomic inserts of more than 300 kb are available.

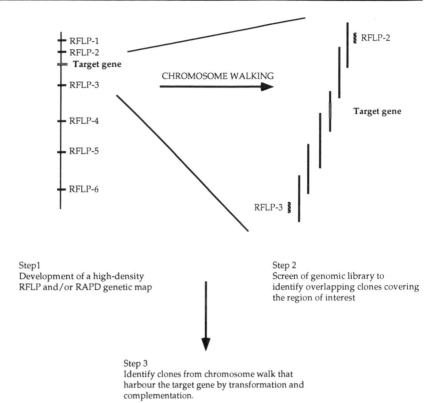

RFLP-1
RFLP-2
Target gene
RFLP-3

CHROMOSOME WALKING

RFLP-4

RFLP-5

RFLP-6

RFLP-2

Target gene

RFLP-3

Step1
Development of a high-density
RFLP and/or RAPD genetic map

Step 2
Screen of genomic library to
identify overlapping clones covering
the region of interest

Fig. 8.5. Major steps involved in map-based cloning.

Step 3
Identify clones from chromosome walk that
harbour the target gene by transformation and
complementation.

The last step in map-based cloning is pinpointing the clone harbouring the target gene among all the overlapping clones identified during chromosome walking. This step can be accomplished for dominant genes by transformation and complementation of recessive individuals for the phenotype confered by the target gene with the clones derived from the chromosome walk. Obviously, the more tightly linked the flanking markers to the target gene, the lower the number of clones that have to be tested. The clone containing the target gene can be identified because it will produce transgenic individuals expressing the appropriate phenotype. This identification can be used provided that the introduced gene is functionally expressed. If transformation is not feasible or transforming sequences are not expressed, other more complex identification strategies can be used (Kerem *et al.*, 1989; Riordan *et al.*, 1989). Map-based cloning has been successfully applied for the isolation of genes responsible for human hereditary diseases, including cystic fibrosis (Rommens *et al.*, 1989) and muscular dystrophy (Kenwrick *et al.*, 1987), and is now being applied in several laboratories for the isolation of plant pest resistance genes. Further aspects of map-based cloning are dealt with in section 19.6.2.

8.5.6 Antisense RNA inactivation

Alteration of gene expression can also be achieved using antisense mRNA technology (van der Krol *et al.*, 1988). This allows the identification of genes whose function is unknown and is achieved by analysis of the altered phenotype produced by the presence of an antisense mRNA of the unknown gene in transgenic plants.

For this procedure it is necessary first to isolate a cDNA encoding an unknown function and then to construct a chimeric gene that directs the production of the antisense strand of mRNA. Afterwards this chimeric gene is introduced into plant cells by gene transfer procedures (Fromm *et al.*, 1986; Klein *et al.*, 1987; Shaw, 1988; Valvekens *et al.*, 1988). The hypothesis is that once inside the cells the antisense mRNA produced by the inserted foreign gene will hybridize with the sense mRNA normally present in the cell and arrest its translation or make it more prone to degradation by ribonucleases. Whatever the mechanism of inactivation, the result is the lack of the corresponding gene product. Antisense mRNA analysis has been used to show that a tomato cDNA (TOM 13) encodes a polypeptide involved in the conversion of 1-aminocyclopropane-1-carboxylic acid to ethylene by the ethylene-forming enzyme (ACC-oxidase) (Hamilton *et al.*, 1990).

8.6 FUTURE PROSPECTS

The technical advances that allow the isolation and characterization of genes have dramatically changed the pace of generating knowledge about the basic biochemical, physiological and molecular events that control plant development and have paved the way for the design of transgenic plants that have great potential in agricultural applications. We are also beginning to witness a major advance in the understanding of complex physiological processes such as drought resistance, carbon assimilation and productivity, that could not be foreseen a decade ago. Strategies such as gene tagging and map-based cloning will facilitate this task by allowing the cloning and characterization of genes that are crucial for these processes and for which only the phenotype produced by the expression of underlying genes is known.

Another important outcome of gene isolation techniques will be the understanding of the mechanisms of plant disease resistance and the functions of disease resistance genes. Although it is out of the scope of this book, it will also be interesting to closely follow the advances on the function of genes encoding transacting factors that regulate gene expression and their potential use to induce dramatic changes in the developmental patterns of transgenic plants. The great success of antisense technology in

plant systems should also prove to be an extremely powerful tool for the study of plant development and to produce transgenic plants from which undesirable agronomic traits have been removed. An illustrative example of the power of this technology is the production of transgenic tomatoes that lack the rate-limiting enzyme in the biosynthetic pathway of ethylene and whose fruits do not ripen and therefore have a much longer life (Oeller *et al.*, 1991). These experiments raise the possibility that fruit ripening might someday be controllable on a commercial scale, thereby preventing spoilage.

In conclusion we dare to say that given the technology available the future of plant sciences and its application for agronomic purposes will be limited only by the creativity of plant scientists.

REFERENCES

Bartels, D., Schneider, K., Terstappen, G., Piatkowski, D. and Salamini, F. (1990), Molecular cloning of abscisic acid-modulated genes which are induced during desiccation of the resurrection plant *Ceratostigma plantagineum. Planta*, **181**, 27–34.

Botstein, D., White R.L., Skolnick, M. and Davis R. (1980), Construction of a genetic linkage map in man using restriction fragment length polymorphism. *Am. J. Hum. Genet.*, **32**, 314–331.

Burke, D.T., Carle, G.F. and Olson, M.V. (1987), Cloning of large segments of exogenous DNA into yeast by means of Artificial Chromosome Vectors. *Science*, **236**, 806–812.

Campbell, A. (1971), Genetic structure. In *The Bacteriophage Lambda*, Hershey, A.D. (ed.), Cold Spring Harbor, New York, pp. 13–25.

Campbell, W.H. and Gowri, G. (1990), Codon usage in higher plants, green algae and cyanobacteria. *Plant Physiol.*, **92**, 1–11.

Dale, J.W. (1987), Cloning in bacteriophage lambda. In *Techniques in Molecular Biology*, Vol. 2, Walker, J.M. and Gaastra, W. (eds), Croom Helm, London.

Dellaporta, S.L., Wood, J. and Hicks J.B. (1985), Maize DNA miniprep. In *Molecular Biology of Plants: A Laboratory Course Manual*, Malmberg, R., Messing, J. and Sussex, I. (eds), Cold Spring Harbor Press, New York, pp. 36–37.

Devereux, J., Haeberli, P. and Smithies, O. (1984), A

comprehensive set of sequence analysis programs for the VAX. *Nucleic. Acids Res.*, **12**, 387–395.

Döring, H.P. and Starlinger, P. (1986), Molecular genetics of transposable elements in plants. *Ann. Rev. Genet.*, **20**, 175–200.

D'Ovidio, R., Tanzarella, O.A. and Porceddu, E. (1990), Rapid and efficient detection of genetic polymorphism in wheat through amplification by polymerase chain reaction. *Plant Mol. Biol.*, **15**, 169–171.

Draper, J. and Scott, R. (1988), The isolation of plant nucleic acids. In *Plant Genetic Transformation and Gene Expression. A Laboratory Manual*, Draper, J., Scott, R., Armitage, P. and Walden, R. (eds), Oxford, pp. 199–236.

Evans, I.M., Gatehouse, L.N., Gatehouse, J.A., Robinson, N.J. and Croy, R.R.D. (1990), A gene from pea (*Pisum sativum* L.) with Thomology to metallothionin genes. *FEBS Lett.*, **262**, 29–32.

Freeman, J., Marquez, A.J., Wallsgrove, R.M., Saarelainen, R. and Forde, B.G. (1990), Molecular analysis of barley mutants deficient in chloroplast glutamine synthetase. *Plant Mol. Biol.*, **14**, 297–311.

Friedman, K.D., Rosen, N.L., Newman, P.J. and Montgomery, R.R. (1988), Enzymatic amplification of specific cDNA inserts from λ gt11 libraries. *Nucleic. Acids Res.*, **16**, 8713–8718.

Frohman, M.A., Dush, M.K. and Martin, G.R. (1988), Rapid production of full-length cDNAs from rare transcripts: amplification using single-specific oligonucleotide primers. *Proc. Natl. Acad. Sci. USA*, **85**, 8998–9002.

Fromm, M.E., Taylor, L.P. and Walbot, V. (1986), Stable transformation of maize after gene transfer by electroporation. *Nature*, **319**, 791–793.

Gautier, M.F., Alary, R. and Jourdier, P. (1990), Cloning and characterization of a cDNA encoding the wheat (*Triticum durum* Desf.) CM 16 protein. *Plant Mol. Biol.*, **14**, 313–322.

Gierl, A. and Saedler, H. (1989), Maize transposable elements. *Ann. Rev. Genet.*, **23**, 121–139.

Hake, S., Vollbrech, E. and Freeling, M. (1989), Cloning Knotted, the dominant morphological mutant in maize using DS2 as a transposon tag. *EMBO J.*, **8**, 15–22.

Hamilton, A.J., Lycett, G.W. and Grierson, D. (1990), Antisense gene that inhibits synthesis of the hormone ethylene in transgenic plants. *Nature*, **346**, 284–287.

Herman, P.L. and Marks, M.D. (1989), Trichome development in *Arabidopsis thaliana*. II. Isolation

and complementation of the GLABROUS 1 gene. *Plant Cell*, **1**, 1051–1055.

Higuchi, R., Beroldingen, C.H. von, Sensabangh, G.F. and Erlich, H. (1988), DNA typing from single hairs. *Nature*, **332**, 543–546.

Huynh, T.V., Young, R.A. and Davis, R.W. (1985), Constructing and screening cDNA libraries in lambda gt10 and gt11. In *DNA Cloning: A Practical Approach*, Vol. I, Glover, D.M. (ed.), IRL Press, Oxford, pp. 49–78.

Jansson, S. and Gustafsson, P. (1990), Type I and type II genes for the chlorophyll a/b-binding protein in the gymnosperm *Pinus sylvestris* (Scots pine): cDNA cloning and sequence analysis. *Plant Mol. Biol.*, **14**, 287–296.

Jeffreys, A.J., Wilson, V., Neumann, R. and Keyte, J. (1988), Amplification of human minisatellites by polymerase chain reaction: towards DNA fingerprinting of single cells. *Nucleic Acids Res.*, **16**, 10953–10960.

Karn, J., Matthes, H.W.D., Gait, M.J. and Brenner, S. (1984), A new selective phage cloning vector L 2001 with sites for *Xba*I, *Bam*HI, *Hind*III, *Eco*RI, *Sst*I and *Xho*I. *Gene*, **32**, 217–221.

Katsube, T., Kazuta, Y., Mori, H., Nakano, K., Tanizawa, K. and Fukui, T. (1990), UDP-Glucose pyrophosphorylase from potato tuber: cDNA cloning and sequencing. *J. Biochem.*, **108**, 321–326.

Kenwrick, S., Patterson, M., Speer, A., Fischbeck, K. and Davies, K. (1987), Molecular analysis of the Duchenne muscular dystrophy region using pulsed field gel electrophoresis. *Cell*, **48**, 351–357.

Kerem, B.S., Rommens, J.M., Buchanan, J.A., Markiewicz, D., Cox, T.K., Chakravarti, A., Buchwald, M. and Tsui, L.C. (1989), Identification of the cystic fibrosis gene: genetic analysis. *Science*, **245**, 1073–1080.

Kiss, G.B., Vincze, E., Végh, Z., Tóth, G. and Soós, J. (1990), Identification and cDNA cloning of a new nodule-specific gene, Nms-25 (nodulin-25) of *Medicago sativa*. *Plant Mol. Biol.*, **14**, 467–475.

Klein, T.M., Wolf, E.D., Wu, R. and Sanford, J.C. (1987), High velocity microprojectiles for delivering nucleic acids into living cells. *Nature*, **327**, 70–73.

Ko, K., Granell, A., Bennett, J. and Cashmore, A. (1990), Isolation and characterization of cDNAs from *Lycopersicum esculentum* and *Arabidopsis thaliana* encoding the 33 KDa protein of the photosystem II-associated oxygen-evolving complex. *Plant Mol. Biol.*, **14**, 217–227.

Koncz, C., Mayerhofer, R., Koncz-Kalman, Z., Nawrath, Ch., Reiss, B., Redei, G.P. and Schell,

J. (1990), Isolation of a gene encoding a novel chloroplast protein by T-DNA tagging in *Arabidopsis thaliana*. *EMBO J.*, **9**, 1337–1346.

Kurkela, S. and Franck, M. (1990), Cloning and characterization of a cold and ABA-inducible *Arabidopsis* gene. *Plant Mol. Biol.*, **15**, 137–144.

Landegren, U., Kaiser, R., Caskey, C.T. and Hood, L. (1988), DNA Diagnostics – molecular techniques and automation. *Science*, **242**, 229–237.

Lee, L.L., Wu, Y., Gibbs, R., Cook, R.G., Muzny, D.M. and Caskey, C.T. (1988), Generation of cDNA probes directed by amino acid sequence: cloning of ureate oxidase. *Science*, **239**, 1288–1291.

Logemann, J., Schell, J. and Willmitzer, L. (1987), Improvement method for the isolation of RNA from plant tissues. *Anal. Biochem.*, **163**, 16–20.

Loh, E.Y., Elliot, J.F., Cwirla, S., Lanier, L.L. and Davis, M.M. (1988), Polymerase chain reaction with single sided specificity: analysis of T cell receptor 5 chain, *Science*, **243**, 217–220.

Lozoya, E., Hoffmann, H., Douglas, C., Schulz, W., Scheel, D. and Hahlbrock, K. (1988), Primary structure and catalytic properties of isoenzymes encoded by the two 4-coumarate: ligase genes in parsley. *Eur. J. Biochem.*, **176**, 661–667.

Mack, D.H. and Sninsky, J.J. (1988), A sensitive method for the identification of uncharacterized viruses related to known virus groups. *Proc. Natl. Acad. Sci. USA*, **85**, 6977–6981.

Maniatis, T., Fritsch, E.F. and Sambrook, K.J. (eds) (1989), *Molecular Cloning – A Laboratory Manual*, Vol. I, II and III, 2nd edn, Cold Spring Harbor Laboratory Press, New York.

Marchuk, D. and Collins, F.S. (1988), pYAC-RC, a yeast artificial chromosome vector for cloning DNA cut with infrequently cutting restriction endonucleases. *Nucleic Acids Res.*, **16**, 7738–7743.

Marks, M.D. and Feldman, K.A. (1989), Trichome development in *Arabidopsis thaliana*. I. T-DNA tagging of the GLABROUS 1 gene. *Plant Cell*, **1**, 1043–1050.

Murray, E.E., Lotzer, J. and Eberle, M. (1989), Codon usage in plant genes. *Nucleic Acids Res.*, **17**, 477–498.

Oeller, P.W., Min-Wong, L., Taylor, L.P., Pike, D.A. and Theologist, A. (1991), Reversible inhibition of tomato fruit senescence by antisense RNA. *Science*, **254**, 437–439.

Old, R.W. and Primrose, S.B. (1989), *Principles of Gene Manipulation. An Introduction to Genetic Engineering*, 4th edn., Blackwell Scientific Publications, Oxford.

Paabo, S., Gifford, J.A. and Wilson, A. (1988), Mitochondrial DNA sequences from a 7,000-year old brain. *Nucleic Acids Res.*, **16**, 9775–9787.

Pelham, H.R.B. and Jackson, R.J. (1976), An efficient mRNA dependent translation system from reticulocyte lysates. *Eur. J. Biochem.*, **67**, 247–252.

Perrey, R., Warskulat, U. and Wink, M. (1990), Molecular cloning of a lupin-specific gene from a cDNA library of suspension-cultured cell of *Lupinus polyphyllus*. *Plant Mol. Biol.*, **15**, 175–176.

Riordan, J.R., Rommens, J.M., Kerem, B.S., Alon, N., Rozmahel, R., Grezelczak, Z., Zielenski, J., Lok, S., Plavsic, N., Chou, J.L., Drumm, M.L., Iannuzi, M.C., Collins, F.S. and Tsui, L.C. (1989), Identification of the cystic fibrosis gene: cloning and characterization of complementary DNA. *Science*, **245**, 1066–1072.

Rommens, J.M., Iannuzi, M.C., Kerem, B., Drumm, M.L., Melmer, G., Dean, M., Rozmahel, R., Cole, J.L., Kennedy, D., Hidaka, N., Zsiga, M., Buchwald, M., Riordan, J.R., Tsui, L.C. and Collins, F.S. (1989), Identification of the cystic fibrosis gene: chromosome walking and jumping. *Science*, **245**, 1059–1065.

Saghai-Maroof, M.A., Loliman, K.M., Jorgensen, R.A. and Allard, R.W. (1984), Ribosomal DNA spacer length polymorphisms in barley: Mendelian inheritance, chromosomal location and population dynamics. *Proc. Natl. Acad. Sci. USA*, **81**, 8014–8018.

Saiki, R.K., Gelfand, H.D., Stoffel, S., Scharf, S.J., Higuchi, R., Horn, G.T., Mullis, K.B. and Erlich, H.A. (1988), Primer-directed enzymatic amplification of DNA with a thermostable DNA polymerase. *Science*, **239**, 487–491.

Saiki, R.K., Scharf, S., Faloona, F., Mullis, K., Horn, G.T., Erlich, H.A. and Arnheim, N. (1985), Enzymatic amplification of B-globin genomic sequences and restriction site analysis for diagnosis of sickle cell anemia. *Science*, **230**, 1350–1354.

Schlessinger, D. (1990), Yeast artificial chromosomes: tools for mapping and analysis of complex genomes. *Trends Genet.*, **6**, 248–258.

Shaw, C.H. (ed.) (1988), *Plant Molecular Biology. A Practical Approach*, IRL Press, Oxford.

Shibata, D.K., Martin, J.W. and Arheim, N. (1988), Use of PCR to isolate DNA from paraffin embedded tissues. *Cancer Res.*, **48**, 4564–4569.

Short, J.M., Fernandez, J.M., Sorge, J.A. and Huse, W.D. (1988), L. Zap. A bacteriophage 1 expression vector with *in vivo* excision properties. *Nucleic Acids Res.*, **16**, 7583–7600.

Steffan, R.S. and Atlas, R.A. (1988), DNA Amplification to enhance detection of genetically, engineered bacteria in environmental samples. *App. Environ. Microbiol.*, **54**, 2185–2191.

Stoflet, E.S., Koeberl, D.D., Sarkar, G. and Sommer, S.S. (1988), Genomic amplification with transcript sequencing. *Science*, **239**, 491–494.

Tanksley, S.D., Young, N.D., Paterson, A.H. and Bonierbale, M.W. (1989), RFLP mapping in plant breeding: New tools for an old science. *Bio/Technology*, **7**, 257–264.

Tood, J.A., Bell, J.I. and McDevitt, H.O. (1987), HLA-DQB gene contributes to suceptibility and resistance to insulin-dependent diabetes mellitus. *Nature*, **329**, 599–604.

Triglia, T., Peterson, M.G. and Kemp, D.J. (1988), A procedure for *in vitro* amplification of DNA segments that lie outside the boundaries of known sequences. *Nucleic Acids Res.*, **16**, 8180–8186.

Valvekens, D., Montagu, M. van and Lijsebettens, M. van (1988), *Agrobacterium tumefaciens*-mediated transformation of *Arabidopsis thaliana* root explants by using kanamycin selection. *Proc. Natl. Acad. Sci. USA*, **85**, 5536–5544.

Van der Krol, A.R., Mol, J.N.M. and Stuitje, A.R. (1988), Antisense genes in plants. An overview. *Gene*, **72**, 45–50.

Vosberg, H.P. (1989), The polymerase chain reaction and improved method for the analysis of nucleic acids. *Hum. Genet.*, **83**, 1–12.

Welsh, J. and McClellan, M. (1990), Fingerprinting genomes using PCR with arbitrary primers. *Nucleic Acids Res.*, **18**, 7213–7218.

Wesetilnyk, E.A. and Hanson, A.D. (1990), Molecular cloning of a plant betaine-aldheyde dehydrogenase, an enzyme implicated in adaptation to salinity and drought. *Proc. Natl. Acad. Sci. USA*, **87**, 2745–2749.

Williams, J.G.K., Kubelik, A.R., Livak, K.J., Rafalski, J.A. and Tingey, S.V. (1990), DNA polymorphisms amplified by arbitrary primers are useful as genetic markers. *Nucleic Acids Res.*, **18**, 6531–6535.

Wong, C., Dowling, C.E., Saiki, R.K., Higuchi, R.G., Erlich, H.A. and Kazazian, H.H. (1987), Characterization of B-thalassaemia mutations using direct genomic sequencing of amplified single copy DNA. *Nature*, **330**, 384–386

Wrischnik, L.A., Higuchi, K.G., Stoneking, M., Erlich, H.A., Arnbeim, A. and Nilson, A.C. (1988), Length mutation in human mitochondrial DNA: direct sequencing of enzymatically amplified DNA. *Nucleic Acids Res.*, **15**, 529–542.

9

Gene transfer to plants: approaches and available techniques

I. Potrykus

9.1 INTRODUCTION

Gene transfer to plants is hampered by a variety of biological problems: the plant cell wall is a perfect barrier and trap for DNA molecules; egg cells, sperm cells and zygotes, are virtually inaccessible; proembryos are extremely small and enclosed within solid tissues; the tiny cells of hidden meristems which contribute to the 'germline' may not be competent for the integration of functional genes. There are no known retroviruses which could help systemically to spread and integrate foreign DNA. We have available only one functional biological vector system which, however, does not work with the agronomically most important groups or varieties of crop plants. Regeneration of transgenic plants mostly depends on 'totipotent' somatic cells, and, although some plant cells are totipotent, the majority, probably, are not. Surprisingly, and despite this long (incomplete) list of problems interfering with the regeneration of transgenic plants, the production of transgenic plants is efficient and routine. Unfortunately, however, this is restricted to some selected 'model plant' species and varieties. Efficient application of gene technology to plant breeding is still limited by the fact that routine and efficient procedures for the recovery of sufficient numbers of independent transgenic

Plant Breeding: Principles and prospects. Edited by M.D. Hayward, N.O. Bosemark and I. Romagosa. Published in 1993 by Chapman & Hall, London. ISBN 0 412 43390 7.

plants which retain their varietal identity, are still missing. There is, however, hope that with continuing efforts to develop novel methods, or improve existing ones, gene technology will become a routine procedure in future plant breeding. The success achieved so far may be considered small, if measured against the ideal situation: easy gene transfer to any given variety of any desired crop plant. The progress made since the recovery of the first transgenic plants in 1984 is, however, impressive (for a review see Gasser and Fraley, 1989).

9.1.1 Recovery of transgenic crop plants

Transgenic plants, in which the foreign gene is inherited according to Mendelian rules have been recovered not only from 'model plants' such as tobacco (*Nicotiana tabacum*), petunia (*P. hybrida*), and *Arabidopsis* (*A. thaliana*), but also from an impressive list of crop plants (Table 9.1). The genes transferred include not only 'model genes' such as those governing resistance to antibiotics or colour markers, but also agronomically interesting traits such as virus resistance, insect resistance, herbicide resistance, male sterility, fruit ripening, etc.

9.1.2 Assessment of the literature on gene transfer methods

The literature describing methods for the production of transgenic plants is rather confusing

Table 9.1. Incomplete list of transgenic crop plants*

Crop	Reference
Rice (*Oryza sativa*)	Shimamoto *et al.* (1989), Datta *et al.* (1990), Christou *et al.* (1991)
Maize (*Zea mays*)	Donn *et al.* (1990), Fromm *et al.* (1990), Gordon-Kamm *et al.* (1990)
Wheat (*Triticum aestivum*)	Vasil and Vasil (1992)
Soybean (*Glycine max*)	Hinchee *et al.* (1988), Christou *et al.* (1990), Christou (1992)
Cotton (*Gossypium hirsutum*)	Firoozabady *et al.* (1987), Umbeck *et al.* (1987), Perlak *et al.* (1990)
Potato (*Solanum tuberosum*)	DeGreef *et al.* (1989), Hoekema *et al.* (1989), Lawson *et al.* (1990)
Tomato (*Lycopersicon esculentum*)	Tumer *et al.* (1987), Smith *et al.* (1988, 1990)
Rapeseed (*Brassica napus*)	Guerche *et al.* (1987), Pua *et al.* (1987), Radke *et al.* (1988)
Tall fescue (*Festuca arundinacea*)	Wang *et al.* (1992)
Orchardgrass (*Dactylis glomerata*)	Horn *et al.* (1988)
Alfalfa (*Medicago sativa*)	Shahin *et al.* (1986), D'Halluin *et al.* (1990), Hill *et al.* (1991)
Lettuce (*Lactuca sativa*)	Michelmore *et al.* (1987), Chupeau *et al.* (1989)
Pea (*Pisum sativum*)	Puonti-Kaerlas *et al.* (1990)
Cabbage (*Brassica oleracea*)	Toriyama *et al.* (1991)
Grapevine (*Vitis rupestris, V. vinifera*)	Mullins *et al.* (1990)
Papaya (*Carica papaya*)	Fitsch *et al.* (1990)
Plum (*Prunus domestica*)	Mante *et al.* (1991)
Poplar (*Populus alba* × *grandidentata*)	Filatti *et al.* (1987), Pythoud *et al.* (1987)
Apple (*Malus pumila*)	James *et al.* (1989), Lambert and Tepfer (1991)
Walnut (*Juglans regia*)	McGranahan *et al.* (1988)
Carnation (*Dianthus caryophyllus*)	Lu *et al.* (1991)
Petunia (*Petunia hybrida*)	Van der Krol *et al.* (1988)

* See also Table 10.2

and misleading, not only for the inexperienced reader but also for experienced scientists working in the field. Many authors, editors and referees have been carried away with optimism and have been misled by indicative evidence and artifacts such that there are nearly more claims for effective methods than there are transgenic plants. On the basis of this experience we consider it mandatory to establish proof before trusting and publishing on integrative transformation (Potrykus, 1990a,b, 1991). From the numerous examples where true transgenic plants have been recovered we have a lead to what constitutes 'proof'. Neither genetic, phenotypic, nor physical data alone are acceptable. Proof for integrative transformation requires: (a) controls for treatment and analysis; (b) tight correlation between treatment and predicted results; (c) tight correlation between physical (e.g. Southern blot) and phenotypic (e.g. enzyme assay) data; (d) complete Southern analysis containing the predicted signal in high-molecular-weight DNA, and showing hybrid fragments between host DNA and the foreign gene, the presence of the complete gene, together with evidence for the absence of contaminating DNA fragments; (e) data that allow discrimination between false positives and correct transformants; (f) correlation of physical and phenotypic evidence with transmission to sexual offspring; (g) molecular analysis of offspring populations. Based on these proofs of the numerous methods applied to gene transfer only four have been successful. These are: (a) *Agrobacterium*-mediated gene transfer (e.g. Hooykaas, 1989; Klee and Rogers, 1989; Binns, 1990); (b) protoplast-based direct gene transfer (e.g. Potrykus *et al.*, 1985a,b; Davey *et al.*, 1989; Paszkowski *et al.*, 1989); (c) microinjection (e.g. Neuhaus *et al.*, 1987; Neuhaus and Spangenberg, 1990; Schnorf *et al.*, 1991); (d) biolistics (e.g. Klein *et al.* 1988a, b; McCabe *et al.*, 1988; Sanford, 1988). Each of these methods has its specific merits and drawbacks, and there is, so far, no single method which has the potential to solve every practical problem (Potrykus, 1990a, 1991).

9.1.3 Unsuccessful approaches

If one trusts indicative evidence then there are numerous different methods to produce transgenic plants, including extremely simple ones, such as: (a) incubation in DNA of dry seeds or embryos (e.g. Töpfer *et al.*, 1989); (b) incubation in DNA of turgescent tissue or cells; (c) pollen transformation (e.g. Hess, 1987); (d) pollen tube pathway (e.g. Luo and Wu, 1988); (e) macroinjection (e.g. De la Peña *et al.*, 1987); (f) electroporation of tissues (e.g. Dekeyser *et al.*, 1990); (g) electrophoresis (e.g. Ahokas, 1989); (h) liposome fusion with tissues (e.g. Gad *et al.*, 1990); (i) liposome injection (e.g. Lucas *et al.*, 1990); (j) microlaser treatment (e.g. Weber *et al.*, 1990); (k) viral vectors (e.g. Fütterer *et al.*, 1990); (l) agroinfection (e.g. Grimsley, 1990); (m) ultrasonication (e.g. Zhang *et al.*, 1991), etc. As, however, no transgenic plants could be recovered from these methods there must be something wrong with trusting indicative evidence, and hence the advice to accept only if complete proof is presented (for a more detailed assessment of the approaches mentioned above see Potrykus, 1990a,b, 1991).

9.2 BIOLOGY OF GENE TRANSFER PROTOCOLS

Some biological parameters will affect the delivery and fate of foreign genes in cells. Consideration of these parameters may help in understanding the problems of some approaches and help in designing better experiments. Not all plant cells are totipotent and plants differ in their capacity to respond to triggers: they are 'competent' for specific internal or environmental factors. Transgenic plants can be regenerated only from cells competent both for regeneration and integrative transformation. Plant tissues are mixed populations of cells with competence for many different responses. Considering the states of competence essential for recovery of transgenic plants the following situation has to be considered. A very small (and varying) minority of cells in plant tissues will be 'competent for both transformation and regeneration'. Others will be competent for transformation or regeneration. A larger fraction of cells will be 'potentially competent', which means that given the correct treatment they will have the potential to shift to the competent state. A variable portion of the cells will not even be potentially competent but will be 'non-competent'. The relative composition of cell populations in tissue is determined by the species, the genotype, the type of organ, the developmental state of the organ and tissue areas within the organ, and even by the individual history of the experimental plant. The most effective trigger for shifting cells potentially competent for regeneration into the competent state is mechanical wounding. The 'wound response' (Kahl, 1982) is probably the biological basis for proliferation and regeneration from somatic cells. Plant species differ in their wound response, as do different cells of the same plant. Graminaceous plant species, especially the cereals and maize (and probably grain legumes) have only a very rudimentary wound response. For some genotypes it is possible to proliferate cells competent for regeneration under experimental conditions that maintain this state ('embryogenic suspensions') (Vasil, 1987). Such cell cultures contain cells competent for regeneration and competent for integrative transformation. Plant cells are efficient barriers and traps for DNA molecules of the size of a functional gene. So far genes can be transported into walled plant cells, only with the help of *Agrobacterium*, viruses, microinjection, and biolistics. Thus production of transgenic plants requires efficient gene transfer into cells competent (simultaneous or consecutive) for both integrative transformation and regeneration. Competence for integrative transformation has obviously little relation to competence for transient expression. Non-viral DNA can integrate into the host genome; its presence in the cells does not, however, guarantee integration. It does not travel from cell to cell and is restricted to the cell to which it has been delivered. Viral DNA does not integrate into the host genome even if present at very high copy number. DNA, like viral RNA, moves from cell to cell and can spread systemically throughout the plant. It is, however, excluded from the meristems and the 'germline'. The above statements do not exclude the possibility of very rare exceptions!

9.2.1 *Agrobacterium*-mediated gene transfer

There is no question that *Agrobacterium tumefaciens* and *A. rhizogenes* provide excellent vector systems for the production of transgenic plants and from their application there is an ever increasing list of novel transgenic species. Numerous excellent reviews (e.g. Fraley *et al.*, 1986; Klee *et al.*, 1987; Hooykaas, 1989) and laboratory manuals (e.g. Draper *et al.*, 1988; Gelvin *et al.*, 1989/90) have been published.

The molecular analysis of the events leading finally to the transfer of the T-DNA to host plant cells is close to completion and elegant protocols have been worked out to use this biological vector for basic and applied research. It is obvious that the process requires a co-ordinated interaction between the bacterium and the plant host. Unfortunately, even though we know so much about the contribution from the bacterial partner, we know little about the interaction with the plant cell. This is, of course far more difficult to study. As long as it is possible to work with plant systems which co-operate, this is no serious problem. Unfortunately, however, experience indicates that numerous (especially economically important) plants are intractable. In these cases more knowledge on the biological basis on the plant side is required. Braun (1952) collected very valuable information on the importance of a state of competence in the plant cell for tumour transformation to occur, and on the same line, Binns (1990) and Binns and Thomashow (1988) reviewed our knowledge of the biology of the 'host range limitations'.

Variation in competence of the plant cell is probably the reason why some plants could be transformed only with great difficulty and after screening of numerous plant genotypes with a wide range of *Agrobacterium* strains (e.g. Hinchee *et al.*, 1988), or during a short developmental window of specific organs, or could not be transformed at all. The common denominator may be found in the phenomenon of 'wound response' (Kahl, 1982). Plants and

tissues differ in their wound response and only those with a pronounced wound response develop larger populations of wound-adjacent competent cells for regeneration and transformation. Plants that are recalcitrant to transformation with *Agrobacterium* probably do not express the appropriate wound response. This is the likely reason for the complete failure to transform cereal plants with *Agrobacterium*, despite the enormous effort which has been invested into this approach. Most experiments have not been published because they were negative; some promising data have been presented at international meetings but since they were not accompanied by sufficient proof they must be considered artifactual.

Transformation of 'monocots' (e.g. Hooykaas-VanSlogteren *et al.*, 1984; Bytebier *et al.*, 1987) is of no importance in this context because cereals are not difficult to transform because they are monocots, but because they do not have the proper wound response. Monocots with a wound response (e.g. *Asparagus*) are as easy to transform as dicots with a wound response. Similarly, dicots without a proper wound response (e.g. grain legumes?) are probably as difficult to transform as cereals. The report on transformation of maize seedlings (Graves and Goldmann, 1986) does not present proof and did not lead to transgenic offspring; the data are, therefore, considered artifacts. Why is it probably impossible to transform cereal plants with *Agrobacterium*? Wounding of differentiated cereal tissues does not lead to wound response-induced dedifferentiation of wound-adjacent cells and accumulation of competent cells. Instead wound-adjacent cells accumulate phenols and die. Although *Agrobacterium* is obviously very efficient at transferring T-DNA into wound-adjacent cells in cereals (e.g. Grimsley *et al.*, 1987) even integration of this T-DNA does not lead to transgenic clones and plants. It is not easy to understand why experiments with meristems (e.g. leaf base and split shoot tip) which can form proliferating cultures *in vitro*, did not yield transgenic clones and plants. A possible explanation may be the fact that cereal cell cultures are not the consequence of proliferating wound meristems but are based on adventitious or axillary meristems

(King *et al.*, 1978). Wounding plus *in vitro* culture, does not lead to many competent cells, but to a few meristem initials that proliferate as meristems. Meristematic cells may not be competent for transformation (Potrykus, 1990a,b). Of the many attempts to transform cereal cell cultures with *Agrobacterium* (an approach, where one could envisage success), only one has been reported successful (Raineri *et al.*, 1990). Unfortunately the data presented do not provide proof as defined above.

The key problem in *Agrobacterium*-mediated transformation of cereals and other recalcitrant plants is, probably, neither *Agrobacterium* (it transfers T-DNA to cereals), nor the host range (cereals are probably included), but rather the availability and accessibility of cells competent for integrative transformation and regeneration.

9.2.2 Protoplasts and direct gene transfer

Protoplasts can be considered ideal experimental systems for gene transfer. The freely accessible plasmalemma guarantees that DNA can reach and enter each and every protoplast in a given population and at concentrations that can be regulated experimentally. The enzymatic (or mechanical) isolation constitutes the wound response, which shifts potentially competent cells into the competent state, thus increasing the proportion of cells competent for regeneration and possible integrative transformation. The foreign genes reach every competent cell, thus increasing the chance for recovery of transgenic plants from a given population. This method of gene transfer does not require any biological vector; the DNA uptake is a straightforward physical process, thus circumventing any possible host range problem. We developed 'direct gene transfer' as an alternative to *Agrobacterium* because of the foreseeable difficulties with cereals (Paszkowski *et al.*, 1984).

DNA uptake can be promoted by various treatments including polyethylene glycol and/or electroporation (Shillito *et al.*, 1985; Negrutiu *et al.*, 1987). Integrative transformation can be very efficient (Negrutiu *et al.*, 1987) and leads to

stable inheritance of predominantly single gene loci of the foreign gene (Potrykus *et al.*, 1985a,b). Co-transformation efficiently transfers non-selectable genes (Schocher *et al.*, 1986), whilst homologous recombination enables gene targeting (Paszkowski *et al.*, 1988). A barrier to gene transfer has, so far, not been detected: virtually every protoplast system has proven transformable if challenged in the appropriate manner, although with different efficiencies. So, are there no problems with the recovery of transgenic plants from protoplasts? Unfortunately, there are many, all related to plant regeneration. Although there has been exciting progress in this respect (Roest and Gilissen, 1989) and further progress can be foreseen, plant regeneration from protoplasts will probably always be a delicate process and will depend upon parameters which are not under experimental control, e.g. species and genotype-dependent competence for wound response and regeneration (Potrykus and Shillito, 1986). However, important progress has recently been achieved with the recovery of transgenic plants from important crops which could not be transformed with *Agrobacterium*: several laboratories have reported transgenic *japonica*-type rice (e.g. Shimamoto *et al.*, 1989), *indica*-type rice (Datta *et al.*, 1990), maize (Donn *et al.*, 1990), and wheat (Vasil and Vasil, 1992). However, despite these achievements application of gene technology to cereals and grasses is not yet routine.

9.2.3 Biolistics or particle gun

Acceleration of heavy microparticles coated with DNA has been developed into a technique to carry genes into virtually all types of cells and tissues (Klein *et al.*, 1988a,b; McCabe *et al.*, 1988; Sanford, 1988). No other gene transfer approach since the early *Agrobacterium*-mediated gene transfer experiments has met with so much enthusiasm and investment in experimentation and manpower. This is understandable because there was the expectation that this technique would solve all gene transfer problems. Indeed the biolistic approach has advantages and potential for general applicability: (a) it is easy to handle; (b) one shot can lead to multiple hits (transfer of genes into many cells); (c) cells survive the intrusion of particles; (d) the genes coated on the particle resume biological activity; (e) target cells can be as different as pollen, cultured cells, cells in differentiated tissues and meristems; (f) they can be located at the surface or in deeper layers of organs; (g) the method depends on physical parameters only, etc. Thus the method allows the transport of genes into many cells at nearly any desired position in an experimental system without too much manual effort.

The enormous investment into this technique has paid off and transgenic plants have been recovered which would have been difficult to produce by other methods (Sanford, 1992). The first transgenic soybean plants were reported simultaneously via *Agrobacterium* (Hinchee *et al.*, 1988) and biolistics (McCabe *et al.*, 1988); however biolistics has by now become far more successful than *Agrobacterium* in this crop (Christou *et al.*, 1990; Christou, 1992). The real breakthrough for biolistics came with the recovery of fertile transgenic maize (Fromm *et al.*, 1990; Gordon-Kamm *et al.*, 1990). This success with maize will guarantee support for further investment into this technique. An interesting question, however, despite this success, is: why, with all the advantages listed above, is this technique so inefficient in yielding stable integrative events, especially in such ideal experimental systems such as embryogenic suspensions? If one compares the number of fertile transgenic maize plants recovered from biolistic treatment in large-scale experiments of embryogenic suspensions with those recovered from a comparable small-scale experiment on direct gene transfer to protoplasts isolated from embryogenic suspensions (Donn, *et al.*, 1990) then the low frequency from the biolistic experiments is rather surprising. If one relates this to the experimental effort and man-years, the difference is even more striking. It is also interesting to note that transgenic maize, and possibly wheat, has been recovered from biolistic treatment of embryogenic cell cultures only.

Biolistics will certainly contribute in future to the production of transgenic plants, but it will probably not be the efficient and generally applicable method as was hoped; this is not only due to technical problems, but probably more to biological problems. Plants which are difficult to transform by *Agrobacterium* probably have only very few competent cells, the particle has to reach these rare cells by a random hit, and the DNA has to integrate into the genome of these cells. Considering the low conversion rate of transient events (hit) to stable integrative events in biolistic systems, integrative transformation in recalcitrant transformation systems must be expected to be very rare events (Potrykus, 1991). A great advantage of the biolistic technique is in its application to studies of transient gene expression in differentiated tissues (Goff *et al.*, 1990). This technique has little competition. Another area where biolistics has advantages is the transformation of tissues *in situ*. Recovery of transgenic soybean from biolistic treatment of seedlings (Christou *et al.*, 1990) is an excellent example of technical development in the right direction. The transgenic plants recovered from these experiments, probably derive from adventitious meristems induced from differentiated cells in the apex below the shoot meristem. This requires competence for this process *in vitro* and will, therefore, again be genotype dependent. Transformation of the cells of shoot meristems, however, with subsequent regeneration of plants by simple and straightforward germination, should open the route for a truly genotype independent transformation procedure.

9.2.4 Microtargeting

Cells in shoot meristems should be transformable by *Agrobacterium*, microinjection and biolistics. All attempts to use *Agrobacterium* for meristem transformation have so far failed, thus meristematic cells have an extremely low competence for acceptance of T-DNA (Potrykus, 1990a,b, 1991). Microinjection has yielded rare transgenic events (G. Neuhaus, personal communication), but could not, so far, be developed to a routine procedure. The problem with traditional biolistic devices, especially for cereal meristems, is that the meristems are very tiny structures (ca 100 μm) and the particles hitting large target areas at random, obviously very seldom hit a meristem. There is, however, a novel device which allows for the targeting of populations of microprojectiles to individual shoot meristems of cereal seedlings (Sautter *et al.*, 1991). This device, which combines advantages of the microinjection technique (the capacity to predict the postion of DNA-delivery) with advantages of the biolistic technique (achieving multiple hits with one event) has specifically been developed for biolistic targeting of shoot meristems. It fulfils all the criteria that are necessary for the predictable and routine acceleration of particle populations into the cells of any given shoot meristem under conditions that will not interfere with subsequent development of the meristem.

As shoot meristems are not only tiny structures but are also composed of very small and delicate cells, it was mandatory during the course of development of the microtargeter to overcome a series of technical problems: (a) the particles must be accelerated to target areas of variable size ranging between 100 and 200 μm; (b) the particles of any given population must be of identical size; (c) they must arrive at the target as individual particles; (d) their penetration into the target must be predictable and variable; (e) the particles have to carry DNA which is freely accessible within the cell. The latter was achieved by developing a method for the preparation of gold particles from solutions of gold salts by the additon of photographic developer. This procedure allows the production of homogenous particles of any desired size. Clumping of particles is prevented by avoiding any kind of macroprojectile. Instead, an aerosol is created from suspensions of gold particles in DNA-containing water. The aerosol is then accelerated in a 'Pitot's' tube by the same gas pressure pulse that created the aerosol. Acceleration occurs in a capillary, the length of which determines the impulse of the particles and the width of the target area. Accurate aiming is achieved with the help of cross-lines in the microscope. As each physical parameter of the system is variable within a wide range, the microtargeting system can therefore, easily be adjusted and optimized to suit the specific

requirements of shoot meristems from different plant species. Many particles can be targeted to a single meristem, thus allowing for the delivery of DNA to statistically every cell. Treated shoot meristems regenerate with high frequency to fertile plants. Data on integrative transformation in the cells of shoot meristems are not yet available, however, integrative transformation in competent cell cultures is very high (Sautter *et al.*, 1991).

9.2.5 Microinjection into zygotic and microspore-derived proembryos

Microinjection uses microcapillaries and microscopic devices to deliver DNA into defined cells in such a way that the injected cell survives and proliferates (Neuhaus and Spangenberg, 1990). This technique has produced transgenic plants from protoplasts (Schnorf *et al.*, 1991) and transgenic chimeras from microspore-derived proembryos (Neuhaus *et al.*, 1987). As biolistics, microinjection delivers DNA into walled plant cells. In comparison with biolistics, it has the disadvantage that only one cell receives DNA per injection and handling requires more skill and instrumentation. It also has advantages: (a) the quantity of DNA delivered can be optimized; (b) the experimenter can decide into which cell to deliver the DNA; (c) delivery is precise and predictable, even into the cell nucleus, and is under visual control; (d) cells of small structures (e.g. microspores and few-celled proembryos which are not available in large quantities as required for the random-targeting biolistic technique) can be precisely targeted; (e) defined microinjected units can be microcultured; (f) in combination with protocols for the culture of zygotic proembryos, microinjection should offer an approach to transformation which should be open for every species and variety with sexual propagation.

On the assumption that few-celled zygotic proembryos contain competent cells, our group has established plant regeneration from isolated zygotic proembryos of maize (*Zea mays*), wheat

(*Triticum aestivum*), rice (*Oryza sativa*), barley (*Hordeum vulgare*), soybean (*Glycine max*), cotton (*Gossypium* hybrid), sunflower (*Helianthus annuus*), tobacco (*Nicotiana tabacum*) and *Arabidopsis thaliana*. Following multiple microinjections with proven marker genes putative primary transgenic chimeras and sexual offspring have been analysed for the presence of the foreign genes (G. Neuhaus, G. Spangenberg, S.K. Datta, personal communication). So far we have only indicative evidence for transgenic chimeras. As we have as yet no proof for transgenic offspring, these may be artifacts. Gene transfer to structures consisting of more than one cell can at the best produce transgenic chimeras. Therefore two interpretations of the data are possible: (a) larger experiments will increase the chance for transmission of the transgene to the offspring (as exemplified with the biolistic approach); or (b) meristematic cells have little competence for integrative transformation.

9.2.6 Other approaches

There are two recent approaches for which the data available are not yet sufficient to constitute proof for integrative transformation, but which otherwise look promising: ultrasonication and electroporation into tissues.

Ultrasonication of DNA into leaf pieces of tobacco (Zhang *et al.*, 1991) has produced phenotypic data which present strong indicative evidence for integrative transformation. Molecular evidence for the presence and activity of the foreign gene, and correlation of this physical evidence with the phenotypic evidence is, however, still missing. As phenotypic evidence has been misleading in numerous previous approaches, it would be premature to conclude that ultrasonication is a functional technique. There are also difficulties in understanding how ultrasonication could transport DNA into walled plant cells.

Electroporation into tissues has earlier yielded indicative evidence for transient expression (Dekeyser *et al.*, 1990). Electroporation into disintegrated cell cultures in sugar cane (Molina *et al.*, 1992) has now yielded indicative evidence

for integrative transformation. Again, the data do not constitute proof as defined above and further analysis and better data will be required to determine whether or not electroporation can indeed lead to uptake of DNA into walled plant cells.

9.3 CONCLUDING REMARKS

Transgenic crop plants can be produced with the help of five techniques. These are: (a) *Agrobacterium*-mediated gene transfer; (b) direct DNA uptake into protoplasts; (c) biolistics; (d) microinjection; (e) microtargeting. The method of choice for species and varieties, where plant regeneration from differentiated tissues is routine and efficient, is *Agrobacterium*. For cases where totipotent protoplasts are readily available, or where there is a realistic chance for the development of cell culture systems which will yield totipotent protoplasts, direct gene transfer is reliable and efficient. Biolistics still requires large-scale experiments, and in part depends upon the establishment of 'embryogenic suspensions', but has proven functional in a series of 'recalcitrant' cases. It has advantages where adventitious meristems develop from surface cells, and where it is relatively easy to prepare large quantities of this material. *Agrobacterium*, which also should be able to interact with such surface cells, is probably not functional because there is no wound response, and because meristem initials do not readily interact with it. Microinjection which, in theory could be taken as an ideal gene transfer system for zygotic proembryos, still has to demonstrate that it can be applied routinely. Microtargeting, the ideal procedure for DNA delivery into shoot meristems, has the potential to develop into a generally applicable method, if the cells of shoot meristems are competent for integrative transformation. Ultrasonication and electroporation into tissues may turn out to represent further alternatives and one day someone may even discover a procedure which opens the ideal pathway via pollen. In general, there is good reason to be optimistic that gene technology will become an integral procedure of plant breeding in the foreseeable future.

REFERENCES

Ahokas, H. (1989), Transfection of germinating barley seed electrophoretically with exogenous DNA. *Theor. Appl. Genet.*, **77**, 469–472.

Binns, A.N. (1990), *Agrobacterium*-mediated gene delivery and the biology of host range limitations. *Physiol. Plant.*, **79**, 135–139.

Binns, A.N. and Thomashow, M.F. (1988), Cell biology of *Agrobacterium* infection and transformation of plants. *Annu. Rev. Microbiol.*, **42**, 575–606.

Braun, A.C. (1952), Conditioning of the host cell as a factor in the transformation process in crown gall. *Growth*, **16**, 65–74.

Bytebier, B., Deboeck, F., DeGreve, H., VanMontagu, M. and Hernalsteens, J.P. (1987), T-DNA organization in tumor cultures and transgenic plants of the monocotyledon *Asparagus officinalis*. *Proc. Natl. Acad. Sci. USA*, **84**, 5345–5349.

Christou, P. (1992), Commercial production of transgenic soybean and rice plants from elite cultivars using electric discharge particle acceleration. In *Advances in Gene Technology: Feeding the World of the 21st Century*, Proc. 1992 Miami Biotechnology Winter Symposium, Whelan, W.J., Ahmad, F., Bialy, H., Black, S., King, M.L., Rabin, M.B., Solomonson, L.P. and Vasil, I.K. (eds), p. 73.

Christou, P., Ford, T.L. and Kofron, M. (1991), Production of transgenic rice (*Oryza sativa* L.) plants from agronomically important *Indica* and *Japonica* varieties via electric discharge particle acceleration of exogeneous DNA into immature zygotic embryos. *Bio/Technology*, **9**, 957–962.

Christou, P., McCabe, D.E., Martinell, B.J. and Swain, W.F. (1990), Soybean genetic transformation – commercial production of transgenic plants. *Trends Biotechnol.*, **8**, 145–151.

Chupeau, M.C., Bellini, C., Guerche, P., Maisonneuve, B., Vastra, G. and Chupeau, Y. (1989), Transgenic plants of lettuce (*Lactuca sativa*) obtained through electroporation of protoplasts. *Bio/Technology*, **7**, 503–508.

Datta, S.K., Peterhans, A., Datta, K. and Potrykus, I. (1990), Genetically engineered *Indica*-rice recovered from protoplast. *Bio/Technology*, **8**, 736–740.

Davey, M.R., Rech, E.L. and Mulligan, B.J. (1989), Direct DNA transfer to plant cells. *Plant Mol. Biol.*, **13**, 273–285.

De Greef, W., Delon, R., De Block, M., Leemans, J. and Botterman, J. (1989), Evaluation of herbicide resistance in transgenic crops under field conditions. *Bio/Technology*, **7**, 61–64.

De la Peña, A., Lörz, H. and Schell, J. (1987), Transgenic plants obtained by injecting DNA into floral tillers. *Nature*, **325**, 274–276.

Dekeyser, R.A., Claes, B., DeRycke, R.M.U., Habets, M.E., VanMontagu, M.C. and Caplan, A.B. (1990), Transient gene expression in intact and organized rice tissue. *Plant Cell*, **2**, 591–602.

D'Halluin, K., Botterman, J. and DeGreef, W. (1990), Engineering of herbicide-resistant alfalfa and evaluation under field conditions. *Crop Science*, **30**, 866–870.

Donn, G., Nilges, M. and Moroz, S. (1990), Stable transformation of maize with a chimaeric, modified phosphinotrycin-acyltransferase gene from *Streptomyces viridochromogenes*. In *Progress in Plant Cellular and Molecular Biology*, Nijkamp, H.J.J., Van der Plas, L.H.W. and Aartrijk, J. (eds), Kluwer, Dordrecht, pp. 53.

Draper, J., Scott, R., Armitage, P. and Walden, R. (eds) (1988), *Plant Genetic Transformation and Gene Expression. A Laboratory Manual*, Blackwell Scientific Publ., Oxford.

Filatti, J.J., Selmer, J., McCown, B., Haissig, B. and Comai, L. (1987), *Agrobacterium*-mediated transformation and regeneration of *Populus*. *Mol. Gen. Genet.*, **206**, 192–199.

Firoozabady, E., DeBoer, D.L., Merlo, D.J., Halk, E.L., Amerson, L.N., Rashka, K.E. and Murray, E.E. (1987), Transformation of cotton (*Gossypium hirsutum* L.) by *Agrobacterium tumefaciens* and regeneration of transgenic plants. *Plant Mol. Biol.*, **10**, 105–116.

Fitsch, M.M.M., Manshardt, R.M., Gonsalves, D., Slighton, J.L. and Sanford, J. (1990), Stable tranformation of papaya via microprojectile bombardment. *Plant Cell Reports*, **9**, 189–194.

Fraley, R.T., Rogers, S.G. and Horsch, R.B. (1986), Genetic transformation in higher plants. *CRC Crit. Rev. Plant Sci.*, **4**, 1–46.

Fromm, M.E., Morrish, F., Armstrong, C., Williams, R., Thomas, J. and Klein, T.M. (1990), Inheritance and expression of chimeric genes in the progeny of transgenic maize plants. *Bio/Technology*, **8**, 833–839.

Fütterer, J., Bonneville, J.M. and Hohn, T. (1990), Cauliflower mosaic virus as a gene expression vector for plants. *Physiol. Plant.*, **79**, 154–157.

Gad, A.E., Rosenberg, N. and Altmann, A. (1990), Liposome-mediated gene delivery into plant cells. *Physiol. Plant.*, **79**, 177–183.

Gasser, C.S. and Fraley, R.T. (1989), Genetically

engineering plants for crop improvement. *Science*, **244**, 1293–1299.

Gelvin, S.B., Schilperoort, R.A. and Verma, D.P.S. (1989/1990), *Plant Molecular Biology Manual*, Kluwer, Dordrecht.

Goff, S.A., Klein, T.M., Roth, B.A., Fromm, M.E., Cone, K.C., Radicella, J.P. and Chandler, V. (1990), Transactivation of anthocyanin biosynthesis genes following transfer of regulatory genes into maize tissue. *EMBO J.*, **9**, 2517–2522.

Gordon-Kamm, W.J., Spencer, T.M., Mangano, L.M., Adams, T.R., Daines, R.J., Start, W.G., O'Brien, J.V., Chambers, S.A., Adams, W.R. jr., Willets, N.G., Rice, T.B., Mackey, C.J., Krueger, R.W., Kausch, A.P. and Lemaux, P.G. (1990), Transformation of maize cells and regeneration of fertile transgenic plants. *The Plant Cell*, **2**, 603–618.

Graves, A.C.F. and Goldmann, S.L. (1986), The transformation of *Zea mays* seedlings via the *Agrobacterium tumefaciens*: detection of T-DNA specific enzyme activities. *Plant Mol. Biol.*, **7**, 43–50.

Grimsley, N.H. (1990), Agroinfection. *Physiol. Plant.*, **79**, 147–153.

Grimsley, N.H., Hohn, T., Davies, J.W. and Hohn, B. (1987), *Agrobacterium*-mediated delivery of infectious maize streak virus into maize plants. *Nature*, **325**, 177–179.

Guerche, P., Jouanin, L., Tepfer, D. and Pelletier, G. (1987), Genetic transformation of oilseed rape (*Brassica napus*) by the of *Agrobacterium rhizogenes* and analysis of inheritance of the transformed phenotype. *Mol. Gen. Genet.*, **206**, 382–386.

Hess, D. (1987), Pollen based techniques in genetic manipulation. *Int. Rev. Cytol.*, **107**, 169–190.

Hill, K.K., Jarvis-Eagan, N., Halk, E.L., Krahn, K.J., Liao, L.W., Mathewson, R.S., Merlo, D.J., Nelson, S.E., Rashka, K.E. and Loesch-Fries, L.S. (1991), The development of virus-resistant alfalfa, *Medicago sativa* L. *Bio/Technology*, **9**, 373–377.

Hinchee, M.A.W., Connor-Ward, D.V., Newell, C.A., McDonnell, R.E., Sato, S.J., Gasser, C.S., Fischhoff, D.A., Re, D.B., Fraley, R.T. and Horsch, R.B. (1988), Production of transgenic soybean plants using *Agrobacterium*-mediated DNA transfer. *Bio/Technology*, **6**, 915–922.

Hoekema, A., Huisman, M.J., Molendijk, L., Van den Elzen, P. and Cornelissen, B.J.C. (1989), The genetic engineering of two commercial potato cultivars for resistance to potato virus X. *Bio/Technology*, **7**, 273–278.

Hooykaas, P.J.J. (1989), Transformation of plant cells via *Agrobacterium*. *Plant Mol. Biol.*, **13**, 327–336.

Hooykaas-VanSlogteren, G.M.S., Hooykaas, P.J.J. and Schilperoort, R.A. (1984), Expression of Ti-plasmid genes in monocotyledonous plants infected with *Agrobacterium tumefaciens*. *Nature*, **317**, 763–764.

Horn, M.E., Shillito, R.D., Conger, B.V. and Harms, C.T. (1988), Transgenic plants of orchardgrass (*Dactylis glomerata* L.) from protoplasts. *Plant Cell Rep.*, **7**, 469–472.

James, D.J., Passey, A.J., Barbara, D.J. and Bevan, M. (1989), Genetic transformation of apple (*Malus pumila* Mill.) using a disarmed Ti-binary vector. *Plant Cell Rep.*, **7**, 658–661.

Kahl, G. (1982), Molecular biology of wound healing: the conditioning phenomenon. In *Molecular Biology of Plant Tumors*, Kahl, G. and Schell, J. (eds), Academic Press, New York, pp. 211–267.

King, J., Potrykus, I. and Thomas, E. (1978), *In vitro* genetics of cereals: problems and perspectives. *Physiol. Veg.*, **16** 81–399.

Klee, H.J., Horsch, R. and Rogers, S. (1987), *Agrobacterium*-mediated plant transformation and its further applications to plant biology. *Ann. Rev. Plant Physiol.*, **38**, 467–486.

Klee, H.J. and Rogers, S.G. (1989), Plant gene vectors and transformation: plant transformation systems based on the use of *Agrobacterium tumefaciens*. In *Cell Culture and Somatic Cell Genetics*, Vol. 6, *Molecular Biology of Plant Nuclear Genes*, Schell, J. and Vasil, I.K. (eds), Academic Press, San Diego, pp. 2–25.

Klein, T.M., Fromm, M.E., Gradziel, T., Sanford, J.C. (1988a), Factors influencing gene delivery into *Zea mays* cells by high velocity microprojectiles. *Bio/Technology*, **6**, 923–926.

Klein, T.M., Fromm, M.E., Weissinger, A., Tomes, D., Schaaf, S., Sleeten, M. and Sanford, J.C. (1988b), Transfer of foreign genes into intact maize cells using high velocity microprojectiles. *Proc. Natl. Acad. Sci. USA*, **85**, 4305–4309.

Lambert, C. and Tepfer, D. (1991), Use of *Agrobacterium rhizogenes* to create chimeric apple trees through genetic grafting. *Bio/Technoloy*, **9**, 80–83.

Lawson, C., Kaniewski, W., Haley, L., Rozman, R., Newell, C., Sanders, P. and Tumer, N.E. (1990), Engineering resistance to mixed virus infection in a commercial potato cultivar: Resistance to potato virus X and potato virus Y in transgenic Russet Burbank. *Bio/Technology*, **8**, 127–134.

Lu, C.Y., Nugent, G., Wardley-Richardson, T., Chandler, S.F., Young, R. and Dalling, M.J. (1991), *Agrobacterium*-mediated transformation of carnation (*Dianthus caryophillus* L.) *Bio/Technology*, **9**, 864–868.

Lucas, W.J., Lansing, A., De Wet, J.R. and Walbot, V. (1990), Introduction of foreign DNA into walled plant cells via liposomes injected into the vacuole: a preliminary study. *Physiol. Plant.*, **79**, 184–189.

Luo, Z. and Wu, R. (1988), A simple method for transformation of rice via the pollen tube pathway. *Plant Mol. Biol. Rep.*, **6**, 165–174.

Mante, S., Morgens, P.H., Scorza, R., Cordts, J.M. and Callahan, A.M. (1991), *Agrobacterium*-mediated transformation of plum (*Prunus domestica* L.) hypocotyl slices and regeneration of transgenic plants. *Bio/Technology*, **9**, 853–857.

McCabe, D.E., Swain, W.F., Martinell, B.J. and Christou, P. (1988), Stable transformation of soybean (*Glycine max*) by particle acceleration. *Bio/Technology*, **6**, 923–926.

McGranahan, G.H., Leslie, C.A., Uratsu, S.L., Martin, L.A. and Dandekar, A.M. (1988), *Agrobacterium*-mediated transformation of walnut somatic embryos and regeneration of transgenic plants. *Bio/Technology*, **6**, 800–804.

Michelmore, R.W., Marsh, E., Seely, S. and Landry, B. (1987), Transformation of lettuce (*Lactuca sativa*) mediated by *Agrobacterium tumefaciens*. *Plant Cell Rep.*, **6**, 439–442.

Molina, P., Arencibia, A., Gutiérrez, C., Fuentes, A., Menéndez, E., Grenidge, V., De la Riva, G. and Selman-Houssein, G. (1992), The use of intact cell electroporation for the transformation and regeneration of sugar cane. In *Advances in Gene Technology; Feeding the World in the 21st Century*, Proc. 1992 Miami Bio/Technoloy Winter Symposium, Whelan, W.J., Ahmad, F., Bialy, H., Black, S., King, M.L., Rabin, M.B., Solomonson, L.P. and Vasil, I.K. (eds), p. 60.

Mullins, M.G., Tang, F.C.A. and Facciotti, D. (1990), *Agrobacterium*-mediated genetic transformation of grapevines: transgenic plants of *Vitis rupestris* Scheele and buds of *Vitis vinifera* L. *Bio/Technology*, **8**, 1041–1045.

Negrutiu, I., Shillito, R.D., Potrykus, I., Biasini, G. and Sala, F. (1987), Hybrid genes in the analysis of transformation conditions. I. Setting up a simple method for direct gene transfer to protoplasts. *Plant Mol. Biol.*, **8**, 363–373.

Neuhaus, G. and Spangenberg, G. (1990), Plant transformation by microinjection techniques. *Physiol. Plant.*, **79**, 213–217.

Neuhaus, G., Spangenberg, G., Mittelsten Scheid, O. and Schweiger, H.G. (1987), Transgenic rape seed plants obtained by microinjection into microspore-derived proembryos. *Theor. Appl. Genet.*, **75**, 30–36.

Paszkowski, J., Baur, M., Bogucki, A. and Potrykus, I. (1988), Gene targeting in plants. *EMBO J.*, **7**, 4021–4027.

Paszkowski, J., Saul, M.W. and Potrykus, I. (1989), Plant gene vectors and genetic transformation: DNA-mediated direct gene transfer to plants. In *Cell Culture and Somatic Cell Genetics of Plants*, Vol. 6, *Molecular Biology of Plant Nuclear Genes*, Schell, J. and Vasil, I.K. (eds), Academic Press, San Diego, pp. 52–68.

Paszkowski, J., Shillito, R.D., Saul, M.W., Mandak, V., Hohn, T. and Potrykus, I. (1984), Direct gene transfer to plants. *EMBO J.*, **3**, 2712–2722.

Perlak, F.J., Deaton, R.W., Armstrong, T.A., Fuchs, R.L., Sims, S.R., Greenplate, J.T. and Fischhoff, D.A. (1990), Insect resistant cotton plants. *Bio/Technology*, **8**, 939–943.

Potrykus, I. (1990a), Gene transfer to cereals: an assessment. *Bio/Technology*, **8**, 535–542.

Potrykus, I. (1990b), Gene transfer to plants: assessment and perspectives. *Physiol. Plant.*, **79**, 125–134.

Potrykus, I. (1991), Gene transfer to plants: assessment of published approaches and results. *Annu. Rev. Plant Physiol. Plant Mol. Biol.*, **42**, 205–225.

Potrykus, I., Paszkowski, J., Saul, M.W., Petruska, J. and Shillito, R.D. (1985a), Molecular and general genetics of a hybrid foreign gene introduced into tobacco by direct gene transfer. *Mol. Gen. Genet.*, **199**, 167–177.

Potrykus, I. and Shillito, R.D. (1986), Protoplasts: isolation, culture and plant regeneration. *Methods Enzymol.*, **118**, 549–578.

Potrykus, I., Shillito, R.D., Saul, M.W. and Paszkowski, J. (1985b), Direct gene transfer: state of the art and future potential. *Plant. Mol. Biol. Rep.*, **3**, 117–128.

Pua, E.C., Mehra-Palta, A., Nagy, F. and Chua, N.H. (1987), Trangenic plants of *Brassica napus* L. *Bio/Technology*, **5**, 815–817.

Puonti-Kaerlas, J., Eriksson, T. and Engström, P. (1990), Production of transgenic pea (*Pisum sativum* L.) plants by *Agrobacterium tumefaciens*-mediated gene transfer. *Theor. Appl. Genet.*, **80**, 246–252.

Pythoud, F., Sinkar, V.P., Nester, E.W. and Gordon, M.P. (1987), Increased virulence of *Agrobacterium rhizogenes* conferred by the vir region of pTiBo542: application to genetic engineering of poplar. *Bio/Technology*, **5**, 1323–1327.

Radke, S.E., Andrews, B.M., Moloney, M.M., Crouch, M.L., Kridl, J.C. and Knauf, V.C.

(1988), Transformation of *Brassica napus* L. using *Agrobacterium tumefaciens*: developmentally regulated gene expression of a reintroduced napin gene. *Theor. Appl. Genet.*, **75**, 685–694.

Raineri, D.M., Bottino, P., Gordon, M.P. and Nester, E.W. (1990), *Agrobacterium*-mediated transformation of rice (*Oryza sativa* L.). *Bio/Technology*, **8**, 33–38.

Roest, S. and Gilissen, L.J.W. (1989), Plant regeneration from protoplasts: a literature review. *Acta Bot. Neerl.*, **38**, 1–23.

Sanford, J.C. (1988), The biolistic process a new concept in gene transfer and biological delivery. *Trends Biotechnol.*, **6**, 299–302.

Sanford, J.C. (1992), The biolistic process – a simple tool for transforming diverse crop species. In *Advances in Gene Technology: Feeding the World in the 21st Century*, Proc. 1992 Miami BioTechnology Winter Symposium, Whelan, W.J., Ahmad, F., Bialy, H., Black, S., King, M.L., Rabin, M.B., Solomonson, L.P. and Vasil, I.K. (eds), p. 4.

Sautter, C., Waldner, H., Neuhaus-Url, G., Galli, A., Neuhaus, G. and Potrykus, I. (1991), Microtargeting: high efficiency gene transfer using a novel approach for the acceleration of microprojectiles. *Bio/Technology*, **9**, 1080–1085.

Schnorf, M., Neuhaus-Url, G., Galli, A., Iida, S., Potrykus, I. and Neuhaus, G. (1991), An improved approach for transformation of plant cells by microinjection: molecular and genetic analysis. *Transgenic Res.*, **1**, 23–30.

Schocher, R.J., Shillito, R.D., Saul, M.W., Paszkowski, J. and Potrykus, I. (1986), Cotransformation of unlinked foreign genes into plants by direct gene transfer. *Bio/Technology*, **4**, 1093–1096.

Shahin, E.A., Spielmann, A., Sukhapinda, K., Simpson, R.B. and Yashar, M. (1986), Transformation of cultivated alfalfa using disarmed *Agrobacterium tumefaciens*. *Crop Sci.*, **26**, 1235–1239.

Shillito, R.D., Saul, M.W., Paszowski, J., Müller, M. and Potrykus, I. (1985), High frequency direct gene transfer to plants. *Bio/Technology*, **3**, 1099–1103.

Shimamoto, K., Terada, R., Izawa, T. and Fujimoto, N. (1989), Fertile rice plants regenerated from transformed protoplasts. *Nature*, **338**, 274–276.

Smith, C.J.S., Watson, C.F., Morris, P.C., Bird, C.R., Seymour, G.B., Gray, J.E., Arnold, C., Tucker, G.A., Schuch, W., Harding, S. and Grierson, D. (1990), Inheritance and effect on ripening of antisense polygalacturonase genes in transgenic tomatoes. *Plant Mol. Biol.*, **14**, 369–379.

Smith, C.J.S., Watson, C.F., Ray, J., Bird, C.R., Morris, P.C., Schuch, W. and Grierson, D. (1988), Antisense RNA inhibiton of polygalacturonase gene expression in transgenic tomatoes. *Nature*, **334**, 724–726.

Töpfer, R., Gronenborn, B., Schell, J. and Steinbiss, H.H. (1989), Uptake and transient expression of chimeric genes in seed-derived embryos. *Plant Cell*, **1**, 133–139.

Toriyama, K., Stein, J.C., Nasrallah, M.E. and Nasrallah, J.B. (1991), Transformation of *Brassica oleracea* with an *S*-locus gene from *B. campestris* changes the self-incompatibility phenotype. *Theor. Appl. Genet.*, **81**, 769–776.

Tumer, N.E., O'Connell, K.M., Nelson, R.S., Sanders, P.R., Beachy, R.N., Fraley, R.T. and Shah, D.M. (1987), Expression of alfalfa mosaic coat protein gene confers cross protection in transgenic tobacco and tomato plants. *EMBO J.*, **6**, 1181–1188.

Umbeck, P., Johnson, G., Barton, K.A. and Swain, W.F. (1987), Genetically transformed cotton (*Gossypium hirsutum* L.) plants. *Bio/Technology*, **5**, 263–266.

Van der Krol, A., Lenting, P.E., Veenstra, J., Van der Meer, I.M., Koes, R.E., Gerats, A.G.M., Mol, J.N.M. and Stuitje, A.R. (1988), An anti-sense chalcone synthase gene in transgenic plants inhibits flower pigmentation. *Nature*, **333**, 866–869.

Vasil, I.K. (1987), Developing cell and tissue culture systems for the improvement of cereals and grass crops. *J. Plant Physiol.*, **128**, 193–218.

Vasil, I.K. and Vasil, V. (1992), Cell culture and genetic manipulation of cereals. In *Advances in Gene Technology: Feeding the World in the 21st Century*, Proc. of the Miami Bio/Technology Winter Symposium, Whelan, W.J., Ahmad, F., Bialy, H., Black, S., King, M.L., Rabin, M.B., Solomonson, L.P. and Vasil, I.K. (eds), p. 33.

Wang, Z., Takamizo, T., Iglesias, V., Osuski, M., Nagel, J., Potrykus, I. and Spangenberg, G. (1992), Transgenic plants of tall fescue (*Festuca arundinacea* Schreb.) obtained by direct gene transfer to protoplasts. *Biotechnology*, **10**, 691–696.

Weber, G., Monajembashi, S., Wolfrum, J. and Greulich, K.O. (1990), Genetic changes induced in higher plant cells by laser microbeam. *Physiol. Plant.*, **79**, 190–193.

Zhang, L.J., Cheng, L.M., Xu, N., Zhao, N.M., Li, C.G., Yuan, J. and Jia, S.R. (1991), Efficient transformation of tobacco by ultrasonication. *Bio/Technology*, **9**, 996–997.

10

The role of gene technology in plant breeding

F. Salamini and M. Motto

10.1 INTRODUCTION

Plant breeders have made spectacular progress in the improvement of many crop species over the last 60 years. In wheat, rice and maize for example about 50% of the increase in the yield over the period 1930–1975 has been attributed to improved varieties; the remainder derives from greater and more efficient use of fertilizers and crop management. Recent advances in genetics, cell and molecular biology, particularly recombinant-DNA technology, provide new opportunities for manipulation of the genome to meet the varying demands of breeders. Therefore, it is expected that the new plant technologies will have a significant impact on crop improvement by supplementing the current activities of plant breeders in expanding and diversifying the gene pool of crop plants, by introducing specific genes not available in the sexually compatible gene pool and by shortening the time required for the production of new varieties and hybrids. This chapter focuses on the relevance of gene technology to the improvement of crop plants.

Plant Breeding: Principles and prospects. Edited by M.D. Hayward, N.O. Bosemark and I. Romagosa. Published in 1993 by Chapman & Hall, London. ISBN 0 412 43390 7.

10.2 GENETIC ENGINEERING OF PLANTS

Crop improvement by genetic engineering is based on the molecular manipulation of relevant genes and availability of vectors for the transformation of the plant cell. Gene technology already offers such methods for isolating, manipulating and obtaining the expression of the plant genes in specific tissues and at the desired level (Chapters 8 and 9 and Weising *et al.*, 1988). Table 10.1 lists chimeric and intact genes that have been expressed in transformed cells or regenerated plants. The list includes genes under the control of T-DNA, viral and plant promoters. Relevant progress has also been made in the development of gene transfer systems for higher plants. The ability to introduce foreign genes into plant cells and to regenerate viable, fertile plants expands our understanding of plant biology and provides an unparalleled opportunity to modify and improve crop plants. Table 10.2 (see however Chapter 9) is a list of species for which the production of transgenic plants has been reported.

The precise characterization of relevant agronomic traits is an essential step to successful use of genetic engineering; traits controlled by one of few Mendelian genes are the best candidates for molecular manipulations. The ideal trait depends on a specific and well characterized product of one or a few structural genes. Unfortunately, many important agronomic characters like yield are not easily related to defined genes. They cannot be easily manipu-

Table 10.1. Chimeric and intact genes in transformed cells or regenerated plants

Promoter	Construction of promoter	Specificity of promoter	Location of sequences responsible for promoter activation	References
Maize zein	3.1 kb genomic clone of zein. Ti inserted	Constitutive in in sunflower	n.d.[a]	Matzke *et al.* (1984)
Bean β-phaseolin	3.8 kb genomic clone of phaseolin. Ti inserted	Seed specific in tobacco	n.d.	Sengupta-Gopalan *et al.* (1985)
Soybean β-conglycin	12 kb genomic clone of conglycinin. Ti inserted	Seed specific in petunia	n.d.	Lawton *et al.* (1987); Beachy *et al.* (1985a)
Pea small subunit of RUBP carboxylase (*pSS*)	Genomic clone of pSS *pSS* − *CAT* − *NOS*[b]	Leaf specific, light regulated in tobacco and	from −35 to −2	Broglie *et al.* (1984); Herrera-Estrella *et al.* (1984); Morelli *et al.* (1985)
Pea *Cab-1* gene	*Cab-1* − *NPTII* − *OCS*[b]	Leaf specific light inducible in tobacco	n.d.	Green *et al.* (1987)
Soybean small subunit of RUBP carboxylase (*SbSS*)	*SbSS* − *NPTII* − *OCS*[b]	Leaf specific, light regulated in soybean protoplasts	n.d.	Facciotti *et al.* (1985)
Wheat *Cab-1*	*Cab-1* − *NPTII* − *NOS*[b]	Leaf specific, phytocrome regulated in tobacco	from −357 to −89	Nagy *et al.* (1987)
Antirrhinum majus chalcone synthase (*Chs*)	*Chs* − *NPTII* − *Chs*[b] (from parsley)	Leaf specific, light induced in tobacco	from −661 to −564	Kaulen *et al.* (1986)
Maize *Adhl*	*Adhl* − *CAT* − *NOS*[b]	Anaerobic regulated in	−247 upstream the start site	Ellis *et al.* (1987); Howard *et al.* (1987)
Heat-inducible gene from *Drosophila*	*Hsp70* − *NPTII* − *OCS*[b]	Heat inducible in tobacco	n.d.	Spena *et al.* (1985); Spena and Schell (1987)
Potato poteinase inhibitor II gene		Leaf specific, wound-induced in tobacco	n.d.	Sanchez-Serrano *et al.* (1987)
Cauliflower mosaic virus	*35S* − *NPTII*	Constitutive in tobacco	from −105 to −46	Sanders *et al.* (1987)
Nopaline synthase gene	*NOS* − *CAT* − *NOS*	Stable and transient expression in tobacco	from −97 to −63	Ebert *et al.* (1987)
T-DNA encoded gene 5	Gene 5 promoter − *OCS*	Specific expression in tobacco stem	n.d.	Koncz and Schell (1986)
Heat inducible gene from soybean	Deletions analysis	Stable transformant in sunflower tumour	to −95	Baumann *et al.* (1987)

[a] n.d. = not determined.
[b] The origin of the components of the chimeric constructions (promoter, coding sequences and 3′ regulatory signals) are as specified following the order given.

Table 10.2. Species for which the production of transgenic plants have been reported[a]*

Plant species	Method (reference)[b]
Herbaceous dicots	
Petunia	At (1)
Tomato	At (2)
Potato	At (3)
Tobacco	At (4), FP (5), PG (6)
Arabidopsis	At (7)
Lettuce	At (8)
Sunflower	At (9)
Oilseed rape	At (10), MI (11)
Flax	At (12)
Cotton	At (13)
Sugarbeet	At (14)
Celery	At (15)
Soybean	At (16), PG (17)
Alfalfa	At (18)
Medicago varia	At (19)
Lotus	At (20)
Vigna aconitifolia	At (21)
Cucumber	Ar (22)
Carrot	Ar (23)
Cauliflower	Ar (24)
Horseradish	Ar (25)
Morning glory	Ar (26)
Woody dicots	
Poplar	At (27)
Walnut	At (28)
Apple	At (29)
Monocots	
Asparagus	At (30)
Rice	FP (31)
Maize	FP (32), PG (33)
Orchard grass	FP (34)
Rye	IR (35)

[a] At, *Agrobacterium tumefaciens*; Ar, *Agrobacterium rhizogenes*; FP, free DNA introduction into protoplasts; PG, particle gun; MI, microinjection; IR, injection of reproductive organs
[b] (1) De Block *et al.* (1984), *EMBO J.*, 31, 681; Horsh *et al.* (1984), *Science*, 223, 496; (2) McCormick *et al.* (1986), *Plant Cell Rep.*, 5, 81; (3) Ooms *et al.* (1987), *Theor. Appl. Genet.*, 73, 744; Sheerman and Bevan (1988), *Plant Cell Rep.*, 7, 47; (4) Horsh *et al.* (1984), *Science*, 223, 496; (5) Paszkowski *et al.* (1984), *EMBO J.*, 3, 2717; (6) Klein *et al.* (1988), *Proc. Natl. Acad. Sci. USA*, 85, 8502; (7) Lloyd *et al.* (1986), *Science*, 234, 464; (8) Michelmore *et al.* (1987), *Plant Cell Rep.*, 6, 439; (9) Evrette *et al.* (1987), *Bio/Technology*, 5, 1201; (10) Fry *et al.* (1987), *Plant Cell Rep.*, 6, 321; Pua *et al.* (1987), *Bio/Technology*, 5, 815; (11) Neuhaus *et al.* (1987), *Theor. Appl. Genet.*, 75, 30; (12) Basoran *et al.* (1987), *Plant Cell Rep.*, 6, 396; Jordan and McHughen (1988), *Plant Cell Rep.*, 7, 281; (13) Firoozabady *et al.* (1987), *Plant Mol. Biol.*, 10, 105; Umbeck *et al.* (1987), *Bio/Technology*, 5, 263; (14) Barnason and Fry, pers. comm.; (15) Catlin *et al.* (1988), *Plant Cell Rep.*, 8, 100; (16) Hinchee *et al.* (1988), *Bio/Technology*, 6, 915; (17) McCabe *et al.* (1988), *Bio/Technology*, 6, 923; (18) Shahin *et al.* (1986), *Crop Sci.*, 26, 1235; (19) Deak *et al.* (1986), *Plant Cell Rep.*, 5, 97; (20) Hernalsteems *et al.* (1984), *EMBO J.*, 3, 3039; (21) Köhler *et al.* (1987), *EMBO J.*, 6, 313; (22) Trulson *et al.* (1986), *Theor. Appl. Genet.*, 73, 11; (23) David *et al.* (1984), *Bio/Technology*, 2, 73; (24) David and Tempé (1988), *Plant Cell Rep.*, 7, 88; (25) Noda *et al.* (1987), *Plant Cell Rep.*, 6, 283; (26) Tepfer (1987) *Cell*, 37, 959; (27), Fillatti *et al.* (1987), *Mol. Gen. Genet.*, 206, 192; Pythoud *et al.* (1987), *Bio/Technology*, 5, 1323; (28) McGranahan *et al.* (1988), *Bio/Technology*, 6, 800; (29) James *et al.* (1989), *Plant Cell Rep.*, 7, 658; (30) Shäfer *et al.* (1987), *Nature*, 327, 529; (31) Toriyama *et al.* (1988), *Bio/Technology*, 6, 1072; Datta *et al.* (1990), *Bio/Technology*, 8, 736; (32) Rhodes *et al.* (1988), *Science*, 240, 204; (33) Gordon-Kamm *et al.* (1990), *Plant Cell*, 2, 603; (34) Horn *et al.* (1989), *Plant Cell Rep.*, 7, 469; (35) De la Peña *et al.* (1987), *Nature*, 325, 274.

* See also Table 9.1

lated using genetic engineering until a better understanding of their genetic control becomes available.

The traits currently amenable to alteration by genetic engineering are plant storage proteins and others which result from the transfer of genes coding new or altered enzymes. The seed storage proteins of cereals and legumes and tuber proteins of potato have been characterized precisely and studied in detail (for a review see Shotwell and Larkins, 1989) and are now the subject of many genetic engineering studies. The improvement of specific plant traits by the use of single genes is limited due to the current lack of cloned genes resulting in useful phenotypes. This is the case for all genes conferring species-specific resistance to diseases. Although these genes exist and have a dramatic effect in excluding compatible interactions among the majority of plants and pathogenic micro-organisms (Vanderplank, 1982), their cloning is proving to be difficult. A preliminary approach to the solution of this problem will be the isolation of alleles conferring disease resistance to parasites which have established a gene-to-gene relation with a plant. Again, such genes have not yet been isolated, although strategies have been designed (Austin *et al.*, 1986; Pryor, 1987).

10.3 MOLECULAR INTERVENTIONS AND BREEDING

The long term goals of biotechnology in crop improvement are the same as those of conventional plant breeding: the creation of new plant varieties with increased vigour and yield, the incorporation of traits for added product value and the development of disease-, herbicide- and pesticide-resistant crops. Moreover, it should be clear that advances made possible through recombinant DNA technology must ultimately be integrated into classical breeding programmes. In the following sections examples of molecular interventions directed to the creation of plants with new properties are described. The cases reported show that plant genetic engineering is a mature technology which already offers original and effective solutions to agronomic problems. According to Gasser and Fraley (1989), genetically engineered rice, maize, soybean, cotton, oilseed rape, sugar beet and alfalfa cultivars are expected to be commercialized between 1993 and 2000.

10.3.1 Use of single genes

(a) Herbicide resistance

Herbicides play an important role in modern agriculture. They permit economic weed control and increase the efficiency of crop production. A number of new herbicides combine high effectivity with lower toxicity to animals and rapid degradation after application. However, they often lack selectivity, a limit to their use in growing crops. Advances in genetic engineering promise to expand the possible applications of broad-spectrum herbicides by introducing genes for resistance into susceptible crops (for reviews see Botterman and Leemans, 1988; Mazur and Falco, 1989).

Resistance can be achieved by at least three different mechanisms: overproduction of a herbicide-sensitive biochemical target (Shah *et al.*, 1986); structural alteration of a biochemical target resulting in reduced herbicide affinity

(Chaleff and Ray, 1984; Comai *et al.*, 1985; Shaner and Anderson, 1985); and detoxification-degradation of the herbicide before it reaches the biochemical target inside the plant cell (De Block *et al.*, 1987).

L-Phosphinothricin and bialaphos
Bialaphos, a tripeptide antibiotic produced by fermentation of *Streptomyces hygroscopicus*, and glucophosinate, the ammonium salt of the chemically synthesized phosphinothricin (PPT), are non-selective herbicides and competitive inhibitors of glutamine synthase (GS) (Leason *et al.*, 1982). The inhibition of GS causes a rapid accumulation of ammonia which leads to the death of the plants. Donn *et al.* (1984) reported the isolation of alfalfa cell lines that were 20- to 100-fold more resistant to PPT than wild-type cells. Resistance to PPT was a consequence of a 4 to 11-fold amplification of a GS gene. The increased synthesis of the enzyme was sufficient to overcome the toxic effects of the inhibitor. De Block *et al.* (1986,1987) proposed an alternative strategy by expressing in the plant an enzyme that detoxifies PPT. The *bar* gene conferring resistance to PPT was isolated from *S. hygroscopicus* (Murakami *et al.*, 1986). It encodes phosphinothricin acetyltransferase (PAT), an enzyme that inactivates the herbicidal compound PPT by acetylation of the free NH_2 group of the molecule (Thompson *et al.*, 1987). The transgenic plants are protected against the applications of glucophosinate and bialaphos under greenhouse and field conditions (Plate 10.1) (De Block *et al.*, 1986, 1987; De Greef *et al.*, 1989). More recently cells from embryogenic maize suspension cultures were transformed with the *bar* gene using microprojectile bombardment (Gordon-Kamm *et al.*, 1990). Transgenic maize plants expressing the *bar* gene are resistant to application of bialaphos.

Glyphosate
Glyphosate, N-phosphonomethyl-glycine, is a broad-spectrum herbicide which interferes with the aromatic amino acid biosynthesis by inhibiting the enzyme 5-enol-pyruvylshikimate-3-phosphate synthase (EPSP), normally expressed in the chloroplast of plants (Steinrücken and Amrhein, 1980; Smart *et al.*, 1985).

Glyphosate is not toxic to animals and is rapidly degraded by soil micro-organisms. Tolerance to this compound is achieved by the overproduction of the target enzyme or by engineering plants with an altered enzyme. Both types of mutations have been selected in *Salmonella typhimurium* (Comai *et al.*, 1983) and the gene *aroA* encoding EPSP was cloned and sequenced (Stalker *et al.*, 1985). This mutant *aroA* gene under the control of the T-DNA mannopine synthase promoter was transferred into tobacco (Comai, *et al.*, 1985) and tomato (Fillatti *et al.*, 1987) plants. Transgenic plants expressing the bacterial EPSP synthase in the cytoplasm show tolerance to glyphosate, indicating that expression in the same cell compartment as the original plant enzyme is not necessary to confer tolerance. The *aroA* gene product is also effective when targeted to the chloroplast. Thus, a gene fusion between the chloroplast transit peptide sequence of a petunia cDNA clone and an *Escherichia coli* mutant *aroA* EPSP gene yielded transgenic tobacco plants that are more tolerant to glyphosate than plants overexpressing the wildtype EPSP gene (Della Cioppa *et al.*, 1987). A higher level of herbicide tolerance has been achieved by accumulating EPSP in the chloroplast (Shah *et al.*, 1986). A cDNA clone encoding EPSP synthase was isolated from a cDNA library of a glyphosate-tolerant *Petunia hybrida* cell line (Steinrücken *et al.*, 1986). The complete EPSP synthase gene of *Petunia* under control of the cauliflower mosaic virus (CaMV) 35S promoter was used for *Agrobacterium*-mediated transformation of *Petunia* cells. Transformed leaf discs produced large quantities of callus tissue and transgenic plants resistant to 4 times the glyphosate dose required to kill wildtype *Petunia* plants tolerant to glyphosate. This higher level of glyphosate tolerance results from the overproduction of EPSP synthase.

Sulphonylureas and imidazolinones

Sulphonylureas block branched chain amino acid biosynthesis inhibiting the metabolic routes leading to valine, leucine and isoleucine (Ray, 1984). The target is the enzyme acetolactate synthase (ALS) (La Rossa and Schloss, 1984; Ray, 1984). An unrelated class of herbicides, the imidazolinones, also inhibits ALS

(Shaner and Anderson, 1985). Mutations conferring resistance to sulphonylureas and/or imidazolinones have been isolated in plants and other organisms. Selection of plant tissues for resistance to this class of herbicides has not yet yielded mutants with overproduction of the target enzyme. A simple explanation would be the high probability of selecting resistance mutations due to alterations of the ALS protein (Comai and Stalker, 1986).

Tobacco mutants resistant to sulphonylureas have been isolated from cell cultures by selecting cell lines growing in the presence of sulphometuron methyl or chlorsulphuron (Chaleff and Ray, 1984). In protoplast regenerated tobacco plants (Chaleff *et al.*, 1987), dominant nuclear mutations at two unlinked loci cosegregated with a herbicide-insensitive ALS enzyme. In mutant *Arabidopsis* plants, a single dominant nuclear mutation at one locus confers resistance (Haughn *et al.*, 1988). The ALS genes are located in the nucleus and include a sequence encoding for a chloroplast transit peptide at their *N*-terminal (Mazur *et al.*, 1987). Transgenic tobacco plants expressing a mutant ALS gene from tobacco or *Arabidopsis* tolerate herbicide concentrations four times that of typical field application rates (Haughn *et al.*, 1988; Lee *et al.*, 1988). Tolerance to imidazolinones has been induced in maize cell lines selected from cell culture. The resistant phenotype was correlated with the presence of an altered ALS enzyme, which was also expressed at the whole plant level (Shaner and Anderson, 1985). Plants homozygous for the resistance gene are tolerant to application of imidazolinone herbicides.

Atrazine

Atrazine and simazine are the best characterized herbicides with respect to their mode of action. This class of compounds inhibits electron transport through photosystem II (PSII) by binding to chloroplast thylakoid membranes. The herbicides bind to a 32 kDa thylakoid membrane polypeptide encoded by the chloroplast *psbA* gene (Zurawski *et al.*, 1982). A potential source of atrazine resistance is chloroplasts from naturally occurring resistant weed biotypes or microbial species. Resistance is maternally inherited in all atrazine-resistant weeds which also carry the highly conserved *psbA* genes (Zurawski *et al.*, 1982). A single amino acid substitution (serine to glycine) at position 264 in the 32 kDa protein results in decreased herbicide binding (Hirschberg and McIntosh, 1983). Transfer of resistant chloroplasts or development of a chloroplast transformation system are possible ways to obtain atrazine resistance in valuable crops. Resistant chloroplasts of *Brassica campestris* have been introduced by standard genetic backcrosses into crops of commercial importance such as oilseed rape, rutabaga and Chinese cabbage. More recently, chloroplasts of a terburtryn-resistant mutant of *Nicotiana plumbaginifolia* have been transferred into the *N. tabacum* nuclear background by protoplast fusion (Menczel *et al.*, 1986). The regenerated plants were resistant to high levels of atrazine (10 kg/ha).

Detoxifying enzymes

Enzymes involved in herbicide detoxification in tolerant crop species have been studied. The genes encoding such proteins are very useful, because they inactivate the herbicides before they are able to inhibit the target enzyme. Examples of such systems are described in the review by Comai and Stalker (1986). One such conjugative enzyme is glutathione-*S*-transferase (GST) which is responsible for the modification of the triazine herbicides by glutathione (Frear and Swanson, 1970). GSTs are multifunctional enzymes detected in at least 17 plant species. cDNA clones corresponding to genes GSTI and GSTII of maize have been isolated (Wiegand *et al.*, 1986). The availability of cloned probes established that the treatment of maize seeds with safeners (herbicide protectant chemicals) results in three- to four-fold activation of GST mRNA in etiolated tissue. Similar results have been obtained in metribuzin-tolerant tomatoes, in which detoxification involves enhanced activity of the enzyme *N*-glucosyltransferase, and tolerance is governed by a single locus (Comai and Stalker, 1986). Furthermore, mixed-function oxidases are involved in the detoxification of 2,4-

dichlorophenoxyacetic acid (2,4-D) in pea (*Pisum sativum*) and of dicamba in tolerant barley (Comai and Stalker, 1986).

Several soil microorganisms involved in herbicide degradation have also been characterized as potential sources of herbicide-resistance genes. A successful example is the engineering of resistance to the benzonitrile herbicide bromoxynil. Bromoxynil (3,5-dibromo-4-hydroxybenzonitrile) is a potent photosynthetic inhibitor in plants. The gene *bxn*, encoding a specific nitrilase, was cloned from the natural soil bacterium *Klebsiella ozaenae*. The *bxn* gene under control of a light-regulated promoter was transferred to tobacco plants to which it conferred resistance to high levels of a commercial formulation of bromoxynil (Stalker *et al.*, 1988). Transgenic tobacco plants resistant to 2,4-D were recently produced through genetic engineering of *tfdA* gene (Lyon *et al.*, 1989; Streber and Willmitzer, 1989) of the soil bacterium *Alcaligenes eutrophus*, encoding the first 2,4-D dichlorophenoxyacetate monoxigenase (DPAM) enzyme involved in a 2,4-D degradative pathway. Tobacco plants expressing this enzyme exhibited resistance when sprayed with levels of herbicide up to 8 times the field application rates (Lyon *et al.*, 1989; Streber and Willmitzer, 1989). The introduction of the gene for DPAM into broad-leaved crop plants may eventually allow 2,4-D to be used as an inexpensive post-emergence herbicide in economically important dicot crops.

(b) Virus resistance

The pathogenic effects of viruses on plant growth, development, and symptom production are described in detail in the specific literature. However, the mechanisms by which plant viruses cause diseases are poorly understood at the molecular level. Virus resistance could provide significant yield protection in important crops such as vegetables, maize, rice, soybean and tropical trees. The approaches to create virus-resistant plants rely on the cloning of viral sequence and on their expression in transgenic plants. Different viral DNA sequences appear to function in transgenic plants to confer resistance to virus infection, e.g. viral coat protein (CP) genes, viral anti-sense RNAs, and viral satellite RNAs.

Resistance based on the expression of viral coat protein

The practice of cross protection has been used for many years to reduce crop losses due to viral infection (Jokusch, 1968). In general, plants infected with one strain of virus exhibit reduced or delayed symptom development in response to a second strain. Abel *et al.* (1986) and Bevan *et al.* (1985) reported on the expression of tobacco mosaic virus (TMV) coat protein gene driven by the CaMV 35S promoter in transformed tobacco plants. When the transgenic plants were inoculated with varying concentrations of TMV, delay in symptom development was observed; moreover, up to 50% of the plants did not show any symptoms at all during the duration of the experiment. This approach produced similar results in transgenic tomato, tobacco, and potato plants against a broad spectrum of plant viruses, including alfalfa mosaic virus, cucumber mosaic virus, potato virus X, and potato virus Y (Tumer *et al.*, 1987; Van Dun *et al.*, 1987; Cuozzo *et al.*, 1988; Hemenway *et al.*, 1988; Hoekama *et al.*, 1989; Kaniewski *et al.*, 1990). One mechanism of coat protein-mediated cross protection involves the uncoating of virus particles before translation and replication (Register and Beachy, 1988).

Transgenic tomatoes carrying the TMV coat protein gene have been evaluated in greenhouse and field tests and shown to be highly resistant to TMV infection (Nelson *et al.*, 1988). The transgenic plants showed no yield loss after virus inoculation, whereas the yield was reduced 23% to 69% in control plants. The level of coat protein in the engineered plants (from 0.01% to 0.5% of the total proteins) is well below the level reported in plants infected with this endemic virus. This should eventually facilitate registration and commercialization of virus-resistant cultivars.

Transgenic plants that express viral antisense RNA

Genes that express antisense RNA are regulatory genes. They direct the synthesis of RNA that by itself exhibits negative control on the expression of other genes at the level of transcription or translation (Weintraub *et al.*, 1985; Green *et al.*, 1986). The antisense RNA contains base sequences complementary to the target ('sense') RNA transcript (e.g. RNA primers of replication, mRNAs), and the formation of RNA–RNA hybrids block the function of sense RNA (section 8.5.6). Antisense RNA may also interfere with mRNA activity, presumably by hybridizing to sequences at or close to the ribosome binding site. The term micRNA (mRNA-interfering complementary RNA) has been proposed for this specific regulatory RNA species. These regulatory genes have been initially discovered in prokaryotes, where they control a diversity of molecular events such as plasmid replication and incompatibility, transposition and phage replication. Recently, complementary mRNA molecules that are transcribed from opposite strands of the same DNA sequence and overlap at their 3′ untranslated region, have been identified in eukaryotes (Spencer *et al.*, 1986; Williams and Fried, 1986). It has been postulated that these transcripts form double-stranded RNA in the nucleus thereby preventing the transport of sense (messenger) RNA into the cytoplasm.

Although claims have been made that the expression of parts of a viral genome in the form of antisense RNA confers disease resistance, experimental data have not yet been presented. On the contrary, expression of TMV antisense RNA in transgenic tobacco plants did not protect against TMV infection (Beachy *et al.*, 1987), whereas TMV CP-expressing plants showed delayed symptom development. Previous experiments, however, on the *in vitro* translation of TMV RNA show that oligodeoxy-nucleotides covering only a portion of the 5′ region of TMV RNA can block its translation (Beachy *et al.*, 1985b). Additional experiments are needed in order to define the effective and optimal antisense RNA structure to be used in transformation experiments of crops.

Transgenic plants that express viral satellite RNA

Satellite RNAs are species of RNA associated with some strains of certain plant RNA viruses. They are not necessary for virus replication

Plate 10.1. Field trial 'Basta' resistant spring oilseed rape: forefront = wildtype cultivar 'Drakar'; back = doubled haploids of cultivar 'Drakar' expressing the *bar* gene (courtesy of J. Leemans, Plant Genetic Systems N.V., Gent, Belgium).

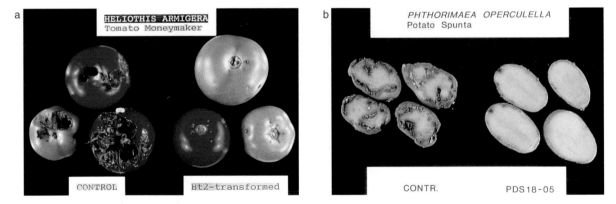

Plate 10.2. (a) Bt tomato: *Heliothis armigera* = tomato fruit worm (a European pest). (b) Bt potato: *Phthorimaea operculella* = potato tuber moth (a common storage pest in the Mediterranean climate) (courtesy of J. Leemans, Plant Genetic Systems N.V., Gent, Belgium).

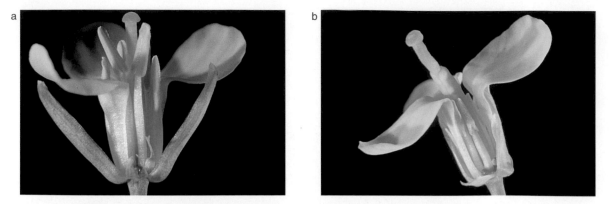

Plate 10.3. Oilseed rape cultivar 'Drakar' flowers from untransformed plants (a), and plants transformed with a *TA29–RNase* gene (b) (courtesy of J. Leemans, Plant Genetic Systems N.V., Gent, Belgium).

Plate 10.4. The transgenic petunia plants were classified according to their phenotype into three groups. (a) Red plants with uniform coloration of the whole flower. (b) White plants with no expression of *Al* cDNA (the coloration seen is due to small amounts of cyanidin production and corresponds to the background coloration observed in the recipient plant RL01). (c, d) Variegated plants with pelargonidin pigmentation in some floral cells. Among the plants of the variegated phenotype the number of *Al*-expressing floral cells varied from a few *Al*-expressing cells (c, region is marked by an arrow) up to about 80% of all floral cells (d) (courtesy of H. Saedler, Max-Planck-Institut für Züchtunsforschung, Köln, Germany).

(Murant and Mayo, 1982; Francki, 1985), do not exhibit any base sequence homology to the genomes of either virus or plant host, and are replicated by the helper virus-coded RNA-dependent RNA polymerase. One aspect of their parasitic association with plant viruses is the fact that they can affect disease and symptom development in virus-infected plants. In cases where the presence of satellite RNA attenuates the disease caused by the virus, satellite RNAs have been used as a biological control in crop protection, e.g. in China in order to protect pepper and tomato from CMV (Po *et al.*, 1987). The satellite RNA of cucumber mosaic virus (CMV) (Harrison *et al.*, 1987) and the satellite RNA of tobacco ringspot virus (TobRV) (Gerlach *et al.*, 1987) confer resistance to infection by satellite-free helper viruses CMV and TobRV, respectively. Transgenic plants expressing satellite RNA present a unique approach towards gentically engineered, virus-resistant crops, in that the protective mechanism seems to be triggered to high efficiency only when the viral pathogen is invading the host plant.

(c) Insect resistance

Chemical control of insects, in addition to being expensive, frequently results in negative environmental effects, therefore, the development of insect-resistant plants is an application of genetic engineering with important implications for both the seed and the agrochemical industries. Plants producing their own protective proteins increase the selectivity of control and reduce damage to non-target insect populations. Progress in engineering insect resistance in transgenic plants has been achieved through the use of the insect control protein genes of *Bacillus thuringiensis* (Bt). Bt toxins differ in their spectrum of insecticidal activity. Most strains of Bt are active against Lepidoptera, but some are specific for Diptera and Coleoptera. The insect toxicity of Bt resides in a large protein with no toxicity to beneficial insects, animals and humans. The mode of action of Bt toxins is exerted by disruption of ion transport across brush border membranes of susceptible insects (Sacchi *et al.*, 1986). Several genes encoding

lepidoptera-type toxins have been isolated. One such gene from Bt ssp. *kurstaki HD-1* (Watrud *et al.*, 1985), contains an open reading frame of 3468 bp encoding a protein of 1156 amino acids. Truncated forms of the gene still encode active toxins as in the case of the *N*-terminal half (646 amino acids) of the Bt *kurstaki* protein. Chimeric Bt *kurstaki* genes with the CaMV 35S promoter and a sequence coding for an active truncated variant as well as the full length gene, have been constructed and expressed in tomato plants (Fischhoff *et al.*, 1987). The chimeric truncated gene was transcribed into mRNA of low abundancy, a finding indicating an unusual instability of the chimeric transcript in the plant cell. The insecticidal protein was nevertheless expressed at a level sufficient to kill larvae of *Manduca sexta*, *Heliothis virescens* and *Heliothis zea*. Progenies of transgenic plants show that the Bt *kurstaki* gene segregates as a single dominant Mendelian marker. Its activity was detectable not only in foliage but also in fruits of transgenic tomato plants.

A second toxin gene has been cloned from the Bt strain *berliner* 1715 (Höfte *et al.*, 1986). It produces a Bt2 protein, 1155 amino acids long, which generates a smaller polypeptide with full toxic activity. Vaeck *et al.* (1987) have used chimeric genes containing the constitutive 2' promoter of the *Agrobacterium* T_R-DNA, the entire coding sequence of Bt2 as well as the truncated toxin gene, to transform tobacco plants. Other constructs contained a chimeric toxin-NPTII gene. The Bt-NPTII fusions were particularly useful because they allowed the selection of transformants with levels of toxin sufficiently high to be insecticidal. In transformed plants their insect toxicity was directly correlated with the level of kanamycin resistance, the selectable marker used in the transformation experiments. A correlation was moreover found between the quantity of the toxin and the insecticidal activity of transgenic plants. Transgenic tobacco plants were protected from feeding damage by larvae of *Manduca sexta*.

The level of insect control observed in the field tests with tobacco, potato and tomato plants has been excellent; in one such test tomato and potato plants containing the Bt gene suffered no agronomic damage (Plate 10.2) under conditions that led to total defoliation of control plants (Delannay *et al.*, 1989). The excellent insect control observed under field conditions indicates that this technology already has commercial application. Market opportunities for caterpillar resistance concern leafy vegetable crops, cotton and maize. Crop targets for beetle resistance are potato and cotton. To protect plants fully against other and more resistant lepidopteran-like Noctuidae, higher levels of Bt toxin will be required. This might be achieved by using chimeric genes containing a strong plant-specific promoter, adapting the bacterial genes to higher levels of expression by optimizing their codon usage in the plant and synthesizing toxin genes coding for proteins with a higher specific activity against specific insects.

Other mechanisms exist which confer to plants field resistance to a wide range of pests. One such mechanism is based on the trypsin inhibitors of cowpea (*Vigna unguiculata*) which have been isolated by Gatehouse *et al.* (1980). These inhibitors are antimetabolic agents against the bruchid beetle *Callosobruchus maculatus* and other insects of the genera *Heliothis*, *Spodoptera*, *Diabrotica* and *Tribolium*. A cowpea trypsin inhibitor gene has been cloned by Hilder *et al.* (1987) using a synthetic oligonucleotide probe. Constructs containing the CaMV 35S promoter and a full length cDNA clone 550 bp long were mobilized into *Agrobacterium* and used to transform leaf discs of tobacco. Transgenic plants contained 3–7 copies of the constructs and a variable level of inhibitors. Western blotting of plant extracts showed that a polypeptide was produced and processed corresponding to the isoinhibitor from cowpea seed. The bioassay for insecticidal activity was done with young tobacco plants infected with emerged larvae of the lepidoptera *Heliothis virescens*. Insect survival and plant damage were decreased on transgenic tobacco compared to the controls, although considerable variability among plants was observed.

(d) Protein quality

It has been recognized for many years that seed proteins are deficient in amino acids that are essential for human and livestock nutrition (Nelson, 1979). In general, cereals are most limited in lysine and legumes in the sulphur amino acids methionine and cysteine. For wheat, in particular, it would be desirable to improve gluten elasticity in many varieties, and this trait is strongly influenced by one group of storage proteins. There has been only limited progress by plant breeders to increase the content of essential amino acids in crops, primarily because genes encoding storage proteins rich in such amino acids are rare. Genetic engineering of genes encoding better seed storage proteins should provide a rather straightforward solution to the problem. To date, work has focused on the isolation and characterization of cDNA and genomic clones for the major seed proteins (Table 10.3). These studies have established that storage proteins are the products of multigene families, often comprising ten or more members tightly linked at a given locus (for a review see Shotwell and Larkins, 1989).

The manipulation of seed quality through genetic engineering faces several obstacles. The addition of one gene to plant genomes modified by transformation may not be very effective in improving the plant phenotype due to the expression of the rest of the multigene family. Recently, Otani, (cited in Scheidegger, 1990) incorporated multiple codons for lysine and tryptophan in a 19 kDa zein cDNA by site-directed mutagenesis. Although the engineered zein was corrected in its deficiency in essential amino acids, the protein was not stably incorporated into its normal cell compartment in the endosperm of maize. The transformation of crops with synthetic or natural genes coding for peptides rich in essential amino acids offers an alternative solution. Recently, a synthetic gene coding for a polypeptide rich in essential amino acids was introduced into potato (Yang *et al.*, 1989). Also, the search for genetic situations leading to a high level of expression of genes coding for proteins useful nutritionally or technologicaly has been proposed. One example is

Table 10.3. The state of research on storage proteins of the major cereal species

Crop	Major storage proteins	State of research				Mutants		References
		Genes cloned[a]	Genes expressed in transgenic plants[a]	Protein made[a]	Modifications desired[b]	Type	Cloned[a]	
Maize	Zeins 21 kDa	+	+		Lys	Regulatory non-allelic to zeins e.g. *02, 07, f12*		Heidecker and Messing (1986)
	19 kDa	+	+					Hoffman *et al.* (1987)
	15 kDa	+					+	Schmidt *et al.* (1987)
	10 kDa	+		+				Motto *et al.* (1988)
Sorghum	Kafirins	(+)	(+)	(+)	Lys	High-lysine		De Rose *et al.* (1988)
Barley	Hordeins B	+	+		Lys	Regulatory also *B-hor* deletion		Shewry *et al.* (1988)
	C	+						Marris *et al.* (1988)
	D	+						
Wheat	Gliadins	+			Lys			Shewry *et al.* (1988)
		+						Payne (1986)
	Glutenins[c]							
	HMW	+	+	(+)	Elasticity			Colot *et al.* (1987)
	LMW	+	+					Robert *et al.* (1988)

[a] +, results available; (+), preliminary results.
[b] Lys, increased lysine; Elasticity, increased elasticity.
[c] HMW, high molecular weight; LMW, low molecular weight.

the high-molecular weight glutenin alleles which are credited with the ability to improve the bread-making quality of wheat (Payne *et al.*, 1984).

Studies on regulatory mutants, which affect whole groups of storage protein genes, are of particular interest. Such mutants are known in barley, sorghum and maize. Progress has been made in maize where clones for one such locus, *opaque-2* (*O2*), have been obtained via transposon tagging (Schmidt *et al.*, 1987; Motto *et al.*, 1988). *O2* regulates the synthesis of one of the major maize storage protein families, the 22 kDa zeins, at the level of transcription (Hartings *et al.*, 1989). The feasibility of using cloned regulatory sequences, such as those of *O2*, to modulate the expression of prolamine genes in cereals can be tested in the future. If feasible, a new tool may soon become available to breed cereals with a better protein quality.

(e) Hybrid breeding

Several possibilities are emerging for the control of plant reproduction either through the manipulation of incompatibility or by inducing male sterility. As these systems are considered in some detail in Chapters 13 and 14, we will restrict ourselves to dealing with those which involve gene isolation and manipulation.

The cloning of alleles responsible for self-incompatibility (Nasrallah *et al.*, 1985; Anderson, *et al.*, 1986) may lead in the future to the breeding of self-compatible genotypes in self-incompatible species. The strategy is based on the duplication of the *S* locus which leads to self-compatibility, as already known from studies on the effect of polyploidization or of cytological duplication of the *S* gene (De Nettancourt, 1977). From this point of view, interesting findings have been reported recently in *Nicotiana alata* which has a gametophytic system of self-incompatiliblity. A diploid plant having the genotype *S1S2* at the incompatibility gene locus will produce *S1* pollen and *S2* pollen. The *S1* pollen will not function on *S1S1* and *S1S2* plants. Similarly, *S2* pollen function is negated by an *S2* allele in the female genotype. The *S* alleles code for glycoproteins (Anderson *et al.*, 1989) present in the stylar tissue which were recently shown to be ribonuclease (S-RNase) (McClure *et al.*, 1989). This finding led to speculation that the S-RNases may be the killer molecules of the natural self-incompatibility system. More recently, as detailed in Chapter 13, McClure *et al.* (1990) have demonstrated that RNA is degraded in incompatible pollen tubes, but not in compatible ones. RNA degradation seems to be restricted to ribosomal RNA, with poly(A)+ RNAs and tRNAs being unaffected. New discoveries in this field may lead to more advanced uses of *S*-based incompatibility in hybrid breeding.

Two systems for inducing male sterility have been developed. The first (Spena *et al.*, 1987; Schmülling *et al.*, 1988), used a molecular construct based on the gene *rol C* of *A. rhizogenes* under the control of the CaMV 35S promoter, flanked by a selectable marker. Tobacco plants transgenic for *rol C* are sterile and phenotypically different from the wild-type. The possibility of selecting in favour of a flanking marker (for example using a gene conferring herbicide resistance) allows the isolation of homozygous male sterile populations to be used in hybrid seed production. Genes transformed into agricultural plants must have a stability of 10^{-3} to be properly utilized in hybrid seed production.

A novel genetic method for inducing pollen sterility, which promises to be applicable to many plant species, has been reported by Mariani *et al.* (1990). The basis of the method is a gene construct that couples a promoter with a highly tissue-specific gene expression to a bacterial or fungal *RNase* gene. The promoter *TA29* (Goldberg, 1988) ensures that the coding sequence to which it is attached is transcribed only in the tapetum at a particular stage of anther development (see Chapter 14). It was first demonstrated that the *TA29* promoter retains its specificity in transgenic tobacco plants. The *TA29* promoter was linked to a synthetic *Aspergillus oryzae T1 RNase* gene and to an *RNase* gene from *Bacillus amyloliquefaciens* (*barnase*). The expression of *RNase barnase* leads to tapetum ablation and the anthers of transgenic tobacco plants did not produce functional pollen. Apart from being male sterile the plants

were phenotypically normal, confirming the fidelity and specificity of the promoter in the constructs. The same specificity of expression and the same level of sterility was observed when the gene constructs were introduced into oilseed rape plants (Plate 10.3). Field trials have shown the stability of the male sterility of the male sterile phenotype and the normal agronomic performance of male sterile lines. The approach has been applied successfully to other crops including cabbage, chicory, cotton and maize (J. Leemans, personal communication).

The ability of the *TA29-RNase* gene to induce male sterility provides a new strategy for the production of hybrid crop plants. Transferring this dominant male sterility gene to plants such as maize will enable hybrids to be produced without detasselling. By coupling the chimeric *TA29-RNase* gene to a dominant herbicide gene (for example, *bar*) breeding systems can be devised to select for uniform populations of male sterile plants. In crop plants where fruits are not the harvested product (for example, lettuce, carrot, cabbage), male sterile plants can be crossed with any pollinator line to produce hybrid seeds. Antisense RNA technology (Izant and Weintraub, 1985) and the existence of *barstar*, a specific protein inhibitor of *barnase* (Hartley, 1988), should facilitate the development of strategies for male fertility restoration, a requirement for hybrid varieties in crops such as wheat, rice and maize. Oilseed rape pollinator lines capable of producing fully fertile hybrid seeds on the male sterile plants have been obtained by expressing *barstar* in their anthers. Hybrid progeny express both *barnase* and *barstar* proteins in the tapetum which form a molecular complex resulting in the full restoration of fertility (J. Leemans, personal communication).

10.3.2 Genetic engineering of simple biochemical pathways

Plants may be defective for groups of chemical reactions leading to the synthesis of specific end products. In the simplest case, only one missing gene can lead to the desired phenotype. This is the case in the experiment des-

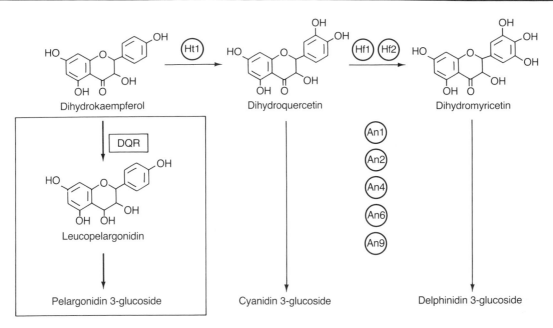

Fig. 10.1. Pathway diagram. Due to mutations in flavonoid 3'-hydroxylase (Ht1) and flavanoid 3',5'-hydroxylase (Hf1, Hf2) dihydrokaempferol is only poorly hydroxylated to dihydroquercetin (DQR) and dihydromyricetin in RLO1 petunia mutants. Dihydrokaempferol is taken as a susbstrate by DQR and converted into lecucopelargonidin which is further processed into pelargonidin-3-glycosides leading to the brick-red colour.

cribed by Meyer *et al.* (1987) in *Petunia*. In this plant, cyanidin and delphinidin-derivatives, but no pelargonidin-derivatives, are produced by the anthocyanin biosynthetic pathway. This is due to the substrate specificity of dihydroflavonol reductase of *Petunia* which cannot reduce dihydrokaempherol. A *Petunia* mutant accumulates this compound and shows no pigmentation (Fig. 10.1). When this genotype was transformed with the *A1* gene of *Zea mays*, which encodes for dihydroflavonol 4-reductase (DFR), the intermediate for pelargonidin biosynthesis was provided, resulting in the synthesis of pelargonidin to produce brick-red flowers. Thus the maize DFR gene product created a novel pigmentation pathway in *Petunia* (Plate 10.4). A second successful approach to manipulate the floral pigmentation pathway was developed independently by Van der Krol *et al.* (1988, 1990) and Napoli *et al.* (1990). Rather than creating a novel side branch, these workers interrupted an existing pathway in petunia by expressing a gene in the inverse orientation (antisense) or by introducing additional copies of a defined pigmentation gene. The antisense of the chalcone synthase gene (CHS) or extra copies of the genes dihydrofolate reductase (DFR) or CHS driven by the CaMV 35S promoter, resulted in an effective reduction of CHS expression. As a consequence, a reduction in floral pigmentation was observed either evenly distributed (up to full white) or variegated in sector or ring patterns (Van der Krol *et al.*, 1988, 1990; Napoli *et al.*, 1990).

To modify the photosynthetic efficiency of agricultural plants is a more complex situation. Possible avenues suggested include increasing the level of carotenoids for obtaining superior pigment/protein complexes in the chloroplast (Sandmann and Bramley, 1985), manipulation of components of the PSII which seems to play a regulatory role (Trebst, 1987), to use in C_3 plants a triose phosphate translocator cloned from C_4 plants (Flügge and Heldt, 1986), to clone and use genes involved in the regulatory mechanisms of leaf senescence (Woodson, 1987), to alter the RuBP-C/RuBP-0 ratio (Kung and Marsho, 1976) or to use genetic engineering to assemble an efficient CO_2 refixation system (Nasyrov, 1978). The last proposal is aimed at the reduction of photorespiration in C_3 plants, by introducing into them a simple biochemical pathway. In its latest version (Kreuzaler, unpublished) the new pathway is based on the enzymes phosphoenolphosphate (PEP) carboxylase and PEP carboxykinase and on a triose-phosphate translocator active at the level of the chloroplast membrane.

The form and habit of a plant, as well as the shape and size of its flowers and other organs, are important traits in determining the commercial value of crop plants. Genes controlling such characteristics can be useful to industry, but their isolation and successful exploitation is a long term goal because of a lack of sufficient information. Flower morphology, for example, is controlled by many genes. The isolation of such genes could facilitate the engineering of the shape of a flower. Mutants in floral morphology have been identified and genetically characterized in ornamental plant species such as *Petunia hybrida* and *Antirrhinum majus* but also in cereal species. Moreover, homeotic flower mutants of *Arabidopsis* and *Antirrhinum* have provided fertile ground for studying flower morphogenesis and isolating the genes involved (Schwarz-Sommer *et al.*, 1990; Coen and Meyerowitz, 1991). The protein encoded by the *Arabidopsis* and *Antirrhinum* homeotic genes *agamous*, *deficiens*, and *floricaula* showed homology to transcription factors. In maize the developmental gene *Knotted-1* (*Kn1*) also encodes a homeodomain-containing protein (Vollbrecht *et al.*, 1991). Sequence comparisons strongly suggest that *Kn1* acts as a transcriptional factor. The *Kn1* homeobox may permit the isolation of genes that, like animal and fungal counterparts, regulated cell fate determination. These findings may provide conceptual understanding of the molecular basis of flower induction and development which may lead to future applications in plant breeding.

10.4 MOLECULAR BIOLOGY AND THE IMPROVEMENT OF TRAITS BASED ON MANY GENES

When the trait of a plant to be improved is dependent on a set of complex metabolic reactions, and particularly if these must still be in part or completely discovered and characterized, the application of molecular approaches to plant breeding has less likelihood of being successful. Nevertheless the feasibility of modifying some of the complex attributes of the plant, like the reaction to biotic and abiotic stresses, has been recently considered. The strategy here is to reveal differential gene actions in contrasting stress situations; to isolate the corresponding genes; to prove their direct role in the phenomenon by testing their effect in transgenic plants, and later to use them directly to modify a crop or as probes to monitor the level of stress.

10.4.1 Disease resistance

Most of the diseases of crop plants are caused by bacteria and fungi. However, limited success in engineering resistance to fungal diseases has been reported and genetically engineered resistance to fungal pathogens and to bacteria remains in the early research stages. The resistance of a plant to potentially pathogenic micro-organisms is based on multiple biochemical factors. Particularly important are the phytoalexins, antimicrobial molecules that are not present in healthy plants but accumulate in response to microbial infection (Collinge and Slusarenko, 1987). Plants may also possess antimicrobial proteins such as chitinases (Schlumbaum *et al.*, 1986) and thionins (Bohlmann *et al.*, 1988) which are potent inhibitors of fungal growth. In addition, there are indications that a 30 kDa ribosome-inactivating protein (RIP) of barley may possess antifungal activity (Leah *et al.*, 1991). By sequence analysis and *in vitro* assays, Maddaloni *et al.* (1991) have also reported that the maize b-32 protein, a 32 kDa cytosolic albumin that accumulates in the endosperm of developing kernels under the control of the regulatory gene *O2*, is a RIP. RIPs make up a large and widely distributed class of plant cytotoxic proteins that irreversibly inactivate eukaryotic ribosomes by depurination of a specific 28 rRNA adenine residue that is essential for translational elongation (Endo *et al.*, 1988). Several different RIPs such as ricin, gelonin, and α-trichosanthin have been tested for use as cell-specific immunotoxins as well as antiviral and insecticidal agents (Coleman and Roberts, 1982; Gatehouse *et al.*, 1990; Leah *et al.*, 1991). However, the evidence that genes conferring disease resistance produce particular compounds in specific tissues is not available even for a single case of host-pathogen interaction (Bennetzen, 1984). This is why improving disease resistance using molecular biology techniques is still not practically feasible. One approach which is actively followed, at present, is directed to understanding the race-specific resistance phenomenon and the molecular cloning of the interacting plant and pathogen genes. Attempts to clone a plant disease-resistance gene by the technique of transposon tagging has been adopted for the isolation of the *Rp1* and *Hm1* genes of maize specifying resistance to the leaf rust fungus *Puccinia sorghi* and to Northern leaf spot induced by *Helminthosporium carbonum*. Recently, Johal and Briggs (cited in Wessler and Hake, 1990), by using the *Mu* transposable element, were able to tag the *Hm1* gene.

10.4.2 Stress tolerance

Despite continuing efforts to improve cultivation practices, crop plants will always be subject to a variety of environmental extremes. In even the most productive agricultural regions, drought and temperature stresses occur throughout the growing season, resulting in injury and reduced plant yield (see Chapter 24). Alteration of gene expression during environmental stress (summarized in Sachs and Ho, 1986) has permitted the isolation of stress-related genes. A summary of such genes already cloned and partially characterized is reported in Table 10.4. Considerable research

effort is still needed to define the role of stress proteins in the acquisition of stress tolerance, including the clarification of the structural and enzymatic functions that these proteins may possess. Also the mechanisms regulating the expression of stress genes in response to different environmental stimuli should be better understood. Although the molecular basis of stress phenomena is at present poorly understood, the molecular approach to this problem has already revealed enough relevant information to permit the monitoring of level of plant tolerance to cold, heat, drought and salts.

Several model systems are being developed in order to understand mechanisms by which plants resist desiccation. These include the study of events taking place in the developing cereal embryo during its acquisition of desiccation tolerance (Bartels *et al.*, 1988) and the isolation of proteins and genes induced by desiccation in resurrection plants such as *Ceraterostigma plantagineum* (Bartels *et al.*, 1990). Resurrection plants can revive from an air-dry condition, and desiccation tolerance is not restricted to seeds, but is also expressed in whole plants.

10.4.3 Diagnostics

The use of molecular probes to help breeding programmes has been frequently proposed. They can specifically monitor the level of stress or disease reactions as well as the presence of pathogenic organisms in the breeding material like potato spindle tuber viroid (PSTV) in potato (Owens and Diener, 1981). A particular case is represented by the probes revealing RFLP alleles linked to agronomically important genes as considered in Chapter 19.

10.4.4 Improved fertilizer efficiency

One of the most publicized potential applications for biotechnology in agriculture is the genetical modification of plants to provide them with their own capacity of fixing nitrogen which is presented in detail in Chapter 26.

Table 10.4. Characterized genes involved in abiotic stress reactions

Stress imposed	Plant	Gene affected	Pathways gene action enzymatic function	Reference[a]
Cold	Tomato	Thiol protease	Degradation of proteins	1
	Alfalfa	An ABA induced gene		2
Heat = high temperature	Soybean	Groups of genes	Exact function and	3
	Pea	68–104 kDa ubiquitous group	identity unclear Hsp. allows cells to	
	Tobacco	20–33 kDa	establish thermo-	
	Tomato		tolerance	
	Maize	small hsp.		4
	Barley	15–18 kDa		5
Anaerobic	Maize	Alcohol dehydrogenase 20 anaerobic polypeptides	Ethanolic fermentation	4
	Rice	Glucose-6-phosphate isomerase	Glycolytic enzymes	
	Sorghum	Fructose-1, 6-diphosphate, Aldolase		6
	Pea	Sucrose synthase,		
	Carrot	Pyruvate decarboxylase		
	Barley	Lactate dehydrogenase		
Heavy metal	Soybean	Cd-binding proteins		7
	Datura innoxia	Cd-metal-binding proteins		8
	Mimulus	Cu-binding proteins		4
	Soybean	As, Cd induce some heat shock genes		9
Salt	*Nicotiana tabacum* (cell culture)	26 kDa protein unique in salt-adapted cells holologous to members proteinase inhibitors		10 11 12
	Barley seedlings	2 new classes of mRNA induced upon 2% NaCl-treatment		13
Water stress Related to water stress and high endogenous ABA-levels	Rice (embryo cont. half seeds)	21 kDa polypeptide (glycin-rich)		14
	Maize (embryo)	15.4 kDa polypeptide (glycin-rich)		15
	Barley	22 kDa		16

[a] (1) Schaffer and Fischer (1988), *Plant Physiol.*, 87, 431–436; (2) Mohapatra *et al.* (1988), *Plant Physiol.*, 87, 468–473; (3) Schlesinger *et al.* (1982), Heat shock from bacteria to man. N.Y. Cold Spring Harbor La.; (4) Sachs and Ho (1986), *Annu. Rev. Plant Physiol.*, 37, 363–376; (5) Key *et al.* (1987), In *Plant Molecular Biology*, Wettstein, D.v. and Chua, N.H. (eds), Plenum Press, New York; (6) Dennis *et al.* (1987), In *Plant Molecular Biology*, Wettstein, D.v. and Chua, N.H. (eds), Plenum Press, New York, pp. 407–417; (7) Fujita and Kawanishi (1987), *Plant Cell Physiol.*, 28, 379–382; (8) Jackson *et al.* (1984), *Plant Physiol.*, 75, 914–918; (9) Edelman *et al.* (1988), *Plant Physiol.* 86, 1048–1056; (10) Singh *et al.* (1987), *Proc. Natl. Acad. Sci. USA*, 84, 739–743; (11) Singh *et al.* (1987), *Plant Physiol.*, 5, 529–536; Richardson *et al.* (1987), *Nature*, 327, 432–434; (13) Ramagopal (1987), *Proc. Natl. Acad. Sci. USA*, 84, 94–98; (14) Mundy and Chua (1988), *EMBO J.*, 7, 2279–2286; (15) Gomez *et al.* (1988), *Nature*, 334, 262–264; (16) Chandler *et al.* (1988), *J. Cell Biol. Suppl.* 12C Abrs. #LO33, 143.

10.5 PROSPECTS

In view of the continuing need for increasing global food production, particularly in developing countries, a science-based, high-yielding, sustainable agriculture must be the key to meeting the required increase. Although plant genes were first isolated only some fourteen years ago and transfer of exogenous DNA into tobacco cells first demonstrated some ten years ago, the application and extension of biotechnology to agricultural problems has already led to the field-testing of genetically modified crop plants. In particular, it is expected that plant biotechnology will contribute significantly to the next 'green revolution'. In fact, the new techniques hold great promise for crop production and protection. Increased productivity through enhanced efficiency of photosynthesis and other physiological processes, improved nutritional and other qualities, and increased resistance or tolerance to biotic and abiotic factors are anticipated through the use of biotechnology. These techniques have already been successfully applied in several areas and in the future will certainly supplement and complement conventional breeding in plant improvement to meet the increasing demands of food production. With the routine transformation of rice and the more recent evidence of transformation of maize the possibility of the world's major cereal crops being modified for improved nutritional value or resistance traits is now feasible in the next few years. In addition, the increasing number of cloned plant genes and the increasing sophistication of our knowledge of the principal developmental and biochemical pathways in plants may permit us to engineer crop plants with higher yields and with less detrimental impact on the environment than now occurs in our current high input agricultural systems. Furthermore, the genetic engineering of plants may also offer significant potential for the agrochemical, food processing, speciality chemical, and pharmaceutical industries in developing new products and manufacturing processes (see Chapter 29 and Gasser and Fraley, 1989).

REFERENCES

Abel, P.P., Nelson, R.S., De, B., Hoffmann, N., Rogers, S.G., Fraley, R.T. and Beachy, R.N. (1986), Delay of disease development in transgenic plants that express the tobacco mosaic virus coat protein gene. *Science*, **232**, 738–743.

Anderson, M.A., Cornish, E.C., Mau, S.L., Williams, E.G., Hoggart, R.M., Atkinson, A., Bonig, I., Grego, B., Simpson, R.J., Roche, P., Haley, J., Penschow, J., Niall, H., Tregear, G., Coghlan, J., Crawford, R. and Clarke, A.E. (1986), Cloning of cDNA for a stylar glycoprotein associated with expression of self-incompatibility in *Nicotiana alata. Nature*, **321**, 38–44.

Anderson, M.A., McFadden, G.I., Bernatzky, R., Atkinson, A., Orpin, T., Dedman, H., Tregear, G., Fernley, R. and Clarke, A.E. (1989), Sequence variability of three alleles of the self-incompatibility gene of *Nicotiana alata. Plant Cell*, **1**, 483–491.

Austin, R.B., Flavell, R.B., Henson, J.E. and Lowe, H.J.B. (1986), *Molecular Biology and Crop Improvement*, Cambridge University Press, Cambridge.

Bartels, D., Schneider, K., Terstappen, G., Piatkowski, D. and Salamini, F. (1990), Molecular cloning of abscisic acid-modulated genes which are induced during desiccation of the resurrection plant *Ceraterostigma plantagineum. Planta*, **181**, 27–34.

Bartels, D., Singh, M. and Salamini, F. (1988), Onset of desiccation tolerance during development of the barley embryo. *Planta*, **75**, 485–492.

Baumann, G., Raschke, E., Bevan, M. and Schöffl, F. (1987), Functional analysis of sequences required for transcriptional activation of a soybean heat shock gene in transgenic tobacco plants. *EMBO J.*, **6**, 1161–1166.

Beachy, R.N., Chen, Z.L., Horsch, R.B., Rogers, S.G., Hoffmann, N.J. and Fraley, R.T. (1985a), Accumulation and assembly of soybean B-conglycinin in seeds of transformed *Petunia* plants. *EMBO J.*, **4**, 3047–3053.

Beachy, R.N., Rogers, S.G. and Fraley, R.T. (1985b), Genetic transformation to confer resistance to plant virus disease. In *Biotechnology in Plant Science. Relevant to Agriculture in the Eighties*, Zaitlin, M., Day, P. and Hollaender, A. (eds), Academic Press, Orlando, pp. 229–247.

Beachy, R.N., Stark, D.M., Deom, C.M. and Fraley, R.T. (1987), Expression of sequences of tobacco mosaic virus in transgenic plants and their role in disease resistance. In *Tailoring Genes for Crop Improvement. An Agricultural Perspective*, Bruening, G., Harada, J. and Kosuge, T. (eds), Plenum Press, New York, pp. 169–180.

Bennetzen, J.L. (1984), Genetic engineering for improved disease resistance. In *Applications of Genetic Engineering to Crop Improvement*, Collins, G.B. and Petolino, J.G. (eds), M. Nijhoff/W. Junk Publishers, Dordrecht, Boston, Lancaster, pp. 491–523.

Bevan, M.W., Mason, S.E. and Goelet, P. (1985), Expression of tobacco mosaic virus coat protein by a cauliflower mosaic virus promoter in plants transformed by *Agrobacterium. EMBO J.*, **4**, 1921–1926.

Bohlmann, H., Clausen, S., Behnke, S., Giese, H., Hiller, C., Reinmann-Phillip, U., Schrader, G., Barkholt, V. and Apel, K. (1988), Leaf-specific thionins of barley – a novel class of cell wall proteins toxic to plant-pathogenic fungi and possibly involved in the defence mechanism of plants. *EMBO J.*, **7**, 1559–1565.

Botterman, J. and Leemans, J. (1988), Engineering of herbicide resistance. *Trends Genet.*, **4**, 219–222.

Broglie, R., Coruzzi, G., Fraley, R.T., Rogers, S.G., Horsch, R.B., Niedermeyer, J.G., Fink, C.L., Flick, J.S. and Chua, N.H. (1984), Light-regulated expression of a pea ribulose-1,5-biphosphate carboxylase small subunit gene in transformed plant cells. *Science*, **224**, 838–843.

Chaleff, R.S. and Ray, T.B. (1984), Herbicide-resistant mutants from tobacco cell cultures. *Science*, **223**, 1148–1151.

Chaleff, R.S., Sebastian, S.A., Creason, G.L., Mazur, B.J., Falco, S.C., Ray, T.B., Mauvais, C.J. and Yadav, M.B. (1987), Developing plant varieties resistance to sulfonylurea herbicides. In *Molecular Strategies for Crop Protection*, Arntzen, C.J. and Rayan, C. (eds), A.R. Liss, New York, pp. 415–425.

Coen, E.S. and Meyerowitz, E.M. (1991), The war of the whorls: genetic interactions controlling flower development. *Nature*, **353**, 31–37.

Coleman, W.H. and Roberts, W.K. (1982), Inhibitors of animal cell-free protein synthesis from grains. *Biochim. Biophys. Acta*, **696**, 239–244.

Collinge, D.B. and Slusarenko, A.J. (1987), Plant gene expression in response to pathogens. *Plant Mol. Biol.*, **9**, 389–410.

Colot, V., Robert, L.S., Kavanagh, T.A., Bevan, M.W. and Thompson, R.D. (1987), Localization of sequences in wheat endosperm protein genes which confer tissue-specific expression in tobacco. *EMBO J.*, **6**, 3559–3564.

Comai, L., Facciotti, D., Hiatt, W.R., Thompson, G., Rose, R.E. and Stalker, D.M. (1985), Expression in plants of a mutant *aroA* gene from *Salmonella typhimurium* confers tolerance to glyphosate. *Nature*, **317**, 741–744.

Comai, L., Sen, L.C. and Stalker, D.M. (1983), An altered *aroA* gene product confers resistance to the herbicide glyphosate. *Science*, **221**, 370–371.

Comai, L, and Stalker, D. (1986), Mechanism of action on herbicides and their molecular manipulation. *Oxford Survey of Plant Molecular and Cell Biology*, **3**, 166–195.

Cuozzo, M., O'Connell, K.M., Kaniewski, W., Fang, R.X., Chua, N.H. and Tumer, N.E. (1988), Viral protein in transgenic tobacco plants expressing the cucumber mosaic virus coat protein or its antisense RNA. *Biotechnology*, **6**, 549–557.

De Block, M., Botterman, J., Vandewiele, M., Dockx, J., Thoen, C., Gossele, V., Rao Movva, N., Thompson, C., Van Montagu, M. and Leemans, J. (1987), Engineering herbicide resistance in plants by expression of a detoxifying enzyme. *EMBO J.*, **6**, 2513–2518.

De Block, M., De Brouwer, D. and Tenning, P. (1986), Transformation of *Brassica napus* and *Brassica oleracea* using *Agrobacterium tumefaciens* and the expression of the *bar* and *neo* genes in the transgenic plants. *Plant Physiol.*, **91**, 694–701.

De Greef, W., Delon, R., De Block, M., Leemans, J. and Botterman, J. (1989), Evaluation of herbicide resistance in transgenic crops under field conditions. *Bio/Technology*, **7**, 61–64.

De Nettancourt, D. (1977), *Incompatiblity in Angiosperms*, Springer-Verlag, Berlin, Heidelberg, New York.

De Rose, R.T., Antony J.L. and Hall, T.C. (1988), Expression of kafirin protein from an unmodified monocot gene sequence in transgenic tobacco seed. *J. Cell. Biochem. Suppl.*, **12**, 179 (Abstr. L 307).

Delannay, X., La Vallee, B.J., Proksch, R.K., Fuchs, R.L., Sims, S.R., Greenplate, J.T., Marrone, P.G., Dodson, R.B., Augustine, J.J., Layton, J.G. and Fischhoff, D.A. (1989), Field performance of transgenic tomato plants expressing the *Bacillus thuringiensis* var. *kurstaki* insect control protein. *Bio/Technology*, **7**, 1265–1269.

Della Cioppa, G., Bauer, S.C., Taylor, M.L., Rochester, D.E., Klein, B.K., Shah, D.M., Fraley, R.T. and Kishore, G.M. (1987), Targeting a herbicide-resistance enzyme from *Escherichia coli* from chloroplasts from higher plants. *Bio/Technology*, **5**, 579–584.

Donn, G., Tischer, E., Smith, J.A. and Goodman, H.M. (1984), Herbicide-resistance alfalfa cells: an example of gene amplification in plants. *J. Mol. Appl. Genet.*, **2**, 621–635.

Ebert, P.R., Ha, S.B. and An, G. (1987), Identification of an essential upstream element in the nopaline synthase promoter by stable and transient assays. *Proc. Natl. Acad. Sci. USA*, **84**, 5745–5749.

Ellis, J.G., Llewellyn, D.J., Dennis, E.S. and Peacock, W.J. (1987), Maize *Adh1* promoter sequences control anaerobic regulation: addition of upstream promoter elements from constitutive genes is necessary for expression in tobacco. *EMBO J.*, **6**, 11–16.

Endo, Y., Tsurugi, K. and Ebert, R.F. (1988), The mechanism of action of barley toxin: a type 1 ribosome-inactivating protein with RNA N-glycosidase activity. *Biochim. Biophys. Acta*, **954**, 224–226.

Facciotti, D., O'Neal, J.K., Lee, S. and Shewmaker, C.K. (1985), Light-inducible expression of a chimeric gene in soybean tissue transformed with *Agrobacterium*. *Bio/Technology*, **3**, 241–246.

Fillatti, J.J., Kiser, J., Rose, R. and Comai, L. (1987), Efficient transfer of a glyphosate tolerance gene into tomato using a binary *Agrobacterium tumefaciens* vector. *Bio/Technology*, **5**, 726–730.

Fischhoff, D.A., Bowdish, K.S., Perlak, F.J., Marrone, P.G., McCormick, S.M., Niedermayer, J.G., Dean, D.A., Kusano-Kretzmer, K., Mayer, E.J., Rochester, D.E., Rogers, S.G. and Fraley, R.T. (1987), Insect tolerant transgenic tomato plants. *Bio/Technology*, **5**, 807–813.

Flügge, I.U. and Heldt, H.W. (1986), Chloroplast phosphate-triose phosphate-phosphoglycerate translocator: its identification, isolation, and reconstitution. *Methods Enzymol.*, **125**, 716–730.

Francki, R.I.B. (1985), Plant virus satellites. *Annu. Rev. Microbiol.*, **39**, 151–174.

Frear, D.S. and Swanson, H.R. (1970), Biosynthesis of S-(4-ethylamino-6-isopropylamino-2-S-triazino) glutathione: partial purification and properties of a glutathione S-transferase from corn. *Phytochemistry*, **9**, 2123–2132.

Gasser, C.S. and Fraley, R.T. (1989), Genetically engineering plants for crop improvement. *Science*, **244**, 1293–1299.

Gatehouse, A.M.R., Barbieri, L., Stirpe, F. and Croy, R.R.D. (1990), Effects of ribosome inactivating proteins on insect development-differences between *Lepidoptera* and *Coleoptera*. *Entomol. Exp. Appl.*, **54**, 43–51.

Gatehouse, A.M.R., Gatehouse, J.A. and Boulter, D. (1980), Isolation and characterization of trypsin inhibitors from cowpea (*Vigna unguiculata*). *Phytochemistry*, **19**, 751–756.

Gerlach, W.L., Llewellyn, D. and Haseloff, J. (1987), Construction of a plant disease resistance gene from the satellite RNA of tobacco ringspot virus. *Nature*, **328**, 802–805.

Goldberg, R.B. (1988), Plants: novel developmental processes. *Science*, **240**, 1460–1467.

Gordon-Kamm, W., Spencer, T.M., Mangano, M.L., Adams, T.R., Daines, R.J., Start, W.G., O'Brien, J.V., Chambers, S.A., Adams, W.R., Willetts, N.G., Rice, T.B., Mackey, C.J., Krueger, R.W., Kausch, A.P. and Lemaux, P.G. (1990), Transformation of maize cells and regeneration of fertile transgenic plants. *Plant Cell*, **2**, 603–618.

Green, P.J., Kay, S.A. and Chua, N.H. (1987), Sequence-specific interactions of a pea nuclear factor with light-responsive elements upstream of the *rbcS-3A* gene. *EMBO J.*, **6**, 2543–2549.

Green, P.J., Pines, O. and Inouye, M. (1986), The role of antisense RNA in gene regulation. *Annu. Rev. Biochem.*, **55**, 569–597.

Harrison, B.D., Mayo, M.A. and Baulcombe, D.C. (1987), Virus resistance in transgenic plants that express cucumber mosaic virus satellite RNA. *Nature*, **328**, 799–802.

Hartings, H., Maddaloni, M., Lazzaroni, N., Di Fonzo, N., Motto, M., Salamini, F. and Thompson, R. (1989), The *O2* gene which regulates zein deposition in maize endosperm encodes a protein with structural homologies to transcriptional activators. *EMBO J.*, **8**, 2795–2801.

Hartley, R.W. (1988), Barnase and barstar expression of its cloned inhibitor permits expression of a cloned ribonuclease. *J. Mol. Biol.*, **202**, 913–915.

Haughn, G.W., Smith, J., Mazur, B. and Somerville, C. (1988), Transformation with a mutant *Arabidopsis* acetolactate synthase gene renders tobacco resistant to sulfonylurea herbicides. *Mol. Gen. Genet.*, **211**, 266–271.

Heidecker, G. and Messing, J. (1986), Structural analysis of plant genes. *Annu. Rev. Plant Physiol.*, **37**, 439–466.

Hemenway, C., Fang, R.X., Kaniewski, W.K., Chua, N.H. and Tumer, N.E. (1988), Analysis of the mechanism of protection in transgenic plants expressing the potato virus X coat protein or its antisense RNA. *EMBO J.*, **7**, 1273–1280.

Herrera-Estrella, L., De Block, M., Messens, E., Harnalsteens, J.P., Van Montagu, M. and Schell, J. (1984), Chimeric genes as dominant selectable markers in plant cells. *EMBO J.*, **2**, 987–995.

Hilder, V.A., Gatehouse, A.M.R., Sheerman, S.E., Barker, R.F. and Boulter, D. (1987), A novel mechanism of insect resistance engineered into tobacco. *Nature*, **330**, 160–163.

Hirschberg, J. and McIntosh, L. (1983), Molecular basis of herbicide resistance in *Amaranthus hybridus*. *Science*, **222**, 1346–1349.

Hoekama, A., Huisman, M.J., Molendijk, L., Van den Elzen, P.J.M. and Cornelissen, B.J.C. (1989), The genetic engineering of two commercial potato cultivars for resistance to potato virus X. *Bio/Technology*, **7**, 273–278.

Hoffman, L.M., Donaldson, D.D., Bookland, R., Rashka, K. and Hermano, E.M. (1987), Synthesis and protein body deposition of maize 15 kD zein in transgenic tobacco seeds. *EMBO J.*, **6**, 3213–3221.

Höfte, H., De Greve, H., Seurinck, J., Jansens, S., Mahillon, J., Ampe, C., Vandekerckhove, J., Vanderbruggen, H., Van Montagu, M., Zabeau, M. and Veack, M. (1986), Structural and functional analysis of a cloned delta endotoxin of *Bacillus thuringiensis berlinger* 1715. *Eur. J. Biochem.*, **161**, 273–280.

Howard, E.A., Walker, J.C., Dennis, E.S. and Peacock, W.J. (1987), Regulated expression of an alcohol deydrogenase 1 chimeric gene introduced into maize protoplasts. *Planta*, **170**, 535–540.

Izant, J.G. and Weintraub, H. (1985), Constitutive and conditional suppression of exogenous and endogenous genes by anti-sense RNA. *Science*, **229**, 345–352.

Jokusch, H. (1968), Two mutants of tobacco mosaic virus temperature sensitive in two different functions. *Virology*, **35**, 94–101.

Kaniewski, W., Lawson, C., Sammons, B., Haley, L., Hart, J., Delannay, X. and Tumer, N.E. (1990), Field resistance of transgenic russett burbank potato to effects on infection by potato virus X and potato virus Y. *Bio/Technology*, **8**, 750–754.

Kaulen, H., Schell, J. and Kreuzaler, F. (1986), Light-induced expression of the chimeric chalcone synthase-NPTII gene in tobacco cells. *EMBO J.*, **5**, 1–8.

Koncz, C. and Schell, J. (1986), The promoter of T_L-DNA gene 5 controls the tissue-specific expression of chimeric genes carried by a novel type of *Agrobacterium* binary vector. *Mol. Gen. Genet.*, **204**, 383–396.

Kung, S.D. and Marsho, T.V. (1976), Regulation of RuDP carboxylase/oxygenase activity and its relationship to plant photorespiration. *Nature*, **259**, 325–326.

La Rossa, R.A. and Schloss, J.V. (1984), The sulfonylurea herbicide sulfometuron methyl is an extremely potent and selective inhibitor of aceto-lactate synthase in *Salmonella typhimurium*. *J. Biol. Chem.*, **259**, 8753–8758.

Lawton, M.A., Tierney, M.A., Nakamura, I., Anderson, E., Komeda, Y., Dube, P., Hoffman, N., Fraley, R.T, and Beachy, R.N. (1987), Expression of a soybean B-conglycinin gene under the control of the cauliflower mosaic virus 35S and 19S promoters in transformed *Petunia* tissues. *Plant Mol. Biol.*, **9**, 315–324.

Leah, R., Tommerup, H., Svensen, I. and Mundy, J. (1991), Biochemical and molecular characterization of three barley seed proteins with antifungal properties. *J. Biol. Chem.*, **266**, 1564–1573.

Leason, M., Cunliffe, D., Parkin, D., Lea, P.J. and Miflin, B.J. (1982), Inhibition of pea leaf glutamine synthetase by methionine sulphoximine, phosphinothricin and other glutamate analogues. *Phytochemisty*, **21**, 855–857.

Lee, K.Y., Townsend, J., Tepperman, J., Black, M., Chui, C.F., Mazur, B., Dunsmuir, P. and Bedbrook, J. (1988), The molecular basis of sulfonylurea herbicide resistance in higher plants. *EMBO J.*, **7**, 1241–1248.

Lyon, B.R., Llewellyn, D.J., Huppatz, J.L., Dennis, E.S and Peacock, W.J. (1989), Expression of a bacterial gene in transgenic tobacco plants confers resistance to the herbicide 2,4-dichlorophenoxyacetic acid. *Plant Mol. Biol.*, **13**, 533–540.

Maddaloni, M., Barbieri, L., Lohmer, S., Motto, M., Salamini, F. and Thompson, R. (1991), Characterization of an endosperm-specific developmentally regulated protein synthesis inhibitor from maize seeds. *J. Genet. Breed.*, **445**, 337–380.

Mariani, C., De Beuckeleer, M., Truettner, J., Leemans, J. and Goldberg, R.B. (1990), Induction of male sterility in plants by a chimaeric ribonuclease gene. *Nature*, **347**, 737–741.

Marris, C., Gallois, P., Copley, J. and Kreis, M. (1988), The 5' flanking regions of a barley hordein gene controls tissue and developmental specific CAT expression in tobacco plants. *Plant Mol. Biol.*, **10**, 359–366.

Matzke, M.A., Susani, M., Binns, A.N., Lewis, E.D., Rubenstein, I. and Matzke, A.J.M. (1984), Transcription of a zein gene introduced into sunflower using a Ti plasmid vector. *EMBO J.*, **3**, 1525–1531.

Mazur, B.J. and Falco, S.C. (1989), The development of herbicide resistant crops. *Annu. Rev. Plant Physiol. Plant Mol. Biol.*, **40**, 441–470.

Mazur, B.J., Chui, C.F. and Smith, J.K. (1987), Isolation and characterization of plant genes coding for acetolactate synthase, the target enzyme for two classes of herbicides *Plant Physiol.*, **85**, 1110–1117.

McClure, B., Haring, V., Ebert, P.R., Anderson, M.A., Simpson, R.J., Sakiyama, F. and Clarke A.E. (1989), Style self-incompatibility gene products of *Nicotiana alata* are ribonucleases. *Nature*, **342**, 955–957.

McClure, B.A., Gray, J.E., Anderson, M.A. and Clarke, A. (1990), Self-incompatibility in *Nicotiana alata* involves degradation of pollen rRNA. *Nature*, **347**, 757–759.

Menczel, L., Polsby, L.S., Steinback, K.E. and Maliga, P. (1986), Fusion-mediated transfer of triazine-resistant chloroplasts. Characterization of *Nicotiana tabacum* cybrid plants. *Mol. Gen. Genet.*, **205**, 201–205.

Meyer, P., Heidmann, I., Forkmann, G. and Saedler, H. (1987), A new petunia flower colour generated by transformation of a mutant with a maize gene. *Nature*, **330**, 677–678.

Morelli, G., Nagy, F., Fraley, R.T., Rogers, S.G. and Chua, N.H. (1985), A short conserved sequence is involved in the light-inducibility of a gene encoding ribulose 1,5-biphosphate carboxylase small subunit of pea. *Nature*, **315**, 200–204.

Motto, M., Maddaloni, M., Ponziani, G., Brembilla, M., Marotta, R., Di Fonzo, N., Soave, C., Thompson, R.D. and Salamini, F. (1988), Molecular cloning of the *O2-m5* allele of *Zea mays* using transposon marking. *Mol. Gen. Genet.*, **212**, 488–494.

Murakami, T., Anzai, H., Imai, S., Satoh, A., Nagaoka, K. and Thompson, C.J. (1986), The bialaphos biosynthetic genes of *Streptomyces hygroscopicus*: molecular cloning and characterization of the gene cluster. *Mol. Gen. Genet.*, **205**, 42–50.

Murant, A.F. and Mayo, M.A. (1982), Satellites of plant viruses. *Ann. Rev. Phytopathol.*, **20**, 4970.

Nagy, F., Bouty, M., Hsu, M-Y, Wong, M. and Chua, N.H. (1987), The proximal region of the wheat *Cab-1* gene contains a 268-bp enhancer-like sequence for phytochrome repsonse. *EMBO J.*, **6**, 2537–2542.

Napoli, C., Lemieux, C. and Jorgensen, R. (1990), Introduction of a chimeric chalcone synthase gene into petunia results in reversible co-suppression of homologous genes *in trans*. *Plant Cell*, **2**, 279–289.

Nasrallah, J.B., Kao, T.H., Goldberg, M.I. and Nasrallah, M.E. (1985), A cDNA clone encoding an *S*-locus-specific glycoprotein from *Brassica oleracea*. *Nature*, **318**, 263–267.

Nasyrov, Y.F. (1978), Genetic control of photosynthesis and improving crop productivity. *Annu. Rev. Plant Physiol.*, **29**, 215–237.

Nelson, O.E. (1979), Genetic modification of protein quality in plants. *Adv. Agron.*, **21**, 171–194.

Nelson, R.S., McCormick, S.M., Delannay, X., Dubé, P., Layton, J., Anderson, E.J., Kaniewska, M., Proksch, R.K., Horsch, R.B., Rogers, S.G., Fraley, R.T. and Beachy, R.N. (1988), Virus tolerance, plant growth, and field performance of transgenic tomato plants expressing coat protein from tobacco mosaic virus. *Bio/Technology*, **6**, 403–409.

Owens, R.A. and Diener, T.O. (1981), Sensitive and rapid diagnosis of potato spindle tuber viroid disease by nucleic acid hybridization. *Science*, **213**, 670–672.

Payne, P.I. (1986), Genetics of wheat storage proteins and the effect of allelic variation on breadmaking quality. *Annu. Rev. Plant Physiol.*, **38**, 141–153.

Payne, P.J., Holt, L.M., Jackson, E.A. and Law, C.N. (1984), Wheat storage protein: their genetics and their potential for manipulation by plant breeding. *Philos. Trans. R. Soc. Lond.*, **304**, 359–371.

Po, T., Zhang, X.H., Qui, B.S. and Wu, G.S. (1987), Satellite RNA for the control of plant diseases caused by cucumber mosaic virus. *Ann. Appl. Biol.*, **111**, 143–152.

Pryor, T. (1987), The origin and structure of fungal disease resistance genes in plants. *Trends Genet.*, **3**, 157–161.

Ray, T.B. (1984), Site of action of chlorsulfuron. Inhibition of valine and isoleucine biosynthesis in plants. *Plant Physiol.*, **75**, 827–831.

Register, J.C. and Beachy, R.N. (1988), Resistance to TMV in transgenic plants results from interference with an early event in infection. *Virology*, **166**, 524–532.

Robert, L., Thompson, R.D. and Flavell, R.B. (1989), Tissue specific expression of a wheat high-molecular weight glutenin gene in transgenic tobacco. *Plant Cell*, **1**, 569–578.

Sacchi, V.F., Parenti, P., Hanozet, G.M., Giordanda, B., Lutly, P. and Wolfersberger, M.G. (1986), *Bacillus thuringiensis* toxin inhibits K^+-gradient-dependent amino acid transport across the brush border membrane of *Pieris brassicae* midgut cells. *FEBS Lett.*, **204**, 213–218.

Sachs, M.M. and Ho, T.H.D. (1986), Alteration of gene expression during environmental stress in plants. *Annu. Rev. Plant Physiol.*, **37**, 363–376.

Sanchez-Serrano, J.J., Keil, M., O'Connor, A., Schell, J. and Willmitzer, L. (1987), Wound-induced expression of a potato proteinase inhibitor II gene in transgenic tobacco 1 plants. *EMBO J.*, **6**, 303–306.

Sanders, P.R., Winter, J.A., Barnason, A.R., Rogers, S.G. and Fraley, R.T. (1987), Comparison of cauliflower mosaic virus 35S and nopaline synthase promoters in transgenic plants. *Nucleic. Acids Res.*, **15**, 1543–1558.

Sandmann, G. and Bramley, P.M. (1985), Carotinoid biosynthesis by *Aphanocapsa* homogenates coupled to a phytoene-generating system from *Phycomyces blakesleeanus*. *Planta*, **164**, 259–263.

Scheidegger, A. (1990), Plant biotechnology goes commercial in Japan. *Trends Biotechnol.*, **8**, 197–198.

Schlumbaum, A., Mauch, F., Vögeli, V. and Boller, T. (1986), Plant chitinases are potent inhibitors of fungal growth. *Nature*, **324**, 365–367.

Schmidt, R.J., Burr, F.A. and Burr, B. (1987), Transposon tagging and molecular analysis of the maize regulatory locus *opaque-2*. *Science*, **238**, 960–963.

Schmülling, T., Schell, J. and Spena, A. (1988), Single genes from *A. rhizogenes* influence plant development. *EMBO J.*, 7, 2621–2629.

Schwarz-Sommer, Z., Huijser, P., Nacken, W., Saedler, H. and Sommer, H. (1990), Genetic control of flower development by homeotic genes in *Antirrhinum majus*. *Science*, **250**, 931–936.

Sengupta-Gopalan, C., Reichert, N.A., Barker, R.F., Hall, T.C. and Kemp, J.D. (1985), Developmentally regulated expression of the bean B-phaseolin gene in tobacco seed. *Proc. Natl. Acad. Sci. USA*, **82**, 3320–3324.

Shah, D.P., Horsch, R.B., Klee, H.I., Kishore, G.M., Winter, J.A., Tumer, N.E., Hironaka, C.M., Sanders, P.R., Gasser, C.S., Aykent, S., Siegel, N.R., Rogers, S.G. and Fraley, R.T. (1986), Engineering herbicide tolerance in transgenic plants. *Science*, **233**, 478–481.

Shaner, D.L. and Anderson, P.C. (1985), Mechanism of action of the imidazolinones and cell culture selection of tolerant maize. In *Biotechnology in Plant Science. Relevance to Agriculture in the Eighties*, Zaitlin, M., Day, P. and Hollaender, A. (eds), Academic Press, New York, pp. 287–299.

Shewry, P.R., Tatham, A.S., Field, J.M., Forde, B.G., Clark, J., Gallois, P., Marris, P., Haiford, N., Forde, J. and Kreis, M. (1988), The structure of barley and wheat prolamins and their genes. *Biochem. Physiol. Pflanz.*, **183**, 117–127.

Shotwell, M.A. and Larkins, B.A. (1989), The biochemistry and molecular biology of seed storage proteins. *Biochem. Plants*, **15**, 297–345.

Smart, C.C., Johänning, G., Müller, G. and Amrhein,

N. (1985), Selective overproduction of 5-enol-pyruvylshikimic acid 3-phosphate synthase in a plant cell culture which tolerates high doses of the herbicide glyphosphate. *J. Biol. Chem.*, **260**, 16338–16346.

Spena, A., Hain, R., Ziervogel, U., Saedler, H. and Schell, J. (1985), Construction of a heat-inducible gene for plants. Demonstration of heat-inducible activity of the *Drosophila* hsp70 promoter in plants. *EMBO J.*, **4**, 2739–2743.

Spena, A. and Schell, J. (1987), The expression of a heat-inducible chimeric gene in transgenic tobacco plants. *Mol. Gen. Genet.*, **206**, 436–440.

Spena, A., Shmülling, T., Koncz, C. and Schell, J.S. (1987), Independent and synergistic activity of *rol A, B* and *C* loci in stimulating abnormal growth in plants. *EMBO J.*, **6**, 3891–3899.

Spencer, C.A., Gietz, R.D. and Hodgetts, R.B. (1986), Overlapping transcription units in the dopa decarboxylase region of *Drosophila*. *Nature*, **322**, 279–281.

Stalker, D.M., Hiatt, W.R. and Comai, L. (1985), A single amino acid substitution in the enzyme 5-enolpyruvylshikimate-3-phosphate synthase confers resistance to the herbicide glyphosate. *J. Biol. Chem.*, **260**, 4724–4728.

Stalker, D.M., McBride, K.E. and Malyi, L.D. (1988), Herbicide resistance in transgenic plants expressing a bacterial detoxification gene. *Science*, **242**, 419–423.

Steinrücken, H. and Amrhein, N. (1980), The herbicide glyphosate is a potent inhibitor of 5-enolpyruvylshikimic acid-3-phosphate synthase. *Biochem. Biophys. Res. Commun.*, **94**, 1207–1212.

Steinrücken, H., Schulz, A., Amrhein, N., Porter, C.A. and Fraley, R.T. (1986), Overproduction of 5-enolpyruvylshikimate-3-phosphate synthase in a glyphosate-tolerant *Petunia hybrida* cell line. *Arch. Biochem. Biophys.*, **244**, 169–178.

Streber, W.R. and Willmitzer, L. (1989), Transgenic tobacco plants expressing a bacterial detoxifying enzyme are resistant to 2,4-D. *Bio/Technology*, **7**, 811–816.

Thompson, C., Movva, N., Tizard, R., Crameri, R., Davies, J., Lauwereys, M. and Botterman, J. (1987), Characterization of the herbicide-resistance gene *bar* from *Streptomyces hygroscopicus*. *EMBO J.*, **6**, 2519–2523.

Trebst, A. (1987), The three-dimensional structure of the herbicide binding niche on the reaction center polypeptides of photosystem II. *Z. Naturforsch.*, **42c**, 742–750.

Tumer, N.E., O'Connell, K.M., Nelson, R.S., Saunders, P.R., Beachy, R.N., Fraley, R.T. and

Shah, D.M. (1987), Expression of alfalfa mosaic virus coat protein gene confers cross-protection in transgenic tobacco and tomato plants. *EMBO J.*, **6**, 1181–1188.

Vaeck, M., Reynaerts, A., Höfte, H., Jansens, S., De Beuckeleer, M., Dean, C., Zabeau, M., Van Montagu, M. and Leemans, J. (1987), Transgenic plants protected from insect attack. *Nature*, **328**, 33–37.

Van der Krol, A.R., Lenting, P.E., Veenstra, J., Van der Meer, I.M., Koes, R.E., Gerats, A.G.M., Mol, J.N.M. and Stuitje, A.R. (1988), An anti-sense chalcone synthase gene in transgenic plants inhibits flower pigmentation. *Nature*, **333**, 866–869.

Van der Krol, A.R., Mur, L.A., Beld, M., Mol, J.N.M. and Stuitje, A.R. (1990), Flavonoid genes in petunia: addition of a limited number of gene copies may lead to a suppression of gene expression. *Plant Cell*, **2**, 291–299.

Van Dun, C.M.P., Bol, J.F. and Van Vloten-Doting, L. (1987), Expression of alfalfa mosaic virus rattle and tobacco virus coat protein genes in transgenic tobacco plants. *Virology*, **159**, 299–305.

Vanderplank, J.E. (1982), *Host–Pathogen Interactions in Plant Disease*, Academic Press, New York, London.

Vollbrecht, E., Veit, B., Sinha, N. and Hake, S. (1991), The developmental gene *Knotted-1* is a member of a maize homeobox gene family. *Nature*, **350**, 241–243.

Walrud, L.S., Perlak, F.J., Tran, M.T., Kusano, K., Mayer, E.J., Miller-Wideman, M.A., Obukowicz, M.G., Nelson, D.R., Kreitinger, J.P. and Kaufman, R.J. (1985), Cloning of the *Bacillus thuringiensis* subsp. *kurstaki* delta endotoxin gene into *Pseudomonas fluorescens*: molecular biology and ecology of an engineered microbial pesti-cide. In *Engineered Organisms in the Environment*, Halvarson, H.O., Pramer, D. and Rogul, M. (eds), American Society for Microbiology, Washington, pp. 40–46.

Weintraub, H., Izant, J.G. and Harland, R.M. (1985), Anti-sense RNA as a molecular tool for genetic analysis. *Trends Genet.*, **1**, 22–25.

Weising, K., Schell, J. and Kahl, G. (1988), Foreign genes in plants: transfer, structure, expression and applications. *Annu. Rev. Genet.*, **22**, 421–477.

Wessler, S. and Hake, S. (1990), Maize harvest. *Plant Cell*, **2**, 495–499.

Wiegand, R.C., Shah, D.M., Mozer, T.J., Harding, E.I., Diaz-Collier, J., Saunders, C., Jaworski, E.G. and Tiemeier, D.C. (1986), Messenger RNA encoding a glutathione-S-transferase responsible for herbicide tolerance in maize is induced in response to safener treatment. *Plant Mol. Biol.*, **7**, 235–243.

Williams, T. and Fried, M. (1986), A mouse locus at which transcription from both DNA strands produces mRNA complementary at their 3' ends. *Nature*, **322**, 275–279.

Woodson, W.R. (1987), Changes in protein and mRNA populations during the senescence of carnation petals. *Physiol. Plant.*, **71**, 495–502.

Yang, M.S., Espinoza, N.O., Nagpala, P.G., Dodds, J.H., White, F.F., Schnorr, K.L. and Jaynes, J.M. (1989), Expression of a synthetic gene for improved protein quality in transformed potato plants. *Plant Sci.*, **64**, 99–111.

Zurawski, G., Bohnert, H.J., Whitfield, P.R. and Bottomlay, W. (1982), Nucleotide sequence of the gene for the M 32,000 thylakoid membrane protein from *Spinacia oleracea* and *Nicotiana debneyi* predicts a totally conserved primary translation product of M 38,950. *Proc. Natl. Acad. Sci. USA*, **79**, 7699–7703.

Part Four

Assessment of Variation

11

Biometrical genetics in breeding

M.J. Kearsey

11.1 NATURE AND SIGNIFICANCE OF QUANTITATIVE TRAITS

Plant breeders are concerned with a wide range of traits in their crop plants. Some of these, such as incompatibility or certain types of disease resistance, may be controlled by just one or two genes whose effects are relatively large. Many, and possibly most traits, however, are complex, being under the control of a number of genes as well as being considerably influenced by the environment. Characters such as yield, quality and maturity date are typical examples. These characters generally show no clear discontinuity between genotypes, the range of appearance of one overlapping that of others so extensively as to give the impression of a continuous distribution.

In general it is assumed that several genes are responsible for the variation in these traits, such genes being referred to as polygenes or quantitative trait loci (QTLs). However, we need not exclude the possibility that only one or a few genes are involved in any particular instance providing the phenotypic differences between the genotypes at those genetic loci are small relative to the non-genetic, environmental, variation within them. Indeed, as we shall see, much of the theoretical framework to the subject has been developed from single gene models.

There is no reason to believe that the genes involved in such traits are in any way different

Plant Breeding: Principles and prospects. Edited by M.D. Hayward, N.O. Bosemark and I. Romagosa. Published in 1993 by Chapman & Hall, London. ISBN 0 412 43390 7.

from the so-called major genes of classical genetics. Any one gene may have many allelic states. Some of the alleles of a given gene may have such marked effects as to be clearly recognizable as a classical major mutant. Other alleles, though potentially separable at the DNA level, may well cause only minor differences at the level of the external phenotype. Thus one allele at a locus involved with growth hormone production could be inactive and result in a dwarf plant, whilst others may simply reduce or increase height by a few percent.

The genes underlying the extensive electrophoretic polymorphism identified by population geneticists have just the properties expected to produce quantitative traits. Thus they have two or more alleles at a locus at high frequencies in natural populations and the differences between the alleles in the biochemical properties of their proteins are small, i.e. they may differ slightly in pH or temperature optimum or K_m (Place and Powers, 1979). Such loci also have non-functional alleles. We should not, however, give the impression that quantitative variation is solely due to minor allelic variation in structural genes; regulatory genes, too, no doubt also contribute to this variation. Indeed, with complex traits such as yield, virtually all the genes of the plant could have an effect, so the final phenotypic variation could be due to small allelic variants at a large number of loci whose immediate function is far removed from the trait itself.

We thus expect the polygenes to show all the typical properties of chromosomal genes both in terms of action and in transmission through meiosis. The area of genetics concerned with the inheritance of these quantitative traits is called 'Biometrical Genetics'. For a more comprehensive account of the theory, the reader should consult Mather and Jinks (1982) and Falconer (1989).

11.2 IMPORTANT QUESTIONS FOR BREEDING

Plant breeders are principally interested in improving the genetic potential of their crop in order to maximize economic gain per unit of input. To this end they require information that will enable them to identify desirable genotypes efficiently, select them and concentrate their genes in a strain, line or variety that is commercially acceptable.

At the start of a breeding programme one has either a population or collection of lines to be used as the base material together with the target traits to be improved. Clearly an important initial question is exactly what type of cultivar one is aiming to produce; an inbred line, an F_1 hybrid, a synthetic population, etc.? The answer to this depends both on the biology, particularly the breeding system of the plant, and, more significantly, on the nature of the genetical control of the characters to be improved. The breeding system can to a large extent be altered whereas the genetic control of traits can not; one can force natural outbreeders such as tobacco and tomatoes to inbreed and inbreeders like barley to outbreed, but it is not easy to change the action and interaction of the polygenes. For this reason, decisions on the type of cultivar should be largely determined by the genetic architecture of the trait, in particular the nature and extent of dominance and gene interaction, and only rarely on the breeding system.

Given a decision on the type of cultivar, it is then necessary to consider the optimum improvement strategy. Should one go for mass selection, or family selection? If family selection, then what type of families, inbred, outbred, full-sib or half-sib? Should one inbreed by single seed descent or use pedigree selection? Should one select at every generation in an inbreeding programme or only at the end? Such decisions, again, require a detailed knowledge of the extent to which the character is genetically, as opposed to environmentally, controlled.

The choice of which particular genotypes to use as parents also creates difficulties. Thus, individual phenotypes may be poor guides to genotypic values because of the influence of the environment. On the other hand, two inbred lines which look similar may in fact be very different genetically and, thus, be capable of yielding useful transgressive segregants. How can we predict the distribution of new, recombinant inbred lines and calculate the proportion which are better than some established, reference variety? If we were to cross these lines, what would their F_1s look like? How could we predict the yield of two and three-way crosses? It is thus essential to have a basic, simple methodology that will provide the breeder with adequate information to make informed choices between genotypes.

Once individuals have been chosen, it is necessary to look at the likely response one will obtain from each cycle of selection and, if possible, predict the likely long- and short-term limits to continued selection. The rate of response will depend on the nature and extent of the genetical variation underlying the trait and on the selection procedure. The limits to response are largely influenced by the number of polygenes responsible for the genetical variation and their linkage relationships. Thus, information on the number and chromosomal distribution of the genes would be essential. Where inbred or hybrid cultivars are the long-term objective, it is important to be able to predict the likely performance of second cycle inbreds and hybrids from information based on material produced from simple crosses of existing inbreds.

Breeders regularly have to improve two or more traits simultaneously, such as yield and quality or height and resistance to lodging. Often such traits show a correlated response so that although yield may improve, quality may decline. It is necessary, therefore, to understand the genetic basis for these correlations before one can predict how the problem can be ameliorated or overcome; i.e. are they due to gene linkage or pleiotropy?

We see that within a breeding programme aimed at quantitative traits, there are many levels at which it is important to have a knowledge of the nature and extent of their genetical and environmental control. Various levels of understanding may be required: (a) is the character inherited? (b) how much variation is genetic? (c) what is the nature of the genetic variation? (d) how is the genetic variation organized? We will now turn to consider ways in which this information can be obtained although in a short chapter we can do no more than consider the basic principles.

11.3 BASIC FEATURES AND PROBLEMS OF POLYGENIC TRAITS

Mendel, and most other geneticists since, have studied major gene differences. In such situations there are few genotypes whereas the phenotypic differences between them are clear and relatively unambiguous. With quantitative traits, on the other hand, there are many genotypes and there are no clear phenotypic differences between them. There is, in fact, a continuous range of variation as illustrated by the F_2 in Fig. 11.1. This leads to a great many

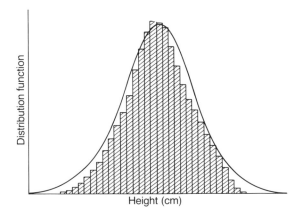

Fig. 11.1. Typical distribution of individuals in a population for a quantitative trait, plant height.

difficulties when studying such traits as the following list shows.

1. Many different genotypes can have the same phenotype. Consider the case of k genes, all having an equal effect on a character. Let us further suppose that there are two alleles and, hence three genotypes, at each locus and that they exhibit co-dominance, i.e. neither allele is dominant. There will thus be a total of 3^k genotypes. For instance, with $k = 3$ we have the following genotypes and phenotypes given that each A, B and C allele adds one unit to the phenotype:

No.	Genotype	Phenotype
(1)	*AABBCC*	6
(2)	*AABBCc*	5
(3)	*AABBcc*	4
(4)	*AABbcc*	5
(5)	*AABbCc*	4
(6)	*AABbcc*	3
(7)	*AAbbCC*	4
.	.	.
.	.	.
.	.	.
(27)	*aabbcc*	0

Although we have 27 genotypes, many have the same phenotype and hence there are only seven phenotypes (0, 1, 2, 3, 4, 5, 6). Thus No(2) looks like No(4), No(3) looks like No(7), etc. Therefore, there is no strict one to one relationship between genotype and phenotype, i.e.,

1 gene	$\longrightarrow$	3 genotypes = 3 phenotypes
2 genes	$\longrightarrow$	9 genotypes = 5 phenotypes
3 genes	$\longrightarrow$	27 genotypes = 7 phenotypes
n genes	$\longrightarrow$	3^n genotypes = $2n + 1$ phenotypes

2. Dominance (allelic-interaction). As with classical genetics, dominance can obscure the true genotype. But the situation can be more complex since both the direction and amount of dominance can vary from gene to gene. That is, A could be dominant to a while b is dominant to B. This will be discussed later.

3. Environmental variation. Not all differences between plants are due to genes and in fact the major part may be due to the environ-

ment, particularly in an agricultural context. Moreover, the genotype and environment may interact to further obscure the genetical effects.

4. Epistasis (non-allelic interaction). When we consider several genes acting together it is possible that the effect of the genotype at one locus may depend on the genotype at another. Consider the two situations in Table 11.1 which shows a trait controlled by

Table 11.1. The effect of the genotype at the *B/b* locus on genotypic differences at the *A/a* locus in (I) the absence and (II) the presence of epistasis. The numbers in the table represent genotypic values in arbitrary units.

(I) *No epistasis i.e. additive*

	aa	*Aa*	*AA*	Genotypic difference *AA − aa*
bb	1	2	3	3 − 1 = 2
Bb	2	3	4	4 − 2 = 2
BB	3	4	5	5 − 3 = 2

(II) *Epistasis i.e. non-additive*

	aa	*Aa*	*AA*	*AA − aa*
bb	1	4	9	9 − 1 = 8
Bb	4	9	16	16 − 4 = 12
BB	9	16	25	25 − 9 = 16

two genes, *A/a*, *B/b*. Table 11.1(I) shows that the change in phenotype when an *AA* genotype is replaced with an *aa* genotype is the same whatever the genotype at the *B/b* locus, i.e. there is no interaction. In Table 11.1(II) on the other hand, the effect of replacing *AA* with *aa* varies considerably with the *B/b* locus, i.e. there is gene interaction. This will cause problems if we wish to make predictions, e.g. of response to selection.

It would appear that there are many difficulties in studying the genetics of quantitative traits. However, because of their widespread occurrence and practical importance, much effort has been devoted to developing methods for their analysis. The following is intended as a basic introduction to these methodologies, illustrating the approach through realistic though genetically simple situations.

11.4 SIMPLE MODELS TO MEANS

In order to make some progress, it is necessary to consider the likely properties of the genes we wish to study. Unfortunately there is not a great deal known about them, so we will assume at the start that they share the properties of any other gene. Thus: (i) polygenes on the same chromosome will show linkage to each other and to other genes; (ii) polygenes on different chromosomes will assort independently at meiosis; (iii) any two alleles may show complete dominance through to no dominance (co-dominance) with respect to each other, but not overdominance, i.e. the heterozygote *Aa* lies in the range $AA \geqslant Aa \geqslant aa$; additionally, (iv) the individual genic effects are small compared to the environmental.

Let us start with two different inbred, homozygous, parental lines, P_1 and P_2, from which we can produce the F_1 and F_2 generations. If we were to grow a very large number of plants of each of these four families in a suitably randomized and replicated trial and score them for some character such as height, we might obtain data similar to that in Fig. 11.1 and Table 11.2.

Table 11.2. Means and variances of the basic generations for height from a trial with complete individual randomization

Generation	Number of Individuals	Mean	Variance
P_1	50	69.6	48.6
P_2	50	68.4	40.1
F_1	100	89.6	53.0
F_2	200	79.5	97.3
B_1	200	77.4	66.5
B_2	200	78.3	84.6

We see that these generations differ in their means (average scores) and their variances. Thus, although the two parents have similar average scores ($P_1 = 69.6$; $P_2 = 68.4$), the F_1 is considerably higher scoring (89.6) and the F_2 is in between (79.5). If we look at the variances we see that the parents and F_1 all have a similar variance (48.675 on average) while the F_2

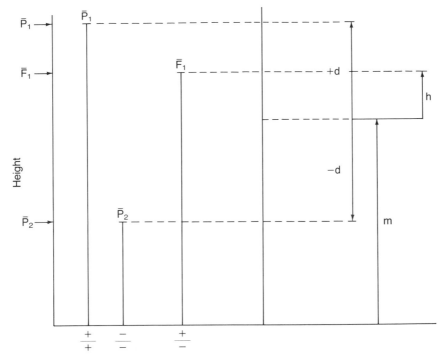

Fig. 11.2. Average phenotypes of two parental inbred lines ($\bar{P}_1$ and $\bar{P}_2$) and their $\bar{F}_1$ for a quantitative trait together with the basic genetic model assuming the means are controlled by a single gene.

variance (97.3) is much larger. How can we set about constructing genetical and environmental models to these means and variances?

Consider two lines which differ by a single gene for height (Fig. 11.2). We will use the convention that P_1 is always the highest scoring parent and refer to the allele which has the increasing (+) effect on the trait as A and the alternative, decreasing (−), allele as a, i.e.,

$$P_1 = AA = +/+$$
$$P_2 = aa = -/-$$
$$F_1 = Aa = +/-$$

The average score of the two parents is used as a reference point, m, and this parameter reflects the combined effects of all those other genes which the two parents share in common. All generation means can then be expressed as deviations from this as shown in Fig. 11.2. Thus,

$$\bar{P}_1 = m + d$$
$$\bar{P}_2 = m - d$$

$$\bar{F}_1 = m + h$$

where the right-hand sides of these equations are the genetic values of each genotype. We could estimate m, d, and h by making comparisons among the means as follows:

$$m = \frac{\bar{P}_1 + \bar{P}_2}{2} = \text{mid-parent}$$

$$d = \frac{\bar{P}_1 - \bar{P}_2}{2}$$

$$h = \bar{F}_1 - m$$

The ratio h/d (the dominance ratio) gives a measure of the degree of dominance of the alleles at that locus, i.e.,

$$h/d = 1 \quad A \text{ completely dominant to } a$$
$$h/d = -1 \quad a \text{ completely dominant to } A$$
$$h/d = 0 \quad \text{no dominance (co-dominance)}$$

i.e. h can take sign, either + or − and, in the

absence of overdominance, h/d lies between $+1.0$ and -1.0.

Now consider two lines differing at two genes (g_1, g_2) and their F_1. There are two possibilities, I and II as follows:

	P_1		P_2		F_1	
	g_1	g_2	g_1	g_2	g_1	g_2
(I)	$+$	$+$	$-$	$-$	$+$	$+$
	$+$	$+$	$-$	$-$	$-$	$-$
	$m + d_1 + d_2$		$m - d_1 - d_2$		$m + h_1 + h_2$	
(II)	$+$	$-$	$-$	$+$	$+$	$-$
	$+$	$-$	$-$	$+$	$-$	$+$
	$m + d_1 - d_2$		$m - d_1 + d_2$		$m + h_1 + h_2$	

The genetic values given assume that the genes behave in an additive manner and do not interact. From the above we notice three things: (1) the F_1s are the same in both I and II; (2) the mid-parents are the same in both I and II; (3) the difference between the two parents varies. Indeed if $d_1 = d_2$ then in the second case both parents have the same phenotype.

Thus we have to take account of how the genes are distributed in the two parents. For example, consider two parents (P_1 and P_2) differing at three loci. There are several possibilities:

	P_1			P_2			F_1		
	g_1	g_2	g_3	g_1	g_2	g_3	g_1	g_2	g_2
(I)	$+$	$+$	$+$	$-$	$-$	$-$	$+$	$+$	$+$
	$+$	$+$	$+$	$-$	$-$	$-$	$-$	$-$	$-$
	$m + d_1 + d_2 + d_3$			$m - d_1 - d_2 - d_3$			$m + h_1 + h_2 + h_3$		
(II)	$+$	$+$	$-$	$-$	$-$	$+$	$+$	$+$	$-$
	$+$	$+$	$-$	$-$	$-$	$+$	$+$	$+$	$-$
	$m + d_1 + d_2 - d_3$			$m - d_1 - d_2 + d_3$			$m + h_1 + h_2 - h_3$		
(III)	$+$	$-$	$+$	$-$	$+$	$-$	$+$	$-$	$+$
	$+$	$-$	$+$	$-$	$+$	$-$	$-$	$+$	$-$
	$m + d_1 - d_2 + d_3$			$m - d_1 + d_2 - d_3$			$m + h_1 + h_2 + h_3$		
(IV)	$-$	$+$	$+$	$+$	$-$	$-$	$-$	$+$	$+$
	$-$	$+$	$+$	$+$	$-$	$-$	$+$	$-$	$-$
	$m - d_1 + d_2 + d_3$			$m + d_1 - d_2 - d_3$			$m + h_1 + h_2 + h_3$		

Clearly there are many possible arrangements and thus the scores of P_1 and P_2 depend not only on the d's but on how the genes are distributed. So in general P_1 and P_2 are defined as follows:

$$P_1 = m + [d]$$
$$P_2 = m - [d]$$

where $[d] = r_d\Sigma d$ and r_d is a measure of how the genes are distributed.

Thus, if P_1 has all $+/+$ loci, and hence P_2 has all $-/-$ loci, then $r_d = 1.0$. However, if P_1 has half the $+/+$ loci and P_2 has the other half $r_d = 0.0$, or if $q =$ proportion of $+$ alleles in P_1 $r_d = (2q - 1)$; i.e.: if $q = 1$, $r_d = 1$; if $q = 0.5$, $r_d = 0$. The mid parent:

$$\frac{\bar{P}_1 + \bar{P}_2}{2} = \frac{(m+[d]) + (m-[d])}{2} = m$$

$$F_1 = m + h_1 + h_2 + \ldots + h_k = m + \Sigma h = m + [h]$$

However, with ambidirectional dominance, the hs can take sign.

Let us now consider the F_2:

	F_1		
	$+/-$ (or A/a)		
	self		
	F_2		
	$+/+$	$+/-$	$-/-$
or	A/A	A/a	a/a
frequency	$\frac{1}{4}$	$\frac{1}{2}$	$\frac{1}{4}$
genetic value	$m + d$	$m + h$	$m - d$

i.e. mean of F_2, $(\bar{F}_2) = \frac{1}{4}(m+d) + \frac{1}{2}(m+h) + \frac{1}{4}(m-d) = m + \frac{1}{2}h$

Now for many genes, providing they do not interact:

$$\bar{F}_2 = m + \frac{1}{2}h_1 + \frac{1}{2}h_2 + \ldots + \frac{1}{2}h_k$$
$$= m + \frac{1}{2}\Sigma h = m + \frac{1}{2}[h]$$

Similarly for the F_3:

	F_2		
	$\frac{1}{4}AA$	$\frac{1}{2}Aa$	$\frac{1}{4}aa$
F_3	AA	Aa	aa
frequency	$\frac{3}{8}$	$\frac{1}{4}$	$\frac{3}{8}$
genetic value	$m + d$	$m + h$	$m - d$

Therefore $\bar{F}_3 = \frac{3}{8}(m+d) + \frac{1}{4}(m+h) + \frac{3}{8}(m-d) = m + \frac{1}{4}h$, or for many genes:

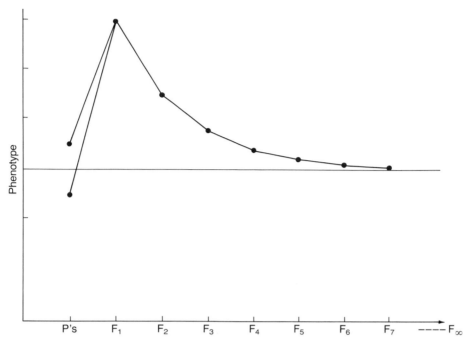

Fig. 11.3. Mean scores of parents, F_1 and subsequent selfed generations assuming a simple additive dominance model of genetical control to illustrate hybrid superiority followed by inbreeding depression.

$$\bar{F}_3 = m + \tfrac{1}{4}h_1 + \tfrac{1}{4}h_2 + \ldots + \tfrac{1}{4}h_k$$
$$= m + \tfrac{1}{4}\Sigma h = m + \tfrac{1}{4}[h]$$

As we go down the selfing series $F_4, F_5, \ldots$, F_∞ the mean depends on the proportion of loci heterozygous (all in F_1, $\tfrac{1}{2}$ in F_2, $\tfrac{1}{4}$ in F_3, etc.). Thus the coefficient of $[h]$ halves each generation and thus the mean changes as shown in Fig. 11.3, providing there is no epistasis. This decline in the mean is often referred to as 'inbreeding depression'.

Before leaving generation means, there are two further generations we need to consider, namely the backcrosses of the F_1 to both parents, B_1 and B_2. The means of these generations are derived in Table 11.6 in the same way as before giving:

$$\bar{B}_1 = m + \tfrac{1}{2}[d] + \tfrac{1}{2}[h]$$
$$\bar{B}_2 = m - \tfrac{1}{2}[d] + \tfrac{1}{2}[h]$$

The data illustrated in Table 11.2 show considerable hybrid vigour (heterosis), with the F_1 much higher yielding than either parent. Using simply the means of the parents and F_1

we can obtain the following parameter estimates:

$$m = \tfrac{1}{2}(\bar{P}_1 + \bar{P}_2) = 69.00$$
$$[d] = \tfrac{1}{2}(\bar{P}_1 - \bar{P}_2) = 0.60$$
$$[h] = \bar{F}_1 - m = 20.60$$

Thus $[h]$ is large and positive, whereas $[d]$ is almost zero. Since the parents must differ genetically or there could be no heterosis, this implies that the coefficient of gene dispersion, r_d, is almost zero.

We cannot deduce anything about the amount of dominance since the ratio $[h]/[d]$ (= $[\Sigma h]/[r_d \Sigma d]$) is inflated by the gene dispersion, r_d. We could, however, use the estimated parameters to predict other generations such as the mean of the recombinant, F_∞ inbred lines, but in order to do this with any confidence we need to be sure that the genes are behaving additively and do not interact. How can we test for this?

The obvious way is to predict the means of

generations that are known, since in these cases we are able to test the predictions. For example, with the data in Table 11.2, we could predict the F_2 and backcross means as follows, where ϵF_2 refers to the expected or predicted value of the F_2, etc.

$$\epsilon F_2 = m + \tfrac{1}{2}[h] = 69.0 + 20.6/2 = 79.30$$

while observed $\bar{F}_2 = 79.54$

Similarly:

$$\epsilon \bar{B}_1 = m + \tfrac{1}{2}[d] + \tfrac{1}{2}[h] = 79.60$$

observed $\bar{\beta}_1 = 77.40$

$$\epsilon \bar{B}_2 = m - \tfrac{1}{2}[d] + \tfrac{1}{2}[h] = 79.00$$

observed $\bar{B}_2 = 78.30$

Clearly, in this example, we can predict other generation means fairly accurately. However we need a test to see if the agreement between the observed and expected generation means is adequate given the experimental error or whether the differences are significant as could occur with epistasis. Such tests of gene interaction can be achieved by 't' tests as follows using the formulae given above to estimate m, $[d]$ and $[h]$. Thus for the F_2:

$$\begin{aligned}\epsilon F_2 &= m + \tfrac{1}{2}[h] \\ &= \tfrac{1}{2}(\bar{P}_1 + \bar{P}_2) + \tfrac{1}{2}\{\bar{F}_1 - \tfrac{1}{2}(\bar{P}_1 + \bar{P}_2)\} \\ &= \tfrac{1}{4}\bar{P}_1 + \tfrac{1}{2}\bar{F}_1 + \tfrac{1}{4}\bar{P}_2\end{aligned}$$

We can now compare the observed F_2 mean with its predicted value (ϵF_2):

$$\begin{aligned}\text{obs}\bar{F}_2 - \epsilon F_2 &= \bar{F}_2 - (\tfrac{1}{4}\bar{P}_1 + \tfrac{1}{2}\bar{F}_1 + \tfrac{1}{4}\bar{P}_2) \\ 4\text{obs}\bar{F}_2 - 4\epsilon F_2 &= 4\bar{F}_2 - (\bar{P}_1 + 2\bar{F}_1 + \bar{P}_2) \\ &= 4\bar{F}_2 - \bar{P}_1 - 2\bar{F}_1 - \bar{P}_2 = C\end{aligned}$$

C should equal 0, apart from experimental error. In our case:

$$C = (4 \times 79.5) - 69.6 - (2 \times 89.6) - 68.4 = 0.8$$

But what is the standard error of C?
If $Q = k_1 a + k_2 b + k_3 c + \dots$

where k_1, k_2, etc. are constants, and a, b, c, etc. are independent, then the variance of Q (S_Q^2) is,

$$S_Q^2 = k_1^2 s_a^2 + k_2^2 s_b^2 + k_3^2 s_c^2 + \dots$$

Then if $C = 4\bar{F}_2 - \bar{P}_1 - 2\bar{F}_1 - \bar{P}_2$

$$S_c^2 = 4^2 S_{\bar{F}_2}^2 + (-1)^2 S_{\bar{P}_1}^2 + (-2)^2 S_{\bar{F}_1}^2 + (-1)^2 S_{\bar{P}}^2$$

where $S_{\bar{F}_2}^2 = S_{F_2}^2/n$, etc.
Therefore,

$$\begin{aligned}S_c^2 &= 16(97.3/200) + 1(48.6/50) + \\ &\quad 4(53.0/100) + 1(40.1/50) = 11.678\end{aligned}$$

or,

$$S_C = \sqrt{11.678} = 3.4173$$
$$'t' = C/S_c = 0.8/3.4173 = 0.234$$

The degrees of freedom (df) of 't' are equal to the total of the df of s_a^2, s_b^2, etc. In the present case the df $\{\Sigma(n-1)\}$ are 396 and thus 't' is not significant. There is thus no reason to reject the null hypothesis, and we can therefore conclude that there is no evidence for epistasis. This test is called by Mather and Jinks (1982) the 'C scaling test'.

Similar tests exist for backcrosses, i.e.,

$$\begin{aligned}A &= 2\bar{B}_1 - \bar{P}_1 - \bar{F}_1 \\ B &= 2\bar{B}_2 - \bar{P}_2 - \bar{F}_1\end{aligned}$$

The best method of testing for epistasis involves using the statistical estimation procedure of weighted least squares. In this method, m, $[d]$ and $[h]$ are estimated simultaneously from all 6 means using as weights the amount of information on each mean which is the reciprocal of the variance of the generation mean ($1/s^2$). This is easily carried out by computer software and provides estimates of parameters which best fit all generations simultaneously, and then tests the goodness of the fit by a χ^2 (for the theory see Mather and Jinks, 1982).

11.5 VARIATION AND COVARIATION

So far, we have only considered models to generation means. However, experience shows that if we score a number of individuals from a particular family for any character they will vary for genetical and environmental reasons. Let us now turn to methods of explaining and interpreting this variation.

11.5.1 Environmental variation

Consider first a family produced either by selfing an inbred line or by crossing two inbred lines, i.e. a P_1, P_2, or F_1 family. Since the individuals within such families are all genetically identical, any variation between them in their phenotype cannot be genetical and it is conventional to refer to such variation as the environmental variation within families (E_w). Such variation will occur even between individuals raised close together in the same plot and is often the major source of variation in any experiment, frequently dwarfing the genetical and macroenvironmental variation such as that due to blocks or sites.

Because E_w is generally such a large component it is worth exploring its underlying causes. There are in fact several possible factors which contribute to it, not all of which are environmental in the strict sense. In order to illustrate all these factors we will use disease susceptibility, measured as the proportion of leaf affected by a pathogen, as an example. Let us suppose that a leaf is taken from every plant and 100 random sectors viewed down the microscope and recorded as infected or not infected.

Since the trait is measured as a proportion (p) it is expected to show binomial 'sampling' variance ($p[1-p]/n$, where $n = 100$) and so the plants will vary from this cause. They will also vary due to 'scoring' errors as the recorder may make mistakes or have to make subjective judgements on whether a sector is infected or not. Leaves on the same plant may differ as they were formed at different times and hence show 'developmental variation' and thus different leaves from the same plant would not have the same degree of infection. This type of variation will cause leaves from different plants to vary also.

The final factor contributing to E_w is true microenvironmental variation. Every plant occupies a separate, potentially different, part of the environment and so may be affected by different levels of moisture, fertility, etc. In those few experimental situations where the components of E_w have been analysed it is frequently found that these true microenvironmental effects are comparatively small contribu-

tors relative to those described previously. However, since binomial, scoring and developmental variation are not due to the environment as such, they cannot be reduced or eliminated by environmental control such as with the use of constant environment facilities. Moreover the actual microenvironmental factors themselves may be very difficult to identify. It therefore follows that E_w will always be present, will frequently be the major source of variation and will be very difficult, if not impossible, to reduce by expensive and sophisticated environmental control systems. In practice, it is far easier and cheaper to reduce the environmental variance on family means by increasing the level of replication.

We saw earlier that in order to make predictions about the mean performance of genotypes we need to know that the genes do not interact. When we wish to predict variances we similarly have to assume that there is no genotype × environmental interaction (G×E). In particular we need to be sure that the environmental variation, E_w, is constant for all genotypes. When we have generations like those illustrated in Table 11.2, we can test for G×E by comparing the value of E_w obtained from the two parents and the F_1 (48.6, 40.1, 53.0) by some appropriate test such as those of Bartlett (1937) or Levene (1960). Preferably these variances should be homogeneous and if not, the heterogeneity should be small. A common cause of heterogeneity is that the variance is positively correlated with the mean.

There may also be environmental variation between families (E_b) which can arise from a number of causes. Some may arise from the design structure such as plots and blocks whereas others may be due to environmental factors, albeit at one generation removed. For example, if we raise two families of P_1, produced by selfing different parent plants, they may differ phenotypically simply because their parents were different owing to the position or time in which they were grown to produce the seed. A common situation in plant breeding is to raise together families of different ages or produced in different sites from seed removed from store. If such effects can produce measurable differences between genetically identical families they will also produce differences between genetically different families and hence be misinterpreted as genetical differences.

Although environmental variation has been developed above in terms of non-segregating families, i.e. inbred parents and F_1s, it will also exist in the same forms in segregating families. Thus E_w will always appear in formulae for variation within families. However, it will be assumed that the investigator has been careful to design the experiment so that E_b is zero.

11.5.2 Genetical variation

As can be seen from the data in Table 11.2, the F_2 generation is more variable than the parents or F_1 and this is of course due to segregation of alleles for which the F_1 was heterozygous and therefore were different in P_1 and P_2. Using the data from Table 11.2, we can see that the environmental variation, E_w, is 48.675 on average, whereas the F_2 variance is 97.3. It follows therefore that the genetical variance of the F_2 is 48.625 (i.e. 97.3–48.675).

The proportion of the variation between individuals in a population or family that arises from genetic segregation is known as the 'broad heritability' of the trait. In our example the genetic variance of the population, which happens to be an F_2, is 48.625 out of a total phenotypic variance of 97.3 and thus the heritability is approximately 0.5; i.e. 50% of the variation observed in this F_2 is due to genes and the remaining 50% is due to environment.

It is important to emphasize that the heritability relates to the genetical control of the variation and not to the extent to which the overall mean performance is determined by genes. There may well be a large number of genes which control a character such as flowering time, but every plant may be the same genotype for most of them and hence there will not be any genetical variation, mutation apart, arising from these genes. Plant breeders are concerned with those loci which do vary in the population because the genotypes at these loci can respond to selection and the heritability refers to these alone.

Different types of family will contain different genotypes in different frequencies and thus we might expect their heritabilities to differ. How, then, can we model the variation in different types of segregating family or population? As with generation means, it is convenient to start with the case of a single gene and then extend the methodology to many genes.

Let the expected genetical variation of an F_2 at a single segregating locus be σ_g^2. Now, from statistical theory:

$$\sigma_g^2 = \epsilon(g_i - \bar{g})^2 = \Sigma a_i(g_i - \bar{g})^2$$

where g_i and a_i are the genetic value and frequency of the ith genotype respectively. Table 11.3(I) illustrates the application of this formula to the F_2.

Table 11.3. The derivation of the genetical variance of an F_2 (see text for explanation of methods I and II)

Genotype	AA	Aa	aa	Mean
Frequency (a_i)	¼	½	¼	
Genetic value (g_i)	$m + d$	$m + h$	$m - d$	$m + \frac{1}{2}h$
$(g_i - \bar{g})$	$(d - \frac{1}{2}h)$	$(\frac{1}{2}h)$	$(-d - \frac{1}{2}h)$	0
g_i' $(= g_i - m)$	d	h	$-d$	$\frac{1}{2}h$

(I) $\sigma_g^2 = \Sigma a_i(g_i - \bar{g})^2$
$= \frac{1}{4}(d - \frac{1}{2}h)^2 + \frac{1}{2}(\frac{1}{2}h)^2 + \frac{1}{4}(-d - \frac{1}{2}h)^2$
$= \frac{1}{4}(d^2 + \frac{1}{4}h^2 - dh) + \frac{1}{2}(\frac{1}{2}h)^2 + \frac{1}{4}(d^2 + \frac{1}{4}h^2 + dh)$
$= \frac{1}{2}d^2 + \frac{1}{4}h^2$

(II) $\sigma_g^2 = \Sigma a_i g_i'^2 - (\Sigma a_i g_i')^2$
$= \frac{1}{4}(d)^2 + \frac{1}{2}(h)^2 + \frac{1}{4}(-d)^2 - (\frac{1}{2}h)^2$
$= \frac{1}{2}d^2 + \frac{1}{4}h^2$

However this method of derivation is rather tedious in complex cases and can readily be simplified by noting that:

$$\Sigma a_i(g_i - \bar{g})^2 = \Sigma a_i g_i^2 + \Sigma a_i \bar{g}^2 - 2\Sigma a_i g_i \bar{g}$$
$$= \Sigma a_i g_i^2 - (\Sigma a_i g_i)^2$$

But g_i has a constant term, m, which does not affect the variance and this can be subtracted before approaching the derivation. Thus if $g_i' = g_i - m$, then,

$$\sigma_g^2 = \Sigma a_i g_i'^2 - (\Sigma a_i g_i')^2$$

The application of this formula is shown in

Table 11.3(II) and we see that as before the genetical variance of an F_2 is:

$$\sigma_g^2 = \frac{1}{2}d^2 + \frac{1}{4}h^2$$

Now for many independent genes, i.e. genes which are independent in action (i.e. they do not interact) and in transmission (i.e. they are not linked), their combined variance is the sum of the variances due to segregation of individual genes. Thus,

$$\sigma_g^2 = (\frac{1}{2}d_1^2 + \frac{1}{4}h_1^2) + (\frac{1}{2}d_2^2 + \frac{1}{4}h_2^2) + \dots \text{ etc.}$$
$$= \frac{1}{2}\Sigma d^2 + \frac{1}{4}\Sigma h^2$$
$$= \frac{1}{2}D + \frac{1}{4}H$$

where $D = \Sigma d^2$ and $H = \Sigma h^2$.

The total phenotypic variation of an F_2 also includes the environmental variation, E_w, therefore,

$$\sigma_p^2 = \frac{1}{2}D + \frac{1}{4}H + E_w$$

From this we see that the genetical variance of an F_2 population is made up of additive (D) and dominance (H) variation and that the coefficient of the additive is twice as large as that of the dominance component. Thus even with complete dominance, additive variation will dominate the genetical variation. We can also note two further features: (i) D, unlike $[d]$, is independent of r_d, the measure of gene dispersion in the parents; (ii) H is independent of the signs of h_i.

If the amount of dominance is constant at all loci, i.e. $h_i/d_i = f$, then $h_i = fd_i$ and thus $H = \Sigma h^2 = f\Sigma d^2$. Therefore, the average amount of dominance (the dominance ratio) can be estimated as $f = \sqrt{(H/D)}$.

However, armed only with the variances within the non-segregating generations (E_w) and the F_2 variance we do not have enough statistics to estimate D and H. In order to do this we would need the variances of other generations such as F_3s or backcrosses. So let us now extend this approach to the F_3 generation.

Suppose we score a total of n F_2 individuals, self them to produce n F_3 families and grow r individual plants from each family in a completely randomized trial which are also scored for the same trait. We thus have the data shown in Table 11.4(I).

Table 11.4. Outline of F_3 data (I) and the corresponding anova (II)

(I) *Outline of F_3 data*

F_2	X_1	X_2	X_3	$\dots$	X_n
F_3 family mean	Y_1	Y_2	Y_3	$\dots$	Y_n
Variance within F_3 family	s_1^2	s_2^2	s_3^2	$\dots$	s_n^2

(II) *Anova of F_3 data*

Item	df	MS	e.m.s.
Between F_3 families	$n-1$	MS_1	$\sigma_{2F3}^2 + r\sigma_{1F3}^2$
Within F_3 families	$n(r-1)$	MS_2	σ_{2F3}^2

Each F_3 family will have a mean Y_i and a variance s_i^2 and we can thus calculate the variance between F_3 family means and the average variance within the n families. We can also calculate the covariance between the F_3 family means (Y_i) and the F_2 parents (X_i). Thus the F_3 provides two variances and one covariance. The two sources of variance are most easily derived from the analysis of variance of the F_3 data as shown in Table 11.4(II).

The component σ_{2F3}^2 is the expected variance within F_3 families, while σ_{1F3}^2 is the expected variance between F_3 family means. These σ^2 can be derived algebraically using the approach in Table 11.3(II) the basic information being illustrated in Table 11.5. We see that the three types of F_2 individual (AA, Aa, aa) produce three types of F_3 families which correspond to P_1, F_2 and P_2, the means and variances of which have been derived previously and are shown as Y_i and σ_i^2 respectively in Table 11.5

We can use this table to derive the two σ^2s:

$$\sigma_{2F3}^2 = \Sigma a_i \sigma_i^2$$
$$= \frac{1}{4}(0) + \frac{1}{2}(\frac{1}{2}d^2 + \frac{1}{4}h^2) + \frac{1}{4}(0)$$
$$= \frac{1}{4}d^2 + \frac{1}{8}h^2$$

this is simply half the F_2 variance.

Making the same assumptions of independence as before, we can extend this to many loci to obtain:

$$\sigma_{2F3}^2 = \frac{1}{4}D + \frac{1}{8}H$$

Similarly:

$$\sigma_{1F3}^2 = \Sigma a_i Y_i^2 - (\Sigma a_i Y_i)^2$$

Table 11.5. Derivation of expected mean squares for F_3 families.

F_2 genotype	AA	Aa	aa	Mean
Frequency a_i	¼	½	¼	
genetic value X_1	$m + d$	$m + h$	$m - d$	$m + \frac{1}{2}h$
F_3 family	AA	$AA : Aa : aa$	aa	Mean
Frequency within family	1	¼ : ½ : ¼	1	
Family mean Y_i	$m + d$	$m + \frac{1}{2}h$	$m - d$	$m + \frac{1}{4}h$
Genetical variance within family (σ_i^2)	0	$\frac{1}{2}d^2 + \frac{1}{4}h^2$	0	$\frac{1}{4}d^2 + \frac{1}{8}h^2$

Again, ignoring the term 'm' as it is a constant and cannot affect the variance, we obtain:

$$\sigma^2_{1F3} = \frac{1}{4}(d)^2 + \frac{1}{2}(\frac{1}{2}h)^2 +$$
$$\frac{1}{4}(-d)^2 - (\frac{1}{4}h)^2$$
$$= \frac{1}{2}d^2 + \frac{1}{16}h^2$$
$$= \frac{1}{2}D + \frac{1}{16}H$$

for many independent genes.

Finally, we can calculate the covariance between the F_2 individuals and the F_3 family means, cov(X,Y):

$$\text{cov}(X,Y) = \Sigma a_i X_i Y_i - (\Sigma a_i X_i)(\Sigma a_i Y_i)$$
$$= \frac{1}{4}(d \cdot d) + \frac{1}{2}(h \cdot \frac{1}{2}h) +$$
$$\frac{1}{4}[(-d)(-d)] - (\frac{1}{2}h)(\frac{1}{4}h)$$
$$= \frac{1}{2}d^2 + \frac{1}{8}h^2$$
$$= \frac{1}{2}D + \frac{1}{8}H$$

for many independent genes.

So far we have considered the genetical components of the σ^2. Providing the F_2 and F_3 were independently randomized there should be no environmental component to the covariance or to σ^2_{1F3}. We must, however, add E_w to σ^2_{2F3} (i.e. $\sigma^2_{2F3} = \frac{1}{4}D + \frac{1}{8}H + E_w$).

We can see that with the F_3s alone we have two statistics (MS_1 and MS_2) but three unknown parameters (D, H and E_w) and thus have insufficient information from which to estimate them. If the parents or F_1 are also included in the same experiment, they will yield independent estimates of E_w and hence D and H too are estimable. The inclusion of the F_2 will similarly provide another statistic to allow all three parameters to be estimated. The covariance between F_2 and F_3 does, in principle, provide an additional, useful statistic as we saw above, but in practice its value is often diminished through the effects of genotype × environment interaction.

This is a particular problem with plants where the effects of different seasons and locations can alter phenotypes considerably and obviously particular F_3 families have to be raised in a different season to the F_2 parents that produced them. The covariance should, therefore, only be included with caution.

Clearly we can pursue this approach down through the F_4 to F_∞ generations and develop new statistics which are different functions of D, H and E_w. In a similar way we can derive the genetical variance within first backcross families as shown in Table 11.6. The two backcross variances contain an additional term, F ($= \Sigma dh$), the size and sign of which will depend on the distribution and direction of dominant alleles in the two parents. Thus if P_1 has most dominant alleles, F will be positive and hence the variance within B_1 will be less than that within B_2. Conversely if F is negative, P_2 has most dominant alleles. If $F = 0$ then both parents carry the same number of dominant alleles.

Thus it is possible to estimate the additive, dominance and environmental contributions to the variability of the trait as shown in Table 11.7. From these estimates we can calculate the dominance ratio, broad and narrow heritability, the latter being that proportion of the phenotypic variation which is due to the additive effect of genes.

11.6 INFORMATION FROM POPULATIONS

Breeders are not always in the enviable position of starting with two inbred lines and the generations derived from them. Instead they

Table 11.6. Derivation of mean and variance of first backcrosses.

B_1	$F_1 \times P_1$		B_2	$F_1 \times P_2$	
	$Aa \times AA$			$Aa \times aa$	
Progeny	AA	Aa		Aa	aa
Frequency	½	½		½	½
Genetic value	$m+d$	$m+h$		$m+h$	$m-d$

$$\bar{B}_1 = \tfrac{1}{2}(m+d) + \tfrac{1}{2}(m+h)$$
$$= m + \tfrac{1}{2}d + \tfrac{1}{2}h$$
$$= m + \tfrac{1}{2}[d] + \tfrac{1}{2}[h]$$
$$\sigma^2_{B1} = \tfrac{1}{2}d^2 + \tfrac{1}{2}h^2 - (\tfrac{1}{2}d+\tfrac{1}{2}h)^2$$
$$= \tfrac{1}{4}d_2 + \tfrac{1}{4}h_2 - \tfrac{1}{2}dh$$
$$= \tfrac{1}{4}D + \tfrac{1}{4}H - \tfrac{1}{2}F$$

$$\bar{B}_2 = \tfrac{1}{2}(m+h) + \tfrac{1}{2}(m-d)$$
$$= m - \tfrac{1}{2}d + \tfrac{1}{2}h$$
$$= m - \tfrac{1}{2}[d] + \tfrac{1}{2}[h]$$
$$\sigma^2_{B2} = \tfrac{1}{2}d^2 + \tfrac{1}{2}h^2 - (\tfrac{1}{2}h-\tfrac{1}{2}d)^2$$
$$= \tfrac{1}{4}d_2 + \tfrac{1}{4}h_2 + \tfrac{1}{2}dh$$
$$= \tfrac{1}{4}D + \tfrac{1}{4}H + \tfrac{1}{2}F$$

Table 11.7. Estimates of components of variances. Data from Table 11.2.

(I) *Data*

Generation	Variance	Model							
		D		H		F		E_W	
P_1	48.6							1.0	
P_2	40.1							1.0	
F_1	53.0							1.0	
F_2	97.3	1/2	+	1/4			+	1.0	
B_1	66.5	1/4	+	1/4	−	1/2	+	1.0	
B_2	84.6	1/4	+	1/4	+	1/2	+	1.0	

(II) *Parameter estimates*

$D = 87.08$
$H = 20.42$
$F = 18.1$
$E_W = 48.675$

Heritability (broad)	$=$	$\tfrac{1}{2}D/(\tfrac{1}{2}D+\tfrac{1}{4}H+E_W)$	$=$	0.50
Heritability (narrow)	$=$	$(\tfrac{1}{2}D+\tfrac{1}{4}H/(\tfrac{1}{2}D+\tfrac{1}{4}H+E_W)$	$=$	0.45
Dominance ratio	$=$	$\sqrt{(H/D)}$	$=$	0.48

may simply have a population. How then can they proceed?

It is first necessary to construct a number of families of known degree of relationship such as full-sibs (FS) or half-sibs (HS) since without a family structure it is not possible to construct any sort of genetical model. Because of the importance of such types of analysis many different designs exist (see Mather and Jinks, 1982), however, we only have space to introduce the principles here and this will again be done using an F_2 population in order to keep the algebra simple.

The simplest type of family structure consists of full-sibs. Parents are chosen from the population at random and crossed together in pairs to produce a number (n) of FS families. These families are referred to by Mather and Jinks (1982) as 'biparental progenies' or 'BIPS' for short. If r progeny are raised and scored from every family in a suitably randomized and replicated experiment, we can examine the variation within and between these FS families in the same way as with F_3 families earlier. This yields the anova shown in Table 11.8.

In order to interpret the σ^2 in this anova, we need to refer to Table 11.9. Along the top and sides are given the various types of male and

Table 11.8.   Anova of BIP families.

Item	df	MS	e.m.s.
Between families	$n-1$	MS_1	$\sigma^2_w + r\sigma^2_B$
Within families	$n(r-1)$	MS_2	σ^2_w
Total	$nr-1$		

female parent from our F_2 population with respect to a single locus, together with their respective frequencies. Mating these at random produces the various families in the body of the table, P_1, P_2, F_1, etc. All of these are familiar from the selfing and backcrossing series, so their means and variances shown in Table 11.9 should present no difficulties. The frequencies of the various families are obtained simply by multiplying the frequencies of the respective parents, e.g. the frequency of family P_1 is obtained as $\frac{1}{4} \times \frac{1}{4}$. Since some of the families (B_1, B_2, F_1) appear more than once in the table, we can summarize the table in terms of the six basic generations whose means, variances and frequencies are as shown at the bottom of Table 11.9.

Before we consider the variances, it is worth noting that the overall mean of the population ($m + \frac{1}{2}h$) is the same as that for an F_2. This is as expected, as random mating an F_2 population does not change the genotype frequencies at a given locus and the population mean is simply a function of these frequencies.

As before σ^2_w is the true variance within families and thus we can derive the expected genetic component of σ^2_w by summing the variances at the bottom of Table 11.9 according to their frequencies, i.e.,

$$\sigma^2_w = \frac{1}{16}[0] + \frac{1}{4}[\frac{1}{4}(d-h)^2] + \frac{1}{8}[0] + \frac{1}{4}[\frac{1}{2}d^2+\frac{1}{4}h^2] + \frac{1}{4}[\frac{1}{4}(d+h)^2]+\frac{1}{16}[0]$$
$$= \frac{1}{4}d^2 + \frac{3}{16}h^2$$

Similarly σ^2_B is the true variance of family means and its genetic component can be derived from the means given in Table 11.9 as:

$$\sigma^2_B = \frac{1}{16}[d]^2 + \frac{1}{4}[\frac{1}{2}(d+h)]^2 + \frac{1}{8}[h]^2 + \frac{1}{4}[\frac{1}{2}h]^2 + \frac{1}{4}[\frac{1}{2}(-d+h)]^2 + \frac{1}{16}[-d]^2-[\frac{1}{2}h]^2$$
$$= \frac{1}{4}d^2+\frac{1}{16}h^2$$

Providing the genes concerned are in linkage equilibrium we can obtain the combined genetical component of these σ^2 by summing over all loci as shown below.

$$\sigma^2_w = \frac{1}{4}\Sigma d^2 + \frac{3}{16}\Sigma h^2 = \frac{1}{4}D + \frac{3}{16}H$$
$$\sigma^2_B = \frac{1}{4}\Sigma d^2 + \frac{1}{16}\Sigma h^2 = \frac{1}{4}D + \frac{1}{16}H$$

Finally we must add in the environmental component, E_w to produce the following complete expectations:

$$\sigma^2_B = \frac{1}{4}D + \frac{1}{16}H$$
$$\sigma^2_w = \frac{1}{4}D + \frac{3}{16}H + E_w$$

Let us now turn to the interpretation of such an analysis. Significant differences between families prove that there is genetical variation and the intraclass correlation, t_{FS}, is a measure of one half the heritability of the trait, albeit the mean of the broad and narrow heritability. Thus,

$$t_{FS} = \sigma^2_B/(\sigma^2_B + \sigma^2_w)$$
$$= (\frac{1}{4}D+\frac{1}{16}H)/(\frac{1}{2}D+\frac{1}{4}H+E_w)$$

or

$$2t_{FS} = (\frac{1}{2}D+\frac{1}{8}H)/(\frac{1}{2}D+\frac{1}{4}H+E_w)$$
$$= (h^2_b+h^2_n)/2$$

Unless we are prepared to assume that there is no dominance, we are unable to estimate D, H and E_w as there are only two statistics.

This analysis is illustrated in Table 11.10(I) for an example with 50 families of 20 individuals each.

If we have scored the phenotypes of the parents we can obtain additional information from the covariance of the FS family means onto the scores of their mid-parents. It is reasonable to expect that the more highly heritable the trait, the more the progeny will resemble their parents and hence the greater the covariance. Let us now quantify this by developing the covariance algebraically.

The bottom line of Table 11.9 gives the average score of the parents for each of the six types of FS family. For example the parents of B_1 are AA and Aa with genetic values d and h, hence the mid-parent is $\frac{1}{2}(d+h)$. To calculate the covariance we simply multiply the corresponding values of the mid-parent values (x_i)

Table 11.9. Family structure for random mating populations

Female		Male		
	Genotype frequency	AA 1/4	AA 1/2	aa 1/4
Genotype frequency	AA 1/4	P_1 1/16	B_1 1/8	F_1 1/16
Genotype frequency	Aa 1/2	B_1 1/8	F_2 1/4	B_2 1/8
Genotype frequency	aa 1/4	F_1 1/16	B_2 1/8	P_2 1/16

	Type of family					
	P_1	B_1	F_1	F_2	B_2	P_2
Frequency (a_i)	$1/16$	$1/4$	$1/8$	$1/4$	$1/4$	$1/16$
Mean (Y_i)	$m+d$	$m+\frac{1}{2}(d+h)$	$m+h$	$m+\frac{1}{2}h$	$m+\frac{1}{2}(-d+h)$	$m-d$
Genetic variance	0	$\frac{1}{4}(d-h)^2$	0	$\frac{1}{2}d^2+\frac{1}{4}h^2$	$\frac{1}{4}(d+h)^2$	0
Mid-parent value (X_i)	$m+d$	$m+\frac{1}{2}(d+h)$	m	$m+h$	$m+\frac{1}{2}(-d+h)$	$m-d$

Table 11.10. Anova of FS and parental data from a trial of 50 biparental families (BIPs) of size $r = 20$

Item	df	MS	F	P
(I) *Anova of FS*				
Between families	49	576.68	7.54	<<0.001
Within families	950	76.52		

$$t_{FS} = \sigma_B^2/(\sigma_B^2+\sigma_W^2)$$

$$= 25.008/(25.008+76.52) = 0.245$$

$$h^2 = 2t_{FS} = 0.49$$

Item	df	MS	F	P
(II) *Regression anova of FS means onto mid-parental values*				
Regression	1	207.05	8.24	0.006
Remainder	48	25.12		
Total	49	28.83		

$$b_{yx} = 0.49$$

Item	df	MS	F	P
(III) *Combined regression and BIP analysis*				
Regression	1	4140.9	8.24	0.006
Remainder	48	502.4	6.57	<<0.001
Within	950	76.5		

with their progeny means (Y_i) and their frequency (a_i):

$$cov(x,y) = \Sigma(a_iX_iY_i) - (\Sigma a_iX_i)(\Sigma a_iY_i) = \frac{1}{4}D$$

This covariance, therefore, is a simple function of the additive genetic variance and has no contributions from non-additive or environmental influences. This latter assumes that the parents and their offspring are separately randomized with respect to their environment and hence their environmental deviations are not correlated.

If we plot the family means against their mid-parental scores, we obtain the relationship between offspring and parents shown in Fig. 11.4. Clearly we would expect that the closer the linear relationship is to unity the greater the genetical control. The slope of this regression (b_{yx}) is estimated as: $cov(x,y)/var(x)$.

Now the X_is are the mid-parent scores. The variance of the individual parental scores is the phenotypic variance of the population ($Vp = \frac{1}{2}D + \frac{1}{4}H + E_w$), thus since the parents were chosen as pairs at random, the variance of the means of pairs of parents is one half the variance of individual parents (recall $Vx = Vx/n$), i.e. $\frac{1}{2}Vp$.

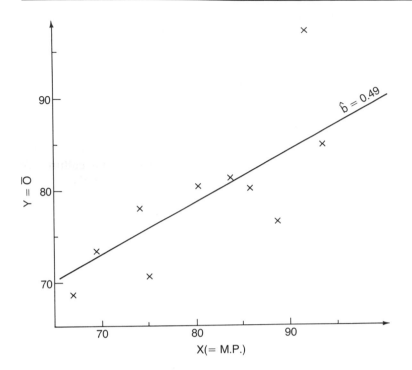

Fig. 11.4. Regression of biparental (full-sib) family means onto mid-parental scores.

Hence, $b_{yx} = (\frac{1}{4}D)/[\frac{1}{2}(\frac{1}{2}D+\frac{1}{4}H+E_w)] = (\frac{1}{2}D)/(\frac{1}{2}D+\frac{1}{4}H+E_w)$ which is the narrow heritability.

Thus, the standard regression approach to mid-parent offspring data yields a direct estimate of the narrow heritability ($b_{yx} = h^2_n$).

Such data, therefore, are amenable to standard regression analysis to provide an anova to test the significance of the regression, together with an estimate of σ^2_B. This is illustrated in Table 11.10(II).

The inclusion of the parental data has created an additional statistic over that available from the progeny alone and thus we could attempt to estimate D, H and E_w. Moreover the two analyses in Table 11.10(I, II) can be combined if we note that the total MS for the regression anova (28.83) corresponds to the MS of the 'Between FS families' item in Table 11.10(I); the apparent numerical difference is simply due to the fact that the MS in the regression analysis is based on family means and hence is a factor of $1/r$ of the other. If we multiply all the MS in Table 11.10(II) by 20, we can then combine the two analyses as shown in Table 11.10(III).

The original FS analysis partitioned the total offspring variation into that within and between FS families, and as we saw, the expected variance of family means is a function of $\frac{1}{4}D + 1/16H$. The regression analysis has further partitioned this variance into the regression onto the mid-parent (the covariance term $\frac{1}{4}D$) and thus the remainder must be the $1/16H$. So this combined analysis provides a test of significance for non-additive variation, H.

As we have seen, the narrow heritability can be estimated directly from the regression of offspring onto mid-parent values. This is an important conclusion, because we now have a method of predicting the performance of offspring from the parents using the standard linear regression equation, $y = a + bx$. The narrow heritability (b) is thus central to predicting the response to selection as will be discussed more fully in Chapter 18.

The use of data from parents and offspring does create the same difficulties raised in the discussion of F_3 data earlier, namely that of G×E interaction. It is thus preferable to estimate heritabilities from contemporary relatives grown in the same trial. We saw above that the intraclass correlation coefficient, t_{FS}, from bips estimates the heritability from contemporary relatives, although it is slightly biased by non-additive variation. Other designs, using mixtures of FS and HS progeny, overcome this problem and yield direct estimates of additive genetical variance (Kearsey, 1965; Kearsey and Jinks, 1968).

11.7 PREDICTION METHODS

We have shown how it is possible to construct genetical models to means and variances for a quantitative trait and how to estimate significant parameters. Let us now turn to see how we can use these parameters to make plant breeding decisions. We will pursue this by looking at specific questions.

11.7.1 Is there genetical variation that can be utilized?

Providing the two parents differ or the F_1 differs from the parents for the traits of interest, there must be genetic variation present in the cross. Similarly, if we can demonstrate that there are significant differences between F_3 or other families (FS or HS) or that the F_2 variance is greater than that of the non-segregating generations, then we have proved that genetic differences exist to be exploited.

The broad heritability of the F_2 will indicate how large this genetic variability is relative to environmental effects while the narrow heritability will show how easy it will be to select for and fix the genetical variation which has been revealed. A small value for the heritability does not necessarily mean that there is little to be gained by selection but simply that gains may be hard won. The use of heritability in this context will be discussed more fully in Chapter 18.

Typically heritabilities are less than 0.5 and frequently much less. This implies that selection in an F_2 generation as part of an inbreeding programme has nothing to commend it since most of the variation selected for is not due to genes and thus is not fixable.

11.7.2 What inbred lines can we obtain from the cross?

In a typical plant breeding programme a number of recombinant inbred lines would be produced from an F_1 in order to identify useful transgressive segregants. These lines might be produced by single seed descent (SSD), pedigree methods or by producing doubled haploids from microspore or ovule culture. To what extent is it possible to predict the range of performance of these lines?

Starting as before with a single gene model, half the recombinant inbred lines will be AA and the other half aa. Thus:

genotype	AA	aa	mean
frequency	½	½	
genetic value	$m + d$	$m - d$	m

The genetical variance of line means will thus be:

$$\sigma^2_1 = \tfrac{1}{2}d^2 + \tfrac{1}{2}(-d)^2 = d^2$$

As before, for many independent genes, this becomes:

$$\sigma^2_1 = \Sigma d^2 = D$$

Thus the distribution of inbred line means is expected to be normally distributed with mean m and variance D, $N(m,D)$, and hence should be predictable from information derived from the early generations (Jinks and Perkins, 1972). For example, using the information from the cross illustrated in Table 11.2, we have seen that $m = 69.0$, $D = 87.08$ (see Table 11.7). We could calculate the proportion of lines exceeding the F_1, for example, using the normal deviate as follows.

$$z = (\bar{F}_1 - m)/\sqrt{D}$$
$$= 20.6/9.33 = 2.21$$

and this corresponds with a proportion of 0.014 in the tail of the distribution, i.e. 1.4% of the derived inbreds should exceed the F_1.

If we had several crosses, we could use the same approach to ask which of them will give the best yield of lines exceeding some existing, target variety. In this way we could identify the most promising crosses at an early stage in the breeding programme.

11.7.3 What second cycle hybrids can be obtained from the inbreds in section 11.7.2?

It may be of interest to know what new F_1 hybrids could be produced by intercrossing the inbreds from section 11.7.2. As with the inbreds, we need to derive the expected distribution of all possible hybrids which can be obtained from such random intercrossing.

If we take the single gene model and consider what happens when we randomly mate the inbred lines, we obtain the following:

♀ \ ♂		AA	aa
		½	½
AA	½	(¼) AA	(¼) Aa
aa	½	(¼) Aa	(¼) aa

Thus we have three types of family, AA, Aa and aa with frequencies ¼, ½ and ¼. These genotypes and frequencies are, of course, those of an F_2. However, we have re-created an F_2 population in which, instead of each individual being unique, it is represented by a family of genetically identical individuals whose true mean is not affected by the environment. Thus, the expected overall mean and variance of these second cycle hybrids is simply the mean and genetic variance of an F_2, i.e. $(m + \frac{1}{2}h)$ and $(\frac{1}{2}D + \frac{1}{4}H)$ and can readily be obtained from the F_2. This distribution will be approximately normally distributed providing there are several genes segregating and so it is possible to make predictions about second cycle hybrids as illustrated above for the recombinant inbred lines (Toledo *et al.*, 1984).

Thus,

$$z = (\bar{F}_1 - \bar{F}_2)/\sqrt{(\frac{1}{2}D + \frac{1}{4}H)}$$
$$= 10.1/\sqrt{48.645} = 1.45$$

This would mean that approximately 7% of second cycle hybrids would exceed the first F_1.

This is higher than the proportion of inbreds above, but could be more expensive to produce.

11.7.4 What is the best inbred cultivar that can potentially be produced from the cross?

If the initial F_1 shows heterosis it is likely that the breeder will aim towards a hybrid variety. However, as we saw in section 11.7.2 it is possible to check to see what proportion, if any, of recombinant inbred lines would exceed the F_1. If this proportion is reasonably large then it may well be worth trying to produce such inbreds. However, with a very large number of genes controlling the variation in the cross, such inbreds may be too few to consider. Nonetheless, it is reasonable to enquire what the possible maximum performance of an inbred variety might be.

The best inbred line (P_{max}) will have all the favourable alleles for the desired trait. That is, its mean will be:

$$\bar{P}_{max} = m + \Sigma d$$

As we have seen, it is possible to estimate m but not Σd; all we have is $[d] = r\Sigma d$. If, as earlier, we assume that the dominance ratio (f) is approximately constant at all loci, i.e.,

$$h_i/d_i = f \text{ or } d_i = h_i/f$$

then,

$$\Sigma d = \Sigma h/f$$

But we can estimate Σh (= $[h]$) and f (= $\sqrt{\{H/D\}}$), therefore,

$$\bar{P}_{max} = m + [h]/\sqrt{\{H/D\}}$$

Using the data and parameters above:

$$\bar{P}_{max} = 69.0 + 20.6/0.48 = 112$$

which is considerably higher than the original F_1 at 89.6. This makes it clear that unless the dominance ratio is greater than one, i.e. overdominance, it should be possible to produce an inbred line which is at least as good, if not considerably better, than the original, heterotic F_1. Current data on a range of organisms confirms this to be true (Jinks, 1981, 1983).

11.7.5 What rate of response might we get to selection?

Since this topic will be considered at some length in Chapter 18, suffice it to say that the response to individual selection is proportional to the narrow heritability, h^2_n (Section 11.6). Thus if we select parents whose score for the trait in question is S (Selection) units above the mean of the population then their progeny would be expected to deviate by R (Response) units from the mean, where:

$$R = S \cdot h^2_n$$

It is possible to see from these five examples that the relatively simple information about the genetic and environmental control of a quantitative character which can be obtained from the early generations of a cross, has very useful predictive potential in a breeding context. The reliability of the predictions will, of course, depend on the reliability of the parameter estimates used to make them. Variances, in particular, are unreliable as they have large standard errors, while those parameters based on generation means can be estimated with reasonable precision. It is up to the breeder and biometrical geneticist to devise sufficiently efficient and reliable estimates in a given context.

11.8 COMPLICATIONS: LINKAGE, EPISTASIS AND G×E

Although we have assumed that linkage, epistasis and genotype × environment interaction are absent in developing genetic models to means and variances it is unlikely that these assumptions will generally hold. With a large number of polymorphic genes controlling the trait, it is to be expected that some at least are linked. Even if gene action were additive at one level, there is no reason to expect that it would be at another. For example, if gene action were additive in its effect on cell diameter, it would not be additive in terms of cell volume since volume is a function of the diameter cubed. Different genotypes, too, could differ in their sensitivity to the environment and this is just one example of genotype × environment interaction.

In so far as we simply wish to manipulate the genetical variation in breeding programmes, the scale of measurement is not very important providing we can make useful decisions and predictions, i.e. we need an additive genetical model. We could therefore change the scale of measurement such as by taking square roots or logs of the original data, provided this removed complications such as epistasis and resulted in a scale on which gene action was effectively additive.

As we have seen earlier it is possible to construct tests for epistasis and interaction between the genotype and microenvironment and so discover whether these are important factors. If they are, we could explore alternative scales by transforming the original individual measurements and then repeating the scaling tests. Very often a simple transformation, such as square roots, will minimize both epistasis and G×E, but it may be that a transformation which reduces one problem will exacerbate another. Systematic approaches have been used successfully to identify optimum or compromise scales in some crops (Smith *et al.*, 1985).

The alternative approach is to accept that epistasis or G×E exists and introduce other parameters into the model to allow for them. This approach may in many cases be relatively easy to achieve providing the interactions are not too complicated. However they do require additional parameters which in turn require more information in terms of additional, independent statistics (Mather and Jinks, 1982).

Linkage, like epistasis, causes non-additivity, but at the variance level, and so will affect predictions. However it can be shown that the biases due to linkage are relatively unimportant in predicting over the short term. This is because linked genes behave not like independent genes but as 'compound' genes or 'effective factors'. It is these 'effective factors' which are being manipulated during selection and it is these same factors which cause the additive genetic variance. So although estimates of *D* and *H* may differ from those that would be obtained if all the genes were unlinked, they are the relevant values for predicting selection response, etc. (Jinks and Pooni, 1982; Kearsey, 1985).

Although the basic concepts have been developed using generations derived by selfing from an F_1 between two inbred lines, they are not restricted to such a special case. Very often the breeder starts with a population about which there is no information on allele frequencies and selfing may be difficult if not impossible. Therefore, a great deal of attention has been paid to methods of obtaining corresponding genetical information from such material. However, because one has less information about the material at the start such as number of alleles at a locus, allele frequencies, and genotype frequencies, it is inevitable that the methods are less powerful when applied to such populations and one is reduced to asking simpler questions. However, providing the population is random mating with two alleles at each locus, models can be constructed which are essentially identical to those for populations with equal allele frequencies. Thus the formulae derived for BIPS and other designs based on F_2s are transferable to the more general cases.

11.9 SUMMARY AND FUTURE PROSPECTS

This chapter has attempted to illustrate simply the basic methods underlying the analysis of polygenic traits in crop plants. We have seen how, by the use of a combination of generation means and variances, it is possible to obtain a variety of information about the genetical and environmental control of agronomic traits. This information in its turn can be used to make important decisions about the long-term strategies and short-term tactics of a plant breeding programme (see Chapter 18).

There are many other biometrical techniques and applications currently available which we have not had time to consider here. For example the use of diallel crosses and triple testcrosses, the analysis of genetical and environmental correlations and their use in selection indices, as well as the methods for counting the number of factors controlling particular traits.

There is considerable interest at present in the use of molecular markers, particularly RFLPs and isoenzymes, to dissect quantitative traits and to provide useful markers for selecting important QTLs with poor penetrance (Tanksley *et al.*, 1989; Lande and Thompson, 1990). Several of these topics are developed in Chapter 19.

REFERENCES

Bartlett, M.S. (1937), Some examples of statistical methods of research in agriculture and applied biology. *J. R. Statist. Soc. Suppl.*, **4**, 137–183.

Falconer, D.S. (1989), *Introduction to Quantitative Genetics*, 3rd edn, Longman, Burnt Mill.

Jinks, J.L. (1981), The genetic framework of plant breeding. *Philos. Trans. R. Soc. London B*, **292**, 407–419.

Jinks, J.L. (1983), Biometrical genetics of heterosis. In *Heterosis*, Frankel, R. (ed.), *Monographs on Theoretical and Applied Genetics*, Vol. 6, Springer–Verlag, Berlin, pp. 1–46.

Jinks, J.L. and Perkins, J.M. (1972), Predicting the range of inbred lines. *Heredity*, **28**, 399–403.

Jinks, J.L. and Pooni, H.S. (1982), Predicting the properties of pure breeding lines extractable from a cross in the presence of linkage. *Heredity*, **49**, 265–270.

Kearsey, M.J. (1965), Biometrical analysis of a random mating population; a comparison of five experimental designs. *Heredity*, **20**, 205–235.

Kearsey, M.J. (1985), The effect of linkage on additive genetic variance with inbreeding an F_2. *Heredity*, **55**, 139–143.

Kearsey, M.J. and Jinks, J.L. (1968), A general method of detecting additive, dominance and epistatic variation for metrical traits. *Heredity*, **23**, 403–409.

Lande, R. and Thompson, R. (1990), Efficiency of marker assisted selection in the improvement of quantitative traits. *Genetics*, **124**, 743–756.

Levene, H. (1960), Robust tests for equality of variance. In *Contributions to Probability and Statistics*, Olkin, I., Ghurye, S.G., Hoeffding, W., Madow, W.G. and Mann, H.B. (eds), Stanford University. Press, Stanford, pp. 278–292.

Mather, K. and Jinks, J.L. (1982), *Biometrical Genetics*, 3rd edn, Chapman and Hall, London.

Place, A.R. and Powers, D.A. (1979), Genetic variation and relative catalytic efficiences of lactate dehydrogenase allozymes in *Fundulus heteroclitus. Proc. Natl. Acad Sci. USA*, **5**(76), 2354–2358.

Smith, B.M., Rogers, W.J. and Kearsey, M.J. (1985), The feasibility of producing inbred rather than F1 hybrid cultivars in Brussels sprouts; a preliminary genetical analysis and choice of material for inbreeding. *Ann. Appl. Biol.*, **107**, 87–99.

Tanksley, S.D., Young, N.D., Paterson, A.H. and Bonierbale, M.W. (1989), RFLP mapping in plant breeding: new tools for an old science. *Biotechnology*, **7**, 257–264.

Toledo, J.F.F. de, Pooni, H.S. and Jinks, J.L. (1984), Predicting the properties of second cycle hybrids produced by intercrossing random samples of recombinant inbred lines. *Heredity*, **53**, 283–292.

12

Biochemical characterization of populations

M. Pérez de la Vega

12.1 INTRODUCTION

The word 'population' is sometimes a misleading term for cultivated plants; this word will be used in this chapter as a general reference for those groups of individuals which, from an agricultural point of view, ought to be more properly named as cultivars, lines or even accessions of a collection. The most inclusive definition for population is a community of individuals which share a common gene pool. This definition means that, for instance, 'Chinese Spring' wheat, 'Tender Green' bean or 'Northern Brewer' hop can be referred to the same respective populations no matter where they are grown. Another useful definition is that a population is a group of individuals of the same species sharing the same territory, which in agriculture can mean a field or a collection or harvesting area. The first definition is more suited to cultivars, pure lines and clones, while the latter is better for local and land races and their collection samples.

The characterization of the genetic variability of a population is a relevant task since genetic diversity within populations and within species determines the rates of adaptive evolution and the extent of response in traditional crop improvement (Chapters 2 and 18). The usual approach to characterization and evaluation of

Plant Breeding: Principles and prospects. Edited by M.D. Hayward, N.O. Bosemark and I. Romagosa. Published in 1993 by Chapman & Hall, London. ISBN 0 412 43390 7.

populations involves cultivation of subsamples and their morphological and agronomic description. Considered as a whole, numerous morphological characteristics are difficult to comprehend in terms of patterns of variation in populations. For this reason, numerical taxonomic techniques are needed to simplify and handle these complex patterns of variation. Traits of agronomic interest like vigour, disease resistance and cold tolerance, usually involve high genotype x environment interactions. Reliable estimates of population values for such traits can thus only be obtained from replicated multi-environment trials or from measurements taken under standardized conditions.

Morphological and agronomic evaluation of population variability may be supplemented and generally surpassed by more direct study of the genome by means of the analysis of biochemical markers. These markers, in particular isozymes, have been extremely useful in improving our knowledge of the genetic composition of populations and for determining the magnitude of various evolutionary forces involved in molding the genetic architecture of plant populations. Already, studies of DNA polymorphism form a further and definitive step in obtaining such knowledge.

Biochemical markers are less affected, if at all, by environmental factors and numerous data can be handled and statistically analysed in terms of patterns of genetic variation, at least those traits like isozyme and DNA polymorphisms whose genetic control is easily understood. Several kinds of biochemical markers have been used for the characterization of plant populations. These markers can be grouped in three classes: (1) a heterogeneous pool of biochemical compounds including phenolics, alkaloids, cyanogens and non-protein amino acids (in the present chapter they will be designated as low-molecular-weight markers); (2) proteins, including both enzymes and storage proteins; (3) DNA markers, including fragments of variable length obtained by digestion with restriction enzymes (restriction fragment length polymorphism, RFLP) by PCR techniques and base sequences.

12.2 BIOCHEMICAL PROCEDURES

Since in population studies a high number of individuals have to be analysed, simple reliable techniques are required. For low-molecular-weight and protein markers, one-dimensional and two-dimensional chromatographic techniques can be used. Usually, two-dimensional techniques have a higher resolution, therefore allowing a clear identification of different phenotypes. However, only a single sample per chromatographic plate or gel can be studied. Since they are also more complicated and time consuming than the one-sample system of analysis, these disadvantages make them unsuitable for general purpose population analysis. One-dimesional techniques allow for the study of several samples simultaneously and, although the resolution is lower, they are faster and more reliable. Therefore, one-dimensional methods continue to be the most widely used for varietal identification and population studies.

12.2.1 Low-molecular-weight markers

The first group includes a series of very heterogeneous biochemical compounds, products of the secondary metabolism of plants. These compounds have a relatively low molecular weight and can be grouped into phenolics, flavonoids, anthocyanins, non-protein amino acids, cyanogens, glycosides, polyacetylenes,

alkaloids and peptides. Most of these compounds have particular characteristics and confer adaptive properties to the plant. Many are toxic to animals, participate in resistance to pests and diseases, frequently as phytoalexins, or give a characteristic odour and/or taste to plants. Many of them having several of these properties. Phenolic compounds are the plant chemicals which have been related most commonly to resistances, sometimes together with peroxidases and polyphenol oxidases. It is of interest to ascertain which populations carry substances involved in resistance responses against pests and diseases, and the frequency and level of expression among individuals. However, many of these substances, are toxic to animals or give an undesirable taste or odour to edible plants. In such cases, the identification of individuals or populations with low levels of a given substance is a task for plant breeders.

Since these substances are chemically very heterogeneous, there are several techniques for their identification and analysis at a population level (McKee, 1973). The most suitable techniques are the chromatographic ones: the classical thin-layer, or the newer liquid, chromatography (high pressure or flow pressure, HPLC or FPLC). Organic solvents are usually required to extract the compounds from ground plant material, and special care has to be taken to avoid their oxidization or degradation. High purity and expensive solvents are required for HPLC analyses. For population studies an inconvenience of these techniques is that the extraction and purification of samples prior to the chromatography are rather arduous and time consuming. Chromatography itself can also be a very slow process, in particular two-dimensional thin-layer. An advantage of these techniques is that they can yield not only qualitative but also quantitative data, which is very important when toxic and other substances are analysed.

12.2.2 Protein electrophoresis

Protein electrophoretic techniques have become the most widely used techniques in biochemical characterization of plant populations, and have been used in more species, populations and samples than any other technique. Both enzymatic and non-enzymatic proteins, generally seed storage proteins, are analysed for this purpose. A wide number of electrophoretic techniques are available which, in general, are relatively inexpensive and fast, so that population samples can be analysed at a reasonable cost and time. These techniques will be briefly described here, although other special purpose techniques will also be mentioned.

Electrophoresis is a technique for separating molecules based on their differential mobility through a solvent in an electric field. Molecules and solvent (a buffer) are generally imbibed in an inert supporting medium in which the molecules can move. For biological polymers, the suitable supporting media are gels made with starch, polyacrylamide or agarose in a buffer solution, and the most widely used technique is zone electrophoresis, which hereafter will be designated only as electrophoresis. An additional advantage of the gels is that they are not only supporting media but they also act as a molecular sieve, therefore molecules are also separated as a result of their size and shape in addition to the main factor determining mobility, the net electric charge.

Electrophoresis can be carried out in gels formed in tubes or slabs, the latter either horizontal or vertical. The most general supports for protein electrophoresis are polyacrylamide and starch. Nucleic acids are resolved using agarose or polyacrylamide gels when respectively restriction fragments or sequences have to be fractioned. The characteristics of starch gels depend upon the starch batch and the method used to manufacture the gel, therefore it is not possible to determine pore size. On the other hand, pore size in a polyacrylamide gel may be predetermined either by adjusting the total percentage of polyacrylamide dissolved in the buffer, or the amount of crosslinker. Furthermore, when there is a wide range in the molecular weights of the material under study, pore gradient gels can be prepared. A slightly different technique is isoelectric focusing (IEF), also called electric focusing. This is a method in which proteins are separated in a pH gradient according to their isoelectric points, that is, the pH at which their net electric charge is zero. The pH gradient is generated by low-molecular-weight synthetic polyampholytes added to a polyacrylamide gel.

An alternative to one-dimensional methods previously described is two-dimensional gel electrophoresis. This technique has a higher resolution in separating different proteins and it is used to separate complex mixtures of proteins into many more components than is possible in conventional one-dimensional electrophoresis. Each dimension separates proteins according to different properties, the first dimension is run in a tube gel, which is then layered horizontally onto the top of a polyacrylamide or starch slab.

(a) Isozyme electrophoresis

In 1959, Markert and Moller introduced the term 'isozyme' to define each one of the possibly many multiple molecular forms of an enzyme. The different molecular forms of an enzyme, which are coded by different alleles of the same locus, are called allozymes, now reserving the term isozyme for enzyme forms which catalyse the same reaction but are coded by more than one locus. Throughout this chapter the term isozyme will be used in its broadest sense. Several compilations of recipes and methods have been published (Brewer and Sing, 1970; Shaw and Prasad, 1970; Shields *et al.*, 1983; Vallejos, 1983; Kephart, 1990). In all methods the isozymes are observed after enzyme system specific staining based on their own catalytic properties, thus electrophoretic conditions have to be non-denaturing. A detailed description of the procedure for starch gel and vertical polyacrylamide gel (PAGE) electrophoresis for isozyme analysis is presented by Shields *et al.* (1983).

(b) Storage protein electrophoresis

Population electrophoretic studies using non-enzymatic proteins of plants have been carried out mainly with seed storage proteins, although other storage proteins such as tuber proteins have been used. Most electrophoretic techniques yield a high number of discrete bands on gels, usually greater than that obtained with any single isozyme pattern. Thus, polyacrylamide gel electrophoresis (PAGE) of storage proteins has been considered the most powerful single system for identifying genetically uniform plant varieties (Nielsen and Johansen, 1986).

As for isozymes, there are techniques that need only a small amount of tissue to obtain sufficient proteins for analysis. For those species with relatively large seeds, a part of the endosperm or the cotyledons can be used to extract proteins and the rest of the seed saved for further use, or alternatively halves of seeds can be differentially extracted. The extracts, sometimes without any further purification, are inserted in polyacrylamide gels. PAGE can be conducted with or without the inclusion of sodium dodecyl sulphate (SDS). The advantage of SDS–PAGE is that proteins are separated by their molecular weight and it has become one of the most widely used techniques for seed storage protein separation. Since storage proteins lack enzymatic activity they are revealed on gels by means of general protein staining techniques. Greater resolution can be achieved with two-dimensional electrophoresis (2D–PAGE) and with isoelectric focusing (IEF). Two-dimensional electrophoresis offers a higher resolution but its use is generally reserved for special studies of a given protein component. In spite of the greater band resolution of the IEF technique, it has not been used as much as PAGE, probably because of its higher cost. In addition to electrophoresis, the reversed-phase high-performance liquid chromatography (RP-HPLC technique) has been proposed for analyses of cereal proteins (Smith, 1986), used in varietal studies of zeins, corn seed storage proteins (Smith, 1989) and glutelins of sorghum (Smith and Smith, 1988).

12.2.3 Electrophoresis of nucleic acids

The analysis of nucleic acids using electrophoresis has undergone an extraordinary development. It is now possible to apply these techniques to the separation of fragments as long as chromosomes or as short as a few tens of nucleotides. Their precision allows for the separation of nucleotide sequences differing in length by a single nucleotide. For the characterization and the estimation of genetic variability in populations, electrophoresis of nucleic acids is a more arduous and expensive technique than the electrophoresis of proteins, at least when crude or low purified protein extracts are used. Therefore, they have not been replaced by nucleic acid techniques and both, protein and nucleic acid electrophoresis, are complementary techniques and probably will be for a long time.

Nucleic acids have a uniform density of electric charge along the molecules and the electrophoretic separation is based on their length. Electrophoretic techniques can be applied to RNA, genomic DNA and cDNA (complementary DNA obtained from an RNA template, normally messenger RNA, by means of reverse transcriptase). The most common techniques at the population level are the analyses of restriction fragment length polymorphisms (RFLP) and the sequencing of genomic DNA and cDNA. In both cases it is necessary to obtain a given DNA segment in the right amount (micrograms or even nanograms) and purity level to be used as probe or to be sequenced. Prior to this, the segment has to be cloned in an adequate vector, either a plasmid or a virus, then replicated, usually into bacteria, excised from the cloning vector and purified (Chapter 8). Recently the polymerase chain reaction (PCR) technique allows for the *in vitro* multiplication of a DNA sequence without the cloning step (Innis *et al.*, 1990). Nucleic acid manipulation is also possible thanks to the use of restriction endonucleases. Each restriction enzyme cleaves genomic DNA into discrete segments which can be cloned and isolated. There are a number of recently published books describing cloning and other nucleic acid manipulation techniques (Sambrook *et al.*, 1989; Schuler and Zielinski, 1989), hereafter the chapter will focus on the use of these techniques in population studies.

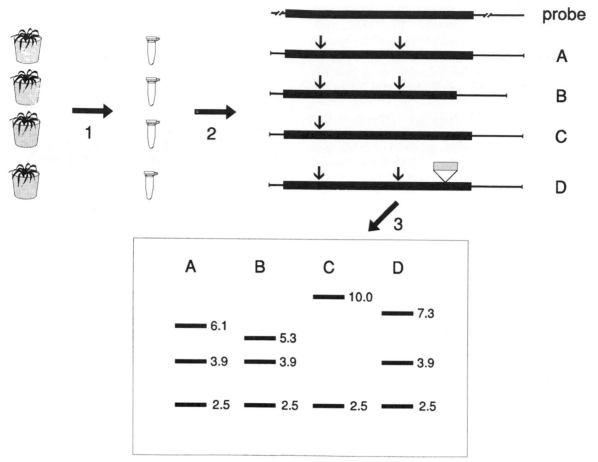

Fig. 12.1. Outline of the RFLP method. (1) DNA extraction from individual plants; (2) digestion of DNA by means of restriction enzymes; (3) electrophoresis in agarose gels. Changes in the position and/or the number of fragments in gels revealed by a given probe indicate the changes in relation to a 'standard' sequence (A). Deletions (B) and insertions (D) change the mobility of affected segments. Base changes that affect restriction enzyme target points (C) change the number of fragments and their relative mobility. Numbers indicate hypothetical fragment length in base pairs × 100.

(a) Restriction fragment length polymorphisms

This technique determines the differences in length of DNA segments and the loss or gain of target points for a given restriction enzyme. Changes in the electrophoretic mobility of a DNA fragment generated by a restriction enzyme indicate differences in the length of a restriction fragment due to insertion or deletion. Changes in the number of fragments together with the appearance of new ones indicate the loss or gain of restriction sites for

such an enzyme, caused by base substitution or by insertion or deletion, whose end points fall within the site (Fig. 12.1).

An outline of the technique would be:

1. Total genomic DNA is isolated from different individuals of populations. A few leaves or even a piece of a leaf is usually enough to provide sufficient amounts of DNA, thus the technique is non-destructive and plants can be saved.

2. DNA samples are digested with a restriction enzyme; small portions of the DNA of each

plant can be digested by different endonucleases and electrophoresed separately.

3. The fragments obtained are separated by electrophoresis in a gel, usually agarose. Samples are located in a row of small wells near one end of the gel and then electrophoresed for several hours. Since the genome of higher plants ranges roughly from 10^9 to 10^{11} base pairs, a given restriction enzyme generates a huge number of discrete fragments resulting in a practically continuous trace of different length DNA fragments. As for isozyme electrophoresis, the answer is to use a method which will reveal only a set of fragments. This can be achieved by the use of specific probes.

4. After electrophoresis, the DNA fragments are denatured and blotted (Southern blotting) to a support membrane where they are firmly placed in the same order as in the gel.

5. Denatured fragments on the membrane are hybridized with a labeled single-strand DNA probe. Labeling has been traditionally achieved by means of radioactive nucleotides, but non-radioactive methods are now available. There are two basic types of DNA sequences that are useful as probes. One is cDNA, which represents coding sequences. The other type of probe is made from genomic clones: total nuclear DNA (coding and non-coding) is digested with a restriction enzyme and the fragments obtained are cloned in a vector.

6. If radioactive probes are used, an autoradiography of the membrane will reveal the set of fragments complementary to the probe. With non-radioactive probes, antibodies against the modified nucleotides and a coupled enzymatic reaction are used to show up the set of fragments directly on the membrane. In both cases, the result is a pattern of discrete bands for each individual of the population.

(b) Random amplified polymorphic DNA

The polymerase chain reaction (PCR) amplifies by several orders of magnitude discrete sequences between the annealing sites of short DNA primers by means of a thermoresistant DNA polymerase and repeated cycles of denaturing, annealing and polymerization. In population studies, genomic DNA is used to amplify one or several sequences, then all the DNA is loaded on agarose or polyacrylamide gels, electrophoresed, and the amplified sequences are observed as discrete bands directly on gels. When the distance between the primer annealing sites varies between individuals the banding pattern generated by the PCR is a genetic polymorphism which can be used in the same way as other genetic markers in population studies.

Primers are short single-stranded DNA segments, usually 15–20 nucleotides in length, although primers as short as 5 nucleotides in length can be used. Two kinds of primers are used, those that target known sequences and those that amplify anonymous unknown sequences. Primers with a known target usually generate one or few discrete bands (segments generated by restriction enzymes are sometimes used to generate additional length polymorphisms from these PCR products), while short primers of arbitrary sequence generate a complex but characteristic spectrum of DNA products, random amplified polymorphic DNA (RAPDs). These patterns, or DNA amplification fingerprints, are being rapidly introduced among the methods for population identification (Caetano-Anollés *et al.*, 1991; Weining and Langridge, 1991). Further details of PCR and the use of RAPDs are presented in sections 8.4 and 19.7.1.

(c) DNA sequencing

Several methods have been described for determining the sequence of nucleotides in DNA. All these methods rely on the use of restriction enzymes to provide specific points of reference for sequence determination. In general, DNA sequences are determined in single-stranded fragments of 100 to a few hundred nucleotides electrophoresed in polyacrylamide gels, either conventional vertical slab gels or the latest capillary gels (Swerdlow and Gesteland, 1990). Large sequences are formed from smaller overlapping sequences. Sequencing techniques require single-stranded DNA radioactively labelled to detect by autoradiography the ladders of fragments differing from each other by a single nucleotide in length. All sequencing techniques create single-stranded fragments that begin at a unique point and terminate at specific bases (G, C, etc.) at different distances from the unique point. The principal difference among sequencing methods is the way in which the DNA is labelled and the set of fragments which are generated (Sambrook *et al.*, 1989).

12.3 APPLICATIONS OF MOLECULAR MARKERS

12.3.1 Low-molecular-weight markers

The advantage of this group of heterogeneous chemical compounds for population and varietal identification is the astonishing number of different molecules found in plants, which makes them very suitable for chemical characterization of populations. Their disadvantage is that the genetic control of their production is generally unknown. Data obtained from the analysis of these products have to be analysed like morphometric data and the methods of numerical taxonomy applied. Either way, determining their presence–absence is a prime objective if they are resistance factors or toxic substances. This group of markers is currently less used than proteins and DNA in population studies. A review of the literature from 1985 to 1990 showed that phenolics, anthocyanins or flavonoids were studied in 31 works distributed among crop groups as follows: cereals 6, legumes 2, ornamentals 6, trees and shrubs (mostly fruit trees) 14, cotton 2 and tomato 1. In 11 of them the relationship between phenolics and resistance to pests and pathogens was investigated.

12.3.2 Isozymes

Unlike low molecular weight markers, the genetic basis of isozyme and storage protein variability is one of the reasons why these markers became as widely used as they are for population characterization. Although the characterization and identification can be carried out on the basis of phenotypic differences of electrophoretic band patterns, the advantage of isozymes is that these patterns can usually be interpreted in terms of loci and alleles. The phenotypic approach to characterizing and comparing populations has been criticized because it does not provide genetic information (Crawford, 1983; Simpson and Withers, 1986). Therefore, isozymes are ideal genetic markers when estimating genetic variability and characterizing plant populations. The advantages of isozymes over other biochemical markers are: (a) allelic expression is generally codominant, free of epistatic interactions and usually unchanged by environmental effects; (b) alleles of different loci are generally distinguishable; (c) enzymatic systems to be studied are usually chosen for technical reasons independent of their level of genetic variability; as a result of this they can represent a random sample of the genome; (d) allelic differences are always detected as mobility differences, independent of the functioning and level of variability of each enzyme system.

Two main drawbacks have been used in arguments against the use of isozymes, and proteins in general, in population and evolutionary studies: (1) not all the genetic changes occurring at the DNA level are detected at the protein level (changes in introns and flanking sequences, synonymous codon changes, changes in amino acid without change in the net electric charge of the protein); (2) only one set of structural genes of organisms are represented in these proteins and this set may not be representative of the whole genome. These disadvantages are certainly limitations to the value of proteins in evolutionary studies, but they are of little relevance to the characterization of populations. It is worth remembering that populations can be characterized only on the basis of phenotypes, morphological or biochemical, and that proteins represent the viable introduction of genotypes and allelic frequencies for the characterization and discrimination of populations, species and higher taxa. Until DNA techniques were introduced in population and evolutionary studies, protein electrophoresis was the most suitable technique, and it is still a very powerful tool.

A limiting characteristic of isozyme systems in population studies is the relatively low number which can actually be observed by gel electrophoresis and specific staining. Of the more than 3000 enzymes known in plants, only about 60 have been assayed for isozyme polymorphism (Vallejos (1983) listed 57 different isozyme systems). In spite of its limitations, starch gel electrophoresis is the most commonly used method for the isozymatic characterization of populations since it is the most versatile and simple electrophoretic technique and, as was pointed out earlier, simple but reliable techniques represent a better ratio of work/results for population studies. Usually starch gels are wide enough to slice them in several thin slabs, staining each one for a different isozyme system. This represents an easy method of obtaining simultaneous data for several loci, from each of the plants analysed. This multilocus information gives better data on the genetic structure of populations, and is of particular relevance for inbreeding species where this structure can be preserved and plays an important role in the adaptation of populations (Allard, 1988; Pérez de la Vega *et al.*, 1991).

In plant studies caution is often necessary due to the polyploid nature of many plant species. In such species, due to the duplication of genes, formal genetic studies are needed to distinguish between homologous and homeologous alleles, and therefore to obtain a full genetic understanding of the results.

Isozymes have been used as markers in formal genetic studies, genetic mapping, population and evolutionary studies, relationship with quantitative traits, identification of cultivars and varieties, genetic resources management, and breeding. Several hundred plant species have been the subject of isozyme studies, including all major families from horsetails and ferns to gymnosperms and angiosperms. Not surprisingly most of the studies have been devoted to crop species and their wild relatives, forest trees and weeds. There are numerous reviews on plant isozymes, but of those dealing with crop (including forest trees) population genetic variability and/or charac-

terization, the following are of importance: Tanksley and Orton (1983), Nielsen (1984), Simpson and Withers (1986), Kahler and Price (1987), Chapman (1989), Hamrick and Godt (1990), Muona (1990). Of these, the review by Simpson and Withers is particularly devoted to plant characterization by isozyme electrophoresis.

Cereals have received far more research effort than any other crop and related species group, although the amount of work on fruit tree characterization is increasing. Maize, one of the pioneer species in population and enzymatic studies, is the most widely examined single species (Table 12.1). The number of samples analysed ranges from 1 to 757, the average being 82.6. The lower numbers are often directed to objectives other than the characterization of populations (e.g., mating system analysis). Those works assaying a high number of samples are usually devoted to comparisons of several species of a genera, the whole collection of a country, or a world-wide collection. The numbers of isozyme systems and loci also vary widely; the averages and ranges of loci and isozyme systems are respectively 16.7 (1 to 70) and 8.4 (1 to 42), although these figures are approximate since some papers only report the polymorphic loci whereas others indicate both polymorphic and monomorphic ones.

The usefulness of isozymes in the biochemical characterization of populations relies on the amount of genetic variability within and between them. Of course, this is also true for other genetic markers, both biochemical and morphological. Among the factors determining the extent of genetic variability are the mode of reproduction and breeding system (Chapter 2 and Hamrick and Godt, 1990). Thus variability within populations increases steadily from clones, genetic pure lines, 'natural' populations of inbreeding species, to outbreeding species. This variability determines the number of individuals and enzyme systems needed to ascertain the degree of uniformity and to distinguish between populations. The number of individuals and loci required to characterize populations increases in proportion to the variability within populations. A few plants of asexually propagated clones could be enough,

Table 12.1. Genera (crops species, wild relatives and weeds) in which biochemical markers have been recently[a] used in population comparative studies[b]

Crops	Isozymes				Storage proteins		Crops	Isozymes			
	NW	NS	NI	NL	NW	NS		NW	NS	NI	NL
Cereals and related species							Trees and shrubs				
Avena	3	64	9	15	1	9	Castanea	1	2	14	—
Dactylis	1	5	7	9			Citrus	1	15	6	7
Festuca	2	13	20	—	1	36	Coffea	1	1	5	—
Hordeum	12	92	21	39	4	19	Malus	2	25	3	5
Lolium	2	33	5	6	2	11	Persea	1	2	4	—
Oryza	3	5	14	24			Prunus	6	131	9	12
Panicum	1	110	11	19							
Pennisetum	2	128	12	48	1	150	Horticulturals				
Saccharum					1	10	and ornamentals				
Secale	5	11	7	14	3	26	Beta	1	394	1	3
Setaria	2	400	—	22			Brassica	1	20	8	—
Sorgum	1	34	14	26			Cucumis	1	757	12	18
Tripsacum					1	51	Cucurbita	2	27	6	12
Triticum	5	52	16	48	10	154	Dianthus	1	100	7	10
Zea	9	92	12	23	1	397	Lactuca	1	31	42	70
Zizinia	1	33	11	17			Raphanus	1	28	8	9
Legumes							Other crops				
Glycine	1	12	17	42			Amaranthus	1	72	9	9
Lens	1	67	9	15			Chenopodium	2	60	6	21
Medicago	1	27	3	5			Fagopyrum	2	36	12	19
Phaseolus	2	73	14	22	2	40	Gossypium	2	153	12	24
Vicia					1	24	Helianthus	1	11	—	10

[a] Data from 1985.

[b] NW = number of published works 1985–1990; NS = number (or average) of samples analysed; NI = Maximum number of isozyme systems described; NL = maximum number of loci.

pure lines obtained by selection can be identified with a few tens of plants, the lowest number accepted as useful is ten. For populations of self-pollinated species such as land races or natural populations of wild plants between 50 and 100 plants would be needed. Frequently, the characterization is based on the allelic or genotypic frequencies of these normally variable populations. A better approach is to characterize these populations by means of multilocus genotype frequencies (Pérez de la Vega *et al.*, 1991).

Outbreeding populations are characterized by means of allelic frequencies at several loci. The number of individuals analysed should be greater than 100. A statistical method of clustering, on the basis of allelic frequencies, can be used to relate the populations. This system of identification and classification has been successful in several cases. It is sometimes difficult, however, to distinguish between populations of outbreeders such as rye although collected from far and wide and the number of individuals analysed was large, and even when populations were identified as belonging to different *Secale* species (Pérez de la Vega and Allard, 1984; Vences *et al.*, 1987). Nevertheless, Adam *et al.* (1987) were able to distinguish between commercial varieties of rye with samples of 200 individuals by means of a likelihood–ratio test.

The next issue to be dealt with is 'how many enzyme systems should be analysed?'. Although there are theoretical estimations from the population genetics approach, experimental evidence shows that in most studies the number of systems surveyed is close to 10, which could represent about 15 polymorphic loci (Table 12.1). In most instances this number of loci is enough to identify populations, especially if multilocus genotypes are scored. Some studies rely on a low number of highly polymorphic loci (esterases, phosphatases and peroxidases are habitually the most polymorphic systems in plants), but it is difficult to distinguish even between cultivars of inbreeding species with one or two systems.

In addition to the characterization and identification of cultivars, varieties, and natural populations, isozyme electrophoresis also is used for the following.

1. Varietal uniformity assessment. One of the tasks in commercial crop production is to maintain cultivars with a minimum level of off-types. Electrophoresis is a rapid method for detecting admixtures with other cultivars, gene flow from nearby fields, and other kinds of contamination (Arús, 1983). Large samples have to be analysed if the frequency of an alien genotype is low.

2. Phylogenetic studies. One of the most important applications of isozymes has been the inference of phylogenetic relationships between species and even higher taxa. The number of papers devoted to this kind of studies is too high to enumerate here (Crawford, 1983; Simpson and Withers, 1986).

3. Estimation of mating system and selection parameters. Mating system and selection are two main factors determining plant population genetic structure (Chapter 2), so it is natural that a great effort has been devoted to estimating their effect on populations of wild and cultivated plants. Data on how to estimate these parameters and reviews on empirical works can be found in recently published books (Tanksley and Orton, 1983; Brown *et al.*, 1990; Weir, 1990).

4. Evaluation of somaclonal variation. Isozymes and storage proteins are among the genetic characteristics used to ascertain somaclonal variation arising during *in vitro* plant tissue culture and to evaluate its potential in plant breeding (Evans and Sharp, 1986).

5. Plant genetic resources conservation and management, and the identification of

centres of domestication and diversity. One of the objectives to which much research has been directed is a better management of resources through the characterization of germplasm accessions and the evaluation of genetic variability in wild populations as potential sources of new desirable genes for breeding purposes (Chapter 3; Brown *et al.*, 1989, 1990; Tanksley and Orton, 1983).

6. Genetic linkage mapping. Tanksley (1983) has stressed the utility of molecular markers in building tight linkage maps and their use in manipulating desired loci in plant breeding. Their relevance is based on the possible application of easily observable markers tightly linked to genes of particular importance, such as resistance genes or quantitative trait loci (QTL) (Chapter 19). Genetic linkage between isozyme genes and resistance and the association of isozyme patterns with resistance have been found in several crop species. Likewise, the development of multilocus associations among virulence and other traits, including isozyme genes, within haplotypes is especially important (Allard, 1990).

12.3.3 Storage proteins

Storage proteins possess a high level of polymorphism, and are very stable. Environmental factors have little or no effect on their presence in mature seeds, although they may affect the amount of protein stored. Like isozymes, they are generally inherited codominantly. Therefore electrophoretic patterns of mature seed proteins are excellent criteria for characterizing and identifying plant populations and cultivars, either alone or together with other markers. Table 12.1 summarizes some recent works devoted to varietal identification. Of course, most analyses of storage proteins in crop species have been developed for those species (cereals and legumes) in which the seed, or its products, is the edible part, but seed proteins have also been used to identify pasture cultivars (Gardiner and Forde, 1988).

Special care has to be taken in the interpretation of differences in electrophoretic band width and intensity when individual seed analyses are carried out. Differences can be due to a different weight/volume ratio in the extraction procedure and/or to different seed protein concentrations among seeds, due to micro-environmental factors during seed formation. Another disadvantage of storage proteins is the difficult genetic interpretation of results if no previous formal genetic studies on their control have been carried out. Information is now available on the genetic control of seed proteins for several of the main crop species such as barley, wheat, maize, soybean, bean, etc. (Gepts, 1990). A difficulty in interpreting the genetics of seed proteins is the multigenic nature of these proteins as has been demonstrated in several plant species (e.g., wheat, maize, pea, bean). Usually, each protein class is coded for by a complex locus, gene cluster, or a family of tandemly arranged sequences of DNA. Heterogeneity in electrophoretic patterns and/or band intensity could be due to: (a) different number of copies in clusters; (b) heterogeneity among sequences within a family; (c) different products of the post-translational processing of proteins in the pathway to their deposition in the protein bodies. Due to this multigenic nature, storage proteins are far more suited to identifying varieties of clonally propagated or inbreeding crops, with low or no internal variability, than to identifying those of outbreeding species which are genetically heterogeneous. In these latter species, and only if the genetic control is known, can the electrophoretic data be transformed in allelic frequencies and proper comparisons carried out.

An additional relevance of the analyses of seed storage proteins in breeding programmes is the relationships between some particular alleles and desirable qualities of seed and seed products. Some alleles of wheat glutenins confer a better bread baking and pasta quality (Payne *et al.*, 1979; Carrillo *et al.*, 1990); in several crop species particular proteins improve their nutritional value due to a better amino acid balance; others confer resistance to pest and diseases (Riggs *et al.*, 1983; Osborn *et al.*, 1988).

This highlights the importance of these proteins in breeding programmes.

12.3.4 DNA polymorphisms

Molecular biology has provided new more powerful tools for studying genetic variation with a greater resolution than all previous experimental methods, including protein electrophoresis. However, current knowledge of molecular diversity in plant populations and species is still scarce (Clegg, 1990). It is limited to a few species and a few genes which frequently are multicopy genes. The main reason, in addition to the relative novelty of the techniques, is that DNA methods are time-consuming and relatively expensive, which means that to date sample sizes analysed have been small from the point of view of population studies, and therefore the ability to test hypotheses statistically has been limited. But the circumstances are rapidly changing and more and more data are accumulating.

The main advantage of nucleic acid techniques is that they give information of any kind of sequence, not only of isozymes or highly expressed non-isozymatic proteins such as storage proteins. Now it is possible to gain information about the whole genome and any of its components, surpassing the limitations of protein electrophoresis, which only gives information on translated sequences. Molecular genetic methods have overcome two main limitations of protein electrophoresis since the detection of variation is not limited to coding regions and all categories of mutational events can, in principle, be detected.

The information obtained with the RFLP technique depends upon both the number of probes available and the number of restriction enzymes used. Each different probe hybridizes with a different set of genomic DNA fragments and each enzyme excises a segment of genomic DNA at different points. Restriction sites can be assigned to genetic loci, and data interpreted in genetic terms. Most of the population analyses are carried out with anonymous probes obtained from genomic DNA, repetitive or single copy, or from messenger RNAs copied to cDNA. Each random genomic or cDNA probe may represent a single locus. The unbiased selection and the assignment to loci allow a direct estimation of level of the genetic variation in populations and species.

Many of the non-coding fragments cloned from genomic DNA are selectively neutral and represent sequences that are more rapidly diverging than cDNA clones. Some lie in regions that are highly polymorphic and therefore they are suitable to distinguish closely related varieties and cultivars. On the other hand, cDNA probes represent fairly well conserved sequences and this enables them to be used as markers across less related taxonomic groups. Several mathematical models have been published that can be used to analyse RFLP data at a population and evolutionary level (Nei, 1987; Nei and Miller, 1990; Weir, 1990).

A major advantage of sequencing techniques is that they let us know precisely the number and kind of genetic changes at the nucleotide level, being closer than ever to the absolute measurement of genetic variability within and between populations and species.

Sequencing techniques are expensive, laborious and time consuming. These are the reasons why, in spite of their power and ability to give genetic information and the continuous increase in the number of plant genes sequenced, they have been used in only a few population studies (Clegg, 1990); that is to say, the same gene has been sequenced in a significant number of individuals of a population or species. The polymerase chain reaction (PCR) represents a very promising technique in molecular DNA analysis devoted to detect multiple DNA sequence variation (Helentjaris *et al.*, 1989; Shattuck-Eidens *et al.*, 1990). Either way, up to now, RFLPs are the most effective DNA technique in identifying population or varieties, although the increasing use of RAPDs will undoubtedly influence this situation. The advantages of RAPDs on RFLPS are that RAPDs are generated by a faster and easier technique and within a few years they will probably contribute many more markers than RFLPs to population and evolutionary studies. Polymorphisms generated by PCR have already demon-strated their usefulness in identifying cereals and legumes (Caetano-Anollés *et al.*, 1991; Weining and Langridge, 1991).

12.3.5 RFLPs in plant population studies

In spite of the short period during which RFLP analysis has been used, its importance in the assessment of plant diversity, population characterization and plant breeding has already been stressed (Chapter 19; Helentjaris *et al.*, 1985; Beckmann and Soller, 1986). While the limited sample of individuals assayed in DNA analysis is a disadvantage, the advantage is that the number of loci that can be detected is many orders of magnitude greater than for isozymes. More than 1000 RFLP markers have been mapped on the tomato genome (Tanksley, personal communication to senior editor), whereas the maximum number of isozyme loci available for study is much lower; 70 is the highest number of loci reported in a single study (Table 12.1). Another advantage is that the variation in DNA sequences is several times higher than protein sequences.

An additional advantage of eukaryotic organisms is that not only the nuclear genome but also organelle genomes can be analysed by RFLP or sequencing techniques. In plants, mitochondrial and chloroplast genomes are being used in population and evolutionary studies (Palmer, 1987). The chloroplast genome in vascular plants is similar in size, gene content and gene order, and the nucleotide sequences of chloroplast genes are highly conserved (Sugiura, 1989), but RFLP analyses can characterize some plant populations or races on the basis of particular cpDNA variants. On the other hand, the plant mitochondrial genome has evolved rapidly in structure, but slowly in sequence, it is a highly plastic entity that is capable of tolerating considerable structural variation with little or no effect on function (Gray, 1989). Thus, again RFLPs could be used to characterize plant populations or races. A recent review on organelle polymorphism in plant populations has been published by Clegg (1990).

Two particular uses of organelle DNA variants are of great interest in plant breeding and population characterization:

1. Determining the causes of male sterility. Mitochondrial DNA variants and mitochondrial plasmid-like DNAs have been related to male sterility in plant species. The relationships between male sterility and varietal changes in mtDNA have been studied in cotton, rice, tobacco, beans and maize (Chapter 14).

2. Determining the maternal ancestor. Since in most plants species organelles are transmitted only by the egg cell, DNA sequences are not changed by recombination among different molecules, and particular variants can be used to ascertain the maternal parent of hybrid populations and species. Clegg (1990) has pointed out the potential in determining plant population structure and gene flow by the combined study of maternally and paternally transmitted genetic markers.

The majority of population plant molecular diversity surveys have focused on the nuclear encoded ribosomal RNA (rDNA) gene family. Several papers on length variation (RFLP) of rDNA intergenic spacers in crop plants and their wild relatives have appeared over the last few years (Rogers and Bendich, 1987). These authors listed 57 species belonging to 32 genera in which data on ribosomal spacer length were known. Among them, population and/or evolutionary studies had been carried out on species of 18 genera. From 1987 more than 15 papers on population variability or phylogenetic relationships studies using rDNA spacer polymorphism have been published; the genera investigated were *Aegilops*, *Brassica*, *Clematis*, *Cucumis*, *Glycine*, *Hordeum*, *Oryza*, *Pennisetum*, *Raphanus*, *Secale* and *Triticum*. Recently, Rocheford *et al.* (1990) have related ribosomal intergenic spacer variability to grain yield in maize. The reasons for the extended use of rDNA spacer polymorphism are the multicopy nature of rDNA that eases the level of detection, and the advantage of the different probes available. However, the frequently heterogeneous nature of rDNA at an individual plant

level makes the precise analysis of rDNA variation difficult at a population level. An individual plant can be heterogeneous with respect to rDNA for three reasons: (1) there are two or more length variants within a single rDNA cluster; (2) different clusters have different length variants; (3) heterozygosity, that is, different length variants on homologous chromosomes.

RFLPs other than rDNA are being increasingly used in comparative studies between related species and between cultivars of crop species. Over the last few years data on several genera have accumulated. Song *et al.* (1990) analysed 38 *Brassica* accessions by means of 33 random nuclear DNA sequences, the *cruciferin* gene and 12 clones of cpDNA. Relationships between *Glycine* species and varieties using anonymous genomic DNA and the urease gene probe have been established by Menancio *et al.* (1990), and cultivars of *G. max* have been studied by Apuya *et al.* (1988) and Keim *et al.* (1989). Cultivars of *Hordeum vulgare* were characterized by means of *Hor* loci probes (Bunce *et al.*, 1986). Accessions of *Lens* species were analysed on the basis of low-copy-number fragments (Havey and Muelhbauer, 1989). Gebhard *et al.* (1989) identified $2n$ breeding lines and $4n$ varieties of potato with RFLP fingerprinting. Harbed *et al.* (1986) used DNA restriction fragment variations of the high molecular weight (HMW) glutenins as criteria to distinguish *Triticum aestivum* cultivars; and Anderson and Greene (1989) compared the sequence of HMW glutenin genes of different genomes and cultivars of *T. aestivum*.

As RFLPs are of extraordinary utility in population and variety identification, they will undoubtedly play a significant role in plant improvement programmes. The inclusion in breeding programmes of a high number of genetic molecular markers, isozymes and RFLPs, has shown that both are effective in identifying and locating quantitative trait loci (QTLs) (Chapter 19). Several papers have described the theoretical rationale of the use of mapped markers (Paterson *et al.*, 1988; Bernatzky and Tanksley, 1989; Young and Tanksley, 1989; Melchinger, 1990; Stuber, 1990). So far, most of the experimental work using

RFLPs, and also isozymes, in relation to QTLs and resistance genes has been done on maize and tomato (Bernatzky and Tanksley, 1989; Stuber 1990). In tomato the association of RFLPs with genes resistant to insects and tobacco mosaic virus, or to genes determining water use efficiency or ripening has been established.

As Bernatzky and Tanksley (1989) pointed out, the use of DNA markers in germplasm and population studies is now limited by cost and the relatively high level of technical equipment required. However, technical advances are making these methods more widely accessible. Isozyme techniques, however, are cheaper and simpler and, although they are limited in the level of polymorphism and the number of detectable loci, they have provided and can still provide a good first estimation of genetic variability. Therefore a rational use of both types of analyses should yield the best results for the biochemical characterization of populations, and the understanding of genetic variability in populations, species and higher taxa.

ACKNOWLEDGEMENTS

I thank Dr Carrillo, Dr Ruiz and Dr García for helpful comments, Dr Arús for providing some references, and C. Polanco for drawing of the figure.

REFERENCES

Adam, D., Simonsen, V. and Loeschcke, V. (1987), Allozyme variation in rye, *Secale cereale* L. 2. Commercial varieties. *Theor. Appl. Genet.*, **74**, 560–565.

Allard, R.W. (1988), Genetic changes associated with the evolution of adaptedness in cultivated plants and their wild relatives. *J. Hered.*, **79**, 225–238.

Allard, R.W. (1990), The genetics of host-pathogen coevolution: Implications for genetic resource conservation. *J. Hered.*, **81**, 1–6.

Anderson, O.D. and Greene, F.C. (1989), The characterization and comparative analysis of high-molecular-weight glutenin genes from genomes A and B of hexaploid bread wheat. *Theor. Appl. Genet.*, **77**, 689–700.

Apuya, N.R., Frazier, B.L., Keim, P., Jill Roth, E. and

Lark, K.G. (1988), Restriction fragment length polymorphisms as genetic markers in soybean *Glycine max* (L.) Merril. *Theor. Appl. Genet.*, **75**, 889–901.

Arús, P. (1983), Genetic purity of commercial seed lots. In *Isozymes in Plant Genetics and Breeding*, Part A, Tanksley, S.D. and Orton, T.J. (eds), Elsevier, Amsterdam, pp. 415–423.

Beckmann, J.S. and Soller, M. (1986), Restriction fragment length polymorphisms in plant genetic improvement. *Oxford Surveys Plant Mol. Cell Biol.*, **3**, 196–250.

Bernatzky, R. and Tanksley, S.D. (1989), Restriction fragments as molecular markers for germplasm evaluation and utilisation. In *The Use of Plant Genetic Resources*, Brown, A.H.D., Frankel, O.H., Marshall, D.R. and Willians, J.T. (eds), Cambridge University Press, Cambridge, pp. 353–362.

Brewer, G.J. and Sing, C.F. (1970), *An Introduction to Isozyme Techniques*, Academic Press, New York.

Brown, A.H.D., Frankel, O.H., Marshall, D.R. and Willians, J.T. (1989), *The Use of Plant Genetic Resources*, Cambridge University Press, Cambridge.

Brown, A.H.D., Clegg, M.T., Kahler, A.L. and Weir, B.S. (1990), *Plant Population Genetics, Breeding and Genetic Resources*, Sinauer Assoc., Sunderland, Massachusetts.

Bunce, N.A.C., Forde, B.G., Kreis. M. and Shewry, P.R. (1986), DNA restriction fragment length polymorphism at hordein loci: application to identifying and fingerprinting barley cultivars. *Seed Sci. Technol.*, **14**, 419–429.

Caetano-Anollés, G., Bassam, B.J. and Gresshoff, P.M. (1991), DNA amplification fingerprinting using very short arbitrary oligonucleotide primers. *Bio/Technology*, **9**, 553–557.

Carrillo, J.M., Vázquez, J.F. and Orellana, J. (1990), Relationships between gluten strength and glutenin proteins in durum wheat cultivars. *Plant Breeding*, **104**, 325–333.

Chapman, C.G.D. (1989), Collection strategies for the wild relatives of field crops. In *The Use of Plant Genetic Resources*, Brown, A.H.D., Marshall, D.R., Frankel, O.H. and Williams, J.T. (eds), Cambridge University Press, Cambridge, pp. 263–279.

Clegg, M.T. (1990), Molecular diversity in plant population. In *Plant Population Genetics, Breeding, and Genetic Resources*, Brown, A.H.D., Clegg, M.T., Kahler, A.L. and Weir, B.S. (eds), Sinauer Associates, Sunderland, Massachusetts, pp. 98–115.

Crawford, D.J. (1983), Phylogenetic and systematic inferences from electrophoretic studies. In *Isozymes in Plant Genetics and Breeding*, Part A, Tanksley, S.D. and Orton, T.J. (eds), Elsevier, Amsterdam, pp. 257–287.

Evans, D.A. and Sharp, W.R. (1986), Somaclonal and gametoclonal variation. In *Handbook of Plant Cell Culture*, Vol. 4, Evans, D.A., Sharp, W.R. and Ammirato, P.V. (eds), Macmillan Publ. Co. New York, pp. 97–132.

Gardiner, S.E. and Forde, M.B. (1988), Identification of cultivars and species of pasture legumes by sodium dodecylsulphate polyacrylamide gel electrophoresis of seed proteins. *Plant Var. Seeds*, **1**, 13–26.

Gebhardt, C., Blomendahl, C., Schachtschabel, V., Debener, T., Salamini, F. and Ritter, E. (1989), Identification of 2*n* breeding lines and 4*n* varieties of potato (*Solanum tuberosum*, ssp. *tuberosum*) with RFLP-fingerprints. *Theor. Appl. Genet.*, **78**, 16–22.

Gepts, P. (1990), Genetic diversity of seed storage proteins in plants. In *Plant Population Genetics, Breeding, and Genetic Resources*, Brown, A.H.D., Clegg, M.T., Kahler, A.L. and Weir, B.S. (eds), Sinauer Associates, Sunderland, Massachusetts, pp. 64–82.

Gray, M.W. (1989), Origin and evolution of mitochondrial DNA. *Annu. Rev. Cell Biol.*, **5**, 25–50.

Hamrick, J.L. and Godt, M.J. (1990), Allozyme diversity in plant species. In *Plant Population Genetics, Breeding, and Genetic Resources*, Brown, A.H.D., Clegg, M.T., Kahler, A.L. and Weir, B.S. (eds), Sinauer Associates, Sunderland, Massachusetts, pp. 43–63.

Harbed, N.P., Bartels, D. and Thompson, R.D. (1986), DNA restriction fragment variation in the gene family encoding high molecular weight (HMW) glutenin subunits of wheat. *Biochem. Genet.*, **24**, 579–596.

Havey, M.J. and Muelhbauer, F.J. (1989), Variability for restriction fragment lengths and phylogenies in lentil. *Theor. Appl. Genet.*, **77**, 839–943.

Helentjaris, T., King, G., Slocum, M., Sidestrang, C. and Wegman, S. (1985), Restriction fragment polymorphisms as probes for plant diversity and their development as tools for applied plant breeding. *Plant Mol. Biol.*, **5**, 109–118.

Helentjaris, T., Shattuck-Eidens, D. and Bell. R. (1989), Use of PCR and direct sequencing to compare DNA sequences from homologous regions for several cultivars. *Maize Genet. Coop. News Lett.*, **63**, 109–110.

Innis, M.A., Gelfand, D.H., Sninsky, J.J. and White, T.J. (1990), *PCR Protocols. A Guide to Methods and Applications*, Academic Press, San Diego.

Kahler, A.L. and Price, S.C. (1987), Isozymes in population genetics, systematics and evolution. In *Grass Systematics and Evolution*, Sodestrom, T.R., Hilu, K.W., Campbell, C.S. and Barkworth, M.E. (eds), Smithsonian Institution Press, Washington DC, pp. 97–106.

Keim, P., Shoemaker, R.C. and Palmer, R.G. (1989), Restriction fragment length polymorphism diversity in soybean. *Theor. Appl. Genet.*, **77**, 786–792.

Kephart, S.R. (1990), Starch gel electrophoresis of plant isozymes: a comparative analysis of techniques. *Am. J. Bot.*, **77**, 693–712.

Markert, C.L. and Moller, F. (1959), Multiple forms of enzymes: tissue ontogenetics, and species specific patterns. *Proc. Natl. Acad. Sci. USA*, **47**, 753–763.

McKee, G.W. (1973), Chemical and biochemical techniques for varietal identifications. *Seed Sci.-Technol.*, **1**, 181–199.

Melchinger, A.E. (1990), Use of molecular markers in breeding for oligogenic disease resitance. *Plant Breeding*, **104**, 1–19.

Menancio, D.I., Hepburn, A.G. and Hymowitz, J. (1990), Restriction fragment length polymorphism (RFLP) of wild perennial relatives of soybean. *Theor. Appl. Genet.*, **79**, 235–240.

Muona, O. (1990), Population genetics in forest tree improvement. In Plant Population Genetics, Breeding, and Genetic Resources, Brown, A.H.D., Clegg, M.T., Kahler, A.L. and Weir, B.S. (eds), Sinauer Associates, Sunderland, Massachusetts, pp. 282–298.

Nei, M. (1987), *Molecular Evolutionary Genetics*, Columbia University Press, New York.

Nei, M. and Miller, J.C. (1990), A simple method for estimating average number of nucleotide substitutions within and between populations from restriction data. *Genetics*, **125**, 873–879.

Nielsen, G. (1984), The use of isozymes as probes to identify and label plant varieties and cultivars. In *Isozymes: Current Topics in Biological and Medical Research*, Rattazzi, M.C., Scandalios, J.G. and Whitt, G.S. (eds), Alan R. Liss, New York, pp. 1–32.

Nielsen, G. and Johansen, H.B. (1986), Proposal for the identification of barley varieties based on the genotypes for 2 hordein and 39 isoenzyme loci of 47 reference varieties. *Euphytica*, **35**, 717–728.

Osborn, T.C., Alexander, D.C., Sun, S.S.M., Cardona, C. and Bliss, F.A. (1988), Insecticidal activity and lectin homology of arcelin seed protein. *Science*, **240**, 207–210.

Palmer, J.D. (1987), Chloroplast DNA evolution and biosystematic uses of chloroplast DNA variation. *Am. Natur.*, **103**, S6–S29.

Paterson, A.H., Lander, E.S., Hewitt, J.D., Peterson, S., Lincoln, S.E. and Tanksley, S.D. (1988), Resolution of quantitative traits into Mendelian factors by using a complete linkage map of restriction fragment length polymorphisms. *Nature*, **335**, 721–726.

Payne, P.I., Corfield, K.G. and Blackman, J.A. (1979), Identification of a high molecular weight subunit of glutenin whose presence correlates with breadmaking quality in wheats of related pedigree. *Theor. Appl. Genet.*, **55**, 153–157.

Pérez de la Vega, M. and Allard, R.W. (1984), Mating system and genetic polymorphism in populations of *Secale cereale* and *S. vavilovii. Can. J. Genet. Cytol.*, **26**, 308–317.

Pérez de la Vega, M., García, P. and Allard, R.W. (1991), Multilocus genetic structure of ancestral Spanish and colonial Californian populations of *Avena barbata. Proc. Natl. Acad. Sci. USA*, **88**, 1202–1206.

Riggs, T.J., Sanada, M., Morgan, A.G. and Smith, D.B. (1983), Use of acid gel electrophoresis in the characterization of 'B' hordein protein in relation to malting quality and mildew resistance of barley. *J. Sci. Food Agric.*, **34**, 576–586.

Rocheford, T.R., Osterman, J.C. and Gardner, C.O. (1990), Variation in the ribosomal DNA intergenic spacer of a maize population mass-selected for high grain yield. *Theor. Appl. Genet.*, **79**, 793–800.

Rogers, S.O. and Bendich, A.J. (1987), Ribosomal RNA genes in plants: variability in copy number and in the intergenic spacer. *Plant Mol. Biol.*, **9**, 509–520.

Sambrook, J., Fritsch, E.F. and Maniatis, T. (1989), *Molecular Cloning. A Laboratory Manual*, Cold Spring Harbor Lab. Press, Cold Spring Harbor, New York.

Schuler, M.A. and Zielinski, R.E. (1989), *Methods in Plant Molecular Biology*, Academic Press, San Diego.

Shattuck-Eidens, D.M., Bell, R.N., Neuhausen, S.L. and Helentjaris, T. (1990), DNA sequence variation within maize and melon: Observations from polymerase chain reaction amplification and direct sequencing. *Genetics*, **126**, 207–217.

Shaw, C.R. and Prasad, R. (1970), Starch gel electrophoresis of enzymes. A compilation of recipes. *Biochem. Genet.*, **4**, 297–320.

Shields, C.R., Orton, T.J. and Stuber, C.W. (1983), An outline of general resources needs and procedures for the electrophoresis separation of active enzymes from plant tissues. In *Isozymes in Plant Genetics and Breeding*, Part A, Tanksley, S.D. and Orton, T.J. (eds), Elsevier, Amsterdam, pp. 443–468.

Simpson, M.J.A. and Withers, L.A. (1986), *Characterization of Plant Genetic Resources Using Isozyme Electrophoresis: A Guide to the Literature*, Int. Board for Plant Genetic Resources, Rome.

Smith, J.S.C. (1986), Biochemical fingerprints of cultivars using reversed-phase high-performance liquid chromatography and isozyme electrophoresis: a review. *Seed Sci. Technol.*, **14**, 753–768.

Smith, J.S.C. (1989), The characterization and assessment of genetic diversity among maize (*Zea mays* L.) hybrids that are widely grown in France: chromatographic data and isozymatic data. *Euphytica*, **43**, 73–85.

Smith, J.S.C. and Smith, O.S. (1988), Variation in kafirin and alcohol-soluble glutelin chromatograms of sorghum inbred lines revealed by reversed-phase high-performance liquid chromatography. *Theor. Appl. Genet.*, **76**, 97–107.

Song, K., Osborn, T.C. and Williams, P.H. (1990), *Brassica* taxonomy based on nuclear restriction fragment length polymorphisms (RFLPs). 3. Genome relationships in *Brassica* and related genera and the origin of *B. oleracea* and *B. rapa* (syn. *campestris*). *Theor. Appl. Genet.*, **79**, 497–506.

Stuber, C.W. (1990), Molecular markers in the manipulation of quantitative characters. In *Plant Population Genetics, Breeding, and Genetic Resources*, Brown, A.H.D., Clegg, M.T., Kahler, A.L. and Weir, B.S. (eds), Sinauer Associates, Sunderland, Massachusetts, pp. 334–350.

Sugiura, M. (1989), The chloroplast chromosomes in land plants. *Annu. Rev. Cell Biol.*, **5**, 51–70.

Swerdlow, H. and Gesteland, R. (1990), Capillary gel electrophoresis for rapid, high resolution DNA sequencing. *Nucleic. Acids Res.*, **18**, 1415–1419.

Tanksley, S.D. (1983), Molecular markers in plant breeding. *Plant Mol. Biol. Rept.*, **1**, 3–8.

Tanksley, S.D. and Orton, T.J. (1983), *Isozymes in Plant Genetics and Breeding*, Parts A and B, Elsevier, Amsterdam.

Vallejos, C.E. (1983), Enzyme activity staining. In *Isozymes in Plant Genetics and Breeding*, Part A, Tanksley, S.D. and Orton, T.J. (eds), Elsevier, Amsterdam, pp. 469–516.

Vences, F.J., Vaquero, F., García, P. and Pérez de la Vega, M. (1987), Further studies on phylogenetic relationships in *Secale*: On the origin of its species. *Plant Breeding*, **98**, 281–291.

Weining, S. and Langridge, P. (1991), Identification and mapping of polymorphisms in cereals based on the polymerase chain reaction. *Theor. Appl. Genet.*, **82**, 209–216.

Weir, B.S. (1990), *Genetic Data Analysis*, Sinauer Associates, Sunderland, Massachusetts.

Young, N.D. and Tanksley, S.D. (1989), Restriction fragment length polymorphism maps and the concept of graphical genotypes. *Theor. Appl. Genet.*, **77**, 95–101.

Part Five

Manipulation of Genetic Systems

13

Self- and cross-incompatibility systems

D. de Nettancourt

13.1 INTRODUCTION

Sexual incompatibility is a widespread phenomenon in flowering plants which restricts inbreeding within populations (self-incompatibility: SI) and also seems to contribute, through the establishment of unilateral barriers between self-incompatible and self-compatible populations, to the reinforcement of speciation (unilateral incompatibility: UI). Darlington and Mather (1949) estimated that half the species in angiosperms display SI; the generalization of the rule proposed by Lewis and Crowe (1958) that SI × SC species are cross-incompatible would imply that UI operates, on average, in more than one interspecific pollination out of four.

The occurrence of self-incompatibility in many genera which contribute to agriculture and its influence upon plant breeding strategies explain why so much is known to date on its classification and inheritance (Fig. 13.1) and on its distribution (Table 13.1). 'Sporophytic incompatibility', in which the incompatibility phenotype in the pollen is determined by the genotype of the pollen producing plant, occurs abundantly in the Cruciferae and the Compositae and has been particularly well studied in the genus *Brassica*. In certain species, sporophytic incompatibility is associated with floral

polymorphism which reinforces the outbreeding potential of the self-incompatible plant. 'Gametophytic incompatibility', where the genotype of the individual microspore determines the phenotype of the pollen, is typical of the Leguminosae and the Solanaceae (monofactorial control), the Graminae (bifactorial) and the Chenopodiaceae and Ranunculaceae (polyfactorial). Gametophytic incompatibility is characterized by very large polyallelic series at the locus which govern the pollen-pistil relationship. On the other hand, the difficulty in elucidating the molecular biology of the complex, interrelating, cellular and subcellular structures operating in anthers, pollen, stigmata and styles accounts for our relative lack of knowledge on the products of incompatibility genes, the nature of the recognition and rejection mechanisms and a possible relationship between the biochemical control of SI and that of UI.

It is the purpose of the present Chapter to describe the modifications of sexual incompatibility systems, which are possible in the current state of knowledge, and to discuss briefly the plant breeding achievements and innovations which could result from further progress in our understanding of sexual incompatibility.

13.2 ELIMINATION OF THE SELF-INCOMPATIBILITY BARRIERS

Different types of manipulations can be exerted by man on sexual incompatibility. Some of

Plant Breeding: Principles and prospects. Edited by M.D. Hayward, N.O. Bosemark and I. Romagosa. Published in 1993 by Chapman & Hall, London. ISBN 0 412 43390 7.

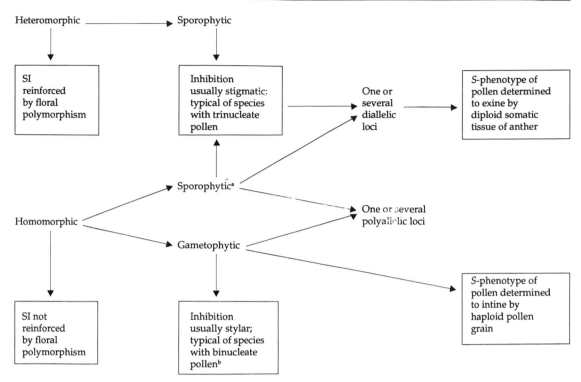

Fig. 13.1. Classification and inheritance of self-incompatibility in the angiosperms. [a]Complemented in certain species by genes which are gametophytic in action (Lewis *et al.*, 1988). [b]Major exception, the grasses where pollen is trinucleate and SI is stigmatic and gametophytic.

them, relatively simple in their means and exploited by plant breeders, lead to the breakdown, either temporary or permanent, of the self-incompatibility mechanism. The techniques available, and their applications in agriculture for the transformation of breeding behaviour or the exploitation of hybrid vigour, are well known (for a review see de Nettancourt, 1977) and it should be sufficient to review here some of their most essential features.

13.2.1 Physiological inhibition

The breakdown of self-incompatibility can be obtained, in several families of plants, through the pollination of buds which do not express the incompatibility phenotype of the mature flower or by means of inhibitors (high temperature, radiations or chemicals preventing RNA or protein syntheses) applied before pollination to the site of the reaction (stigma or style). Such effects are remarkable, from a theoretical point of view, because they contribute to the evidence suggesting that SI does not result from the absence of a stimulation of pollen growth between unlike gene products in pollen and style but from a rejection reaction between like gene products which can be prevented by inhibitors of metabolic activity (for a discussion see Lewis, 1979).

The physiological inhibition of SI is also important for plant breeding because it can be extensively exploited for the creation of the *S*-homozygous lines needed for F_1-hybrid seed production. Lewis (1949) has provided a summary-outline of the method and of its current applications for the exploitation of hybrid vigour through the production of double-

Table 13.1. Self-incompatibility systems recorded in some of the genera which play an important role in plant breeding sciences (from de Nettancourt, 1977). The list of references does not always include the names of the scientists having identified the systems but simply provides a basic reference from which more specific information may be obtained

Genus	Self-incompatibility system recorded	Reference
Beta	Homomorphic, polyfactorial (at least 4 loci), probably gametophytic	Lundqvist *et al.* (1973)
Brassica	Homomorphic, monofactorial, sporophytic	Thompson (1957)
Coffea	Homomorphic, gametophytic	Devreux *et al.* (1959)
Dactylis	Homomorphic, bifactorial, gametophytic	Lundqvist (1965)
Fagopyrum	Heteromorphic, monofactorial, sporophytic	Dahlgreen (1922)
Festuca	Homomorphic, bifactorial, gametophytic	Lundqvist (1961)
Helianthus	Homomorphic, presumably sporophytic	Kinman (1963)
Hordeum	Homomorphic, bifactorial, gametophytic	Lundqvist (1962)
Linum	Heteromorphic, monofactorial, sporophytic	Lewis (1943)
Lotus	Homomorphic, monofactorial, gametophytic	Bubar (1958)
Lycopersicon	Homomorphic, monofactorial, gametophytic	Lamm (1950)
Medicago	Homomorphic, monofactorial, gametophytic	Duvick (1966)
Nicotiana	Homomorphic, monofactorial, gametophytic	East and Mangelsdorf (1925)
Phalaris	Homomorphic, bifactorial, gametophytic	Hayman (1956)
Prunus	Homomorphic, monofactorial, gametophytic	Crane and Brown (1937)
Pyrethrum	Homomorphic, sporophytic	Brewer (1968)
Pyrus	Homomorphic, gametophytic	Lewis and Modlibowska (1942)
Raphanus	Homomorphic, monofactorial, sporophytic	Sampson (1957)
Secale	Homomorphic, bifactorial, gametophytic	Lundqvist (1956)
Solanum	Homomorphic, bifactorial, gametophytic	Pandey (1962)
Theobroma	Homomorphic, monofactorial, gametophytic	Pushkarnath (1942)
Trifolium	Homomorphic, sporophytic	Cope (1962)
	Homomorphic, monofactorial, gametophytic	Duvick (1966)

cross hybrids in cabbages, fodder kale, Brussels sprouts and cocoa.

In addition to the effects of bud pollination and of inhibitors on the promotion of selfing in self-incompatible species, a few reports have been made on the influence of compatible pollen upon the growth of self-incompatible pollen tubes. This influence, often referred to as a 'mentor pollen effect', has been occasionally observed in the case of sporophytic self-incompatibility and, even more rarely, in gametophytic systems (Miri and Bubar, 1966). It occurs when mixtures of compatible and self-incompatible pollen are applied on the same stigma. The technique generally requires, in order to prevent competition between compatible and incompatible pollen, that the compatible pollen be killed by methanol or very high radiation doses (Chapter 5).

13.2.2 Permanent breakdown of sexual incompatibility

(a) Mutations conferring self-compatibility

The sporophytic system of self-incompatibility is not adapted to mutation studies: changes at the locus (usually known as the *S*-locus) or loci, governing SI, cannot be detected unless they are established within the individual anther throughout the majority of tapetal cells which determine the incompatibility phenotype of the pollen. Mutations directly induced in the PMC or in the microspores are not expressed by the pollen grain and are not selected, upon selfing, by the pollen sieve.

In contrast, gametophytic SI, particularly when governed by a single gene, is well suited for mutation studies and large-scale screening tests leading to the recovery of SC mutants. In practice, any fertile pollen grain expressing a self-compatibility mutation will reach the ovule and transmit the mutated character to the next generation. Millions of pollen grains can be submitted, after mutagenic treatment of meiotic buds or microspores, to such a natural screening mechanism. It is probable that most self-compatibility mutants result from the deletion or inactivation of the part of the self-incompatibility locus which governs production in the pollen grain of the incompatibility proteins. Cases are known, however, of mutations which do not lead to modifications of the pollen phenotype but to alterations restricted to the pistil which has lost the capacity to identify and reject self-pollen. It is from the study of SC mutations that the model of a tripartite SI locus, with one pollen activity part, a pistil activity part and a specificity segment common to pollen and pistil, was established by Lewis (1949, 1960).

Self-compatibility mutations have been induced, usually by radiations, in many genera (for a review see de Nettancourt, 1977) such as *Oenothera*, *Prunus*, *Nicotiana*, *Petunia* and *Trifolium*. Van Gastel and de Nettancourt (cited in de Nettancourt, 1977) have worked out the theory for obtaining double-cross hybrids through the use of plants heterozygous for a permanent self-compatibility mutation at the monofactorial gametophytic locus of self-incompatibility. The system proposed replaces almost entirely the low-seed-yielding and tedious procedure of bud pollination or pistil treatment by the simple self-pollination of self-compatible inbreds, which are *S*-heterozygous and carry a pollen-part mutation within one of their two *S*-alleles. *S*-heterozygotes selected after several generations of inbreeding can then be submitted to the effects of an inbreeding technique (bud pollination or pistil treatment) known to induce temporary self-compatibility. Self-incompatible homozygous lines can, thereafter, be recovered from the progenies resulting from such techniques and made available to the breeder for testing combining abilities and producing F$_1$ hybrid seed. Sweet-cherry mutants combining the self-compatibility char-

acter and a modified growth habit, which facilitates harvesting, are cultivated in Canada. On the whole, however, only little use has been made so far of the ease with which it is possible to transform certain SI cultivars into self-pollinators (see Lewis, 1979).

(b) The generation of new self-incompatibility alleles

While heteromorphic self-incompatibility is usually governed by one or several diallelic loci, very high numbers of different alleles have been estimated to operate at the incompatibility loci of many species with homomorphic incompatibility. This is particularly the case in the monofactorial sporophytic system of *Brassica* and in gametophytic incompatibility. One would expect the *S*-locus, on the basis of such large allelic series, to constitute choice material for the induction of so-called 'constructive mutations' leading to the transformation of a given *S*-allele with a specificity S_x into a new *S*-allele with a specificity S_y. Surprisingly enough, none of the numerous experiments undertaken on many different test species with acute X-rays, chronic gamma-rays, fast neutrons and EMS ever led to the generation of new incompatibility specificities detectable as functional *S*-alleles in the progenies of treated plants. Such negative results contrast sharply with the relative facility with which new *S*-alleles can be detected in advanced progenies of plants submitted to several generations of selfing by means of bud pollinations or treatments with inhibitors (see de Nettancourt, 1977); they cast doubts on the efficiency of artificial mutagens and imply that a complex polygenic system regulates and intensifies the apparition in inbred backgrounds of new alleles at the self-incompatibility locus.

(c) The effects of polyploidy

The tetraploid relatives or artificially induced tetraploids of diploid species with gametophytic monofactorial self-incompatibility usually display a self-compatible phenotype. The capacity of such tetraploids to accomplish self-fertilization does not result from any change in the pistil, which maintains its aptitude to reject incompatible pollen, but from the loss, in the diploid pollen grains, of the incompatible phenotype. Lewis (1947, 1949) has shown that the production of compatible pollen by tetraploids is limited to S-heterozygous plants ($S_xS_xS_yS_y$) and that only the grains which carry different alleles (S_xS_y) are compatible. A somewhat analogous situation is found in certain pollen-part mutants which produce haploid pollen bearing a small centric fragment. It is possible that in such mutants and in S-heterozygous diploid pollen, competitive interaction between different S-alleles prevents the production in sufficient quantities of any of the two S-proteins specified by the S-alleles. However, no evidence has ever been provided in favour of the hypothesis that the self-incompatibility locus is localized on the centric fragments which characterize certain SC mutants.

13.3 OVERCOMING INTERSPECIFIC INCOMPATIBILITY

At a time when molecular techniques for the transfer between species of plasmids, chromosomes and individual genes are proving their efficiency and their usefulness for plant breeding, only little needs to be said here on more conventional approaches involving the use of mutagens or of mentor pollen for breaking down reproductive barriers (for further discussion see Chapters 4 and 5).

13.3.1 Induced mutations

The earliest and most concerted efforts to remove interspecific incompatibility by means of induced mutations are certainly those of Davies and Wall (1961) who attempted to bypass, after radiation exposure of male and female gametes, incompatibility or post-zygotic isolation between species of the genera *Trifolium*, *Brassica*, *Vicia*, *Hordeum*, *Antirrhinum* and *Lycopersicum*. The only instance for which a clear-cut effect of irradiation could be found

concerned the cross *Brassica oleracea* × *B. nigra*, where reproductive isolation does not result from incompatibility relations, but from embryo abortion. More promising effects of irradiation on interspecific incompatibility have been reported by Pandey (1974) who found that X-irradiation of the anthers of the self-incompatible *Nicotiana glauca* enabled the pollen of this species to be accepted by the otherwise incompatible pistil of *N. forgetiana*. Pandey listed several arguments in support of his conviction that this radiation effect resulted from a mutation of the S-gene complex.

13.3.2 Mentor pollen effects

The most impressive application of such effects for plant breeding purposes was made by Knox *et al.* (1972) who mixed killed compatible pollen from the pistillate partner with live incompatible pollen for producing large quantities of interspecific hybrids in *Populus*. It was shown that the mentor effect could also be induced by adding to the incompatible pollen proteins extracted from the wall of compatible pollen. Since the mentor proteins are probably exine-held, it is likely that mentor substances originate from the tapetum during pollen development. They should therefore be essentially restricted to sporophytic systems.

13.3.3 Bud pollinations and inhibitors

No report appears to have been made of the efficiency of bud pollination or inhibitors for breaking interspecific incompatibility. The contrast with the situation concerning self-incompatibility is particularly striking and suggests that UI results from an absence of complementary stimulation rather than from an oppositional inhibition to pollen germination and pollen growth. The observation substantiates the opinion of Hogenboom (1975) that interspecific incompatibility is a process, basically distinct from self-incompatibility, which reflects the incompleteness of the pollen-pistil relationship.

13.4 THE MOLECULAR BIOLOGY OF INCOMPATIBILITY: STATE OF RESEARCH AND EXPECTATIONS FOR PLANT BREEDING

Early discoveries on the biochemistry of incompatibility have been reviewed by de Nettancourt (1977). It should therefore be sufficient to mention here, as examples of pioneering work, the contributions to pollen serology of Lewis (1952), Mäkinen and Lewis (1962), Linskens (1965), the presentations by Lewis (1965) of the dimer hypothesis of self-incompatibility and of the tripartite structure model of the *S*-locus, and the work of Heslop-Harrison *et al.* (1974, 1975), of Knox *et al.* (1972) and of Dickinson and Lewis (1973) on the determination, from tapetum to exine, of pollen incompatibility in sporophytic systems.

Current research on the molecular biology of incompatibility concentrates on the identification of *S*-gene products, the nature of recognition and rejection reactions and the characterization of the SI gene. Breakthroughs in some of these areas should, in the long term, be essential for the development of new tools and approaches in plant breeding.

13.4.1 Identification of *S*-gene products

Such identification could, in certain cases, facilitate a systematic analysis of the conditions favouring outbreeding in pseudo-compatible crops or permit the rapid acquisition of knowledge on the distribution of different *S*-genotypes in populations of perennials. It would in addition render possible further progress in our understanding of the system which, as indicated below, are prerequisites to new developments in plant breeding.

There is a strong presumption, first suggested by Nasrallah and Wallace (1967) and confirmed by several authors (Nasrallah *et al.*, 1970, 1972, 1982; Nishio and Hinata, 1977, 1982; Hinata and Nishio, 1978; Gaude, 1990), that the incompatibility molecules operating on the female side in *Brassica* are families of *S*-specific glycoproteins. *S*-specific glycoproteins have

also been identified in the gametophytic systems of *Nicotiana* and *Lycopersicum* by Clarke and collaborators (see Anderson *et al.*, 1986). Dickinson (1990) reviewed the evidence indicating that *S*-specificity is associated in glycoproteins to a variable region of the polypeptide chain and not to the glycosyl group. The possibility remains, however, that these specific glycoproteins do not correspond to the products of the *S*-genes but to those of associated loci closely linked to *S*. The confusion does not appear likely, however, because the glycoproteins considered as SI products constitute a large portion of the total proteins produced by the female tissue (Gaude, 1990) and because bioassays confirm the involvement of these glycoproteins in gametophytic and sporophytic self-incompatibility (for a discussion and references see Dickinson, 1990). Mc Clure *et al.* (1989) have reported that the glycoproteins, corresponding to five different *S*-alleles, isolated from style extracts of *N. alata*, are ribonucleases.

More uncertainty characterizes the identification of incompatibility proteins in the pollen and there is little biochemical evidence in support of the hypothesis that SI results, initially, from the presence of identical components in pollen and pistil. Only trace amounts of RNase activity could be detected by McClure *et al.* (1989) in the pollen of SI plants (*N. alata*) where a specific relationship had been found in the style between different ribonucleases and different *S*-alleles. The fact that mutations towards new *S*-specificities would rapidly lead to a breakdown of the system if different specificity parts operated independently in pollen and pistil (Lewis, 1960) and the observation that a change in specificity results from the generation of a new allele which is expressed in pollen and pistil (de Nettancourt *et al.*, 1971) certainly suggest, however, that the same genetic element governs the presence in pollen and pistil of the specific, but possibly very discrete, substances involved in the recognition reaction. The elegant works of Knox and of Heslop-Harrison and co-workers (see de Nettancourt, 1977) show that the best place for searching for such *S*-specific male determinants is probably in the tapetum and the pollen exine of sporophytic SI species. Indeed, very recently, Guilluy *et al.* (1990) have been able to detect in *Brassica*, through the use of PCR amplification, the presence in anthers of mRNA homologous to stigmatic *S*-specific glycoproteins of the same plant. This work, which also revealed the presence of *S*-specific substances in vegetative tissue, seems to confirm the hypothesis that the products of the *S*-gene in pollen and pistil are identical.

13.4.2 Nature of the recognition and rejection reactions

The importance of research in this area for plant breeding is particularly obvious since it should yield knowledge facilitating the control and exploitation of important mechanisms for outbreeding and reproductive isolation in flowering plants. Obviously, no important information on the process which enables pollen and pistil to recognize their respective *S*-genotypes will be available before the components of the reaction are identified. Such components correspond perhaps to short sequences of nucleotides or amino acids, associated in the pistil to members of a polymorphic family of glycoproteins, and dispersed in the intine or the exine, of the pollen grain. How these molecules, identical, complementary or homologous, recognize each other and meet to trigger a reaction which will prevent or stop the germination or growth of the incompatible pollen is not known. It does appear, however, that the recognition can be extremely rapid and leads, in many species, to a deposition, as a consequence or cause of inhibition, of callosic occlusions below the pollen and within the stigma papillae or, in the case of stylar incompatibility, to the accumulation around the pollen tube apex of a callose-rich inner wall and its subsequent opening in the conductive tissue.

It has recently been reported by McClure *et al.* (1990) for *N. alata*, that *S*-allele specific degradation of pollen RNA occurs *in vivo* after incompatible pollinations; they concluded, in view of the *S*-specificity of stylar RNases, that the gametophytic self-incompatibility system in *N. alata* acts through a cytotoxic mechanism directed against pollen RNA. As pointed out by the authors, a major unresolved question is the nature of allelic specificity in SI. Specificity does not appear to reside in the substrate for *S*-RNases (since they degrade RNA from many different sources) nor to be related to cleavage patterns. McClure *et al.* (1990), who agree that the degradation of rRNA could be a secondary consequence of some other inhibitory mechanism, are testing the hypothesis that the *S*-allele specificity of interaction between pollen tube and style resides in the mechanism by which the *S*-RNase gains access to the cytoplasmic compartment of the pollen tube. The implications of this work are wide since anther specific expression of a ribonuclease gene has recently been reported to cause male sterility when introduced into tobacco and oilseed rape by means of recombinant DNA techniques (Chapters 10 and 14 and Mariani *et al.*, 1990).

13.4.3 Characterization of the *S*-gene

Achievements in this area have been summarized by Dickinson (1990) and Gaude and Dumas (1990). In several laboratories, after the pioneering work by Nasrallah and collaborators in 1985, cDNA clones coding for glycoproteins segregating with specific *S*-alleles in *Brassica*, *Nicotiana* or *Lycopersicum* have been isolated in various parts of the world. In *Brassica*, comparisons of amino acid sequences suggest only very small differences in nucleotide composition between different *S*-alleles. More allelic variability was found in *Nicotiana* (Anderson *et al.*, 1989). In both the gametophytic and the sporophytic systems, this variability is concentrated in regions which probably control *S*-specificity.

The time is becoming ripe, although many mysteries remain and very little is known on the genetic determinant operating in SI pollen, to proceed to the transfer of SI in SC species. A first attempt in this direction has been reported by Moore and Nasrallah (1990) who introduced the *S*-gene of *B. oleracea* into the amphidiploid, self-compatible, *Nicotiana tabacum*. Evidence of the expression of the *S*-gene was found at the RNA and protein levels in the style of transformed tobacco plants.

The prospects for plant breeding of such manipulations are most exciting; at the same time, however, it must be remembered that the requirements for the expression of SI are not necessarily satisfied by the genetic background of SC species (Martin, 1961, 1964, 1968; de Nettancourt, 1977). The expression of SI in self-fertilizers will possibly require a considerable amount of genetic adjustment and modification.

REFERENCES

Anderson, M.A., Cornish, E.C., Mau, S.L., Williams, E.G., Hoggart, R., Atkinson, A., Bonig, I., Grego, B., Simpson, R., Roche, P.J., Haley, J.D., Penschow, J.D., Niall, H.D., Tregear, G.W., Coghlan, J.P., Crawford, R.J. and Clarke, A.E. (1986), Cloning of cDNA for a stylar glycoprotein associated with expression of self-incompatibility in *Nicotiana alata*. *Nature*, **321**, 38–44.

Anderson, M.A., Ebert, P.R., Altschuler, M. and Clarke, A.E. (1989), Molecular genetics of self-incompatibility in flowering plants. In *Plant Reproduction: From Floral Induction to Pollination*, Vol. I, Lord, E. and Bernier, G. (eds), The American Society of Plant Physiologists Symposium Series.

Brewer, J.G. (1968), Flowering and seed-setting in Pyrethrum (*Chrysanthemum cinerariaefolium* Vis.), a review. *Pyrethrum Post*, **9**(4), 18–21.

Bubar, J.S. (1958), An association between variability in ovule, development within ovaries and self-incompatibility in *Lotus* (Leguminosae). *Can. J. Bot.*, **36**, 14–22.

Cope, F.W. (1962), The mechanism of pollen incompatibility in *Theobroma cacao*. *Heredity*, **17**, 157–182.

Crane, M.B. and Brown, A. (1937), Incompatibility and sterility in the sweet cherry, *Prunus avium* L. *J. Pomol.*, **15**, 86.

Dahlgreen, K.V.O. (1922), Vererbung der Heterostylie bei *Fagopyrum* (nebst einigen Bemerkungen über *Pulmonaria*). *Hereditas*, **3**, 91–99.

Darlington, C.D. and Mather, K. (1949), *The Elements of Genetics*, Allen and Unwin Ltd, London.

Davies, D.R. and Wall, E.T. (1961), *Gamma Radiation and Interspecific Incompatibility in Plants. Effects of Ionizing Radiations on Seeds*, IAEA, Vienna, pp. 73–87.

de Nettancourt, D., Ecochard, R., Perquin, M.D.G., van der Drift, T. and Westerhof, M. (1971), The generation of new S-alleles at the incompatibility locus of *L. peruvianum* Mill. *Theor. Appl. Genet.*, **41**, 120–129.

de Nettancourt, D. (1977), *Incompatibility in Angiosperms*, Springer-Verlag, Berlin.

Devreux, M., Vallaeys, G., Pochet, P. and Gilles, A. (1959), Recherches sur l'autostérilité du caféier Robusta (*Coffea canephora* Pierre). *Publ. INEAC*, **78**, 44.

Dickinson, H.G. (1990), Self-incompatibility in flowering plants. *Bioessays*, **12**, 155–161.

Dickinson, H.G. and Lewis, D. (1973), The formation of the tryphine coating the pollen grains of *Raphanus*, and its properties relating to the self-incompatibility system. *Proc. R. Soc. London, Ser. B*, **184**, 148–165.

Duvick, D.N. (1966), Influence of morphology and sterility on breeding methodology. In *Plant Breeding*, Frey, K.J. (ed.), Iowa State University Press, Ames, pp. 85–138.

East, E.M. and Mangelsdorf, A.J. (1925), A new interpretation of the hereditary behaviour of self-sterile plants. *Proc. Natl. Acad. Sci. USA*, **11**, 166–171.

Gaude, T. (1990), *Etude de la reconnaissance pollen-stigmate chez les plantes à fleur: détermination des facteurs de reconnaissance pollinique chez Brassica oleracea*. Thèse de doctorat d'état, Université Claude Bernard.

Gaude, T. and Dumas, C. (1990), Molecular aspects of the self-incompatibility systems of *Brassica* and *Nicotiana*. *Bot. Acta*, **103**, 323–326.

Guilluy, C.M., Gaude, T., Digonnet-Kerhoas, C., Chaboud, A., Heizman, Ph. and Dumas, C. (1990), New data and concepts in angiosperm fertilization. In *Mechanism of Fertilization*, Nato AS Series, **45**, 253–270.

Hayman, D.L. (1956), The genetic control of incompatibility in *Phalaris coerulescens*. *Aust. J. Biol. Sci.*, **9**, 321–331.

Heslop-Harrison, J., Knox, R.B. and Heslop-Harrison, Y. (1974), Pollen-wall proteins: exine-held fractions associated with the incompatibility response in Cruciferae. *Theor. Appl. Genet.*, **44**, 133–137.

Heslop-Harrison, J., Knox, R.B., Heslop-Harrison, Y. and Mattsson, O. (1975), Pollen wall proteins: emission and role in incompatibility responses. In *The Biology of Male Gamete*, Duckett, J.G. and Racey, P.A. (eds). *Biol. J. Linn. Soc.* Suppl. 1, Vol. VII, pp. 189–202.

Hinata, K. and Nishio, T. (1978), S-allele specificity of stigma proteins in *Brassica oleracea* and *B. campestris*. *Heredity*, **41**, 93–100.

Hogenboom, N.G. (1975), Incompatibility and incongruity: two different mechanisms for the non-functioning of intimate partner relationships. *Proc. R. Soc. London. Ser. B*, **188**, 361–375.

Kinman, M.L. (1963), Current status of sunflower production and research, and the possibilities of this crop as an oilseed in the United States. Presented before the *Oilseed and Peanut Research and Marketing Advisory Committee*, Peoria, Ill. Feb. 7 (cited in Duvick, 1966).

Knox, R.B., Willing, R. and Ashford, A.E. (1972), Role of pollen-wall proteins as recognition substances in interspecific incompatibility in poplars. *Nature*, **237**, 381–383.

Lamm, R. (1950), Self-incompatibility in *Lycopersicum peruvianum* Mill. *Hereditas*, **36**, 509–511.

Lewis, D. (1943), The physiology of incompatibility in plants. II. *Linum grandiflorum. Ann. Bot. II*, **7**, 115–122.

Lewis, D. (1947), Competition and dominance of incompatibility alleles in diploid pollen. *Heredity*, **1**, 85–108.

Lewis, D. (1949), Structure of the incompatibility gene. II. Induced mutation rate. *Heredity* **3**, 339–355.

Lewis, D. (1952), Serological reactions of pollen incompatibility substances. *Proc. R. Soc. London, Ser. B*, **140**, 127–135.

Lewis, D. (1960), Genetic control of specificity and activity of the *S* antigen in plants. *Proc. R. Soc. London, Ser. B*, **151**, 468–477.

Lewis, D. (1965), A protein dimer hypothesis on incompatibility. Proc. 11th Int. Congr. Genet. The Hague, 1963. In *Genetics Today*, Vol. 3, Geerts, S.J. (ed.), Pergamon Press, London, pp. 656–663.

Lewis, D. (1979), Sexual incompatibility in plants. In *Studies in Biology*, No. 110, Edward Arnold, London, pp. 1–60.

Lewis, D. and Crowe, L.K. (1958), Unilateral incompatibility in flowering plants. *Heredity*, **12**, 233–256.

Lewis, D. and Modlibowska, I. (1942), Genetical studies in pears. IV. Pollen-tube growth and incompatibility. *J. Genet.*, **43**, 211–222.

Linskens, H.F. (1965), Biochemistry of incompatibility. Proc. 11th Int. Congr. Genet. The Hague, 1963. In *Genetics Today*, Vol. 3, Geerts, S.J. (ed.), Pergamon Press, London, pp. 621–636.

Lundqvist, A. (1956), Self-incompatibility in rye. I. Genetic control in the diploid. *Hereditas*, **42**, 293–348.

Lundqvist, A. (1961), A rapid method for the analysis of incompatibilities in grasses. *Hereditas*, **47**, 705–707.

Lundqvist, A. (1962), Self-incompatibility in diploid *Hordeum bulbosum* L. *Hereditas*, **48**, 138–152.

Lundqvist, A. (1965), The genetics of incompatibility. Proc. 11th Int. Congr. Genet. The Hague, 1963. In *Genetics Today*, Vol. 3, Geerts, S.J. (ed.), Pergamon Press, London, pp. 637–647.

Lundqvist, A., Osterbye, U., Larsen, K. and Linde-Laursen, I. (1973), Complex self-incompatibility systems in *Ranunculus acris* L. and *Beta vulgaris* L. *Hereditas*, **74**, 161–168.

Mäkinen, Y.L.A. and Lewis, D. (1962), Immunological analysis of incompatibility (S) proteins and of cross-reacting material in a self-compatible mutant of *Oenothera organensis*. *Genet. Res.*, **3**, 352–363.

Mariani, C., DeBeuckeler, M., Truettner, J., Leemans, J. and Goldberg, R.B. (1990), Induction of male sterility in plant by a chimeric ribonuclease gene. *Nature*, **347**, 737–741.

Martin, F.W. (1961), The inheritance of self-incompatibility in hybrids of *Lycopersicum esculentum* Mill. × *L. chilense* DUN. *Genetics*, **46**, 1443–1454.

Martin, F.W. (1964), The inheritance of unilateral incompatibility in *Lycopersicum hirsutum*. *Genetics*, **50**, 459–469.

Martin, F.W. (1968), The behaviour of *Lycopersicum* incompatibility alleles in an alien genetic milieu. *Genetics*, **60**, 101–109.

Mc Clure, B.A., Gray, J.E., Anderson, M.A. and Clarke, A.E. (1990), Self-incompatibility in *Nicotiana alata* involves degradation of pollen rRNA. *Nature*, **347**, 757–760.

Mc Clure, B.A., Haring, V., Ebert, P.R., Anderson, M.A., Simpson, R.J., Sakiyama, F. and Clarke, A.E. (1989), Style self-incompatibility gene products of *Nicotiana alata* are ribonucleases. *Nature*, **342**, 955–957.

Miri, R.K. and Bubar, J.S. (1966), Self-incompatibility as an out-crossing mechanism in birdsfoot trefoil (*Lotus corniculatus*). *Can. J. Plant Sci.*, **46**, 411–418.

Moore, H.M. and Nasrallah, M.E. (1990), A *Brassica* self-incompatibility gene is expressed in the stylar transmitting tissue of trans-genic tobacco. *Plant Cell*, **2**, 29–38.

Nasrallah, M.E., Barber, J.T. and Wallace, D.H. (1970), Self-incompatibility proteins in plants: detection, genetics, and possible mode of action. *Heredity*, **25**, 23–27.

Nasrallah, J.B., Kao, T.H., Chen, C.H., Goldberg, M.L. and Nasrallah, M.E. (1987), Amino acid sequence of glycoproteins encoded by three alleles of the S-locus of *Brassica oleracea*. *Nature*, **326**, 617–619.

Nasrallah, J.B., Kao, T.H., Goldberg, M.L. and Nasrallah, M.E. (1985), A cDNA clone encoding an S-locus specific glycoprotein from *Brassica oleracea*. *Nature*, **318**, 263–267.

Nasrallah, M.E. and Wallace, D.H. (1967), Immuno-genetics of self-incompatibility in *Brassica oleracea* L. *Heredity*, **22**, 519–527.

Nasrallah, M.E., Wallace, D.H. and Savo, R.M. (1972), Genotype, protein, phenotype relationships in self-incompatibility of *Brassica*. *Genet. Res.*, **20**, 151–160.

Nishio, T. and Hinata, K. (1977), Analysis of S-specific proteins in stigma of *Brassica oleracea* L. by isoelectric focusing. *Heredity*, **38**, 391–396.

Nishio, T. and Hinata, K. (1982), Comparative studies on S-glycoproteins purified from different S-genotypes in self-incompatible *Brassica* species. I. Purification and chemical properties. *Genetics*, **100**, 641–647.

Pandey, K.K. (1962), Interspecific incompatibility in *Solanum* species. *Am. J. Bot.*, **49**, 874–882.

Pandey, K.K. (1974), Elimination of heterozygosity and efficiency of genetic systems. *Theor. Appl. Genet.*, **44**, 199–205.

Pushkarnath, M. (1942), Studies in sterility in potatoes. I. The genetics of self- and cross-incompatibilities. *Ind. J. Genet.*, **2**, 11.

Sampson, D.R. (1957), The genetics of self-incompatibility in the radish. *J. Hered.*, **48**, 26–29.

Thompson, K.F. (1957), Self-incompatibility in marrow-stem kale, *Brassica oleracea* var. *acephala*. I. Demonstration of sporophytic system. *J. Genet.*, **55**, 45–60.

14
Male sterility

J.M. Lasa and N.O. Bosemark

14.1 INTRODUCTION

Male sterility in plants implies an inability to produce or to release functional pollen, and is the result of failure of formation or development of functional stamens, microspores or gametes. As a consequence male sterility is sometimes divided into: (a) 'Pollen sterility' in which male sterile individuals differ from normal only in the absence or extreme scarcity of functional pollen grains; (b) 'Structural or staminal male sterility' in which male flowers or stamens are malformed and non-functional or completely absent; (c) 'Functional male sterility' in which perfectly good and viable pollen is trapped in indehiscent anthers and thus prevented from functioning. Of the three types of male sterility mentioned, pollen sterility is by far the most common and the only one that has played a major role in plant breeding and hybrid seed production. The reason for the absence of pollen in this type of male sterility is a breakdown of microsporogenesis, usually caused by an abnormal or malfunctioning tapetum, the innermost layer of the anther wall.

Based on its inheritance or origin, male sterility may be further divided into: (1) 'Nuclear male sterility' (NMS), also called 'genic', 'genetic' or 'Mendelian', in which the male sterility is governed solely by one or more nuclear genes; (2) 'Cytoplasmic male sterility' (CMS) in which male sterility comes about as a result of the combined action of nuclear genes and genic or structural changes in the cytoplasmic organellar genomes resulting in what is often referred to as 'sterile cytoplasm' (S) as opposed to normal 'fertile cytoplasm' (N); (3) 'Non-genetic, chemically induced male sterility' which results from the application of specific chemicals referred to as gametocides or chemical hybridizing agents (CHA).

Recent reviews on male sterility and its use in plant breeding can be found in Hanson and Conde (1985), Driscoll (1986), Kaul (1988) and Rao *et al.* (1990).

14.2 NUCLEAR MALE STERILITY

Nuclear male sterility, originating through spontaneous mutation, is a common phenomenon in nature. That the frequency of such mutations is high is evidenced by the fact that in species like maize, tomato and barley, which have been intensively studied, more than 40 non-allelic male sterility genes have been described (Kaul, 1988). As pointed out by Duvick (1966), nuclear male sterility can probably be found in all diploid species. Mutations resulting in nuclear male sterility have also been induced in several important crop plants. Both ionizing radiation and chemical mutagens such as ethyl methane sulphonate (EMS) and ethylene imine (EI) have been used (Kaul, 1988). Spontaneous nuclear male sterility is usually controlled by a single recessive gene. This does not permit the creation of a population where all plants are male sterile. The highest proportion of male steriles that can be realized is 50%, which is obtained in the backcross *msms* × *MSms*. Less

Plant Breeding: Principles and prospects. Edited by M.D. Hayward, N.O. Bosemark and I. Romagosa. Published in 1993 by Chapman & Hall, London. ISBN 0 412 43390 7.

frequently, nuclear male sterility has been found to be controlled by more than one recessive gene (Driscoll, 1986), by polygenes (Athwal *et al.*, 1967) or by dominant genes (Mathias, 1985).

14.3 CYTOPLASMIC MALE STERILITY (CMS)

Based on its origin, CMS may be divided into autoplasmic CMS and alloplasmic CMS (Lacadena, 1968). Autoplasmic CMS would then refer to those cases where CMS has arisen within a species as a result of spontaneous mutational changes in the cytoplasm, most likely in the mitochondrial genome. Alloplasmic CMS, on the other hand, would comprise such cases where CMS has arisen from intergeneric, interspecific or occasionally intraspecific crosses and where the male sterility can be interpreted as being due to incompatibility or poor co-operation between the nuclear genome of one species and the organellar genome of another. This category also includes CMS in products of interspecific protoplast fusion.

Of the 140 species with CMS known in 1972, Laser and Lersten (1972) classified 56% as being spontaneous autoplasmic in origin. Although, as pointed out by Belhassen *et al.* (1991), the categorization of male sterility is frequently based on very limited evidence, it is interesting to note that spontaneous autoplasmic CMS has been estimated to be only one quarter as frequent as spontaneous NMS (Kaul, 1988). The reason for this latter difference may be partly a greater genetic stability of the organellar genomes, partly the ability of nuclear genes to compensate for changes in the organellar genomes as evidenced by the occurrence of CMS restorer genotypes. The nuclear genetic control of CMS is predominantly governed by one or more recessive genes, but dominant genes (Gabelman, 1956) as well as polygenes (Bradner and Childers, 1968), have been reported to be involved in CMS.

As to the cause of CMS several theories have been presented. Thus Atanasoff (1964) suggested that viral infection may be the cause of CMS. Although at the time a plausible explana-tion for autoplasmic CMS, compatible with published reports of graft transmission of CMS in petunia (Frankel, 1962), recent experiments have shown that the virus infection hypothesis is not correct and that male sterility in petunia, in common with CMS in other species, is associated with a variant mitochondrial DNA arrangement (Evenor *et al.*, 1988). Although probably a rare exception, it should be mentioned here that the CMS '447' of *Vicia faba* has been shown to be linked to the presence of cytoplasmic virus-like particles, and that the frequent instability of this CMS is strongly related to the content of these particles (Dulieu *et al.*, 1988; Berthaut *et al.*, 1991).

That the mitochondria may be the site of the changes in the cytoplasm giving rise to CMS was first proposed by Rhoades (1950). However, only much later, when molecular techniques had become available, was it possible to demonstrate that mitochondrial DNA (mtDNA) from fertile and male sterile cytoplasms in maize show different restriction endonuclease digestion patterns, which strongly supported the hypothesis of a mitochondrial origin of CMS (Levings and Pring, 1977; Pring and Levings, 1978). Since then, similar differences between the mtDNA from fertile and CMS plants have been found in many species including sugar beet (Powling, 1982), oilseed rape (Vedel *et al.*, 1982), sorghum (Conde *et al.*, 1982), petunia (Kool *et al.*, 1983), sunflower (Siculella and Palmer, 1988), rice (Kadowaki and Harada, 1989) and carrot (Pingitore *et al.*, 1989).

It is now becoming clear that the different mtDNA restriction endonuclease digestion patterns are reflections of aberrant intra- or intermolecular DNA recombination events in the mitochondrial genome which have either modified existing genes or created new genes some of which are more or less related to the male sterile phenotype (Leaver *et al.*, 1988; Makaroff and Palmer, 1988; Fauron and Havlik, 1989; Leaver, 1989; Halldén *et al.*, 1990). Also the analyses of the mitochondrial and chloroplast genomes of cytoplasmic male sterile somatic hybrids in the genera *Nicotiana*, *Petunia* and *Brassica* (Belliard *et al.*, 1979; Boeshore *et al.*, 1983; Fluhr *et al.*, 1983; Clark *et al.*, 1986; Barsby *et al.*, 1987) and evidence from studies of fertile revertants of the maize T and S male sterile cytoplasms (Umbeck and Gengenbach, 1983; Schardl *et al.*, 1985), strongly support the conclusion that CMS is due to a malfunction of the mitochondria and that the association of CMS with the mitochondrial genome is a general phenomenon. However, conclusive evidence, linking CMS to the mitochondrial genome still exists only for a few species and it cannot be excluded that occasionally CMS is related to changes in the chloroplast DNA (cpDNA). Thus, it has been proposed that cpDNA at least contributes to CMS in *Nicotiana* (Frankel *et al.*, 1979), *Gossypium* (Galun and Wilkins, 1989) and *Sorghum* (Chen *et al.*, 1990).

14.4 CHEMICALLY INDUCED MALE STERILITY

In several economically important crops attempts to develop hybrid varieties have been hampered by lack of a suitable and well functioning male sterility system. In other crops the cost of hybrid breeding based on cytoplasmic male sterility may be prohibitive. As a consequence several chemical companies have developed so called gametocides or chemical hybridizing agents (CHA), which when sprayed on plants at a certain stage before flowering render the plants male sterile. A successful CHA should produce completely male sterile plants without affecting female fertility, require a minimum number of applications and be effective over a wide range of environmental conditions (Brears and Bingham, 1989).

The major problems with the CHAs produced so far have been lack of effective male sterilization, partial female sterility and phytotoxicity. Another limiting factor has been the growth stage interval in which the CHA must be applied to be effective. If this interval is too narrow, adverse weather conditions may prevent a timely application of the chemical (McRae, 1985). Since the first attempts to use maleic hydrazide as a chemical sterilant 35 years ago, many different substances have been synthesized and a great deal of work and money has been invested in CHAs. However,

as yet no entirely satisfactory product has appeared on the market (Brears and Bingham, 1989).

14.5 USE OF MALE STERILITY IN PLANT BREEDING

Male sterility plays an important role in plant breeding, firstly in the production of hybrid seed, and secondly as a plant breeding tool facilitating population improvement, backcrossing, interspecific hybridization and other intermediate breeding procedures.

14.5.1 Hybrid seed production

Following the success of hybrid maize, hybrid breeding methods are now successfully employed in sorghum, sugar beet, sunflower, tomato, carrot, onion and several other field, vegetable, and ornamental crops. To utilize hybrid vigour effectively it must be possible to produce hybrid seed in such quantities that the F_1 can be grown directly by the farmer. To the maize breeder this does not present a problem since in this monoecious species the male inflorescence can easily be removed without damaging the female flowers. However, apart from a few vegetable and flower species, in which both the seed production per plant and the value of the seed are very high, maize is the only important agricultural crop that can be manually emasculated on a sufficiently large scale to permit production of hybrid seed for commercial use. As a consequence in the last 50 years plant breeders have introduced different kinds of male sterility into plant populations and used them to circumvent the restrictions to large scale controlled hybridization imposed by flower morphology and breeding systems.

(a) Nuclear male sterility (NMS)

Recessive NMS

The fact that NMS does not permit the production of a uniformly male sterile population seriously limits its use in hybrid seed production. To permit the use of recessive NMS in large-scale hybrid seed production it is necessary to rogue out 50% male fertile plants from the backcross *msms* × *Msms*. Since safe classification of fertile *versus* male sterile plants cannot usually be made until shortly before flowering, such roguing becomes far too difficult and costly in most crops to permit economic production of large quantities of hybrid seed. Several solutions to this problem have been proposed comprising: (1) vegetative propagation; (2) marker assisted selection; (3) cytogenetic methods; (4) temporary restoration of fertility; (5) use of functional male sterility.

Vegetative propagation of MS plants to be used as females in hybrid seed production has so far been used mainly in ornamental plants where the price of commercial seed is high (Reimann-Philipp, 1983). However, with the development of cheap and efficient methods of micropropagation, hybrid seed production based on vegetatively propagated heterozygous male sterile genotypes is likely to increase in importance, especially among vegetable crops (see Chapter 16).

More that 60 years ago Singleton and Jones (1930) suggested that a seed colour gene could be used as a marker for a linked *ms* gene in maize. With such a marker, seeds resulting in fertile plants could be removed before sowing, permitting a virtually pure stand of male sterile plants to be produced. Since then several similar systems of marker-assisted selection have been developed, involving seed characters such as white endosperm in maize (Galinat, 1975, 1976) and shrunken endosperm in barley (Ulrich and Eslick, 1977) and seedling characters such as short leaf in sorghum (Webster, 1977) and the woolly character in tomato (Durand, 1981). However, sufficiently tight linkages have not been found between markers and *ms* genes for any of these systems to be used on a significant scale. An exception may be a recent tomato line from Purdue University, USA, possessing a seedling marker closely linked to a recessive male sterility (Anonymous, 1991b).

Recently, Jorgensen (1987) proposed that tight linkages may be synthesized through genetic transformation, which would allow the introduction of an appropriate marker gene to a random location in each of a large number of plants carrying a suitable *ms* gene. Standard linkage analysis of the progenies of independent transgenic plants would then be used to identify sufficiently tight linkage between the marker and the *ms* gene. The method should be applicable to any hybrid breeding programme provided a reliable *ms* gene, a suitable marker and an efficient transformation–regeneration system are available.

Cytogenetic methods combine *ms* genes with chromosomal selective mechanisms to produce pure male sterile progenies. In the 'Balanced Tertiary Trisomic' (BTT) system, originally proposed by Ramage and Wiebe (Ramage, 1965), this is achieved through an extra translocated chromosome, which carries a dominant *Ms* gene closely linked to the translocation breakpoint, while the two normal chromosomes carry the corresponding recessive *ms* gene. In a modification of the original BTT system the extra chromosome carries the dominant alleles of the *ms* locus and a recessive lethal marker gene closely linked to each other and the translocation breakpoint (Ramage and Scoles, 1981). Upon selfing such trisomic plants, the extra chromosome is transmitted through the egg cell, resulting in viable and fertile offspring and maintenance of the line. When the trisomics are used as pollinators on *msms* females, the extra chromosome is not transmitted and the offspring is thus 100% male sterile. The BTT system was developed to produce hybrid barley varieties but has been used only to a limited extent (Ramage, 1983). In an alternative system developed for maize (Patterson, 1975), a recessive *ms* gene is linked in repulsion with a chromosome deficiency. Since male gametes carrying the deficiency do not function, the effect becomes the same as with the BTT-system. Even more complicated is the XYZ-system in wheat (Driscoll, 1972, 1985), which involves an alien chromosome addition. Neither of the two last mentioned systems have been put to practical use.

Under certain circumstances it is possible to breed pure nuclear male sterile lines via self-pollination. However, this requires that the plants are pollen sterile under certain conditions but not under other. Thus, Hansche and Gabelman (1963) found partial male sterility in carrots, which has a differential penetrance in different environments, and which thus makes it possible to increase seed in one location and grow male sterile blocks for hybrid seed production in another. Similarly, Yang *et al.* (1988) described the use of a partial male sterility, sensitive to photoperiod, for commercial production of hybrid rice. However, if the environmental conditions promoting pollen fertility are present for a few days during the critical period of pollen development, there is a risk of break down of sterility in the hybrid seed production fields.

Also several chemicals have been used to restore temporarily the fertility in various kinds of male sterile plants. Thus, gibberellic acid (GA3) has been reported to restore pollen fertility in the 'corn grass' mutant in maize (Nickerson, 1960), in stamenless mutants in tomato (Phatak *et al.*, 1966; Sauhney and Greyson, 1973) as well as in a nuclear male sterile in barley (Kasembe, 1967). An advantage of chemical reversion of male sterility is that it is not necessary to achieve complete fertility restoration. Side effects of the chemicals on the plants are also not a serious problem since the treated plants are not used to grow the hybrid seed. Because of the possibilities offered by temporary chemical fertility restoration in hybrid seed production, several authors have recommended intensified research in this field (Driscoll and Barlow, 1976; Hockett *et al.*, 1978; Foster, 1981; and McRae, 1985). Recently, Pareddy (1990) reported on the effect of 12 different plant growth regulators (PGR), including GA3, on development and reversion of two kinds of NMS in maize *in vitro*. Although the cultured tassel meristems developed normally on a liquid Murashige-Skoog medium and expressed male sterile characteristics similar to those *in vivo*, none of the tested PGRs reverted the NMS tassels to fertility *in vitro*. It was concluded that in addition to PGRs, nutritional and environmental factors may be involved in restoration of NMS.

Attempts to use functional male sterility in hybrid seed production have also been made. Thus, Roever (1948) described a recessive gene in tomato which produces normal pollen in anthers that do not dehisce; because of this, natural selfing is prevented. However, this mutant can be selfed by manually inducing dehiscence of the anthers. The male sterile mutant can thus be easily maintained and used for hybrid seed production without emasculation. Similar mutants have been found in maize (Horowitz and Obregon, 1951), in eggplant (Jasmin, 1954) and in beans (Wyatt, 1983). Although some of the mutants mentioned seem to have potential for hybrid seed production, so far they appear not to have been used commercially. However, very recently a method for production of hybrid tomato seed, involving a functional male sterile maternal line and a thiamine-dependant male line, was described (Barabás, 1991). The system is said to have been introduced into Hungarian tomato breeding.

Dominant NMS

Dominant nuclear MS genes have been described in several crop species including carrot, cotton, wheat and oil seed rape (Mathias, 1985). Although dominant NMS may be used to facilitate crossing in recurrent selection programmes, its inheritance normally prevents its use in hybrid seed production. However, recently, scientists at Plant Genetic Systems (PGS), in Belgium, and University of California, USA, have transformed tobacco and oilseed rape plants with a chimeric gene consisting of an anther tapetum specific gene (*TA29*) and a ribonuclease gene (*RNase*). Expression of the chimeric gene selectively destroys the anther tapetal cells and results in male sterile plants. Female fertility is not affected and transformed plants are normal in all other respects. By linking the *TA29-RNase* gene to a dominant herbicide resistance gene, one can use the corresponding herbicide to produce uniform populations of male sterile plants (Mariani *et al.*, 1990). The same group has also identified a restorer gene (*barstar*) that inhibits the effect of the *RNase* gene (Mariani *et al.*, 1992), thus permitting the use of this dominant NMS system for hybrid seed production in the type of scheme shown in Fig. 14.1.

Two more organizations, Paladin Hybrids, Canada, and ICI, UK, have filed patents on

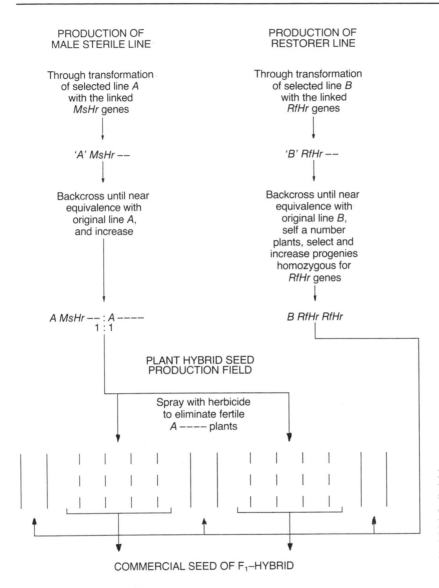

PRODUCTION OF
MALE STERILE LINE

Through transformation
of selected line *A*
with the linked
MsHr genes

'A' *MsHr* — —

Backcross until near
equivalence with
original line *A*,
and increase

A MsHr — — : *A* — — — —
1 : 1

PRODUCTION OF
RESTORER LINE

Through transformation
of selected line *B*
with the linked
RfHr genes

'B' *RfHr* — —

Backcross until near
equivalence with
original line *B*,
self a number
plants, select and
increase progenies
homozygous for
RfHr genes

B RfHr RfHr

PLANT HYBRID SEED
PRODUCTION FIELD

Spray with herbicide
to eliminate fertile
A — — — — plants

COMMERCIAL SEED OF F_1–HYBRID

Fig. 14.1. Steps in a hybrid seed production system based on the PGS-dominant nuclear *Ms* gene linked to a dominant herbicide resistance (*Hr*) gene.

systems of genetically engineered dominant nuclear male sterility (Anonymous, 1991a; Cutler, 1991). A fourth system, under development at the USDA Agricultural Research Service, University of California Plant Gene Expression Centre, Albany, aims to produce plants with pollen that self-destructs when required. To achieve this, plants are transformed with a chimeric gene consisting of a pollen specific promoter (*LAT59*) and a gene (*fms2*) that converts indole acetamide (IAM) into indole acetic acid (IAA). If the system works as expected, plants carrying the *LAT59-fms2* gene, when sprayed with IAM, will selectively process this substance into IAA at concentrations sufficiently high to kill the pollen and render the plant male sterile (McCormick *et al.*, 1989; Wood, 1990). Although this latter system avoids both the problem of elimination of fertile pollen and the restoration of fertility in F_1 hybrids, as with CHAs, a possible limiting factor would be the need for a chemical application with all the associated problems, such as weather conditions and the span of flowering.

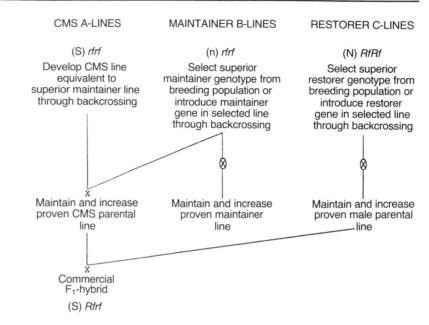

Fig. 14.2. Steps in the development of a hybrid seed production system based on cytoplasmic male sterility. Please, note that for ease of tracing the *Rf* gene in backcrossings, restorer lines are usually developed in S-cytoplasm, not in N-cytoplasm as shown here.

There is reason to believe, however, that several alternative systems of engineered NMS will be developed in the coming years.

(b) Cytoplasmic male sterility (CMS)

Cytoplasmic male sterility is the most important system used in hybrid seed production and so far, with few exceptions, the only one by which hybrid seed can be produced both effectively and economically. The inheritance of CMS may be illustrated in onion where the nuclear control is exercised by only one recessive gene. As may be seen in Fig. 14.2, it is only the combination of 'sterile' cytoplasm and homozygosity for the recessive gene *rf*, (S)*rfrf*, that results in male sterility. A genotype of the constitution (N)*rfrf* is called a maintainer since a male sterile plant will produce a uniformly male sterile offspring only when pollinated by plants of this genotype. A genotype that masks the expression of the CMS trait and which, when used as a pollinator on a CMS female, restores the pollen fertility of the progeny is called a restorer. As in the case of CMS in onion, the restorer gene may be the dominant allele of the recessive maintainer gene, and the restorer genotype thus (N)-*RfRf*. However, full restoration frequently requires the involvement of other nuclear genes

and may even be accompanied by changes in the mitochondrial genome (Mackenzie and Chase, 1990). Hybrid breeding and seed production, as illustrated in Fig. 14.2, thus requires the following materials and procedures.

Maintainer genotypes (N)rfrf

If one wants to use CMS in hybrid breeding, one must first find or introduce maintainer genotypes in one's own breeding material and then through crossing and backcrossing transfer their nuclear genotypes into sterile cytoplasm. Maintainer genotypes, often called B-lines, look precisely the same as any normal fertile genotype with N-cytoplasm. To identify maintainer genotypes it is thus necessary to testcross fertile plants individually to CMS-plants and to classify the progenies for male sterility. If a particular testcross progeny consists of only 100% male sterile plants, the pollinator plant used in that testcross was of maintainer genotype. The genotype of the maintainer plant is normally preserved via self-pollination, which is also part of the maintainer line development whenever the objective is to develop an inbred CMS line.

The frequency of maintainer genotypes varies considerably from species to species as well as between different sterile cytoplasms within a species. Thus, in maize 70% of all inbred lines are maintainers for the T cytoplasm and 40% for the S cytoplasm (Duvick, 1965). As a comparison, in most sugar beet populations only 2–5% of the plants are maintainers for the CMS discovered by Owen (1942, 1945).

Equivalent CMS lines (S)rfrf

Once a new maintainer line has been evaluated and its breeding value proven, its nuclear genotype will have to be transferred to the sterile cytoplasm through crossing and repeated backcrossing. The resultant CMS line (A-line) will be largely isogenic to the maintainer line except for the organellar genomes. It can easily be propagated by further crossing to its maintainer counterpart (B-line).

Restorer genotypes

In biennial crops such as onion and sugar beet, which are grown for their vegetative parts only, male fertility of the hybrid crop is not required. However, for crops like sorghum and sunflower, where the harvested material is the seed, the pollinator parent used on the CMS female parent, must be homozygous for the required restorer genes to ensure pollen fertility and seed set in the resultant hybrid. As with maintainers, restorers have to be identified through testcrossing with CMS plants and subsequent classification of the progenies for pollen fertility. Restorers are usually developed in sterile cytoplasm, inbred and selected for agronomic characteristics and combining ability in the same way as are the maintainers.

Once good CMS and restorer lines, with acceptable vigour and superior combining ability, are available, commercial seed production should present no major problems provided the crop is a normally cross pollinating species. However, hybrid breeding based on CMS is frequently tedious and costly and sometimes impractical for one or several of the following causes.

1. Maintainer and/or restorer genotypes are too scarce in the breeding populations to permit direct isolation, and the corresponding genes may thus have to be introduced into contrasting populations prior to selection and line development. In certain cases of alloplasmic CMS, e.g. the 'Ogura' CMS in oilseed rape, both the CMS and the restorer genes may have to be introduced from a species belonging to another genus (Rouselle and Dosba, 1985; Pellan-Delourme and Renard, 1988). In other cases full and reliable restoration may require the assemblage of several genes with a complementary effect, as is the case in wheat with the *timopheevi* CMS (Wilson and Driscoll, 1983).

2. Dependence of maintenance as well as restoration on environmental conditions, especially temperature and genetic background (in onion: Jones and Clarke, 1943; in petunia: Izhar, 1978; in rye: Scoles and Evans, 1979; in chives: Tatlioglu, 1985; and in oilseed rape with 'Polima' CMS: Fan *et al.*, 1986).

3. Association of CMS with negative traits, e.g. chlorophyll deficiency at low temperatures and flower malformations in oilseed rape with 'Ogura' CMS (see Chapter 7 and Rousele, 1982), or extreme susceptibility to a special race of pathogen as in the case of the 'Texas' CMS in maize (Chapter 25; Hooker *et al.*, 1970; Pring and Lonsdale, 1989).

4. Finally, but not dependent on the male sterility system employed, there are the cases where hybrid seed production has turned out to be impractical and uneconomical because of problems in seed production caused by flower morphology and restricted pollen dispersal. However, these problems have been experienced mainly in strictly self–pollinating species like wheat, barley, faba beans and soybeans.

Some of the problems mentioned above may be overcome or obviated by proper planning of the work and/or with the help of *in vitro* culture techniques. Thus, the following section will discuss methods of recurrent selection employing NMS which may also facilitate the development of inbred maintainer and restorer genotypes. Also the rather drastic defects associated with the transfer of the 'Ogura' CMS to oilseed rape have been cured through protoplast fusion and organelle exchange (Chapter 7 and Pelletier *et al.*, 1983). The problem of environmental sensitivity experienced with some crops and CMS sources may be more difficult to ameliorate. However if it is impossible to overcome by stringent selection of maintainers and restorers, an alternative CMS source or an entirely new hybridizing system may be required.

We may conclude that although it is time consuming and costly to initiate a CMS-based hybrid breeding programme, once established it can also be very efficient and reliable, as is the case in sorghum, sugar beet, sunflower and rye. Concerns over the vulnerability resulting from the massive use of a single sterile cytoplasm should not be dismissed but should be judged from case to case. However, where several CMS mutants of different origin exist, it may be advisable to safeguard hybrid seed production through adapting a couple of alternative sterile cytoplasms to commercial use, and eventually even to use them on a limited scale.

(c) Chemical hybridizing agents

When using CMS for hybrid seed production, development of maintainers, equivalent CMS lines and restorers will have to precede evaluation for hybrid performance. With a CHA large number of testcrosses for combining ability can be made, treating one of the parents of a potential hybrid with the CHA. If the performance of the resultant F_1 hybrid is good enough, commercial production is just a matter of scaling up. Thus, CHAs are interesting both as a breeding tool in search for good combiners and as a means of large-scale hybrid seed production.

Over the last 30 years a large number of CHAs have been produced. With few exceptions these chemicals have also damaged female fertility or had other weakness. While such products may still be useful for intermediate breeding purposes, where only limited seed quantities are required, they do not yet permit the economic production of large quantities of hybrid seed with high and reliable germination. However, in recent years new and much improved products have been developed and are being tested by some of the major chemical companies involved in plant breeding. If and when a highly efficient and safe CHA becomes available, in those crops in which it may be used, it is likely to have a major impact on variety development.

14.5.2 Intermediate breeding procedures

In addition to hybrid seed production, male sterility is very useful in plant breeding programmes through facilitating backcrosses, testcrossing for combining ability, and interspecific, and intergeneric hybridization. Although it is generally accepted that a wide gene pool is a prerequisite for succesful breeding work, breeders usually use only a very limited amount of the available genetic variation. Thus, it has been estimated that 90% of the US maize breeding effort is based on only 3 out of 130 racial complexes (Brown, 1975), and that as little as 2% of the genetic variability is used in the breeding of barley (Persson, 1984). Whereas in cross-pollinated species new genetic variability can easily be introduced into breeding populations and these be continously improved through various methods of recurrent selection, in self-pollinated crops the use of such methods has been restricted by the problem of making the large number of crosses among selected genotypes required in each cycle of selection. However, as was first suggested by Gilmore (1964), nuclear male sterility may be used to make recurrent selection also available to breeders of self-pollinated crops. The American barley breeder R.F. Eslick named this process 'Male Sterile Facilitated Recurrent Selection' (MSFRS) (Ramage, 1975). According to Ramage (1987) MSFRS, in its simplest form, involves: (a) selecting plants, both male fertile and male sterile, from a population segregating for the desired characters and male sterility; (b) intercrossing the selected plants; (c) bulking the crossed seed, and growing and harvesting the F_1 generation. The resulting F_2 generation provides the population from which the next cycle of selection is made. New sources of germplasm may be introduced into the population in any cycle by crossing them onto selected male sterile plants. This procedure has been applied in breeding sorghum (Doggett and

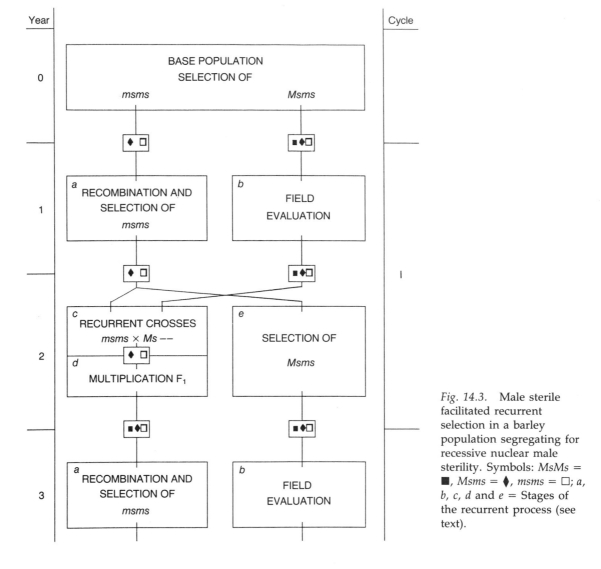

Fig. 14.3. Male sterile facilitated recurrent selection in a barley population segregating for recessive nuclear male sterility. Symbols: *MsMs* = ■, *Msms* = ◆, *msms* = □; *a*, *b*, *c*, *d* and *e* = Stages of the recurrent process (see text).

Eberhart, 1968; Doggett, 1972), barley (Jain and Sunneson, 1963; Eslick, 1977; Ramage, 1979), sugarbeet (Bosemark, 1971), soybean (Brim and Stuber, 1973; Burton *et al.*, 1990), wheat (Driscoll, 1977), sunflower (Anand, 1977), and rice (Fujimaki, 1979), and it has been proposed in some other crops like faba beans (Duc *et al.*, 1985).

Depending on the crop and the breeding objective MSFRS may be applied in different ways. A system currently in use at the Estación Experimental de Aula Dei, Zaragoza, Spain, for the development of new barley varieties for semiarid conditions, is shown in Fig. 14.3. The base population is formed from crosses of selected genotypes with a NMS donor line. In the multiple F_2-generation, 25% of the plants will be male sterile, and by harvesting only such plants, in the next two generations of recombination, a 1:1 equilibrium of male sterile to fertile segregants, is reached. From this population, male steriles and fertiles are chosen (*Year 0*, Fig. 14.3), to initiate the recurrent selection process, which consists of:

Year 1 (Fig. 14.3), allow recombination among progenies of male steriles to take place; and select superior male sterile plants (Plot *a*). Concurrently (Plot *b*), carry out field evaluation of the progenies of the selected fertile plants.

Year 2 (Fig.14.3), in Plot *c*, male sterile segregants from selected male steriles in Plot *a*, are crossed with superior field evaluated lines from Plot *b* and, in the same year, multiplied to obtain segregation for male sterility (Plot *d*). Concurrently (Plot *e*), superior fertile plants are selected among the progenies of the male sterile plants from Plot *a*.

New variability may be introduced by crosses between male steriles from the population and selected germplasms, but only after field evaluation of the crosses. Varieties are developed from superior progenies (Plot *b*), through pedigree selection, SSD or haplo-diploidization techniques.

In the same way as incorporation of NMS permits breeders of self-pollinated crops to exploit cross-pollinated crop breeding methodology, it permits breeders of cross-pollinated crops with hybrid breeding programmes to introduce genes for obligate self-fertility into their breeding populations, thereby facilitating both recurrent selection with S_1-progeny testing, testcrossing, introduction of exotic germplasm and inbred line development. An example of such a system as developed for sugar beet is shown in Fig. 14.4 (Bosemark, 1971, 1993). In *year 1* recombination and production of half-sib families is obtained by harvesting seed from selected male sterile segregants. In *year 2* S_1 seed is obtained from selected fertile plants in the half-sib families. In *year 3* the S_1 families are evaluated in a replicated yield trial. Selected lines are recombined in *year 4*(1) to form the half-sib families for the next cycle of selection. Since the regular segregation of male steriles in such a population makes it easy to introduce maintainer and restorer genes, it becomes natural to develop genetically contrasting populations that are uniformly of maintainer, or uniformly of restorer genotype, or which have a high frequency of the required genes in question. This will alleviate the problems when restorers or maintainers are scarce, as mentioned earlier.

Systems of MSFRS may also be used to upgrade and adapt populations incorporating exotic material and to introduce new genetic variability into the elite breeding populations in a controlled manner (Bosemark, 1989).

14.6 CONCLUSIONS AND PERSPECTIVES

Increasing interest in both hybrid breeding and efficient methods of population improvement has made pollination control via male sterility an important objective in many plant breeding programmes. Beside searching for spontaneous cases of male sterility, breeders have used mutagens to induce both NMS and CMS and wide hybridization and *in vitro* techniques to construct new alloplasmic CMS. To this has been added in the last couple of years the possibilities to use recombinant DNA techniques to engineer entirely new systems of dominant NMS. If these systems, and those which will no doubt rapidly follow, yield what they promise, all the more or less complicated systems developed to facilitate the use of recessive NMS in large-scale hybrid seed production, will soon become history. Not even the elegant system proposed by Jorgensen (1987) will be competitive because of the uncertainty and time involved in constructing a leak-proof marker-linked NMS system.

Where good and proven systems of CMS exist, as in the crops mentioned earlier, these will most likely remain in use since the immediate benefit of introducing a new system would be relatively small compared to the cost. Similarly in population improvement with self-pollinating species, where the objective is not to develop inbred lines for hybrid seed production, the use of existing sources of recessive NMS will continue. However, in those crops where hybrids are not already produced, the introduction of a CMS system, even if readily available, may well be more costly and involve more risk than the introduction of an engineered male sterility system such as that developed by Plant Genetic Systems, which would be possible to use not only in hybrid seed production but also in population improvement.

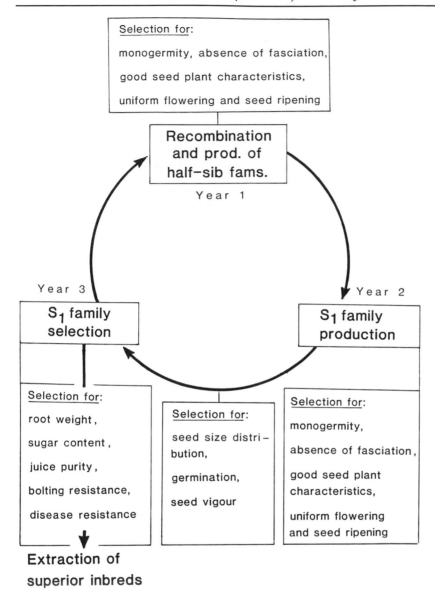

Selection for:

monogermity, absence of fasciation,

good seed plant characteristics,

uniform flowering and seed ripening

Recombination and prod. of half—sib fams.

Year 1

Year 3

S$_1$ family selection

Year 2

S$_1$ family production

Selection for:

root weight,

sugar content,

juice purity,

bolting resistance,

disease resistance

Selection for:

seed size distri—bution,

germination,

seed vigour

Selection for:

monogermity,

absence of fasciation,

good seed plant characteristics,

uniform flowering and seed ripening

Extraction of superior inbreds

Fig. 14.4. Male sterile facilitated recurrent S$_1$ selection in a self-fertile, monogerm, maintainer ($S^F S^F$, *mm*, *xxzz*) sugar beet population segregating for the recessive nuclear male sterility gene a_1. The figure also illustrates the characters that may be selected for in each of the three years of a selection cycle.

Finally, it should be emphasized that further changes may be brought about should a CHA that fulfils all requirements of function and safety, become available. Although less likely to be effective in crops with a long flowering period coupled with substantial vegetative growth such as sugar beet and forages, in several other crops including maize, the major hybrid crop, such a CHA would have important consequences, both for the seed industry and for the farmers.

REFERENCES

Anand, I.J. (1977), Breeding methodology for population improvement in sunflower. In *Proc. 3rd Int. Congress of SABRAO, Plant Breed. Papers. 2. Oilseed Breed.*, Canberra, pp. 1–46.

Anonymous (1991a), Increasing interest in male sterility. *Agrow*, **133**, 19.

Anonymous (1991b), Purdue releases new male sterile tomato breeding system for joint licensing. *Ag. Biotech. News*, **8**(4), 19.

Atanasoff, D. (1964), Viruses and cytoplasmic heredity. *Z. Pflanzenzücht.*, **51**, 197–214.

Athwal, D.S., Phul, P.S. and Minocha, J.L. (1967), Genetic male sterility in wheat. *Euphytica*, **16**, 354–360.

Barabás, Z. (1991), Hybrid seed using nutritional mutants. *Euphytica*, **53**, 67–72.

Barsby, T.L., Yarrow, S.A., Kemble, R.J. and Grant, J. (1987), The transfer of cytoplasmic male sterility to winter-type oilseed rape (*Brassica napus* L.) by protoplast fusion. *Plant Sci.*, **53**, 243–248.

Belhassen, E., Dommée, B., Atlan, A., Gouyon, P.H., Pomente, D., Assouad, M.W. and Couvet, D. (1991), Complex determination of male sterility in *Thymus vulgaris* L.: genetic and molecular analysis. *Theor. Appl. Genet.*, **82**, 137–143.

Belliard, G., Vedel, F. and Pelletier, G. (1979), Mitochondrial recombination in cytoplasmic hybrids of *Nicotiana tabacum* by protoplast fusion. *Nature*, **281**, 401–403.

Berthaut, N., Duc, G., Dulieu, P., Boudon-Padieu, E. and Dulieu, H. (1991), Application d'une méthode immunoenzymatique à la sélection pour la stabilité de la stérilité mâle cytoplasmique chez la féverole (*Vicia faba* L.). *Agronomie*, **11**, 83–92.

Boeshore, M.L., Lifshitz, I., Hanson, M.R. and Izhar, S. (1983), Novel composition of mitochondrial genomes in *Petunia* somatic hybrids derived from cytoplasmic male sterile and fertile plants. *Mol. Gen. Genet.*, **190**, 459–467.

Bosemark, N.O. (1971), Use of Mendelian male sterility in recurrent selection and hybrid breeding in beets. *Eucarpia Fodder Crops Section*, Report of Meeting in Lusignan, pp. 127–136.

Bosemark, N.O. (1989), Prospects for beet breeding and use of genetic resources. In *Report of an International Beta Genetic Resources Workshop*, Wageningen, IBPGR, Rome, pp. 89–90.

Bosemark, N.O. (1993), Genetics and breeding. In *The Sugar Beet Crop*, Cooke, D.A. and Scott, R.K. (eds), Chapman and Hall, London.

Bradner, N.R. and Childers, W.R. (1968), Cytoplasmic male sterility in alfalfa. *Can. J. Plant Sci.*, **48**, 111–112.

Brears, T. and Bingham, J. (1989), Exploitation of heterosis in hybrid wheat using gametocides. *Vortr. Pflanzenzüchtg.*, **16**, 397–409.

Brim, C.A. and Stuber, C.W. (1973), Application of genetic male sterility to recurrent selection schemes in soybeans. *Crop Sci.*, **13**, 528–530.

Brown, W.L. (1975), A broader germplasm base in corn and *Sorghum*. *Proc. Corn. Sorghum Res. Conf.*, **30**, 81.

Burton, J.W., Koinange, E.M.K. and Brim, C.A. (1990), Recurrent selfed progeny selection for yield in soybean using genetic male sterility. *Crop Sci.*, **30**, 1222–1226.

Chen, Z., Liang, G.H., Muthukrishnan, S. and Kofoid, K.D. (1990), Chloroplast DNA polymorphism in fertile and male-sterile cytoplasms of sorghum (*Sorghum bicolor* (L.) Moench). *Theor. Appl. Genet.*, **80**, 727–731.

Clark, E., Schnabelrauch, L., Hanson, M.R. and Sink, K.C. (1986), Differential fate of plastid and mitochondrial genomes in *Petunia* somatic hybrids. *Theor. Appl. Genet.*, **72**, 748–752.

Conde, M.F., Pring, D.R., Schertz, K.F. and Ross, W.M. (1982), Correlation of mitochondrial DNA restriction endonuclease patterns with sterility expression in six male-sterile sorghum cytoplasms. *Crop Sci.*, **22**, 536–539.

Cutler, K. (1991), Canadian seed firm offers 'Male sterility technology for hire'. *Seed Industry*, **42**(6), 4–21.

Doggett, H. (1972), Recurrent selection in *Sorghum* populations. *Heredity*, **28**, 9–29.

Doggett, H. and Eberhart, S.A. (1968), Recurrent selection in *Sorghum*. *Crop Sci.*, **8**, 119–121.

Driscoll, C.J. (1972), XYZ system of production of hybrid wheat. *Crop Sci.*, **12**, 516–517.

Driscoll, C.J. (1977), Reorganizing existing variability in wheat and related species. In *Proc. 3rd Int. Congr. of SABRAO*, Canberra.

Driscoll, C.J. (1985), Modified XYZ system of producing hybrid wheat. *Crop Sci.*, **25**, 1115–1116.

Driscoll, C.J. (1986), Nuclear male sterility system in seed production of hybrid varieties. *C.R.C. Critical Rev. Plant Sci.*, **3**, 227–256.

Driscoll, C.J. and Barlow, K.K. (1976), Male sterility in plants – Induction, isolation and utilization. In *Induced mutations in cross breeding*, IAEA, Vienna, pp. 123–131.

Duc, G., Le Guen, J., Picard, J. and Ber Thelem, P. (1985), Proposed use for a newly discovered dominant male sterile allele for breeding purposes in *Vicia faba*. *FABIS Newslett.*, **12**, 8–10.

Dulieu, P., Pémir, F., Dulieu, H. and Gautheron, D.C. (1988), Purification of virus-like particles from male-sterile *Vicia faba* and detection by ELISA in crude leaf extracts. *Plant Sci.*, **56**, 9–14.

Durand, Y. (1981), Relationships between the marker genes *aa* and *wo* and the male sterility gene ms_{35}. In *Proc. Meet. Eucarpia Tomato Working Group*, Avignon, Philouze, J. (ed.), pp. 225–228.

Duvick, D.N. (1965), Cytoplasmic pollen sterility in corn. *Adv. Genet.*, **13**, 1–56.

Duvick, D.N. (1966), Influence of morphology and sterility on breeding methodology. In *Plant Breeding*, Frey, K.J. (ed.) Iowa State Univ. Press, Ames, pp. 85–138.

Eslick, R.F. (1977), Male sterile facilitated recurrent selection – advantages and disadvantages. *Proc. 4th Regional Winter Cereal Workshop (Barley)*, II, 84–91.

Evenor, D., Barjoseph, M. and Izhar, S. (1988), Attempts to detect extra genomial factors in cytoplasmic male-sterile petunia lines. *Theor. Appl. Genet.*, **76**, 455–458.

Fan, Z., Stefansson, B.R. and Sernyk, J.L. (1986), Maintainers and restorers for three male-sterility-inducing cytoplasms in rape (*Brassica napus* L.). *Can. J. Plant Sci.*, **66**, 229–234.

Fauron, C. and Havlik, M. (1989), The maize mitochondrial genome of the normal type and the cytoplasmic male sterile type T have very different organization. *Curr. Genet.*, **15**, 149–154.

Fluhr, R., Aviv, D., Edelman, M. and Galun, E. (1983), Cybrids containing mixed and sorted-out chloroplasts following interspecific somatic fusions in *Nicotiana*. *Theor. Appl. Genet.*, **65**, 289–294.

Foster, C.A. (1981), Breeding F₁ hybrid cereals – research results and development prospects. In *Opportunities for Manipulation of Cereal Productivity. British Plant Growth Regulator Group*, Monograph 7, pp. 24–35.

Frankel, R. (1962), Further evidence of graft-induced transmission to progeny of cytoplasmic male sterility in petunia. *Genetics*, **47**, 641–646.

Frankel, R., Scowcroft, W.R. and Whitfeld, P.R. (1979), Chloroplast DNA variation in isonuclear male-sterile lines of *Nicotiana*. *Mol. Gen. Genet.*, **169**, 129–135.

Fujimaki, H. (1979), Recurrent selection by using genetic male sterility for rice improvement. *JARQ*, **13**, 153–156.

Gabelman, W.H. (1956), Male sterility in vegetable breeding. In *Brookhaven Symp. Biol*, No. 9, *Genet. Plant Breed.*, pp. 113–122.

Galau, G.A. and Wilkins, T.A. (1989), Alloplasmic male sterility in AD allotetraploid *Gossypium hirsutum* upon replacement of its resident A cytoplasm with that of D species *G. harknesii*. *Theor. Appl. Genet.*, **78**, 23–30.

Galinat, W.C. (1975), Use of male sterile 1 gene to eliminate detasseling in production of hybrid seed of bicolor sweet corn. *J. Hered.*, **66**, 387–388.

Galinat, W.C. (1976), Genetic systems for the production of hybrid corn seed without detasseling. *Maize Genet. Coop. Newslett.*, **50**, 72.

Gilmore, E.C. Jr (1964), Suggested method of using reciprocal recurrent selection in some naturally self pollinated species. *Crop Sci.*, **4**, 323–325.

Halldén, C., Lind, C., Säll, T., Bosemark, N.O. and Bengtsson, B.O. (1990), Cytoplasmic male sterility in *Beta* is associated with structural rearrangements of the mitochondrial DNA and is not due to interspecific organelle transfer. *J. Mol. Evol.*, **31**, 365–372.

Hansche, P.E. and Gabelman, W.H. (1963), Phenotypic stability of pollen sterile carrots, *Daucus carota* L. *Proc. Am. Soc. Hort. Sci.*, **82**, 341–350.

Hanson, M.R. and Conde, M.F. (1985), Functioning and variation of cytoplasmic genomes: Lessons from cytoplasmic-nuclear interactions affecting male fertility in plants. *Int. Rev. Cytol.*, **94**, 213–267.

Hockett, E.A., Baenziger, P.S. and Steffens, G.L. (1978), A proposal for increased research on chemical induction of fertility in genetic male sterile barley. *Euphytica*, **27**, 109–111.

Hooker, A.L., Smith, D.R., Lim, S.M. and Beckett, J.B. (1970), Reaction of corn seedlings with male sterile cytoplasm to *Helminthosporium maydis*. *Plant Dis. Rep.*, **54**, 708–712.

Horowitz, S. and Obregon, P. (1951), The character 'indehiscent anthers' in maize and its possible use in the production of hybrids. *Agron. Trop. Rev. Inst. Nac. Agric., Venezuela*, **1**, 77. Plant Breed. Abstr., 22, abstr. 353, 1952.

Izhar, S. (1978), Cytoplasmic male sterility in *Petunia*. III. Genetic control of microsporogenesis and male fertility restoration. *J. Hered.*, **69**, 22–26.

Jain, S.K. and Suneson, C.A. (1963), Male sterility for increased outbreeding in populations of barley. *Nature*, **199**, 407–408.

Jasmin, J.J. (1954), Male sterility in *Solanum melongena* L.: preliminary report on a functional type of male sterility in egg plants. *Proc. Am. Soc. Hort. Sci.*, **63**, 443.

Jones, H.A. and Clark, A.E. (1943), Inheritance of male sterility in the onion and the production of hybrid seed. *Proc. Am. Soc. Hort. Sci.*, **43**, 189–194.

Jorgensen, R.A. (1987), A hybrid seed production method based on synthesis of novel linkages between marker and male-sterile genes. *Crop Sci.*, **27**, 806–810.

Kadowaki, K. and Harada, K. (1989), Differential organization of mitochondrial genes in rice with normal and male-sterile cytoplasms. *Jpn J. Breed.*, **39**, 179–186.

Kasembe, J.N.R. (1967), Phenotypic restoration of fertility in a male sterile mutant by treatment with gibberellic acid. *Nature*, **215**, 668.

Kaul, M.L.H. (1988), *Male Sterility in Higher Plants, Monogr. Theor. Appl. Genet.*, No. 10, Springer-Verlag, Berlin, Heidelberg, New York.

Kool, A.J., Bot, P.V.M. and Marrewijk, G.A.M. (1983), Analysis of cytoplasmic variations associated with cytoplasmic male sterility in *Petunia hybrida* plants. In *10th Eucarpia Cong.*, Wageningen, abstr., p. 115.

Lacadena, J.R. (1968), Hybrid wheat VII. Test on the transmission of cytoplasmic male sterility in wheat by embryo-endosperm grafting. *Euphytica*, **17**, 439–444.

Laser, K.D. and Lersten, N.R. (1972), Anatomy and cytology of microsporogenesis in cytoplasmic male sterile angiosperms. *Bot. Rev.*, **38**, 425–454.

Leaver, C.J. (1989), Structure and function of the mitochondrial genome in relation to cytoplasmic male sterility. *Vortr. Planzenzüchtg.*, **16**, 379–390.

Leaver, C.J., Isaac, P.G., Small, I.D., Bailey-Serres, J., Liddell, A.D. and Hawkesford, M.J. (1988), Mitochondrial genome diversity and cytoplasmic male sterility in higher plants. *Philos. Trans. Roy. Soc. London B*, **319**, 165–176.

Levings, C.S. III and Pring, D.R. (1977), Diversity of mitochondrial genomes among normal cytoplasms of maize. *J. Hered.*, **68**, 350–354.

Mackenzie, S.A. and Chase, C.D. (1990), Fertility restoration is associated with loss of a portion of the mitochondrial genome in cytoplasmic male-sterile common bean. *Plant Cell*, **2**, 905–912.

Makaroff, C.A. and Palmer, J.D. (1988), Mitochondrial DNA rearrangements and transcriptional alterations in the male-sterile cytoplasm of Ogura radish. *Mol. Cell Biol.*, **8**, 1474–1480.

Mariani, C., De Beuckeleer, M., Truettner, J., Leemans, J. and Goldberg, R.B. (1990), Induction of male sterility in plants by a chimaeric ribonuclease gene. *Nature*, **347**, 737–741.

Mariani, C., Gossele, V., De Beuckeleer, M., De Block, M., Goldberg, R.B., De Greef, W. and Leemans, J. (1992) A chimaeric ribonuclease-inhibitor gene restores fertility to male sterile plants. *Nature*, **357**, 384–387.

Mathias, R. (1985), A new dominant gene for male sterility in rapeseed, *Brassica napus* L. *Z. Pflanzenzücht.*, **94**, 170–173.

McCormick, S., Twell, D., Wing, R., Ursin, V., Yamaguchi, J. and Zarabell, S. (1989), Anther-specific genes: Molecular characterization and promoter analysis in transgenic plants. In *Plant Reproduction: From Floral Induction to Pollination*, Lord, E. and Bemier, G. (eds), Am. Soc. Plant Physiol. Symp. Series, Vol. I, pp. 128–135.

McRae, D.H. (1985), Advances in chemical hybridization. *Plant Breeding Rev.* **3**, 169–191.

Nickerson, N.H. (1960), Sustained treatment with gibberellic acid of maize plants carrying one of the dominant genes teopod and corn grass. *Am. J. Bot.*, **47**, 809–815.

Owen, F.V. (1942), Male sterility in sugar beets produced by complementary effects of cytoplasmic and Mendelian inheritance. *Am. J. Bot.*, **29**, 692.

Owen, F.V. (1945), Cytoplasmically inherited male sterility in sugar beets. *J. Agric. Res.*, **71**, 423–440.

Pareddy, D.R. (1990), Studies on development and attempted chemical reversion of cultured tassels of two genic male steriles of maize (*ms14* and *ms24*). *Maydica*, **35**, 203–208.

Patterson, E.B. (1975), An alternative approach for producing corn hybrids. *Illinois Res.*, **17(3)**, 16–18.

Pellan-Delourme, R. and Renard, M. (1988), Cytoplasmic male sterility in rapeseed (*Brassica napus* L.): female fertility of restored rapeseed with 'Ogura' and cybrids cytoplasms. *Genome*, **30**, 234–238.

Pelletier, G., Primard, C., Vedel, F., Chetrit, P., Remy, R., Rousselle, P. and Renard, M. (1983), Intergeneric cytoplasmic hybridization in Cruciferae by protoplast fusion. *Mol. Gen. Genet.*, **191**, 244–250.

Persson, G. (1984), Methods for the genetic improvement of quality and quantity in barley. In *Cereal Grain Protein Improvement*, IAEA, Vienna.

Phatak, S.C., Wittwer, S.H., Honma, S. and Bukovac, M.J. (1966), Gibberellin induced anther and pollen development in a stamenless tomato. *Nature*, **209**, 635–636.

Pingitore, M., Matthews, B. and Bottino, P.J. (1989), Analysis of the mitochondrial genome of *Daucus carota* with male sterile and male fertile cytoplasm. *J. Hered.*, **80**, 143–145.

Powling, A. (1982), Restriction endonuclease analysis of mitochondrial DNA from sugarbeet with normal and male-sterile cytoplasms. *Heredity*, **49**, 117–120.

Pring, D.R. and Levings, C.S. III (1978), Heterogeneity of maize cytoplasmic genomes among male sterile-cytoplasms. *Genetics*, **89**, 121–136.

Pring, D.R. and Lonsdale, D.M. (1989), Cytoplasmic male sterility and maternal inheritance of disease susceptibility in maize. *Annu. Rev. Phytopathol.*, **27**, 483–502.

Ramage, R.T. (1965), Balanced tertiary trisomics for use in hybrid seed production. *Crop Sci.*, **5**, 177–178.

Ramage, R.T. (1975), Techniques for producing hybrid barley. *Barley Newslett.*, **18**, 62–65.

Ramage, R.T. (1979), Male sterile facilitated recurrent selection. *Barley Newslett.*, **22**, 56–59.

Ramage, R.T. (1983), Heterosis and hybrid seed production in barley. In *Heterosis. Reappraisal of Theory and Practice, Monogr. Theor. Appl. Genet.*, No. 6, Frankel, R. (ed.), Springer-Verlag, Berlin, pp. 71–93.

Ramage, R.T. (1987), A history of barley breeding methods. *Plant Breed. Rev.*, **5**, 95–138.

Ramage, R.T. and Scoles, M.P. (1981), Allele tests and BTT establishment of seedling lethal mutants on chromosome 2. *Barley Genet. Newslett.*, **11**, 35.

Rao, M.K., Devi, K.U. and Arundhati, A. (1990), Applications of genic male sterility in plant breeding. *Plant Breeding*, **105**, 1–25.

Reimann-Philipp, R. (1983), Heterosis in ornamentals. In *Heterosis. Reappraisal of Theory and Practice, Monogr. Theor. Appl. Genet.*, No. 6, Frankel, R. (ed.), Springer-Verlag, Berlin, pp. 234–259.

Rhoades, M.M. (1950), Gene induced mutation of a heritable cytoplasmic factor producing male sterility in maize. *Proc. Natl. Acad. Sci. USA*, **36**, 634–635.

Roever, W.E. (1948), A promising type of male sterility for use in hybrid tomato seed production. *Science*, **107**, 506.

Rouselle, P. (1982), Premiers résultats d'un programme d'introduction de l'androstérilité 'Ogura' du radis chez la colza. *Agronomie*, **2**, 859–864.

Rouselle, P. and Dosba, F. (1985), Restauration de la fertilité pour l'androstérilité génocytoplasmique chez la colza (*Brassica napus* L.). Utilisation des *Raphano-Brassica*. *Agronomie*, **5**, 431–437.

Sawhney, V.K. and Greyson, R.I. (1973), Morphogenesis of the stamenless-2 mutant in tomato. II. Modification of sex organs in the mutant and normal flowers by plant hormones. *Can. J. Bot.*, **51**, 2473–2479.

Schardl, C.L., Pring, D.R. and Lonsdale, D.M. (1985), Mitochondrial DNA rearrangements associated with fertile revertants of S-type male-sterile maize. *Cell*, **43**, 361–368.

Scoles, G.J. and Evans, L.E. (1979), The genetics of

fertility restoration in cytoplasmic male-sterile rye. *Can. J. Genet. Cytol.*, **21**, 417–422.

Siculella, L. and Palmer, J.D. (1988), Physical and gene organization of mitochondrial DNA in fertile and male-sterile sunflower. CMS-associated alterations in structure and transcription of the *atpA* gene. *Nucleic Acids Res.*, **16**, 3787–3799.

Singleton, W.R. and Jones, D.F. (1930), Heritable characters of maize. XXXV. Male sterile. *J. Hered.*, **21**, 266–268.

Tatlioglu, T. (1985), Influence of temperature on the expression of cytoplasmic male sterility in chives (*Allium schoenoprasum* L.). *Z. Pflanzenzüchtg.*, **94**, 156–161.

Ulrich, S.E. and Eslich, R.F. (1977), Inheritance of the shrunken endosperm character, *sex3C* of Bomi Riso mutant 1508 and its association with lysine content. *Barley Genet. Newslett.*, **7**, 66–73.

Umbeck, P.F. and Gengenbach, B.G. (1983), Reversion of male-sterile T-cytoplasm maize to male fertility in tissue culture. *Crop Sci.*, **23**, 584–588.

Vedel, F., Mathieu, C., Lebacq, P., Ambard-Bretteville, F. and Remy, R. (1982), Comparative macromolecular analysis of the cytoplams of normal and cytoplasmic male sterile *Brassica napus*. *Theor. Appl. Genet.*, **62**, 255–262.

Webster, O.J. (1977), *Sorghum* studies in Arizona. *Sorghum Newslett.*, **20**, 81.

Wilson, P. and Driscoll, C.J. (1983), Hybrid wheat. In *Heterosis. Reappraisal of Theory and Practice, Monogr. Theor. Appl. Genet.*, No. 6, Frankel, R. (ed.), Springer-Verlag, Berlin, pp. 94–123.

Wood, M. (1990), Lethal genes may wipe out unwanted pollen. *Agricultural Research*, **38**(8), 22–23.

Wyatt, J.E. (1983), Functional male sterility in beans, *Phaseolus vulgaris* L. *Plant Breed. Abstr.*, **53**, abstr. 9769.

Yang, R.C., Wang, N.Y., Liang, K.J., Cheng, Q.H. and Li, W.M. (1988), 5460ps: *indica* photosensitive genic male-sterile rice. *Int. Rice Res. Newslett.*, **13**(6), 6–7.

15

Apomixis

A.P.M. den Nijs and G.E. van Dijk

15.1 INTRODUCTION

Most plants reproduce sexually through seeds. A zygote is formed by fusion of reduced female and male gametes (amphimixis) and develops into an embryo. Plants of some species reproduce through seeds having an embryo which is formed without reduction of the chromosome number and fertilization. Such vegetative or asexual reproduction by means of seeds is called apomixis.

Apomixis is found mostly in polyploid species of the families Gramineae, Rosaceae and Asteraceae and often confers fertility to hybrid genotypes which otherwise would have been sterile. Most successful apomictic species are facultative with sexual reproduction and apomixis being in equilibrium. Such species usually comprise sexual and apomictic entities, often with several ploidy levels and are called agamospecies or agamocomplexes.

Clausen (1961) explained the evolutionary adaptability and multitude of micro-species recognizable in successful agamocomplexes by their dual ability to occasionally sidestep sexual reproduction and to multiply the successful combinations asexually.

Apomixis occurs in numerous species of agricultural value. Among the grasses it prevails in polyploid species of *Paspalum, Panicum,*

Poa, Bothriochloa, Dichanthium, Eragrostis, Pennisetum and *Cenchrus* (Bashaw, 1975). Other important genera include *Citrus* and *Rubus*, and crops as diverse as sugar beet, apple, pearl millet, wheat and maize rank apomictic species among their wild relatives.

Apomixis is potentially a powerful breeding tool to fix heterosis, but it has mostly been regarded as a barrier for the breeder because recombination by means of crossing is very difficult or even impossible. During the last two decades however, this opinion has lost ground as more detailed studies of apomictic species revealed exceptional sexual genotypes. Harlan *et al.* (1964) demonstrated that in the prairie-grass *Dichanthium* apomixis was relatively simply inherited and Taliaferro and Bashaw (1966) proposed a model for the genetic basis of apomixis in the forage crop buffelgrass (*Pennisetum ciliare* now called *Cenchrus ciliaris*). These authors proposed a specific breeding method for this species. Since then breeders have become more and more convinced that apomixis is not a 'dead end'. Breeding methods for apomictic crops were reviewed by Bashaw (1980), Hanna and Bashaw (1987) and Van Dijk (1991). Savidan (1990) concluded that 'transfer of apomixis to wheat, rice, maize, sorghum and millets is currently being attempted'. Potato and sugar beet can be added to this list. New techniques such as tissue culture, somatic hybridization, protoplast regeneration and genetic manipulation, but also isozyme and RFLP (restriction fragment length polymorphisms) analysis of progenies, may contribute to the breeder's ability to handle apomixis.

Plant Breeding: Principles and prospects. Edited by M.D. Hayward, N.O. Bosemark and I. Romagosa. Published in 1993 by Chapman & Hall, London. ISBN 0 412 43390 7.

15.2 TYPES OF APOMIXIS AND DEFINITIONS

In a broad sense apomixis may include asexual reproduction through purely vegetative organs such as rhizomes, stolons, bulbils, etc., as well as many modern biotechnological methods of asexual reproduction. In this chapter the term will be restricted to the asexual formation of seed according to the definitions commonly used by plant breeders (Bashaw, 1980; Van Dijk, 1991). In this sense apomixis is synonymous with agamospermy: seed formation without fertilization of the egg cell. The main ways in which the embryos can be formed are shown in Fig. 15.1 (Grant, 1971). The normal sexual life cycle is included for comparison.

The routes can be summarized as follows:

Diplospory: a non-reduced embryo-sac develops from an archespore cell (embryo-sac mother cell) through omission or restitution of meiosis; the egg cell develops parthenogenetically into an embryo, or another cell of the embryo-sac divides and develops into an embryo. The latter route is called apogamety.

2. Apospory: the non-reduced embryo-sac develops from a somatic cell of the nucellus or the integument instead of the embryo-sac mother cell.

3. Adventitious or nucellar embryony: the embryo develops directly from the sporophytic tissue, without formation of a gametophyte. This most closely equals pure asexual reproduction.

The underlying processes of agamospermy,

A. NORMAL SEXUAL LIFE CYCLE

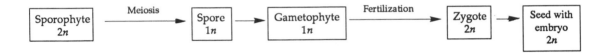

B. GAMETOPHYTYC APOMIXIS

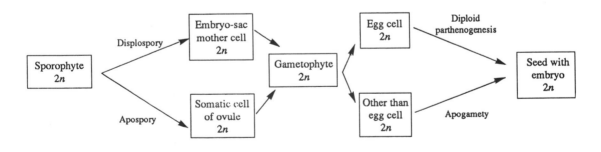

C. ADVENTITIOUS EMBRYONY

Fig. 15.1. Modes of agamospermous reproduction. A normal sexual life cycle is shown for comparison (after Grant, 1971).

non-reduction and parthenogenesis have been recognized since the excellent pioneering treatise by Gustafsson (1946). A recent classification with emphasis on the embryological development was proposed by Naumova (1990). The occurrence of types of apomixis in several plant families is given in Table 15.1. It should perhaps

Table 15.1. Numbers of species possessing various types of apomixis (after Nygren, 1967).

Family	Apospory	Diplospory	Adventitious embryony	Total
Asteraceae	18	51		69
Gramineae	68	27		95
Liliaceae		1	6	7
Rosaceae	65	3		68
Rutaceae	2		5	7
Urticaceae	2	7		9
28 other families	5	15	33	53
Total	160	104	44	308

be noted, that an extensive survey of the plant kingdom for apomixis has yet to be completed.

Special attention must be paid to the effect of pollen on apomixis. If no fertilization of the central nucleus is necessary for seed development, apomixis is called autonomous and the resulting endosperm will have double the somatic or the somatic chromosome number (Asker, 1980).

In many apomictic species pollination is obligatory for the formation of endosperm and the development of the egg cell: this is called pseudogamy. Although the egg is not fertilized, the central nucleus, or one or two of the unreduced polar nuclei are. Endosperm resulting from a fertilized central nucleus should be 5*n*, but, for example in pseudogamous *Ranunculus auricomus* all combinations of 1 to 3 polar nuclei and 1 to 2 sperm cells were found (Rutishauser, 1969).

Pollen development in apomictic plants can be as regular as in sexual plants giving (poly) haploid pollen. In many species, however, meiosis is irregular resulting in aneuploid sporad cells. This may be due to the polyploid and often aneuploid nature of most apomicts. The pollen quality of pseudogamous species is better than that of species with autonomous apomixis (Nygren, 1967).

A special form of apomixis is haploid parthenogenesis. In this case a normal, reduced egg cell or other nucleus in the embryo-sac develops parthenogenetically into an embryo and seed, giving a plant with half the chromosome number of the parent. Haploid parthenogenesis can be autonomous or it can be induced, for example by pollination. In rare cases the nucleus of one of the sperm cells of the pollen tube can displace the egg cell in an embryo-sac yielding a haploid with the nucleus of the male partner and the cytoplasm of the female. This form of androgenesis is considered in Chapter 17 and could be a useful technique for quickly transferring nuclei into different cytoplasms.

15.3 CHARACTERS OF APOMICTIC PLANTS

Most apomictic species are polyploids with exceptions for the genus *Citrus* and some species of *Potentilla*. According to Stebbins (1971) both auto- and allopolyploids are found, the latter being more common, but recent evidence shows more and more autopolyploids among the apomictic species. Apomixis is a means to maintain genotypes with irregular chromosome numbers, and aneuploidy is often found. In some species a wide range of chromosome numbers occurs, e.g. in *Poa pratensis* it varies from 2*n* = 28 to 2*n* = 154. From crossing experiments it has also become clear that apomictic plants are usually very heterozygous (Clausen, 1952, 1961).

As apomixis equals vegetative reproduction by seed, the offspring will be uniform and identical to the mother plant. In obligate apomicts this holds true for almost all offspring, the only deviating types being rare 3*n* plants due to fertilization of an unreduced egg cell (Bashaw, 1980).

Facultative apomicts combine apomictic and sexual reproduction. The sexual reproduction is of the cross-fertilizing type, which is also often true for related species. The combined sexual and apomictic reproduction results in mixed progeny of maternal and aberrant types (Table 15.2).

Plants originating from regular fertilization of

Table 15.2. Four modes of reproduction in a facultative diploid apomict and the resulting offspring (after Asker, 1979)

Egg cells	Offspring formed by	
	Parthenogenesis	Fertilization
Unreduced	Uniform, maternal	Variable 'autotriploids' ($3x$)
Reduced	(Poly)haploids	Variable 'diploids' ($2x$)

reduced or unreduced gametes are expected to be $2n$ or $3n$, respectively. However, many deviations may occur because of cytological irregularities both on the male and the female side. In the offspring of the pseudogamous aposporous *Poa ampla* ($2n = 63$) Clausen (1961) found 90–95% maternal types with 63 chromosomes whereas the remainder consisted of aberrants, with chromosome numbers varying from $2n = 56$ to $2n = 147$. In facultative apomicts the uniformity of the offspring depends on the degree of apomictic reproduction in the parent plant.

In facultative aposporous apomicts embryosacs of meiotic and mitotic origin may occur in the same nucellus. In many apomictic species high frequencies of multiple or twin seedlings occur. In obligate apomicts multiple seedlings are always of apomictic origin and identical to the mother plant, whereas in facultative apomicts non-identical twins are possible.

Apomictic reproduction can be recognized because the offspring of a single mother plant is often more uniform than expected and similar to the mother plant. Such offspring is often called maternal offspring. The percentage of uniform maternal offspring indicates the degree of apomixis in the parent. When pollen parents seem to have no or little influence on the progeny, apomixis is also very likely, especially when the female plant has recessive characters. If self-fertilization is possible, lack of inbreeding depression in the progeny is another indication of apomixis.

An enhanced frequency of multiple seedlings also indicates the possible occurrence of apomixis. Van Dijk (1979) found 1–2% twin seedlings in the apomictic grass *Poa pratensis*,

compared with only 0.04% in the cross-fertilizing amphimictic species *Lolium perenne* and *Phleum pratense*. Bashaw (1980) even reported up to 22% identical twins in buffelgrass (*Cenchrus ciliaris*). On the other hand, Gröber *et al.* (1978) could not detect a reliable correlation between the frequency of polyembryony and the degree of apomixis in their *Poa pratensis* breeding materials. An odd chromosome number ($3n$, aneuploid) combined with normal seed set is also typical of apomictic reproduction.

Although several offspring characters may indicate the occurrence of apomixis, embryological studies of young ovules are required to confirm the mechanism of apomictic behaviour. The occurrence of multiple embryo-sacs in the young ovule is an important criterion. In some grasses the aposporous ones are not full-grown but 4-nucleate in contrast to the 8-nucleate sexual ones (Bashaw, 1980). In diplosporous apomicts cytological study should identify whether the first division of a megaspore mother cell is meiotic or mitotic. Such studies are difficult because of the early stage of development of the ovule when meiosis occurs (Bashaw, 1980), but they have become more efficient with the help of new clearing techniques (Young *et al.*, 1979). The occurrence of aposporous embryo-sacs is not in itself a guarantee for completely or partly apomictic progeny, because such embryo-sacs may fail to develop into seeds (Grazi *et al.*, 1961). The absence of two types of embryo-sacs on the other hand is no guarantee for a completely sexual offspring. Therefore, embryological studies may reveal the mechanisms of apomixis, whereas progeny tests are required to determine the degree of apomixis.

15.4 GENETIC BASIS OF APOMIXIS

For genetic studies apomictic genotypes are usually crossed with related sexual forms. As a first approach one can then distinguish between maternal and aberrant offspring plants, looking at apomixis 'as a whole' (Asker, 1979). One of the earliest observations from crosses between apomictic and sexual forms was a disturbance of the apomictic reproduction. This led to the assumption of a special physiological condition for apomictic reproduction. In *Poa*, Müntzing (1940) believed that apomixis is 'due to a rather delicate genetic balance' and 'this balance may be upset in various ways, by crosses with other types or merely by a quantitative change in chromosome number either in a plus or minus direction'. In *Potentilla argentea* and some other species, chromosome doubling by colchicine has shifted the apomictic mode of reproduction towards sexuality (Asker, 1979). Several authors emphasize that a study of the underlying processes, especially non-reduction, parthenogenesis and pseudogamy is needed for a full understanding of the inheritance of apomixis (Asker, 1980).

From a study of apomixis in the rubber-plant guayule (*Parthenium argentatum*) Powers (1945) reasoned that at least three sets of genes are necessary for apomictic reproduction, namely for (1) failure of reduction in number of chromosomes; (2) failure of fertilization; (3) development of non-reduced unfertilized egg-cells. Although this model seems very fanciful, the material has never allowed it to be tested (Nogler, 1984). Gerstel *et al.* (1953) suggested that apomixis in *Parthenium argentatum* is based on at least four recessive genes, a minimum of two being concerned with the process of meiotic reduction and two more with the requirement of fertilization (Table 15.3).

A species in which much research has revealed comparatively little about the inheritance of apomixis is the aposporous facultative apomictic grass *Poa pratensis*. Åkerberg and Nygren (1959) stated that many factors are involved and any disturbance will destroy the system, whereas Grazi *et al.* (1961) believed the clarification of the genetic background to be extremely difficult, if not entirely impossible, because of the high chromosome number and a very irregular meiosis, which leads to the production of functioning gametes with variable chromosome number. In early studies apomixis was described as a recessive character, but Almgård (1966) described a dominant pattern of apospory in crosses between *Poa pratensis* and the sexual species *P. longifolia*. Funk and Han (1967) also found apomixis to be dominant in *P. pratensis* and presumed that it is controlled by two or more dominant complementary genes plus modifying factors.

Mutants for one of the component processes of apomixis may be useful for genetic analysis. Matzk (1991) isolated mutants in *Poa pratensis* that are incapable of parthenogenesis. In crosses with this material parthenogenesis appeared to be dominant over sexuality. Non-parthenogenetic genotypes were assumed to be nulliplex at one locus, while parthenogenetic genotypes have one or more dominant alleles (Table 15.3).

In species of the agamo-complex involving

Table 15.3. Genetic base of apomixis in some crop species

Species	Number of loci	Gene action	Reference
Bothriochloa spp.	2	Dominant	Harlan *et al.* (1964)
Cenchrus ciliaris	2	Epistasis	Taliaferro and Bashaw (1966)
	3	Epistasis, incomplete penetrance	Gustine *et al.* (1989)
Citrus spp.	Few	Dominant, nucellar embryony	Cameron and Soost (1982)
Panicum maximum	2	Recessive	Hanna *et al.* (1973)
	1	Dominant	Savidan (1983)
Parthenium	3		Powers (1945)
argentatum	4	2 for non-reduction	Gerstel *et al.* (1953)
Paspalum notatum	Few	Recessive	Burton and Forbes (1961)
Poa pratensis	Many	Recessive	Åkerberg and Nygren (1959) Grazi *et al.* (1961)
	2 +	Dominant, modifiers	Funk and Han (1967)
	1	Dominant, only parthenogenesis	Matzk (1990)

Bothriochloa and *Dichanthium* Harlan *et al.* (1964) found apomixis to be dominant over sexuality and concluded that two genes were involved. The apomictic plants were heterozygous, the sexual ones homozygous recessive for the character. However, these conclusions did not hold true for the *Dichanthium* types investigated by D'Cruz and Reddy (1971), who concluded that apomixis is associated with polyploidy, and depends upon a complex genetic balance for its expression. Similarly, in *Paspalum notatum* Burton and Forbes (1961) explained the inheritance of obligate apomixis in this tetraploid species by assuming a few recessive genes for apomixis.

More detailed knowledge about the genetic background of obligate apomixis was derived from research in *Cenchrus ciliaris*. After discovering a sexual plant, Taliaferro and Bashaw (1966) used this plant for selfing and for crossing experiments. The offspring of the selfed plant consisted of completely sexual plants and obligate apomicts in the ratio 13:3. In crosses with apomictic plants the sexual plant gave a ratio of 5 sexual:3 apomictic plants. The data were explained by assuming 2 genes and epistasis, the sexual parent being *AaBb* and the apomictic *Aabb*. Apospory was considered due to the action of the dominant gene *A*, which could be suppressed by a second gene *B*. The double recessive *aabb* reproduced sexually. Recently in this species facultative apomixis was discovered to be caused by a third gene, epistatic to gene *A* and having a lower penetrance than gene *B* (Gustine *et al.*, 1989). The authors initiated a programme to isolate the *A* and *B* genes, via their protein products. Total pistil proteins were isolated from progenies differing for reproductive behaviour following selfing of a sexual type *AaBb*. These two kinds of progeny were compared in two-dimensional electropherograms, but no systematic differences in proteins were correlated with sexual *versus* apomictic embryo-sac formation (Gustine *et al.*, 1989).

In *Panicum maximum* also both sexual plants and aposporous apomicts are found and here apomixis can be obligate or facultative. Hanna *et al.* (1973) studied sexual × apomictic hybrids and explained the results on the basis of a

model involving at least two loci with apomixis as the recessive character, *aabb*, *Aabb* and *aaBb* being apomicts, plants with both *A* and *B* being sexual. It should be noted, however, that *P. maximum* is in fact a tetraploid. Savidan (1983) explained segregation data in *Panicum maximum* by designating sexual tetraploids as *aaaa*, and all the apomicts he tested as *Aaaa*; apomixis was facultative with a low percentage of sexuality. The inheritance of the percentage sexual reproduction could not be clarified, but it was proposed that a single dominant gene *A* controlled both apospory and parthenogenesis as one process (Table 15.3).

The inheritance of nucellar or adventitious embryony in common *Citrus* species appeared to be rather simple (Cameron and Soost, 1982). Crosses between strictly sexual parents as a rule yielded sexually reproducing hybrids whereas crosses between sexual and nucellar parents usually produced hybrids of both types. Parents that reproduced by nucellar embryony sometimes produced offspring that was completely sexual. Nucellar embryony appears to be dominant over sexual embryony and very few genes may be involved. Strictly sexual plants appear to be homozygous recessive for these genes (Table 15.3).

In addition to the data from crop species, information is available from many studies of wild species of, for example the genera *Hieracium*, *Taraxacum*, *Ranunculus* and *Potentilla* (Asker, 1979; Nogler, 1984). At first sight, also in nature examples of a simple genetic base of apomixis are rather the exception. However, Nogler (1984) discusses the monogenic inheritance of aposporous *Ranunculus auricomus* and concludes from cytogenetic investigations that such simple inheritance may be common for apospory.

A cytoplasmic basis for the inheritance of apomixis was postulated by Lehnhardt and Nitzsche (1988). The authors perceive a parallel distribution pattern throughout the plant kingdom of both cytoplasmic inheritance and apomixis, and feel that none of the different proposed nuclear genetic systems are adequate to explain the inheritance of apomixis. However, there are no indications in the literature for cytoplasmic influence on apomixis.

It is clear, that a better insight into the apomictic phenomena and their genetic background is needed to improve practical breeding programmes. Although apomixis is clearly under genetic control, genetical research indicates a wide range of behaviour from simple monogenic control to unexplainable genetic situations. The conflicting data from different experiments indicate that various genetic mechanisms for apomixis may exist even within the same species, but they also highlight the need for a critical re-appraisal of the earlier studies with help of more powerful analytical methods (Nogler, 1984; Savidan, 1990).

15.5 ENVIRONMENTAL INFLUENCE

For any particular clone of a facultative apomictic species, grown in a controlled environment, the proportion of sexual to apomictic seeds is fairly constant (Stebbins, 1971). However, environmental factors have been reported to affect the degree of sexuality. In *Dichanthium aristatum* a short day treatment (8 h) during the development of the inflorescences resulted in 79% of the embryosacs being aposporous, which was significantly higher than the 47% in 16 h (Knox and Heslop-Harrison, 1963). The effects on the composition of the progeny were not given. Williamson (1981) likewise reported an influence of the light regime during the period just prior to ear emergence until anthesis in *Poa ampla* and *Poa pratensis*. In contrast to *Dichanthium*, short-day treatment in *Poa ampla* promoted sexuality compared to a 20 h day length regime.

Increased sexuality in a *Poa pratensis* plant occurred when it flowered in early spring in the greenhouse compared to the same genotype flowering in the open (Grazi *et al.*, 1961). Han (1971) observed sexual embryos in panicles of *Poa pratensis* after a temperature shock by hot water just before flowering, whereas gibberellic acid, day length, light intensity, nitrogen-dose and the position of the flower had no effect. Plants grown in glasshouses gave more sexual offspring than plants grown in the field, which facilitates hybridization programmes (Section 15.6).

In seedlots of *Poa pratensis* cultivars, grown at contrasting sites, the rate of aberrants was higher when the flowering period was longer (Hovin *et al.*, 1976), but the authors judged that the negative effect on the uniformity of the offspring was limited. However, seed multiplication in unusually hot summers often reveals too much heterogeneity in the offspring of commercial *Poa* varieties (Vijn, personal communication).

In *Malus* species a considerable influence of the time of year on expression of apomixis was reported by Schmidt (1977), with temperature being presumably the most important factor. Also, a facultative apomictic line of *Cenchrus ciliaris* was influenced by the environment: sexuality was greater at the end of the season than during summer (Sherwood *et al.*, 1980). However, no single factor could be identified which was responsible for this effect. Applications of plant growth hormones had different effects. Maleic hydrazide shifted the ratio towards higher frequency of apospory. In *in vitro* experiments gibberellic acid increased the ratio of reduced to unreduced embryo-sacs (Gustine *et al.*, 1989).

In conclusion it seems that the environment has a certain influence, but a possible shift in the ratio of sexual *versus* apomictic reproduction depends as much on the strength of the apomictic genotype as on the environmental conditions.

15.6 BREEDING APOMICTIC CROPS

Apomixis offers many advantages to the breeder. A single outstanding plant of an obligate apomict is sufficient for a cultivar and cultivars are uniform, stable and easy to describe. Heterosis and epistatic gene effects can be fixed. For seed production the isolation distances between cultivars can be reduced. Finally, viral infections can be eliminated from the plants e.g. in *Citrus*. It seems obvious, that these advantages weigh most heavily in less-developed countries where breeding and seed production often lack technological sophistication.

However, for the plant breeder apomixis creates the problem of generating new variation for selection to act upon. Therefore different approaches to breeding apomicts have been developed.

15.6.1 Ecotype selection

With high levels of variation in the species, cultivars can be developed by collection and ecotype screening. Outstanding genotypes can be multiplied for commercial use. Collecting new genotypes is easy where ecotypes have evolved that are adapted to various natural and man-made environments. As an example, in 1936 Joseph Valentine observed an outstanding plant of *Poa pratensis* on a golf course of the Merion Golf Club (USA) and this single plant was the basis of the variety 'Merion', successful for many years in parts of the USA, Canada and Western-Europe (Van Dijk, 1991).

The production of cultivars from ecotypes comes to its natural end when the best types have been collected. This is also illustrated in *Poa pratensis*, in which, in 1977, 12 out of 24 applications for registration of new cultivars in the Netherlands were rejected because of lack of distinctness from existing cultivars, in spite of the fact that the breeders collected both in Europe and in the United States. Some successful ecotypes seem to have a wide distribution and reappear regularly in breeders collections (Duyvendak and Luesink, 1979).

A further source of variation is the aberrant plants occurring in the offspring of apomicts. In obligate apomicts only rare $3n$ plants are found which maintain apomictic reproduction. In facultative apomicts the aberrants are more varied and some of them may well have originated by hybridization. Unfortunately, most of the aberrants are weaker than the parent plant. Duich and Musser (1959) found 2.8–4.2% aberrants in cv. 'Merion' of *Poa pratensis*, 19% of them (less than 1% of the total offspring) being equal or superior to the original cultivar.

Although many varieties have been produced by collection and ecotype selection, further breeding needs hybridization and it is at this point that apomixis presents a barrier.

15.6.2 Hybridization of obligate apomicts

Obligate apomixis precludes hybridization until exceptional sexual types are discovered. This is illustrated by the breeding history of *Cenchrus ciliaris*, which is used as a fodder grass in parts of the USA (Taliaferro and Bashaw, 1966). Before 1958 new variation was sought in introductions from abroad and by mutagenic treatments, but all attempts failed. In 1958 a sexual plant was discovered and this plant made it possible to elucidate the genetic background of apomixis (section 15.4) and to start a breeding programme by selfing sexual plants and crossing sexuals and apomicts. Both the S_1 and hybrid offspring contained sexual and apomictic plants. Selfing and backcrossing sexual hybrids to the male parent were used to recover recessive traits (Fig. 15.2). The apomicts can be evaluated, while with the sexuals the breeding programme can be continued. This programme could be efficient because the genetic basis of apomictic reproduction was known, the apomictic plants breed true and can be recognized early (Bashaw, 1975).

Fertilization of an unreduced (aposporous) egg sometimes occurs in obligate apomicts, but until recently it was considered to occur too infrequently for practical plant breeding. However, Bashaw and Hignight (1990) proved that in *Cenchrus ciliaris* a sufficient number of these BIII 'triploid' hybrids ($2n+n$) with useful fertility could be obtained. This could be used as an alternative means of gene transfer in obligate apomicts.

Like in *Cenchrus*, the improvement of *Paspalum notatum* was unsuccessful until the obligate apomictic reproduction could be circumvented. Burton and Forbes (1961) doubled the chromosome number of sexual diploid accessions by colchicine treatment of seeds and crossed these new tetraploids with the apomictic tetraploids ($2n = 40$). Both obligate apomictic and com-

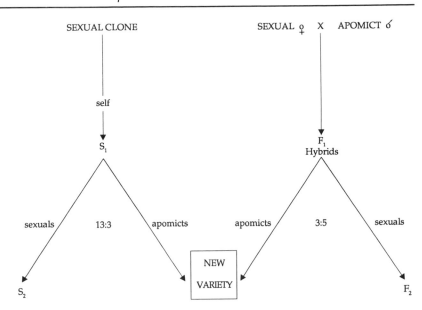

Fig. 15.2. Breeding of *Cenchrus ciliaris* based on a sexual individual (Taliaferro and Bashaw, 1966). With the sexual S_2 and F_2 plants the scheme can be repeated.

pletely sexual F_1 hybrids were recovered. Recently also sexual accessions were discovered in the apomictic forage grass *Brachiaria decumbens*, which enabled the start of a hybridization programme for improvement of nutritive value (Do Valle *et al.*, 1989).

15.6.3 Hybridization of facultative apomicts

Although the distinction between facultative and obligate apomixis is important for breeding work, it must be remembered that a facultative apomict with a very low tendency to sexuality approaches the obligate types. More detailed study often reveals facultative apomixis in hitherto obligate apomictic species, e.g. in *Cenchrus ciliaris* (Sherwood *et al.*, 1980) and in *Eragrostis curvula* (Voigt and Burson, 1983). By crossing tetraploid sexuals with apomicts the latter authors obtained segregating progenies with about 40% apomictic plants. Promising apomicts with less then 13% sexuality could be selected and their progeny were considered sufficiently uniform to become a cultivar.

In *Panicum maximum* most biotypes are tetraploids and facultative apomicts, but obligate apomixis also occurs. Sexual tetraploids were

obtained from colchicine-treated natural sexual diploids (Savidan, 1982). A breeding scheme of recurrent crosses between sexuals × apomicts at the tetraploid level extended the sexual gene pool and produced new apomictic hybrids. The programme was facilitated by a high recovery rate of apomicts and by the low degree of sexuality in the apomictic plants (Savidan, 1983). Several measures for further improving the efficiency of the scheme have been described (Savidan *et al.*, 1989).

Guayule is a facultative apomictic tetraploid species producing rubber. Breeding for increased rubber content has proved to be very difficult and breeding at the diploid level is advocated as an attractive way to make quick progress as in potato and alfalfa. Doubling of the chromosome number of selected diploids should then yield new elite tetraploids, which should be apomictic. Tetraploid guayule was recently synthesized *de novo* from diploid sexual genotypes (Hashemi *et al.*, 1989). Unlike the natural tetraploids, these colchicine-derived artificial tetraploids reproduce sexually, having a predominantly normal meiosis. Apomixis must be restored by recurrent hybridization with apomictic types as in *Panicum*.

Hybridization breeding of *Poa pratensis*, a facultative pseudogamous apomict, has proved to be complicated. The species has a wide distribution and is used both as a forage and as turfgrass. It may have originated through species crossing and allopolyploidy, combined with a high degree of autopolyploidy (Gustafsson and Gadd, 1965). Viability is barely affected by small variations in chromosome number and aneuploidy is common. The number of chromosomes is in the range $2n = 28$–154, with most ecotypes having 45–90. Within the species genotypes occur that are nearly 100% apomictic as well as some that are completely sexual, but a high level of apomixis is common. The genetic control of apomixis has not been elucidated, although various conflicting suggestions have been put forward (section 15.4). Åkerberg and Nygren (1959) felt that crossing and recombination should certainly be used in breeding, but without a thorough knowledge of the highly heterozygous parents, it would be no more than a gamble. Crossing of apomictic plants in a glasshouse was initiated in the seventies in the USA (Pepin and Funk, 1971). The percentage aberrants in the progenies depended on the sexuality of the female parent whereas their quality depended on the cross combination and the degree of transgressive segregation that occurred. Many hybrids were 'triploids' and a high level of apomixis could be retained. Funk and Han (1967) reported that about 14% of the hybrid plants were predominantly apomictic, from one combination even 86%. Hintzen (1979) found in F_1 progenies 3–58% aberrants with an average of 17%. About one-third of these crosses produced superior apomictic plants in 0.4–4.5% of the aberrant progeny. However, very few selections reached the variety trials and Hintzen and Van Wijk (1985) concluded that hybridization was no more efficient than ecotype selection.

The success of hybridization breeding depends partly on the degree of sexuality that is acceptable in cultivars. Pepin and Funk (1974) consider 80% apomictic seed to be the minimum, but in many countries registration authorities demand a much higher uniformity (e.g. in the Netherlands 95–98%, according to Hintzen and Van Wijk, 1985).

Poa pratensis offers many possibilities for interspecific hybridization, which may alterna-tively be used to sidestep apomixis. Extensive hybridization experiments involving many *Poa* species have demonstrated that hybrids between two apomictic species usually became sexual, but recombinations of parental characters appeared in the F_2 (Clausen, 1952). *Poa pratensis* can absorb traits from many species, giving rise to apomictic segregants that taxonomically would be classified as *Poa pratensis* (Clausen, 1961). As an example, from crosses between sexual *Poa longifolia* ($2n = 28$, 42) $\times$ *P. pratensis* it appeared that intermediate hybrids were unstable but in the next generation traits could be stably transferred (Almgard, 1966; Williamson and Watson, 1980; Van Dijk and Winkelhorst, 1982). Apomictic reproduction was recovered in further generations; thus Williamson and Watson (1980) found 22% of the hybrid progenies to be apomictic in the third generation, Van Dijk and Winkelhorst (1982) found 0–30% very uniform progenies in advanced backcrosses. The latter authors stressed the twofold use of interspecific crosses in breeding apomicts: the introduction of an entirely sexual phase and the possibility of introgression.

It is important to identify the most promising partner for interspecific hybridization. *Poa palustris* appeared to be a better crossing parent than *Poa longifolia* (Nitzsche, 1983) and Van Dijk (1979) reported a higher recovery of apomixis when *P. chaixii* was used instead of *P. longifolia*. The ideal partner should not only give fertile recognizable hybrids, it should also contain transmissible traits for enrichment of the apomict and allow a rapid recovery of apomixis in the generations following hybridization.

Time-consuming hybridization procedures in *Poa pratensis* are made more efficient if selection can be carried out in early phases of the programme. Isozyme analysis can be used to distinguish maternal from deviating offspring at the seedling stage. Preliminary tests of offspring from controlled crosses between highly apomictic cultivars of *Poa pratensis* indicated that in this species isozymes may be helpful depending on the isozyme type of the parent of the cross (Den Nijs, 1990). However, almost half the offspring which differed from the maternal parent could not be positively identified as a hybrid because no male parent isozyme band(s) were present. This suggests that apart from hybrids other deviating offspring are also produced in these crosses.

In *Citrus*, the nucellar embryony has been a major obstacle to the systematic production of hybrids, because many cultivars produce almost entirely nucellar embryos and there may be several in a single seed (Cameron and Soost, 1982). However, other cultivars form both nucellar and sexual embryos and even purely sexual cultivars have been identified. The latter have been widely used in crosses. Crosses between sexuals and apomicts produce hybrids of both types. *Citrus* species offer one of the rare situations where apomictic breeding can be performed at the diploid level, though tetraploid and triploid cultivars occur.

15.6.4 Induction of mutations

Induction of variation by mutagenic treatment appears at first sight very promising, as useful mutations could directly be fixed by apomixis. Apart from that, sexuality could also be induced. Experiments with *Poa pratensis* (Julén, 1961; Gustafsson and Gadd, 1965) suggested that both targets can be obtained. Treating with X-rays, dry seeds of the cultivar 'Fylking' increased the number of aberrants from 2% to 14% (Julén, 1961). The aberrants were generally weaker than the original cultivar, but some were at least as vigorous. A large number were chimeric and had to be separated into different parts. Julén concluded that it would be possible to induce sexuality by X-rays, make crosses and in later generations select new apomictic types with favourable characters. Indeed Hanson and Juska (1962) irradiated *P. pratensis* cv. 'Merion' with thermal neutrons and selected a mutant with enhanced rust resistance. However, it had lost the resistance to *Drechslera poae*, which made 'Merion' highly appreciated. The success of mutation induction has not been evident (Bashaw *et al.*, 1983) and breeders have turned their attention to other ways of inducing variation such as tissue culture.

15.6.5 Tissue culture

The application of tissue culture techniques as a source of useful genetic variation was recently tested by Taliaferro *et al.* (1989) with the apomictic bluestem grass *Bothriochloa*. Twenty-one variant regenerants (R_1) from callus of four accessions of different *Bothriochloa* species were selected for study of isozymes, chromosome number and reproductive behaviour. The percentage of variant R_1 plants was 1–8% depending on the accession. All R_1-plants differed from the 'parent' accession for both peroxidase and esterase isozyme bands, but there were no differences between the regenerants from one accession, indicating a common origin. The chromosome number of the R_1 deviants was either equal to the 'parent' or higher and the plants appeared to be apomictic. Thus new morphological, biochemical and cytogenetic variation was induced, while apomixis was retained.

In *Poa pratensis* Wu and Jampates (1986) used tissue culture as a source of novel genetic variation. They initiated callus on shoot tip pieces of cultivars 'Baron' and 'Merion' and, after a callus transfer, isolated regenerated shoots which were established in soil. All 58 regenerated plants had the same isozyme genotype and chromosome number as the 'parent'. In field tests of over 500 regenerants from callus cultures of 13 Dutch *P. pratensis* cultivars, only 14 deviating plants were found (Den Nijs, 1990). The progeny of these aberrant plants appeared to be very uniform, so the high apomictic tendency of the 'parent' had not been lost. The experiment did not yield promising new somaclonal variation.

In vitro regeneration of plants from immature endosperm could release variation in pseudogamous apomicts such as *Poa pratensis*, as the endosperm is the product of fertilization. Matzk (1991) succeeded in callus induction and root regeneration from such cultures, but so far no plants have been regenerated (Matzk, personal communication).

Tissue culture of interspecific hybrids can also provide for the introgression of desirable genes from the wild species into the crop due to chromosomal rearrangements. This approach was recently tried in an interspecific *Pennisetum* hybrid involving three species, in which apospory originated from *P. squamulatum* (Ozias-Akins *et al.*, 1989). Differences between regenerants of a single genotype were obtained, which persisted after one cycle of apomictic reproduction. The authors, however, are not convinced that the variation is genetic, since they consider that epigenetic changes in gene expression may persist during aposporic seed formation.

Tissue culture of the displosporous apomictic diploid wild species *Arabis holboellii* Hornem. was used by Lehnhardt and Nitzsche (1988) in attempts to obtain genetic variation for reproductive traits. The experiments aimed at discovering a cytoplasmic basis for the inheritance of apomixis. Among 32 plants regenerated from callus, no significant morphological variants were detected, but all were sterile. This was taken as an indication of cytoplasmic inheritance of apomixis. Since high levels of male and/or female sterility among first generation regenerants from tissue culture have frequently be reported, this claim clearly requires confirmation.

Somatic hybridization (Chapter 7) could also be used in a different way of breeding apomicts. A programme for somatic hybridization of Kentucky bluegrass was initiated at Wageningen, The Netherlands, with the aim of asexually fusing elite apomictic genotypes. Callus cultures, initiated from immature inflorescences, were the source of suspension cultures with shoot forming ability. The latter cultures, in turn, were used to isolate protoplasts, which are necessary for protoplast fusion. Although some of these protoplasts formed microcalli and underwent some morphogenesis, only albino shoots could be regenerated (Van der Valk *et al.*, 1988) and no protoplast fusions have yet been attempted.

15.7 INTRODUCING APOMIXIS IN AMPHIMICTIC CROPS

The advantages of apomictic reproduction have inspired many attempts to introduce this char-acter into amphimictic crops. This could be achieved by interspecific hybridization between the amphimictic crop and related apomicts. The prospects of this approach depend on the availability of such relatives and the genetic basis of the apomixis. Because of the conflicting and incomplete evidence on the genetics in diverse species, there is no consensus about the feasibility of two different breeding approaches: transfer *vs* synthesis. Synthesis would involve combining different mutations for elements of apomixis, e.g. unreduced egg cell development, suppression of recombination (meiotic mutants), and parthenogenesis of the egg cell (while the endosperm may still need to be fertilized to develop). Meiotic mutants are very common in the plant kingdom and have almost always incomplete penetrance, i.e. are expressed in only part of the meiotic cells (Kaul and Murthy, 1985). Such mutants could in principle be employed to synthesize apomixis (Asker, 1980). On the other hand, as argued by Savidan (1990), in some species firm data are available for a single gene regulating apomixis as a process. This would be ideal for transfer of apomixis from species to species provided incongruity barriers can be overcome.

Interspecific hybridization was used as a tool for the introduction of apomixis into pearl millet from *Pennisetum squamulatum* (Dujardin and Hanna, 1989). The latter species is hexaploid, but could be hybridized with tetraploid ($2n = 28$) breeding lines of *P. glaucum*. Among back-cross progenies onto pearl millet apomictic types were obtained, one BC_3 derivative with $2n = 29$ being highly apomictic. It seems likely therefore that gene(s) for apomixis are localized on a small part of the genome of the donor species (Dujardin and Hanna, 1989). Attempts to speed up introgression of these gene(s) via tissue culture are underway (section 15.6.5 and Ozias-Akins *et al.*, 1989).

Attempts to introduce apomixis into the sugar beet, *Beta vulgaris*, have been less successful so far. From crosses with tetraploid apomictic forms of *Beta lomatogona* and *Beta intermedia*, triploid, tetraploid and pentaploid apomictic plants were selected (Cleij *et al.*, 1968, 1976). However, Jassem (1990) recently pointed out, there are still many problems with the facultative nature of the apomixis and no cultivars have been released as yet.

The cross-fertilizer *Zea mays* ($2n = 20$) hybridizes with the facultative pseudogamous apomict *Tripsacum dactyloides* ($2n = 72$). By repeated crossing of the hybrid ($2n = 46$) to *Zea mays*, partially apomictic plants with $2n = 56$ and $2n = 28$ were obtained (De Wet *et al.*, 1973). Some of the $2n = 46$ plants were also apomictic. It was recently confirmed, that in *Tripsacum dactyloides* apomixis is of the diplosporic type (Burson *et al.*, 1990). Diploid (pistillate) plants were completely sexual, two triploid and one tetraploid accessions were obligate pseudogamous apomicts. Although the genetic basis of apomixis is as yet unclear in this species, the attempts to introduce apomixis in maize continue. CIM-MYT has recently started a large crossing programme with many new accessions of *Tripsacum* (Savidan, personal communication).

Interspecific hybridization has also been employed to introduce apomixis in the apple (*Malus* × *domestica*), where apomictic reproduction of rootstocks would be valuable (Schmidt, 1977). Some tetraploid and triploid wild *Malus* species are facultative aposporous apomicts. The balance between apomixis and sexuality appears to be influenced by season, with low temperatures favouring apospory. In some years less than half the offspring was maternal (Schmidt, 1988).

Apomixis would also be very useful for the production of true breeding heterozygous cultivars in self-fertilizing crops. In bread wheat, *Triticum aestivum*, apomixis might be transferred from the related apomictic species *Agropyron scabrum*, which gives fertile hybrids with *Triticum aestivum* (Doll, 1973). Carman has recently obtained apomictic BC_1-progeny of a wheat × *Elymus rectisetus* cross (Savidan, personal communication). Alternatively, Matzk (1982) proposed to synthese apomictic reproduction in important cereal crops, based on mutants for component processes. Parthenogenesis of egg cells, suppressed meiotic reduction division and aposporous embryo-sac formation have all been observed in genotypes of cereal species. However, a major difficulty of this approach will be the requirement for normal development of endosperm, which is the principal commercial component of the harvested seed.

As an example of the 'synthesis' approach, some work on the potato is presented. Since the discovery in the seventies of the potential use of numerically unreduced gametes for potato breeding, proposals have been put forward to introduce apomixis in this crop. Hermsen (1980) suggested that in *Solanum tuberosum* apomictic cultivars could be used for raising potatoes from botanical seeds instead of tubers. Vegetative propagation through apomixis would decrease some of the deleterious effects of using seed tubers, e.g. transmission of diseases and laborious seed tuber production. Therefore the prospects of growing potatoes from botanical seed (True Potato Seed, TPS) are very appealing especially in areas where seed tuber production is difficult. Indeed TPS is being used on a small scale in various potato growing areas. Hermsen (1980) called apomixis still a 'utopian scheme' for potato breeding, but nevertheless presented evidence for the existence of several key elements of the process. Apospory does not exist in *Solanum*. Diplospory may be used, provided that meiotic recombination in the unreduced egg cells is impeded. This goal can be achieved via asynapsis, but this anomaly has so far not been discovered. However, mutants have been discovered for desynapsis (Jongedijk and Ramanna, 1988), and the formation of unreduced female and male gametes appears to be under relatively simple genetic control. Induction of parthenogenesis is routinely performed in potato in e.g. $4x \times 2x$ crosses where special pollinators are used which have the genetic capacity to induce both normal and unreduced egg cells to develop (Hermsen, 1980). It remains a challenge to breeders to combine these different genetic systems effectively to facilitate large-scale production of diplosporic parthenogenetic seeds with or preferably without pollination (Jongedijk, 1985).

15.8 CONCLUDING REMARKS AND PERSPECTIVES

The history of breeding apomicts is rather short compared to that of most other crops. The art of breeding apomictic crops is first to overcome the barriers for hybridization and second to fix the outstanding recombinants again by apomixis. In other words, to alternately release variation and to seal it into new apomictic forms.

Since the pioneering studies by Gustafsson (1946) progress has undoubtedly been made in breeding apomictic crops. This chapter aims to assess the state of the art of this work, but it can and will not claim completeness. Very efficient breeding programmes can be performed in obligate apomicts, provided that sexual or partly sexual cross-compatible plants are available. The breeder of facultative apomicts has more possibilities for starting a hybridization programme. However, the varying levels of apomictic reproduction in the hybrid offspring is a complication and often prevents completion of the breeding programme. Breeding facultative apomicts is easy to start but more difficult to reach fruition. The high potential of apomixis for the fixation of heterosis has also encouraged attempts to introduce it into amphimictic crops.

International collaboration is strengthened by the Apomixis Newsletters and the International Network for Apomixis Research, APONET, recently created by Savidan (ORSTROM/ CIMMYT). Several international agencies promote apomixis research especially for crops of less developed nations.

New breeding tools and biotechnology may be useful for manipulating apomixis, both in apomictic crops and for introducing apomixis into otherwise sexual crops. These include more efficient ways to make successful interspecific crosses e.g. by embryo rescue and somatic embryogenesis in callus derived from single hybrid offspring, but also better cytological and biochemical detection techniques. The Herr-clearing technique for ovules has proved extremely useful (Young *et al.*, 1979). Biochemical or molecular markers for early progeny testing can speed up genetic analysis. Ideally, each genotype should be individually distinguishable. This ideal can be approached by molecular genome analysis through RFLPs. Although RFLP analysis still is a very expensive and laborious technique, it could be useful for appraisal of the reproductive behaviour of putative apomicts. The technique is much more powerful than, for example isozyme analysis. RFLPs could also be used to locate genes for apomixis on the genome and thereafter to try and isolate these genes if close linkages with markers are obtained.

Somatic hybridization could be an ideal tool to combine two elite genotypes into a new apomictic polyploid which is expected to express the good characters of both 'parents'. When asymmetric protoplast fusion becomes possible between distantly related species, novel combinations of apomictic and sexual species may also be obtained. Alternatively, molecular analysis of the process of apomixis may identify the genes responsible and enable their isolation, cloning and use in genetic transformation. Eventually breeders might make crosses and selections, and transform the elite selections into apomictic cultivars. The ultimate breeder's goal, complete mastery over meiosis and fertilization, will then be reached.

ACKNOWLEDGMENTS

The authors gratefully acknowledge helpful criticism by colleagues Drs W. Lange and J. Hoogendoorn and Dr Y. Savidan (ORSTROM/ CIMMYT), which significantly improved the chapter.

REFERENCES

Åkerberg, E. and Nygren, A. (1959), *Poa pratensis, trivalis, palustris, compressa* und verwandte Arten. In *Züchtung der Futterpflanzen*, Kappert, H. and Rudorf, W. (eds), Paul Parey, Berlin, pp. 392–418.

Almgård, G. (1966), Experiments with *Poa*. III. *Lantbr. Högsh. Ann.*, **32**, 3–64.

Asker, S. (1979), Progress in apomixis research. *Hereditas*, **91**, 231–240.

Asker, S. (1980), Gametophytic apomixis: elements and genetic regulation. *Hereditas*, **93**, 277–293.

Bashaw, E.C. (1975), Problems and possibilities of apomixis in the improvement of tropical forage grasses. In *Tropical Forages in Livestock Production Systems*, Am. Soc. Agron. Spec. Pub., **24**, 23–30.

Bashaw, E.C. (1980), Apomixis and its application in

crop improvement. In *Hybridization of Crop Plants*, Fehr, W.R. and Hadley, H.H. (eds), Am. Soc. Agron. and Cr. Sci. Soc., Madison, pp. 45–63.

Bashaw, E.C. and Hignight, K.W. (1990), Gene transfer in apomictic buffelgrass through fertilization of an unreduced egg. *Crop Sci.*, **30**, 571–575.

Bashaw, E.C., Voigt, P.W. and Burson, B.L. (1983), Breeding challenges in apomictic warm season grasses. In *Proc. XIV Int. Grassland Congress*, Lexington, 1981, Smith, J.A. and Hays, V.W. (eds), Westview Press, Boulder, pp. 179–181.

Burson, B.L., Voigt, P.W., Sherman, R.A. and Dewald, C.L. (1990), Apomixis and sexuality in eastern gamagrass. *Crop Sci.*, **30**, 86–89.

Burton, G.W. and Forbes, J. Jr (1961), The genetics and manipulation of obligate apomixis in common bahia grass (*Paspalum notatum* Flugge). In *Proc. VIIIth Int. Grassland Congress*, Reading, Skidmore, C.L. (ed.), Alden Press, Oxford, pp. 66–71.

Cameron, J.W. and Soost, R.K. (1982), Breeding and development. In 75 years of *Citrus* research. *California Agric.*, **36**, 4–6.

Clausen, J. (1952), New bluegrasses by combining and rearranging genomes of contrasting *Poa* species. In *Proc. VIth Int. Grassland Congress*, Pennsylvania, Wagner *et al.* (eds), Pennsylvania State College, Pennsylvania, pp. 216–221.

Clausen, J. (1961), Introgression in polyploid Poas. *Euphytica*, **10**, 87–94.

Cleij, G., de Bock, T.S.M. and Lekkerkerker, B. (1968), Crosses between *Beta intermedia* Buce and *B. vulgaris* L. *Euphytica*, **17**, 11–20.

Cleij, G., de Bock, T.S.M. and Lekkerkerker, B. (1976), Crosses between *Beta vulgaris* L. and *B. lomatogona* F. et M. *Euphytica*, **25**, 539–547.

D'Cruz, R. and Reddy, P.S. (1971), Inheritance of apomixis in *Dichanthium*. *Ind. J. Gen. Pl. Br.*, **31**, 451–460.

Doll, R. (1973), Die Bedeutung der Apomixis für die Pflanzenzüchtung. *Biol. Rundschau*, **11**, 362–365.

Duich, J.M. and Musser, H.B. (1959), The extent of aberrants produced by 'Merion' Kentucky Bluegrass, *Poa pratensis* L., as determined by first and second generation progeny test. *Agron. J.*, **51**, 421–424.

Dujardin, M. and Hanna, W.W. (1989), Developing apomictic pearl millet – characterization of a BC₃ plant. *J. Genet. Breed.*, **43**, 145–151.

Duyvendak, R. and Luesink, B. (1979), Preservation of genetic resources in grasses. In *Broadening*

Genetic Base Crops, Zeven, A.C. and Harten, A.M. van (eds), Pudoc, Wageningen, pp. 67–73.

van Dijk, G.E. (1979), Wild species for the breeding of grasses. In *Broadening Genetic Base Crops*, Zeven, A.C. and Harten, A.M. van (eds), Pudoc, Wageningen, pp. 211–216.

van Dijk, G.E. (1991), Breeding of apomictic crops. In *Advances in Plant Breeding*, Vol. 2, Mandal, A.K., Gangula, P.K. and Bannerjee, S.P. (eds), CBS Publishers, New Delhi, pp. 22–34.

van Dijk, G.E. and Winkelhorst, G.D. (1982), Interspecific crosses as a tool in breeding *Poa pratensis* L. 1. *Poa longifolia* Trin × *P. pratensis* L. *Euphytica*, **31**, 215–223.

Funk, C.R. and Han, S.J. (1967), Recurrent intraspecific hybridization: a proposed method of breeding Kentucky bluegrass, *Poa pratensis*. *N. J. Agric. Expt. Sta.*, Bull. **818**, 3–14.

Gerstel, D.U., Hammond, B.L. and Kidd, C. (1953), An additional note on the inheritance of apomixis in guayule. *Bot. Gaz.*, **115**, 89–93.

Grant, V. (1971), *Plant Speciation*, Columbia University Press, New York.

Grazi, F., Umaerus, M. and Åkerberg, E. (1961), Observations on the mode of reproduction and the embryology of *Poa pratensis*. *Hereditas*, **47**, 489–541.

Gröber, K., Matzk, F. and Zacharias, M. (1978), Untersuchungen zur Entwicklung der apomiktischen Fortpflanzungsweise bei Futtergraesern, III. *Kulturpflanze*, **26**, 303–327.

Gustafsson, Å. (1946), Apomixis in higher plants. Part I. The mechanisms of apomixis. *Lunds Univ. Arsskr.*, N.F. Adv. 2, **42**, 1–66.

Gustafsson, Å. and Gadd, I. (1965), Mutations and crop improvement. 4. *Poa pratensis* L. (Gramineae). *Hereditas*, **53**, 90–102.

Gustine, D.L., Sherwood, R.T. and Gounaris, J. (1989), Regulation of apomixis in buffelgrass. In *Proc. XVIth Int. Grassland Congress*, Nice, Association Française pour la Production Fourragère, pp. 411–412.

Han, S.J. (1971), Effects of genetic and environmental factors on apomixis and the characteristics of nonmaternal plants in Kentucky Bluegrass (*Poa pratensis* L.). *Herb. Abstr.*, **41**, 297.

Hanna, W.W. and Bashaw, E.C. (1987), Apomixis: Its identification and use in plant breeding. *Crop Sci.*, **27**, 1136–1139.

Hanna, W.W. and Dujardin, M. (1982), Apomictic interspecific hybrids between pearl millet and *Pennisetum orientale* L. C. Rich. *Crop Sci.*, **22**, 859–867.

Hanna, W.W., Powell, J.B., Millot, J.C. and Burton,

G.W. (1973), Cytology of obligate sexual plants in *Panicum maximum* Jacq. and their use in controlled hybrids. *Crop Sci.*, **13**, 695–697.

Hanson, A.A. and Juska, F.V. (1962), Induced mutations in Kentucky bluegrass. *Crop Sci.*, **2**, 369–371.

Harlan, J.R., Brooks, M.H., Borgaonkar, D.S. and Wet, J.M.J. de (1964), The nature and inheritance of apomixis in *Bothriochloa* and *Dichanthium*. *Bot. Gaz.*, **125**, 41–46.

Hashemi, A., Estilai, A. and Waines, J.G. (1989), Cytogenetics and reproductive behavior of induced and natural tetraploid guayule (*Parthenium argentatum* Gray). *Genome*, **32**, 1100–1104.

Hermsen, J. (1980), Breeding for apomixis in potato, pursuing a utopian scheme. *Euphytica*, **29**, 595–607.

Hintzen, J.J. (1979), Methods for apomictic species. In *Plant Breeding Perspectives*, Sneep, J. and Hendriksen, A.J.T. (eds), Pudoc, Wageningen, pp. 186–189.

Hintzen, J.J. and van Wijk, A.P.J. (1985), Ecotype breeding and hybridization in Kentucky bluegrass (*Poa pratensis* L.). In *Proc. Vth Int. Turfgrass Res. Conf.*, Avignon, Lemaire, F. (ed.), INRA, Paris, pp. 213–219.

Hovin, A.W., Berg, C.C., Bashaw, E.C., Buckner, R.C., Dewey, D.R., Dunn, G.M., Hoveland, C.S., Rineker, C.M. and Wood, G.M. (1976), Effects of geographic origin and seed production environments on apomixis in Kentucky bluegrass. *Crop Sci.*, **16**, 635–638.

Jassem, B. (1990), Apomixis in the genus *Beta*. Review. *Apomixis Newsl.*, **2**, 7–23.

Jongedijk, E. (1985), The pattern of megasporogenesis and megagametogenesis in diploid *Solanum* species hybrids; its relevance to the origin of 2n-eggs and the induction of apomixis. *Euphytica*, **34**, 599–611.

Jongedijk, E. and Ramanna, M.S. (1988), Synaptic mutants in potato, *Solanum tuberosum* L. I. Expression and identity of genes for desynapsis. *Genome*, **30**, 664–670.

Julén, G., 1961. The effect of X-rays on the apomixis in *Poa pratensis*. In *Effects of Ionizing Radiation on Seeds*, IAEA, Paul Gerin Publ., Vienna, pp. 527–532.

Kaul, M.L.H. and Murthy, T.G.K. (1985), Mutant genes affecting meiosis in higher plants. *Theor. Appl. Genet.*, **70**, 449–466.

Knox, R.B. and Heslop-Harrison, J. (1963), Experimental control of aposporous apomixis in a grass of the *Andropogonea*. *Bot. Notiser*, **116**, 127–141.

Lehnhardt, B. and Nitzsche, W. (1988), Loss of apomixis in *Arabis holboellii* Hornem. when regenerated in tissue culture. *Angew. Botanik*, **62**, 225–232.

Matzk, F. (1982), Vorstellungen ueber potentielle Wege zur Apomixis bei Getreide. *Arch. Zuecht. forsch.*, Berlin, **12**, 182–195.

Matzk, F. (1991), Studies towards novel methods in breeding of the apomictic species *Poa pratensis* L. In *Fodder crops breeding: achievements, novel strategies and biotechnology*, Nijs, A.P.M. den and Elgersma, A. (eds), Pudoc, Wageningen, pp. 191–192.

Müntzing, A. (1940), Further studies on apomixis and sexuality in *Poa*. *Hereditas*, **26**, 115–190.

Naumova, T. (1990), Apomixis and amphimixis in angiosperms: classification. *Apomixis Newsl.*, **2**, 33–38.

Nitzsche, W. (1983), Interspecific hybrids between apomict forms of *Poa palustris* L. × *Poa pratensis* L. *Proc. 14th Int. Grassl. Congr. Lexington, Ky*, pp. 155–157.

den Nijs, A.P.M. (1990), Experimenting with apomixis and sexuality in *Poa pratensis*. *Apomixis Newsl.*, **2**, 52–54.

Nogler, G.A. (1984), Gametophytic apomixis. In *Embryology of angiosperms*, Johri, B.M. (ed.), Springer, Berlin, pp. 475–518.

Nygren, A. (1967), Apomixis in the angiosperms. In *Handbuch der Pflanzenphysiologie B 18*, Ruhland, W. (ed.), Springer, Berlin, pp. 551–596.

Ozias-Akins, P., Dujardin, M., Hanna, W.W. and Vasil, I.K. (1989), Quantitative variation recovered from tissue cultures of an apomictic, interspecific *Pennisetum* hybrid. *Maydica*, **34**, 123–132.

Pepin, G.W. and Funk, C.R. (1971), Intraspecific hybridization as a method of breeding Kentucky bluegrass (*Poa pratensis* L.) for turf. *Crop Sci.*, **11**, 445–448.

Pepin, G.W. and Funk, C.R. (1974), Evaluation of turf, reproductive, and disease-response characteristics in crossed and selfed progenies of Kentucky bluegrass. *Crop Sci.*, **14**, 356–359.

Powers, L. (1945), Fertilization without reduction in guayule (*Parthenium argentatum* L.) and a hypothesis as to the evolution of apomixis and polyploidy. *Genetics*, **30**, 332–346.

Rutishauser, A. (1969), *Embryologie und Fortpflanzungsbiologie der Angiospermen*, Springer Verlag, Wien.

Savidan, Y.H. (1982), Embryological analysis of facultative apomixis in *Panicum maximum* Jacq. *Crop Sci.*, **22**, 467–469.

Savidan, Y.H. (1983), Genetics and utilization of apomixis for the improvement of guineagrass (*Panicum maximum* Jacq.). In *Proc. XIV Int. Grassland Congress*, Lexington, 1981, Smith, J.A. and Hays, V.W. (eds), Westview Press, Boulder, pp. 182–184.

Savidan, Y.H. (1990), Apomixis and its use in plant breeding. The genetic control of apomixis. *Apomixis Newsl.*, **2**, 24–26.

Savidan, Y.H., Jank, L., Costa, J.C.G. and do Valle, C.B. (1989), Breeding *Panicum maximum* in Brazil. 1. Genetic resources, modes of reproduction and breeding procedures. *Euphytica*, **41**, 107–112.

Schmidt, H. (1977), Contributions on the breeding of apomictic apple stocks. 4. On the inheritance of apomixis. *Z. Pfl. Zücht.*, **78**, 3–12.

Schmidt, H. (1988), Criteria and procedures for evaluating apomictic rootstocks for apple. *Hort-Science*, **23**, 104–107.

Sherwood, R.T., Yound, B.A. and Bashaw, E.C. (1980), Facultative apomixis in buffelgrass. *Crop Sci.*, **20**, 375–379.

Stebbins, G.L. (1971), *Chromosomal Evolution in Higher Plants*, Edward Arnold, London.

Taliaferro, C.M. and Bashaw, E.C. (1966), Inheritance and control of obligate apomixis in breeding buffelgrass, *Pennisetum ciliare. Crop Sci.*, **6**, 473–476.

Taliaferro, C.M., Dabo, S.M., Mitchell, E.D., Johnson, B.B. and Metzinger, B.D. (1989), Morphologic, cytogenetic and enzymatic variation in tissue culture regenerated plants of apomictic old-world blue stem grasses (*Bothriochloa* sp.). *Plant Cell, Tissue and Organ Culture*, **19**, 257–266.

van der Valk, P., Zaal, M.A.C.M. and Creemers-Molenaar, J. (1988), Regeneration of albino plantlets from suspension culture derived protoplasts of *Poa pratensis* L. (Kentucky bluegrass). *Euphytica Supplement*, 169–176.

do Valle, C.B., Savidan, Y.H. and Jank, L. (1989), Apomixis and sexuality in *Brachiaria decumbens* Stapf. In *Proc. XVIth Int. Grassland Congress*, Nice, Association Française pour la Production Fourragère, pp. 407–308.

Voigt, P.W. and Burson, B.L. (1983), Breeding of apomictic *Eragrostis curvula*. In *Proc. XIV Int. Grassland Congress*, Lexington, 1981, Smith, J.A. and Hays, V.W. (eds), Westview Press, Boulder, pp. 160–163.

de Wet, J.M.J., Harlan, J.R., Engle, L.M. and Grant, C.A. (1973), Breeding behaviour of maize-tripsacum hybrids. *Crop Sci.*, **13**, 254–256.

Williamson, C.J. (1981), The influence of light regimes during floral development on apomictic seed production and on variability in resulting seedling progenies of *Poa ampla* and *P. pratensis*. *New Phytol.*, **87**, 769–783.

Williamson, C.J. and Watson, P.J. (1980), Production and description of interspecific hybrids between *P. pratensis* and *P. longifolia*. *Euphytica*, **29**, 715–725.

Wu, L. and Jampates, R. (1986), Chromosome number and isoenzyme variation in Kentucky bluegrass cultivars and plants regenerated from tissue culture. *Cytologia*, **51**, 125–132.

Young, B.A., Sherwood, R.T. and Bashaw, E.C. (1979), Cleared-pistil and thick sectioning techniques for detecting aposporous apomixis in grasses. *Can. J. Bot.*, **57**, 1668–1672.

16

Micropropagation and somatic embryogenesis

C.H. Bornman

16.1 INTRODUCTION

Due both to ease of production and handling as well as the potential for specialized treatment, seed is the preferred vehicle for the propagation and cultivation of most agronomic and forest tree species (Gray, 1990). However, a glance at the list of the world's top crops (Witt, 1985) reveals that 10 of the 30 crops with an annual production of between 10 and 450 million metric tonnes are vegetatively propagated. For a variety of reasons, including genetic self-incompatibility, uniformly homogeneous seed can either not be produced or utilized for a number of agronomical and horticultural crops (Redenbaugh, 1990a,b). In these cases, plants are propagated vegetatively, such as sugarcane and fruit crops, or by genetically non-uniform seed, as for example lucerne (alfalfa) and forage grasses.

Interest in the application of tissue culture techniques as an alternative means of asexual propagation was greatly stimulated by the work of Morel (1960, 1965), who demonstrated that orchids could be multiplied rapidly by shoot meristem culture *in vitro*. Morel first multiplied tropical orchids through the division of protocorms, differentiated structures that develop naturally on orchid embryos, as well as from *Cymbidium* shoot tip cultures *in vitro*. The

Plant Breeding: Principles and prospects. Edited by M.D. Hayward, N.O. Bosemark and I. Romagosa. Published in 1993 by Chapman & Hall, London. ISBN 0 412 43390 7.

protocorms could be sectioned and subcultured, each section regenerating a new protocorm from which an orchid plantlet developed within weeks. This technique found almost immediate practical application in the early 1970s and today many commercial laboratories in Europe, North America and South-East Asia annually produce millions of orchid plants at low cost (Boxus, 1987). The world-wide production of micropropagated plants (mostly ornamental) is estimated at between 50–100 million, but the market potential is probably 5–10 times these numbers.

Micropropagation has been used to only a very limited extent for agronomic crops, most of which are seed-propagated, but quite extensively for horticultural and ornamental species, which are normally propagated vegetatively. Seeds, generally, are an inexpensive commodity, so that their substitution by vegetative propagules would be limited by the latter's cost effectiveness (Redenbaugh, 1990a). As agronomic crops are grown extensively, an alternative system to seed would demand a scale-up in micropropagation technology to a level that at the very least would have to approach seed efficiency. Until relatively recently, axillary shoot production represented the most advanced micropropagation technology but, especially since the mid-1980s, interest in somatic embryogenesis has increased dramatically (Fujii *et al.*, 1987; McKersie *et al.*, 1989; Becwar *et al.*, 1989; Gray, 1990; Redenbaugh, 1990a; Webster *et al.*, 1990). Today, production of somatic embryos *in vitro* and their conversion into artificial or synthetic seed would appear to offer the only suitable micropropagation potential as an efficient system of vegetative propagation. According to Gray (1987), the ability to propagate outstanding individuals vegetatively, but at seed efficiencies, would revolutionize production of both seed and vegetatively propagated crops. Crop uniformity in terms of maturation, harvest, yield, etc., would no longer be limited to self-pollinated crops or F_1 hybrid cultivars such as maize. Redenbaugh (1990c) holds out the possibility of capitalizing on variability rather than homozygosity across a given population, such as with variety development in lucerne. Elite plants with novel gene combinations, but which are not true-breeding, could be used as an explant source for artificial seed production. For vegetatively propagated crops, propagation and planting efficiencies could be increased; and in the case of seed propagated crops, new hybrids could be used immediately, obviating the time-consuming and cost-intensive development of parental inbreds. In the case of both types of crops, individual plants transformed by genetic engineering could be increased, and perhaps even utilized, without sexual recombination (Gray, 1990). However, despite much progress in other areas of plant biotechnology, mass clonal propagation *in vitro* has so far seen limited large-scale commercial application, not least as a result of the inherent high costs of the technique, coupled with the relatively low value of individual plants of most species. In this review, the term 'micropropagation' is used mainly in reference to mass production of clonal plants for commercial purposes, thus excluding the limited-scale regeneration of plants by embryo rescue or from ovules, microspores and transgenic tissues.

Sensu stricto, micropropagation, that is clonal multiplication *in vitro*, encompasses three types of vegetative propagation: (1) axillary shoot production from tissues containing pre-existing meristems; (2) adventitious shoot production following induction of adventitious meristems; (3) somatic embryogenesis. A potential fourth type of propagation exists for multiplication and regeneration with protoplasts and single cells as starting points, culminating with the production of either adventitious shoots or somatic embryos as in (2) and (3) above. Each of these methods of propagation has its advantages and disadvantages (Table 16.1).

16.2 MICROPROPAGATION

In meristem (or meristem-tip) culture the explant, up to 1 mm in length, may consist of the shoot apical meristem or, more frequently, of the apical dome plus one or more subjacent leaf primordia. A major advantage of utilizing such a small explant is the potential for eliminating pathogenic organisms that may be present in the donor plant. When elimination of viral or mycoplasmal infection is an objective, meristems of the smallest possible size that are still capable of regeneration are used. Apical domes smaller than 0.25 mm in height are difficult to grow and unlikely to root, whereas those larger than 0.75 mm may produce plants that are still infected. An advantage of meristem-tip culture is that plantlet production from adventitious meristems is avoided, thus capturing and maintaining the inherent genetic stability (Grout, 1990).

The technique of shoot-tip culture is more expedient for large-scale micropropagation. In addition to the shoot apex, a relatively large number of developing leaves and leaf primordia are present. Depending upon species, shoot tip explants vary in length from 1 to 10 mm. Development of the explant *in vitro* is confined to the outgrowth of an organized shoot, again without an intermediate adventitious stage. As explant source, terminal buds appear to have a stronger potential for growth than lateral buds. However, in order to accelerate propagation rates, axillary buds that develop in the leaf axils of the expanding plants *in vitro* may be also used as secondary propagules (Hu and Wang, 1983; Grout, 1990). These buds are excised and subcultured as nodal segments. Under appropriate culture conditions the growth of the new shoot does not involve adventitious organogenesis, thus presumably minimizing problems related to genetic instability and somaclonal variation that are often associated with the latter. Adventitious meristems may also be

Table 16.1. Summary of the advantages and limitations of different methods of micropropagation

Plant regeneration via	Advantages	Limitations
1. *Axillary shoot production*		
Shoot-tips	Preformed incipient shoot Easy dissection High survival and growth rates Genetically stable Phenotypically homogeneous Wide species applicability	Staged process involving explant establishment, multiplication of propagules, rooting, hardening-off and transplantation Initial low multiplication rate Unsuitable for virus elimination Sometimes diminishing organogenic capacity
Axillary (and terminal) buds	As above	As above, plus difficult to desinfest
Meristem-tips	Preformed meristem Genetically stable Phenotypically homogeneous Useful for pathogen elimination	Difficult dissection Low survival rate Less applicable than shoot-tip culture
2. *Adventitious shoot production*		
Organogenesis	Initial high multiplication rate Positive somaclonal variation	Staged process, as above under 1, but additionally involving callus induction, meristem induction and organ differentiation Less applicable than shoot-tip culture Negative somaclonal variation
3. *Somatic embryo-genesis*	Developmentally analogous to zygotic embryos Continuous production	Precocious germination as a result of asynchronous development
Synthetic seed	May: circumvent long breeding cycles circumvent self-incompatibility decrease hybrid seed cost be used in production of hybrids be used for germplasm conservation eliminate need for inbreds	High relative cost of production

deliberately induced via an organogenic phase in an unspecialized tissue such as callus, or in specialized tissues such as those comprising the epidermis and subepidermis of explants of varying origin, for example, cotyledons, hypocotyls, petioles and leaf bases.

Embryos arising from somatic cells are termed somatic embryos or embryoids. Although the organizational development of somatic embryos is more variable than that of zygotic embryos, there is a remarkable structural and physiological resemblance between them. In contrast to a bud, a fully developed somatic embryo possesses a primordial hypocotyl-root axis, the apex and base of which are terminated by a shoot apical meristem and a root apical meristem, respectively. Early research dealt mainly with induction of somatic embryos in different species (Ammirato, 1983), but now, in addition to the problems of synchronous embryo development and maturation, attention is turning to the potential use of the somatic embryo as an artificial or synthetic seed. Synthetic seed is seen as an alternative to greenhouse propagation through stem cuttings and micropropagation of shoots and buds *in vitro*.

The potential fourth type of micropropagation referred to earlier, involving protoplasts as a starting point, is limited by the longer period required for plant regeneration as well as by the risk of sacrificing clonal fidelity as a result of the generation of genetic variation following the mass regeneration of plants. This variation, of protoclonal or somaclonal origin, may express itself in the form of either deleterious or beneficial traits in the regenerated plants (Brown *et al.*, 1986). Furthermore, fostering the generation of variation, protoplast fusion is currently used in somatic and gametosomatic hybridization and is fast becoming an adjunct to conventional plant breeding (see Chapter 7 and Power and Davey, 1990).

16.2.1 Axillary shoot production through enhanced release of axillary buds

As pointed out by Hu and Wang (1983), the merit of using the enhanced release of axillary buds as a means of regeneration is that an incipient shoot has already been differentiated *in vivo*. The risk for genetic change is therefore minimized and all that remains is establishment of the complete plant by elongation of a stem and development of a root system. When it is necessary to maintain a chimeral plant genetic structure (Grout, 1990), this type of culture preserves the precise arrangement of the apical cell layers. In a typical chimera, the epi- and subepidermal layers of the meristem are of differing genetic background. It is the contribution in a particular arrangement to the leaf and floral organs that produces the desired characteristics. In shoots of adventitious origin there is a risk that the original chimeral pattern is disturbed. Axillary shoot production has been used extensively for ornamental and herbaceous horticultural crops. According to Hu and Wang (1983), this success is due in part to the weak apical dominance and strong root regenerating capacity of many herbaceous plants. There are numerous examples (Hu and Wang, 1983) of plants produced by axillary shoot culture. Some representative genera are: *Anthurium, Chrysanthemum, Dianthus, Fuchsia, Gerbera, Gloxinia, Phlox, Saxifraga, Allium, Asparagus, Rheum* and *Fragaria*.

16.2.2 Adventitious shoot production through organogenesis

Adventitious shoots arise from adventitious meristems and usually involve an intermediate callus phase. Callus cultures, which can be maintained for extended periods by intermittent subculture, are convenient for long-term maintenance of cell lines. They may be used as the tissue source from which cell suspensions are derived or, if embryogenic, for the sustained production of somatic embryos. Callus tissue is the form of culture in which adventitious meristems usually are induced and from which plant regeneration is initiated. However, the frequency of genetic changes, in the form of aneuploidy and polyploidy, resulting from mitotic abnormalities, is often increased in this type of micropropagation. Micropropagation of axillary and adventitious shoots involves a number of distinct developmental stages (Murashige, 1974): (1) explant establishment; (2) propagule multiplication; (3) rooting; (4) hardening-off and transplantation. Each stage usually involves different chemical and physical environments. Stage (4), strictly, is a macropropagation stage and in many cases even stage (3) has to be carried out *ex vitro*. With few exceptions, of which potato and, to a lesser degree, sugarcane are perhaps the most notable, micropropagation has had limited application for agronomic crops (Redenbaugh, 1990a). Micropropagation of woody plants has lagged far behind that of herbaceous, ornamental plants. As a technique, micropropagation has an inherent high cost that can only be offset when the value of the individual plants is high.

16.2.3 Somatic embryogenesis

A particular type of differentiation that may be observed from immature embryos and other tissues cultured *in vitro* on defined media or from cells cultured in suspension, is the development of embryo-like structures that appear to recapitulate the typical stages in the embryogenesis of a zygote (Raghavan, 1983). However, as these embryo-like structures are formed from sporophytic or somatic cells, in contrast to gametophytic or gametic cells, they are referred to as somatic embryos or embryoids and the process by which they arise as somatic embryogenesis. Somatic embryogenesis was first described in carrot (*Daucus carota*) by Reinert (1958, 1959) and Steward *et al.* (1958). Induction, culture, developmental physiology and biochemistry of somatic embryogenesis has been studied most comprehensively in the Apiaceae and, during the past decade, in the Pinaceae. An impressive number of over 200 species (Ammirato, 1983) have produced somatic embryos in culture and in addition to the two families aforementioned, much work has also focused on species in the Fabaceae, Poaceae and Solanaceae.

The natural occurrence of embryos, developed from cells that are not the products of gametic fusion, are well known in the form of adventive or asexual embryogenesis. In members of the Rutaceae, in particular *Citrus*, cells of the nucellus frequently develop into embryos. Cells of the embryo sac such as the egg, synergids or antipodals, may also develop into embryos with the gametic chromosome number (Ammirato, 1983). Adventive cells may also originate from within anthers or from endospermal cells (Chapter 15 and Tisserat *et al.*, 1979).

Although microspores and egg cells are programmed for terminal differentiation, culture *in vitro* of anthers and ovules at certain stages of development may give rise to the differentiation of haploid embryos in processes known as androgenesis and gynogenesis, respectively (Chapter 17). It is recognized that the transformation of microspores and egg cells into embryos strictly does not fall within the ambit of somatic embryogenesis (Raghavan, 1983). Nevertheless, it is important to point out that in both somatic and gametic embryogenesis cells that have not embarked upon a pathway of irreversible differentiation within an already committed plan, exhibit their inherent genetic potential in the same way and to the same extent as do those of a zygote.

(a) Somatic embryogenesis as compared with zygotic embryo development

The principal difference between somatic and zygotic embryogenesis lies in the respective methods by which they are initiated. The zygotic embryo develops from a fertilized egg, the resulting plant containing a potentially different meiotic recombination of genes (Gray, 1990). As they arise from the cells of a single individual, somatic embryo-derived plants constitute a clone. Both embryo types share similar gross ontogenies (Fig. 16.1), passing through: globular, heart-shaped, torpedo and cotyledonary stages in the case of dicotyledons; globular, scutellar and coleoptilar stages for monocotyledons (Gray and Conger, 1985); and embryonal-suspensor mass, globular, torpedo and cotyledonary for conifers (Becwar *et al.*, 1989).

Somatic embryos usually develop asynchronously from a cell mass comprising embryogenic tissue and multiple stages of development may therefore be represented in a single culture, with the result that embryos bypass maturation, become disorganized, and form new embryogenic cells and tissue. Cotyledons and radicles may also develop at different rates, resulting in precocious germination. Often, somatic embryos display structural abnormalities such as poorly developed cotyledons and hypocotyl-root axes, additional apical meristems, and additional or insufficient cotyledons. Somatic embryos typically lack a

Fig. 16.1. Comparison of zygotic and somatic embryogenesis. After formation of a zygote from the fertilized egg, zygotic embryos pass through distinct, sequential and synchronous developmental stages to the mature embryo. In somatic embryogenesis, after a proembryonal stage, somatic embryos are formed following an asynchronous developmental stage. Maturation is often incomplete. Embryos in various stages of development may dedifferentiate into embryogenic cells which, in turn, give rise to proembryonal tissue, in a constant cycle of 'repetitive' embryogensis. Alternatively, with time embryogenic cells may progressively lose their embryogenicity and revert to callus cells or die.

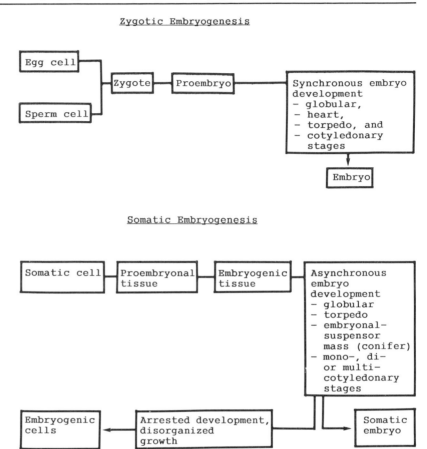

quiescent phase, as a consequence of which they continue to grow, germinate, become disorganized or die. A quiescent or maturation phase may be induced by careful desiccation (Roberts *et al.*, 1990) or by treatment with abscisic acid, and/or high levels of sucrose (Kim and Janick, 1990). Lack of synchronous development and even maturation are severe limitations to the use of somatic embryos as artificial or synthetic seed.

16.3 CULTURE MEDIA AND EXPLANT SOURCES IN MICROPROPAGATION

16.3.1 Axillary and adventitious shoot production

The most commonly used culture media are based on the well-established formulations of

Heller (1953), Murashige and Skoog (1962), White (1963), Gamborg *et al.* (1968) and Schenk and Hildebrandt (1972). Sucrose is the most common carbon source. Of the phytohormones known to control various stages of growth and development, namely auxins, gibberellins, cytokinins, abscisic acid (ABA) and ethylene, only auxin and cytokinin are routinely incorporated into the culture medium. The most commonly used auxins are 2,4-dichlorophenoxyacetic acid (2,4-D), naphthaleneacetic acid (NAA) and indoleacetic acid (IAA). Benzyladenine and furfurylaminopurine (kinetin) are the most frequently employed cytokinins. Careful manipulation of the relative levels and ratios of auxin and cytokinin make it possible to control organogenesis, growth and development. Evans *et al.* (1983) and George *et al.* (1987) have provided extensive reviews and lists of plant species that have been cultured and media that have been used in plant tissue culture.

16.3.2 Somatic embryogenesis

Somatic embryos have been induced and grown on a range of media (Ammirato, 1983), from relatively dilute media such as White's (1963) to the high salt media of Murashige and Skoog (1962), Gamborg *et al.* (1968) and Schenk and Hildebrandt (1972). The presence of auxin is usually critical for embryo initiation, and the lowering of auxin or its absence is necessary for embryo development. Abscisic acid has been found to be essential for embryo maturation, preventing precocious germination and repetitive embryogenesis. Embryonic, meristematic and reproductive tissues have a propensity for embryogenic growth. Immature embryos are often used as explant source, as for example in *Lolium* (Dale, 1980), *Pennisetum* (Vasil and Vasil, 1980), *Triticum* (Ozias-Akins and Vasil, 1982), *Zea* (Lu *et al.*, 1982), *Dactylis* (Gray and Conger, 1985), *Picea* (Hakman *et al.*, 1985), but just about any part of the plant body taken at the time of development has been used: excised embryo, hypocotyl root, petiole, intercalary meristem, immature inflorescence, floral or reproductive tissue, and cultured cells and protoplasts.

16.4 MICROPROPAGATION AS EXEMPLIFIED BY POTATO (*Solanum* spp.)

The culture of meristems from potato shoots and tuber sprouts has several important applications in potato production: virus eradication, germplasm conservation and exchange, and potato tuber seed production through propagation *in vitro*. Meristem-tip culture, sometimes in combination with thermotherapy and chemotherapy, is used to eliminate viruses such as PVA, PVG, PVM, PVS, PVX, PVY, leaf-roll virus, paracrinkle virus and potato spindle tuber viroid (Miller and Lipschutz, 1984). Following indexing of the meristem-derived plantlets using serological and other tests, the virus-free plants are then increased by clonal multiplication.

Traditional methods of potato tuber seed increase are based on low propagation ratios that vary from 1:3 to 1:15, 1 tuber yielding 3–15 seed tubers (Bryan *et al.*, 1981). However, using one or a combination of macro- and micro-propagation methods, multiplication ratios of 1:50 to 1:several thousand cuttings per year can be produced, each cutting yielding 5 or more seed tubers. Wooster and Dixon (1985) found that micropropagation using single node explants not only greatly increased the rate of seed tuber multiplication as compared with stem cutting propagation, but allowed entry of new cultivars to statutory trials two years after the initiation of stocks rather than after the more usual three or four. However, although successful, the method was judged as labour intensive and therefore expensive. Several micropropagation techniques are in use, some of which are illustrated in Fig. 16.2.

16.4.1 Meristem-tip culture

A meristem tip (0.075–0.1 mm) with one subjacent leaf primordium (Fig. 16.2, method A) is dissected from a tuber sprout and cultured on an agar-solidified nutrient medium for ca 2–3 months to obtain rooted plantlets ca 30 mm long.

16.4.2 Axillary node and shoot production

The meristem-derived plantlets, after pathogen testing, can then either be placed as nodal explant segments or intact in a horizontal position on a medium with very low (0.005 μM NAA) auxin levels to encourage rooting of a number of axillary shoots (Fig. 16.2, method B). This step is repeated by pressing the newly formed and developing axillary shoots back on to the medium until a mass of axillary shoots is obtained. Assuming that 5 nodes were available from an axillary shoot, Grout (1990) could achieve 5, 25, 125, 625, 3125 and 15 625 nodes for further subculture after 3, 4, 5, 6, 7 and 8

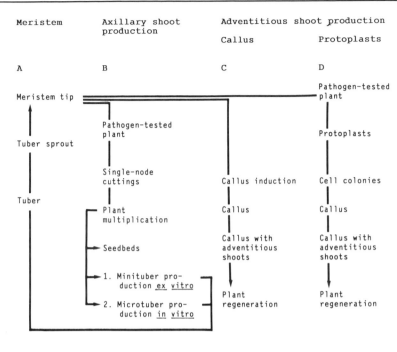

Meristem	Axillary shoot production	Adventitious shoot production	
		Callus	Protoplasts
A	B	C	D

Fig. 16.2. Four methods of micropropagation as applied to potato (*Solanum* spp.): A, meristem tip; B, axillary shoot regeneration; C, adventitious shoot regeneration via callus; and D, adventitious shoot regeneration via protoplasts. Genetic variation increases progressively from axillary-bud to protoplast-regenerated plants, from left to right (Schilde-Rentschler and Schmiediche, 1984).

months from initiation of the culture, respectively. Axillary plants produced *in vitro* can be bulked up to provide nodal cuttings that are rooted in soil and transplanted to the field for commercial seed potato and tuber production. This type of production has proved highly successful in Dalat, Vietnam where millions of rooted plantlets are sold by micropropagator farmers to farming co-operatives for transplanting directly to the field (van Uyen, 1984). The procedure is summarized in Table 16.2. Micropropagated axillary plants can also be used for greenhouse production of minitubers and *in vitro* production of microtubers (Fig. 16.2, method B).

16.4.3 Minituber production

If placed in fine sand with the leaf above and the axillary bud below the surface, leaf-bud cuttings obtained from axillary shoots derived axenically *in vitro* will produce minitubers (>1

cm) that are capable of yielding up to 1 kg of normal tubers when planted in the field (Tovar *et al.*, 1985). Minitubers (Fig. 16.2, method B) may be used as part of an integrated rapid multiplication scheme or as foundation potato tuber seed.

16.4.4 Microtuber production

Tuberization *in vitro* (Fig. 16.2, method B) can be induced by transferring harvested axillary shoots to a cytokinin containing culture medium and growing the cultures under an 8 h photoperiod of low irradiation at ca 20°C. Compared with the production of *in vitro* plantlets for export, microtubers (up to 1 cm in diameter and 50 mg fresh weight) have the following advantages (Schilde-Rentschler and Schmiediche, 1984; Tovar *et al.*, 1985): (1) they can be produced in large quantities regardless of season; (2) they can be stored for several

Table 16.2. Rapid potato multiplication procedure used by farmers in Vietnam (van Uyen, 1984)

No. of plants	Procedure
1	*in vitro* plantlet in test tube (from research station) multiplied twice over a 2-month period, 5 cuttings per tube
25	*in vitro* plantlets grown for 1 month in test tubes
100	*in vitro* nodal (cuttings rooted in soil beds) 2,500 cuttings/m^2
1 000	apical cuttings rooted in soil beds at 2 500 cuttings/m^2 harvested weekly for 2 months
10 000	apical cuttings rooted in banana leaf pots after 15–20 days transplanted to the field
Tubers	large-sized used for consumption small-sized used for seed for further multiplication

months; (3) transfers to fresh media are unnecessary; (4) shipment is facilitated; (5) strictest quarantine standards can be met.

16.4.5 Adventitious shoot production via callus and single-cell cultures

Several different protocols have been established for the isolation, culture and regeneration of plants from callus (Fig. 16.2, method C) induced on various plant parts, as well as from protoplasts (Fig. 16.2, method D) (see Miller and Lipschutz, 1984). Such regenerants, because of the length of time spent in the callus phase, may be subject to somatic mutations. While this may not be a problem for ornamental plants in which genetic change of an important character is easily observed, in crop plants this may not be so.

16.5 SYNTHETIC SEED

Synthetic or artificial seed (see Redenbaugh, 1993) consists of somatic embryos enclosed by a protective coating. Two types of synthetic seed have been developed (Fujii *et al.*, 1987): (1) hydrated and (2) desiccated. Hydrated artificial seeds are somatic embryos encapsulated in a hydrated gel, such as calcium alginate (Figs 16.3 and 16.4). Alginate is non-toxic and the gelling process is not temperature-dependent. Embryos, at the appropriate stage of development, are mixed with sodium alginate and dropped into a bath of calcium salt to form calcium alginate capsules with diameters 4–6 mm. The capsules are then delivered directly into the soil by fluid drilling. However, very few rates of conversion, that is, production of a whole plant from a true somatic embryo, have been reported (Fujii *et al.*, 1987; Redenbaugh *et al.*, 1987b). For albuminous seed crops, such as celery, an artificial endosperm may be necessary and it is therefore desirable that coating gels retain compounds such as sucrose.

Desiccated artificial seeds are produced by either coating somatic embryos with a water-

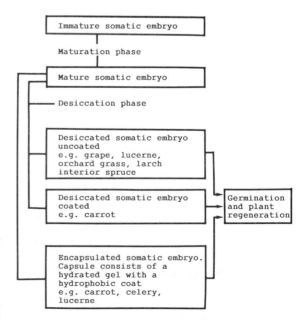

Fig. 16.3. Use of the somatic embryo as artificial seed.

Fig. 16.4. Sugar beet somatic embryos encapsulated in calcium alginate. Although somatic embryogenesis is poorly developed in sugar beet, these embryoids, derived from immature zygotic embryos 18 days postanthesis, are able to germinate *in vitro* and develop into plantlets at a conversion rate of ca 5%. The capsules are approximately 7.5 mm in diameter.

soluble resin, polyoxyethylene glycol, and allowing them to dry, or desiccation without coating. The desiccation process itself may damage the embryos. However, treatment of somatic embryos of *Medicago sativa* (McKersie *et al.*, 1990) and Interior spruce (Roberts *et al.*, 1990) at the early cotyledonary stage of development with abscisic acid or sucrose, increases tolerance to desiccation. *Medicago* somatic embryos can then be dried to 15% or less moisture content and stored at ambient room temperature and humidity without loss of viability.

16.5.1 Applications of synthetic seed

Potential application of synthetic seed will vary by crop (Table 16.3), but most probably will be determined by the need for improvement in the production of the crop, as well as in the ability to produce somatic embryos of that crop. In many cases, somatic embryos can be produced for species for which there is no apparent need to replace true seed. In other cases, where true seed cannot be used, or where there is a need to replace cumbersome vegetative propagation, somatic embryos either cannot be produced efficiently or cannot be produced at all.

(a) Seed-propagated, predominantly outbreeding crops

Lucerne (*Medicago sativa* L.) and orchardgrass (*Dactylis glomerata* L.) are examples of seed-propagated self-incompatible crops. Lucerne is a naturally cross-pollinated, autotetraploid species that suffers from severe inbreeding depression, the commercial cultivars of which derive from synthetic populations. As it is presently impossible to establish true breeding lines, breeder's seed has to be produced by crossing several different genotypes that are then planted to produce foundation seed (McKersie *et al.*, 1990; Redenbaugh, 1990a). The foundation seed, in turn, is planted for the production of certified seed but, as a result of a decrease in heterozygosity, the certified seed generation may yield 10–15% less than the breeder seed generation. Several schemes for hybrid seed production have been proposed, but chiefly as a result of the high cost of parental line propagation, none have been commercially successful. As production of parental lines and control of pollen is not technically feasible in lucerne, Redenbaugh (1990a) has suggested that artificial seed technology could potentially fill the void. By propagating male- and female-sterile or self-sterile parental lines through synthetic seeds, a pollination control system can be introduced in the plant population. Regeneration of lucerne plants via somatic embryogenesis is summarized in Fig. 16.5.

Somatic embryogeny in lucerne is well characterized and embryo-to-plant conversion frequencies as high as 90% can be obtained under laboratory conditions (Redenbaugh, 1990c). Field conversion frequencies, however, are still low, of the order of 15–25%. Low unit (per plant) value and current low cost of lucerne

Table 16.3. Potential application of synthetic seed technology to 13 selected crops. In part, based on Gray (1990) and Redenbaugh (1990a)

Crop	Commercial propagation[a]	Relative seed cost[b]	Somatic embryos		Synthetic seed	
			Obtained[c]	Quality and efficiency of system[d]	Relative need[b]	Potential application[e]
Barley	1	3	6	8	3	10
Conifer	1	3	6	8[g]	4	15
Cotton	1	4	6	8	4	15
Grape	2	na	6	9	4	11
Lucerne	1	3	6	9	5	11, 15
Maize	1	4	6	8	3	10, 13
Orchard grass	1	3	6	9	4	11
Soybean	1	4	6	8	4	15
Sugar beet	1	4	7[f]	–	3	10, 13
Sugarcane	2	na	7[f]	–	5	12
Tomato	1	5	7	–	5	14
Watermelon (seedless)	1	5	7	–	5	14
Wheat	1	3	6	8	3	10

[a] 1: seed; 2: vegetative.
[b] 3: low; 4: moderate; 5: high; na: not applicable.
[c] 6: yes; 7: no.
[d] 8: poorly-developed; 9: well-developed.
[e] 10: limited usefulness because existing methods are effective; 11: circumvention of self-incompatibility; 12: decrease of labour cost; 13: elimination of need for inbreds; 14: decrease of hybrid seed cost; 15: mass production of hybrids.
[f] Although reported, development of somatic embryogenesis lags far behind that of other systems.
[g] Exceptions are larch and some species of spruce.

seed limit application of synthetic seed technology. Redenbaugh *et al.* (1987a) have estimated the cost of a synthetic propagule at US$0.00026.

However, this must be put against the low conversion rates in the field.

Conifers are also planted by seed, although in

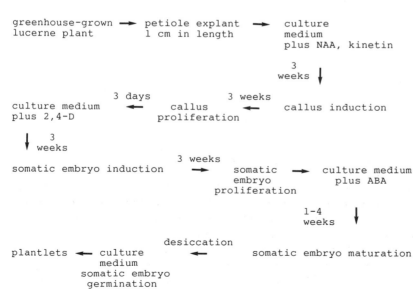

Fig. 16.5. Regeneration of lucerne plants via somatic embryogenesis. Total regeneration time, including a number of subcultures to media of varying composition, is from 15 to 25 weeks.

some cases their numbers are supplemented to a very limited extent by micropropagation (*Pinus radiata*) and rooted cuttings (*Picea abies* and *Cryptomeria japonica*). They are highly heterozygous and improvement by conventional breeding is time-consuming. Synthetic seed offers, at least theoretically, the possibility of cloning superior trees at a cost that should be more economical than the production of rooted cuttings. Well developed somatic embryo systems exist only for Norway spruce (*Picea abies* L.), Interior spruce (*Picea engelmanii* Parry + *P. glauca* (Moench) Voss) and larch (*Larix* spp.).

Tropical crops propagated by seed include cacao (*Theobroma cacao*), coconut (*Cocos nucifera*), oil palm (*Elaeis guineensis*) and coffee (*Coffea* spp.). A characteristic of the seeds of many tropical species such as *Citrus* spp., coconut, cacao and mango is the short duration of their viability. A problem associated with this characteristic is the difficulty in storing seeds at temperatures below 10°C and desiccation below 20–40% RH (Bewley and Black, 1985) generally is damaging. Although vegetative multiplication by cuttings, budding and grafting is used, mainly for breeding purposes and to preserve germplasm, coffee propagation is mainly through seed. However, somatic embryogeny in coffee is more advanced than for most tropical crops and costs of synthetic seed propagules, even if high, are expected to be competitive because of the high per unit plant value.

(b) Hybrid seed

Hybrid seed is difficult to produce for certain seed-propagated crops such as cotton (*Gossypium hirsutum* L.) and soybean (*Glycine max* Merrill.) due to problems with flower abscission in the former and cleistogamy in the latter. Seed of most existing cultivars is derived from self-pollination. Small numbers of hybrids could be produced by hand-pollination and subsequent mass production through artificial seed, thus exploiting hybrid vigour at the production level. For crops such as tomato (*Lycopersicon esculentum* Mill.) and seedless watermelon (*Citrullus lanatus* [Thunb.] Matsum and Nakai),

Table 16.4. Comparison of positive (+) and negative (−) aspects of clonal and seed propagation systems based on Fujii *et al.* (1987)

Greenhouse cuttings
+ Genetic uniformity maintained
− Rooting required before field planting
− Low multiplication rate; limited by size of donor plant
− Low volume, small-scale propagation method
− High cost per plantlet

Micropropagation
+ Genetic uniformity maintained
− Acclimation required before field planting
− Relatively low multiplication rate
− Relatively low volume, small-scale propagation method
− Very high cost per plantlet

Synthetic seeds
+ Genetic uniformity maintained
+ Direct delivery of propagules to field
+ Rapid multiplication rate
+ High volume, medium-scale propagation method
− High cost per plantlet

True seeds
− Genetic uniformity may not be maintained
+ Direct delivery to field
+ High volume, large-scale multiplication
+ Very rapid multiplication rate
+ Low cost per plantlet

the high cost of hybrid seed is offset by the value of the resulting crop (Gray, 1990). Synthetic seed may be less expensive than conventionally produced hybrid seed, but somatic embryogenesis has not been achieved for either of these species.

(c) Vegetatively-propagated crops

Two examples of outbred, clonally propagated, self-incompatible crops that show inbreeding depression are grape (*Vitis* spp.) and sugarcane (*Saccharum* spp.). Developmental costs of synthetic seed are presently difficult to justify, because existing methods of propagation (cuttings for grape and stem cuttings or seed pieces for sugarcane) tend to be cost effective. In the case of sugarcane, the large land area required for seedcane production on the one hand and the cost of labour involved in micropropagation on the other are factors that should favour development of synthetic seed technology. However, no efficient system for the production of somatic embryos in sugarcane exists. According to Gray (1990) use of artificial seeds could be advantageous for germplasm conservation of grape, where a well developed somatic embryo system exists.

16.5.2 Limiting problems

The difficulty in producing large numbers of high quality, vigorous somatic embryos limits the application of artificial seed technology. To compete commercially with true seed (Table 16.4) or efficient micro- and macropropagation systems, somatic embryos must be able to germinate rapidly and uniformly, and develop into plants at rates and frequencies approaching those of true seed. Production of artificial seed requires careful control of a number of stages, from the initial choice of explant, to induction, growth, development and maturation of the somatic embryo. Recently, research has started to shift toward improving embryo quality, as applications for using somatic embryos for vegetative propagation are becoming better identified (McKersie *et al.*, 1990; Redenbaugh, 1990a,b,c; Webster *et al.*, 1990).

16.6 PROSPECTS

In the short term, the use of micropropagation for various horticultural, especially ornamental, crops and its current limited use for agronomic crops is not expected to change significantly. However, emphasis on the development of cost-effective micropropagation techniques is expected to increase greatly. The technique with the greatest potential is somatic embryogenesis which, in the form of synthetic seed, may allow the extensive scale-up in micropropagation technology required for the production of cultivars that cannot be produced commercially

by other methods. With somatic embryogenesis having been reported in nearly all the major monocotyledonous and dicotyledonous crops, as well as in a number of gymnospermous species, future research should focus on increasing the efficiency of conversion of somatic embryos to plants for such species where an alternative to existing systems of reproduction is required.

REFERENCES

Ammirato, P.V. (1983), Embryogenesis. In *Handbook of Plant Cell Culture*, Vol. 1, *Techniques for Propagation and Breeding*, Evans, D.A., Sharp, W.R., Ammirato, P.V. and Yamada, Y. (eds), Macmillan, New York, pp. 82–123.

Becwar, M.R., Noland, T.L. and Wycoff, J.L. (1989), Maturation, germination, and conversion of Norway spruce (*Picea abies* L.) somatic embryos to plants. *In Vitro Cell Dev. Biol.*, **25**, 575–580.

Bewley, J. and Black, M. (1985), *Seed Physiology of Development and Germination*, Plenum Press, New York.

Boxus, P. (1987), *In vitro* vegetative propagation of plants. *Nestlé Research News 1986/87*, Nestec Ltd, Vevey, Switzerland, pp. 73–79.

Brown, C., Lucas, J.A., Crute, I.R., Walkey, D.G.A. and Power, J.B. (1986), An assessment of genetic variability in somaclonal lettuce plants (*Lactuca sativa*) and their offspring. *Ann. Appl. Biol.*, **109**, 391–407.

Bryan, J.E., Jackson, M.T. and Nelson Meléndez, G. (1981), *Rapid Multiplication Techniques for Potato*, International Potato Center (CIP), Lima.

Dale, P.J. (1980), Embryoids from cultured immature embryos of *Lolium multiflorum*. *Z. Pflanzenphysiol.*, **100**, 73–77.

Evans, D.A., Sharp, W.R., Ammirato, P.V. and Yamada, Y. (1983), *Handbook of Plant Cell Culture*, Vol. 1, *Techniques for Propagation and Breeding*, Macmillan, New York.

Fujii, J.A., Slade, D.T., Redenbaugh, K. and Walker, K.A. (1987), Artificial seeds for plant propagation. *Tibtech.*, **5**, 335–339.

Gamborg, O.L., Miller, R.A. and Ojima, K. (1968), Nutrient requirements of suspension cultures of soybean root cells. *Exp. Cell. Res.*, **50**, 151–158.

George, E.F., Putlock, D.J.M. and George, H.J. (1987), *Plant Culture Media*, Vol. I, *Formulations and Uses*, Exegetics Ltd, Westbury.

Gray, D.J. (1987), Introduction to the symposium, In

Proc. Symp. Synthetic Seed Technology for the Mass Cloning of Crop Plants: Problems and Perspectives. *HortScience*, **22**, 796.

Gray, D.J. (1990), Synthetic seed for clonal production of crop plants. In *Recent Advances in the Development and Germination of Seeds*, Taylorson, R.B. (ed.), Plenum Press, New York, pp. 29–45.

Gray, D.J. and Conger, B.V. (1985), Somatic embryo ontogeny in tissue cultures of orchardgrass. In *Tissue Culture in Forestry and Agriculture*, Henke, R.R., Hughes, K.W., Constantin, M.J. and Hollaender, A. (eds), Plenum Press, New York, pp. 49–57.

Grout, B.W.W. (1990), Meristem-tip culture. In *Plant Cell and Tissue Culture*, Pollard, J.W. and Walker, J.M. (eds), The Humana Press Inc., Clifton, New Jersey, pp. 81–91.

Hakman, I., Fowke, L.C., von Arnold, S. and Eriksson, T. (1985), The development of somatic embryos in tissue cultures initiated from immature embryos of *Picea abies* (Norway spruce). *Plant Sci.*, **38**, 53–59.

Heller, R. (1953), Recherches sur la nutrition minerale des tissues vegetaux cultives, *in vitro*. *Ann. Sci. Not. Bot. Biol. Veg.*, **14**, 1·223.

Hu, C.Y. and Wang, P.J. (1983), Meristem, shoot tip, and bud cultures. In *Handbook of Plant Cell Culture. Vol. I. Techniques for Propagation and Breeding*, Evans, D.A., Sharp, W.R., Ammirato, P.V. and Yamada, Y. (eds), Macmillan, New York, pp. 177–227.

Kim, Y-H. and Janick, J. (1990), Synthetic seed technology: improving desiccation tolerance of somatic embryos of celery. *Acta Hortic.*, **280**, 23–28.

Lu, C., Vasil, I.K. and Ozias-Akins, P. (1982), Somatic embryogenesis in *Zea mays* L. *Theor. Appl. Genet.*, **62**, 109–112.

McKersie, B.D., Bowley, S.R., Senaratna, T., Brown, D.C.W. and Bewley, J.D. (1989), Application of artificial seed technology in the production of alfalfa (*Medicago sativa* L.). *In Vitro Cell. Dev. Biol.*, **25**, 1183–1188.

McKersie, B.D., Senaratna, T. and Bowley, S.R. (1990), *Artificial Seeds: a Potential Revolution in the Propagation of Plants*, Highlights 2–6, Ontario Ministry of Agriculture and Food, Guelph, Ontario.

Miller, S.A. and Lipschutz, L. (1984), Potato. In *Handbook of Plant Cell Culture*, Vol. 3, Ammirato, P.V., Evans, D.A., Sharp, W.R. and Yamada, Y. (eds), Macmillan, New York, pp. 291–326.

Morel, G.M. (1960), Producing virus-free cymbidiums. *Am. Orchid. Soc. Bull.*, **29**, 495–497.

Morel, G.M. (1965), Clonal propagation of orchids by meristem culture. *Cymbidium Soc. News*, **20**, 3–11.

Murashige, T. (1974), Plant propagation through tissue cultures. *Annu. Rev. Plant Physiol.*, **25**, 135–166.

Murashige, T. and Skoog, F. (1962), A revised medium for rapid growth and bio-assays with tobacco tissue cultures. *Physiol. Plant.*, **15**, 473–497.

Ozias-Akins, P. and Vasil, I.K. (1982), Plant regeneration from cultured immature embryos and inflorescences of *Triticum aestivum* (wheat): evidence for somatic embryogenesis. *Protoplasma*, **110**, 95–105.

Power, J.B. and Davey, M.R. (1990), Protoplasts of higher and lower plants. In *Plant Cell and Tissue Culture*, Pollard, J.W. and Walker, J.M. (eds), Humana Press, Clifton, New Jersey, pp. 237–259.

Raghavan, V. (1983), Biochemistry of somatic embryogenesis. In *Handbook of Plant Cell Culture*, Vol. 1., *Techniques for Propagation and Breeding*, Evans, D.A., Sharp, W.R., Ammirato, P.V. and Yamada, Y. (eds), Macmillan, New York, pp. 655–671.

Redenbaugh, K. (1990a), Applications of micropropagation for agronomic crops. In *Micropropagation*, Debergh, P. and Zimmerman, R.H. (eds), Kluwer Academic Publishers, Dordrecht and Boston, pp. 285–309.

Redenbaugh, K. (1990b), Application of artificial seed to tropical crops. *HortScience*, **25**, 251–255.

Redenbaugh, K. (1990c), Coating and field trialing somatic embryos. In *Seeds: Genesis of Artificial and Natural Forms*, International Symposium (Le Biopole Vegetal), Amiens, France, pp. 134–141.

Redenbaugh, K., Slade, D., Viss, P. and Fujii, J.A. (1987a), Encapsulation of somatic embryos in synthetic seed coats. *HortScience*, **22**, 803–809.

Redenbaugh, K., Viss, P., Slade, D. and Fujii, J.A. (1987b), Scale-up: artificial seeds. In *Plant Cell and Tissue Culture*, Green, C., Somers, D., Hackett, W. and Biesboer, D. (eds), Liss, New York, pp. 473–493.

Redenbaugh, K. (ed.) (1993) Synseeds: Applications of Synthetic Seeds to Crop Improvement. CRC Press, Boca Raton, FL, pp. 1–481.

Reinert, J.M. (1958), Morphogenese und ihre Kontrolle an Gewebekulturen aus Karotten. *Naturwissenschaften*, **45**, 344–345.

Reinert, J.M. (1959), Über die Kontrolle der Morphogenese und die Induktion von Adventiveembryonen an Gewebekulturen aus Karotten. *Planta*, **58**, 318–333.

Roberts, D.R., Sutton, B.C.S. and Flinn, B.S. (1990),
    Synchronous and high frequency germination of
    interior spruce somatic embryos following partial
    drying at high relative humidity. *Can. J. Bot.*, **68**,
    1086–1090.

Schenk, R.U. and Hildebrandt, A.C. (1972), Medium
    and techniques for induction and growth of
    monocotyledonous and dicotyledonous plant
    cell cultures. *Can. J. Bot.*, **50**, 199–204.

Schilde-Rentschler, L. and Schmiediche, P.E. (1984),
    Tissue culture: past, present, and future. *CIP
    Circ.*, **12**, 1–6.

Steward, F.C., Mapes, M.O. and Mears, K. (1958),
    Growth and organized development of cultured
    cells. II. Organization in cultures grown from
    freely suspended cells. *Am. J. Bot.*, **45**, 705–
    708.

Tisserat, B., Esan, B.B. and Murashige, T. (1979),
    Somatic embryogenesis in angiosperms. *Hortic.
    Rev.*, **1**, 1–78.

Tovar, P., Estrada, R., Schilde-Rentschler, L. and

Dodds, J.H. (1985), Induction and use of *in vitro*
    potato tubers. *CIP Circ.*, **13**, 1–5.

van Uyen, N. (1984), A new use for tissue culture and
    rapid multiplication: potato production by Viet-
    namese farmers. *CIP Circ.*, **12**, 7–9.

Vasil, V. and Vasil, I.K. (1980), Isolation and culture
    of cereal protoplasts. II. Embryogenesis and
    plantlet formation from protoplasts of *Pennisetum
    americanum*. *Theor. Appl. Genet.*, **56**, 97–100.

Webster, F.B., Roberts, D.R., McInnis, S.M. and
    Sutton, B.C.S. (1990), Propagation of interior
    spruce by somatic embryogenesis. *Can. J. For.
    Res.*, **20**, 1759–1765.

White, P.R. (1963), *A Handbook of Plant Tissue Culture*,
    Jaques Cottell, Pennsylvania.

Witt, S.C. (1985), *Biotechnology and Genetic Diversity*,
    California Agricultural Lands Project, Center for
    Science Information, San Francisco.

Wooster, P. and Dixon, T.J. (1985), Application of
    micropropagation to potato seed production. *J.
    Natl. Inst. Agric. Bot.*, **17**, 99–106.

17

Andro- and parthenogenesis

B. Foroughi-Wehr and G. Wenzel

17.1 INTRODUCTION

The invention of nature to double the genome when evolution jumped from lower to higher plants guaranteed genetic security and stability; it reduced, however, the chance of a quick adaptation to new conditions created, for example, by a new environment or by the higher variability of harmful micro-organisms still having the flexible haploid complement. Thus, a plant breeder who is dependent on genetic variability seeks to circumvent diploidy, at least in several steps of a breeding programme. Of course, the final product must be diploidized again, to produce doubled haploid lines (DH) with the necessary stability for the new genotype.

It is no longer a question that the most appropriate starting material for the production of haploids are the gametes. Consequently, two principal methods exist for the induction of plants with a simpler genome: either starting from the female or the male gametophyte, which means the *in vitro* regeneration of pollen or the induction of haploid tissue from the embryo sac, normally the egg cell.

Both processes occur eventually in nature, but since the seeds of such haploids are generally absent altogether or are at least sterile, they have no regular chance to produce a subsequent generation. Haploids from the female gametophyte, which are called parthenogenetic haploids, are more frequent in

Plant Breeding: Principles and prospects. Edited by M.D. Hayward, N.O. Bosemark and I. Romagosa. Published in 1993 by Chapman & Hall, London. ISBN 0 412 43390 7.

nature than haploids originating from the male gametophyte, the androgenetic process of haploid production. The first spontaneous haploids found and described by Blakeslee (1922) were parthenogenetic ones. Usually they develop from one of the synergids together with the normal diploid embryo, resulting in twins, one of which is haploid while the other is diploid. For practical breeding purposes this spontaneous haploidization rate is much too low. Androgenesis may also appear naturally, but then a microspore or a sperm nucleus develops without fertilization on or in the embryo sac, an event even rarer than natural parthenogenesis.

Due to the low rate of spontaneous haploid formation, procedures for increasing haploid production have been developed. Embryo sac androgenesis can be increased genetically in maize using the specific gene *ig* together with a marker gene *R-navajo* (Kermicle, 1969), but this method proved to be of limited practical utility. In barley, haploids can be obtained by special methods, such as the incorporation of the *hap*-gene (Hagberg and Hagberg, 1980).

A breakthrough has been the successful induction of haploids from young anthers cultured *in vitro* in 1964 (Guha and Maheshwari, 1966). It turned out that the pollen grain, the nucleus of which has normally already undergone an unequal division, is too late a stage for the desired somatic development and that the uninucleate condition of the microspore is needed. The development of the microspore to a haploid sporophyte, is called 'microspore androgenesis' or simply 'androgenesis'. This androgenesis occurs only under *in vitro* conditions, thus a prerequisite for

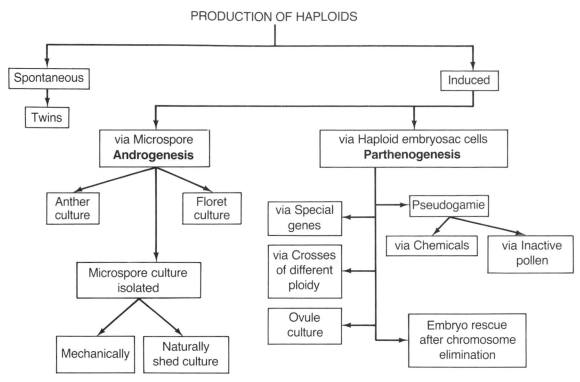

Fig. 17.1. Schematic summary of procedures resulting in the formation of haploid plants.

androgenetic haploids was the ability to handle cells *in vitro*. The same was true for the parthenogenetic method of haploid induction. Here, techniques such as embryo culture or the more sophisticated isolation of unfertilized egg cells and their regeneration *in vitro* had to be made reproducible, before the artificial induction of parthenogenetic haploids attained an acceptable frequency. The main processes for haploid induction are shown in Fig. 17.1.

Since the first successes in *in vitro* regeneration of haploids, great research effort has been directed towards improving the technique for economically important plants such as cereals, vegetable crops, and sugar beet. Today, the methodology has reached a stage where private laboratories are adopting this technique at their own economic risk.

The principal ideas of how to make use of the simpler gametophytic genome in the basic and applied sciences are more than 50 years old, and very few new concepts have been added

since Blakeslee, Belling, Farnham and Bergner itemized them in 1922. The main advantage of using haploids is the rapid and complete homozygosity of the offspring, because phenotypic selection for quantitatively, and particularly for qualitatively, inherited characters is much easier and, therefore, breeding is more efficient. In tetraploids like *Medicago* and potato, haploidization is used to produce diploids. Experiments are in progress to test whether breeding at the diploid level is of commercial interest. In several tetraploid crops it has become possible, by this approach, to combine wild species with the cultivated forms at the diploid level.

A rather recent application of DH lines is their use in genome mapping. In a population of DH lines the identification of phenotypic markers is more efficient, as intermediate expressions due to heterozygosity are excluded. At the plant level as well as at the marker level (today mainly by the use of RFLPs, discussed in

Chapters 12 and 19) a gene will segregate in a 1:1 ratio. This is of particular importance when a polygenically controlled character is being mapped.

Here, we present general principles and the up-to-date knowledge in haploid production and use, with emphasis on important crops. For further information, see the reviews of Wenzel and Foroughi-Wehr (1984) and Dunwell (1985).

17.2 PARTHENOGENESIS

17.2.1 *In vitro* production of haploid plants from unpollinated ovaries or ovules

In several crops it is possible to induce parthenogenesis by *in vitro* culture of unpollinated ovaries and ovules (Yang and Zhou, 1982). The sporophytic megagametophyte differentiates into an embryo sac containing the egg cell, synergids and antipodals. These haploid embryo sac cells are able to develop a haploid embryo or callus *in vitro*. The culture of unfertilized ovules or ovaries in many different plant species has been attempted, but growth in the majority stopped at the callus stage, and in only some did development lead to the production of haploid green plants (Table 17.1).

17.2.2 Methods and factors influencing culture results

Many of the regular plant tissue culture methods are applicable to ovule or ovary culture. As in all tissue culture procedures, the regeneration response depends on the genotype. In five wheat genotypes, for example, the percentage of ovules producing haploid plants ranged from 1.3 to 10.9% (Zhu *et al.*, 1981). Among 50 breeding lines of onions, four were found to be notably more productive (Keller, 1990). The developmental stage of the gynogenetic cells inside the ovule cannot be determined at the time of inoculation. Some investigators tried to find a correlation between the size of ovules and the response (D'Halluin and Keimer, 1986). In addition, the pollen developmental stage was used as an indirect criterion for identifying the embryo sacs appropriate for haploid induction. It has been confirmed that for most plant species it is possible to induce embryos from ovules over a broad range of growth stages.

From one ovule several embryos can be induced in barley (San Noeum, 1976) and sugar beets (D'Halluin and Keimer, 1986). However, usually only one embryogenetic structure develops at the micropylar end of the embryo sacs (Zhou and Yang, 1981; Hosemanns and Bossoutrot, 1983). Figure 17.2 shows an opened ovule of wheat with a developing embryo.

Table 17.1. Parthenogenetic haploid production

Species	Reference
Hordeum vulgare	San Noeum (1976)
Triticum aestivum	Zhu and Wu (1979)
Nicotiana tabacum	Zhu and Wu (1979)
Oryza sativa	Asselin de Beauville (1980)
Gerbera jamesonii	Sitbon (1981)
Zea mays	Ao *et al.* (1982)
Citrus sinensis	Kochba and Spiegel-Roy (1982)
Beta vulgaris	Hosemanns and Bossoutrot (1983)
Helianthus annuus	Cai and Zhou (1984)
Cucurbita pepo	de Vaulx and Chambonnet (1986)
Hevea brasiliensis	Chen *et al.* (1988)
Allium cepa	Muren (1989)

Fig. 17.2. Parthenogenesis demonstrated on the *in vitro* development of an unfertilized wheat embryo, 3 weeks after isolation from the floret.

One of the major problems in producing gynogenetic plants is how to preclude a somatic origin of the embryogenic tissue. Calli or embryos may be produced simultaneously from the gametophytic and the sporophytic cell. If the haploid chromosome number can be counted exactly in the regenerants, a clear discrimination is possible. But if both haploid as well as diploid plants are produced, the origin of the non-haploids, whether they have doubled spontaneously or are of somatic origin, can only be identified in the offspring.

17.2.3 Use of parthenogenesis in other crops

Other techniques for the induction of parthenogenetic procedures include pollination with irradiated (100 krad) pollen or with chemically (basic toluidine-blue) treated pollen, for example, in *Populus* (Illies, 1974; Wang *et al.*, 1975), and sugar beet (Yüce, 1973). Due to the low success rate, such approaches have not attained practical importance.

The situation is different in potato. The cross between 4x *Solanum tuberosum* with 48 chromosomes and 2x *S. phureja* with 24 chromosomes leads to parthenogenetic haploids of *S. tuberosum* ($2n = 2x = 24$). The reason for the haploid formation is that the second microspore mitosis is blocked in special *S. phureja* genotypes, resulting in only one sperm nucleus, which can either fertilize the embryo sac nucleus or the egg cell. In the first case, the egg cell remains unfertilized giving rise to a parthenogenetic haploid seed. From successful pollinations, up to 40% haploid seeds can be obtained. The selection of the dihaploids (2x), is facilitated because, due to the incompatibility between endosperm and embryo, no viable triploids are formed. This embryo/endosperm block does not work when, in a subsequent combination, 2x *S. tuberosum* is pollinated with *S. phureja*. As a consequence, huge numbers of diploid hybrid seeds must be screened for the very small fraction of monoploids ($2n = x = 12$). Because of this laborious selection, the androgenetic tech-

nique is much more efficient for this second step of ploidy reduction from 2x to 1x.

17.2.4 Value of parthenogenesis for breeding

Up till now the practical application of parthenogenesis has been limited by the low frequency of haploid induction in most species, except for potato. Only in plant species where anther culture has been unsuccessful as in sugar beet (Speckmann *et al.*, 1986) or onion (Muren, 1989), is haploid parthenogenesis from unpollinated ovules or ovaries of practical interest.

17.2.5 Embryo rescue after wide hybridization

In many interspecific and intergeneric crosses fertilization occurs normally and a hybrid embryo arises, however, because of the failure of normal endosperm formation the embryos degenerate. The embryos of such unsuccessful crosses are often capable of growing into normal hybrid plants on an endosperm replacement (an artificial medium *in vitro*). Since Laibach (1929) obtained hybrid plants from the interspecific cross *Linum perenne* × *L. austriacum* by growing the excised embryo *in vitro*, this technique has been practised for many wide crosses within both dicots and monocots. The embryos were either rescued from the ovule several days after pollination, then followed by *in vitro* culture, or were cultured *in vitro in ovulo* for a certain period and the embryo then isolated. The latter method is used if the incompatibility of the partners leads only to a very small undifferentiated pro-embryo, too small for isolation. This has been tested, for example in interspecific hybrids of *Medicago* species (McCoy and Smith, 1986) and of *Allium* species (Gonzales and Ford-Lloyd, 1987). Depending on the chromosome number of the parents, hybrid plants were obtained with euploid or aneuploid chromosome number. The resulting hybrids were mostly intermediate in morphological characteristics showing the presence of both parental genomes.

17.2.6 Haploid production by chromosome elimination

A special case of incompatibility in wide crosses is chromosome incompatibility. In some inter-specific crosses, the phenomenon of chromosome elimination occurs during early embryo development. Due to a mechanism not yet understood, the chromosomes of one parent are partially or completely lost during the early divisions of the young hybrid embryo. In nature, such a loss of chromosomes, normally coupled to a failure of endosperm development, leads to sterility. If all chromosomes of one parent are lost the embryo rescue results in haploid plants. For barley, this system was studied by Kasha and Kao (1970) and Jensen (1975) in crosses of *Hordeum vulgare* × *H. bulbosum*. They demonstrated that by crossing diploid barley with diploid *H. bulbosum* high frequencies of barley monoploids could be produced. Although normal fertilization takes place, the *H. bulbosum* chromosomes are eliminated during the development of the embryo. Since the endosperm fails to develop, the embryo rescue technique is required to prevent starvation and death of the embryos. Apart from crosses with *H. bulbosum*, haploids of barley have been reported from crosses with rye (*Secale cereale*) (Fedak, 1977) or with *Psathyrostachys fragilis* as well as from crosses with other wild barley species (Jorgensen and von Bothmer, 1988).

Wheat haploids have been produced by fertilizing wheat with *H. bulbosum* (Snape *et al.*, 1979), but the use of this technique is limited to genotypes which carry the rye crossability genes (*Kr* genes) (Sitch *et al.*, 1985). Although considerable effort has been directed towards overcoming these incompatibility barriers and increasing the frequency of fertilization by several pre- and post-pollination treatments, success in wheat is still limited. A similar mechanism of chromosome elimination leads to haploids in wheat crosses with maize (Laurie and Bennett, 1987). This method may be more applicable as maize seems to be insensitive to the crossability genes which determine the frequency of fertilization in wheat × *H. bulbosum* crosses (Snape *et al.*, 1979). The survival rate of the embryos *in vivo* and *in vitro* is low, but can be improved by spikelet culture (Laurie and Bennett, 1988).

The *bulbosum* system in barley has been practised for nearly 20 years. The first doubled haploid variety of barley 'Mingo' was licensed for sale in Canada in 1979 (Ho and Jones, 1980). During the past decade the method has been considerably improved by selecting more compatible *H. bulbosum* genotypes and by optimizing different culture factors. Field evaluation of DH lines showed no significant difference between single seed descent lines and DH lines, for grain yield, heading date, and plant height (Choo, 1988). The *H. bulbosum* method has led to considerable progress in haploid production and to a better understanding of incompatibility reactions. However, since it is possible to make use of the greater potential of the androgenetic system for haploid production, the *bulbosum* method is now of decreasing value for practical plant breeding.

The agronomic performance of DH wheat lines from wheat × *H. bulbosum* crosses (Snape *et al.*, 1988) as well as from wheat × maize crosses (Laurie and Snape, 1990) has been investigated. The results suggest that most DH lines represent the genotype of their wheat parent. But so far progress in wheat breeding via DHs has only been achieved with material derived androgenetically.

17.3 ANDROGENETIC HAPLOID PRODUCTION

The most promising and also most successful method of inducing DH lines today is by microspore androgenesis. Figure 17.3 shows a schematic presentation of anther culture as exemplified by the potato. A microspore, i.e., a pollen in an early developmental stage can be induced *in vitro* to form a plantlet. This technology has been applied successfully to many of species covering a range of genera. The most important progress is in the crop plant species

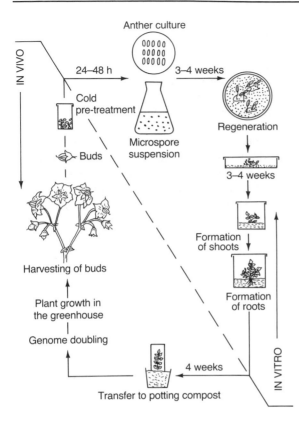

Fig. 17.3. Scheme for the androgenetic induction of haploids in potato.

of the Gramineae and Brassicaceae, where the technique is used nowadays in practical breeding. In spite of this we know very little about the induction mechanisms responsible for transforming a microspore into a sporophyte. The most important factors that influence success are summarized below.

17.3.1 The anther donor plant

(a) Genotype

Success in microspore culture is predominantly dependent on the genotype of the anther donor material. In all plant species examined, there are major differences in anther response between genotypes under given culture conditions. Culture conditions could probably be

optimized for each genotype, as proposed by Dunwell (1981). However, we believe that it is more effective to broaden the genetic base for tissue culture ability by selection and breeding.

In maize, the anther culture process itself was used to select for genetic improvement of *in vitro* response. Two anther-derived plants were crossed to produce an F_1 hybrid. The anther response of the S_1 families from this cross was almost eight times higher than the original cross. The enhanced anther culture responsiveness of commercial germplasm should broaden the practical applicability (Petolino *et al.*, 1988). Barloy *et al.* (1989) concluded from their investigations with different DH maize lines that it is relatively easy to introduce responsiveness into elite lines by repeated backcrossing.

In barley and potato, high-responding genotypes were crossed with agronomically important varieties. The F_1 generation showed an intermediate reaction between both parents, making it likely that a high response in anther culture is heritable (Wenzel and Foroughi-Wehr, 1984). In comparisons of eight varieties of winter barley at the Scottish Crop Research Institute, the high responding genotype 'Igri' (Foroughi-Wehr and Friedt, 1984) produced the highest percentage of green structures (Powell, 1988). Similar results were obtained in a trial among 16 varieties in Denmark (Knudsen *et al.*, 1989). For rye, comparable results were reported (Wenzel *et al.*, 1977); F_1 hybrids from two DH lines showed, as in maize, an increased microspore response (Flehinghaus *et al.*, 1991). In wheat, in a set of crosses to non-responsive varieties, the good regeneration capacity of the variety 'Centurk' could be transferred to the F_1. Reciprocal crosses indicated negligible cytoplasmic effects (Bullock *et al.*, 1982). Similar results have been obtained for potato (Wenzel, 1980; Uhrig and Salamini, 1987) and *Brassica napus* (Keller, 1984; Siebel and Pauls, 1989).

Genetic analysis of *in vitro* culture response is rare. Henry and de Buyser (1985) stated that haploid production in wheat is controlled by at least three different and independently inherited traits: embryo induction, regeneration ability, and the ratio of green to albino plants. They suggested that regeneration ability could be linked to the 1BL–1RS translocation. The results with 45 winter wheat cultivars obtained by Müller *et al.* (1989) confirmed the higher regeneration frequency in 1BL–1RS translocation lines. The results were verified in other investigations (Foroughi-Wehr and Zeller, 1990), but it could be assumed that, in addition to the 1BL–1RS translocations, there are other genetic systems which influence microspore regeneration in wheat. Higgins and Mathias (1987) analysed the chromosomal effect of regeneration capacity from callus cultures and reported significantly increased morphogenesis of different substitution lines of *4B* chromosomes in 'Chinese Spring' donor material.

(b) Growth of donor plants

Even if the donor plant genotype is excellent, the response of microspores in culture is dependent on their condition. Dunwell and Perry (1973) first pointed to the influence of the photoperiod and light regime on the androgenetic response in anther culture of tobacco. In *Brassica napus*, pollen embryogenesis can be improved by preconditioning the donor plants at low temperature (Keller and Stringam, 1978). In cereals, an influence of the growing conditions of the donor material is known for wheat, barley, rice and rye. Even the very responsive variety 'Igri' regenerates no green plants at all if the growing conditions of the donor plants are not satisfactory. Critical environmental factors include light intensity, photoperiod, temperature, nutrition, and CO_2 concentration. Also pest control measures may have a detrimental effect on microspore development. In most plant species, field-grown material produced during the normal growing season was superior to greenhouse-grown material.

17.3.2 Anther culture

(a) Developmental stage of the anthers

For most plant species, flower buds, inflorescences, or spikes harvested at the early or mid-uninucleate stage of microspore development are most responsive *in vitro*. As soon as starch deposition has begun, no sporophytic development and, subsequently, no macroscopic structures are formed. On the other hand, very young microspores, tetrads, or spores immediately after release from the tetrads are ineffective. The pollen stage in the anther is either determined by rapid staining with acetocarmine or, if the exine is too thick, as in potato or rapeseed, by fixation in Carnoy's solution overnight and with subsequent staining. The size of the flower buds that contain anthers in the right stage depends, to some extent, on their position in the inflorescence, on the genotype, and on the plant age. In cereals, by checking the stage of one anther this can be used as a marker for the estimation of microspore development in the remaining florets from the same variety during the same season. Pollen in florets from the middle of the inflorescence are further developed than those in apical and basal florets. Quite often, anthers from the lower and the upper third of the ear are at the correct stage for inducing sporophytic development, when central florets show the first microspore mitosis.

After surface sterilization of the buds or ears, the anthers are removed aseptically and planted on a liquid or solid culture medium. The induction of uninucleate microspores to form sporophytic embryoids or calli has been studied in detail by Sunderland in *H. vulgare* (Sunderland and Dunwell, 1977; Sunderland *et al.*, 1979; Sunderland and Evans, 1980) and in wheat by Krüger (1987). It was found that the three different pathways of androgenetic microspore development elaborated in *Datura* (Sunderland, 1974) are transferable to other plant species. Here, however, multicellular microspores develop along all pathways, even if the vegetative or the generative nucleus starts to divide first (Sunderland and Evans, 1980). In wheat, the generative nucleus degenerates during the *in vitro* development and is not involved in formation of microstructures. During establishment of an anther culture procedure, it is recommended that some anthers be sacrificed after 5 days and up to 3 weeks of culture. These should be checked microscopically for multicellular structures which are a reliable proof of the desired reaction.

The culture of entire florets, tassels, or parts of inflorescences and the consequent induction of haploid plants has been reported in barley (Wilson, 1977), rice (Kim and Raghavan, 1988), *Festuca arundinacea* (Kasperbauer *et al.*, 1980), and maize (Pareddy and Greyson, 1985). The results are highly genotype dependent and generally less successful than anther or microspore culture.

(b) Pretreatment of anthers

Certain pretreatments of the inflorescences can have a positive effect on the development of the microspore. The most effective technique used in anther culture, especially for the cereals, is cold pretreatment. Periods from 72 h up to 4 weeks at 4–10°C are recommended (barley: Sunderland *et al.*, 1981; potato: Wenzel and Uhrig, 1981; maize: Genovesi and Collins, 1982; wheat: Lazar *et al.*, 1985; rice: Chen *et al.*, 1981; rye: Wenzel *et al.*, 1977). The genus *Brassica* is an exception as the culture effect is better after an incubation of excised inflorescences (in water) at 45°C for 3 hours prior to anther culture (Keller, 1984). An influence on the ploidy level of regenerated rice plants by cold pretreatment was also reported (Yang *et al.*, 1985).

The mechanisms underlying the positive effect on callus or plantlet formation are unknown. As a result of cold storage, weak or non-viable anthers and microspores are killed; they then become dark brown in colour. At low temperature, increased numbers of microspores are arrested during the first mitosis, as starch production is blocked. Thus, the spikes are enriched in vigorous material, which increases the possibility of subsequent development. The cold pretreatment retards ageing of the anther wall, and in particular the degradation of the tapetum matrix, allowing a higher proportion of microspores to change their developmental pattern from gametophytic to sporophytic. Cold pretreatment may also provide an unspecific shock, resulting in the establishment of endogenous cellular conditions that favour the desired development. This is quite probable, as other shocks such as pretreatment with high temperature in *Brassica* (Keller, 1984), spraying with ethrel (Bennett and Hughes, 1972), centri-fugation (Nitsch, 1974), and irradiation with 1 krad X-ray in rye (Stolarz, 1974) and barley have stimulatory effects.

On the other hand, several studies showed a high frequency of callus or embryoid formation for good responding genotypes such as cv. 'Igri' in barley, without any pretreatment (Powell, 1988). In wheat a negative effect of low temperature on the anther response has also been reported (Marsolais *et al.*, 1984). If the donor plants grow under optimal conditions, the promoting effect of pretreatment is minimized and the whole procedure may be omitted.

(c) Culture medium

For programmed culture success understanding of plant–medium interaction is required, but still in its infancy. Consequently there exists no general logical approach to a best culture medium. Trial and error are the dominant features in composing media. In general, a first medium which shall induce the sporophytic development of the gametophytic tissue needs some auxins, whereas media for plantlet regeneration demand cytokinins. There is a list of several hundred specialized media which are elaborated on a wide range of plants but normally only on a few genotypes. Today two defined media (Murashige and Skoog, 1962; Chu, 1978) and the undefined potato extract medium (Chuang *et al.*, 1978) are most widely used, and can be recommended for first trials on new species. Beside inorganic compounds, the type of carbohydrates responsible for energy delivery, and the physical nature of the medium (liquid, solid, or semisolid) are parameters of central importance.

Liquid media are preferably used for pollen, embryoid, or callus induction. Kao (1981) reported a considerable increase in the frequency of pollen regeneration by adding 'Ficoll' to liquid medium. 'Ficoll' prevents small structures such as callus or embryoids from sinking into the medium and dying through anaerobiosis. Figure 17.4 shows the embryoid development from one anther of potato in 0.5 ml of such a liquid culture medium. For solid media agar used to be the only gelling agent. In principle it should have no nutritional effect.

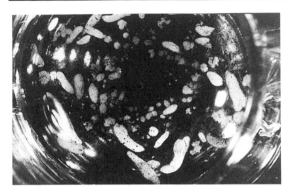

Fig. 17.4. Embryoids growing from shed pollen culture of 2*x* *Solanum tuberosum* microspores. The picture shows embryoids growing from one anther in one well of a multiplate with holes of 1 cm in diameter (Zitzlsperger, unpublished).

Results are more reproducible by using the inorganic gelling agent gelrite. Recently agar has been replaced by starch, which combine the solidifying characteristic with nutritional influences. Sorvari (1986) successfully used barley starch for barley anther culture. In wheat, wheat starch was most responsive for anther culture (Foroughi-Wehr and Zeller, 1990), and in rye, rye starch is under trial. Starch is not only cheaper than agar but is also broken down by enzymes into sugars, and some recent improvements are due to different carbohydrate sources. In most culture media sucrose is recommended. Since Raquin (1983) found that maltose improves embryo induction and development in anther culture of petunia, more attention has been given to this sugar.

In summary, the available literature does not allow recommendation of any anther culture medium for general applicability. The concentration and composition of sugars, as well as the type of gelling agent are presently the compounds which have the greatest influence on medium quality (Foroughi-Wehr, 1993). For practical application, where F_1 hybrids are used containing heterozygous microspore populations, a satisfactory situation has not been reached.

(d) Culture conditions

The effects of incubation conditions such as light (quantity and quality) and temperature have been investigated in a number of studies. Usually the anthers were cultured in the dark in a range between 20 and 30°C. Regenerated plantlets were transferred to light with the intensity of 3000–10000 lux. There is a strong correlation between light intensity and temperature, in low temperature conditions the light intensity can also be low and *vice versa*. In *Datura innoxia*, microspore embryoids are not formed at 20°C or lower (Sopory and Maheshwari, 1976). Ouyang *et al.* (1983) showed that in wheat the optimal culture temperature ranged from 26 to 30°C, varying with genotypes. For field-grown material, the optimal culture temperature was generally 2°C higher than for greenhouse-grown material (Ouyang *et al.*, 1987). The orientation of the anthers on the medium with regard to the loculi has a positive effect on response in *Datura* (Sopory and Maheshwari, 1976), barley (Hunter, 1985; Shannon *et al.*, 1985), and maize (Tsay *et al.*, 1986).

17.3.3 Isolated microspore culture

To date, the culture of isolated microspores is possible in many plants, such as potato (Uhrig, 1985), rapeseed (Lichter, 1982), tobacco (Nitsch, 1974), *Datura* (Nitsch and Norreel, 1973), barley (Sunderland and Xu, 1982; Datta and Wenzel, 1988; Kuhlmann, 1991), wheat (Wei, 1982; Datta and Wenzel, 1987), rice (Chen *et al.*, 1981), and maize (Coumans *et al.*, 1989; Pescitelli *et al.*, 1989). Figure 17.5 shows the development of isolated microspores of the barley variety 'Igri' after mechanical isolation. Microspores in uninucleate stage are either isolated mechanically from the anther or shed initially into liquid culture media. To improve the efficiency of isolated pollen culture, it proved to be beneficial to condition the liquid culture medium by the addition of anthers or ovules (Köhler and Wenzel, 1985). Siebel and Pauls (1989) compared anther culture and microspore culture in *Brassica napus* and demonstrated a much higher efficiency of embryo production in microspore culture. There were no differences between the two populations with respect to agronomic characters in the field. Isolated microspore

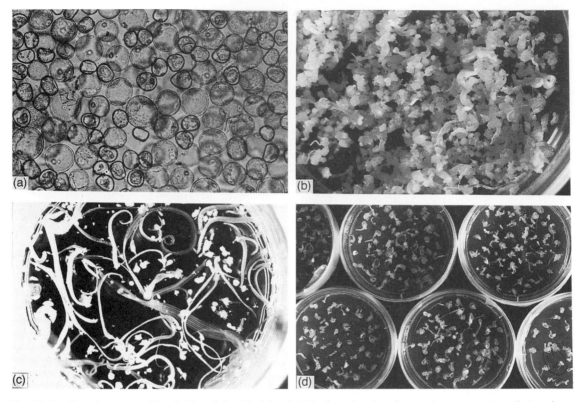

Fig. 17.5. Development of haploid and doubled haploid barley plantlets from microspores in isolation from the anther. (a) Microspore in liquid culture medium 2 days after mechanical isolation from the anther; (b) regeneration of embryoid-like structures 5 weeks later; (c) plantlet formation still in liquid culture; (d) further growth on solid medium (Kuhlmann, 1991).

cultures offer a more effective system of regenerating a random sample from the microspore population than does anther culture. In addition, these cultures are a good source for *in vitro* selection (Chapter 20) and gene transfer (Kuhlmann, 1991).

17.3.4 The problem of albino plants after androgenesis

The occurrence of albino plantlets is a serious problem in androgenetic plants of cereals. In haploids of gynogenetic origin albinos occur only occasionally. Also in dicots albino plantlets appear sporadically after microspore regeneration. The primary cause of albinism is not understood. Detailed investigations of proplastids and of the plastid genome of the regener-

ated albino plantlets showed some deletions of the membrane structures as well as different blockages in the plastid development (Chen *et al.*, 1988). In wheat, comparison of RFLP analyses of chloroplast DNA from green and albino tissue showed that in the latter there was a wide range of cpDNAs (Day and Ellis, 1984). Whether this deletion of the plastid genome exists in the immature microspore or whether it is altered during the *in vitro* phase is not known. It is certain that the frequency of albino plants is related to different factors such as the stage of pollen (wheat: He and Ouyang, 1983), the culture conditions (rice: Genovesi and Magill, 1979; barley: Kuhlmann and Foroughi-Wehr, 1989), or the genotype (rice: Wang *et al.*, 1978). Tuvesson *et al.* (1989) pointed out that nuclear genes in wheat are responsible for albinism. On the other hand, the maternal

Table 17.2. The ploidy level and the number of plants setting seed of DH-progenies from six different F$_1$-crosses of winter barley

| F$_1$-cross | Ploidy level | | | | | |
| | 2n = x = 7 | | 2n = 2x = 14 | | 2n = 4x = 28 | |
	Seed set	*Sterile*	*Seed set*	*Sterile*	*Seed set*	*Sterile*
1	11	3	14	1	1	–
2	22	10	88	1	9	2
3	16	6	15	1	2	–
4	45	15	207	4	17	
5	67	35	164	–	26	–
6	10	10	44	1	9	–
Total	171	79	532	8	64	4

inheritance of chloroplasts suggests their elimination or possible alteration during pollen formation (Vaughn *et al.*, 1980). Nuclear gene changes, as well as cytoplasmic ones, are possible reasons for albinism but an additional influence of physiological factors cannot be excluded. The dramatic decrease of albino plantlet formation with improved culture techniques, supports this view.

17.3.5 Ploidy level and genome doubling

Plants developed from microspores or ovaries are either haploid, diploid, or tetraploid; a few are aneuploid or mixoploid. In barley, potato, rapeseed, and rye the majority (ca 70%) of the plants developed are diploid, whereas in maize (Nitsch *et al.*, 1982) and in wheat, most of the regenerants remain haploid. Various factors influence the ploidy level of the regenerated plants. In *Brassica napus*, this character affects to be under genetic control (Chuong *et al.*, 1988) and the proportion of diploids rises with increasing duration of the culture period (Siebel and Pauls, 1989). The same is true for potatoes, where monoploid ($n = x = 12$) formation, dependent on the genotype, ranged from 0 to 20% (J. Zitzlperger, personal communication). Factors such as the pollen stage at the time of inoculation, the hormones in the media or the culture conditions may also influence the ploidy level of the plants.

The chromosome number of most haploid plants may be doubled by colchicine, as described by Jensen (1975); different species may require different concentrations. In potato and maize, the efficiency of doubling haploid plants is very low. But Wan *et al.* (1989) used colchicine treatment of an embryogenic haploid maize callus initiated from anther culture to produce fertile doubled haploid inbred lines at high frequency. In potatoes the return from the $2x$ to the practically important $4x$ level is not only possible via meiotic doubling but also via protoplast fusion. This strategy is described in Chapter 7.

In rice and barley the direct production of doubled haploids by androgenesis has been improved to the extent that colchicine treatment of the remaining haploids can be omitted. Under field conditions seed set could be found in more than 50% of haploid barley plants without colchicine treatment. Table 17.2 shows the ploidy level of the root tip cells and the seed set of the mature plants of the DH offspring of six F$_1$-crosses. It can also be seen that at the diplod level a few plants are sterile; chromosomal mutations may be the reason. The high occurrence of haploid plants setting seed, implies the production of unreduced gametes on the haploid cell level.

17.4 HAPLOIDS IN PLANT BREEDING

Since improved doubled haploid production via anther culture has been developed this method has been increasingly applied to practical breeding programmes in potato, rapeseed, and the cereals. Of particular interest in barley is the use of doubled haploids in the incorporation of resistance to the soil-borne barley yellow mosaic virus (BaYMV) (Foroughi-Wehr and Friedt, 1984). In wheat (De Buyser *et al.*, 1987) and rice (Huang *et al.*, 1988), new cultivars have been developed via this technique. It is generally accepted today that a stepwise reduction in the ploidy level of potato offers the advantage of a simpler inheritance and also a better chance to combine qualitatively inherited characters (Wenzel *et al.*, 1979). Data now available demonstrate that field resistance to potato virus Y and potato leaf roll virus is maintained and expressed at all ploidy levels which means that both monogenically and polygenically inherited resistances are maintained during the successive haploidization process. When a genome has passed the monohaploid level, an additive character can be transferred efficiently to the next generation via classical crossing. Similar results have been obtained for polygenic nematode resistance (Uhrig and Wenzel, 1987). In a vegetatively propagated crop, such as potato, the F_1 hybrid can be propagated as a new cultivar.

In *Brassica napus*, quality characteristics are emphasized in breeding techniques with haploids (Keller, 1984; Siebel and Pauls, 1989). With this species, vigour of microspores was positively correlated with high glucosinolate content in the androgenetic regenerants. More than 70% of androgenetic plantlets tested had a glucosinolate content much higher than did the original material (Hoffmann *et al.*, 1982). Since the aim of breeding was for low glucosinolate content, spontaneous selection for this character worked against the programme. Development of microspores with a high level of glucosinolates, being chemically quite similar to phytohormones, may have been favoured. Very low or very high glucosinolate levels are diverse qualities needed either for human consumption or for industry. A simple screening for extremes would increase breeding flexibility at an early stage.

Since regeneration is no longer a main obstacle in using DHs, other questions are gaining importance. Snape and Simpson (1981) have examined the genetic effects when starting from different filial generations. The question asked was: can DHs be used more economically in an F_1 generation, eg., a preselected F_2 or F_3, or should the haploid step be combined with other breeding techniques? It could be demonstrated that one haploid step followed by selection in the greenhouse and in the field during the first androgenetic (A_1) and two subsequent selfed generations (A_2, A_3) is the most efficient procedure, if characters from related varieties are to be combined. Haploids can also increase the efficiency for the most complex breeding problem: the transfer of quantitatively inherited traits from distant genotypes to cultivars. For such complex breeding programmes a combination of recurrent selection, which guarantees variability, and haploid selection, which guarantees secure selection, has been developed. This breeding technique, called recurrent selection alternating with haploid steps, is shown in Fig. 17.6. The degree of relationship of the two parents determines the number of backcrosses needed. The necessary offspring size of DH population for quantitative characters depends on the degree of linkage and on the number of genes involved. The main advantage of using this method is to speed up the breeding process and to make selection of quantitatively inherited traits more effective and reliable (Foroughi-Wehr and Wenzel, 1990).

17.5 PROSPECTS

The use of haploids is a common step in modern breeding programmes. It is irrelevant whether these haploids are produced parthenogenetically or androgenetically. Normally, parthenogenesis is laborious as it demands greenhouse work and, additionally, *in vitro* culture. An exception is potato, where no *in vitro* culture step is necessary. In most cases, microspore androgenesis has been improved so much that the number of green haploid plants necessary for applied work (normally around 100 per donor genotype) can be produced. This is particulary true for most cereals and for

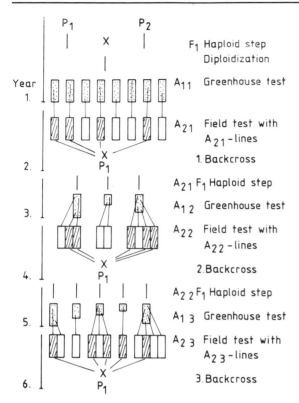

Fig. 17.6. Recurrent selection alternating with haploid steps, a breeding scheme for the introduction of mono- or polygenically inherited traits from distantly related genotypes (Foroughi and Wenzel, 1990).

Brassicaceae. There are strong genotypic differences but, in principle, the procedure is economically effective. In crops where there are still difficulties, e.g., legumes, a concentrated input might solve the problem. Sugar beet, however, is an example of a failure of androgenesis, and parthenogenesis is presently the only promising procedure available.

In those crops where the technical problems of haploid induction are solved, the question of their incorporation into breeding schemes is presently under investigation. When the aim is rapid incorporation of a monogenic trait or combination of monogenic traits, the time gain is beneficial, but much more important are strategies which allow the combination of quantitatively inherited characters. Here, haploids have their biggest potential, as they allow effective selection of polygenic traits in a very early generation. That an A_1 is equivalent to a F_n is probably the most far-reaching advantage of parthenogenesis and androgenesis.

REFERENCES

Ao, G.M., Zhao, S.X. and Li, G.H. (1982), *In vitro* induction of haploid plantlets from unpollinated ovaries of corn (*Zea mays* L.). *Acta Genet. Sin.*, **9**, 281–283.

Asselin de Beauville, M. (1980), Obtention d'haploides *in vitro* á partir d'ovaires non fecondés de Riz, *Oryza sativa* L. *C. R. Acad. Sci.*, **296 D**, 489–492.

Barloy, D., Denis, L. and Beckert, M. (1989), Comparison of the aptitude for anther culture in some androgenetic doubled haploid maize lines. *Maydica*, **34**, 303–308.

Bennett, M.D. and Hughes, W.G. (1972), Additional mitosis in wheat pollen induced by ethrel. *Nature*, **240**, 566–568.

Blakeslee, A.F., Belling, J., Farnham M.E. and Bergner, A.D. (1922), A haploid mutant in the Jimson weed, *Datura stramonium*. *Science*, **55**, 646–647.

Bullock, W.P., Baenziger, P.S., Schaeffer, G.W. and Bottino, P.J. (1982), Anther culture of wheat (*Triticum aestivum* L.) F_1's and their reciprocal crosses. *Theor. Appl. Genet.*, **62**, 155–159.

Cai, D.T. and Zhou, C. (1984), *In vitro* induction of haploid embryoids and plantlets from unpollinated young florets and ovules of *Helianthus annuus* L. *Kexue Tongbao*, **29**, 680–682.

Chen, Y., Zuo, Q., Li, S., Lu, D. and Zheng, S. (1981), Green plant regenerated from isolated rice pollen grains *in vitro* and the induction factors. *Acta Genet. Sin.*, **8**, 158–163.

Chen, Y.R., Lin, T.L. and Huang, Y.F. (1988), Variations in the plastids of albino plants derived from rice anther culture. *Proc. Natl. Sci. Counc. B. Roc.*, **12**, 62–68.

Chen, Z.H., Li, W.B., Zhang, L.H., Xu, X.E. and Zhang, S.J. (1988), Production of haploid plantlets in cultures of unpollinated ovules of *Hevea brasiliensis* Muell. In *Arg. Somatic Cell Genetics of Woody Plants*, Proc. IUFRO Working Party, Grosshandsdorf, 1987, Ahuja, M.R. (ed.), pp. 39–44.

Choo, T.M. (1988), Cross prediction in barley using doubled-haploid lines. *Genome*, **30**, 366–371.

Chu, C.C. (1978), The N6 medium and its applications to anther culture of cereal crops. In *Proc.*

Symp. Plant Tissue Culture, Science Press, Peking, pp. 43–45.

Chuang, C.C., Ouyang, T.W., Chia, H., Chou, S.M. and Ching, C.K. (1978), A set of potato media for wheat anther culture. In *Proc. Symp. Plant Tissue Culture*, Science Press, Peking, pp. 51–56.

Chuong, P.V., Deslauriers, C., Kott, L.S. and Beversdorf, W.D. (1988), Effects of donor genotype and bud sampling in microspore culture of *Brassica napus*. *Can. J. Bot.*, **66**, 1653–1657.

Coumans, M.P., Sohota, S. and Swanson, E.B. (1989), Plant development from isolated microspore of *Zea mays* L. *Plant Cell Rep.*, **7**, 618–621.

Datta, S.K. and Wenzel, G. (1987), Isolated microspore derived plant formation via embryogenesis in *Triticum aestivum* L. *Plant Sci.*, **48**, 49–54.

Datta, S.K. and Wenzel, G. (1988), Single microspore derived embryogenesis and plant formation in barley (*Hordeum vulgare* L.). *Arch. f. Züchtungsforschung.*, **18**, 125–131.

Day, A. and Ellis, T.H.N. (1984), Chloroplast DNA deletions associated with wheat plants regeneration from pollen: Possible basis for maternal inheritance of chloroplasts. *Cell*, **39**, 359–366.

De Buyser, J., Henry, Y., Lonnet, P., Hertzog, R. and Hespel, A. (1987), 'Florin': A doubled haploid wheat variety developed by the anther culture method. *Plant Breeding*, **98**, 53–56.

D'Halluin, K. and Keimer, B. (1986), Production of haploid sugarbeets (*Beta vulgaris* L.) by ovule culture. In *Genetic Manipulation in Plant Breeding*, Horn, W., Jensen, C.J., Odenbach, W. and Schieder, O. (eds), Walter de Gruyter, Berlin, pp. 307–309.

Dunwell, J.M. (1981), Influence of genotype and environment on growth of barley embryos *in vitro*. *Ann. Bot. (London) (N.S.)*, **48**, 535–542.

Dunwell, J.M. (1985), Anther and ovary culture. In *Cereal Tissue and Cell Culture*, Bright, S.W.J. and Jones, M.J.K. (eds), Martinus Nijhoff, Dordrecht, pp. 1–44.

Dunwell, J.M. and Perry, E. (1973), The influence of *in vivo* growth conditions of *N. tabacum* plants on the *in vitro* embryogenetic potential of their anthers. *John Innes Inst. Rep.*, **64**, 69–70.

Fedak, G. (1977), Haploids from barley × rye crosses. *Can. J. Genet. Cytol.*, **19**, 15–24.

Flehinghaus, T., Deimling, S. and Geiger, H.H. (1991) Methodical improvements in rye anther culture. *Plant Cell Rep.* **10**, 397–400.

Foroughi-Wehr, B. (1993) Protokoll für die effiziente Produktion von doppelhaploiden Gerste- und Weizenpflanzen. *Nachrichtenbl. deut. Pflanzenschutzd*, in press.

Foroughi-Wehr, B. and Friedt, W. (1984), Rapid production of recombinant barley yellow mosaic

virus resistant *Hordeum vulgare* lines by anther culture. *Theor. Appl. Genet.*, **67**, 377–382.

Foroughi-Wehr, B. and Wenzel, G. (1990), Recurrent selection alternating with haploid steps a rapid breeding procedure for combining agronomic traits in inbreeders. *Theor. Appl. Genet.*, **80**, 564–568.

Foroughi-Wehr, B. and Zeller, F.J. (1990), *In vitro* microspore reaction of different German wheat cultivars. *Theor. Appl. Genet.*, **79**, 77–80.

Genovesi, A.D. and Collins, G.B. (1982), *In vitro* production of haploid plants of corn via anther culture. *Crop Sci.*, **22**, 1137–1144.

Genovesi, A.D. and Magill, C.W. (1979), Improved rate of callus and green plant production from rice anther culture following cold shock. *Crop Sci.*, **19**, 662–664.

Gonzales, L.G. and Ford-Lloyd, B.V. (1987), Facilitation of wide-crossing through embryo rescue and pollen storage in interspecific hybridization of cultivated *Allium* species. *Plant Breeding*, **4**, 318–322.

Guha, S. and Maheshwari, S.C. (1966), Cell division and differentiation of embryos in the pollen grains of *Datura in vitro*. *Nature*, **212**, 97.

Hagberg, A. and Hagberg, G. (1980), High frequency of spontaneous haploids in the progeny of an induced mutation in barley. *Hereditas*, **93**, 341–343.

He, D.G. and Ouyang, J.W. (1983), Response of wheat anthers at different development stages to *in vitro* cultures. *Ann. Rep. Inst. Genet. Acad. Sinica*, Peking, 1982, p. 30.

Henry, Y. and de Buyser, J. (1985), Effect of the 1B/1R translocation on anther culture ability in wheat (*Triticum aestivum* L.). *Plant Cell Rep.*, **4**, 307–310.

Higgins, P. and Mathias, R.J. (1987), The effect of the 4B chromosomes of hexaploid wheat on the growth and regeneration of callus cultures. *Theor. Appl. Genet.*, **74**, 439–444.

Ho, K.M. and Jones, G.E. (1980), 'Mingo' barley. *Can. J. Plant Sci.*, **60**, 279–280.

Hoffmann, F., Thomas, E. and Wenzel, G. (1982), Anther culture as a breeding tool in rape. II. Progeny analysis of androgenetic lines and of induced mutants from haploid cultures. *Theor. Appl. Genet.*, **61**, 225–232.

Hosemanns, D. and Bossoutrot, D. (1983), Induction of haploid plants from *in vitro* culture of unpollinated beet ovules (*Beta vulgaris* L.). *Plant Breeding*, **91**, 74–77.

Huang, C.S., Tsay, H.S., Chern, C.G., Chen, C.C., Yeh, C.C. and Tseng, T.H. (1988), Japonica rice breeding using anther culture. *J. Agric. Res. China*, **37**, 1–8.

Hunter, C.P. (1985), The effect of anther orientation on the production of microspore-derived embryoids and plants of *Hordeum vulgare* cv. 'Sarbarlis'. *Plant Cell Rep.*, **4**, 267–268.

Illies, Z.M. (1974), Induction of haploid parthenogenesis in aspen by postpollination treatment with toluidine-blue. *Silvae Genet.*, **23**, 167–226.

Jensen, C.J. (1975), Barley monoploids and doubled monoploids: technique and experience. In *Barley Genetics III*, Gaul, H. (ed.), Karl Thiemig, München, pp. 316–345.

Jorgensen, R.B. and Bothmer, R. von (1988), Haploids of *Hordeum vulgare* and *H. marinum* from crosses between the two species. *Hereditas*, **108**, 207–212.

Kao, K.N. (1981), Plant formation from barley anther cultures with Ficoll-media. *Z. Pflanzenphysiol.*, **103**, 437–443.

Kasha, K.J. and Kao, K.N. (1970), High frequency haploid production in barley (*Hordeum vulgare* L.). *Nature*, **225**, 874–876.

Kasperbauer, M.J., Buckner, R.C. and Springer, W.D. (1980), Haploid by anther-panicle culture of tall fescue *Festuca arundinacea*. *Crop Sci.*, **20**, 103–107.

Keller, J. (1990), Haploids from unpollinated ovaries of *Allium cepa* single plant screening, haploid determination, and long term storage. In *Progress in Plant Cellular and Molecular Biology*, Nijkamp, H.J.J., Plas, L.H.W. van der and Aartrijk, J. van (eds), Kluwer Academic Publishers, Amsterdam, pp. 275–279.

Keller, W.A. (1984), Anther culture of *Brassica*. In *Cell Culture and Somatic Cell Genetics of Plants*, Vasil, I.K. (ed.), Academic Press, Orlando, pp. 302–310.

Keller, W.A. and Stringam, G.R. (1978), Production and utilization of microspore-derived haploid plants. In *Frontiers of Plant Tissue Culture*, Thorpe, T. (ed.), University of Calgary Press, Calgary, pp. 113–122.

Kermicle, J.L. (1969), Androgenesis conditioned by a mutation in maize. *Science*, **166**, 1422–1424.

Kim, M.Z. and Raghavan, V. (1988), Induction of pollen plantlets in rice by spikelet culture. *Plant Cell Rep.*, **7**, 560–563.

Knudsen, S., Due, I.K. and Andersen, S.B. (1989), Components of response in barley anther culture. *Plant Breeding*, **103**, 241–146.

Kochba, J. and Spiegel-Roy, P. (1982), Progress in selection for sodium chloride, 2,4-D dichlorophenoxyacetic acid (2,4-D) and streptomycin tolerance in *Citrus sinensis* ovular callus lines. In *Induced Mutations in Vegetatively Propagated Plants II*, Coimbatore, 11–15 Feb. 1980, pp. 77–89.

Köhler, F. and Wenzel, G. (1985), Regeneration of isolated barley microspores in conditioned media and trials to characterize the responsible factor. *J. Plant Physiol.*, **121**, 181–191.

Krüger, H.-U. (1987), Zytologische Charakterisierung androgenetischer Entwicklungsphasen bei Weizen (*Triticum aestivum* L.). *Arch. Züchtungsforch.*, **17**, 297–307.

Kuhlmann, U. (1991), *Mikrosporenregeneration und Untersuchungen zur Transformation bei Gerste.* Dissertation, TU München.

Kuhlmann, U. and Foroughi-Wehr, B. (1989), Production of doubled haploid lines in frequencies sufficient for barley breeding programs. *Plant Cell Rep.*, **8**, 78–81.

Laibach, F. (1929), Das Taubwerden von Bastardsamen und die künstliche Aufzucht früh absterbender Bastardembryonen. *Z. Bot.*, **17**, 417–459.

Laurie, D.A. and Bennett, M.D. (1987), The effect of the crossability loci Kr1 and Kr2 on fertilization frequency in hexaploid wheat × maize crosses. *Theor. Appl. Genet.*, **73**, 403–409.

Laurie, D.A. and Bennett, M.D. (1988), The production of haploid wheat plants from wheat × maize crosses. *Theor. Appl. Genet.*, **76**, 303–397.

Laurie, D.A. and Snape, J.W. (1990), The agronomic performance of wheat doubled haploid lines derived from wheat × maize crosses. *Theor. Appl. Genet.*, **79**, 813–816.

Lazar, M.D., Schaeffer, G.W. and Baenziger, P.S. (1985), The physical environment in relation to high frequency callus and plantlet development in anther culture of wheat (*Triticum aestivum* L.) cv. 'Chris'. *J. Plant Physiol.*, **121**, 103–109.

Lichter, R. (1982), Induction of haploid plants from isolated pollen of *Brassica napus*. *Z. Pflanzenphysiol.*, **105**, 427–434.

Marsolais, A.A., Seguin–Swarts, G. and Kasha, K.J. (1984), The influence of anther cold pretreatments and donor plant genotype on *in vitro* androgenesis in wheat (*Triticum aestivum* L.). *Plant Cell Tissue Organ Culture*, **3**, 69–79.

McCoy, T.J. and Smith, L.Y. (1986), Interspecific hybridization of perennial *Medicago* species using ovule-embryo culture. *Theor. Appl. Genet.*, **71**, 772–783.

Müller, G., Borschel, H., Vahl, U., Wiberg, A., Härtel, H. and Damisch, W. (1989), Die Nutzung der Antherenkulturmethode im Zuchtprozeß von Winterweizen. I. Zur Androgenesefähigkeit von 1B-1R-Weizen-Roggen-Translokationsformen. *Plant Breeding*, **102**, 196–207.

Murashige, T. and Skoog, F. (1962), A revised medium for rapid growth and bio assays with tobacco tissue cultures. *Physiol. Plant.*, **15**, 473–497.

Muren, R.C. (1989), Haploid plant induction from unpollinated ovaries in onion. *HortScience*, **24**, 833–834.

Nitsch, C. (1974), La culture de pollen isolé sur milieu synthetique. *C. R. Acad. Sci.*, **278 D**, 1031–1034.

Nitsch, C., Andersen, S., Godars, M., Neuffer, M.G. and Sheridan, W.F. (1982), Production of haploid plants of *Zea mays* and *Pennisetum* through androgenesis. In *Variability in Plants Regenerated from Tissue Culture*, Earle, E.D. and Demarly, Y.Y. (eds), Preager, New York, pp. 69–91.

Nitsch, C. and Norreel, B. (1973), Effet d'un choc thermique sur le pouvoir embryogene du pollen de *Datura innoxia* cultivé dans l'anthère ou isolé de l'anthère. *C. R. Acad. Sci.*, **276 D**, 303–306.

Ouyang, J.W., He, D.G., Feng, G.H. and Jia, S.E. (1987), The response of anther culture to culture temperature varies with growth conditions of anther-donor plants. *Plant Sci.*, **49**, 145–148.

Ouyang, J.W., Zhou, S.M. and Jia, S.E. (1983), The response of anther culture to culture temperature in *Triticum aestivum*. *Theor. Appl. Genet.*, **66**, 101–109.

Pareddy, D.R. and Greyson, R.J. (1985), *In vitro* culture of immature tassels of an inbred field variety of *Zea mays* cv. 'Oh43'. *Plant Cell Tissue Organ Culture*, **5**, 119–128.

Pescitelli, S.M., Mitchell, J.C., Jones, A.M., Pareddy, D.R. and Petolini, J.F. (1989), High frequency androgenesis from isolated microspore of maize. *Plant Cell Rep.*, **7**, 6773–6776.

Petolino, J.F., Jones, A.M. and Thompson, S.A. (1988), Selection for increased anther culture response in maize. *Theor. Appl. Genet.*, **76**, 157–159.

Powell, W. (1988), The influence of genotype and temperature pre-treatment on anther response in barley (*Hordeum vulgare* L.). *Plant Cell Tissue Organ Culture*, **12**, 291–297.

Raquin, C. (1983), Utilization of different sugars as carbon sources for *in vitro* cultures of *Petunia*. *Z. Pflanzenphysiol.*, **111**, 453–457.

San Noeum, L.H. (1976), Haploides d'*Hordeum vulgare* L. par culture *in vitro* d'ovaires non fécondés. *Ann. Amélior. Plantes*, **26**, 751–754.

Shannon, P.R.M., Nicholson, A.E., Dunwell, J.M. and Davies, D.R. (1985), Effect of anther orientation on microspore-callus production in barley (*Hordeum vulgare* L.). *Plant Cell Tissue Organ Culture*, **4**, 271–280.

Siebel, J. and Pauls, K.J. (1989), A comparison of anther and microspore culture as a breeding tool in *Brassica napus*. *Theor. Appl. Genet.*, **78**, 473–479.

Sitbon, M. (1981), Production of haploid *Gerbera jamesonii* plants by *in vitro* culture of unfertilized ovules. *Agronomie*, **1**, 807–817.

Sitch, L.A., Snape, J.W. and Firman, S.J. (1985), Intrachromosomal mapping of crossability genes in wheat (*Triticum aestivum*). *Theor. Appl. Genet.*, **70**, 309–314.

Snape, J.W., Chapman, V., Moss, J., Blanchard, C.E. and Miller, T.E. (1979), The crossabilities of wheat varieties with *Hordeum bulbosum*. *Heredity*, **42**, 291–298.

Snape, J.P. and Simpson, E. (1981), The genetical expectations of doubled haploid lines derived from different filial generations. *Theor. Appl. Genet.*, **60**, 123–128.

Snape, J.W., Sitch, L.A., Simpson, E. and Parker, B.B. (1988), Tests for the presence of gametoclonal variation in barley and wheat doubled haploids produced using the *Hordeum bulbosum* system. *Theor. Appl. Genet.*, **75**, 509–513.

Sopory, S.K. and Maheshwari, S.C. (1976), Development of pollen embryoids in anther cultures of *Datura innoxia*. I. General observation and effects of physical factors. *J. Exp. Bot.*, **27**, 49–57.

Sorvari, S. (1986), Comparison of anther cultures of barley cultivars in barley-starch and agar gelatinized media. *Ann. Agr. Fenn.*, **25**, 249–254.

Speckmann, G.J., Van Geyt, J.P.C. and Jacobs, M. (1986), The induction of haploids of sugar beet (*Beta vulgaris* L.) using anther and free pollen culture or ovule and ovary culture. In *Genetic Manipulation in Plant Breeding*, Horn, W., Jensen, C.J., Odenbach, W. and Schieder, O. (eds), Walter de Gruyter, Berlin, pp. 351–353.

Stolarz, A. (1974), The induction of androgenesis in pollen grains of *Secale cereale* L. Strzekeciriskie Jare *in vitro* conditions. *Hodowla Rosl.*, *Aklim. Nasienn.*, **18**, 217–220.

Sunderland, N. (1974), Anther culture as a means of haploid induction. In *Haploids in Higher Plants: Application and Potential*, Kasha, K.J. (ed.), University of Guelph, Guelph, pp. 91–122.

Sunderland, N. and Dunwell, J.M. (1977), Anther and pollen culture. In *Plant Tissue and Cell Culture.*, Street, H.E. (ed.), University of California Press, Berkeley, pp. 223–264.

Sunderland, N. and Evans L.J. (1980), Multicellular pollen formation in cultured barley anthers. II. The A-pathway, B-pathway and C-pathway. *J. Exp. Bot.*, **31**, 501–514.

Sunderland, N. and Xu, Z.H. (1982), Shed pollen

culture in *Hordeum vulgare. J. Exp. Bot.*, **33**, 1086–1095.

Sunderland, N., Xu, Z.H. and Huang, B. (1981), Recent advances in barley anther culture. In *Barley Genetics IV*, Proc. Fourth Int. Barley Genetics Symp., Edinburgh University Press, Edinburgh, pp. 699–703.

Sunderland, N., Roberts, M., Evans, L.J. and Wildon, D.C. (1979), Multicellular pollen formation in cultured barley anthers. I. Independent division of the generative and vegetative cells. *J. Exp. Bot.*, **30**, 1133–1144.

Tsay, H.S., Miao, S.H. and Widholm, J.M. (1986), Factors affecting haploid plant regeneration from maize anther culture. *J. Plant Physiol.*, **126**, 33–40.

Tuvesson, I.K.D., Pedersen, S. and Andersen, S.B. (1989), Nuclear genes affecting albinism in wheat (*Triticum aestivum* L.) anther culture. *Theor. Appl. Genet.*, **78**, 879–88.

Uhrig, H. (1985), Genetic selection and liquid medium conditions improve the yield of androgenetic plants from diploid potatoes. *Theor. Appl. Genet.*, **71**, 455–460.

Uhrig, H. and Salamini, F. (1987), Dihaploid plant production from 4*x*-genotypes of potato by the use of efficient anther plants producing tetraploid strains (4*x* EAPP–clones) – proposal of a breeding methodology. *Plant Breeding*, **98**, 228–235.

Uhrig, H. and Wenzel, G. (1987), Breeding for virus and nematode resistance in potato through microspore culture. In *Biotechnology in Agriculture and Forestry Vol. III*, Bajaj, Y.P.S. (ed.), Springer Verlag, Berlin, pp. 346–357.

Vasil, I.K. (1980), Androgenetic haploids. *Int. Rev. Cytol.*, Suppl. 11B, 195–217.

Vaughn, K.C., DeBonte, L.R., Wilson, K.G. and Schaeffer, G.W. (1980), Organelle alteration as a mechanism for maternal inheritance. *Science*, **206**, 196–198.

de Vaulx, R.D. and Chambonnet, D. (1986), Obtention of embryos and plants from *in vitro* culture of unfertilized ovules of *Cucurbita pepo*. In *Genetic Manipulation in Plant Breeding*, Horn, W., Jensen, C.J., Odenbach, W. and Schieder, O. (eds), Walter de Gruyter, Berlin, 295–297.

Wan, Y., Petolino, J.F. and Widholm, J.M. (1989), Efficient production of doubled haploid plants through colchicine treatment of anther-derived maize callus. *Theor. Appl. Genet.*, **77**, 889–892.

Wang, C., Sun, C.S., Chu, C.C. and Wu, S.C. (1978), Studies on the albino pollen plantlets of rice. In *Proc. Symp. Plant Tissue Culture*, Science Press, Peking, pp. 149–159.

Wang, Ch., Sun, S. and Chu, Z.-Ch. (1975), The induction of *Populus* pollen-plants. *Acta Bot. Sin.*, **17**, 56–59.

Wei, Z.M. (1982), Pollen callus culture of *Triticum aestivum*. *Theor. Appl. Genet.*, **63**, 71–73.

Wenzel, G. (1980), Recent progress in microspore culture of crop plants. In *The Plant Genome*, Davies, D.R. and Hopwood, D.A. (eds), John Innes Charity, Norwich, pp. 185–196.

Wenzel, G. and Foroughi-Wehr, B. (1984), Anther culture in cereal and grasses. In *Cell Culture and Somatic Cell Genetics of Plants*, Vasil, I.K. (ed.), Academic Press Inc., pp. 311–327.

Wenzel, G. and Uhrig, H. (1981), Breeding for nematode and virus resistance in potato via anther culture. *Theor. Appl. Genet.*, **59**, 333–340.

Wenzel, G., Hoffmann, F. and Thomas, E. (1977), Increased induction and chromosome doubling of androgenetic haploid rye. *Theor. Appl. Genet.*, **51**, 81–86.

Wenzel, G., Schieder, O., Przewozny, T., Sopory, S.K. and Melchers, G. (1979), Comparison of single cell culture dederived *Solanum tuberosum* L. plants and a model for their application in breeding programs. *Theor. Appl. Genet.*, **55**, 49–55.

Wilson, H.M. (1977), Culture of whole barley spikes stimulates high frequencies of pollen calluses in individual anthers. *Plant Sci. Lett.*, **9**, 233–238.

Yang, H.Y. and Zhou, C. (1982), *In vitro* induction of haploid plants from unpollinated ovaries and ovules. *Theor. Appl. Genet.*, **63**, 97–104.

Yang, J.S., Sohn, J.K. and Chung, G.S. (1985), Effects of low temperature treatment on culturability and ploidy level in rice anther culture. *Korean J. Breed.*, **17**, 344–347.

Yüce, S. (1973), *Haploide bei der Zuckerrübe.*, Dissertation, University Gießen.

Zhou, C. and Yang, H.Y. (1981), Induction of haploid rice plantlets by ovary culture. *Plant Sci. Lett.*, **20**, 231–237.

Zhu, Z. and Wu, H. (1979), *In vitro* production of haploid plantlets from the unpollinated ovaries of *Triticum aestivum* and *Nicotiana tabacum*. *Acta Genet. Sin.*, **5**, 181–183.

Zhu, Z.C., Wu, H.S., An, Q.K. and Liu, Z.Y. (1981), Induction of haploid plantlets from unpollinated ovaries of *Triticum aestivum* culture *in vitro*. *Acta Genet. Sin.*, **8**, 386–389.

Part Six

Selection Methods

18

Selection strategies and choice of breeding methods

J. Moreno-González and J.I. Cubero

18.1 INTRODUCTION

The shift from a hunting and gathering economy to agriculture was possible because natural and artificial modifications in wild species allowed for a lasting interaction between biological organisms and man in such a way that they became dependent on each other. The placing of plants and animals in a disturbed environment provided the selective pressure required to produce new forms, well differentiated from the wild stock. This was named automatic selection by Harlan (1975) who described the conditions for its existence in crop plants. Very briefly, harvesting and repeated sowing of the harvested material (not collecting any more from the wild) was a powerful filter for selecting mutants of agricultural value such as absence of dormancy, synchronous flowering and maturity, large seeds, hard spikes in cereals, indehiscence in legumes and better regeneration ability in vegetatively propagated crops.

Automatic selection was unconsciously practised for millennia. Most of our present crops and landraces were domesticated and developed by this effective method. However, selection is consciously and 'methodically' practised nowadays, not only when a wild plant needs to be domesticated but also when some specific traits are sought (e.g., photoperiod insensitivity in wheat, uniform maturity in tomato, resistance to low temperatures in winter cereals, etc.). Ancient farmers only practised mass or bulk techniques of selection. They left very few descriptions of their selection criteria. Roman agronomists (Columela, around 60 AD) recommended picking of the biggest wheat ears and the largest kernels for getting a good harvest the next season.

Systematic selection of plants was carried out by agricultural experimental stations of Europe and America long before Mendel (1866) established the basis of genetics. The scientific basis of plant breeding was established soon after the rediscovery of Mendel's work in 1900. Johannsen (1903) correctly interpreted the results of selection in a self-fertilized population of beans by distinguishing selection within a homozygous line from selection among segregating homozygous genotypes in the original population, which permitted him to clarify the differences between environmental and genetic effects. Resistance to yellow rust (*Puccinia striiformis*) in wheat was demonstrated by Biffen (1905) to be controlled by a single recessive factor inherited in a Mendelian manner. The methods for developing and producing maize inbred lines and hybrids, which Shull (1909) described, constitutes the basis of commercial seed production in many crops.

More complicated genetic models were later postulated to explain the behaviour and the inheritance of traits that show continuous

Plant Breeding: Principles and prospects. Edited by M.D. Hayward, N.O. Bosemark and I. Romagosa. Published in 1993 by Chapman & Hall, London. ISBN 0 412 43390 7.

Table 18.1. Estimates of total genetic gains and regression coefficients for maize, soybean and wheat grain yield when testing hybrids or varieties from different eras from five authors' studies (adapted from the references)

Reference	Crop	Genetic gains		Time period (year)	Testing characteristics		
		Total kg/ha	Regression coefficient kg/ha yr		Varieties (no.)	Densities (no.)	Location
Russell (1984)	Maize	2700	54	1930–80	24	3	Iowa
Duvick (1977)	Maize	2060	52	1930–70	69	3	Iowa
Tollenar (1989)	Maize	3880	130	1959–88	9	4	Ontario
Cox et al. (1988)	Wheat	1106	16.3	1918–87	38		Kansas
Spetch and Williams (1984)	Soybean	1410	18.8	1902–77	240		Nebraska

variation and are influenced by environmental conditions. The application of mathematics to these models by Fisher (1918), Wright (1921), Haldane (1924) and Kempthorne (1969) led to the establishment of the theoretical basis of quantitative genetics. Quantitative genetic theory was first used by animal breeders (Lush, 1931, 1933, 1947; Dickerson and Hazel, 1944) as a tool to overcome the inherent limitations due to long generation intervals and low flexibility in the mating systems of most animal species under selection. Later on, quantitative genetics provided the necessary knowledge to plant breeding researchers (Harrington, 1937; Harlan *et al.*, 1940; Immer, 1941; Hull 1945; Sprague, 1946; Comstock *et al.*, 1949; Lonnquist, 1964) to formulate alternative breeding strategies and selection methods that can cope with different breeding situations. It also provided the basis for estimation of genetic parameters and prediction of genetic gains.

The choice of a selection method will determine the effectiveness of reaching a given goal, even sacrificing the purity of the method to be applied.

The objectives of the present Chapter are: (1) to discuss genetic concepts that may help breeders to choose appropriate breeding strategies for different situations; (2) to discuss the different forms of plant material available and their possible use in breeding programmes; (3) to describe and to compare different selection schemes for developing useful cultivars.

18.2 MODERN CROP CULTIVARS

Most varieties used in modern agriculture were developed by applying classical plant breeding methods to appropriate genetic materials. These methods are called classical to differentiate them from others such as those exploiting genetic engineering, molecular genetics or biotechnology. Significant genetic improvement of many crop cultivars has been achieved by classical methods during the last 50–70 years. The genetic change has been assessed by comparing commercial varieties from different eras in common experiments. The studies of Russell (1984), Duvick (1977) and Tollenar (1989) showed, by applying linear regression analysis to performance on to the year of release, that the average annual genetic gain of hybrid maize grain yield during the last 50–60 years was about 75 kg/ha (Table 18.1). The work of Cox *et al.* (1988) showed that the average annual genetic gain in grain yield for red winter wheat was 16.2 kg/ha or about 1.1% per year, over the last 70 years (Table 18.1). Significant genetic gains over the years have also been reported for other crops such as cotton, soybean and sorghum (Fehr, 1984).

18.3 FACTORS AFFECTING THE CHOICE OF BREEDING STRATEGY

A breeding programme set up to improve certain characteristics in a crop cannot be considered as an isolated research activity, but needs to be linked to other areas of research on that crop and related crops. In general, plant breeding programmes are costly and lengthy. Results of these programmes, often condensed in an improved variety, take several years to reach fruition. Thus, before starting a breeding programme on a specific crop, it is necessary to analyse the problem in depth to minimize the risks of wrong choices and unexpected results. The analysis should focus on the objectives and goals of the work as well as the methods to be used to reach these goals.

18.3.1 Strategic perspective

Some of the strategic perspectives to be taken into account when planning a breeding programme include the following.

1. What will be the impact of a plant breeding programme on the agricultural development of a region? Will it result in significant changes in the acreage of this crop or related crops? As an example, the introduction of dwarf and semi-dwarf varieties of wheat from the breeding programme of the International Center for Maize and Wheat Improvement (CIMMYT) significantly increased the acreage, the yield and the total production of this crop in India, Pakistan, and Middle East, which simultaneously caused the displacement of other varieties and crops from their original growing areas. A similar pattern, although not so striking, occurred with the rice breeding programme from the International Rice Research Institute (IRRI) in the Far East (Swaminathan, 1986).

2. What will be the economical, social and environmental impact derived from the proposed breeding programme? Will the farming conditions be changed in a certain area? The uniformity and homogeneity of varieties derived from a modern breeding programme will allow better mechanization and commercialization of the crop, but at the same time they may also mean a change from crop diversity to a monoculture. In addition, new varieties will usually need better growing conditions and larger amount of inputs such as fertilizers, irrigation, chemicals and weed control. These are important factors that interact with regional agricultural and social policies and environmental planning.

3. The objectives of the breeding programme should be realistic. Since the results of the programme will become available only after several years, the breeder should foresee under what environmental conditions the improved variety will be grown. For example, how much nitrogen will farmers use for the improved crop 10 to 15 years hence? The testing environments should be similar to the future growing environments. The problem becomes more complex for horticultural products, because the breeder needs to know in advance what will be the taste of the consumers when the variety will be released. Therefore, a breeding programme cannot be designed independent of other research activities such as cultural practices, agroindustrial changes and commercialization. It needs to be integrated in to an 'agricultural package'.

4. It should be realized that plant breeding is both a science and an art. Books contain descriptions of methods, but they rarely contain the biological, statistical and logistic details to be taken into account before deciding the main guidelines of a breeding programme. Scientific knowledge is necessary, but experience and a well-trained team of collaborators are also essential to achieve success in this field. As an example, the scientific basis and the description of a suitable ideotype to be selected are not usually found in books.

5. The amount of resources allocated to the breeding programme should be adjusted to the magnitude of the programme and the expected results (Chapter 30).

18.3.2 Factors affecting the selection methods

(a) Reproductive system of the plants

Breeding methods need to generate quickly and easily many distinct genotypes for evaluation.

Knowledge of the reproductive system is essential to develop efficient techniques of outcrossing and selfing in the breeding material. Such mechanisms determine not only the choice of the breeding method but also the seed production methods of the commercial varieties.

In general, self-pollinating (autogamous) plants cannot be easily outcrossed to other plants. Their natural sexual barriers to outcrossing, either morphological or functional, will have to be overcome by artificial techniques such as emasculation, or genetic male sterility. In addition, autogamous plants are homozygous. Therefore, they do not make use of the dominance effects of genes at individual loci. Usually, varieties of autogamous plants are pure lines, or multiline varieties whose seed is commercially maintained and produced by self-pollination. However, commercial hybrids can also be produced in certain self-pollinating crops (wheat, tomato, tobacco, cotton, etc.).

As cross-pollinating (allogamous) plants naturally cross, it is generally fairly easy to make controlled crosses either by hand or by exposing female plants to the desired pollen. Unwanted and uncontrolled crosses can be prevented by bagging the flowers of the female plants. Heterozygosity is a common feature in allogamous plants, thus dominance effects may also contribute to the performance of the traits in these plants. From a theoretical point of view, dominant unfavourable alleles for a trait tend to be eliminated from the genome because of natural or artificial selection against them when they are expressed in both the homozygous and heterozygous state; consequently, recessive favourable alleles tend to be fixed. On the contrary, recessive unfavourable alleles are not easily eliminated because they may escape selection when they are in the heterozygous state; consequently, dominant favourable alleles are also not easily fixed and are found segregating in populations. These reasonings are in agreement with the complementary theory to explain heterosis in plants, which requires a large proportion of recessive favourable alleles to be fixed in both parents and a large proportion of dominant favourable alleles to be dispersed in both parents (Chapter 2;

Bruce, 1910; Jones, 1917, 1945, 1958; Falconer, 1981).

Breeding methods of vegetatively propagated plants generate a large amount of variability, which may be in the form of homozygous or heterozygous genotypes, for evaluation in a segregating population. Selection of desirable genotypes under correctly defined environmental conditions will identify the best clones for use as commercial varieties. Because each commercial variety is vegetatively propagated from a single selected genotype, then the additive, dominance and epistatic effects are all inherited in that variety.

Environmental factors may occasionally interact with the reproductive system of the plants as in facultative apomixis and percentage of partial allogamy in some crops. As an example, landraces and cultivars of faba bean (a partial allogamous plant) from Central Europe frequently show outcrossing rates of 70% or more. On the other side, faba bean behaves as a strict selfer in Southern Sudan. The photoperiod, the temperature, the insect population and some other circumstances can explain the shift.

Since the methods and the breeding products are different for autogamous and allogamous plants one may ask, what are the acceptable limits of outcrossing for autogamy and allogamy in selection programmes? A maximum rate of 5% outcrossing may be accepted with selection methods set up for selfers. For allogamous species, there is more flexibility since any inbred progeny are not good competitors and most of the selected plants in a population will thus come from outcrossing. Therefore, an outcrossing rate of 75–80% is sufficient for handling selection material as if it were fully allogamous. However, these are approximate figures that depend on the traits being selected.

(b) Heterosis

Heterosis is a common phenomenon in allogamous species and has also been reported in some self-pollinating crops such as soybean (Hillsman and Carter, 1981; Nelson and Bernard, 1984), tomato, wheat, etc. The level of heterosis present for the trait to be selected must be taken into account when considering the choice of selection method. Heterosis was first successfully used in the development of hybrids in maize (Shull, 1909). Commercial production of hybrid seed is now being widely used in several crops, both allogamous and autogamous. A principle of commercial seed production is that it must be simple and cost effective; thus higher seed production costs must be compensated for by a higher yield in the hybrids. It is easy to produce hybrid seed in maize by planting alternate rows of males and females which are emasculated. It is also easy in asparagus because of the dioecious differentiation of male and female plants. In other crops, such as beet, sunflower, pepper, wheat, rice, etc., large-scale emasculation is not possible and it is necessary to introduce one or another kind of male sterility (Chapter 14). The high production costs of tomato hybrid seed based on hand emasculation are acceptable only because of the high price and the small quantities of hybrid seed required.

(c) Cytogenetic structure

The ploidy level can influence the expression of a particular character in such a way that it can affect the breeding strategy. Tetraploidy is commonly used in forage crops to increase dry matter yield and expression of other favourable traits (e.g., red clover, perennial and italian rye grass, orchard grass, etc.).

Selection for certain traits of commercial triploid and tetraploid crops can be more effective if performed at the diploid level before returning to their polyploid level (e.g., bolting resistance in sugar beet, disease resistance and combining ability in potato, etc.). As discussed in Chapter 6, alloploidy is useful for either creating new species such as triticale (wheat × rye) and tritordeum (wheat × barley) or transferring genes from the wild to the cultivated genotypes. Induced autopolyploidy can also influence other plant mechanisms such as the incompatibility system in species with gametophytic incompatibility (see Chapter 13).

(d) Differentiation between qualitative and quantitative traits

Selection methods are different when applied to qualitative and quantitative traits. Qualitative traits show simple Mendelian inheritance, controlled by one or two genes. They show visual or measurable differences between classes of the segregating genotypes. Some examples of qualitative traits are kernel, flower and leaf colours, plant size (dwarf *vs* normal type), specific resistance to diseases, amino acid content in kernels, fruit shape (pointed *vs* inverted-blunt in *Capsicum*), etc. Introduction of desirable qualitative characteristics in a target cultivar is usually carried out by backcross selection procedures.

As considered in Chapters 2 and 11, quantitative traits show continuous variation in their phenotypic expression and segregants cannot be grouped into distinct classes. Such traits have a complex inheritance and are controlled by many genes with small effects that are influenced by the environment and interaction with other loci. Some examples of quantitative traits are grain yield, total dry matter, grain protein content, seed oil content, general resistance to diseases, etc. The breeding methods used in quantitative traits are cyclic or recurrent. They are based on selecting in each cycle genotypes that show favourable combinations of genes. If many genes are involved, it is very unlikely to get at random an optimum gene combination in only one cycle of selection. Thus, it will be necessary to recombine the selections to increase the probability of getting more favourable combinations in the next cycles. In this way, recurrent selection procedures will lead to a small, but steady genetic gain to the trait in each cycle.

(e) Gene action

If the alleles at an individual locus have exclusively additive effects for the three possible genotypes, the heterozygote and the two homozygotes, then they are said to have additive gene action. If the alleles have a supplementary dominance effect in the heterozygote, then they contribute to the dominance variance and it can be said that they have dominance gene action. The following model was proposed (Falconer, 1981) to interpret mathematically the action of the genes: genotypes *BB*, *Bb* and *bb* have genotypic values *a*, *d* and *–a*, respectively; where *B* and *b* refer to the favourable and unfavourable alleles, respectively; *a* and *d* are the additive and dominance effects, respectively. The same model was proposed by Mather and Jinks (1971) but with a different notation, where *d* and *h* are the corresponding additive and dominance values, respectively: the Falconer notation will be used in this chapter. If gene action is additive, then *d* is zero. On the other hand, if there is dominance gene action, then *d* will be different from zero. The more the *d* value deviates from zero the more important the dominance gene action will be.

The kind of breeding procedure chosen will depend on the prevalence of one or the other type of gene action in the breeding material under selection. If additive effects are the most important, the selection product will be a variety, pure line or strain to be improved within a population. Intrapopulation selection methods will be effective in accumulating favourable alleles in breeding materials with additive gene action. Unfavourable alleles can be eliminated when they show up in both the homozygous and the heterozygous state. Self-pollinating crops have generally a prevalence of additive effects; e.g., the additive model accounted for almost the whole genetic variance in some soybeans studies (Horner and Weber, 1956; Gates *et al.*, 1960; Brim and Cockerham, 1961). If dominance effects are important, hybrid varieties will show the highest performance. The most appropriate selection methods under these circumstances will be those that take advantage of heterosis and the general and specific combining abilities. Interpopulation selection methods will be effective in accumulating favourable gene combinations in both the homozygous and the heterozygous states. Recessive unfavourable alleles can only be eliminated when they show up in the homozygous state. Therefore, selfing that increases homozygosity and allows elimination of unfavourable recessive alleles, alternating with crossing that recovers heterozygosity and allows expression of heterosis should be a good breeding strategy in this case.

(f) Heritability

The knowledge of the phenotypic variance of a trait and its separation into genetic and environmental components is useful for helping breeders to design an effective selection method. The heritability parameter (h^2) has been mathematically defined as the ratio between the additive variance (σ_A^2) and the phenotypic variance (σ_P^2) (Chapter 11):

$$h^2 = \frac{\sigma_A^2}{\sigma_P^2} \qquad (18.1)$$

where,

$$\sigma_P^2 = \sigma_G^2 + \frac{\sigma_{G\times E}^2}{e} + \frac{\sigma_E^2}{r\times e} \qquad (18.2)$$

where σ_G^2, $\sigma_{G\times E}^2$ and σ_E^2 are genetic, genetic × environmental and environmental (experimental error) variances, respectively; *e* and *r* are numbers of environments and replications, respectively. Heritability has also been defined as the linear regression coefficient of mid-parents on offspring and it can be used for predicting expected genetic gain. It means that only a part of the total apparent variation existing in a generation is genetically transmitted to the next one. Heritability is a useful parameter for comparing and improving the efficiency of selection methods. Reduction of the environmental error and genetic × environmental components in the phenotypic variance (Equation 18.2) can be accomplished by improving the experimental techniques (it reduces σ_E^2) and by increasing the number of replications (*r*) and the environments (*e*) in the selection experiments. Selection units can only be replicated if they are families but not if individual plants unless they can be clonally propagated. Thus, for traits with low heritability the genetic gain will increase if they are selected on a family basis.

(g) Classical vs special methods

A decision has to be made to choose either a classical (selection, testing, crosses, selfing,

Table 18.2. Germplasm sources of maize inbred line development in 36 seed companies from a 1984 survey in the US Corn Belt (adapted from Darrah and Zuber, 1986)

Source	Unweighted crude data (%)	Weighted for company production (%)
Single crosses	41.8	20.1
Three-way or double crosses	4.4	3.9
Backcrosses	20.1	4.6
Local varieties	1.7	0.1
Exotics or semiexotics	9.3	4.5
Synthetics	11.2	5.9
Recurrent improved populations	7.8	7.3
Other	3.6	53.6

polyploidy, mutation) or a special (tissue culture, genetic engineering, genetic markers) method, or to combine both kinds of methods, which is the most likely avenue for future plant breeding. The choice will depend not only on the required time to reach the objectives but also on the available facilities, researcher experience and training.

18.4 PLANT MATERIALS

Choice of the right breeding material either for immediate use, or for long-term population improvement programmes, is a crucial and important decision that may determine the success and the results of any breeding programme in the future. The results of a study showed that only 17 good maize inbred lines were developed by the pedigree method from 1521 samples of open-pollinated varieties (Moreno-González, 1988). Furthermore, six of the lines traced back to the same local variety 'Aranga' which was sampled in different farms of the region in four different years. It seems that the 'Aranga' variety has a favourable gene pool for line development and its choice among the others is not by chance. Darrah and Zuber (1986) summarized a survey of 36 maize seed companies from the US Corn Belt about the use of maize germplasm. They reported that the major source of germplasm for development of inbred lines was the single crosses (Table 18.2) that seem to contain the most favourable gene combinations. Therefore, a careful choice of the breeding material is an important matter to

consider before starting a breeding programme. Plant material can be first classified into adapted materials, exotic materials and wild relatives.

18.4.1 Adapted plant materials

Adapted plant materials are those that can be immediately cultivated in the region or area for which the breeding programme was designed. Depending on the degree of selection undergone, different types can be distinguished.

1. Landraces are local varieties developed by farmers over many generations of selection without applying plant breeding procedures. It was the material first used in the early stages of plant breeding. Many of the landraces, which were used in the past, are today conserved in gene banks and are still useful for new breeding programmes. Some examples of this material are the well-known local varieties of maize 'Reid Yellow Dent' and 'Lancaster Sure Crop' (Hallauer, 1990), the wheat cultivars 'Turkey' and 'Kharkof' (Cox *et al.*, 1988), the oats varieties 'Kherson' and 'Red Rustproof' (Poehlman, 1979) and the sunflower cultivar 'Peredovik'.

2. Synthetics or composites are developed by breeders by recombining selected lines, cultivars or local varieties. They mark a second step in the use of breeding material as sources of second cycle lines or cultivars. Some examples are 'Stiff Stalk Synthetic' in maize (Hallauer *et al.*, 1983) and population

'P21' selected from the sunflower cultivar 'Peredovik'.

3. Recurrent selection populations derive from long-term breeding programmes by evaluating genotypes and recombining the selected ones. They should have a high frequency of favourable genes and enough variability for being adapted to a wide range of environments. This allows the derivation of superior lines and commercial cultivars.

4. Elite lines developed from selected breeding material are used as refined and adapted stocks to develop new improved second or third cycle lines by backcrossing.

5. Selected crosses originated from elite lines are widely used to develop new inbred lines by the pedigree method (Table 18.2).

18.4.2 Exotic plant materials

Exotic materials cannot be directly used in the area for which the breeding programme is being carried out. To obtain positive effects from the exotic populations, they need to be selected for adaptability (Crossa *et al.*, 1987; Mungoma and Pollak, 1988). Because of its low grain yield, exotic maize germplasm has not been an important source for new lines in the US Corn Belt, except for specific traits such as resistance to European corn borer from 'Maíz Amargo' from Argentina (Hallauer, 1990). However, many authors have emphasized the possibilities of transferring favourable alleles from exotic to adapted germplasm (Zuber, 1975; Dudley, 1982; Goodman, 1984). Theoretical studies by Crossa (1989) have estimated the probability of fixation of favourable alleles of exotic populations into adapted populations for different population sizes and different segregating generations (F_2, backcrosses).

18.4.3 Wild relatives

Wild relatives are species that either have shared common ancestors with a crop or are themselves the ancestors of the crop, remaining in the wild under natural selection. They can be utilized for transferring useful genes from the wild to adapted crops. There are many reports in the literature that emphasize the success of such transfer. Genes controlling high soluble solids in the wild species *Lycopersicon chmielewskii* have been transferred to the tomato crop *L. esculentum* (Rick, 1974; Osborn *et al.*, 1987). The grassy stunt virus was a serious problem in the rice crop of South and Southeast Asia during the late 1960s to the late 1970s. Resistance to this disease was found in its relative *Oryza nivara* and later incorporated into the high-yielding variety 'IR36' (Hoyt, 1988). Genes for rust resistance of wheat have been found in *Agropyron elongatum*. Results of Nault *et al.* (1982) show that *Zea diploperennis* has genes for immunity to several maize viruses. Because *Z. diploperennis* is interfertile with the maize crop, it offers the potential to contribute to maize genes for immunity or tolerance to virus diseases. Resistance to powdery mildew has been transferred from *Helianthus debilis* to cultivated sunflower (Jan and Chandler, 1985). Genes for *Cercospora* resistance have also been transferred from *Beta maritima* to sugar beet.

18.5 THE ENVIRONMENTS AND THEIR INTERACTIONS WITH THE GENOTYPES

Depending on the size and magnitude of the environments, two types can be distinguished: microenvironments and macroenvironments. Microenvironment is the space surrounding a plant which provides it with nutrients, water, light, temperature and physical space for its growth. Knowledge of the microenvironmental effects is important when selecting individual plants, as in mass selection. For this kind of selection, one of the important factors to be controlled is the plant density and the plant-to-plant distance to minimize the effects of differences in competitiveness among plants. When evaluating genotypes on a plot basis, the environmental variance or experimental error (σ_E^2) from Equation 18.2 has two components:

$$\sigma^2_E = \sigma^2_e + \frac{\sigma^2_{we}}{n} \qquad (18.3)$$

where σ_e^2 is the plot-to-plot environmental

variance due to plot environmental random effects and experimental technique random effects; σ^2_{we} is the plant-to-plant within-plot variance due to plant microenvironmental random effects and within-family genetic effects; and n is the number of plants per plot. When selecting individual plants, n equals one. Thus σ^2_{we} can be further divided as follows:

$$\sigma^2_{we} = \sigma^2_{me} + \sigma^2_{wf} \qquad (18.4)$$

where σ^2_{me} is the plant-to-plant microenvironmental variance and σ^2_{wf} is the within-family genetic variance.

The term macroenvironment refers to the growing conditions given at a specific location in a year, such as temperature, water supply, fertilization, light and soil type. Further on, in this chapter, the term environment will be used instead of macroenvironment.

Evidence of the importance of genotype × environment (G×E) interaction has been presented for many crops. One of the problems that breeders find in attempts to select the best genotypes is the presence of this genotype × environment interaction (Finlay and Wilkinson, 1963; Eberhart and Russell, 1966; Hill and Baylor, 1983; Moreno-González, 1986). Breeders try to develop cultivars not only with high performance but also with good stability over environments. Therefore, knowledge of the magnitude of the G×E interaction is important to make decisions on the choice of the selection scheme for developing new cultivars (Chapter 22 and Chakroun *et al.*, 1990). If interactions are very large, it is advisable to subdivide the area for which the breeding programme was first aimed into subareas, and to apply a specific breeding programme to each of them; costs and expected results should be considered when making the decision.

Different approaches have been suggested to minimize this G×E interaction.

1. Comstock and Moll (1963) showed that increasing the number of environments (locations and years) and replications in the selection experiments will reduce the phenotypic variance and consequently increase the genetic gain. However, this requires considerable effort and cost.

2. Moreno-González (1986) proposed an alternative method of reciprocal recurrent selection in which populations A and B are each selected in a different environment. The method was specially designed for a hybrid programme to be applied to an area where interactions arise from different kinds of environmental factors and proved to be more efficient than the classical reciprocal recurrent selection method when the dominance effects operating in each of the selection environments were divergent.

3. Another approach that may increase the efficiency in selecting genotypes with lower G×E interactions is the improvement in the statistical analysis. The additive main effects and multiplicative interaction (AMMI) model has been shown to give a more predictive accuracy than simple prediction of means over environments and replicates in yield trials (Gauch, 1988; Gauch and Zobel, 1988). As has been suggested AMMI may have a number of promising implications for plant breeding research programmes (see Chapter 22).

An axiom in traditional plant breeding has been to select where the variety is cultivated. This is called *in situ* selection which means to select for the specific conditions of a region or area; e.g. drought, salt and cold tolerance, etc. This selection will identify good regionally adapted genotypes, but it will produce a strong G×E interaction when these genotypes are grown in a wider range of environments. The *in situ* selection will increase the varietal offer for the different regions while maintaining a rich gene reserve in the commercial varieties. The reasoning behind this selection is that genes selected under optimal conditions might not be expressed in a non-optimal environment. In contrast to this, *ex situ* selection is performed in a small number of good environments which were found to be highly correlated with other environments after G×E interaction studies. These high-yielding environments will favour the expression of the full genetic potential of the breeding material. The thought underlying this method is that the best genes will be detected and adaptable varieties could be developed for a large range of diverse environments. Seed companies and international breeding centres will prefer to develop highly adaptable varieties because of a higher efficiency in handling and development of the breeding material, lower costs in seed production and greater commercial impact.

18.6 CONCEPT OF GENETIC GAIN

According to Falconer (1981), response to selection is the difference between the averaged phenotypic values of the progeny from selected parents and the parental generation before selection. Estimates of response to selection or genetic gain are useful for comparing different selection methods and predicting the genetic improvement of a population in a selection programme.

Derivation of the genetic gain is as follows: let X_i ($i = 1, 2, \ldots, n$) be the phenotypic value of the selection unit ('selection units' can be either single plants or families) in the original generation (G_0) and $\bar{X}$ be their averaged mean. Those units with higher phenotypic values, X_s ($s = 1, 2, \ldots, j; j<n$), are selected, $\bar{X}_s$ being their averaged mean. In certain cases, as in mass selection or ear-to-row selection, the selected generation descends directly from the selected units. However, in most of the cases, there is an intermediate generation of recombination (R) before producing the selected generation (S) whose individuals have the phenotypic values Y_js (Fig. 18.1). It can be assumed that each individual from S is related to one individual from the selected units in the G_0 generation according to the following linear relation:

$$Y_j = a + b_{y.x}X_j + \epsilon_j \qquad (18.5)$$

where $b_{y.x}$ is the linear regression coefficient of Y_js on X_js; ϵ_j is the error associated with the phenotypic value of plant j with expected value zero and normally and independently distributed; and a is the intercept that can be estimated from Equation 18.5 by assuming that all selection units from G_0, with phenotypic mean $\bar{X}$, are intermated to produce a generation with the same phenotypic mean value $\bar{X}$. Then:

$$\bar{X} = a + b_{y.x}\bar{X} \qquad (18.6)$$

By substituting a in Equation 18.5, it results:

$$Y_j - \bar{X} = b_{y.x}(X_j - \bar{X}) + \epsilon_j \qquad (18.7)$$

By averaging Equation 18.7 over all individuals of S and their corresponding selection units in G_0, it becomes:

$$\bar{Y} - \bar{X} = b_{y.x}(\bar{X}_s - \bar{X}) \qquad (18.8)$$

where $\bar{Y} - \bar{X}$ is the selection response or genetic gain, which is often designated by $\triangle G$; $\bar{X}_s - \bar{X}$ is the selection differential s, which can be expressed in standardized units $s = i\sigma_x$, where i is the selection intensity and σ_X is the phenotypic standard deviation; and $b_{y.x}$ can be expressed as $\text{cov}(Y,X)/\sigma_X^2$ by elementary statistics. Then, Equation 18.8 becomes the following general equation of genetic gain:

$$\triangle G = i\text{cov}(Y,X)/\sigma_X \qquad (18.9)$$

where $\text{cov}(Y,X)$ is the covariance between the selected units from G_0 and their corresponding relatives from the S selected generation; and σ_X is the phenotypic standard deviation of the G_0 generation. The $\text{cov}(Y,X)$ term depends on the relationship between the selection units evaluated and their corresponding individuals in the selected generation. It can be proved that its value is proportional to the additive genetic variance of the population. Therefore, the larger the genetic variability in the population the greater the genetic progress.

The phenotypic standard deviation (σ_P) can be reduced by improving the experimental technique and by increasing the number of replications and environments in the selection experiments (Equation 18.2).

18.7 POPULATION SIZE AND SELECTION INTENSITY

Because the heritability has been defined as the regression coefficient of parents on offspring, then $h^2 = b_{y.x}$ and Equation 18.9 can also be written in the following way (Chapter 11):

$$\triangle G = i\sigma_x h^2 \qquad (18.10)$$

GENERATION 0 (G₀)

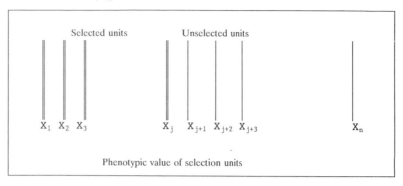

Mean of Generation 0 = $\overline{X}$; Mean of the selected units = $\overline{X}_s$

RECOMBINATION GENERATION (R)

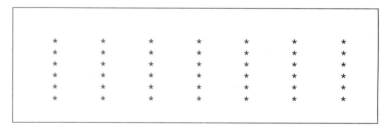

SELECTED GENERATION 1 (S)

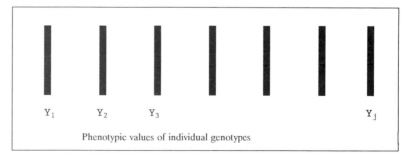

Fig. 18.1. Scheme for breeding a selected generation based on evaluation of selection units with recombination of either the selected units or remnant seed from the selected units.

Mean of selected Generation 1 = $\overline{Y}$

If a normal distribution of phenotypic values of individuals or selection units is assumed in each generation, and a portion Pr of such individuals is selected, then we can establish a correspondence between Pr and i from the Tables in Falconer (1981, Appendix); e.g., for a portion of 1%, 5%, 10%, 15%, 20%, 25% and 30% of selected individuals or units, the corresponding i values are 2.66, 2.06, 1.76, 1.55, 1.40, 1.27, and 1.16, respectively.

From Equation 18.10, it can be seen that as i increases (i.e., as the portion of selected individuals is reduced), the genetic gain will increase. However, a minimum population size of selected individuals should be retained in the recombination generation to avoid fixation of unfavourable genes. This is especially necessary for populations where the frequency of unfavourable alleles is relatively high, as in low-performing populations. In these cases, breeders are recommended to select at least 30 individuals for recombination. Based on the probability of fixation of unfavourable alleles, Rawlings (1980) reported that a population size of 20–30 does not seriously limit the long term selection gain. The probability of fixation of a neutral gene in one generation, $u_1 (p)$, is:

$$u_1 (p) = p^{2N} \qquad (18.11)$$

where p is the frequency of the gene and N is the population size or the number of recombinant individuals. More complicated equations of probability of fixation of genes under selection in a time interval can be found in Crow and Kimura (1970). A useful equation is:

$$u(p) = \frac{1-e^{-Nsp}}{1-e^{-Ns}} \qquad (18.12)$$

where $u(p)$ is the ultimate probability of fixation of a favourable gene, p is the initial gene frequency, s is the selection advantage of the gene and N is the effective population size.

Data adapted from Barker and Curnow (1969) based on Equation 18.12 show that after many generations of selection on 150 unlinked loci at low gene frequency, the ultimate expected genetic gain reached was 11.8%, 47.5%, 73.8% and 92% of the total possible genetic gain when applying population sizes 4, 16, 32 and 64, respectively. After 30 generations of selection, the expected genetic gain relative to the total possible genetic gain was 11.7%, 38.1%, 50.33% and 58.1% for population sizes 4, 16, 32 and 64, respectively (Table 18.3). Thus, the population size (i.e., the number of units to be selected) should be chosen in relation to the length of the selection programme. Once the number of selected units is decided, the selection intensity of each experiment is immediately computed.

Table 18.3. Expected genetic change (%) relative to the the total possible genetic change for different values of the effective population size (N) and after different generations of selection for the genetic model specified below[a] (adapted from Barker and Curnow, 1969)

N	Generations				
	1	5	10	30	limit
1	2.7	5.5	5.7	5.7	5.7
4	1.4	5.1	8.1	11.7	11.8
16	1.4	6.7	13.1	38.1	47.7
32	1.4	7.0	14.4	50.3	73.8
64	1.4	7.2	15.2	58.1	92.0
256	1.4	7.3	15.7	63.1	100.0
∞	1.4	7.3	15.8	63.6	100.0

[a] The assumed genetic model was 150 unlinked loci of equal additive effects under 2.5% selection intensity and heritability 0.2. Gene frequency was 0.1 for 50 loci, 0.2 for other 50 loci and 0.3 for the remaining 50 loci.

18.8 SOME METHODOLOGICAL PROBLEMS

18.8.1 Partial allogamous populations

Independence between the breeding values and dominance deviations of genotypes is assumed in panmictic populations. However, under partial inbreeding (partial allogamy), this assumption does not hold any more and the following expression for the covariance (cov) can be established (Wricke and Weber, 1986) following Falconer's model:

$$\text{cov}(A,D) = 4pq(q-p)F[a+(q-p)d]d \qquad (18.13)$$

where A and D are the breeding values and the dominance deviations of genotypes, respectively; a and d are the additive and the dominance effects; p and q are the gene frequencies of alleles B and b; and F is the coefficient of inbreeding. Except for the cases of non-dominance ($d = 0$) and F_2 populations from the cross of two inbreds ($p = q$), estimation of genetic parameters (heritability, variances, expected response to selection) in partial allogamous populations under selection should take into account the covariance component. Likewise, Kempthorne (1969) derived expressions of change in the population mean in non-random mating populations under selection.

If the objective of the breeding programme is to develop synthetic varieties, then estimation of the general combining ability (GCA) by a topcross or polycross should include the effect of selfing plants. In this case, the concept of general varietal ability (GVA) which consider the inbred progeny is preferred to the GCA in partial allogamous populations (Gallais, 1979).

There is a controversy about whether to select for self-fertility or self-sterility in partial auto-gamous crops. Although self-sterility could seem more attractive because of its similarity with strict outcrossers and possible enhancement of heterosis, poor pollination and seed set of the improved variety has been a problem in some environments, especially for insect pollinated crops. Because of these disadvantages, the tendency nowadays for population improvement and building up synthetic varieties is to use self-fertile lines which can also receive alien pollen while favouring an optimum level of outcrossing in the farmer's fields to increase heterosis.

18.8.2 Autogamous populations

The objective in self-fertile crops is to concentrate different useful genes (yield, resistance, quality, etc.) in the same genotype. Under complete autogamy, inbreeding will fix favourable and unfavourable alleles at the same time. Thus, if in the process an unfavourable allele is fixed in a line its corresponding favourable gene will never be found in the descendants of this line. In order to overcome this problem, two different alternatives to the methods of line development (section 18.10) have been suggested for using recurrent procedures: (1) to make handcrosses among F_2 plants, which requires a huge work; (2) to introduce male sterility into the population. This will allow recombination of genotypes and handling the selfer as an outcrosser (Chapter 14).

18.9 BACKCROSSING

Backcrossing (BC) is used to transfer useful genes from a genetic stock (population, inbred line, individual plants, varieties, wild relatives, etc.) called the donor to a breeding material (inbred line, population, etc.) which it is desired to improve, the recurrent parent. The method consists in crossing the donor and the recurrent parent to make the F_1 generation followed by one or more backcrosses to the recurrent parent. Different types of genes can be transferred: (1) single dominant genes; (2) single recessive genes; (3) polygenes underlying a quantitative trait.

1. Single dominant genes are easy to transfer. Genotypes carrying the target genes can be easily identified in the backcross generations and used to make new backcrosses (Fig. 18.2).
2. Single recessive genes do not show up in the backcross generations. F_2 intermating generations after each two backcrosses are necessary for unmasking the homozygous recessive genotypes to be again backcrossed to the recurrent parent (Fig. 18.3). However, the intermating generations can be avoided and the backcrossing time reduced by using a more complicated procedure. Random BC plants are backcrossed and selfed at the same time, while keeping appropriate registers. Only backcross progenies from plants whose selfed progenies segregated for the useful gene are retained; remaining backcrosses are eliminated (Fig. 18.4). This process may be enhanced by the employment of marker assisted selection procedures as presented in Chapter 19.
3. When transferring polygenic traits, F_2 recombination generations are necessary after the cross and backcross generations to increase their variability and to recombine the selected genotypes. Individuals from the F_2 generations with the highest level of performance are backcrossed to the recurrent parent. If the heritability of the polygenic trait is low, instead of selecting individual plants it is advisable to create S_1 families from the F_2 generations and to

Backcross generation	Donor genotype	Segregating genotype	Recurrent genotype	Contribution of recurrent genotype

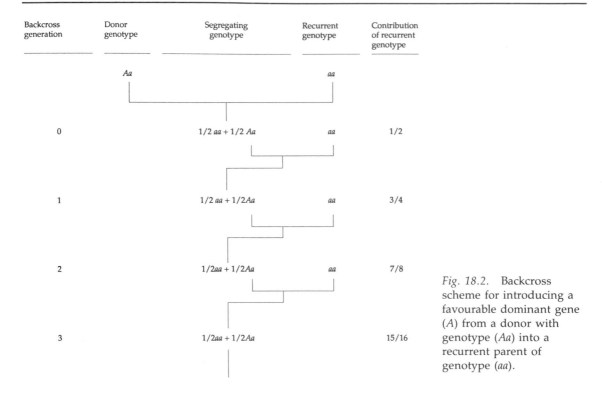

Fig. 18.2. Backcross scheme for introducing a favourable dominant gene (*A*) from a donor with genotype (*Aa*) into a recurrent parent of genotype (*aa*).

evaluate them. Remnant seed from the best families are backcrossed.

The fraction of the recurrent parent (receptor) present in the progency of each successive backcross following the first F_1 cross is given by the expression: $1 - \frac{1}{2}^{n+1}$, where n is the backcross number.

Dudley (1982, 1984a,b) developed a theory of transfer of favourable alleles from inbred lines and populations to elite inbreds or populations. The number of backcrosses to be made depends on the relative performance of the donor compared to the recurrent parent (Chapter 19).

18.10 LINE DEVELOPMENT

Lines can be considered the foundations of many breeding programmes and are necessary for some breeding strategies such as: (a) development and production of hybrids by crossing highly homozygous inbred lines; (b) direct utilization of pure lines or mix-line varieties as farmer seed in self-pollinating crops; (c) development of genetic stocks containing distinct registered and characterized genes for possible use in different breeding situations; e.g., genes for disease resistance, androsterility, food and feed quality, etc.; (d) development of synthetic varieties by intermating of lines for direct use by farmers or for population improvement; (e) sources of second or third cycle lines. Several methods for the development of lines are available.

18.10.1 Pedigree method

The pedigree method has been described in different textbooks (Allard, 1960; Poehlman, 1979; Fehr, 1987). Briefly, it consists in crossing two lines, each providing desirable genes which are to be incorporated into the resulting lines. The F_2 segregating generation is grown and selection is made on selfed plants. Progenies from the selected F_2 plants are grown in individual rows in the F_3 generation and re-

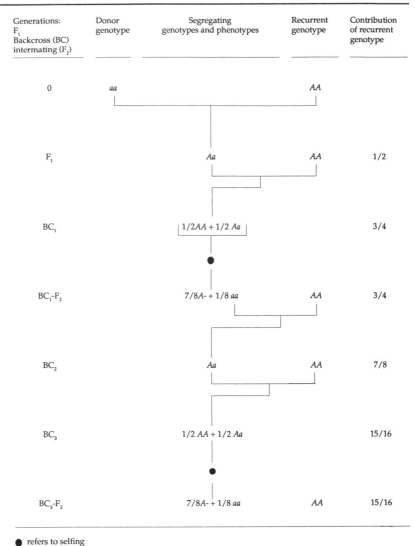

Generations: F₁ Backcross (BC) intermating (F₂)	Donor genotype	Segregating genotypes and phenotypes	Recurrent genotype	Contribution of recurrent genotype
0	aa		AA	
F₁		Aa	AA	1/2
BC₁		1/2AA + 1/2 Aa		3/4
BC₁-F₂		7/8A- + 1/8 aa	AA	3/4
BC₂		Aa	AA	7/8
BC₃		1/2 AA + 1/2 Aa		15/16
BC₃-F₂		7/8A- + 1/8 aa	AA	15/16

Fig. 18.3. Scheme for introducing a favourable recessive gene (*a*) from a donor with genotype (*aa*) into a recurrent parent of genotype (*AA*) by backcrossing and an alternating F₂ intermating generation.

● refers to selfing

selected on the basis of desired characteristics. Additional selection for higher heritability traits (e.g., plant type, earliness, etc. in some crops) may be carried out on individual plants within the selected F₃ progenies. F₄ progeny rows are again grown and the procedure is repeated for two or more generations until uniformity is apparent in individual progenies. Afterwards, replicated testing of lines is carried out for two or more generations. Only the highest yielding lines are retained each year. The method keeps records of each family in each generation in such way that lines can be traced back to establish their relationship.

The pedigree method is widely used for development of pure lines and new elite varieties in self-fertilized crops and for development of inbred lines in cross-fertilized crops such as maize (Hallauer, 1990). It is a simple and effective method of selection. The crucial point for this method is the choice of the parents for the single cross. For self-fertilized crops, one choice could be to cross the best adapted line with a good line having the desired characteristics lacking in the other line. The right choice depends on the knowledge of the material and the intended objective. If the end-product is a hybrid variety, choice of the lines should avoid

Generations: F₁ Backcross (BC) Selfing (S)	Donor genotype	Segregating genotypes and phenotypes	Recurrent genotype	Contribution of recurrent genotype

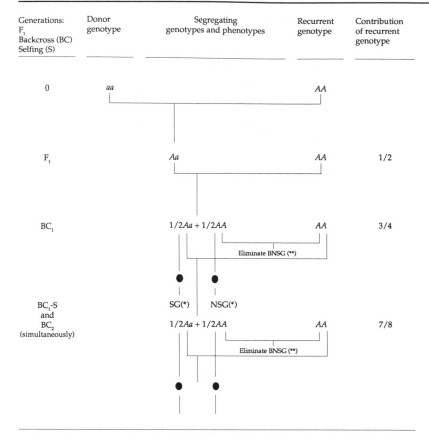

Fig. 18.4. Scheme for introducing a favourable recessive gene (*a*) from a donor with genotype (*aa*) into a recurrent parent of genotype (*AA*) by simultaneous backcrossing and selfing.

(*) SG and NSG refer to the segregating and non-segregating selfed progenies
(**) BNSG refers to backcrosses corresponding to non-segregating selfed progenies
● refers to selfing

destroying any heterotic potential of the line to be improved.

The pedigree method has many variants (Hallauer, 1990). The different protocols depend on the source material and the objectives to be reached. In some instances, an intermediate sibbing generation (recombination among selected sibs within progenies) after the first selfing generations will help to make additional selection within families.

A more general procedure of pedigree line development may use as source breeding material any population, local variety, three-way cross or double cross, instead of the single cross. After obtaining the segregating generation, the remaining steps are the same as the pedigree method described.

18.10.2 Single Seed Descent (SSD)

The procedure first proposed by Goulden (1941) and later described by Brim (1966) consists in advancing the F_2 population without selection for four to five generations in such a way that each F_6 or F_7 seed traces back to a different F_2 plant. Only one seed is retained from each plant in each generation. Afterwards, the seed is increased and replicated tests of the resultant lines are carried out for two or more generations to select the highest-yielding lines.

This method is simpler than the pedigree method. The advancing of the unselected lines can be easily made in a small nursery space, or in the greenhouse during the winter, so shortening the time for line development. However, neither progeny selection to eliminate undesired families, nor additional visual selection for high heritability traits among plants within families can be carried out during this process. Therefore, lines developed with this method are subjected to less visual selection than those resulting from the pedigree method. However, the SSD method could be a good choice when speed and simplicity are required for the breeding programme; e.g., if a large number of single crosses as source material are being developed at the same time, a simple method may be needed.

The advantages of the method are: (1) quick increase of the additive variance among families (Table 18.4); (2) reduction of the breeding space and saving of work for records and notes; (3) suitable for low heritability traits where visual selection is not effective.

Table 18.4. Coefficients of the additive genetic variances among families and within families relative to the the additive genetic variance of the F_2 generation (σ^2_{AF2}) for different selfing generations in the development of pure lines

| Generation | | Coefficient of the σ^2_{AF2} | |
Parent	Family	Among families	Within families
F_2	F_3	1	1/2
F_3	F_4	3/2	1/4
F_4	F_5	7/4	1/8
F_5	F_6	15/8	1/16
F_∞	F_∞	2	0

The SSD method has been used for line development in many self-fertilized crops such as soybean, barley, wheat and oats. It has also been used for accelerating the derivation of F_4 and F_5 lines from planned crosses in recurrent selection programmes in soybean (Piper and Fehr, 1987) and peanut (Monteverde-Penso and Wynne, 1988).

18.10.3 Other methods

Other methods for line development are the bulk method, mass selection and early generation testing. They are useful for self-pollinated crops. The bulk method consists in advancing the F_2 and subsequent generations by harvesting an unselected random bulk sample of seed until homogeneity of lines in the F_5 or F_6 generation is apparent. Thereafter, seed is increased and lines are tested.

Mass selection is similar to the bulk method, but selection for agronomic traits (plant type, maturity, disease tolerance, etc.) is carried out in the F_2 and subsequent generations. After the F_5 or F_6 generation, lines are extracted and tested.

The early generation testing method consists in deriving F_3 or F_4 families from F_2 plants for testing. Ivers and Fehr (1978) evaluated the pedigree method, the single-seed descent method and two methods of early generation testing of lines. Their results showed that no method of inbreeding was superior in effectiveness for identifying high-yielding lines.

18.11 INTRAPOPULATION RECURRENT IMPROVEMENT

The term population improvement generally refers to recurrent or cyclic selection applied to populations. It was suggested as a means for increasing the frequency of favourable genes for quantitative traits (Jenkins, 1940; Comstock *et al.*, 1949). Recurrent selection requires recombination of selected genotypes, therefore it can be easily applied to cross-fertilized crops, but male sterility or artificial emasculation are needed if applied to self-fertilized crops. The improvement aimed at may be either within populations or the cross between two populations.

Improvement within populations is appropriate if: (1) the end product is going to be a population or synthetic variety, as in certain allogamous crops where the higher cost of hybrid seed production is not compensated for by a greater advantage in yield; e.g. forage grasses, forage legumes, etc.; (2) elite pure lines developed from an improved population are needed either for superior farmer cultivars in self-fertilized crops or for parents of hybrids; (3) one wants to develop a mixed-genotype variety of an autogamous crop. According to the kind of selection units used, methods can be classified as individual plant selection or family selection.

18.11.1 Parental control

The genetic gain (Equation 18.9) can be increased if the proportion of alleles that pass to the next generation from selected individuals increases. The control of selection over both parents, females and males, will double the genetic gain as compared to the control over females only. Different situations can control the selection over the two parents: (1) selected individuals are selfed instead of being pollinated from selected and unselected plants; in this case one additional generation is required for recombination; (2) selection is carried out before pollination and recombination is among selected individuals only; (3) recombination is among selected clones.

18.11.2 Individual plant selection

(a) Mass selection

The mass selection method for line development described above, and mainly useful for self-pollinated crops, differs from that which will be described below mass selection in populations. In this method, the selection units are individual plants selected on the basis of their phenotype. Seed from progenies of selected plants as females pollinated by the population at large will form the next generation. In cross-fertilized plants, these progenies are half-sibs. The genetic gain for mass selection ($\triangle G_m$) can be derived from Equation 18.9 in the following way:

$$\text{cov}(X,Y) = \frac{1}{2}\sigma^2_A \qquad (18.14)$$

where $\text{cov}(X,Y)$ is the parent–offspring covariance and σ_A^2 is the additive variance. The phenotypic variance (σ_X^2) includes the total genetic variance (σ_G^2), $\sigma_G^2 = \sigma_A^2 + \sigma_D^2$; the plot-to-plot environmental variance (σ_e^2); the plant-to-plant environmental variance (σ^2_{we}),

$\sigma^2_{we} = \sigma^2_{wf} + \sigma^2_{me}$; and the genotype × environmental variance ($\sigma^2_{G\times E}$), $\sigma^2_{G\times E} = \sigma^2_{A\times E} + \sigma^2_{D\times E}$. The within-family genetic variance, (σ^2_{wf}) for individual plant selection is zero. Then:

$$\triangle G_m = \frac{\frac{1}{2}i\sigma^2_A}{\sqrt{\sigma^2_A + \sigma^2_D + \sigma^2_{A\times E} + \sigma^2_{D\times E} + \sigma^2_e + \sigma^2_{me}}} \qquad (18.15)$$

where σ_D^2, $\sigma^2_{A\times E}$ and $\sigma^2_{D\times E}$ are the dominance, additive × environment and dominance × environment variances, respectively; the other parameters have been previously defined.

Advantages of this method are: (1) simplicity; (2) evaluation of a large number of individuals (plants) in a small field area, which makes it possible to apply a high selection pressure; (3) completion of a cycle of selection per generation.

Mass selection is less efficient for low heritability traits because of the large denominator in Equation 18.15. In fact, Hallauer and Sears (1969) found that the regression coefficients of yield on cycles of selection were not significantly different from zero in two open-pollinated varieties of maize under mass selection.

The 'grid selection', introduced by Gardner (1961), is a modification of mass selection. It divides the area of the field into parts (grids) and selects the best plant in each grid. From the theoretical point of view, this modification reduces the plot-to-plot environmental variance (σ_e^2). Many selection experiments were conducted under Gardner's grid system (Hallauer and Sears, 1969; Verhalen *et al.*, 1975; Gardner, 1978; Cortez-Mendoza and Hallauer, 1979). However, the grid system did not always improve the selection response for low heritability traits.

To increase the efficiency of grid selection, Bos (1983) proposed to adjust the plant yields by 'standardization in each grid' and to apply truncation selection to the adjusted yield in the following way:

$$z_{ij} = \frac{y_{ij} - y_i}{s_i} \qquad (18.16)$$

where z_{ij} and y_{ij} are the adjusted and unadjusted yields of plant j in grid i, respectively; y_i.

is the mean yield of all plants in grid i; and s_i is the phenotypic standard deviation in grid i.

Another modification of mass selection, called the 'honeycomb design', has been developed by Fasoulas (1973). Each single plant is in the centre of a regular hexagon and is compared to the other six equidistant plants. Fasoulas and Zaragotas (1990) introduced a modification in the original method by considering each single plant the centre of a moving circle encompassing another 12, 18, 36 or 54 plants to which the plant in the centre is compared. These modifications try to reduce both the plot-to-plot and the plant-to-plant microenvironmental variances (σ_e^2 and σ_{me}^2) in Equation 18.15. However, Hanson *et al.* (1979) claimed the original method was so complex as to be impractical. Results by Mitchell *et al.* (1982) indicated that Fasoulas method was only marginally superior to mass selection based on unadjusted single plant yields for durum wheat. In contrast, results of Bos (1981) on winter rye were disappointing.

Genetic progress resulting from mass selection for low heritability traits such as grain yield were not always consistent in the reported experiments (Table 18.5). However, mass selection for high heritability traits such as early flowering and other agronomic traits in maize (Genter, 1976a; Cortez-Mendoza and Hallauer, 1979; Troyer and Larkins, 1985; Troyer, 1990) proved to be very efficient.

Mass selection can be used as a recurrent method in cross-fertilized and self-fertilized crops. The self-fertilized crops require selected plants to be male sterile as in sorghum (Atkins, 1980; Lothrop *et al.*, 1985). If a dominant gene (*Ms*) controls male sterility and selection is only practised on male sterile plants, the equilibrium situation will be reached in the second generation with one-half of the plants being of genotype *Ms/ms* and the other half *ms/ms* (Chapter 14).

(b) Phenotypic recurrent selection

The breeding scheme for phenotypic recurrent selection is the same as for mass selection, but the selection control is over both seed and pollen parents. Thus, the genetic gain per cycle in Equation 18.15 is multiplied by two. In some

annual crops such as maize, selected plants are selfed and one additional generation is needed for intermating. If the recombination generation is carried out in the winter nursery, the genetic gain per year will be doubled. In biennial crops like sugar beet where selection is done before flowering, recombination is only among selected individuals and the genetic gain of Equation 18.15 is also doubled. This selection was successfully practised for increasing oil content (Alexander, 1977; Misevic and Alexander, 1989) and for some physiological kernel traits in maize (Hartung *et al.*, 1989).

18.11.3 Family selection

In this group of selection methods, the selection units are not individual plants, but families. The general scheme has three steps: (1) to create a family structure; (2) to evaluate and to select best families by progeny testing; (3) to recombine the selected families. Each step is generally carried out in one generation, but two or more steps may be combined in one generation as in half-sib family selection. Each step may take more than one generation, e.g. S_2 family development will need two generations; the family evaluation is usually carried out in one season but could be done in two to reduce the genetic × year interaction; and one additional generation of recombination could be needed to break up linkages. Different methods of intrapopulation improvement are discussed below.

(a) Half-sib family selection

The general procedure has only two generations: (1) generation 1, half-sib families are created by allowing free pollination of selected female plants; (2) generation 2, seed from generation 1 half-sib families are evaluated for selection in replicated trials and different environments.

Remnant seed from generation 1 is then used for recombining these selected families to form the next cycle. Thus, the creation of families and the recombination steps are both carried out in the same generation 1.

The genetic gain for half-sib family selection

Table 18.5. Maize genetic gains (%) averaged over cycles in the populations *per se* and population topcrosses for several intrapopulation selection methods (adapted from references).

Selection method	Selected population	Selection cycles (no.)	Genetic gain per cycle		Evaluated population topcrosses (no.)	Reference
			Population per se	Population topcrosses		
Mass selection	Krug(M)C6	6	1.5			Hallauer and Sears (1969)
	Iowa Ideal(M)C6	6	1.4			Hallauer and Sears (1969)
	PHCB(M)C9	9	0.9	0.3	12	Genter and Eberhart (1974)
	PHWI(M)C9	9	−0.3	0.2	12	Genter and Eberhart (1974)
	NHC(M)C12	12	0.3	0.0	12	Genter and Eberhart (1974)
	Weighted average		0.63	0.17		
Modified ear-to-row	Hays Golden C10	10	5.4			Compton and Bahadeer (1977)
	Kitale(ETR)C6	6	1.8	−0.7	1	Darrah *et al.* (1978)
	Ecuador573(ETR)C6	6	6.3	−0.7	1	Darrah *et al.* (1978)
	Weighted average		4.66	−0.7		
Full-sib family	Jarvis(FS)C6	6	3.5	2.3	2	Moll and Stuber (1971)
	Ind. Chf.(FS)C6	6	2.8	2.3	2	Moll and Stuber (1971)
	Hybrid (FS)C4	4	8.5			Genter (1976b)
	Jarvis(FS)C10	10	2.9	2.3	4	Moll and Hanson (1984)
	Ind. Chf.(FS)C10	10	1.8	2.0	4	Moll and Hanson (1984)
	Weighted average		3.27	2.18		
S_1 or S_2 family	VLE(S)C2	2	1.1	0.5	9	Genter (1973)
	VCBS(S)C2	2	7.2	3.0	9	Genter (1973)
	Fla. 767(S)C5	5	3.7	2.0	2	Horner *et al.* (1973)
	VCBS(S)C4	4	5.4	3.9	12	Genter and Eberhart (1974)
	BSK(S)C4	4	2.5	−0.6	12	Genter and Eberhart (1974)
	BS13(S)C2	2	4.2[a]	1.5	9	Lamkey and Hallauer (1984)
	BS13(S)C2	2	3.2[a]	1.4	8	Oyervides-Garcia and Hallauer (1986)
	BSSS2(S1)C3	3	7.9	2.2	8	Oyervides-Garcia and Hallauer (1986)
	BSSS2(S1)C2	2	6.9	2.1	8	Oyervides-Garcia and Hallauer (1986)
	BSK(S)C8	8	2.9	1.2	7	Tarner and Smith (1987)
	BSK(S)C8	8	4.0			Rodriguez and Hallauer (1988)
	BS2(S)C4	4	6.0			Rodriguez and Hallauer (1988)
	BS16(S)C3	3	3.2			Rodriguez and Hallauer (1988)
	BS13(S)C4	4	2.8[a]			Rodriguez and Hallauer (1988)
	BS2(S2)C5	5	4.9	4.1	1	Iglesias and Hallauer (1989)
	BSTL(S2)C5	5	1.9	0.7	1	Iglesias and Hallauer (1989)
	BS16(S2)C5	5	−2.9	0.9	1	Iglesias and Hallauer (1989)
	FS8A(S2)C4	4		3.3	3	Horner *et al.* (1989)
	FS8B(S2)C4	4		3.0	3	Horner *et al.* (1989)
	Weighted average		3.50	1.76		

[a] Estimated assuming the same genetic gain for the first seven cycles of the testcross selection method (half-sib families) than for the remaining cycles of the S_2 selection method.

($\triangle G_{HS}$) can be derived from Equation 18.9. The term cov(X,Y) is the covariance between parents and offspring through the remnant seed, which is equivalent to the covariance among half-sibs, that is $\sigma_A^2/4$; the term σ_X is the square root of the phenotypic variance among half-sib families (σ_{HS}). Then:

$$\Delta G_{HS} = \frac{i\frac{1}{4}\sigma_A^2}{\sqrt{\frac{1}{4}\sigma_A^2 + \dfrac{\frac{1}{4}\sigma_{A\times E}^2}{e} + \dfrac{\sigma_e^2}{er} + \dfrac{\frac{3}{4}\sigma_A^2 + \sigma_D^2 + \sigma_{me}^2}{ern}}} \qquad (18.17)$$

where e, r and n are the numbers of environments, replications and plants per plot, respectively; σ_e^2 is the plot-to-plot environmental variance; and σ_{me}^2 is the plant-to-plant microenvironmental variance. Genetic gain per year can be doubled in summer crops if the recombination is carried out in the winter nursery. Half-sib selection, as described herein, is a simple procedure which is especially useful if selfing is difficult (e.g., presence of self-incompatibility, as in rye, rapeseed and sunflower populations) or controlled crosses are hard to make.

The common breeding scheme of perennial forage grasses and legumes follows the half-sib selection pattern. It consists of: (1) generation of half-sib families in a 'polycross mating system' from previously selected clones which are vegetatively maintained; (2) evaluation of the families in replicated rows during two or three years; (3) recombination of the selected clones. In this case, the genetic gain from Equation 18.17 will be multiplied by two, because genes from both selected parents will contribute to the next cycle. However, one additional generation will be necessary for recombination.

If the population to be improved has more than one female inflorescence, such as prolific ears in maize or multiheads in sunflower, it is possible to open-pollinate the first inflorescence and to self the second one. Two testing alternatives are suggested: (1) testing of half-sib families only; (2) simultaneous testing of half-sib and S_1 families to compute a combined selection index (Goulas and Lonnquist, 1977). Genetic gain is doubled as compared to Equation 18.17 for alternative (1) and even greater for alternative (2) because of the additional advantage of the combined selection (Goulas and Lonnquist, 1977); however, one additional generation is required for recombining the remnant selfed seed from the selected plants. The same scheme can be used in single-inflorescence populations, but development of the half-sib families is much more laborious since individual plants are selfed and also used as males to cross each with four to six plants of the same population as tester.

Traditional ear-to-row selection is the simplest scheme of half-sib selection to be conducted in allogamous crops. Each cycle is completed in one generation. Unreplicated half-sib progenies from selected individuals are grown in one environment. The best individuals within the best families are selected to create the progenies of the next cycle. The genetic gain is very much reduced as compared to Equation 18.17. The numerator is divided by 2 because there is only female control and the value assigned to r and e in the denominator is 1. This scheme has been successfully used in the long-term selection experiments of the University of Illinois since 1896 for selecting the high heritable traits of oil and protein content in maize (Dudley et al., 1974)

(b) Modified ear-to-row selection

This method was suggested by Lonnquist (1964) for maize. The three steps, development, evaluation and recombination of families are gathered in one generation. Half-sib families are evaluated in replicated trials. In the same year, an isolated crossing field is planted according to the following lay-out: female rows are each planted with seed from individual half-sib families; alternate male rows are planted with bulked seed from the whole population. Female rows are detasselled before flowering and allowed to be pollinated by the unselected population. At harvest time, the best five to eight ears (depending on the selection intensity) are collected in each female row. Based on results of the testing trials, only ears from the selected families (1/5 to 1/8 of the total depending on the selection intensity) are saved to form the half-sib families of the next generation. So, the number of families is kept equal through selection cycles. The genetic gain ($\triangle G_{mHS}$) has two components one for the selection among families and another for the intra-family selection. It can be expressed in the following way:

$$\triangle G_{mHS} = \frac{\frac{1}{8}i\sigma_A^2}{\sigma_{pHS}} + \frac{\frac{3}{8}i'\sigma_A^2}{\sigma_{we}} \qquad (18.18)$$

where i and i' are the selection intensities among families and within-families, respectively; σ_{pHS} is the phenotypic standard deviation of the half-sib families; and σ_{we} is the square root of the plant-to-plant within-plot variance.

When comparing genetic gains per year the only advantage found in this method over the half-sib family selection method is the additional intra-family selection whose efficiency depends on the control of σ_{we}. However, this advantage disappears if in the half-sib method (1) additional within-family selection is carried out in the generation 1, or (2) recombination is made in the off-season nursery. Compton and Bahadur (1977) and Darrah *et al.* (1978) reported gains in maize yield from 1.8 to 6.3% per cycle when applying the modified ear-to-row selection method to maize populations (Table 18.5).

This procedure has been adapted to perennial forage crops by combining half-sib family selection in replicated rows and within-family individual plant selection in a separated field. Recombination is carried out by a polycross system with clones from individual selected plants in the selected families. Genetic gain is doubled, as compared to Equation 18.18, since there is control of both parents (Hill and Byers, 1979).

(c) Full-sib family selection

Full-sib family selection is similar to the half-sib family selection, but families are created in generation 1 by making crosses between two plants, each from a family selected in the previous generation. The genetic gain ($\triangle G_{FS}$) per cycle is the following:

$$\triangle G_{FS} = \frac{i\sigma_A^2}{2\sigma_{FS}} \qquad (18.19)$$

where σ_{FS} is the phenotypic standard deviation of full-sib families. From the theoretical point of

view, the genetic gain is almost twice as large in this method as in the half-sib selection. If the recombination generation is carried out in a winter nursery, one cycle can be completed each year. In this case, the theoretical genetic gain per year can be comparable to that of other more powerful methods such as S_1 or S_2 family selection. If the selected families are intermated three generations are required per cycle. Full-sib family selection has been used for maize intrapopulation improvement at CIMMYT for many years. Moll and Stuber (1971), Genter (1976b) and Moll and Hanson (1984) reported averaged selection responses of 3.3% per cycle for maize populations *per se* and 2.2% for the population crosses after several cycles of selection (Table 18.5).

(d) Selfed (S_1 or S_2) family selection

Three and four generations per cycle are needed for the S_1 and S_2 family selection methods, respectively. The basic scheme consists in generating S_1 or S_2 families, evaluating selfed families in replicated multienvironment trials and recombining remnant seed from selected families. The theoretical genetic gain is the following:

$$\triangle G_{s1} = \frac{i\sigma_{A1}^2}{\sigma_{pS1}} \qquad (18.20)$$

where σ_{A1}^2 is the additive genetic variance among S_1 families, which is slightly different from the additive genetic variance in a Hardy–Weinberg equilibrium population (σ_A^2) unless there is no dominance or the gene frequency is 0.5 for all alleles; σ_{pS1} is the phenotypic standard deviation among S_1 families. The equation of genetic gain for S_2 families has the same structure as Equation 18.20 but the additive genetic variance among S_2 families is expected to be about 50% larger than among S_1. Consequently, the standard phenotypic deviation for S_2 is expected to be a little larger than for S_1. From the theoretical point of view, both methods have higher expected genetic gain per cycle than those described above. However, if a winter nursery is available and comparisons are made on genetic gain per year, there is no clear advantage of these methods over full-sib selection.

Several maize selection experiments for intra-population improvement by using the S_1 or S_2 family methods have been reported (Genter, 1973; Horner *et al.*, 1973, 1989; Genter and Eberhart, 1974; Lamkey and Hallauer, 1984; Oyervides-Garcia and Hallauer, 1986; Tarner and Smith, 1987; Rodriguez and Hallauer, 1988; Helms *et al.*, 1989; Iglesias and Hallauer, 1989). Averaged genetic gain per cycle for several populations was computed as 3.36% in the populations *per se* and 1.76% in the population crosses (Table 18.5).

Among the recurrent family selection methods, the S_1 seems to be the most suitable for use in self-fertilized crops such as soybean, wheat, barley, etc. The conventional method described by Piper and Fehr (1987) for soybean that resulted in a higher genetic gain when several methods were compared consisted in: (1) deriving 8 Tabout 10 F_4 (S_2) lines from each of 30 F_1 crosses; (2) evaluating Tthe S_2-derived lines at two stages (in the first stage, evaluation of each line was conducted in 12-hill plots, in the second stage the 30% highest yielding lines were evaluated in replicated row plots at three locations); (3) intermating each of the 30 selected lines with other two selected lines to produce 30 crosses for next cycle. Multiple generations of intermating did not enhance the genetic gain in the Piper and Fehr study. A similar procedure was previously described by Miller and Rawlings (1967) for cotton.

(e) Modified S_1 recurrent selection method

A modification of the S_1 selection method (MS_1) has been proposed by Dhillon and Khehra (1989) following a similar pattern to the modified ear-to-row selection proposed by Lonnquist (1964). The evaluation and the recombination phases were combined in the same season, reducing the time interval of the S_1 method to two growing seasons. The S_1 families are evaluated in replicated experiments in different environments. An isolated crossing field is grown in the same season, where plant-to-plant selection is practised within S_1 family female rows, which are detasselled. Alternate male rows are planted with a balanced seed compo-

site from all S_1 families. One cycle of MS_1 recurrent selection could be completed in a year, provided off-season nurseries were available.

Expected genetic gain in the MS_1 method has two components, selection among the S_1 families and selection within the S_1 families. In terms of expected genetic gain, the relative efficiency of the MS_1 method was calculated to be superior to the mass selection, half-sib, modified ear-to-row, S_1, and full-sib (three seasons per cycle) selection methods, and as good or better than the modified full-sib (two seasons per cycle) method (Dhillon and Khehra, 1989).

18.11.4 Relative merits of intrapopulation selection methods

Although the S_1 and S_2 selection methods theoretically have the highest expected genetic gain per cycle for intrapopulation improvement, a large body of experimental evidence has shown that other methods such as full-sib and testcross selections for maize yield resulted in a greater genetic gain for both the populations *per se* and the population crosses (Tables 18.5 and 18.6). Horner *et al.* (1989) explained these results by suggesting that non-additive gene action in the overdominance range is important in those populations because, in the absence of overdominance, the S_2 method is expected to be a better procedure. This explanation is, however, in disagreement with Design III experiments summarized by Gardner (1963) which suggested that the average degree of dominance for corn yield is in the partial dominance range. Although overdominance is an unlikely explanation (Chapter 2) there is no reason to discard it from the general genetic model because it may be present for some loci. Another alternative or complementary explanation to the results from the former selection experiments might be that a more favourable dominance × dominance epistasis is likely to be accumulated in the full-sib and testcross selection methods than with the S_1 or S_2 method. In agreement with this explanation, Moreno-González and Dudley (1981) ventured the hypothesis that selection for yield combin-

Table 18.6. Maize genetic gains (%) averaged over cycles in the populations *per se* and population topcrosses for the testcross selection method (adapted from references).

Selected population[a]	Selection cycles (no.)	Genetic gain per cycle		Evaluated population topcrosses (no.)	Reference
		Population per se	Population topcrosses		
VCBS(HT)C3	3	7.3	3.7	12	Genter and Eberhart (1974)
BSSS(HT)C7	7	−0.5	0.4	12	Genter and Eberhart (1974)
BSSS(HT)C7	7	1.3	3.1	2	Eberhart *et al.* (1973)
Fla. 767(IT)C5	5	4.8	4.4	2	Horner *et al.* (1973)
Fla. 767(PT)C5	5	4.2	2.4	2	Horner *et al.* (1973)
VLE(HT)C2	2	−0.6	0.8	9	Genter (1973)
VCBS(HT)C2	2	1.3	2.3	9	Genter (1973)
Alph(IT)C5	5	4.9	5.0	5	Russell *et al.* (1973)
(WF9xB7)(IT)C5	5	4.0	2.9	5	Russell *et al.* (1973)
FSB(HT)C7	7	2.6	2.6	2	Horner *et al.* (1976)
Kolkmeier(IT)C5	5	1.6	2.9	8	Walejko and Russell (1977)
Lancaster(IT)C5	5	−2.0	1.8	10	Walejko and Russell (1977)
BSK(HIT)C8	8	4.0	1.4	7	Tarner and Smith (1987)
BS12(IT)C7	7	13.0			Rodriguez and Hallauer (1988)
BSK(HIT)C8	8	3.3			Rodriguez and Hallauer (1988)
FS8A(IT)C4	4		4.6	3	Horner *et al.* (1989)
FS8B(IT)C4	4		5.4	3	Horner *et al.* (1989)
Weighted average		3.51	2.30		

[a] Letters HT, IT, PT and HIT between parentheses in the population names refer to hybrid tester, inbred tester, parental population tester and hybrid followed by inbred tester, respectively.

ing ability in some maize inbreds might have offset the diminishing effect of the dominance × dominance epistasis which was consistently detected by generation mean analysis in other unselected maize traits such as plant and ear height.

18.12 TESTCROSS SELECTION

The testcross selection intends to improve not only the population *per se* but also its combining ability. Thus, it can be considered as mid-way between intrapopulation and interpopulation selection. The basic steps of the method consist of: (1) selfing and simultaneously crossing each plant of a population to a tester; (2) evaluating crosses in replicated multienvironment trials; (3) recombining selfed remnant seed corresponding to selected crosses. One modification of this method consists in adding one more step to the procedure; the S_1 families are grown in an isolated crossing field as females and pollinated

with the tester as male thus avoiding handcrossing. The remaining steps are the same.

A critical aspect of this method is the choice of the tester. Broad and narrow genetic base testers are used for improving the general and specific combining abilities of populations, respectively. Therefore, testers can be populations, double and single hybrids and inbred lines. Low-yielding testers have been suggested to increase the gene frequency of favourable dominant alleles in intrapopulation and interpopulation selections (Rawlings and Thompson, 1962; Moreno-González and Grossman, 1976), since they increase the genetic variance of crosses and allow a better discrimination among genotypes. If the tester is the population itself, the method will become a half-sib selection method with three generations per cycle. In this case the genetic gain will be double that of Equation 18.17. This procedure has been used in maize, sunflower and sugar beet. Reported genetic gain in several maize selection experiments that used the testcross method averaged 3.51 and 2.30% per cycle for the population *per se* and the population crosses, respectively (Table 18.6).

18.13 INTERPOPULATION RECURRENT IMPROVEMENT

Interpopulation recurrent selection is appropriate if the final product is a hybrid variety. In this case, the development of genetically divergent populations seems likely to increase the chances of getting elite lines which will yield superior cross progenies. Several breeding schemes based on the procedure proposed by Comstock *et al.* (1949) have been suggested.

18.13.1 Reciprocal Recurrent Selection (RRS)

Reciprocal Recurrent Selection (RRS) was first designed by Comstock *et al.* (1949). It tries to alter two genetically different populations so as to improve their crossbred mean. The RRS method was designed to make use of additive, dominance and overdominance gene action. It is especially efficient when overdominance exists (Moreno-González and Grossman, 1976). It is also efficient for selecting favourable epistatic gene combinations in the population cross. The method consists of the following steps: (1) individual plants from two populations (A and B) are selfed and, at the same time, crossed with 3 to 5 plants from the reciprocal female tester population (B and A); (2) selection in each population is based on the performance of evaluated testcross half-sib families; (3) remnant seed from selected S_1 families are mated at random within A and B to form new A and B populations.

The change in the crossbred mean ($\triangle G$) was calculated by Moreno-González and Grossman (1976). When neglecting second order terms, a useful and simple equation of genetic gain can be written as follows:

$$\triangle G_{(A \times B)} = \frac{i\sigma^2_{A(HSA)}}{4\sigma_{P(HSA)}} + \frac{i'\sigma^2_{A(HSB)}}{4\sigma_{P(HSB)}} \quad (18.21)$$

where i and i' are the selection intensities in

populations A and B, respectively; $\sigma^2_{A(HSA)}/4$, $\sigma^2_{A(HSB)}/4$, $\sigma_{P(HSA)}$ and $\sigma_{P(HSB)}$ are the additive genetic variances and square roots of the phenotypic variances among half-sib families generated from individuals in populations A and B when crossed with tester populations B and A, respectively.

Results of three long-term selection experiments in maize that used the RRS method have been published on different occasions (Table 18.7). Average genetic gain per cycle was 0.78% for the populations *per se*, 3.43% for the hybrid between the reciprocal populations, and 1.57% for the crosses of the populations with different testers. As expected, more favourable genes must have been accumulated in the population hybrid than in the populations *per se* and other population crosses.

18.13.2 Full-Sib Reciprocal Recurrent Selection (FSRRS)

The FSRRS method was designed for maize yield selection by Hallauer and Eberhart (1970). It is a modification of the RRS method by Comstock *et al.* (1949). At least, one prolific population is required. The method consists in generating full-sib families by crossing the top ear of a plant from population A with a plant from B, while at the same time the lower ear is selfed (to be saved as remnant seed). If reciprocal plants from population B have also two ears, their top ear is crossed with its corresponding A plant and the lower ear is selfed. If plants from B have only one ear, they are selfed. Theoretical advantages of this method over the RRS method are (1) only one-half of families are evaluated in each cycle; (2) superior $S_0 \times S_0$ crosses can be advanced in further generations and evaluated as $S_1 \times S_1$, $S_2 \times S_2$, ..., etc. crosses, in such a way that hybrids can be simultaneously developed in a few years, while improving the population cross in the long run; (3) additive genetic variance of full-sib families is double that of half-sib families.

Expected genetic gain for FSRRS ($\triangle G_{FSRRS}$) is as follows:

$$\triangle G_{FSRRS} = \frac{i\sigma^2_A}{2\sigma_{PFS}} \quad (18.22)$$

Table 18.7. Maize genetic gains (%) averaged over cycles in the populations *per se,* the population hybrids and population topcrosses for two reciprocal recurrent selection methods (adapted from references).

Selection method	Selected population	Selection cycles (no.)	Genetic gain per cycle			Reference
			Population per se	Population hybrids	Population topcrosses	
Reciprocal	Jarvis(R)C6	6	1.7	3.0		Moll and Stuber (1971)
recurrent	Ind. Chief(R)C6	6	1.2	3.0		Moll and Stuber (1971)
selection	BSSS(R)C5	5	−0.1	4.1	1.2	Eberhart *et al.* (1973)
	BSCB(R)C5	5	1.0	4.1	0.3	Eberhart *et al.* (1973)
	KitaleII(R)C3	3	−0.1	7.0		Darrah *et al.* (1978)
	Ecuador573(R)C3	3	4.6	7.0		Darrah *et al.* (1978)
	Jarvis(R)C10	10	1.7	2.3	2.8	Moll and Hanson (1984)
	Ind. Chief(R)C10	10	1.2	2.3	1.4	Moll and Hanson (1984)
	BSSS(R)C8	8	2.5		1.3	Oyervides-Garcia and Hallauer (1986)
	BSSS(R)C10	10	1.3			Rodriguez and Hallauer (1988)
	BSCB1(R)C10	10	−1.6			Rodriguez and Hallauer (1988)
	Weighted average		0.78	3.43	1.57	
Full-sib	BS10(FR)C5	5	3.2	2.5		Hallauer (1984)
reciprocal	BS11(FR)C5	5	2.9	2.5		Hallauer (1984)
recurrent	BS10(FR)C7	7	2.0			Rodriguez and Hallauer (1988)
selection	BS11(FR)C7	7	0.8			Rodriguez and Hallauer (1988)
	BS10(FR)C8	8	3.0	6.5	3.0	Eyherabide and Hallauer (1991)
	BS11(FR)C8	8	1.6	6.5	2.2	Eyherabide and Hallauer (1991)
	Weighted average		2.17	4.96	2.6	

where i is the selection intensity, σ_A^2 is the additive genetic variance of the population cross, and σ_{PFS} is the phenotypic standard deviation of the crossed full-sib families.

The FSRRS method has been used in maize and sunflower. Results of genetic progress in two maize populations BS10 and BS11 under the FSRRS method are presented in Table 18.7 (Hallauer, 1984; Rodriguez and Hallauer, 1988; Eyherabide and Hallauer, 1991)). Genetic gain per cycle was 2.17% for the populations *per se,* 4.96% for the population hybrid and 2.6% for the population topcrosses, which is larger than the progress achieved under the RRS method according to published information.

18.13.3 Combined S_2 and Full-Sib Reciprocal Recurrent Selection (CRRS)

This method was designed by Moreno-González and Hallauer (1982). It tries to make

additional S_2 family evaluation of each one of the members that generate the full-sib families in order to select more precisely the best individuals in each population. Four seasons are necessary to complete one cycle: (1) to generate S_1 families in populations A and B; (2) to plant paired A and B S_1 rows, to self plants in each row and cross them onto plants in its reciprocal paired row to obtain S_2 and full-sib families; (3) to evaluate full-sib families and corresponding S_2 progenies; (4) to recombine remnant S_1 seed from selected members within each population.

Moreno-González and Hallauer (1982) showed that this method was more efficient than the FSRRS method under equivalent amount of effort for low heritability of the trait being selected and when few replications are used in the selection experiment. Other advantages of the CRRS method compared to the FSRRS are that additional pressure can be applied to the S_1 families and one-ear plant (non-prolific) populations can be used for the method. Disadvantages of the method compared to FSRRS are that generation of families is a more complex procedure and three times as many families need to be evaluated in the selection trials. No published data, so far, are available for assessment of progress of selection under this method.

18.13.4 Other modifications of the Reciprocal Recurrent Selection

Several modifications of the reciprocal recurrent selection have been proposed. Moreno-González and Grossman (1976) suggested two alternatives to RRS. The first (RRS-I) uses, as the tester of population A, a population (LB) that is derived from B by family selection for low yield. The second (RRS-II) is similar to RRS-I, but also uses, as the tester of B, a population (LA) that is derived from A by family selection for low yield. The choice of the selection method depends on the importance of the gene effects. If overdominance is unimportant, RRS-II should the best selection method, followed by RRS-I and RRS. On the contrary, if overdominance is important, RRS should be preferred to RRS-I followed by RRS-II.

Russell and Eberhart (1975) suggested the use as the tester of population A an inbred line derived from or related to population B; reciprocally, the tester of population B should be an inbred line derived from or related to population A. Besides improving the population cross, selected strains from one population are crossed to selected strains from the other population to identify the best hybrid combination. Thus, this method enables one simultaneously to improve the breeding populations and develop single hybrids.

Another modification of the RRS method in which prolific, two-ear-plant populations are required has been proposed by Paterniani (1967). The recombination and the creation of the crossed half-sib families are carried out in the same generation. Two separate blocks, one for each population, are grown. In the first season the top ear from individual plants in

each population is crossed with bulk pollen from the reciprocal population. At the same time, the second ear is not selfed but open-pollinated within each population. In the second season the reciprocal crosses are evaluated. After evaluation, the open-pollinated seed corresponding to the selected families will form the new cycle within each population. Genetic gain is reduced to one half but the method is simple and only two generations per cycle are required. If a winter nursery is available one cycle per year can be completed.

Paterniani and Vencovsky (1977) proposed the crossing of individual half-sib families as females from one population to the bulk reciprocal population as a male in isolated crossing fields in the first season. Testcrosses are evaluated in the second season and remnant seed from half-sib families corresponding to selected crosses are recombined to create again half-sib families in the third season. Genetic gain is one-half of the original RRS but the method is easy since no hand pollination is involved. At the same time sampling problems are minimized because bulk pollen from the whole reciprocal population instead of from four or five plants is used as tester.

18.14 CONCLUSIONS

A large body of empirical evidence has been accumulated to demonstrate that classical selection methods have been efficient in obtaining genetic progress in cultivated crops through the years. Evidence not only comes from selection experiments, but also from assessment of commercial varieties from different eras in common trials. Data from long term selection experiments indicated that genetic variance was not exhausted and improvement continued after 70 generations of selection for oil and protein in maize ((Dudley *et al.*, 1974). Genetic variance decreased in the maize population BSSS improved after nine cycles of reciprocal recurrent selection but not in the population improved after seven cycles of testcross followed by three cycles of S_2 recurrent selection methods (Helms *et al.*, 1989). It seems that the depletion of genetic variance has not been critical in long-term selected populations in limiting their genetic improvement.

Several factors affect the choice of breeding method. The knowledge of the reproductive system of the crop under selection will indicate the ease with which plants may be outcrossed and selfed. This is important for anticipating the technical difficulties with the method chosen. The kind of gene action (additiveness, dominance or overdominance) prevalent for the traits to be selected will indicate which genetic model to consider in the selection procedure. If additive effects are the most important, the line development or the S_1–S_2 recurrent selection methods will be the most appropriate. If dominance effects, and thereby heterosis, are important and hybrid seed production is easy, hybrids will be the highest performing cultivars and testcross or interpopulation selection methods will be recommended. If dominance is important but hybrid seed production is not feasible, the cultivar will be a pure line for self-pollinated crops and a segregating population (synthetic variety) for cross-pollinated crops. In this case, intrapopulation selection methods will be recommended.

Identification of favourable genetic characteristics in the source plant breeding materials is essential for the success of a breeding programme. These characteristics could be high performance, large genetic variability, and good combining ability of the trait to be selected, favourable genes for specific features such as disease resistances, genes for general adaptability to different environments, easiness of crossing, etc. Selection for qualitative traits will use specific simple breeding methods such as backcrossing or the pedigree method. On the other hand selection for quantitative traits may use any kind of breeding method, however, recurrent selection procedures will be the most efficient.

If heritability for the selected trait is high, selection based on individuals plants or unreplicated families, such as mass selection, simple phenotypic selection or ear-to-row selection can be appropriate. If heritability is low, selection based on replicated family units will be recommended. If genotype × environment (G×E) interaction is important, more than one environment should be used in the selection work. Other approaches to reduce this kind of interaction are specific breeding strategies (such as reciprocal recurrent selection of populations A and B in different environments, Moreno-González 1986) and more powerful data analysis such as AMMI.

18.15 PROSPECTS

Classical breeding methods will continue to be used in the future, unless other methods can surpass the 1% yearly genetic increase that has been achieved in many crops. New techniques have become available for plant breeders in recent years. These techniques will have to show that they can efficiently produce in the future as many and as good products as traditional breeding. The combination of classical and innovative methods is the likely logical approach in the future. The strategy will depend on the crop species and breeding objectives. Some of the prospects that can be visualized are as follows.

1. Enhancement of the present genetic variability in the source populations. Response to selection depends on the genetic variability. Introduction of exotic material after adaptation to the growing conditions will bring in new favourable alleles that might be lost in the current breeding populations. Thus, basic studies on the performance, adaptation and heterotic pattern of the exotic material should be conducted. The diallel mating system among populations may provide the information required.
2. Identification of genes in the wild stock for some special characteristics of the crops like disease resistance, male sterility, food and feed quality, flavours and endurance of horticultural products, exotic types of ornamental plants, etc. Thus, wild relatives of the crops should be widely screened.
3. Identification of quantitative trait loci (QTLs) associated to markers for selecting inbred lines carrying the favourable QTLs and increasing the frequency of these QTLs in populations.

4. Emphasis on breeding programmes to select for products of industrial use. Special kinds of oils, starch, proteins and metabolites are needed in the pharmaceutical, cosmetic, painting, food, plastic and motor industry. Crops with an efficient production of starch and sugar could be used for alcohol transformation as energy source.

5. Better understanding of quantitative genetics. Although quantitative genetics theory has developed considerably in recent years, there are areas where further studies would provide additional knowledge which will result in more efficient breeding methods; e.g., better understanding of heterosis, epistatic effects and overdominance; more efficient methods of estimating the number of effective factors affecting a quantitative trait (QTLs); and mapping of QTL by restricted fragment length polymorphisms (RFLP) or other molecular techniques.

6. Better control of the testing environments will reduce the $G \times E$ interaction. Studies of the effects of competitiveness and selection in stress *vs* non-stress environments on genetic gains are needed.

7. Improvement of data analysis. The use of powerful computers will improve the analysis of experiments and facilitate the identification of the best genotypes without noise.

8. Improvement in the precision of mechanization and automation will ameliorate the experimental field techniques.

9. Utilization of commercial hybrids in some other crops. The discovery of effective male sterility systems either genetic, cytoplasmic or chemical would allow the exploitation of heterosis and dominance effects in some crops. The use of gametocides would facilitate the application of outcrossing methods to autogamous plants and the production of hybrids on a large scale.

10. Tissue culture. Growing the cell culture in a medium containing a selection pressure agent for a specific characteristic (e.g., phytotoxins, herbicides, or other stress agent) will eventually allow for selection of desirable mutant genotypes.

11. Doubled haploid lines. The techniques for development of pure lines from anther culture or interspecific crosses followed by doubling of the chromosome number are available for some crops (rice, rapeseed, tobacco, barley, etc.). Additional selection for desirable mutants can be made in the growing medium, but field testing of the lines should be carried out.

12. Genetic engineering. Genes responsible for useful traits in some plants have been isolated and sequenced. A lot of work in the area is undergoing in the world. Molecular techniques of synthesis, cloning and transferring of genes into the genome of higher plants (tobacco, potato, etc.) are available. They may change the characteristics of a cultivar. However, the transferred gene is not always expressed in the genetic background of a plant and sometimes may negatively affect the expression of other traits. So far, work has focused on major genes, the question arises whether QTLs will be cloned, transferred and expressed. Most of these prospects are discussed in detail in other Chapters.

REFERENCES

Alexander, D.E. (1977), High oil corn. Current status of breeding and utilization. *Ann. Illinois Corn Breeders School*, University of Illinois, Urbana, **13**, 1–9.

Allard, R.W. (1960), *Principles of Plant Breeding*, John Wiley, New York.

Atkins, R.E. (1980), Registration of IAP1R(M)C4 sorghum germplasm. *Crop Sci.*, **20**, 676.

Barker, L.H. and Curnow, R.N. (1969), Choice of population size and use of variation between replicate populations in plant breeding selection programs. *Crop Sci.*, **9**, 555–560.

Biffen, R.H. (1905), Mendel's law of inheritance and wheat breeding. *J. Agr. Sci.*, **1**, 4–48.

Bos, I. (1981), The relative efficiency of honeycomb selection and other procedures for mass selection in winter rye (*Secale cereale* L.). Dr thesis, Agriculture University, Wageningen.

Bos, I. (1983), About the efficiency of grid selection. *Euphytica*, **32**, 885–893.

Brim, C.A. (1966), A modified pedigree method of selection in soybeans. *Crop Sci.*, **6**, 220.

Brim, C.A. and Cockerham, C.C. (1961), Inheritance

of quantitative characters in soybeans. *Crop Sci.*, **1**, 187–190.

Bruce, A.B. (1910), The mendelian theory of heredity and augmentation of vigor. *Science*, **32**, 627–628.

Chakroun, M., Taliaferro, C.M. and McNew, R.W. (1990), Genotype–environment interactions of bermudagrass forage yields. *Crop Sci.*, **30**, 49–53.

Columela, L.I.M. (60 AD), *De los trabajos de campo* (Spanish translation from *De Re Rustica*), Holgado Redondo, A. (ed.), Ministerio de Agricultura, Pesca y Alimentación, Madrid, 1988.

Compton, W.A. and Bahadur, K. (1977), Ten cycles of progress from modified ear-to-row selection in corn. *Crop Sci.*, **17**, 378–380.

Comstock, R.E. and Moll, R.H. (1963), Genotype × environment interactions. In *Statiscal Genetics and Plant Breeding*, Robinson, H.F. and Hanson, W.D. (eds), NAS-NCR publication 982.

Comstock, R.E., Robinson, H.F. and Harvey, P.H. (1949), A breeding procedure designed to make maximum use of both general and specific combining ability. *Agron. J.*, **41**, 360–367.

Cortez-Mendoza, H. and Hallauer, A.R. (1979), Divergent mass selection for ear length in maize. *Crop Sci.*, **19**, 175–178.

Cox, T.S., Shroyer, J.P., Ben-Hui, L., Sears, R.G. and Martin, T.J. (1988), Genetic improvement in agronomic traits of hard red winter wheat cultivars from 1919 to 1987. *Crop Sci.*, **28**, 756–760.

Crossa, J. (1989), Theoretical considerations for the introgression of exotic germplasm into adapted maize populations. *Maydica*, **34**, 53–62.

Crossa, J., Gardner, C.O. and Mumm, R.F. (1987), Heterosis among populations of maize (*Zea mays* L.) with different levels of exotic germplasm. *Theor. Appl. Gen.*, **73**, 445–450.

Crow, J.F. and Kimura, M. (1970), *An Introduction to Population Genetics Theory*, Harper & Row, New York.

Darrah, L.L., Eberhart, S.A. and Penny, L.H. (1978), Six years of maize selection in 'Kitale synthetic II', 'Ecuador 573' and 'Kitale composite A' using methods of the comprehensive breeding system. *Euphytica*, **27**, 191–204.

Darrah, L.L. and Zuber, M.S. (1986), 1985 United States farm maize germplasm base and commercial breding strategies. *Crop Sci.*, **26**, 1109–1113.

Dhillon, B.S. and Khehra, A.S. (1989), Modified S_1 recurrent selection in maize improvement. *Crop Sci.*, **29**, 226–228.

Dickerson, G.E. and Hazel, L.N. (1944), Effectiveness of selection on progeny performance as a supple-

ment to earlier culling in livestock. *J. Agr. Res.*, **49**, 459–476.

Dudley, J.W. (1982), Theory for transfer of alleles. *Crop Sci.*, **22**, 631–637.

Dudley, J.W. (1984a), A method of identifying lines for use in improving parents of a single cross. *Crop Sci.*, **24**, 355–357.

Dudley, J.W. (1984b), A method for identifying populations containing favorable alleles not present in elite germplasm. *Crop Sci.*, **24**, 1053–1054.

Dudley, J.W, Lambert, R.J. and Alexander, D.E. (1974), Seventy generations of selection for oil and protein concentration in the maize kernel. In *Seventy Generations of Selection for Oil and Protein in Maize*, Dudley, J.W. (ed.), Crop Science Society of America Inc., Madison, Wisconsin.

Duvick, D.N. (1977), Genetic rates of gain in hybrid maize yields during the past 40 years. *Maydica*, **22**, 187–196.

Eberhart, S.A., Debela, S. and Hallauer, R.A. (1973), Reciprocal recurrent selection in the BSSS and BSCB1 maize populations and half-sib selection in BSSS. *Crop Sci.*, **13**, 451–456.

Eberhart, S.A. and Russell, W.A. (1966), Stability parameters for comparing varieties. *Crop Sci.*, **6**, 36–40.

Eyherabide, G.H. and Hallauer, A.R. (1991), Reciprocal full-sib recurrent selection in maize. I. Direct and indirect responses. *Crop Sci.*, **31**, 952–959.

Falconer, D.S. (1981), *Introduction to Quantitative Genetics*, 2nd edn., Longman Inc., New York.

Fasoulas, A. (1973), *A New Approach to Breeding Superior Yielding Varieties*, Pub. No. 4, Department of genetics and plant breeding, Aristotelian University of Thessaloniki.

Fasoulas, A. and Zaragotas, D.A. (1990), *New Developments in the Honeycomb Selection Designs*, Pub. No. 12, Department of genetics and plant breeding, Aristotelian University of Thessaloniki.

Fehr, W.R. (ed.) (1984), *Genetic Contribution to Yield Gains of Five Major Crops*, CSSA Spec. Pub. No. 7, CSSA, Madison, Wisconsin.

Fehr, W.R. (1987), *Principles of Cultivar Development*, Vol. 1, *Theory and Technique*, McGraw Hill Inc., New York.

Finlay, K.W. and Wilkinson, G.N. (1963), The analysis of adaptation in a plant-breeding program. *Aust. J. Agr. Res.*, **14**, 742–754.

Fisher, R.A. (1918), Correlations between relatives on the supposition of Mendelian inheritance. *Trans. R. Soc. Edin.*, **52**, 339–433.

Gallais, A. (1979), The concept of varietal ability in plant breeding. *Euphytica*, **28**, 811–823.

Gardner, C.O. (1961), An evaluation of mass selection and seed irradiation with thermal neutrons on yield of corn. *Crop Sci.*, **1**, 241–245.

Gardner, C.O. (1963), Estimates of genetic parameters in cross-fertilizing plants and their implications in plant breeding. In *Statistical Genetics and Plant Breeding*, Robinson, H.F. and Hanson, W.D. (eds), NAS–NRC Publication 982.

Gardner, C.O. (1978), Population improvement in maize. In *Maize Breeding and Genetics*, Walden, D.B. (ed.), Wiley, New York, pp. 207–228.

Gates, C.R., Weber, C.R. and Horner, T.W. (1960), A linkage study of quantitative characters in a soybean cross. *Agron. J.*, **52**, 45–49.

Gauch, H.H. (1988), Model selection and validation for yield trials with interaction. *Biometrics*, **44**, 705–715.

Gauch, H.H. and Zobel, R.W. (1988), Predictive and postdictive success of statistical analyses of yield trials. *Theor. Appl. Genet.*, **76**, 1–10.

Genter, C.F. (1973), Comparisons of S_1 and testcross evaluation after two cycles of recurrent selection in maize. *Crop Sci.*, **13**, 524–527.

Genter, C.F. (1976a), Mass selection in a composite of intercrosses of Mexican races of maize. *Crop Sci.*, **16**, 556–558.

Genter, C.F. (1976b), Recurrent selection for yield in the F_2 of a maize single cross. *Crop Sci.*, **16**, 350–352.

Genter, C.F. and Eberhart, S.A. (1974), Performance of original and advanced maize populations and their diallel crosses. *Crop Sci.*, **14**, 881–885.

Goodman, M.M. (1984), Evaluation of exotics. *Ann. Illinois Corn Breeders School*, University of Illinois, Urbana, **20**, 85–100.

Goulas, C.K. and Lonnquist, J.H. (1977), Comparison of combined half-sib and S_1 family selection with half-sib, S_1, and selection index procedures in maize. *Crop Sci.*, **17**, 754–757.

Goulden, C.H. (1941), Problems in plant selection. In *Proc. 7th International Genetical Congress*, Edinburgh, pp. 132–133.

Haldane, J.B.S. (1924), A mathemathical theory of natural and artificial selection. Part I. *Trans. Camb. Phil. Soc.*, **23**, 19–41.

Hallauer, A.R. (1984), Reciprocal full-sib selection in maize. *Crop Sci.*, **24**, 755–759.

Hallauer, A.R. (1990), Methods used in developing maize inbreds. *Maydica*, **35**, 1–16.

Hallauer, A.R. and Eberhart, S.A. (1970). Reciprocal full-sib selection. *Crop Sci.*, **10**, 315–316.

Hallauer, A.R., Russell, W.A. and Smith, O.S. (1983), Quantitative analysis of Iowa Stiff Synthetic. *Stadler Genet. Symp.*, **15**, 83–104.

Hallauer, A.R. and Sears, J.H. (1969), Mass selection for yield in two varieties of maize. *Crop Sci.*, **9**, 47–50.

Hanson, P.R., Jenkins, G. and Westcott, B. (1979), Early generation selection in a cross of spring barley. *Z. Pflanzenzunchtg.*, **83**, 64–80.

Harlan, J.R. (1975), *Crops and man*, ASA, CSSA Publishers, Madison.

Harlan, H.V., Martin, M.L. and Stevens, H. (1940), *A study of methods in barley breeding*, US Dept. Agric. Tech. Bull. No. 720.

Harrington, J.B. (1937), The mass-pedigree method in the hybridization improvement of cereals. *J. Am. Soc. Agron.*, **29**, 379–384.

Hartung, R.C., Poneleit, C.G. and Cornelius, P.L. (1989), Direct and correlated responses to selection for rate and duration of grain fill in maize. *Crop Sci.*, **29**, 740–745.

Helms, T.C., Hallauer, A.R. and Smith, O.S. (1989), Genetic variability in improved and nonimproved 'Iowa Stiff Synthetics' maize populations. *Crop Sci.*, **29**, 959–962.

Hill, R.R. Jr. and Baylor, J.J. (1983), Genotype × environment interaction analysis in alfalfa. *Crop Sci.*, **23**, 811–815.

Hill, R.R. Jr. and Byers, R.A. (1979), Allocation of resources in selection for resistance to alfalfa blotch leafminer in alfalfa. *Crop Sci.*, **19**, 253–257.

Hillsman, K.J. and Carter, H.W. (1981), Performance of F_1 hybrid soybeans in replicated row trials. *Agron. Abstr. American Society of Agronomy*, Madison, Wisconsin, p. 63.

Horner, E.S., Lundy, H.W., Lutrick, M.C. and Chapman, W.H. (1973), Comparison of three methods of recurrent selection in maize. *Crop Sci.*, **13**, 485–489.

Horner, E.S., Lutrick, M.C., Chapman, W.H. and Martin, F.G. (1976), Effect of recurrent selection for combining ability with a single-cross tester in maize. *Crop Sci.*, **16**, 5–8.

Horner, E.S., Magloire, E. and Morera, J.A. (1989), Comparison of selection for S_2 progeny *vs* testcross performance for population improvement in maize. *Crop Sci.*, **29**, 868–874.

Horner, T.W. and Weber, C.R. (1956), Theoretical and experimental study of self–fertilized populations. *Biometrics*, **12**, 404–414.

Hoyt, E. (1988), *Conserving the wild relatives*, IBPGR/IUCN/WWF, Rome.

Hull, F.H. (1945), Recurrent selection and specific combining ability in corn. *J. Am. Soc. Agron.*, **37**, 134–145.

Iglesias, C.A. and Hallauer, A.R. (1989), S_2 recurrent

selection in maize populations with exotic germ-plasm. *Maydica*, **34**, 133–140.

Immer, F.R. (1941), Relation between yielding ability and homozygosis in barley crosses. *J. Am. Soc. Agron.*, **33**, 200–206.

Ivers, D.R. and Fehr, W.R. (1978), Evaluation of the pure-line method for cultivar development. *Crop Sci.*, **18**, 541–544.

Jan, C.C. and Chandler, J.M. (1985), Transfer of powdery mildew resistance from *Helianthus debilis* Nutt. to cultivated sunflower. *Crop Sci.*, **25**, 664–666.

Jenkins, M.T. (1940), Segregation of genes affecting yield of grain in maize. *J. Am. Soc. Agron.*, **32**, 55–63.

Johansen, W.L. (1903), *Ueber Erblichkeit in Populationen und in reinen Linein*, Gustav Fisher, Jena.

Jones, D.F. (1917), Dominance of linked factors as a means of accounting for heterosis. *Genetics*, **2**, 466–479.

Jones, D.F. (1945), Heterosis resulting from degenerative changes. *Genetics*, **30**, 527–542.

Jones, D.F. (1958), Heterosis and homeostasis in evolution and in applied genetics. *Am. Nat.*, **92**, 321–328.

Kempthorn, O. (1969), *An Introduction to Genetic Statistics*, The Iowa State University Press, Ames.

Lamkey, K.R. and Hallauer, A.R. (1984), Comparison of maize populations improved by recurrent selection. *Maydica*, **29**, 357–374.

Lonnquist, J.H. (1964), Modification of the ear-to-row procedures for the improvement of maize populations. *Crop Sci.*, **4**, 227–228.

Lothrop, J.E., Atkins, R.E. and O.S. Smith, (1985), Variability for yield components in IAP1R grain sorghum random-mating population. I. Means, variances components and heritabilities. *Crop Sci.*, **25**, 235–240.

Lush, J.L. (1931), The number of daughters necessary to prove a sire. *J. Dairy Sci.*, **14**, 209.

Lush, J.L. (1933), The bull index problem in the light of modern genetics. *J. Dairy Sci.*, **16**, 501.

Lush, J.L. (1947), Family merit and individual merit as basis for selection. *Am. Nat.*, **81**, 241.

Mather, K. and Jinks, J.L. (1981), *Biometrical Genetics*, Cornell University Press, New York.

Mendel, G. (1866), Versuche über Pflanzenhybriden. *Verh. Naturfosh. Ver in Brünn*, **4**, 3–47. Translation in Mendel's principles of heredity by Bateson, W. (1909), Cambridge University Press, Cambridge.

Miller, P.A. and Rawlings, J.O. (1967), Selection for increased lint yield and correlated responses in upland cotton, *Gossypium hirsutum* L. *Crop Sci.*, **7**, 637–640.

Misevic, D. and Alexander, D.E. (1989), Twenty-four cycles of phenotypic recurrent selection for percent oil in maize. I. *per se* and test-cross performance. *Crop Sci.*, **29**, 320–324.

Mitchell, J.W., Baker, R.J. and Knott, D.R. (1982), Evaluation of honeycomb selection for single plant yield in durum wheat. *Crop Sci.*, **22**, 840–843.

Moll, R.H. and Hanson, W.D. (1984), Comparisons of effects of intrapopulation *vs* interpopulation selection in maize. *Crop Sci.*, **24**, 1047–1052.

Moll, R.H. and Stuber, C.W. (1971), Comparisons of response to alternative selection procedures initiated with two populations of maize (*Zea mays* L.). *Crop Sci.*, **11**, 706–711.

Monteverde-Penso, E.J. and Wynne, J.C. (1988), Evaluation of three cycles of recurrent selection for fruit yield within a population of Virginia-type peanut. *Crop Sci.*, **28**, 75–78.

Moreno-González, J. (1986), Choice of environments in reciprocal recurrent selection programs. *Theor. Appl. Genet.*, **71**, 652–656.

Moreno-González, J. (1988), Evaluation, development and improvement of source breeding materials for cold regions. In *Maize Breeding and Maize Production*, Euromaize 88, Maize Research Institute of Zemun Polje, Belgrade.

Moreno-González, J. and Dudley, J.W. (1981), Epistasis in related and unrelated maize hybrids determined by three methods. *Crop Sci.*, **21**, 644–651.

Moreno-González, J. and Grossman, M. (1976), Theoretical modifications of reciprocal recurrent selection. *Genetics*, **84**, 95–111.

Moreno-González, J. and Hallauer, A.R. (1982), Combined S_2 and crossbred family selection in full-sib reciprocal selection. *Theor. Appl. Genet.*, **61**, 353–358.

Mungoma, C. and Pollak. L.M. (1988), Heterotic patterns among ten Corn Belt and exotic maize populations. *Crop Sci.*, **28**, 500–504.

Nault, L.R., Gordon, D.T., Dansteegt, V.D. and Iltis, H.H. (1982), Response of annual and perennial teosintes to six maize viruses. *Plant Dis.*, **66**, 61–62.

Nelson, R.L. and Bernard, R.L. (1984), Production and performance of hybrid soybeans. *Crop Sci.*, **24**, 549–553.

Osborn, T.C., Alexander, D.G. and Fobes, J.J. (1987), Identification of restricted fragment length polymorphisms linked to genes controlling soluble solids content in tomato fruit. *Theor. Appl. Genet.*, **73**, 350–356.

Oyervides-Garcia, M. and Hallauer, A.R. (1986),

Selection-induced differences among strains of Iowa Stiff Stalk Synthetic maize. *Crop Sci.*, **26**, 506–511.

Paterniani, E. (1967), Selection among and within half-sib families in a Brazilian population of maize (*Zea mays* L.). *Crop Sci.*, **7**, 212–217.

Paterniani, E. and Vencovsky, R. (1977), Reciprocal recurrent selection in maize (*Zea mays* L.) based on testcrosses of half-sib families. *Maydica*, **22**, 141–152.

Piper, T.E. and Fehr, W.R. (1987), Yield improvement in a soybean population by utilizing alternative strategies of recurrent selection. *Crop Sci.*, **27**, 172–178.

Poehlman, J.M. (1979), *Breeding Field Crops*, 2nd edn., Avi. Pub. Co., Westport.

Rawlings, J.O. (1980), Long and short-term recurrent selection in finite populations – choice of population size. In *World Soybean Research Conference. II Proceedings*, Corbin, F.T. (ed.), Westview Press, Boulder, pp. 201–215.

Rawlings, J.O. and Thompson, D.L. (1962), Performance level as criterion for the choice of maize testers. *Crop Sci.*, **2**, 217–220.

Rick, C.M. (1974), High soluble-solids content in large fruited lines derived from a wild green-fruited species. *Hilgardia*, **42**, 493–510.

Rodriguez, O.A. and Hallauer, A.R. (1988), Effects of recurrent selection in corn populations. *Crop Sci.*, **28**, 796–800.

Russell, W.A. (1984), Agronomic performance of maize cultivars representing different eras of breeding. *Maydica*, **29**, 375–390.

Russell, W.A. and Eberhart, S.A. (1975), Hybrid performance of selected maize lines from reciprocal recurrent and testcross selection programs. *Crop Sci.*, **15**, 1–4.

Russell, W.A., Eberhart, S.A. and Vega, O.U.A. (1973), Recurrent selection for specific combining ability for yield in two maize populations. *Crop Sci.*, **13**, 257–261.

Shull, G.H. (1909), A pure-line method of corn breeding. *Am. Breeder's Assoc. Rep.*, **5**, 51–59.

Specht, J.E. and Williams, J.H. (1984), Contribution of genetic technology to soybean productivity. Retrospect and prospect. In *Genetic Contributions to Yield Gains of Five Major Crop Plants*, Fehr, W.R. (ed.), CSSA special publication No. 7, Madison.

Sprague, G.F. (1946), Early testing of inbred lines of corn. *J. Am. Soc. Agron.*, **38**, 108–117.

Swaminathan, M.S. (1986), *The Green Revolution*, Comm. to the 20th anniversary of CIMMYT, September 22–24, Mexico, D.F.

Tarner, A.H. and Smith, O.S. (1987), Comparison of half-sib recurrent selection in the Krug Yelow Dent maize populations. *Crop Sci.*, **27**, 508–513.

Tollenar, M. (1989), Genetic improvement in grain yield of comercial maize hybrids grown in Ontario from 1959 to 1988. *Crop Sci.*, **29**, 1365–1371.

Troyer, A.F. (1990), Selection for early flowering in corn: three adapted synthetics. *Crop Sci.*, **30**, 896–900.

Troyer, A.F. and Larkins, J.R. (1985), Selection for early flowering in corn: 10 late synthetics. *Crop Sci.*, **25**, 695–697.

Verhalen, L.M., Barker, J.L., McNew, R.W. (1975), Gardner's grid system and plant selection efficiency in cotton. *Crop Sci.*, **15**, 588–591.

Walejko, R.N. and Russell, W.A. (1977), Evaluation of recurrent selection for specific combining ability in two open-pollinated maize cultivars. *Crop Sci.*, **17**, 647–651.

Wricke, G. and Weber, W.E. (1986), *Quantitative genetics in selection in plant breeding*, Walter de Gruyper, Berlin, New York.

Wright, S. (1921), Systems of matings. *Genetics*, **6**, 111–178.

Zuber, M.S. (1975), Corn germplasm base in the US: Is it narrowing, widening or static? *Corn and Sorghum Ind. Res. Conf.*, **30**, 277–286.

19

Marker-assisted selection

P. Arús and J. Moreno-González

19.1 INTRODUCTION

The use of polymorphic single genes to facilitate the process of plant breeding was proposed early in this Century (Sax, 1923). The basic principle is that selection for characters with easily detectable phenotypes can simplify the recovery of genes of interest linked to them and more difficult to score. The first marker loci available were those that have an obvious impact on the morphology of the plant. Genes that affect form, coloration, male sterility or disease resistance among others have been genetically analysed in many plant species. In some well characterized crops like maize, tomato, pea, barley or wheat, tens or even hundreds of such genes have been assigned to different chromosomes (O'Brien, 1990).

Selection with markers depends heavily on the quality of the polymorphisms used. The most important properties for good quality markers are: (1) easy recognition of all possible phenotypes (homo- and heterozygotes) from all different alleles; (2) early expression in the development of the plant; (3) no effects on the plant morphology of alternate alleles at the marker loci; (4) low or null interaction among markers allowing the use of many at the same time in a segregating population. Unfortunately, the general properties of morphological markers are far from ideal: dominance and late expression, deleterious effects, pleiotropy, epi-

stasis and rare polymorphisms are the rule. As a consequence, their use in plant breeding has been very limited.

New sources of high quality genetic markers, based on the identification of polymorphisms in proteins and DNA, have been developed during the last three decades. They have been termed molecular markers (Tanksley, 1983) and include isozymes, RFLPs and RAPDs (for a detailed description see Chapters 8 and 12). These markers have most or all the requisite properties mentioned above and, for this reason, their potential as tools for plant breeding is much greater than that of morphological genes. The objective of this chapter is: (a) to review the use of markers in plant breeding with emphasis on applications leading to the improvement of selection efficiency; and (b) to discuss theoretical models for the analysis and selection of quantitative traits associated with marker loci.

19.2 APPLICATIONS OF UNMAPPED MARKERS

When the phenotype for one or more marker loci is known for two individuals, inferences can be made about the phenotype of the progeny resulting from their cross. Hence, given an individual of dubious origin it is possible to determine by its marker phenotype, whether it may derive from a certain cross or not. Similarly, the paternity of a given individual can be tested knowing the marker phenotype of its putative offspring, and eventually that of the other parent. Many practical problems of plant breeding respond to

Plant Breeding: Principles and prospects. Edited by M.D. Hayward, N.O. Bosemark and I. Romagosa. Published in 1993 by Chapman & Hall, London. ISBN 0 412 43390 7.

situations as simple as these. Marker loci are very efficient tools to solve them or simplify in some way their resolution.

Identification of true hybrid individuals from apomictic seedlings can be done in the progeny of crosses between individuals of the genus *Citrus* with the use of isozyme genes (Torres *et al.*, 1982). True hybrids can also be separated from selfs or other sorts of seed contaminants in F₁ hybrid seed samples with the use of isozyme genes fixed for different alleles in each of the inbred lines (Arús *et al.*, 1982). Isozymes are also excellent selectable markers for early detection of sexual (Parfitt *et al.*, 1985; McGranahan *et al.*, 1986; Chaparro *et al.*, 1989) and parasexual (Lo Schiavo *et al.*, 1983) interspecific hybrids.

One of the most interesting applications of marker loci is in the selection of haploid or double haploid individuals originating from anther or microspore culture (Zamir *et al.*, 1981; Colby and Peirce, 1988). Plants of haploid origin regenerated from microspores of an individual heterozygous for a marker locus will all be homozygotes, whereas individuals arising from diploid anther wall tissue will be heterozygotes. Some of the regenerated diploids are the result of spontaneous chromosome doubling from cells originally haploid, and thus homozygous (Chapter 17). In this case, marker identification is especially effective since methods based on morphological characters associated with haploidy or chromosome counts are useless. The same principle can be used to test whether the segregations observed in haploid progenies derived from heterozygotes at marker loci, segregate according to the expected 1:1 ratio. Orton and Browers (1985) in *Brassica oleracea*, and Hayward *et al.* (1990) in *Lolium perenne* analysed such segregations at four and two isozyme loci respectively. Significant deviations from expected were found for all the loci studied, indicating that the recovery of a random set of gametes cannot be taken for granted. A similar approach was taken by Kling *et al.* (1987) to compare the allele frequencies of three maize populations with those of the random inbred lines derived from them. Significant departures from expected allele frequencies were only found in two of the 10 isozyme loci studied.

Simple methods for estimating the relative proportion of outcrossing and self-fertilization can be devised with one or more molecular markers (Ritland and Jain, 1981; Tanksley, 1984; Miller *et al.*, 1989). The mating system has important implications for the level of genetic variability and its organization in natural populations (Gottlieb, 1981). Knowledge of this information is relevant for the application of markers to more specific uses, like the development of better strategies for germplasm conservation (Brown, 1978), the verification of the accuracy of pedigrees (Smith and Smith, 1989) or the identification of cultivars (Chapter 12 and Bailey, 1983).

19.3 SELECTION FOR MAJOR GENES LINKED TO MARKERS

Major genes responsible for economically important characters are frequent in the plant kingdom (Gottlieb, 1986). Characters like disease resistance, male sterility, self-incompatibility and others related to the shape, colour and architecture of the whole plant, fruits, flowers or leaves are often of mono) or oligogenic nature. Marker loci tightly linked to major genes can be used for selection, sometimes more efficiently than direct selection for the target gene. There are three situations in which marker selection will clearly be more favourable: (1) when the selected character is expressed late in plant development, like fruit and flower features or adult characters in species with a juvenile period; (2) when the expression of the target gene is recessive; (3) when there is a requirement for special operations in order for the gene to be expressed, as in the case of breeding for disease or pest resistance. Additional advantages can be expected from the use of markers in breeding for disease resistance (Koebner and Martin, 1990; Melchinger, 1990). Since selection can be performed without inoculation, errors due to unreliable inoculation methods are avoided and breeding for resistance can be done in areas where field inoculation with the pathogen of interest is not allowed for safety reasons. Moreover, problems in the recognition of the effects of environmentally unstable resistance genes can be eluded.

If one marker is to be used for gene tagging, it is necessary that linkage with the target gene be tight (>5 recombination units or cM) in order to insure that only a minor fraction of the selected individuals will be recombinants. Alternatively, when two flanking markers can be used, it is only required that the interval between them is approximately of 20 cM, since selection for both markers at the same time results in the recovery of the target gene with a probability of at least 99% (Tanksley, 1983).

19.3.1 Isozymes as markers for major genes

A list of isozyme genes that can be used as markers of interest in plant breeding is presented in Table 19.1. Linkage between isozymes and other morphological markers useful for the assignment of the former to pre-existing linkage maps, but judged irrelevant for plant breeding (Tanksley and Rick, 1980; Havey and Muehlbauer, 1989), have been omitted.

Although the list of Table 19.1 is probably incomplete, it is clear that the total number of isozyme-gene linked pairs detected so far is not very large. There is an obvious reason for this: as the number of isozymes is low, the probability of finding one of them linked to an important gene is small. Moreover, inheritance studies including co-segregation of isozymes and major genes are relatively uncommon in the literature. Thus, the potential of isozymes as markers of individual genes is still underexploited, and many more useful linkages are likely to be found in the near future.

In addition to single marker tagging, there are two examples in which interval mapping can be used. In both cases, nuclear male sterility (*ms-10*) in tomato (Tanksley and Zamir, 1988), and annual habit (*Hb*) in celery (Quiros *et al.*, 1987), the flanking markers are an isozyme and a morphological marker. Strong correlations between isozyme phenotypes and metric characters like photosynthetic ability (Weng and Chen, 1989) and seed protein content (Shenoy *et al.*, 1990) have been found in rice. In both cases isozymes account for a considerable part of the total variation, suggesting that a major quantitative trait locus (QTL) may have been localized. The use of markers for the study of

Table 19.1. Major genes of economic interest for which an isozyme marker can be used for gene tagging.

Character	Crop	Isozyme gene	Genetic distance (cM)	References
Nematode resistance (*Mi*)	Tomato	*Aps-1*	>1	Rick and Fobes (1974)
Bean yellow mosaic virus resistance (*Mo*)	Pea	*Pgm-p*	2	Weeden *et al.* (1984)
Male sterility (*ms-10*)	Tomato	*Prx-2*	1.5	Tanksley *et al.* (1984)
Strawbreaker foot rot resistance (*Pch*)	Wheat	*Ep-D1, Ep-1A*	3	McMillin *et al.* (1986) Worland *et al.* (1988)
Incompatibility (*S*)	Apple	*Got-1*	2	Manganaris and Alston (1987)
Enation mosaic virus resistance (*En*)	Pea	*Adh-1*	4	Weeden and Provvidenti (1988)
Yellow mosaic virus resistance (*Ym*)	Barley	*Est-1, Est-2, Est-4*	3.7	Konishi *et al.* (1989)
Rust resistance (*Rph-10*)	Barley	*Est-2*	5	Feuerstein *et al.* (1990)
Rust resistance (*Rph-11*)	Barley	*Acp-3*	7	Feuerstein *et al.* (1990)
Rust resistance (*Rph-11*)	Barley	*Dip-2*	11	Feuerstein *et al.* (1990)
Bitterness (*Bi*)	Watermelon	*Pgm-1*	11.3	Navot *et al.* (1990)
Flesh color (*red*)	Watermelon	*Gdh-2*	12.8	Navot *et al.* (1990)
Sex	Honey locust	*6Pgd-2*	–	Schnabel and Hamrick (1990)

quantitatively inherited characters is considered in detail in section 19.5.

19.3.2 RFLPs as markers for major genes

The limited number of markers inherent in isozymes does not exist with RFLPs. Thus, in species with saturated linkage maps, many if not all the major genes of economic interest can be labelled either by a single RFLP or by a bracket of adequately spaced markers. Selection for many valuable genes at the same time can be facilitated if they are linked to such marker loci. This can, for example, be exploited when resistance to different diseases has to be introduced in a sensitive cultivar (Tanksley *et al.*, 1989), or when various genes for resistance to a specific pathogen have to be accumulated in a certain genotype in order to provide durable resistance (Melchinger, 1990).

19.4 APPLICATION OF MARKER SELECTION IN BACKCROSS BREEDING

The transfer of a gene of interest from a donor to a recipient (recurrent) genotype is commonly done with the backcross method. The objective is to obtain an individual with the same genotype as the recurrent parent (RP), with the exception of one or a few genes provided by the donor parent (DP). Frequently, the RP is a cultivar of good general performance and the DP has a poor agricultural value (individuals from wild species and old or exotic cultivars are often used as DPs) but possesses a valuable character lacking in the RP. The backcross method proceeds as follows: the F_1 and subsequent generations are repeatedly crossed back to the RP, decreasing the proportion of the DP genome by one half per generation. Selection is performed in each backcross generation for the target character and for the good characteristics associated with the recurrent parent. A minimum of five or six backcross generations is required to complete the process (Allard, 1960).

The recovery of the recurrent genotype can be accelerated with the use of molecular markers. If the F_1 is heterozygous for a marker locus, individuals with the RP allele(s) at this marker in the first or subsequent backcross generations will also carry a chromosome segment of the RP tagged by the marker. Using the model developed by Hanson (1959), Tanksley and Rick (1980) predicted that, under some simplifying assumptions, the proportion of the RP genome expected in the first backcross generation after selection for twelve markers (one per chromosome in tomato) was nearly the same as in the third backcross without selection. Experimental results by Tanksley *et al.* (1981) were consistent with theoretical expectations.

Selection for morphological characters can also contribute to the return to the RP genome (Tanksley *et al.*, 1981; Muehlbauer *et al.*, 1988). If screening for markers is done first, selected plants can be reselected later on the basis of the presence of the character to be transferred from the DP and the morphological resemblance to the RP, resulting in a joint use of both kinds of information.

The proportion of the genome that can be monitored with markers depends on their number and position on the chromosomes. For some species, it is possible to use maps of RFLPs which cover almost the entire genome at evenly separated distances of 20–30 cM. Hence, it would be theoretically feasible to select individuals practically identical to the recurrent parent by screening a sufficiently large number of progeny in the first backcross generation. Since this is nearly impossible in practice, because the probability of finding such an individual is extremely low ($1/2^n$, n being the number of loci), selection for markers has to be done in several steps, choosing the individual(s) closer to the recurrent parent in subsequent backcross generations until the return to the recurrent parent is complete. Computer simulations using the tomato as a model showed that by selecting the best plant out of a total of 30 per generation, the whole recurrent genome could be recovered in three generations (Tanksley *et al.*, 1989).

Obviously, the introgressed gene needs a different treatment. If linked to a marker or within a marker bracket, it can be easily selected in the heterozygous state. If unlinked, or of unknown position, marker selection should be relaxed in order to be sure that at least one of the chosen plants carries the appropriate segment of the donor parent. A strategy for that purpose is to select, in addition to the plants chosen for marker resemblance to the RP, *n* heterozygous individuals, one for each of the markers utilized (Tanksley *et al.*, 1981).

This approach to backcross breeding has been termed 'whole genome selection' (Tanksley *et al.*, 1989; Young and anksley, 1989). It results essentially in a saving of time since the number of generations needed to obtain a variety can be drastically reduced. The backcross method is commonly used in annual or biennial species, but rarely applied to fruit trees, forest trees and other woody perennials. This occurs because their long intergeneration period (i.e. juvenility in fruit trees lasts usually from five to seven years (Torres, 1983)) makes it too long a process involving more than two or three generations. The gene pool that tree breeders can use is generally restricted to the cultivated species. Valuable genes for quality, disease resistance or hardiness, present in wild relatives, have been used for rootstock breeding, but are seldom employed for cultivar improvement. Whole genome selection offers an excellent opportunity for the introgression of new genes within a reasonable time span and with population sizes per generation limited to a few hundred individuals. An important effort has been made during the last decade for the development of isozyme markers in woody perennials (Mitton, 1983; Torres, 1983; Hauagge *et al.*, 1987; Lee and Ellstrand, 1987). This effort should continue now with the development of RFLP-saturated maps, which will allow the breeder to realize all the potential of marker-assisted selection in this important group of species.

The same principle used in whole genome selection for the faster recovery of the recurrent parent genome applies to a completely different situation: selection for the target gene implies also the introgression of a chromosome segment of the donor parent. Mathematical models, studied by Stam and Zeven (1981), predict that the size of the introgressed fragment will be considerable for a large number of generations. These authors calculated the average length of this segment to be of approximately 20 cM after 10 backcross generations. Ten generations later, an average of 10 cM will still be retained. Experimental observations support these data: Young and Tanksley (1989) found that the fragment of the donor genome present in a sample of tomato cultivars resistant to tobacco mosaic virus ranged from 4 to 51 cM in length. This variation in length, which reflects different degrees of 'linkage drag' (Brinkman and Frey, 1977), can result in negative effects for the new cultivar, since large DNA fragments of the donor parent are likely to contain genes conferring unexpected and often undesirable characters (Zeven *et al.*, 1983).

Flanking marker loci in the vicinity of the target gene can be used to minimize the linkage drag effect (Young *et al.*, 1988; Young and Tanksley, 1989). Since recombination occurs between the target gene and the markers, individuals carrying the gene of interest and the alleles of the recurrent parent can be recovered after a few backcross generations (Tanksley *et al.*, 1989). The size of the donor linkage block can thus be limited to the distance between the two markers, and the whole operation can be done in a much shorter period than that expected without marker selection.

19.5 MAPPING QUANTITATIVE CHARACTERS

The theoretical bases for identification of quantitative trait loci (QTL) associated with individual marker loci (M) have been studied by several authors (Jayakar, 1970; MacMillan and Robertson, 1974; Soller and Beckmann, 1983; Edwards *et al.*, 1987; Cowen, 1988). Likewise, the use of flanking marker loci for QTL identification has been suggested by Lander and Botstein (1989) and Knapp *et al.* (1990). Experimental studies (Law, 1967; Tanksley *et al.*, 1982; Osborn *et al.*, 1987; Stuber *et al.*, 1987; Paterson *et al.*, 1988) have shown that marker genes are in fact linked to genes controlling quantitative characters in several crop species like tomato, wheat and maize. Based on these results, simple approximate models for the separation of genetic effects are formulated in the following paragraphs.

19.5.1 Genetic models

Different genetic models can be set up for estimating the contribution of the marker-linked loci to the quantitative trait under study.

(a) Individual marker locus

Consider two inbred parental lines P_1 and P_2 with individual marker genotypes MM and mm, respectively and QTL genotypes QQ and qq, respectively. The F_1 cross between P_1 and P_2 has the following chromosome array:

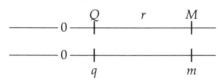

where r is the recombination frequency between Q and M. The kinds of gametes and their frequencies derived from meiosis in the F_1 generation would be:

Gametes	Frequencies
QM	$1/2\ (1-r)$
qm	$1/2\ (1-r)$
Qm	$1/2\ r$
qM	$1/2\ r$

Different kinds of generations and progenies have been studied for mapping the QTLs (Soller and Beckmann, 1983; Cowen, 1988; Knapp *et al.*, 1990). These generations can be F_2 individual plants, lines derived from self-fertilized F_2 plants (S_1), F_1 backcrosses to P_1 (B_1) and P_2 (B_2), S_1 progeny derived from backcrosses (BS_1 and BS_2), doubled haploids (DH), recombinant inbreds (RI) and F_1 testcrosses (F1TC). Since association between markers and QTLs occurs only in generations in linkage disequilibrium, the use of these is an essential requirement for this kind of analysis. The F_2 and backcross generations B_1 and B_2 provide useful information and need a short period of time and little

effort for their development. However, evaluation of genotypes from individual plants in these generations has a large environmental error for traits of low heritability. In this case the selfed progenies S_1, BS_1 and BS_2 are a more appropriate choice.

F_2 generation

Following Soller and Beckmann (1983) and Edwards *et al.* (1987), let us assume that $+a$, d and $-a$ are the genotypic effects of QQ, Qq and qq, respectively. The frequencies and genotypic values of members of an F_2 are shown in Table 19.2. Plants from the F_2 generation can be

Table 19.2. Genotypes, frequencies and genotypic values of individual plants in the F_2 generation

Genotype	Frequency	Genotypic value
$QQMM$	$1/4(1-r)^2$	$+a$
$QQMm$	$1/2r(1-r)$	$+a$
$QQmm$	$1/4r^2$	$+a$
$QqMM$	$1/2r(1-r)$	d
$QqMm$	$1/2(1-2r+2r^2)$	d
$Qqmm$	$1/2r(1-r)$	d
$qqMM$	$1/4r^2$	$-a$
$qqMm$	$1/2r(1-r)$	$-a$
$qqmm$	$1/4(1-r)^2$	$-a$

separated into the three marker classes MM, Mm and mm with the following expected class means:

Marker class	Expected class mean
$MM(F_2)$	$\mu_1 + (1-2r)a + 2r(1-r)d$
$Mm(F_2)$	$\mu_1 + (1-2r+2r^2)d$
$mm(F_2)$	$\mu_1 + (2r-1)a + 2r(1-r)d$

where μ_1 is the genotypic value of the remaining genes contributing to the quantitative trait. The means are adjusted for the relative genotypic frequencies within the marker classes MM, Mm and mm.

Contrasts among marker classes will estimate the QTL effects in the following way:

$$\frac{1}{2}[MM(F22) - mm(F_2)] = a(1-2r) \tag{19.1}$$

$$Mm(F_2) - \frac{1}{2}[MM(F_2) + mm(F_2)] = d(1-2r)^2 \tag{19.2}$$

$$\frac{2Mm(F_2) - MM(F_2) - mm(F_2)}{MM(F_2) - mm(F_2)} = \frac{d}{a}(1-2r) \tag{19.3}$$

Disadvantages of this method are:

1. Estimate of additive a and dominance d effects and degree of dominance d/a are biased downward by the coefficients $(1-2r)$, $(1-2r)^2$ and $(1-2r)$, respectively. Therefore, the higher the recombination frequency between the marker and the QTL, the larger the bias.

2. If several loci with positive and negative effects were associated with the marker locus, then a global confounding QTL effect would be estimated instead of individual effects for each QTL. Equations 19.1 and 19.2 can be more precisely written in the following way:

$$\Sigma a_i \, (1-2r_i) = 1/2[MM(F_2)-mm(F_2)\,] \tag{19.4}$$
$$\Sigma d_i \, (1-2r_i)^2 = Mm(F_2) - 1/2[MM(F_2)+ \\ mm(F_2)] \tag{19.5}$$

where subscript i refers to loci Q_i/q_i ($i = 1, 2, \ldots, n$) associated with the marker gene M/m.

3. If the heritability of the trait under selection is low, phenotypic values of individual plants will have a large environmental error component, therefore a high number of plants need to be scored to detect significant effects.

S₁ generation
Each S_1 line is classified into the same marker class as the F_2 plant from which it was derived. By making the same contrasts as above, the following estimates can be written:

$$\frac{1}{2}[MM(S_1) - mm(S_1)] = a(1-2r) \tag{19.6}$$

$$2Mm(S_1) - MM(S_1) - mm(S_1) \\ = d(1-2r)^2 \tag{19.7}$$

$$\frac{2MM(S_1) - MM(S_1) - mm(S_1)}{\frac{1}{2}[MM(S_1) - mm(S_1)]} = \frac{d}{a}(1-2r) \tag{19.8}$$

Disadvantages (1) and (2) of the F_2 generation

also hold for the S_1. However, because of a lesser environmental error component in the performance of S_1 lines a lower number of F_2 plants need to be scored for similar precision of the estimates.

Backcross generation
A backcross of the F_1 generation to lines P_1 $(QQMM)$ and P_2 $(qqmm)$ produces the genotypes and frequencies expressed in Table 19.3.

Table 19.3. Genotypes and frequencies of the backcrosses to parents P_1 $(QQMM)$ and P_2 $(qqmm)$ for the individual marker model

Genotype	Genotype frequency[a]	
	Backcross to P_1	Backcross to P_2
QQMM	1/2(1−r)	
QQMm	1/2r	
QqMM	1/2r	
QqMm	1/2(1−r)	
QqMm		1/2(1−r)
Qqmm		1/2r
qqMm		1/2r
qqmm		1/2(1−r)

[a] r is the recombination frequency between loci Q/q and M/m.

Separation of plants into marker classes yields the following expected marker means:

Marker class	Expected marker class mean
$MM(B_1)$	$\mu_2 + (1-r)a + rd$
$Mm(B_1)$	$\mu_2 + (1-r)d + ra$
$Mm(B_2)$	$\mu_3 + (1-r)d - ra$
$mm(B_2)$	$\mu_3 + (r-1)a + rd$

where μ_2 and μ_3 are the genotypic values of the remaining genes contributing to the quantitative trait in the backcross to P_1 (B_1) and P_2 (B_2), respectively. Estimates of a and d are as follows:

$$1/2[MM(B_1)-Mm(B_1)+Mm(B_2)-mm(B_2)] \\ = a(1-2r) \tag{19.9}$$
$$1/2[Mm(B_1)+Mm(B_2)-MM(B_1)-mm(B_2)] \\ = d(1-2r) \tag{19.10}$$

If the dominance effect is different from zero,

comparison of Equations 19.2 and 19.10 allows estimation of *r*:

$$r = \frac{1}{2} - \frac{2Mm(F_2) - MM(F_2) - mm(F_2)}{2[Mm(B_1) + Mm(B_2) - MM(B_1) - mm(B_2)]}$$

(19.11)

Selfed backcross progenies
(BS$_1$ and BS$_2$)
Since the selfed progenies from the backcrosses can be replicated, they have less environmental error than their individual parent plants, therefore, they are more appropriate for use in detecting QTL of low heritability. Each selfed backcross progeny is classified into the same marker class as the individual plant from which it was derived. Additive and dominance effects are estimated in this generation by making analogous comparisons as above. If the dominance effect is different from zero, an estimate of *r* is obtained in the same way as in Equation 19.11 but with better precision:

$$r = \frac{1}{2} - \frac{2Mm(S_1) - MM(S_1) - mm(Sd21)}{2[Mm(BS_1) + Mm(BS_2) - MM(BS_1) - mm(BS_2)]}$$

(19.12)

The combining of generations for
individual marker loci
If dominance is different from zero, the bias can be mitigated by using combined generations F_2 (or S_1) and backcrosses (or BS_1, S_2) for estimating $(1-2r)$ (Equations 19.11 and 19.12). The error associated with estimates of gene effects can be reduced by using selfing generations. However, if a correction for the bias $(1-2r)$ is introduced, then an important additional error is associated with these unbiased estimates. Estimation of a global confounding effect from distant and close QTL associated with the marker as indicated in disadvantage (2) is an inherent limitation of the individual marker locus model.

(b) Flanking marker loci

The use of flanking marker loci has been suggested by Lander and Botstein (1989) and Knapp *et al.* (1990) as a means of overcoming some of the limitations and disadvantages of the individual marker models.

The flanking marker loci model takes advantage of the possibility of mapping the genome at high-density intervals (i.e., 5-30 cM) by RFLPs and isozyme markers.

Let us assume two inbred lines P_1 and P_2 with flanking marker genotypes $M_1M_1M_2M_2$ and $m_1m_1m_2m_2$, respectively, and QTL genotypes Q_1Q_1 and q_1q_1, respectively. The F_1 cross between P_1 and P_2 has the following chromosome array:

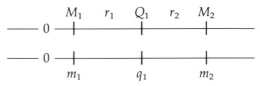

where r_1 and r_2 are the recombination frequencies between loci M_1 and Q_1 and between Q_1 and M_2, respectively. Knapp *et al.* (1990) developed models with flanking markers for the backcross progeny group and F_2 and F_3 progeny for both situations involving double-crossover and no-double-crossover. They pointed out that the no-double-crossover model may be justified because the errors caused by double-crossovers are smaller than the misclassification errors. The chance of a double crossover is at most 1% in the case of a 20 cM RFLP map (Lander and Botstein, 1989). Thus, the no-double-crossover situation will be considered in the flanking marker model developed in this Chapter. The genotypes and frequencies of the backcross generations appear in Table 19.4. Following a similar approach to that of Knapp *et al.* (1990) but using a slightly different strategy, the marker classes and their expected values in the backcross generations are expressed in Table 19.5.

Our model assumes genotypic values $+a$, d and $-a$ for the QTL genotypes QQ, Qq and qq, respectively; $r = r_1 + r_2$ where r (the recombination frequency between M_1 and M_2) can be directly estimated from the marker phenotypes; and $\rho = r_1/r$.

From Table 19.5, unbiased estimates of the additive *a* and dominance *d* effects of the QTL, ρ, μ_4 and μ_5 by contrasting marker classes from the backcross generations B_1 and B_2 were found in the following way:

Table 19.4. Genotypes and frequencies of the backcrosses to parents P$_1$ ($M_1M_1Q_1Q_1M_2M_2$) and P$_2$ ($m_1m_1q_1q_1m_2m_2$) for the flanking marker model with no-double-crossover

Genotype	Genotype frequency[a]	
	Backcross to P$_1$	Backcross to P$_2$
$M_1M_1Q_1Q_1M_2M_2$	$1/2(1-r_1-r_2)$	
$M_1M_1Q_1Q_1M_2m_2$	$1/2r_2$	
$M_1M_1Q_1q_1M_2m_2$	$1/2r_1$	
$M_1m_1Q_1Q_1M_2M_2$	$1/2r_1$	
$M_1m_1Q_1q_1M_2M_2$	$1/2r_2$	
$M_1m_1Q_1q_1M_2m_2$	$1/2(1-r_1-r_2)$	
$M_1m_1Q_1q_1M_2m_2$		$1/2(1-r_1-r_2)$
$M_1m_1Q_1q_1m_2m_2$		$1/2r_2$
$M_1m_1q_1q_1m_2m_2$		$1/2r_1$
$m_1m_1Q_1q_1M_2m_2$		$1/2r_1$
$m_1m_1q_1q_1M_2m_2$		$1/2r_2$
$m_1m_1q_1q_1m_2m_2$		$1/2(1-r_1-r_2)$

[a] r_1 and r_2 are recombination frequencies between loci M_1/m_1 and Q_1/q_1 and Q_1/q_1 and M_2/m_2, respectively.

Table 19.5. Marker classes and their expected values in the first backcross generations B$_1$ and B$_2$ for a flanking marker model with no-double-crossover

Backcross generation	Marker class		Expected marker class value[a]
B$_1$	$M_1M_1M_2M_2$	1	$\theta_1 = \mu_4 + a$
	$M_1M_1M_2m_2$	2	$\theta_2 = \mu_4 + (1-\rho)a + \rho d$
	$M_1m_1M_2M_2$	3	$\theta_3 = \mu_4 + \rho a + (1-\rho)d$
	$M_1m_1M_2m_2$	4	$\theta_4 = \mu_4 + d$
B$_2$	$M_1m_1M_2m_2$	5	$\theta_5 = \mu_5 + d$
	$M_1m_1m_2m_2$	6	$\theta_6 = \mu_5 + (1-\rho)d - \rho a$
	$m_1m_1M_2m_2$	7	$\theta_7 = \mu_5 + \rho d - (1-\rho)a$
	$m_1m_1m_2m_2$	8	$\theta_8 = \mu_5 - a$

[a] $\rho = r_1/r$ where $r = r_1 + r_2$ and r_1, r_2 and r are the recombination frequencies between M_1 and Q_1, Q_1 and M_2 and M_1 and M_2, respectively. μ_4 and μ_5 are the genotypic values of the remaining genes contributing to the quantitative trait in generations BC$_1$ and BC$_2$, respectively.

$$a = (\theta_1 - \theta_4 + \theta_5 - \theta_8)/2 \qquad (19.13)$$
$$d = (\theta_5 - \theta_8 - \theta_1 + \theta_4)/2 \qquad (19.14)$$
$$\rho = (\theta_1 - \theta_2 + \theta_3 - \theta_4)/4(\theta_1 - \theta_4) + (\theta_7 - \theta_8 + \theta_5 - \theta_6)/4(\theta_5 - \theta_8) \qquad (19.15)$$

$$\mu_4 = (\theta_2 + \theta_3 - \theta_5 + \theta_8)/2 \qquad (19.16)$$
$$\mu_5 = (\theta_6 + \theta_7 - \theta_1 + \theta_4)/2 \qquad (19.17)$$

where θ_1, θ_2, θ_3, θ_4, θ_5, θ_6, θ_7 and θ_8 are the means of the flanking marker classes $M_1M_1M_2M_2$ (1), $M_1M_1M_2m_2$ (2), $M_1m_1M_2M_2$ (3), $M_1m_1M_2m_2$ (4) in generation B$_1$ and $M_1m_1M_2m_2$ (5), $M_1m_1m_2m_2$ (6), $m_1m_1M_2m_2$ (7) and $m_1m_1m_2m_2$ (8) in generation B$_2$, respectively; μ_4 and μ_5 are the genotypic values of the remaining loci contributing to the quantitative trait in generations B$_1$ and B$_2$, respectively.

As was seen above, the use of selfed backcross progeny instead of individual plants would improve the precision of the estimates for traits of low heritability.

19.5.2 Strategies for data analysis

(a) Method of comparison of marker class means

The models described above allow estimation of gene effects and recombination frequencies by the method of comparison of marker class means.

In the case of an individual marker, unbiased estimates can be obtained under the following assumptions. (1) A dominance effect is present at the QTL to estimate the bias $(1-2r)$, although this estimate will have a large error; (2) only one QTL is associated with the marker; (3) the QTL segregates independently from the remaining QTLs.

In the case of flanking markers, unbiased estimates can be obtained under the following assumptions. (1) Only one QTL is located between the two markers, otherwise gene effects are estimated as if there was one linked QTL and the ρ estimate is inaccurate; (2) the QTL between the two markers segregates independently from the remaining loci contributing to the quantitative trait, otherwise the method will have an unpredictable bias.

(b) Method of maximum likelihood

Individual marker case

The method of maximum likelihood (ML) for estimating QTL effects linked to an individual marker in a F_2 population was used by Weller (1986). The proposed ML technique assumes normal distributions of the genotypic values of QQ, Qq and qq at the QTL with means μ_{11}, μ_{12} and μ_{22} and variances σ^2_{11}, σ^2_{12} and σ^2_{22}, respectively. An iterative computer algorithm was suggested to estimate the parameters. Starting values of the means, $\bar{X}_{11}$, $\bar{X}_{12}$ and $\bar{X}_{22}$, and the variances, S_{11}, S_{12} and S_{22} were assigned to the distributions of QQ, Qq and qq genotypic values, respectively. A starting value was also assigned to the recombination frequency, r. Likelihood of occurrence of the observed performance of all F_2 individuals based on normal distributions with the starting values assigned to the parameters was calculated with the following likelihood function:

$$L = [\prod_{i=1}^{I} f(MM_i)][\prod_{j=1}^{J} f(Mm_j)][\prod_{k=1}^{K} f(mm_k)] \quad (19.18)$$

where $f(MM_i)$, $f(Mm_j)$ and $f(mm_k)$ are the probability density functions of individuals i, j and k belonging to marker classes MM, Mm and mm, respectively. I, J and K are the number of individuals in marker classes MM, Mm and mm, respectively. The probability density functions of the marker classes are:

$$f(MM) = (1-r)^2 f(QQ) + 2r(1-r)f(Qq) + r^2 f(qq) \quad (19.19)$$
$$f(Mm) = r(1-r)f(QQ) + [1-2r(1-r)]f(Qq) + r(1-r)f(qq) \quad (19.20)$$
$$f(mm) = r^2 f(QQ) + 2r(1-r)f(qq) + (1-r)^2 f(qq) \quad (19.21)$$

where $f(QQ)$, $f(Qq)$ and $f(qq)$ are the probability density functions of genotypes QQ, Qq and qq at the QTL from normal distributions as defined above. After the starting values are tested, other values are sequentially tested following the algorithm described by Weller (1986) until a maximum likelihood is found. Using this technique, Weller (1986) was able accurately to estimate the means and variances, but not the recombination frequency from Monte Carlo simulated data for one QTL. One of the disadvantages of this method is that it does not

yield meaningful results for loci of small effects, unless a large number of individuals are scored.

Flanking markers case

Lander and Botstein (1989) and Knapp *et al.* (1990) suggested the exploitation of the power of high-density linkage maps to detect QTL effects by using techniques of maximum likelihood estimation with missing data. The method is similar to that described above for an individual marker but extended to all flanking markers. For the B_1 generation, following Lander and Botstein's (1989) approach, the linear equation relating phenotypes ϕ_i and genotypes g_i is:

$$\phi_i = \mu + bg_i + \epsilon_i \quad (19.22)$$

where g_i is a dummy variable (1 and 0 for genotypes QQ and Qq, respectively); ϵ is a random normal variable with mean 0 and variance σ^2; μ, b and σ^2 are unknown parameters; μ is the genotypic value of Qq and the remaining loci contributing to the QT; and b is the estimated gene effect when substituting an allele at the QTL. The likelihood function for a genotype (L_{gi}) is the probability density function of a normal distribution:

$$L_{gi}(\mu, b, \sigma^2) = \frac{1}{\sqrt{2\pi\sigma^2}} e^{-\frac{(\phi_i - \mu - bg_i)^2}{2\sigma^2}} \quad (19.23)$$

The likelihood function for all individuals falling in the marker class k (L_k) is:

$$L_k(\mu, b, \sigma^2) = \Pi_i[P_i(1)L_i(1) + P_i(0)L_i(0)] \quad (19.24)$$

where k(1 to 4) refers to the marker class in the BC_1 generation (Table 19.4); $P_i(1)$ and $P_i(0)$ refer to the probability of QQ and Qq occurring in the marker class k, respectively (e.g., $P_i(1) = r_2$ and $P_i(0) = r_1$ for marker class 2, $M_1M_1M_2m_2$, Table 19.4); $L_i(1)$ and $L_i(0)$ are likelihood functions for genotypes QQ and Qq, respectively.

The likelihood function for all scored individuals is:

$$L(\mu, b, \sigma^2) = \Pi_k L_k(\mu, b, \sigma^2) \quad (19.25)$$

Maximum likelihood estimates (MLE) with missing data are obtained with a computer programme that explores different combinations of values for the parameters by using the EM algorithm (Dempster *et al.*, 1977).

Since many flanking markers are tested in the whole set of the B_1 generation, a maximum likelihood is also obtained assuming no QTL effect (i.e., $b = 0$) to avoid false positives. As indicated by Paterson *et al.* (1988) the odds ratio is the chance that the data would arise from a QTL with this effect divided by the chance that they would arise given no linked QTL. The $\log_{10}$ of the odds ratio (LOD) is calculated with the following expression:

$$ LOD = \log_{10}[L(\mu,b,\sigma^2)/L(\mu_0,0,\sigma_0^2)] \quad (19.26) $$

The LOD threshold value for avoiding a false positive with 0.95 probability when testing 60 flanking markers in 1200 cM was estimated to be about 2.4 (Lander and Botstein, 1989).

By using simulated data for a 250 backcross progeny, the method detected the largest 4 out of 5 actual QTL (Lander and Botstein, 1989). Shortcomings of the method are: (1) small gene effects (i.e., less than one third the standard phenotypic deviation) are not easily detected unless a large progeny family is scored; (2) dominance effects are not considered in the method; (3) if a QTL in a flanked region does not segregate independently from others in nearby flanked regions, a larger progeny will be required; (4) two divergent strains, artificially selected for high and low performance, are recommended for estimating large gene effects.

(c) Other methods

Because of the shortcomings of the methods described above, other strategies should be explored to analyse the data. Strategies should be able to fulfil the following requirements: (1) allow unbiased estimates of additive and dominance gene effects to be obtained; (2) to get estimates of ρ (it locates the QTL on the genome between the two flanking markers) within a small surrounding of the real QTL position; (3) to use a limited number of individuals because complete mapping is a complex and expensive procedure; (4) to protect the estimates against large type I and type II errors.

Multiple linear regression for analysing RFLP-associated QTL data was suggested by Cowen (1989) who pointed out that the stepwise regression (Draper and Smith, 1981) seems to be the most reasonable analytic procedure when compared to the backward elimination procedure and the 'linear regression by leaps and bounds' (Romero-Severson *et al.*, 1989). Advantages of the stepwise method are: (1) incorporation and elimination of a variable in the model can be controlled backward and forward by a significant F-value chosen by the researcher; (2) the computer subroutine for this method is available in many statistical packages. However, one disadvantage common to all multiple regression methods is that the estimate of the value of the partial regression coefficient for any predictor depends on the other predictors included in the model when the independent variables are correlated. So far, no conclusive information is available about the relative efficiency of the different methods for statistical analysis of RFLP-associated QTL data. Further research should continue in this field.

19.5.3 Plant breeding applications

QTL mapping is becoming an important tool in plant breeding. Several studies have shown the utility of genetic markers (morphological, isozymes and RFLP) for detecting and locating genes underlying quantitative variation on the tomato (Tanksley and Hewitt, 1988; Paterson *et al.*, 1988; Osborn *et al.*, 1987; Weller, 1987) and maize genomes. These identified regions can be targeted in selection experiments. Various applications of markers to facilitate the selection of quantitative characters are described in the following paragraphs.

(a) Location of QTL on the genome

Results of maize studies (Stuber *et al.*, 1987; Edwards *et al.*, 1987) to locate QTLs associated with grain yield and 24 yield-related traits using 17 to 20 isozyme individual marker loci have shown: (1) association among marker loci and many quantitative traits were highly significant; (2) different genomic regions contributed to yield; (3) types of gene action were predominantly dominant for grain yield, top ear grain weight and ear length, and additive for ear number, kernel row number and second ear grain weight; (4) single and cumulative marker loci effects accounted for between 0.3 and 16% and between 8 and 40% of the phenotypic variation of the traits, respectively.

(b) Development of inbred lines

Marker information can be used to facilitate selection when developing inbred lines from the F_2 by the pedigree method in either self-pollinated or cross-pollinated crops (Dudley, 1991). Consider that the cross $M_iM_iQ_iQ_i \times m_im_iq_iq_i$ ($i = 1, 2, \ldots$) is selfed for many generations to obtain inbred lines. Assume that four favourable unlinked QTLs are to be retained in the lines. If no marker information is available the chance that one inbred carries all four QTLs is $(\frac{1}{2})^4 = 0.0625$. On the other hand, if the QTLs can be followed by markers and no selection for M_iM_i is carried out at any generation, the probability of recovering inbred lines with the favourable genotype Q_iQ_i among the marker phenotypes M_iM_i is given by the following equation (Haldane and Waddington, 1931):

$$p(Q_i\,Q_i/M_i\,M_i) = \frac{1}{1+2r} \qquad (19.27)$$

where $p(Q_iQ_i/M_iM_i)$ is the probability of finding Q_iQ_i on marker M_iM_i. If the recombination value is $r = 0.2$ for each marker-QTL combination, the chance of recovering one inbred with all four QTLs is $(0.7143)^4 = 0.2603$. A large number of individuals need to be grown in each generation because the proportion of individuals with homozygous markers M_iM_i for the four loci is $(0.25)^4 = 0.0039$ in the F_2, and $[\frac{1}{2}(1-1/2^{n+1})]^4$ in the nth generation after the F_2. If selection of M_iM_i individuals is carried out in the F_2, the probability of recovering one inbred carrying the four QTLs is $(1-r)^4 = (0.8)^4 = 0.4069$. In this situation, a large number of individuals have also to be examined in the F_2, but only a few need to be grown in the following selfed generations. Assume now that flanking markers separated at $r = 0.2$ are used to follow the QTL. If one or two individuals

with the four double flanking marker homozygotes $M_iM_iM_jM_j$ are selected in the F_2, the probability of deriving one inbred line that carries the four QTLs is at least $[1-(\frac{1}{2}r)^2]^4 = (0.99)^4 = 0.9606$. However, an F_2 of very large size has to be examined, because the proportion of individuals with the double flanking marker homozygous for the four loci is $[\frac{1}{2}(1-r)]^8 = 0.000655$. These examples illustrate the advantages of using markers for the development of inbred lines, specially when many QTLs have to be fixed.

(c) Improvement of inbred lines

A method for the transfer of favourable alleles affecting a quantitative trait from a donor inbred line (P_w) to a parental elite inbred line (P_1) used in a hybrid ($P_1 \times P_2$) has been proposed by Dudley (1984). This method can be monitored with the aid of marker loci as follows: after the cross $P_1 \times P_w$ and $P_2 \times P_w$, plants or families from appropriate generations (F_2, backcross, or their selfing progenies) are grown and scored for identifying the favourable alleles present in P_w and lacking in P_1 and P_2 that are linked to marker loci. F_2 plants, or better double-haploid lines from F_1 gametes, carrying a high concentration of favourable target alleles with large effects can be identified by their marker phenotype and backcrossed to the original P_1.

When improving second cycle inbred lines by recurrent backcrossing (e.g., $M_iM_iQ_iQ_i$ is the donor and $m_im_iq_iq_i$ the recurrent parent, $i = 1, 2, \ldots$), Dudley (1991) has shown that selection for markers M_im_i in each backcross generation will markedly increase the frequency of favourable genotypes Q_iQ_i in generations beyond BC_1, particulary if the recombination value $r \leq 0.1$. The use of flanking markers will greatly increase the chance of finding favourable QTL alleles in the recovered inbred although it will require a larger number of individuals to be examined in the backcross generations.

(d) Improvement of populations

Synthetic populations are usually at linkage disequilibrium during the first random mating generations after they are formed. In addition, disequilibrium for tightly linked loci may remain for many generations because the rate of decay for approaching equilibrium is $1-r$ (Falconer, 1976). High-density marker mapping will be necessary for the improvement of these populations. Theoretical studies on the improvement of linkage disequilibrium populations using markers have been made by Dudley (1991). Assume that the frequency of four independent marker loci linked with favourable QTLs is 0.25 each in the population. Then, the probability of obtaining an individual homozygous for all four markers is $(0.25)^8 = 0.000015$. If individuals with at least one dose of a marker allele are selected, the probability of obtaining individuals carrying the four markers is $(7/16)^4 = 0.0366$, a much more manageable number. Dudley (1991) concluded that as the number of marker–QTL combinations increase, or as the frequency of each marker–QTL combination in the population decreases, selection to increase the frequency of each marker–QTL combination prior to selection for multiple combinations will be necessary.

(e) Hybrid performance

The genetic distance, as estimated with markers, has been proposed for the establishment of relationships between inbred lines and to predict the performance of their single crosses. Results of Lee *et al.* (1989), Godshalk *et al.* (1990), Melchinger *et al.* (1990) and Dudley *et al.* (1991) indicated that RFLP-based Rodger's genetic distance can be used to allocate maize inbreds to heterotic groups, but are of limited use in predicting heterotic performance of single crosses between unrelated lines. In addition, a large number of markers are required to obtain reliable estimates of genetic distance (Melchinger *et al.*, 1991). The limitations of the Rodgers's genetic distance for hybrid prediction could be explained because it is only based on the difference of allele frequencies at different marker loci, not on the quantitative trait effect associated with the markers. A genetic distance weighted on the relative effect of each QTL-associated marker is suggested as a more reliable predictor of hybrid performance.

19.6 PROSPECTS OF MARKER RESEARCH

The use of marker loci in plant breeding has evolved mainly on two fronts. First, by devising new applications of these tools to unexplored areas. Second, by the improvement of methods in order to obtain either more markers of better quality or more efficient procedures than those previously available. Two recent developments, one concerning a new field of application; marker-aided cloning of genes, and the other a new source of polymorphisms; random amplified polymorphic DNA (RAPD) markers, are summarized below.

19.6.1 Map-based gene cloning

Gene-cloning methods are well developed for those eukaryotic genes whose products (mRNA, proteins) are known (Chapter 8). However, the products of genes involved in the expression of agriculturally important characters such as disease resistance are seldom known. For this reason, it has been necessary to develop new approaches to gene cloning. One approach, known as 'reverse genetics', is based on tight linkage between the gene of interest and RFLP markers (Tanksley *et al.*, 1989; Young, 1990; some of the technical aspects of this process have been considered in section 8.5.5). The first step of this method consists of identifying markers flanking the target gene and tightly linked to it. These markers can be found either by classical linkage analysis in F_2 or backcross populations or, and more efficiently, by using a set of near isogenic lines (NILs) which allow easy detection of markers located in the linkage block of the gene of interest (Young *et al.*, 1988; Muehlbauer *et al.*, 1989; Messeguer *et al.*, 1991). The next step is to move from one of the markers to the other through the target gene by means of a 'chromosome walk'. With this technique, partially overlapping DNA probes covering the entire 'physical' distance between both markers are identified. The final step is the localization of the target gene. If transformation is feasible, a

DNA clone containing the gene can be identified because plants transformed with such a clone will exhibit the phenotype conferred by the gene of interest.

The process of map-based cloning is a difficult one. At the present time none of the genes of interest in plant breeding have been cloned with this procedure. However, the development of methods which allow the separation and cloning of large DNA fragments promise rapid progress towards the isolation of valuable genes. Many genes responsible for disease resistance in plants are genetically well characterized and located on classical or molecular marker maps. Their economical importance is very high and yet virtually nothing is known about their biochemistry. They are natural targets for map-based cloning and some of them, including those which confer resistance to viruses (Young *et al.*, 1988), nematodes (Messeguer *et al.*, 1991) or fungal diseases (Young, 1990; Yu *et al.*, 1991), have started to be studied with this objective. The isolation, cloning and molecular characterization of such genes will provide information of great relevance for the breeding of new and more stable forms of disease resistance in plants.

19.6.2 RAPD markers

Williams *et al.* (1990) and Welsh and McClellan (1990) proposed the use of single short random primers (usually 10-mers) in a polymerase chain reaction (PCR) as a method of generating polymorphic markers (RAPDs). DNA polymorphisms result either from differences in the DNA sequence at primer binding sites or by chromosome changes affecting the amplified regions. After electrophoresis, the amplified fragments can be visualized by fluorescence and the resulting pattern of bands can be interpreted genetically. Technical details of PCR are presented in sections 8.4 and 12.1.3.

RAPD markers have some important advantages over RFLPs. First, the process is simpler and faster, since after DNA extraction only two relatively rapid operations (4–6 hours) are involved: DNA amplification and electrophoretic separation. Second, the method requires less investment in laboratory equipment, is less labour-intensive and is safer since radioactivity is not necessary. Third, it requires small amounts of genomic DNA.

RAPD markers have proven useful for cultivar identification (Hu and Quiros, 1991), development of genome-specific markers (Quiros *et al.*, 1991) and detection of markers tightly linked to genes of interest by means of NIL analysis (Martin *et al.*, 1991). Using RAPD and RFLP markers, Michelmore *et al.* (1991) proposed a simple and effective method for the allocation of markers to specific regions of the chromosome. The procedure, termed 'bulked segregant analysis', consists in the comparison between two pooled DNA samples made up from individuals of a segregating population in linkage disequilibrium. These samples differ for a specific DNA region that contains the gene of interest (i.e., individuals susceptible *vs* resistant for a particular disease), but are random for the rest of the genome. Differences in banding patterns between both samples will reveal the presence of markers located in the region of interest. Bulked segregant analysis may be performed in a segregating population, which is much easier to obtain than a pair of NILs. Moreover, the same population can be used to study any region of the chromosome having a segregating locus as well as for the determination of the genetic distance between markers and target genes.

Results obtained in plants indicate that RAPDs: (1) generally behave as dominant markers (presence *vs* absence of the amplified product); (2) are highly polymorphic and informative (average of 5–10 loci per primer); (3) complement RFLP maps in the sense that RAPDs can be obtained from highly repetitive DNA fragments allowing the access to regions of the chromosome previously not studied; (4) require considerably less time and effort than RFLPs for a similar level of information. While important restrictions to the use of RAPDs derive from their dominant mode of inheritance, these markers are for many other reasons a major advancement in marker research, and an important step forward to the increased utility of markers in practical plant breeding.

REFERENCES

Allard, R.W. (1960), *Principles of Plant Breeding*, John Wiley & Sons, New York.

Arús, P., Tanksley, S.D., Orton, T.J. and Jones, R.A. (1982), Electrophoretic variability as a tool for determining seed purity and for breeding hybrid varieties of *Brassica oleracea*. *Euphytica*, **31**, 417–428.

Bailey, D.C. (1983), Isozymic variation and plant breeder's rights. In *Isozymes in Plant Genetics and Breeding*, Tanksley, S.D. and Orton, T.J. (eds), Elsevier Science Publishers, Amsterdam, pp. 425–440.

Brinkman, M.A. and Frey, K.J. (1977), Yield component analysis of oat isolines that produce different grain yields. *Crop Sci.*, **17**, 165–168.

Brown, A.H.D. (1978), Isozymes, plant population genetic structure and genetic conservation. *Theor. Appl. Genet.*, **52**, 145–157.

Chaparro, J.X., Goldy, R.G., Mowrey, B.D. and Werner, D.J. (1989), Identification of *Vitis vinifera* L. × *Muscadina rotundifolia* Small hybrids by starch gel electrophoresis. *HortScience*, **24**(1), 128–130.

Colby, L.W. and Peirce, L.C. (1988), Using an isozyme marker to identify doubled haploids from anther culture of asparagus. *HortScience*, **23**(4), 761–763.

Cowen, N.M. (1988), The use of replicated progenies in marker-based mapping of QTLs. *Theor. Appl. Genet.*, **75**, 857–862.

Cowen, N.M. (1989), Multiple linear regression analysis of RFLP data sets used in mapping QTLs. In *Development and Application of Molecular Markers to Problems in Plant Genetics*, Helentjaris, T. and Burr, B. (eds), Cold Spring Harbor Laboratory, New York.

Dempster, A.P., Laird, N.M. and Rubin, D.B. (1977), Maximum likelihood from incomplete data via the EM algorithm. *J. R. Stat. Soc.*, **39**, 1–38.

Draper, N.R. and Smith, H. (1981). *Applied Regression Analysis*, 2nd edition, Wiley, New York.

Dudley, J.W. (1984), A method of identifying lines for use in improving parents of a single cross. *Crop Sci.*, **24**, 355–357.

Dudley, J.W. (1991), Use of molecular markers in corn breeding. In *1991 Illinois Corn breeders School*, Department of Agronomy, University of Illinois, Urbana, pp. 120–138.

Dudley, J.W., Saghai Maroof, M.A. and Rufener, G.K. (1991), Molecular markers and grouping of parents in maize breeding programs. *Crop Sci.*, **31**, 718–723.

Edwards, M.D., Stuber, C.W. and Wendel, J.F. (1987), Molecular marker-facilitated investigations of quantitative-trait loci in maize. I. Numbers, genomic distribution and types of gene action. *Genetics*, **116**, 113–125.

Falconer, D.S. (1976), *Introduction to Quantitative Genetics*, Longman, London.

Feuerstein, U., Brown, A.D.H. and Burdon, J.J. (1990), Linkage of rust resistance genes from wild barley (*Hordeum spontaneum*) with isozyme markers. *Plant Breeding*, **104**, 318–324.

Godshalk, E.B., Lee, M. and Lamkey, K.R. (1990), Relationship of restriction fragment length polymorphisms to single-cross hybrid performance of maize. *Theor. Appl. Genet.*, **80**, 273–280.

Gottlieb, L.D. (1981), Electrophoretic evidence and plant populations. *Prog. Phytochem.*, **7**, 1–46.

Gottlieb, L.D. (1986). The genetic basis of plant form. *Philos. Trans. R. Soc. Lond.*, **313**, 197–208.

Haldane, J.B.S. and Waddington, C.H. (1931), Inbreeding and linkage. *Genetics*, **16**, 357–374.

Hanson, W.D. (1959), Early generation analysis of lengths of heterozygous chromosome segments around a locus held heterozygous with backcrossing or selfing. *Genetics*, **44**, 833–837.

Hauagge, R., Kester, D.E. and Assay, R.A. (1987), Isozyme variation among California almond cultivars: I. Inheritance. *J. Amer. Soc. Hort. Sci.*, **112**, 687–693.

Havey, M.J. and Muehlbauer, F.J. (1989), Linkages between restriction fragment length, isozyme and morphological markers in lentil. *Theor. Appl. Genet.*, **77**, 395–401.

Hayward, M.D., Olesen, A., Due, I.K., Jenkins, R. and Morris, P. (1990), Segregation of isozyme marker loci amongst androgenetic plants of *Lolium perenne* L. *Plant Breeding*, **104**, 68–71.

Hu, J. and Quiros, C.F. (1991), Identification of broccoli and cauliflower cultivars with RAPD markers. *Plant Cell Rep.* **10**, 505–511.

Jayakar, S.D. (1970), On the detection and estimation of linkage between a locus influencing a quantitative character and a marker locus. *Biometrics*, **26**, 451–464.

Kling, J.G., Gardner, C.O. and Thomas-Compton, M.A. (1987), Gene frequencies in mass-selected corn populations and in derived random inbred lines. *Crop Sci.*, **27**, 190–194.

Knapp, S.J., Bridges, W.C. Jr and Bikers, D. (1990), Mapping quantitative trait loci using molecular marker linkage maps. *Theor. Appl. Genet.*, **79**, 583–592.

Koebner, R.M.D. and Martin, P.K. (1990), Association of eyespot resistance in wheat cv 'Capelle-

Desprez' with endopeptidase profile. *Plant Breeding*, **104**, 312–317.

Konishi, T., Kawada, N., Yoshida, H. and Sohotme, K. (1989), Linkage relationship between two loci for the barley yellow mosaic resistance of Mokusekko 3 and esterase isozymes in barley (*Hordeum vulgare* L.). *Jpn. J. Breed.*, **39**, 423–430.

Lander, E.S. and Botstein, D. (1989), Mapping mendelian factors underlying quantitative traits using RFLP linkage maps. *Genetics*, **121**, 185–199.

Law, C.N. (1967), The location factors controlling a number of quantitative characters in wheat. *Genetics*, **56**, 445–461.

Lee, J.M. and Ellstrand, N.C. (1987), Inheritance and linkage of isozymes in the cherimoya. *J. Hered.*, **78**, 383–387.

Lee, M., Godshalk, E.B., Lamkey, K.K. and Woodman, W.W. (1989), Association of restriction fragment length polymorphisms among maize inbreds with agronomic performance of their crosses. *Crop Sci.*, **29**, 1067–1071.

Lo Shiavo, F., Giulano, G. and Terzi, M. (1983), Identifying natural and parasexual hybrids. In *Isozymes in Plant Genetics and Breeding*, Tanksley, S.D. and Orton, T.J. (eds), Elsevier Science Publishers, pp. 305–312.

Manganaris, A.G. and Alston, F.H. (1987), Inheritance and linkage relationships of glutamate oxaloacetate transaminase isozymes in apple. 1. The gene GOT-1, a marker for the S incompatibility locus. *Theor. Appl. Genet.*, **74**, 154–161.

Martin, G.B., Williams, J.G.K. and Tanksley, S.D. (1991), Rapid identification of markers linked to a *Pseudomonas* resistance gene in tomato by using random primers and near-isogenic lines. *Proc. Natl. Acad. Sci. USA*, **88**, 2336–2340.

McGranahan, G.H., Tulecke, W., Arulsekar, S. and Hansen, J.J. (1986), Intergeneric hybridization in the Juglandaceae: *Pterocarya* sp. × *Juglans regia*. *J. Am. Soc. Hort. Sci.*, **111**(4), 627–630.

McMillan, I. and Robertson, A. (1974), The power of methods for detection of major genes affecting quantitative characters. *Heredity*, **32**, 349–356.

McMillin, D.E., Allan, R.E. and Roberts, D.E. (1986), Association of an isozyme locus and strawbreaker foot rot resistance derived from *Aegilops venticosa* in wheat. *Theor. Appl. Genet.*, **72**, 743–747.

Melchinger, A.E. (1990), Use of molecular markers in breeding for oligogenic disease resistance. *Plant Breeding*, **104**, 1–19.

Melchinger, A.E., Lee, M., Lamkey, K.R. and Woodman, W.L. (1990), Genetic diversity for restriction fragment length polymorphisms: relation to estimated genetic effects in maize inbreds. *Crop Sci.*, **30**, 1033–1040.

Melchinger, A.E., Messmer, M.M., Lee, M., Woodman, W.L. and Lamkey, K.R. (1991), Diversity and relationship among U.S. maize inbreds revealed by restriction fragment length polymorphisms. *Crop Sci.*, **31**, 669–678.

Messeguer, R., Ganal, M.W. and Tanksley, S.D. (1991), High resolution RFLP map around the root knot nematode resistance gene (Mi) in tomato. *Theor. Appl. Genet.*, **82**, 529–536.

Michelmore, R.W., Paran, I. and Kesseli, R.V. (1991), Identification of markers linked to disease resistance genes by bulked segregant analysis: A rapid method to detect markers in specific genomic regions by using segregating populations. *Proc. Natl. Acad. Sci. USA*, **88**, 9828–9832.

Miller, P.J., Parfitt, D.E. and Weinbaum, S.A. (1989), Outcrossing in peach. *HortScience*, **24**(2), 359–360.

Mitton, J.B. (1983), Conifers. In *Isozymes in Plant Genetics and Breeding*, Part B, Tanksley, S.D. and Orton, T.J. (eds), Elsevier Science Publishers, Amsterdam, pp. 443–472.

Muehlbauer, G.J., Specht, J.E., Staswick, P.E., Graef, G.L. and Thomas-Compton, M.A. (1989), Application of the near-isogenic lines-gene mapping technique to isozyme markers. *Crop Sci.*, **29**, 1548–1553.

Muehlbauer, G.J., Specht, J.E., Thomas-Compton, M.A., Staswick, P.E. and Bernard, R.L. (1988), Near-isogenic lines. A potential resource in the integration of conventional and molecular marker linkage maps. *Crop Sci.*, **28**, 729–735.

Navot, N., Sarfatti, M. and Zamir, D. (1990), Linkage relationships of genes affecting bitterness and flesh colour in watermelon. *J. Hered.*, **81**(2), 162–164.

O'Brien, S.J. (1990), *Genetic Maps*, Book 6, *Plants*, Cold Spring Harbor Laboratory Press, New York.

Orton T.J. and Browers, M.A. (1985), Segregation of genetic markers among plants regenerated from cultured anthers of broccoli (*Brassica oleracea* var 'italica'). *Theor. Appl. Genet.*, **69**, 637–643.

Osborn, T.C., Alexander, D.C. and Fobes, J.F. (1987), Identification of restriction fragment length polymorphisms linked to genes controlling soluble solids content in tomato fruit. *Theor. Appl. Genet.*, **73**, 350–356.

Parfitt, D.E., Arulsekar, S. and Ramming, D.W. (1985), Identification of Plum × Peach hybrids by isozyme analysis. *HortScience*, **20**(2), 246–248.

Paterson, A.H., Lander, E.S. Hewitt, J.D., Paterson, S., Lincoln, S.E. and Tanksley, S.D. (1988),

Resolution of quantitative traits into mendelian factors by using a complete linkage map of restriction fragment length polymorphisms. *Nature*, **335**, 721–726.

Quiros, C.F., Douches, D. and D'Antonio, V. (1987), Inheritance of annual habit in celery: cosegregation with isozyme and anthocyanin markers. *Theor. Appl. Genet.*, **74**, 203–208.

Quiros, C.F., Hu, J., This, P., Chevre, A.M. and Delseny, M. (1991), Development and chromosomal localization of genome-specific markers by polymerase chain reaction in *Brassica*. *Theor. Appl. Genet.*, **82**, 627–631.

Rick, C.M. and Fobes, J.F. (1974), Association of an allozyme with nematode resistance. *Rep. Tomato Genet. Coop.*, **24**, 25.

Ritland, K. and Jain, S.K. (1981), A model for the estimation of outcrossing and gene frequencies using an independent loci. *Heredity*, **47**, 35–52.

Romero-Severson, J., Lotzer, J., Brown, C. and Murray, M. (1989), Use of RFLPs for analysis of quantitative trait loci in maize. In *Developments and Application of Molecular Markers to Problems in Plant Genetics*, Helentjaris, T. and Burr, B. (eds), Cold Spring Harbor Laboratory, New York, pp. 97–102.

Sax, K. (1923), The association of size differences with seed-coat pattern and pigmentation in *Phaseolus vulgaris*. *Genetics*, **8**, 552–560.

Schnabel, A. and Hamrick, J.L. (1990), Nonrandom associations between sex and 6-phosphogluconate dehydrogenase isozyme genotypes in *Gleditsia triacanthos* L. *J. Hered.*, **81**(3), 230–233.

Shenoy, V.V., Seshu, D.V. and Sachan, J.K.S. (1990), *Shikimate dehydrogenase-1²* allozyme as a marker for high seed protein content in rice. *Crop Sci.*, **30**, 937–940.

Smith, J.S.C. and Smith, O.S. (1989), The description and assessment of distances between inbred lines of maize: The utility of morphological, biochemical and genetic descriptors and a scheme for the testing of distinctiveness between inbred lines. *Maydica*, **34**, 151–161.

Soller, M., and Beckmann, J.S. (1983), Genetic polymorphism in varietal identification and genetic improvement. *Theor. Appl. Genet.*, **67**, 25–33.

Stam, P. and Zeven, A.C. (1981), The theoretical proportion of the donor genome in near-isogenic lines of self-fertilizers bred by back-crossing. *Euphytica*, **30**, 227–238.

Stuber, C.W., Edwards, M.D. and Wendel, J.F. (1987), Molecular marker-facilitated investigations of quantitative trait loci in maize. II.

Factors influencing yield and its component traits. *Crop Sci.*, **27**, 639–648.

Tanksley, S.D. (1983), Molecular markers in plant breeding. *Plant Mol. Biol. Reporter*, **1**, 3–8.

Tanksley, S.D. (1984), High rates of cross-pollination in chile pepper. *HortScience*, **19**(4), 580–582.

Tanksley, S.D. and Hewitt, J. (1988), Use of molecular markers in breeding for soluble solids content in tomato. A re-examination. *Theor. Appl. Genet.*, **75**, 811–823.

Tanksley, S.D., Medina-Filho, H. and Rick, C.M. (1981), The effect of isozyme selection on metric characters in an interspecific backcross of tomato – basis of an early screening procedure. *Theor. Appl. Genet.*, **60**, 291–296.

Tanksley, S.D., Medina-Filho, H. and Rick, C.M. (1982), Use of naturally-occurring enzyme variation to detect and map genes controlling quantitative traits in an interspecific backcross of tomato. *Heredity*, **49**, 11–25.

Tanksley, S.D. and Rick, C.M. (1980), Isozymic gene linkage map of the tomato; applications in genetics and breeding. *Theor. Appl. Genet.*, **57**, 161–170.

Tanksley, S.D., Rick, C.M. and Vallejos, C.E. (1984), Tight linkage between a nuclear male-sterile locus and an enzyme marker in tomato. *Theor. Appl. Genet.*, **68**, 109–113.

Tanksley, S.D., Young, N.D., Paterson, A.H. and Bonierbale, M.W. (1989), RFLP mapping in plant breeding: new tools for an old science. *Bio/Technology*, **7**, 257–264.

Tanksley, S.D. and Zamir, D. (1988), Double tagging of a male-sterile gene in tomato using a morphological and enzymatic marker gene. *HortScience*, **23**(2), 387–388.

Torres, A.M. (1983), Fruit trees. In *Isozymes in Plant Genetics and Breeding*, Part B, Tanksley, S.D. and Orton, T.J. (eds), Elsevier Science Publishers, Amsterdam, pp. 401–421.

Torres, A.M., Soost, R.K. and Mau-Lastovicka, T. (1982), Citrus isozymes. Genetics and distinguishing nucellar from zygotic seedlings. *J. Hered.*, **73**, 335–339.

Weeden, N.F. and Provvidenti, R. (1988), A marker locus, *Adh-1*, for resistance to pea enation mosaic virus in *Pisum sativum*. *J. Hered.*, **79**(2), 128–131.

Weeden, N.F., Provvidenti, R. and Marx, G.A. (1984), An isozyme marker for resistance to bean yellow mosaic virus in *Pisum sativum*. *J. Hered.*, **75**, 411–412.

Weller, J.I. (1986), Maximum likelihood techniques

for the mapping and analysis of quantitative trait loci with the aid of genetic markers. *Biometrics*, **42**, 627–640.

Weller, J.I. (1987), Mapping and analysis of quantitative trait loci in *Lycopersicon* (tomato) with the aid of genetic markers using approximate maximum likelihood methods. *Heredity*, **59**, 413–421.

Welsh, J. and McClellan, M. (1990), Fingerprinting genomes using PCR with arbitrary primers. *Nucleic. Acids Res.*, **18**, 7213–7218.

Weng, J. and Chen, C. (1989), Photosynthetic ability, grain yield and an esterase band in rice genotypes. *Euphytica*, **42**, 265–268.

Williams, J.G.K., Kubelik, A.R., Livak, K.J., Rafalski, J.A. and Tingey, S.V. (1990), DNA polymorphisms amplified by arbitrary primers are useful as genetic markers. *Nucleic. Acids Res.*, **18**, 6531–6535.

Worland, A.J., Law, C.N., Hollins, T.W. Koeber, R.M.D. and Guira, A. (1988), Location of a gene for resistance to eyespot (*Pseudocercosporella erpotrichoides*) on chromosome 7D of bread wheat. *Plant Breeding*, **101**, 43–51.

Young, N.D. (1990), Potential applications of map-based cloning to plant pathology. *Physiol. Mol. Plant Pathol.*, **37**, 81–94.

Young, N.D., Zamir, D., Ganal, M.W. and Tanksley, S.D. (1988), Use of isogenic lines and simultaneous probing to identify DNA markers tightly linked to the *Tm-2a* gene in tomato. *Genetics*, **120**, 579–583.

Young, N.D. and Tanksley, S.D. (1989), RFLP analysis of the size of chromosomal segments retained around the *Tm-2* locus of tomato during backcross breeding. *Theor. Appl. Genet.*, **77**, 353–359.

Yu, Z.H., Mackill, D.J., Bonman, J.M. and Tanksley, S.D. (1991), Tagging genes for blast resistance in rice via linkage to RFLP markers. *Theor. Appl. Genet.*, **81**, 471–476.

Zamir, D., Tanksley, S.D. and Jones, R.A. (1981), Genetic analysis of the origin of plants regenerated from anther tissues of *Lycopersicon esculentum*. *Plant Sci. Lett.*, **21**, 223–227.

Zeven, A.C., Knott, D.R. and Johnson, R. (1983), Investigation of linkage drag in near isogenic lines of wheat by testing for seedling reaction to races of stem rust and yellow rust. *Euphytica*, **32**, 319–327.

20

Gametophytic and sporophytic selection

E. Ottaviano and M. Sari-Gorla

20.1 GAMETOPHYTIC SELECTION *VERSUS* SPOROPHYTIC SELECTION

Gametophytic selection as a tool in plant breeding refers to methodologies based on the application of selection pressures to the male gametophytic generation with the purpose of improving traits of the sporophytic progeny. Frequently it is also referred to as pollen selection.

The rationale behind this methodology is strictly related to the main features of the angiosperm life cycle, which, as a consequence of sexual reproduction, is composed of two alternate generations: (1) the gametophyte, which develops from haploid cells (spores) produced by meiosis; (2) the sporophyte which develops from the zygote (diploid) into mature plants carrying female (pistil) and male (anther) structures. For the purposes of this chapter, the development of the male gametophytic generations is of particular importance (Fig. 20.1a,b). Inside the young anther, archesporial cells, differentiated from the meristematic tissue, divide to form the wall layers and a large number of pollen mother cells. These undergo meiosis, producing microspore tetrads. After tetrad release, the microspore develops and undergoes the first haploid mitotic division

Plant Breeding: Principles and prospects. Edited by M.D. Hayward, N.O. Bosemark and I. Romagosa. Published in 1993 by Chapman & Hall, London. ISBN 0 412 43390 7.

forming one vegetative and one generative cell. The generative cell, during development in the anther (trinucleate species) or during pollen tube growth (binucleate species), produces two spermatic cells by means of a second mitotic division. Mature pollen is shed by the anther and, after a short period of life as an independent organism, is deposited on the stigma, where it germinates, producing a pollen tube which, passing through the transmitting tissue of the style, reaches the ovule and discharges the two male gametophytic cells to accomplish the double fertilization.

With reference to the alternating generations, male gametophytic selection (MGS) can be described as a change in gene frequencies resulting from selection pressures applied to the stages of the male gametophytic generation and sporophytic selection (SS) as a change in gene frequencies produced by selection acting on the diploid phase, which begins with the zygote and ends with the completion of meiosis (Fig. 20.2). Selection can take place on both the female and male gametophytic generations. However, because of the peculiar features of the male gametophytic generation, such as the large numbers of individuals competing with each other during development in the anther and during pollen germination and tube growth in the style, and because of the independent stage (the lag period between shedding and style deposition), the phenomenon is much more pronounced in the male than in the female gametophyte selection generation and easier to manipulate for breeding purposes.

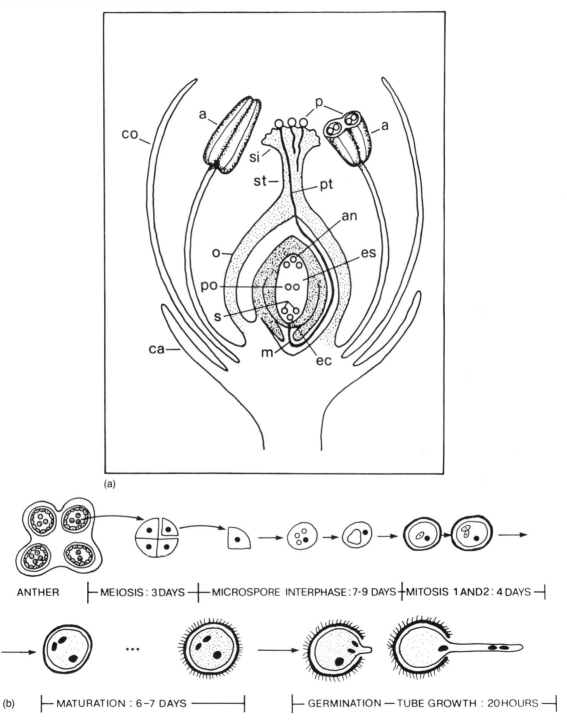

(a)

ANTHER ├─MEIOSIS : 3 DAYS ─┼─MICROSPORE INTERPHASE : 7-9 DAYS ┼MITOSIS 1 AND 2 : 4 DAYS ─┤

(b) ├─MATURATION : 6-7 DAYS ─────────┤ ├─ GERMINATION ─ TUBE GROWTH : 20 HOURS ─┤

Fig. 20.1. (a) Angiosperm floral structure in longitudinal section: a, anther; co, corolla; ca, calyx; p, pollen grains; si, stigma; st, style; o, ovary; po, polar nuclei; s, synergids; an, antipodals; es, embryo sac; pt, pollen tube; ec, egg cell; m, micropyle. (b) Male gametophyte developmental stages from meiosis to fertilization (maize).

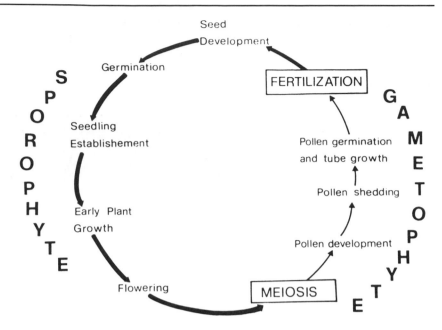

Fig. 20.2. Sporophytic and male gametophytic generations in higher plants.

For genes that are equally expressed in the male gametophytic and in the sporophytic generations, it is to be expected that the efficiency of MGS is much greater than selection applied to the sporophyte, in consequence of the population size and the haploid state. The male gametophytic population is much larger than the sporophytic population. In maize, for example, a single plant produces several million pollen grains and about 10–30 grains are deposited on a single silk. Consequently, a very high intensity of selection can be obtained. The haploid state has a twofold effect. First, it exposes the recessive alleles and, consequently, with an equal coefficient of selection, the evolution rate under MGS is much higher than under SS (Table 20.1). Moreover, if a superior zygote occurs with the frequency of p^2, a superior gamete would be found with the frequency of p, that is, the probability of selecting complex allele combinations is greater in the gametophytic generation than in the sporophytic. The higher probability of selecting superior genotypes is the basic assumption underlying the gametic selection proposal of Stadler (1944). However, that selection refers to field testing of the sporophytic progeny, and so does not derive the advantage of the large

Table 20.1. Evolution of a population under sporophytic, gametophytic or sporophytic and gametophytic selection. The values of the coefficients of selection (s) are assumed equal to 0.1

Gene frequency variation	Number of generations required		
	Sporophytic	Gametophytic	Sporophytic and gametophytic
0.5 – 0.4	9	8	4
0.4 – 0.3	12	8	4
0.3 – 0.2	22	10	7
0.2 – 0.1	57	16	12
0.1 – 0.01	1,236	46	43
0.01 – 0.001	9,034	45	45
Total	10,370	133	116

population size and does not allow the recovery of the superior gametes.

MGS can play a fundamental role in the evolution of a plant population. Differential gene transmission in the gametophytic generation was first envisaged by Mendel and demonstrated by Heribert-Nilsson (1923) and Correns (1928) to be a mechanism which can lead to distorted segregation. The contribution of pollen selection to the rapid evolution of angiosperm species was first indicated by Buchholz (1922). However, in view of the manifold positive aspects of diploidy in higher plant genetic systems and the large body of results of genetic analysis of many traits showing concordance with expectations based on the absence of variability in gametophytic fitness components (pollen viability, germination and tube growth), the classical view was that there is a gradual reduction of the gametophytic generation, with a consequent decrease in gene expression, leading to a specific genetic domain controlling pollen development and function. According to this view, most of the genes expressed in the gametophytic generation are not expressed or do not play important roles in the sporophyte and, in consequence, pollen selection could not have an important effect on the fitness of the sporophytic generation. While in the animal this expectation seems to be confirmed by experimental observations, in angiosperm plants it has never been fulfilled (for a review see Heslop-Harrison, 1980).

During the last two decades, several experimental results have contrasted with the classical view. As will be described later in this chapter, it has been shown that a large portion of the plant genome expressed in the male gametophytic generation, is also expressed in the sporophytic generation, and that selection applied to the pollen population can affect many important traits of the plant progeny. Consequently, basic mechanisms controlling higher plant evolution may be the result of selection acting during the gametophytic generation.

It has been suggested by Mulcahy (1979) and Ottaviano and Sari-Gorla (1979) that pollen selection is responsible for the high evolution rate in angiosperms. In fact, in haploids, since all genetic differences are phenotypically expressed, the evolution rate can reach a maximum value. The main factor limiting a high evolution rate in both diploids and haploids is represented by the population size. A high evolution rate implies a high cost of selection and, when selection in a small population reduces population size, the process can lead to extinction. However, the high cost that this rate implies would be compatible with the gametophytic generation because of the large population size available.

Gametophytic selection can act as a mechanism regulating the genetic structure of populations with regard to the genetic load of quantitative traits, which are generally determined by complex gene combinations (Ottaviano *et al.*, 1988b). For these characters a large load is expected to be produced by genetic recombination. The large population size of the gametophytic generation enables selection to remove a major portion of the recombination products at a cost which is compatible with the biological features of the species. All these mechanisms are expected to act in special selective environments, also in crop plants cultivated in the field and in breeding nurseries. Monoculture in fields, high population density and compressed flowering times produce higher pollination densities and consequently a higher intensity of selection. In breeding nurseries MGS is expected to increase because crosses between divergent genetic material and inbreeding are frequently carried out with a high pollination density. Moreover, chemical treatment and biotic stresses can affect both the sporophytic and the gametophytic generation.

20.2 BIOLOGICAL ASSUMPTION AND EXPERIMENTAL EVIDENCE FOR MGS

20.2.1 Pollen development

The main assumption of MGS is that important sporophytic traits are controlled by genes expressed equally in the male gametophytic generation. Although support for the hypothesis is directly provided by analyses of gene expression and by selection experiments, the framework to which genetic analysis is to be referred and the interpretation of the results of selection experiments are based directly on the physiological and developmental activities of the male gametophyte. Pollen physiology and development have been reviewed recently by Heslop-Harrison (1987) and Mascarenhas (1989), here only the aspects more relevant to MGS are considered.

In the different stages of the male gametophytic phase important physiological processes occur, in fact, although very reduced in morphological complexity, the male gametophyte undergoes a number of discrete differentiation events and has several very specialized functions.

During microspore maturation in the anthers, many complex metabolic pathways are active. Thus, the biosynthesis of the cell wall, which has quite a complex structure, being composed of different layers of material, calls for the production and assembly of several types of polysaccharide molecules. The increase of cytoplasm and the cell division processes imply an active basal metabolism and energy production. In species where the division of the generative cell occurs before the release of pollen from the anthers, the elements necessary for the beginning of pollen function, germination and early tube growth, are already present in the mature grain. The analysis of its content reveals a very large number of different mRNAs, the majority of them represented by a large number of copies. Germination in this case occurs very soon (1–10 minutes) (Hoekstra, 1983) after arrival of the pollen on the stigma, and the maximum tube growth rate is attained almost immediately. For pollen released in the bicellular stage, much of the development is left until after the pollen is rehydrated on the stigma; germination may take up to one hour, and the full tube growth rate is not attained immediately.

With regard to the genetic control of these functions, the main sporophytic role resides in the control of the differentiation processes that lead to meristematic cells inside the anther to develop microsporocytes. The tapetal sporophytic cells are of great importance for microspore nutrition and are responsible for the synthesis of the outermost pollen wall layer (exine). Moreover, the cytoplasm of the microspore is of sporophytic origin, even if sporophytic genes are repressed here. However, after microspore release from the tetrads, many pollen-specific characters are expected to be directly controlled by the pollen genome itself. Thus, the vegetative cell, which constitutes the greater part of the pollen, has an active role in the control of its metabolism, storage of reserve substances, synthesis of the inner layer of the wall (intine), control of opening of germination apertures, and in governing the synthesis of enzymes and other proteins, required by the pollen tube wall.

For a short time mature pollen grains are completely independent of the mother plant, exist as free organisms, and, as such, are directly exposed to the environment; this phase, between anthesis and the deposition on the stigma of a female plant, is metabolically quiescent, particularly in the case of trinucleate pollen. However, mature pollen grains are exposed to environmental effects, which may include stress conditions, such as high or low temperature, humidity, drought or pollutants.

Pollen germination begins with rehydration (during dispersal, pollen has a substantially lower water content than a normal somatic cell), which takes place mainly through water transfer from the stigma. After extrusion of the tube through the germ pore, it penetrates the stigma and grows within the transmitting tissue of the style. The tube growth pattern is different from that of the other types of plant cells and has a very fast and strong elongation, reaching a total length frequently 1000 times the grain diameter. In maize for example, a pollen grain of about 90 μm diameter is able to develop a tube 20 cm long. Thus the reserve substances contained in the pollen grain alone should not be sufficient to furnish the material needed to sustain tube development, involving the synthesis of large amounts of cell wall material and other cell compounds. Moreover, after the initial period of growth, living cytoplasm occupies only a small part of the tube and the tube tip is isolated from the proximal portion by callose plugs that are formed behind the tip where the active cytoplasm is confined. The material of the abandoned wall is not recycled, thus representing a net loss of reserves for the male gametophyte. The pollen protoplasm, on the other hand, takes up nutrients from the style by means of an active pollen-style interaction. Pollen germination on the stigma, tube growth within stylar tissues and fertilization are the result of a complex of interactions at the physiological and molecular level that have to be regulated by genes with tissue-specific expression.

In conclusion, even if the number and the specific physiological role of the genes controlling pollen maturation are unknown, it is reasonable to suppose that complete gametophyte development and function calls for the expression of numerous genes; in fact, chromosomal deficiencies that can be transmitted through the female counterpart (continuously dependent on the sporophyte) are selected against in the male gametophyte (McClintock, 1944; Stadler and Roman, 1948).

20.2.2 Gene expression in the male gametophytic generation and its significance

Direct support for MGS is provided by data showing that a large portion of the plant genome is expressed in the haploid phase, and that genes expressed in pollen are also expressed in the sporophytic generation. The most evident proof of haploid gene expression is furnished by single gene segregation in the pollen population produced by a heterozygous plant (for a review see Ottaviano and Mulcahy, 1989). In maize, F_1 plants from homozygous adh^+ and adh^- (alcohol dehydrogenase activity–inactivity) parental lines produce 50% adh^+ and 50% adh^- pollen. This can be visualized by staining pollen grains with a specific reaction mixture; pollen grains containing the active enzyme stain blue, if there is no enzyme activity, the grains remain uncoloured. In the same species direct visualization of single grain genetic content can be obtained for *ß-gal⁻* (*ß*-galactosidase), for *sp* ('small pollen'), for *waxy* (amylose or amylopectin grain content colours differently when treated with iodine solution), for *am* ('amylose extender') and for *Rf3*, a fertility restoration factor specific for *CMS-S* cytoplasm (Laughnan and Gabay, 1973). Plants with *CMS-S* cytoplasm and heterozygous *Rf3/rf3* produce 50% empty sterile pollen grains and 50% plump, normal grains (Fig. 20.3). Frequently single gene expression is not directly analysable in pollen but can be detected on the basis of its effect on the progeny, since the presence of genes affecting pollen development or function are revealed by distorted Mendelian segregation. For example, 'gametophytic factors' specifically determine pollen fertilization ability; they have been described in several plant species, including maize, tobacco, wheat, tomato, lima bean and barley. In general, the dominant allele, designated *Ga*, at a single Mendelian locus has a marked selective advantage in pollen over its recessive allelic counterpart, *ga*. As a result, *Gaga* plants produce aberrant segregation ratios, with a deficiency of *ga* carrying progeny at the 'gametophytic factor' locus as well at linked loci.

The mode of action of the 'gametophytic factors' varies according to species and loci. In maize, nine different genes have been identified; in some cases *ga*-bearing pollen is able to grow on *GaGa* or *Gaga* silks, but in competition with *Ga* pollen is outgrown and clearly reveals a poor fertilization ability. A particularly strong effect is expressed by the *Gas* allele; on silks carrying the *Gas* allele, *ga* pollen fails to induce seed set even in the absence of competing *Ga* gametophytes. It has been demonstrated that *ga* pollen germinates on *GasGas* styles no less than *Ga* pollen, but its tube growth rate rapidly diminishes and, after about eight hours, ceases (House and Nelson, 1958; Bianchi and Lorenzoni, 1975).

These effects can be caused only by specific interactions between *Ga* controlled functions in the pollen and in the style, but the mechanism and the gene products involved in this phenomenon are completely unknown.

The most dramatic effects of pollen-style interaction are due to gametophytic incompatibility: the growth of the incompatible tube (tube bearing the same *S* allele as the style) is completely blocked in the style (Chapter 13).

In the examples concerning enzymes, the differences between the two segregating pollen types are in terms of presence–absence of enzymatic activity, ability–inability to produce the final product of a metabolic pathway. Investigation of a number of genes is possible by means of isozyme analysis, i.e., forms of the same enzyme having small structural differences are revealed by differences in electrophoretic mobility. Micro-electrophoresis of single grains has revealed the presence of only the fast or the slow form on the zymogram of

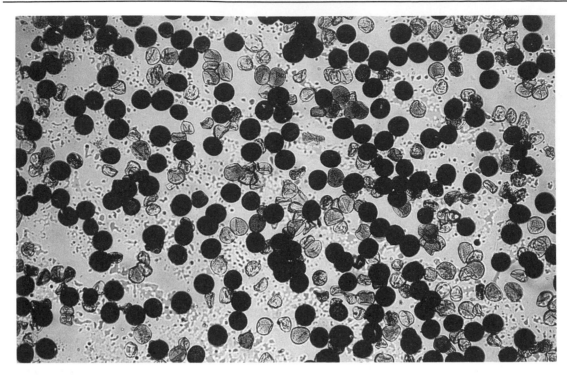

Fig. 20.3. Segregation revealed at pollen level proving post-meiotic gene expression; pollen produced by heterozygous *Rf3/rf3* plants.

pollen produced by hybrid F/S plants (Mulcahy *et al.*, 1979). However, this can also be done by analysing the pollen population as a whole. For dimeric or multimeric enzymes (the active form consists of two, or more, monomeric subunits), the zymogram of F_1 offspring of lines bearing the fast or the slow variant has three bands: the two homodimeric forms and a heterodimeric enzyme, formed by random association of the two monomers. Pollen produced by these plants is expected to show three bands in the case of diploid gene expression, only the two homodimeric bands in the case of haploid transcription, since it occurs when only one allele is present in each cell and thus only monomers of one type can be produced. Since isozyme polymorphism is a widespread phenomenon and the coding genes are randomly located on the chromosomes, this procedure makes it possible to analyse the type of expression of several genes controlling basic metabolic pathways, thus providing a good

sampling, both functional and spatial, of the phenomenon. The studies carried out in different plant species as tomato, maize, barley, *Populus* (e.g. Tanksley *et al.*, 1981; Sari-Gorla *et al.*, 1986) have revealed haplo-diploid expression for the majority of the loci. For the same enzyme, the zymogram of pollen from F/S plants shows a two-banded pattern, which contrasts with the three- (or more, when multimeric) banded pattern of sporophytic tissues. It has been estimated that about 10% of the genes are pollen-specific, 60–75% are expressed both in sporophytic tissues and in pollen, and 10–40% only in the sporophyte, according to the species.

All these studies relate to mature pollen grains, in which, because of the relative dormancy, most of the metabolic processes are reduced to a very low level. Information about gene expression during microsporogenesis, has been obtained by isozyme analyses and by enzyme activity detection in *Brassica* and maize.

These studies, although confined to a limited sample of genes, indicate that in the developing microspore a highly regulated system is operating. Some genes are expressed from the very early stages until maturity, reaching a peak just before the first microspore mitosis, others are phase-specific and are expressed only in the early or only in the later stages of maturation. There is also evidence that some genes reveal sporophytic control in the early stages and gametophytic control in the later stages of maturation (Frova, 1990).

A different way of obtaining a quantitative estimate of the number of genes expressed in pollen is based on RNA analysis. In *Tradescantia* and maize, investigation of the kinetics of hybridization of [³H]cDNA with poly(A) RNA in excess has revealed the presence in mature pollen grains of 20000–24000 different mRNAs (Willing and Mascarenhas, 1984). Studies of heterologous hybridization between cDNA from sporophytic tissues and poly(A) RNA from pollen, and *vice versa*, have indicated that about 10–20% of the sequences expressed in pollen are pollen-specific, whereas a minimum of 65% of the mRNAs in pollen are similar to those found in the sporophyte (Willing *et al.*, 1988). Although this experimental approach does not indicate specific physiological functions and does not discriminate between pre-meiotic (sporophytic) or post-meiotic (gametophytic) transcription, the results obtained show that gene expression in the male gametophytic generation together with the gametophytic–sporophytic overlap involve a large portion of the plant genome.

In recent years, cDNA libraries obtained from poly(A) RNA of mature pollen have been constructed for several plant species (*Tradescantia*, maize, tomato, *Oenothera*). These libraries contain clones that are specific to pollen as well as other clones that are expressed, both in pollen and in vegetative tissues (Mascarenhas, 1989). RNA blot hybridization using pollen-specific cDNA clones as probes allows the transcriptional pattern to be studied during microsporogenesis. A group of genes are active soon after meiosis and mRNAs reach their maximum accumulation by late pollen interphase, then decrease; another set of pollen–specific genes is activated after microspore mitosis and mRNA increases in content up to maturity. Even if the physiological role of the individual gene is unknown, the pattern of accumulation, the kinetics of which is similar to that observed in the case of some enzymes, suggests that they perform a major function during germination and early tube growth.

Information concerning the control of micro-sporogenesis, pollen development and the function of genes expressed in the male gametophytic generation, is provided by mutant analysis. Because of the difficulties of detection and characterization of male gameto-phytic mutants (within species morphological variability of pollen is very small and efficient selection methods for viable gametophytic mutants are difficult to set up), the problem has been approached indirectly by selecting for sporophytic or endosperm mutants and then testing for gametophytic expression. The rationale is based on the observation that components of pollen development and function show a positive correlation with endo-sperm development and that some alleles causing defective endosperm development are expressed in the male gametophyte. In maize, the analysis of the distorted segregations from Mendelian expectation permits testing for gametophytic gene expression and identification of the phase (pre- or post-shedding) in which the gene is expressed. Of 32 defective endosperm mutants, each representing a different gene controlling endosperm development, 22 also affected microspore development or tube growth (Ottaviano *et al.*, 1988a). In *Arabidopsis thaliana* (Meinke, 1985) the gametophytic expression of ten embryo-lethal mutants has been demonstrated.

Most relevant to MGS, whether in throwing light on the evolution of plant adaptation or for plant breeding purposes, is the evidence concerning gene expression induced by environmental stresses. The problem has been studied with regard to heat shock treatments. HSPs (heat shock proteins) are normally found in all sporophytic tissues and a large body of data indicates that they are associated with the acquisition of thermotolerance (Krishnan *et al.*, 1989; Mascarenhas, 1984). In maize, tobacco and *Arabidopsis* genotypes well adapted to a temperate climate this class of induced protein is not synthesized in the pollen grain or germinating pollen. The analysis of developing microspores in maize revealed that they respond to heat shock by producing a set of HSPs; some of these have appeared in all the developmental stages analysed and are common to the sporophyte, whereas others show tissue- and phase-specific expression (Frova, *et al.*, 1989). Similar results have been obtained in *Sorghum*. However, in this species, where material from subtropical climates has been analysed, the ability to synthesize HSPs is retained in germinating pollen, supporting the idea that pollen selection plays an important role in plant adaptation (Frova *et al.*, 1991).

The information reported above largely supports the main assumption of MGS. It also indicates that response to selection is likely to be phase-dependent, according to the type and the amount of gene expression. The possibility of inducing response to specific stresses suggests that efficient methods of breeding for tolerance or resistance to these factors could be developed.

20.3 POLLEN COMPETITIVE ABILITY

The previous section reports data demonstrating the considerable extent both of gametophytic gene expression and of gametophytic–sporophytic genetic overlap. If the genes responsible for these phenomena are directly involved in the control of pollen development and function, one would expect a large amount of genetic variability for pollen competitive ability and a significant portion of this variability should show a genetic correlation with sporophytic traits.

In population genetics pollen competitive ability (PCA) is considered synonymous with pollen fitness. It is the result of several factors controlling: (1) the amount of viable pollen produced by a single plant; (2) pollination ability (the amount of pollen reaching the stigmas); (3) postpollination ability (pollen germination and tube growth in the style

tissues); (4) fertilization ability (Pfahler, 1975). In the context of applied MGS, pollen competitive ability refers to a single male gametophyte (developing microspore, mature pollen grain, germinating pollen grain or growing pollen tube). It is therefore the result of pre-shedding components (viability and competition within the anther) and post-shedding components (viability of the pollen grain during the independent stage, differential pollen germination time, pollen tube growth rate and fertilization ability).

The improvement of pollen competitive ability is of importance in plant breeding for two reasons. First, good pollen quality has a direct effect on female fertility and consequently on seed production. There are data, in maize, indicating that pollen quality (viability and function) shows a large amount of genetic variability in response to environmental stresses, such as high temperatures, and that this has consequences for seed set (Herrero and Johnson, 1980; Schoper *et al.*, 1986). Second, pollen competitive ability due to pollen tube growth rate has been shown to be correlated to different plant traits representing plant vigour (Mulcahy, 1971, 1974; Ottaviano *et al.*, 1980). On the assumption that the correlation is, at least in part, due to the effects of genes expressed in both sporophytic and gametophytic tissues, gametophytic selection could be incorporated into conventional plant breeding programmes to improve plant vigour.

During the last few years a number of studies have been carried out to prove that MGS for PCA has direct effects on both plant fertility and vigour. This selection can take place in all stages of pollen development and function. However, most of the data concern pollen function, i.e., pollen germination and tube growth. The selection derives from the competition between the many pollen grains deposited on the same stigma and between the many pollen tubes growing in the same style. Both the stigma and the stylar tissues act as sieves and selection results from differential germination time on the stigma and tube growth rate in the style. Selection intensity can be modulated either by the amount of pollen deposited on the stigma (pollination intensity) or by varying the dis-

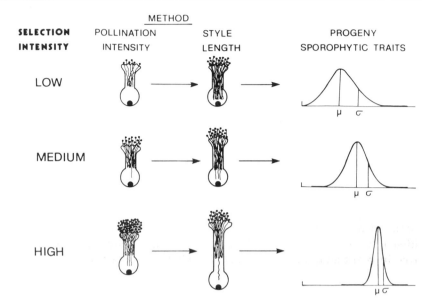

Fig. 20.4. Sporophytic effects of selection for pollen competitive ability. Different selection intensities are obtained by varying the amount of pollen on the stigma or the distance between stigma and ovule.

tance between the stigma and the ovule (Fig. 20.4). To prove that the response is related to genes expressed in the gametophytic phase and is not merely the result of sporophytic variability, selective forces have been applied to the gametophytic population produced by a single heterozygous plant or genetically homogeneous heterozygous families (F_1 progeny of inbred parents or vegetative clonal progeny).

The method based on variation of the distance that the competing pollen tubes have to cover in the style was first suggested by Correns (1928) in *Dianthus chinensis*. Because of the favourable ear structure it has been extensively used in maize to study the gametophytic expression of sporophytic and endosperm mutants or to analyse the variation in pollen competitive ability (Mangelsdorf and Jones, 1926; Jones, 1928; Mulcahy, 1971; Ottaviano *et al.*, 1980). In this species a single ear has hundreds of flowers and the silk (the style and stigma) length varies according to the position of the flower on the ear, increasing from the top to the base, where it can reach a length of about 20 cm. This structure makes it possible to set up MGS experiments and to obtain a quantitative evaluation of the components of pollen competitive ability. Selection experiments exploit the fact that because several pollen tubes are competing within the same silk, the intensity of

gametophytic selection increases from the apex to the base of the ear. Consequently, kernels at the base of the ear are produced after more intense gametophytic selection, whereas those at the apex are produced after less intense selection. Gametophytic competitive ability can be evaluated by a mixed pollination technique. The use of a mixture of two pollen types distinguished by an aleurone colour gene allows estimation of: (1) relative pollen germinability and viability; (2) the relative pollen tube growth rate. This procedure was used in a study of MGS carried out according to a divergent recurrent selection scheme (Ottaviano *et al.*, 1982, 1988b). The results obtained showed a clear response for one of the components of gametophytic competition. Families from the population produced under high selection intensity revealed higher values of the parameter describing pollen tube growth rate. Moreover, correlated responses were obtained for sporophytic traits (kernel weight, seedling growth and root tip growth *in vitro*). Quantitative genetic analysis of pollen tube growth rate and kernel weight has shown that a significant portion (about 20%) of the between-family genetic variability for both gametophytic and sporophytic characters is released by MGS. Since the procedure adopted was strictly based on within plant selection, the response observed demonstrated that the pollen tube growth rate is affected by genes expressed in the post-meiotic phase, and that these genes are also involved in the control of kernel and seedling weight and root tip growth. Data confirming these results have been obtained in maize by Landi *et al.* (1989). The same method adapted to different flower or inflorescence structures has been used to demonstrate the effect of MGS in *Cucurbita pepo* (Stephenson *et al.*, 1988), *Dianthus chinensis* (Mulcahy and Mulcahy, 1975) and *Anchusa officinalis* (McKenna, 1986). The selection method based on pollination intensity (the amount of pollen deposited on the stigma) was first used by Ter-Avanesian (1949, 1978) in *Vigna*, cotton and wheat. The responses were detected both in population variances and in the means for characters expressing plant vigour and fertility. Positive results have also been obtained in *Petunia hybrida* (Mulcahy *et al.*, 1978), *Turnera ulmifolia*, a neotropical weed (McKenna, 1986), in *Cucurbita pepo* and *Lotus corniculatus* (Schlichting *et al.*, 1987) and *Cassia fasciculata* (Lee and Hartgerink, 1986).

Considered as a whole, these studies clearly indicate that major components of PCA are controlled to a significant extent by gametophytic genes which are expressed also in the sporophyte, where they contribute to the determination of plant vigour. The most direct evidence of this is given by the response to selection, applied to the pollen population of a single heterozygous plant, for both pollen and sporophyte qualities. In maize, in particular, the alternative explanation of the observed correlated response, i.e. that it is the result of linkage between genes acting on the two different phases, is ruled out by the structure of the experiment carried out by Ottaviano *et al.* (1988b). Although the physiological role of these genes is still unknown, these results and the information reported in the previous sections of this chapter suggest that several functions common to both plant and pollen growth could be involved. Support for this hypothesis is provided by findings that mutants of genes controlling endosperm development in maize or embryo development in *Arabidopsis* also affect pollen development or pollen tube growth (section 20.2.2). In maize a marked reduction of indoleacetic acid synthesis has been observed in the endosperm of two of these mutants (Torti *et al.*, 1986).

Physiological and developmental considerations and the results of genetic analyses reported in section 20.2 indicate that pollen development and function can be expected to be under the control of tissue specific genes. Indications of the involvement of genes specific to the gametophytic generation have been obtained by means of RFLP analysis based on a set of recombinant inbreds (RIs) derived from the F_1 cross between two divergent inbred lines (Sari-Gorla, 1992). Relative pollen tube growth rate, measured in each RI according to the method based on mixed pollination referred to above, confirmed that the character has the typical distribution of a quantitative character and that a large portion of the variability is genetic in origin ($h^2 = 0.77$). Chromosome localization of quantitative trait loci (QTL) by means of RFLP analysis detected six unambiguously assigned QTLs controlling 60% of the genetic variability. A very important aspect of these results is revealed when they are compared with the chromosomal localization of *Ga* factors (as mentioned earlier, these genes have specific functions expressed in the silk and in the male gametophytic tissue and have major effects on pollen tube growth in the style) obtained by classical mapping procedures. Three of these genes are found in a chromosomal region detected by RFLP analysis. As alleles of these factors having major genetic effects do not segregate in the RI population analysed, it follows that, in a maize population, the *ga* loci are present as isoalleles having small phenotypic effects not detectable by means of standard genetic analysis. In general, genetic dissection by means of RFLP analysis supports the idea that the variability in PCA is the result of the effects of two categories of genes: (1) genes controlling metabolic processes common to pollen and sporophyte; (2) genes specific for pollen development and function.

20.4 BIOTIC AND ABIOTIC STRESSES: POLLEN ASSAY AND SELECTION

Section 20.3 reported experimental results indicating that the efficiency of selection for plant vigour and fertility would be improved if conventional methods of selection also exploited selection pressures applied to the male gametophytic generation. A number of studies carried out during the last decade clearly show that MGS can provide a very efficient method for the improvement of specific traits, such as tolerance to physical and chemical stresses or resistance to diseases. In this respect it is important to distinguish pollen selection in *sensu strictu* from pollen assay. The former implies control of the character by genes expressed in the post-meiotic phase (gametophytic gene expression) and gametophytic–sporophytic genetic overlap. The latter term only requires an assay of the character at the pollen level, regardless of gametophytic or sporophytic control.

The possibility of rapid screening of large sets of genotypes by means of pollen assay has been exploited with regard to a number of traits showing association between pollen and sporophyte. In maize, *in vitro* pollen germination and tube growth have been used to test tolerance to EPTC, a thiocarbamate selective herbicide (Sari-Gorla *et al.*, 1992). A diverse set of inbred lines, chosen in order to represent the most important breeding germplasm, revealed a large amount of genetic variability with regard to EPTC tolerance at the pollen level. Extreme minus and plus variants in the distribution were also the most divergent genotypes in terms of herbicide effects at the seedling stage (Fig. 20.5). This association with herbicide tolerance has been proved also for ethofumesate in sugar beet (Smith, 1986), and for chlorsulfuron in maize (Sari-Gorla *et al.*, 1989). Association between pollen and sporophyte has been reported for resistance to pathotoxins produced by *Helminthosporium maydis* (Laughnan and Gabay, 1973), by *Alternaria brassicola* (Hodgkin and McDonald, 1986) and by *Alternaria alternata* f.sp. *lycopersici* (Bino *et al.*, 1988), and for tolerance to heavy metals (copper and zinc) in

Silene dioica and *Silene alba* (Searcy and Mulcahy, 1985a,b). Association has also been proved for oil quality in *Brassica napus* (Evans *et al.*, 1988), for glucosinolate content (Dungey *et al.*, 1988) and for low temperature tolerance in potato (Kristjansdottir, 1990). In transgenic plants the expression at the pollen level of a chimeric gene conferring resistance to kanamycin has been reported in tomato (Bino *et al.*, 1987).

Although pollen–sporophyte association for useful traits can provide an important tool for rapid and inexpensive screening of large sets of genotypes, a real breakthrough for plant breeding methodology would be achieved if selection could operate at the single pollen grain level within the population of grains produced by segregating plants. Most of the work carried out in this field has aimed at testing this possibility. The criterion generally adopted for this purpose is based on the evaluation of the segregating progeny produced by the male gametophytic population from single heterozygous genotypes (plants or vegetatively cloned families). The treatment (physical or chemical stress) is applied at different stages of the male gametophytic phase, i.e., developing microspores, germinating pollen grains or elongating pollen tubes. A comprehensive study, with checking of correspondence between pollen and sporophyte throughout the generations, has been carried out on EPTC resistance in maize (Sari-Gorla *et al.*, 1992). Two divergent genotypes detected by pollen assay, as mentioned before, were crossed and the herbicide treatment applied during pollen germination on the silks of the recessive parental lines. The backcross progeny recovered showed a significant increase of resistant seedlings in comparison with seedlings from untreated pollen. The pattern of distribution of the backcross progeny, although the character does not show complete penetrance, indicates that resistance is due to a small number (2–3) of genes. Significant response to MGS for herbicide tolerance in maize has also been demonstrated for chlorsulfuron (Sari-Gorla *et al.*, 1989). Here the character shows a typical quantitative distribution. Response to MGS for low temperature tolerance in tomato was demonstrated by Zamir and Gadish (1987) in an interspecific cross between *Lycopersicon*

Fig. 20.5. (a) Pollen of susceptible and tolerant inbred lines of maize (P_1 (left) and P_2 (right)) growing on medium supplemented with herbicide. (b) Seedling response to herbicide treatment of the same inbreds, as in (a), and of their reciprocal F_1 progenies.

hirsutum, a species adapted to low temperatures, and a sensitive species, *L. esculentum*. In the backcross progeny obtained by pollinating *L. esculentum* plants with pollen of the F_1 hybrid, root elongation of seedlings was less inhibited by cold than in the control progeny. The same approach was used to prove the efficiency of MGS for tolerance to salinity in the cross between *Solanum pennellii* (salt tolerant) and *Lycopersicon esculentum* (sensitive) (Sacher *et al.*, 1983). In *Silene dioica* and *Mimulus guttatus* selection for heavy metal tolerance during pollen development was effective in increasing tolerance in the sporophyte (Searcy and Mulcahy, 1985a,b). In potato Pallais *et al.* (1986) detected effects of wet-storage of pollen on the quality of the true seeds.

The efficiency of MGS for tolerance or resistance to biotic or abiotic stresses depends on the development of suitable technologies, the stage in which the treatment reaches the maximum selection intensity and the genetic control of the character. Studies of gene expression have revealed that the amount of genetic variability phenotypically expressed can be expected to be stage-dependent (section 20.2). Although there is no direct information concerning the amount of genetic variability for useful traits expressed in the male gametophytic generation, evidence is indirectly provided by the selection experiments reported above, where the results of pre- and post-shedding treatments were compared. For instance, chlorsulfuron and heavy metal treatments were more effective in pre-pollination stages, whereas selection for tolerance to low temperature in tomato was more effective during the post-pollen shedding stages. Basic mechanisms determining important traits, such as tolerance to temperature stresses and disease resistances, may not be expressed constitutively but induced by the stressing factor, as, for example, the response to temperature shocks and to diseases. In these cases the selection method has to be designed so as to apply the selection pressure when genetic differences are fully expressed.

Experimental evaluation of the efficiency of MGS may suffer from the fact that the method acts on genetic differences expressed at cellular or subcellular levels, whereas the response is evaluated on the developing sporophyte (seedlings, plants), where most of the characters are the result of complex interactions acting at all levels of plant organization. For example, tolerance to high temperature stresses is also the result of leaf morphology, leaf pubescence, epicuticular wax and transpiration intensity. Because of this complexity, which involves the simultaneous segregation of a large number of genes controlling tolerance at different levels of the plant organization, the effect of the selection on components expressed at cellular and subcellular levels cannot be easily detected and evaluated. This problem can be solved by the use of molecular genetic markers (isozymes and RFLPs) which allow detection of chromosome segments carrying genes controlling the character and where response is evaluated as differences in gene frequency. Moreover, the results can be used to identify markers for assisted selection in the following generations. The method was first used by Zamir *et al.* (1982) and Zamir and Vallejos (1983) for response to low temperature stresses in tomato. MGS induced differential transmission of isozyme alleles indicated that the genes are localized on two different chromosome segments. Similar results have been obtained by Sari-Gorla *et al.* (1988) for tolerance to salinity in tomato where MGS applied to the gametophytic population of the F_1 of an interspecific cross between *Solanum pennellii* and *L. esculentum* detected two chromosome segments carrying genes involved in the control of tolerance.

20.5 TECHNOLOGY FOR MGS IN PLANT BREEDING

As mentioned above, MGS can be used to pursue two different objectives: the improvement of fertility related to pollen quality and the improvement of useful traits of the sporophytic generation. Selection for pollen quality is indirectly taken into account in most breeding programmes when the material is evaluated for pollen production and for fertility. Selection pressures could be increased by a direct evaluation of *in vivo* or *in vitro* pollen viability and tube growth, for which several methods have been developed (Stanley and Linskens, 1974).

The usefulness of MGS as a tool to improve sporophytic traits is confirmed by the studies reported in previous sections. Since the success of the method is independent of the genotype and since its large-scale application does not require sophisticated and expensive technologies, it can easily be included in breeding programmes based on conventional methods. The proposal to use MGS as a method in plant breeding has been made only recently and consequently is still in the experimental stage. Only two studies, both carried out in maize, report the results of complete cycles of selection, where the method was integrated into recurrent selection procedures, with the aim of selecting for traits reflecting plant vigour (Ottaviano *et al.*, 1988b; Landi *et al.*, 1989). In the same species it has also been shown that pollen-style interaction, evaluated as pollen competitive ability, can be used as an indicator of combining ability between the pollen and the silk parent for sporophytic traits (Ottaviano *et al.*, 1980).

A very promising use of MGS is in selection for tolerance to biotic and abiotic stresses. In fact the method possesses most of the advantages of *in vitro* selection based on cell or tissue cultures (large population size, controlled environment). It also has a number of further advantages. (1) There is the possibility of exploiting variability produced by genetic recombination, the amount of which can be regulated by the choice of the parental genotypes of the hybrid plants. In *in vitro* selection this feature is found only in procedures based on dihaploid technology (Wenzel, 1989). (2) It does not depend on the solution of problems connected with plant regeneration, including the negative aspects of somaclonal variation.

A comparative review of the results obtained by selection based on *in vitro* cultures of cell and plant tissues and by MGS has been given by Ottaviano *et al.* (1990). The analysis (Table 20.2) indicates that most of the characters that can be selected by *in vitro* culture systems are also suitable for MGS. The more numerous results obtained by means of *in vitro* selection surely reflect the far greater use of *in vitro* technologies

by both private and public institutions. The results of this comparative analysis also suggest a strategy for defining the most promising applications of MGS. Characters which reveal genetic variability in *in vitro* cultures, are likely to be suitable for the MGS approach.

The efficiency of MGS depends on the development of suitable methodologies serving to maximize selection intensity. The main limitations relate to the dosage of selective agents. In fact, the most suitable stages for the application of selective agents is that of microspore development and those of pollen germination and tube elongation. In these stages selective agents, especially in the case of chemicals (herbicides, phytotoxins, antimetabolites), have to reach the gametophyte in spite of the protection provided by the sporophytic tissues of the anther or of the pistil, which can be easily damaged by them. Consequently the treatment has to be applied at low dosages. The possibility of overcoming this difficulty may be provided by different methodologies, based largely on *in vitro* techniques which, in this respect, have been only partially explored.

A very promising methodology could be developed from technologies of *in vitro* culture of microspores producing mature pollen grains which can be used directly for pollination. Very encouraging results in this field have been obtained by Benito Moreno *et al.* (1988) in tobacco; immature pollen was isolated at early binucleate stage and grown in artificial medium until maturation, when it could be used to pollinate emasculated flowers; the flowers set seed and the seed germinated normally.

Floral buds and young inflorescences of a number of species have been cultured *in vitro*, using different systems (stage, media) according to the species involved (Rastogi and Sawhney, 1989); for instance, in maize immature tassels are dissected aseptically at a very early stage, placed in flasks containing artificial medium and incubated in growth chambers for 16–25 days, until the spiklets extrude anthers; these are then excised and dried, to permit the pollen to shed. The application of physical stress or addition of chemicals to the culture medium may lead to selection of resistant pollen grains, to use for *in vivo* pollination.

Table 20.2. *In vitro* and pollen selection for resistance or tolerance to different stress factors (partly based on Ottaviano *et al.*, 1990)

Factor	In vitro *selection* (cells or tissues)		Pollen selection
	Cell lines	Regenerated plants	
Temperature (low)	*Nicotiana* *Capsicum* Carrot		Tomato
Salinity	*Nicotiana* *Datura* Carrot *Medicago* Rice Flax	Tobacco	*L. esculentum* × *S. pennellii*
Drought (water-stress)	Tomato		
Metals			
A1	Tomato	Carrot *Nicotiana*	
Hg	*Petunia* Tobacco		
Zn			*Silene dioica*
Cu			*Mimulus guttatus*
Herbicides			
Glyphosate	*Corydalis* Carrot		
PPT	Alfalfa		
Chlorosulfuron		Tobacco	Maize
Picloram		Tobacco	
Paraquat		Tobacco *L. esculentum* × *peruvianum*	
Amitrole		Tobacco	
2,4-D		*Lotus corniculatus*	
MCPA	Potato	Potato	
Antimetabolites			
Hy-pro		*Solanum*	
MSO		Tobacco	
5-met-try		*Datura*	
Phytotoxins			
Fusarium	Potato	Medicago Barley	Tomato
Helmynthosporium	Maize		
Alternaria	Tobacco	*Brassica*	
Phytophthora	Potato		
Pseudomonas	Tobacco		
Xanthomonas	Peach		
Phoma	*Brassica*		

However, the *in vitro* development of angiosperm flowers, leading to the production of viable pollen from floral primordia, is not always successful, and depends to a large extent on the species (cucumber, tobacco and tomato give better results) and, within species, on the genotype.

More general is the use of cultured spikes or

tassels (wheat, maize) at a later stage, without excision of the organ (Donovan and Lee, 1977). The procedure has been used to select for herbicide (chlorsulfuron) tolerance in maize. Plants about two weeks before anthesis are cut at the base of the immature tassel, sterilized and put into flasks containing nutritive medium. The plants are grown in controlled chambers, adding fresh medium, which is taken up by the growing plant, until anthesis. Completely normal pollen, with respect to both quantity and quality, is produced. During pollen maturation, selection can be applied by adding the selective factor to the medium or by maintaining the plant under stress conditions.

A very simple method, used to select for salt tolerance in tomato (Sacher *et al.*, 1983), is to grow plants in hydroponic culture, in the presence of the selective agent; however, it has to be taken into account that the greater the distance the selective factor is applied from its target (the developing anthers), the less is the probability that it will reach the target; in fact, it is possible that the chemical, or toxin, will be metabolized before reaching the male inflorescence, thus becoming ineffective or producing different effects, such as toxicity to the plant. It may be possible to overcome this problem by direct injection of the selective agent at the base of the tassel.

The mature pollen stage would seem the easiest to manipulate. Pollen can be directly subjected to selective treatment, but in general at this stage pollen is not very sensitive to external agents, since it is relatively quiescent in metabolic terms and very well protected against environmental damage; thus it is necessary for the treatment to operate during the first phase of pollen germination, when there is an open interface with the environment. The possible experimental approaches vary according to the species. In fact pollen of some plants, such as grasses, and trinucleate pollen in general, germinates and grows *in vitro* much less readily than binucleate pollen. In particular, bursting of the grain or the emerging tube tips is frequently observed when pollen is put into a liquid culture medium. Although a high sugar concentration can prevent this bursting, it can also reduce viability and germinability. A common

feature of successful growth of trinucleate pollen is a solid support, obtained by adding agar or agarose to the basal medium.

For pollen which can retain fertilization ability in liquid medium, the treatment is carried out by adding the selective agent to the medium or simply to water, and then spraying the suspension directly on to the stigmas, which are covered with a plastic bag to prevent desiccation. Again, pollen can be soaked in a solution containing the selective agent for a time and then subjected to centrifugation. The supernatant is then eliminated, the pollen is rinsed, recentrifuged, and finally the pellet is used for pollination. The procedure is quite delicate, since pre-germinated pollen may suffer tube rupture, thus failing fertilization. In some species (*Brassica*) it is necessary to remove the stigmas from styles before pollination to obtain a seed set (Hodgkin, 1987).

For pollen types that burst in a liquid environment, pollen can be pre-germinated on solid medium, supplemented with the selective factor and then covered with a membrane. After the very early germination, the membrane is removed and the pollen transferred with a camel hair brush to the styles for pollination. A recently described method for obtaining a homogeneous suspension of germinating grains (Kranz and Lörz, 1990), consists in spraying hydrated pollen onto a thin layer of moisture-delivering emulsion and, when the tubes have emerged, to cover them with mineral oil. The suspension of pollen grain can then be used for pollination. The problem is that not many compounds are soluble in paraffin oil or similar, and so the method has only a limited applicability.

Pollen can be selected during the fertilization process. The selective agent is applied to the style with a camel hair brush or sprayed on the flower which is then enclosed in a bag. About an hour after the treatment, pollination is carried out with fresh pollen, which will be affected during its germination and growth. For plant species with suitable flower morphology an artificial style (or stigma) can be used (Bowman, 1984). The stigma-style tract is eliminated and replaced with an artificial synthetic stigma, made of solid medium, to which the chemical, toxin or other selective factor, is added. Pollen tubes passing through the artificial stigma are filtered by this, thus only tubes insensitive to the selective factor will pass the barrier and achieve fertilization.

20.6 CONCLUDING REMARKS

The evidence concerning pollen gene expression and the genetic overlap accumulated in recent years is of great importance in higher plant biology. Even though some significant studies on differential gene transmission, gametophytic factors and selective fertilization were carried out early in the century, the importance of the phenomenon continued to be underestimated until 15–20 years ago. The great potential importance, for both evolutionary and applied plant sciences, of the demonstration that pollen is able to express genetic variability and that the genes involved function, at least in part, at both the gametophytic and sporophytic phase, has opened a new field of research.

It is now well established that the pollen genome controls, at least in part, its own characteristics, since microspore development, grain germination and tube growth are, to a large extent, under control of the pollen genes. This may be due to the expression of pollen-specific genes, but in part it is due to the effect of genes that are expressed also in sporophytic tissues: in particular, endosperm traits and, in general, traits which are the expression of growth processes, such as root length, appear to have a common genetic basis. It is reasonable to suppose that basic cellular functions are controlled by the same set of genes, expressed in both generations, since most genes coding enzymes of the basic cellular metabolism have been revealed to have haplo-diploid expression. Moreover, also some specific traits of agronomical importance have been shown to be under haplo-diploid control, thus suggesting the suitability of using pollen as a breeding tool.

The main advantage of gametophytic selection is the possibility of its inclusion in a conventional breeding programme; unlike other biotechnological approaches which are very promising but as yet unsuitable for general practical application, the procedure does not call for knowledge of the genetic control of the character, or the use of particularly sophisticated equipment, and its success does not depend on the genetic background of the plant. Furthermore, besides improving plants in terms of specific agronomic qualities, such as stress tolerance, it can also select for general gametophytic fitness. It can be particularly useful when the selected character is under complex genetic control, or when the genes involved are recessive alleles which are particularly rare in the populations; particular genetic combinations can be isolated with greater probability in pollen than in the sporophyte, due to the smaller number of haploid than diploid combinations and the very different sizes of the populations that can be manipulated.

The main limitation is that only traits expressed at the cellular level can be selected in pollen. Furthermore, in the case of a character that has a component expressed at this level, this component has to be important in the determination of the trait if the selection is to be effective. In some cases the application of the selective pressure can be particularly difficult; for instance, resistance to phytotoxins cannot be easily assayed *in vivo*, because the anthers or the pistil would be damaged by the treatment. *In vitro* techniques of microspore culture, pollen germination and fertilization may provide interesting solutions.

A phenomenon calling for further investigation is pollen competition in natural populations and its consequences in the sporophytic progeny. On the basis of the evidence presented in this chapter it is possible to construct precise mathematical models; however, the part played by this phenomenon in plant population evolution has yet to be exactly ascertained. With regard to crop species, the existence of high pollen competition is assured by manual pollination or by the density of pollen shed in cultivated fields, but so far few cases have been extensively studied in natural populations.

A field to a large extent unexplored is that of pollen-style and pollen-pollen interactions; they are an important component of fertilization ability, since the success of a pollen genotype in fertilization is conditioned by the passage through the stylar tissue, which acts as a sieve, and by the pollen types competing in the same style. Although the existence of these genetic interactions has been clearly demonstrated (for instance the *ga* factors described in section 20.2.2), the mechanisms involved are unknown. Some models, based on that relating to gametophytic incompatibility, have been proposed, but no clear indication with regard to this phenomenon is yet available.

REFERENCES

Benito Moreno, R.M., Macke, F., Hauser, M., Alwen, A. and Heberle-Bors, E. (1988), Sporophytes and male gametophytes from *in vitro* cultured immature tobacco pollen. In *Sexual Reproduction in Higher Plants*, Cresti, M., Gori, P., Pacini, E. (eds), Springer-Verlag, Berlin, New York, pp. 137–142.

Bianchi, A. and Lorenzoni, C. (1975), Gametophytic factors in *Zea mays*, In *Gamete Competition in Plants and Animals*, Mulcahy, D.L. (ed.), Elsevier, Amsterdam, pp. 257–264.

Bino, R.J., Franken, J., Witsenboer, H.M.A., Hille, J. and Dons, J.J.M. (1988), Effects of *Alternaria alternata* f.sp. *lycopersici* toxins on pollen. *Theor. Appl. Genet.*, **76**, 204–208.

Bino, R.J., Hille, J. and Franken, J. (1987), Kanamycin resistance during *in vitro* development of pollen from transgenic tomato plants. *Plant Cell Rep.*, **6**, 333–336.

Bowman, R.N. (1984), Experimental non-stigmatic pollinations in *Clarkia unguiculata* Lind (Onagraceae). *Am. J. Bot.*, **71**, 1338–1346.

Buchholz, J.T. (1922), Developmental selection in vascular plants. *Bot. Gaz.*, **73**, 249–286.

Correns, C. (1928), Bestimmung, Vererbung und Verteilung des Geschlechtes bei den hoheren Pflanzen. *Handbuch der Vererbungswissenschaft*, **2**, 1–138.

Donovan, G.R. and Lee, J.W. (1977), The growth of detached wheat heads in liquid culture. *Plant Sci. Lett.*, **9**, 107–113.

Dungey, S.G., Sang, J.P., Rothnie, N.E., Palmer, M.V., Burke, D.G., Knox, R.B., Williams, E.G., Hilliard, E.P. and Salisbury, P.A. (1988), Glucosinolates in the pollen of rape seed and Indian mustard. *Phytochemistry*, **27**, 815–817.

Evans, D.E., Rothnie, N.E., Sang, J.P., Palmer, M.V., Mulcahy, D.L., Singh, M.B. and Knox, R.B. (1988), Correlation between gametophytic (pollen) and sporophytic (seed) generations for polyunsaturated fatty acids in oilseed rape *Brassica napus* L. *Theor. Appl. Genet.*, **76**, 411–419.

Frova, C. (1990), Analysis of gene expression in microspores, pollen and silks of *Zea mays* L. *Sex. Plant Reprod.*, **3**, 200–206.

Frova, C., Taramino, G. and Binelli, G. (1989), Heat-shock proteins during pollen development in maize. *Dev. Genet.*, **10**, 324–332.

Frova, C., Taramino, G. and Ottaviano, E. (1991), Sporophytic and gametophytic heat shock protein synthesis in *Sorghum bicolor*. *Plant Sci.*, **73**, 35–44.

Heribert-Nilsson, N. (1923), Zertationsversuche mit Durchtrennung des Griffels bei *Oenothera Lamarkiana*. *Hereditas*, **4**, 177–190.

Herrero, M.P. and Johnson, R.R. (1980), High temperature stress and pollen viability of maize. *Crop Sci.*, **20**, 796–800.

Heslop-Harrison, J. (1980), The forgotten generation: some thoughts on the genetics and physiology of Angiosperm gametophytes. In *The Plant Genome*, 4th John Innes Symposium, Davies, D.R. and Hopwood, D.A. (eds), John Innes Institute, Norwich, pp. 1–14.

Heslop-Harrison, J. (1987), Pollen germination and pollen tube growth, *Int. Rev. Cytol.*, **107**, 1–70.

Hodgkin, T. (1987), A procedure suitable for *in vitro* pollen selection in *Brassica oleracea*, *Euphytica*, **36**, 153–159.

Hodgkin, T. and MacDonald, M.V. (1986), The effect of a phytotoxin from *Alternaria brassicicola* on *Brassica* pollen. *New Phytol.*, **104**, 631–636.

Hoekstra, F.A. (1983), Physiological evolution in angiosperm pollen: possible role of pollen vigour. In *Pollen: Biology and Implication for Plant Breeding*, Mulcahy, D.L. and Ottaviano, E. (eds), Elsevier Biomedical, New York, pp. 35–41.

House, L.R. and Nelson, O.E. (1958), Tracer study of pollen tube growth in cross-sterile maize. *J. Hered.*, **49**, 18–21.

Jones, D.F. (1928), *Selective Fertilization*, University of Chicago Press, Chicago.

Kranz, E. and Lorz, H. (1990), Micromanipulation and *in vitro* fertilization with single pollen grains of maize. *Sex Plant Reprod.*, **3**, 160–169.

Krishnan, M., Nguyen, H.T. and Burke, J.J. (1989), Heat shock protein synthesis and thermal tolerance in wheat. *Plant Physiol.*, **90**, 140–145.

Kristjansdottir, I.S. (1990), Pollen germination *in vitro* at low temperature in European and Andean tetraploid potatoes. *Theor. Appl. Genet.*, **80**, 139–142.

Landi, P., Frascaroli, E., Tuberosa, R. and Conti, S.

(1989), Comparison between responses to gametophytic and sporophytic recurrent selection in maize (*Zea mays* L.), *Theor. Appl. Genet.*, **77**, 761–767.

Laughnan, J.R. and Gabay, S.J. (1973), Reaction of germinating maize pollen to *Helminthosporium maydis* pathotoxins. *Crop Sci.*, **43**, 681–684.

Lee, T.D. and Hartgerink, A.P. (1986), Pollination intensity, fruit maturation pattern, and offspring quality in *Cassia fasciculata* (Leguminosae). In *Biotechnology and Ecology of Pollen*, Mulcahy, D.L., Bergamini Mulcahy, G. and Ottaviano, E. (eds), Springer-Verlag, New York, pp. 417–422.

Mangelsdorf, P.C. and Jones, D.F. (1926), The expression of mendelian factors in the gametophyte of maize. *Genetics*, **11**, 423–455.

Mascarenhas, J.P. (1984), Molecular mechanisms of heat stress tolerance. In *Applications of Genetic Engineering to Crop Improvement*, Collins, G.B. and Petolino, J.G. (eds), M. Nijhoff/Dr. W. Junk Publ., Dordrecht, pp. 391–425.

Mascarenhas, J.P. (1989), The male gametophyte of flowering plants. *Plant Cell*, **1**, 657–664.

McClintock, B. (1944), The relationship of homozygous deficiency to mutations and allelic series in maize. *Genetics*, **29**, 478–502.

McKenna, M.A. (1986), Heterostyly and microgametophytic selection: The effect of pollen competition on sporophytic vigor in two distylous species. In *Biotechnology and Ecology of Pollen*, Mulcahy, D.L., Bergamini Mulcahy, G. and Ottaviano, E. (eds), Springer-Verlag, New York, pp. 443–448.

Meinke, D.W. (1985), Embryo-lethal mutants of *Arabidopsis thaliana*: analysis of mutants with a wide range of lethal phases. *Theor. Appl. Genet.*, **69**, 543–552.

Mulcahy, D.L. (1971), A correlation between gametophytic and sporophytic characteristics in *Zea mays* L. *Science*, **171**, 1155–1156.

Mulcahy, D.L. (1974), Correlation between speed of pollen tube growth and seedling weight in *Zea mays* L. *Nature*, **249**, 491–492.

Mulcahy, D.L. (1979), The rise of the Angiosperms: a genecological factor. *Science*, **206**, 20–23.

Mulcahy, D.L. and Mulcahy, G.B. (1975), The influence of gametophytic competition on sporophytic quality in *Dianthus chinensis*. *Theor. Appl. Genet.*, **46**, 277–280.

Mulcahy, D.L., Mulcahy, G.B. and Ottaviano, E. (1978), Further evidences that gametophytic selection modifies the genetic quality of the sporophyte. *Soc. Bot. Fr. Actualite Bot.*, **1–2**, 57–60.

Mulcahy, D.L., Mulcahy, G.B. and Robinson, R.W. (1979), Evidence for post-meiotic genetic activity in pollen of *Cucurbita* species. *J. Hered.*, **70**, 365–368.

Ottaviano, E. and Mulcahy, D.L. (1989), Genetics of Angiosperm Pollen, *Adv. Genet.*, **26**, 1–64.

Ottaviano, E., Petroni, D. and Pè, E. (1988a), Gametophytic expression of genes controlling endosperm development in maize. *Theor. Appl. Genet.*, **75**, 252–258.

Ottaviano, E. and Sari Gorla, M. (1979), Genetic variability of male gametophyte in maize. Pollen genotype and pollen-style interaction. In *Israeli–Italian Joint Meeting on Genetics and Breeding of Crop Plants*, Monogr. Genet. Agraria IV, Roma, pp. 89–106.

Ottaviano, E., Sari-Gorla, M. and Mulcahy, D.L. (1980), Pollen tube growth rate in *Zea mays*: implications for genetic improvement of crops. *Science*, **210**, 437–438.

Ottaviano, E, Sari-Gorla, M. and Mulcahy, D.L. (1990), Pollen selection, efficiency and monitoring. In *Isozymes: Structure, Function and Use in Biology and Medicine*, Ogita, Z. and Markert, C.L. (eds), Wiley–Liss Inc., New York, pp. 575–588.

Ottaviano, E., Sari-Gorla, M. and Pè, E. (1982), Male gametophytic selection in maize, *Theor. Appl. Genet.*, **63**, 249–254.

Ottaviano, E., Sari-Gorla, M. and Villa, M. (1988b), Pollen competitive ability in maize: within population variability and response to selection. *Theor. Appl. Genet.*, **76**, 601–608.

Pallais, N., Malagamba, P., Fong, N., Garcia, R. and Shcmiediche, P. (1986), Pollen selection through storage: a tool for improving true potato seed quality? In *Biotechnology and Ecology of Pollen*, Mulcahy, D.L., Bergamini Mulcahy, G. and Ottaviano, E. (eds), Springer-Verlag, New York, pp. 153–158.

Pfahler, P.L. (1975), Factor affecting male transmission in maize (*Zea mays* L.). In *Gamete Competition in Plants and Animals*, Mulcahy, D.L. (ed.), North Holland Publ., Amsterdam, pp. 115–124.

Rastogi, R. and Sawhney, V.K. (1989), *In vitro* development of Angiosperm floral buds and organs. *Plant Cell Tissue Organ Cult.*, **16**, 145–174.

Sacher, R., Mulcahy, D.L. and Staples, R. (1983), Developmental selection for salt tolerance during self pollination of *Licopersicon* × *Solanum* F$_1$ for salt tolerance of F$_2$. In *Pollen: Biology and Implications for Plant Breeding*, Mulcahy, D.L. and Ottaviano, E. (eds), Elsevier Biomedical, New York, pp. 329–334.

Sari-Gorla, M., Frova, C., Binelli, G. and Otta-viano, E. (1986), The extent of gametophytic–sporophytic gene expression in maize. *Theor. Appl. Genet.*, **72**, 42–47.

Sari-Gorla, M., Mulcahy, D.L., Gianfranceschi, L. and Ottaviano, E. (1988), Gametophytic selection for salt tolerance in tomato, *Genet. Agr.*, **42**, 92–93.

Sari-Gorla, M., Ottaviano, E., Ferrario, S., Gianfranceschi, L. and Villa, M. (1992), Herbicide (tiocarbamate) tolerance in maize. Genetics and pollen selection. In *Angiosperm Pollen and Ovules. Basic and Applied Aspects*, Ottaviano, E., Mulcahy, D.L. and Sari-Gorla, M. (eds), Springer-Verlag, New York, pp. 364–369.

Sari-Gorla, M., Ottaviano, E., Frascaroli, E. and Landi, P. (1989), Herbicide-tolerant corn by pollen selection, *Sex Plant Reprod.*, **2**, 65–69.

Sari-Gorla, M., Pè, M.E., Mulcahy, D.L. and Ottaviano, E. (1992) Genetic dissection of pollen competitive ability. *Heredity*, **69**, 423–430.

Schlichting, C.D., Stephenson, A.G., Davis, L.E. and Winsor, J.A. (1987), Pollen competition and offspring variance. *Evol. Trend. Plant.*, **1**, 35–39.

Searcy, K.B. and Mulcahy, D.L. (1985a), Pollen selection and the gametophytic expression of metal tolerance in *Silene dioica* (Caryophyllaceae) and *Mimulus guttatus* (Scrophulariaceae). *Am. J. Bot.*, **72**, 1700–1706.

Searcy, K.B. and Mulcahy, D.L. (1985b), The parallel expression of metal tolerance in pollen and sporophytes of *Silene dioica* (L.) Clairv., *S. alba* (Mill.) Krause and *Mimulus guttatus* DC. *Theor. Appl. Genet.*, **69**, 597–602.

Schoper, J.B., Lambert, R.J. and Vasilas, B.L. (1986), Pollen viability, pollen shedding and combining ability for tassel heat tolerance in maize. *Crop Sci.*, **27**, 27–31.

Smith, G.A. (1986), Sporophytic screening and gametophytic verification of phytotoxin tolerance in sugarbeet (*Beta vulgaris* L.). In *Biotechnology and Ecology of Pollen*, Mulcahy, D.L., Bergamini Mulcahy, G. and Ottaviano, E. (eds), Springer-Verlag, New York, pp. 83–88.

Stadler, L.J. (1944), Gamete selection in corn breeding. *J. Am. Soc. Agron.*, **36**, 988–989.

Stadler, L.J. and Roman, H. (1948), The effect of X-rays upon mutation of a gene A in maize. *Genetics*, **33**, 273–303.

Stanley, R.G. and Linskens, H.F. (1974), *Pollen. Biology Biochemistry Management*, Springer-Verlag, Berlin.

Stephenson, A.G., Winsor, J.A. and Schlichting, C.D. (1988), Evidence for non-random fertilization in the common zucchini, *Cucurbita pepo*. In *Sexual Reproduction in Higher Plants*, Cresti, M., Gori, P. and Pacini, E. (eds), Springer-Verlag, Berlin, pp. 333–338.

Tanksley, S.D., Zamir, D. and Rick, C.M. (1981), Evidence for extensive overlap of sporophytic and gametophytic gene expression in *Lycopersicon esculentum*. *Science*, **213**, 453–455.

Ter-Avanesian, D.V. (1949), The role of the number of pollen grains per flower in plant breeding. *Bull. Appl. Bot. Plant Breed.*, **28**, 19–33 (in Russian).

Ter-Avanesian, D.V. (1978), The effect of varying the number of pollen grains used in fertilization. *Theor. Appl. Genet.*, **52**, 77–79.

Torti, G., Manzocchi, L. and Salamini, F. (1986), Free and bound indole-acetic acid is low in the endosperm of the maize mutant defective endosperm-B18. *Theor. Appl. Genet.*, **72**, 602–605.

Vasil, I.K. (1987), Physiology and culture of pollen. *Int. Rev. Cytol.*, **107**, 127–171.

Wenzel, G. (1989), Biotechnology in agriculture – an overview. In *Biotechnology: a Comprehensive Treatise*, Biotechnology, Vol. 6B, VCH Publ., New York, pp. 771–796.

Willing, R.P., Bashe, D. and Mascarenhas, J.P. (1988), An analysis of the quantity and diversity of messenger RNAs from pollen and shoots of *Zea mays*. *Theor. Appl. Genet.*, **75**, 751–753.

Willing, R.P. and Mascarenhas, J.P. (1984), Analysis of complexity and diversity of mRNAs from pollen shoots of *Tradescantia*. *Plant Physiol.*, **75**, 865–868.

Zamir, D. and Gadish, I. (1987), Pollen selection for low temperature adaptation in tomato. *Theor. Appl. Genet.*, **74**, 545–548.

Zamir, D., Tanksley, S.D. and Jones, A.J. (1982), Haploid selection for low temperature tolerance of tomato pollen. *Genetics*, **101**, 129–137.

Zamir, D. and Vallejos, E.C. (1983), Temperature effects on haploid selection of tomato microspores and pollen grains. In *Pollen: Biology and Implication for Plant Breeding*, Mulcahy, D.L. and Ottaviano, E. (eds), Elsevier Sci. Publ. Co., New York, pp. 335–342.

21

In vitro selection

G. Wenzel and B. Foroughi-Wehr

21.1 INTRODUCTION

Natural selection serves the demands of nature and the plant itself resulting, e.g. in fragile ears, long stolons or toxic seeds to secure survival of the species. At this point man has to interact to select the plant that most serves his needs as food, feed or industrial raw material. Consequently this is the central step of the breeding process. Success depends on the ease and speed with which the superior plants within a segregating population can be identified. In classical plant breeding programmes, selection is carried out on huge populations normally in the field. Field selections are strongly influenced, however, by environmental conditions, and are uncertain and lengthy, especially in the case of breeding for quantitative characteristics with a polygenic background. Such traits usually show only slight changes per selection cycle and it may easily take 10–20 years to improve a desired agronomic trait. For this reason, there was a feeling of euphoria when it became possible to grow plants *in vitro* under artificial conditions in the petri dish. But the realities and challenges of plant biotechnology are still under discussion (Vasil, 1990), and neither positive nor negative results can be generalized.

The possibility of growing plants and even plant cells in the same way as micro-organisms offers a basis for new selection techniques. One advantage of this is that the varying influence of

the environment is circumvented; even more important is the chance to work on huge single-cell populations, each cell of which might be a genetically different individual, as in a population of haploid microspore derived cells. In combination with other biotechnological approaches, particularly genome and gene diagnosis coupled to the polymerase chain reaction (PCR), selections at the DNA level are already possible on small portions of the plant, and can be performed in the petri dish. Consequently *in vitro* techniques represent a qualitative improvement of the selection process in plant breeding, whereas in the field of production and use of variability biotechnology serves only as a quantitative extension of available tools (Fig. 21.1).

At the beginning of this account of *in vitro* selection there should be a caution. After a boom in reports on successful *in vitro* selections, often coupled with the use of somaclonal variation, today in only a few examples is the *in vitro* characteristic expressed under field conditions. Not infrequently, the lack of a profound knowledge of plant physiology has made *in vitro* selection more of an art than a reproducible scientific tool.

21.2 *IN VITRO* SELECTION ON THE SPOROPHYTE

A first advantage of *in vitro* selection is the avoidance of an unpredictable environment. Thus, with plant screening already partially transferred from the field to the greenhouse it was a logical development to proceed to the

Plant Breeding: Principles and prospects. Edited by M.D. Hayward, N.O. Bosemark and I. Romagosa. Published in 1993 by Chapman & Hall, London. ISBN 0 412 43390 7.

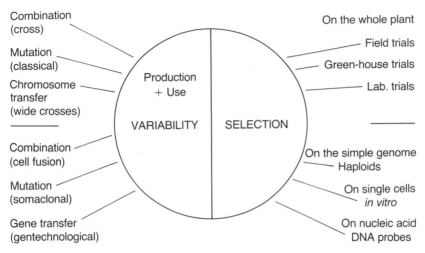

Fig. 21.1. Summary of classical (upper part) and biotechnological methods used in plant breeding.

next step: from the greenhouse to the glass vessel, to *in vitro* selection.

21.2.1 Whole plants or plant organs

Seedlings of heterozygous plant populations can be screened not only in the field but also *in vitro* as was demonstrated by Miedaner *et al.* (1987) for the selection of wheat resistant to *Fusarium culmorum*. In a dual culture system, they were able to select plants which grew normally in the presence of the fungus while susceptible material grew more slowly (Fig. 21.2). For such tests whole plants are not always needed; often it is much more convenient to culture only plant organs; the rest of the plant can be kept *in vivo*. In a number of methods leaf segments are cultured on aqueous agar media, where they stay vigorous long enough to express disease symptoms. Along this line, leaf segments of young rye or barley plants were inoculated with standardized spore suspension of powdery mildew. The infection frequency was measured by counting the lesions of mildew per leaf area (Fig. 21.3), and also spectrophotometrically by determining the turbidity of spore suspensions after washing off the spores from the infected leaves. The most resistant plants were selected and used for further breeding (Lind and Wenzel, 1984).

In *Brassica napus* the sensitivity of seedlings of different cultivars in the presence of a culture filtrate from *Leptosphaeria masculans* was broadly correlated with the known resistance of those cultivars to the pathogen. Thus, the filtrate can also be used for selection in a seed population *in vitro* (Newsholme *et al.*, 1989). Sometimes the culture medium used in such *in vitro* tests on whole plants, seedlings, or embryos influences the reaction to a pathogen. Msikita *et al.* (1990) found that the rate of development and the severity of symptoms of *Pythium aphanidermatum* on axillary embryos of *Cucumis sativus* varied depending on the phytohormones of the MS medium (Murashige and Skoog, 1962). Thus, on media with 2 mg/l BAP and 0.2 mg/l NAA there was less blight than on such with higher or lower phytohormone concentrations. Such early screening systems, though not undifferentiated *in vitro* cultures in the true sense of the word, are of utmost importance in applied breeding. Today, these simple screenings, e.g., for the susceptibility against fungal spores, are the *in vitro* selections which already pay very well in numerous breeding procedures for disease resistance, in contrast to a number of more spectacular but often unreproducible elegant single cell systems.

A similar approach was used to identify tolerance or resistance against inorganic salts. Salt-tolerant sugar beet plants could be developed from a petiole culture excised from germinating mature embryos. Multiple salt as $NaHCO_3$, $NaCl$, $MgSO_4$ and $CaSO_4$ was added

Fig. 21.2. *In vitro* test of wheat cultivars for relative resistance to *Fusarium culmorum* at the whole plant level: (a) less susceptible cultivar, which grows quite normally in the presence of the fungal mycelium; (b) uninfected control of the same cultivar; (c) susceptible cultivar; the mycelium reduces the growth considerably compared to (d) the uninfected control (Miedaner *et al.*, 1987).

to the medium on which 5% of the petioles survived and regenerated. After several *in vitro* cycles the plantlets were transferred to soil amended with salts and seeds were obtained. The progeny also showed salt tolerance. Consequently salt tolerance was declared to be heritable, although a Mendelian ratio of segregation could not be found (Freytag *et al.*, 1990). Selection for salt tolerance is only successful if variability is available in the population or if it arises during the *in vitro* passages. Shoot cultures of sugar beet, tobacco, Chinese cabbage and rapeseed were grown on media with additional salt. There was no evidence for any

increase in salt tolerance in the cultures after up to 24 sub-culture cycles. All selected cultures except those from sugar beet were less vigorous than the control cultures (Chandler *et al.*, 1988). Thus, a general proof for the usefulness of *in vitro* selection at the plant organ level for salt tolerance is still missing.

21.2.2 Callus cultures

Section 21.2.1 described *in vitro* selection on the organized plant. Such plants are normally grown from seeds, which means that in most

Fig. 21.3 *In vitro* test for the level of mildew resistance on 2 cm long leaf pieces of rye. The different pustule development after a uniform inoculation can be numerically evaluated (Lind and Wenzel, 1984).

cases only one plant is available per genotype. This is a severe drawback in selection, particularly for quantitative traits. It would be advantageous first to multiply such plants vegetatively and then to screen a larger group of plants from each genotype. Vegetative *in vitro* multiplication, which is now possible for more than 300 plant species (Vasil and Vasil, 1980), is normally accompanied by callus formation. As the multiplication of such callus is rather fast and proceeds to shoot regeneration, it is logical to try selection directly at the callus level. Regeneration, still difficult in many plant species, has then to be induced in only the few selected genotypes.

(a) Selection of somaclonal variants

Callus cultures not only offer a system for genotype multiplication, they often additionally express spontaneous new variability. By a number of treatments such as irradiation or extended culture times, the amount of new variability can be increased. The use of this spontaneous or induced variability, so called somaclonal variation, has been controversial for the last 10 years. Cell culture was either judged to act as a new mutagen with the known drawbacks of classical mutagenesis (e.g., Hoffmann *et al.*, 1982) or it was viewed as an excellent technique for intracultivar improvement (e.g., Larkin and Scowcroft, 1981). In the latter case, it was anticipated that the genetic changes appearing in plants regenerated from de- and re-differentiated cells during *in vitro* culture are stable, are expressed sexually or are at least vegetatively transmitted. The cause of the somaclonal variation is not understood, but in applied work such understanding might not be a prerequisite for using the procedure. The creation of new variability *per se* is advantageous particularly if the new character does not exist in the natural gene pool. However, the change of one or more characters in an existing cultivar must be detected and thus coupled with efficient *in vitro* selection systems to lead to improved cultivars. Despite this prerequisite the possibility of finding new traits has stimulated tremendous interest in such unconventional breeding methods.

Selection for disease resistance

Variability among plants regenerated from tissue culture was first reported in sugarcane. Heinz *et al.* (1977) found regenerants from calli that were resistant to the Fiji virus disease and later found resistance to *Helminthosporium sacchari*. Recently, a somaclonal variant of *Apium graveolens* resistant to *Fusarium* was obtained from a susceptible cultivar via callus culture. It could be shown that this character was heritable and could be incorporated into breeding lines (Heath-Pagliuso and Rappaport, 1990). Early examples dealing with *in vitro* selection for disease resistance were reviewed by Wenzel (1985) and Daub (1986).

Selection for environmental stress

At the very beginning of the applied use of tissue culture, there was interest in selection for abiotic stresses. Numerous results are summarized by Dix (1980) but, as most of them never reached application, they will not be considered here. Selection for freezing tolerance was carried out in a tissue population derived from callus cultures of the winter wheat variety 'Norstar'. Less freezing tolerant as well as more freezing tolerant somaclones were found, and Lazar *et al.* (1988) stated that this variation was heritable. In vegetatively propagated crops such as ornamentals, spontaneous desirable mutants could be selected from regenerated callus, e.g. in *Gladiolus* (Buiatti *et al.*, 1969) and in *Fuchsia* (Bouharmont and Dabin, 1986). As a physiological understanding of the reasons for a superior reaction of a new somaclone is still missing, approaches using calli and aiming at an agronomically useful improvement need good luck; the positive examples cited above are proof that success is possible.

(b) Selection using a selective agent

Somaclonal variation might become superior to simple mutation breeding when selection pressure is applied during the *in vitro* phase. It should be aimed at specific characters, such as resistance to biotic and abiotic stresses, or changes in developmental patterns. The host pathogen system opened a path for the development of selection methods using the correlation between *in vitro* parameters and *in vivo* behaviour.

Selection for disease resistance

More than 40 years ago Morel (1946) realized the possible advantage of growing pathogens together with their hosts (*Vitis vinifera* and *Plasmopara viticola*) *in vitro*. However, the further development of such dual culture techniques to select *in vitro* for fungal resistance has not been very rapid. Recently, significant differences in the growth of *Phytophthora cinnamomi* on resistant and susceptible pine callus lines were found. A prerequisite for successful experiments is appropriate culture conditions for both the fungus and the callus in one petri dish (Jang and Tainter, 1991). Here, instead of the pathogen, toxins or culture filtrates produced by the pathogen were used for selection. The assumption of this approach is that toxin metabolites in the filtrates produced by the pathogen play a role in pathogenesis, and that they can be used to exert selection pressure for cells that are resistant to the pathogen. Although there is no guarantee that plants regenerated from the resistant calli or cells will also be resistant to the pathogen, this method has been applied to several plant-pathogen systems. Behnke (1979, 1980) reported on a screening procedure in potato callus with culture filtrate of *Phytophthora infestans* and *Fusarium oxysporum*. Alfalfa plants resistant to *Fusarium oxysporum* f. sp. *medicaginis* arose from calli grown in a culture filtrate of the fungus (Arcioni *et al.*, 1987). However, the stability of the resistance in subsequent generations was not confirmed by progeny analysis leaving doubts on the genetic basis and nature of the changes. The same applies to alfalfa genotypes with increased resistance to *Fusarium* selected from suspension cultures (Binarova *et al.*, 1990).

The first experiment where controlled selection for resistance to a pathogen was made with the responsible toxin alone, was performed on maize callus (Gengenbach *et al.*, 1977). However, the reversion from susceptibility to resistance against *Helminthosporium maydis* was associated with genetic changes in the cytoplasm leading to reversion from male sterility to male fertility. Thus, this result was of no practical value. Calli of barley and wheat genotypes were screened for their resistance to purified culture filtrate produced by the fungus *Helminthosporium sativum*. Plants have been regenerated from somaclones surviving the toxin treatment and the first *in vitro*-derived generation produced significantly improved lines (Chawla and Wenzel, 1987). This means that the selection at the *in vitro* level worked (Fig. 21.4) but the correlation with subsequent field tests was unreliable (Wenzel and Foroughi-Wehr, 1990). It was unclear as to whether this discrepancy was due to lack of genetic differences or whether the test procedures were inadequate to detect the slight, probably quantitative alterations in the field. Branchard (1982) also found no correlation between the reaction of barley callus and barley seedlings to *Rhynchosporium*, and suggested that the toxin responsible for the measured reaction may be destroyed by a glucodase manufactured by the callus *in vitro*. Hunold *et al.* (1990) regenerated barley plants after selection using the toxin from *Drechslera teres*. In the regenerated plants the resistance level was improved, but the same was true for the unselected control plants. This may be explained as a somaclonal variation in this trait. Similar results were obtained in wheat tissue culture, where a selection programme was carried out with syringomycin produced by the bacterium *Pseudomonas syringae* (Pauly *et al.*, 1987). Secondary embryoid lines of *Brassica*

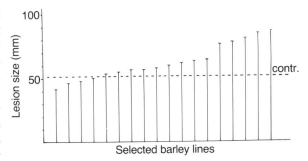

Fig. 21.4. Reaction of 18 barley lines of the sexual progeny of *in vitro*-selected somaclones to the infection with *Helminthosporium sativum*. Unselected control plants produced lesions of 52 mm² (Wenzel and Foroughi-Wehr, 1990).

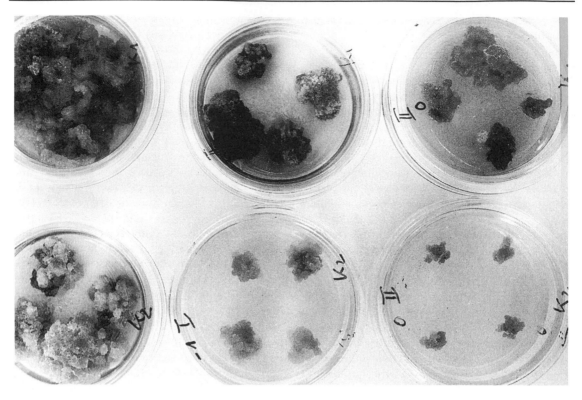

Fig. 21.5. Calli from two dihaploid potato clones on media with two (middle and right) concentrations of semipurified toxin from *Phytophthora infestans* (left, controls). Upper row: cultures from a hypersensitive clone where the calli turned black; lower row: calli from a susceptible clone showing strong growth inhibition (Foroughi-Wehr *et al.*, 1986).

napus differed in sensitivity to a selection medium containing partially purified *Alternaria brassicola* filtrate. Some regenerants were more resistant to the pathogen than seed grown plants of the same cultivar, although there was no correlation between sensitivity to the selection medium and susceptibility to the pathogen (McDonald *et al.*, 1986).

In contrast to these rather negative findings, Helgeson *et al.* proved in 1972 (and this was actually the most positive start signal for this type of experiments) that tobacco calli resistant to *Phytophthora parasitica* can be obtained from resistant plants and that susceptible plants yield only susceptible calli. Also, in tomato, Lai *et al.* (1988) demonstrated that plants could be regenerated from cotyledonary explants resistant to the culture filtrate of *P. infestans*. Furthermore, Deaton *et al.* (1982) showed that resistance to

Phytophthora parasitica is expressed quantitatively in tobacco calli. Experiments with the defined toxin cercosporin, which *Cercospora* species produce in large quantities, did not work, however, since also cultivars resistant to the fungus were killed by the toxin (Daub, 1984). Here, the antibiotic cercosporin is probably used by the fungus as a defensive rather than as an offensive system. Similar problems arise when hypersensitive reactions are to be screened at the cell level. In this type of resistance, the cells in the resistant plant are killed so quickly that the pathogen cannot survive, multiply or spread, thus allowing all other tissue parts to survive. This type of cell interaction requires organized tissue. As Fig. 21.5 demonstrates, calli from *Phytophthora*-hypersensitive potato plants on *Phytophthora*-toxin containing media rapidly turn black and die, whereas calli from other clones, with a non-hypersensitive field resistance, grow more slowly. At present, it is not possible to select *in vitro* for hypersensitivity, one of the most powerful tools in disease resistance strategies. Not being able to use *in vitro* selection in connection with hypersensitivity reduces the usefulness of the tissue culture approach considerably.

Selection for salt tolerance

Selection for tolerance to abiotic stresses involves exposing tissue to the stress for which resistance is desired. Selection on media enriched in NaCl or Na_2SO_4, has proven not to be difficult. In the review of *in vitro* selection for salt tolerance by Raghava Ram and Nabors (1985) cell lines could be selected from several plant species, but plant regeneration was very rare. Such work depends, first of all, on the selection of cell lines exhibiting enhanced resistance to salinity. Callus growth on a medium with a high level of either NaCl or Na_2SO_4 has been reported for sugar beet (Pua and Thorpe, 1986), rapeseed (Mandal *et al.*, 1989), tomato (Hassan and Wilkins, 1988) and potato (Sabbah and Tal, 1990). Several experiments have been reported on the selection of salt-tolerant cell lines of rice (Kim *et al.*, 1988), but most were incapable of regenerating plants, because of the long callus phase. Even when plants could be regenerated from salt-tolerant callus of *japonica* or *indica* rice, proof of the genetic alteration is often still missing (Kim *et al.*, 1988; Subhashini and Reddy, 1989). Vajrabhaya *et al.* (1989) obtained from callus of 450000 embryos of two *indica* rice cultivars one line in which the progenies showed a heritable, salt-tolerant character. In maize, plant regeneration was obtained from a long-term salt-tolerant cell culture, in which tolerance could be preserved *in vitro*, in the absence of NaCl, for 3 months. However, the seed set of the regenerants was very poor. Among the regenerated salt-tolerant plants, four survived in soil and were selfed or crossed with each other. Nevertheless no increased salt-tolerance of the offspring was evident, although more vigour was observed in few lines (Lupotto *et al.*, 1989). Most plants regenerated from NaCl-tolerant cell lines in

alfalfa showed chromosome anomalies. Only one regenerated plant flowered but was sterile so that the heritability of NaCl tolerance could not be tested (McCoy, 1987). Plants were regenerated from a salt-tolerant cell line of *Linum usitatissimum*; the progeny of these plants were superior when grown in saline soil as well as in normal soil. This indicates that the selected trait is the result of a more general mechanism and might be due to an increase in total vigour of the regenerants (McHughen, 1987). Jain *et al.* (1990) concluded from their experiments with *Brassica juncea* that different somaclones may have different salt tolerance mechanisms, because they selected lines that reacted differently in their salt tolerance during vegetative and reproductive phases. One explanation of the salt-resistant cell lines obtained in many species, and the rare development of stable regenerants from such lines, is that this resistance results from a physiological adaptation. Such habituated, epigenetic changes may even obscure the selection of cells with stable genetic mutations. However only the regeneration of fertile plants and the transmission of the resistant phenotype to sexual progeny provides definitive proof of genetic changes which will be useful in the breeding process.

Selection for tolerance to metals

In vitro selections have been performed for resistance to metals such as Cd, Zn or Al. In a first report of this nature, Meredith (1978) used tomato cell cultures in selection experiments for resistance to aluminium. The selected clones remained resistant after 2–4 months of growth in the absence of the stress, but no plants were regenerated.

Selection for temperature stress

Cold-tolerant callus cultures may also be established *in vitro*. However regeneration has been obtained only in a few cases and the correlation between *in vitro* and *in vivo* resistance for this character has not been verified.

Selection for herbicide tolerance

Since the early work by Chaleff and Parsons (1978) on a picloram-resistant tobacco somaclone, numerous reports on tolerance to various herbicides have appeared. Today, there is considerable interest in understanding the mechanisms underlying this tolerance and in using it in combination with gene technology (for detailed information see Chapter 10). In the case of paraquat, enzymes convert the toxins into inactive products, while amplification of target genes confers resistance to glyphosate and L-phosphinothricin (Comai and Stalker, 1986). For a wide range of other herbicides, *in vitro* cultures were used to detect herbicide-tolerant tissue. In most cases, it was possible to regenerate plants expressing some level of tolerance, e.g. in *Linum* for chlorsulfuron (Jordan and McHughen, 1987), where the offspring from seeds also expressed the tolerance. Of particular interest is tolerance in those crops where gene transfer techniques are difficult, e.g. imidazolinone-tolerant maize (Newhouse *et al.*, 1991). In other cases, callus cultures were used to identify herbicide–tolerant mutants and to isolate the mutated gene sequence (Amrhein *et al.*, 1987). In the case of glyphosate, the resistance identified in *Petunia* callus cultures was transferred to several crop species (Horsch *et al.*, 1987). This last example is one of the most important results of applied *in vitro* selection.

21.2.3 Single cell systems

A system that starts with single cells can help circumvent problems of cross-feeding and chimerism, and thereby increase the chances of finding valuable mutants. Furthermore, the number of individuals can be increased to 10^6, which is within the range of the natural mutation rate. Two kinds of single-cell systems of somatic origin are available: the cell suspension culture and the protoplast culture. Here, microbiological procedures can be applied efficiently, because a large number of individuals can be screened in a small area.

(a) Selection for biotic stress resistance

Shepard (1981) postulated that protoplasts are useful in creating new genotypes as he found tremendous variability in potato clones from the tetraploid cultivar 'Russet Burbank' and claimed that up to 30% of the regenerated protoclones were useful variants. Although many of the variants were gross aberrants, some showed improved resistance to *Alternaria solani* (Matern *et al.*, 1978) or *Phytophthora infestans*. However, not all variants were genetically stable, even during vegetative propagation. These may have originated from certain changes other than genetic mutation. From the potato cv. 'Judith', plantlets could be regenerated with increased resistance to *Phytophthora infestans* via protoplast culture (Meulemans and Fouarge, 1986). Clones with significantly enhanced tuber resistance to *Erwinia* have been found in non-selected populations of protoplast-derived regenerants (Taylor *et al.*, 1988).

In experiments with dihaploid potatoes ($2n = 2x = 24$), Wenzel (1980) regenerated more than 3000 clones from protoplasts and found only a few grossly aberrant types, most of which turned out to be aneuploids with no practical value. The remainder were uniform in phenotype and in their total protein pattern. This difference in the behaviour of tetraploid and diploid potatoes shows the strong influence of ploidy level on the survival of variants. Thus, it can be concluded that the degree of variation depends both on the ploidy level and on the genotype. The potato cultivar 'Russet Burbank', which has given a very high frequency of somaclonal variation, for example, also shows a distinctly high frequency of variants in the field. A specific contribution of an *in vitro* culture step to the programmed production of useful new genotypes is consequently still under question.

The single-cell system can be used with more precision in resistance selection when the toxic compounds responsible for the virulence are used to eliminate sensitive protoplasts. Potato protoplast-derived callus tissue was exposed to different inoculum concentrations of *Erwinia carotovora* subsp. *carotovora*. The reaction to the pathogen varied considerably within the callus populations, and callus lines with improved resistance have been identified. However, the number of regenerants was too small for field tests for soft rot, or for making a correlation

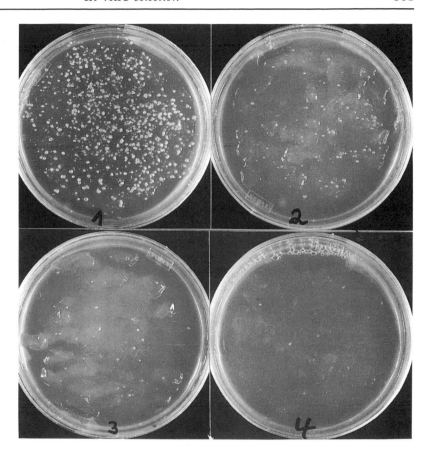

Fig. 21.6. Development of calli from dihaploid potato protoplasts on different levels of semi-purified toxin from *Fusarium sulphureum*: (1) toxin-free control; (2) 0.01; (3) 0.02; (4) 0.03 toxin units (Schuchmann, 1985).

between *in vitro* and *in vivo* reaction (Taylor and Secor, 1990). Potato protoplasts of dihaploid clones were regenerated in the presence of selective concentrations of extracts from *Fusarium sulphureum*, *F. coeruleum* or *Phytophthora infestans* (Fig. 21.6). A significantly higher number of protoclones proved not to be affected by the toxin in secondary callus or in a leaf test, compared to regenerated protoclones that had developed in the absence of the selection pressure (Fig. 21.7). Although it has been proven that *in vitro* selection for fungal resistance works on the *in vitro* level, the correlation with subsequent field tests was unreliable (Wenzel and Foroughi-Wehr, 1990). This may be due not only to the rather artificial *in vitro* screening and the not too sensitive field tests, which only detect differences greater than differences arising from the average amount due to somaclonal variation. One must know more about the actions of the pathogen or the

reactions of the infected plant, in order to develop precise strategies that make possible the detection of useful new genotypes.

For the first screening, it is important to use selective agents that are essentially correlated with the infection process as, for example, the toxin sirodesmin PL produced by *Phoma lingam* in *Brassica napus*. Protoplasts isolated from resistant and susceptible varieties die when treated with the toxin in a concentration higher than 1 μM, whereas in cell aggregates, significant differences between resistant and susceptible varieties were obtained. The same clear correlation between sensitivity to the toxin and susceptibility to the pathogen has been found for intact plants (Sjödin and Glimelius, 1989a). Hammerschlag (1988) induced callus from peach embryos and subjected it to culture filtrate of *Xanthomonas campestris*, the causal agent of leaf spot disease of peach. From a total of four regenerated plants, two were signifi-

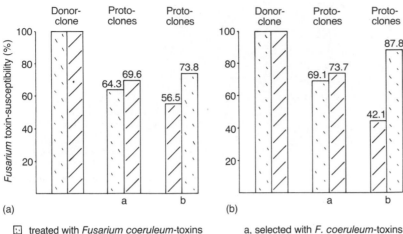

Fig. 21.7. Reaction of secondary callus from selected protoclones of a dihaploid potato clone on medium containing toxins of *Fusarium coeruleum* or *F. sulphureum*: (a) relative size of lesions on leaves; (b) relative increase in callus growth (Wenzel *et al.,* 1987).

⬚ treated with *Fusarium coeruleum*-toxins

⬚ treated with *Fusarium sulphureum*-toxins

a, selected with *F. coeruleum*-toxins

b, selected with *F. sulphureum*-toxins

cantly more resistant to *Xanthomonas* than the donor genotype.

(b) Selection for abiotic stress resistance

Selection for cold tolerance

In potato, frost tolerance could be increased by accumulation of proline in the leaves. From a cell suspension of dihaploid potato clones, somaclones were selected with a proline content up to 25 times higher than the wild type. Frost-killing temperature in the regenerated plants was also higher than in the wild type, but was not always correlated with a higher proline content in the leaves. In contrast to the resistance expressed in leaves, tuber frost tolerance was not improved (Swaaij *et al.,* 1986, 1987).

Selection for Al-tolerance
Selection in cell suspension by the use of media containing Al ions, either as Fe-EDTA or as $Al_2(SO_4)_3$, produced tolerant cell lines of *Daucus carota* (Ojima and Ohira, 1982), *Nicotiana plumbaginifolia* (Conner and Meredith, 1985) and *Solanum tuberosum* (Wersuhn *et al.,* 1988). The regenerated calli remained tolerant on an Al-free medium for several months, and Al-tolerant regenerants were obtained. In *D.*

carota, young seedlings grown from seeds of the selected regenerated plant maintained tolerance. In potato, the regenerants were vegetatively propagated, but here Al-tolerance was not constant; after the first trial, 30% of the clones showed significantly higher Al-tolerance than the control clones, but after further trials, only 5% maintained the new trait constantly and also after subculture in control medium. Conner and Meredith (1985) reported a 50% constancy in *N. plumbaginifolia,* and found a single dominant mutation causing the Al-tolerance.

Selection for herbicide tolerance
A suspension culture of *Brassica napus* cv. 'Jet Neuf' was used to isolate sulphonylurea herbicide resistance. The medium was supplemented with chlorsulfuron in a concentration of 5×10^{-8} M, surviving cells were regenerated and the resistant variants formed callus on the selection medium, indicating that the resistance was not lost during organogenesis (Saxena *et al.,* 1990). Atrazine- and diuron-tolerant calli and plants were recovered from a protoplast culture of *Nicotiana plumbaginifolia* after addition of a plastome mutagen, N-ethyl-N-nitrosourea, to freshly isolated protoplasts. The surviving calli were regenerated into plants and sprayed with the herbicide, some clones were resistant to atrazine and diuron, respectively (Rey *et al.,*

1990). Wersuhn *et al.* (1987) selected potato clones resistant to the herbicides 2-methyl-4-chlorophenoxy acid and sodium-2,2-dichloropropionate. In other cases the regenerated plants often showed abnormalities and the resistance to the herbicide was weak as in the reaction of *Lotus corniculatus* to chlorsulfuron (MacLean and Grant, 1987).

21.3 SELECTION OF CELL FUSION PRODUCTS

Protoplast fusion is a useful technique for somatic combination of heterozygous genomes without recombination. The details of the different methods used and the difficulties to select the hybrids are discussed in Chapter 7. With this technique it is possible to add two genomes completely or to inactivate one fusion partner partially or completely. Such asymmetric hybridization, in combination with *in vitro* selection, makes it possible to transfer single characters from one plant species to another, e.g. to transfer a chromosome fragment from carrot bearing a gene for resistance against methotrexate into *Nicotiana tabacum* protoplasts (Dudits *et al.*, 1987), or kanamycin resistance from *N. plumbaginifoliu* to *N. tabacum* (Bates *et al.*, 1987). Furthermore, it was possible to transfer the resistance against *Phoma lingam* from different resistant *Brassica* species to *Brassica napus* by asymmetric fusion. The donor protoplasts were irradiated with X-rays and, after fusion with the *Brassica napus* protoplasts, selection was applied to the hybrid cell cultures with the toxin, sirodesmin PL, produced by *Phoma lingam* (Sjödin and Glimelius, 1989a,b). Cytoplasmic male sterility was also transferred by asymmetric fusion in *Daucus carota* (Tanno-Suenaga *et al.*, 1988), in *Brassica* species (Pelletier *et al.*, 1988) and in *Nicotiana* (Menczel *et al.*, 1983).

21.4 *IN VITRO* SELECTION IN GAMETOPHYTIC CELLS

Whereas new variability in sporophytic cell populations supposedly comes from sponta-

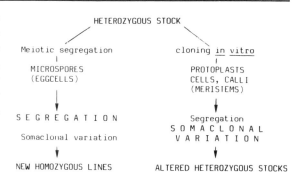

Fig. 21.8. Comparison of the two systems leading to variability as a prerequisite for *in vitro* selection.

neous mutations, which appear at random, gametophytic cell populations express variability that depends on the segregating genotype of the parents (Fig. 21.8). The gametophyte population will contain all parental traits in new combinations, due to recombination during meiosis. Thus, the chances of finding valuable combinations of agronomically important characters are higher than for *in vitro* mutagenesis.

21.4.1 Selection in microspore populations

The most suitable source of heterogeneous cell populations are microspores from heterozygous parents. Regeneration is possible either within the anther (anther culture) or in isolated culture (isolated microspore culture or shed-pollen culture, see Chapter 17). Isolated microspore culture may offer a more effective system of regenerating a random sample from the microspore population than does anther culture. The advantages of using microspores for selection purposes are: (1) a huge population of single haploid cells is available; (2) a lower level of somaclonal variation will result in very few plants with new and mostly undesired characteristics; (3) the selection can be carried out at the earliest developmental or breeding stage; (4) the expression of recessive traits and selected individuals will be homozygous after chromosome doubling. In contrast to selection in somatic cell populations, a desired recessive character is delivered by one of the parents. This means that even if one screening is not sufficient to detect a complex reaction resulting in the phenotypic behaviour being finally expressed in the field, the technique may be sufficient to identify one component of this complex. As long as this component is linked to the whole process, the whole complex is probably transferred; such a selection system may work in segregating homozygous material.

In a combined anther culture and *in vitro* selection approach Ye *et al.* (1987) cultured anthers from a barley F_1 cross between a salt-tolerant and a normal, sensitive cultivar, in liquid media containing up to 0.8% Na_2SO_4. No progeny from the F_1 microspores cultured in the high-salt medium were as susceptible as the susceptible parent. The results indicate that, in these experiments, sensitive microspores were eliminated. Selected lines exhibiting elevated levels of tolerance to salt, were the result of recombination rather than gametoclonal variation. Again, cross-feeding may obscure the effect of selection during dense microspore culture within the anther, as in somatic cell cultures. Therefore, culture in isolation from the anther may be beneficial, particularly when it passes through an embryogenic regeneration pathway. As this regeneration procedure additionally avoids callus formation, the frequency of the normally undesired genetic aberrants (due to gametoclonal variation) will be reduced, resulting in a higher proportion of lines with good agronomic fitness. Microspores or haploid embryos of *Brassica* were screened for their tolerance to imidazolinone or sulphonylurea herbicides (Swanson *et al.*, 1988, 1989). Doubled haploid plants could be brought to maturity and the offspring tolerated at least twice the recommended field rates of the corresponding herbicide (Swanson *et al.*, 1989).

21.4.2 Selection during ovule and embryo culture

Ovule culture is still difficult (see Chapter 17), and no experiments with *in vitro* selection have been made as yet. After fertilization, however, screening of young embryos is possible. This selection closely resembles procedures of seedling selection (Section 21.2.1), but has all the advantages of *in vitro* culture. Each embryo results from a sexual process and represents a different genotype, a striking difference from the selection of somatic cells. In barley and wheat immature embryos were screened on B5 medium (Gamborg *et al.*, 1968) containing fusaric acid. Of 2 mm large wheat embryos (10 days after pollination) 2% could survive on medium with 0.4 mm fusaric acid, whereas in barley 0.2 mm was the critical concentration (Köhler *et al.*, 1986). To detect intravarietal variation in resistance to the herbicide asulam, Collin *et al.* (1986) screened the barley variety 'Midas' using excised embryos. Plants could be selected showing an enhanced resistance compared to the unselected variety. The resistance was stable and transmitted to the progeny.

21.5 *IN VITRO* SELECTION AND MOLECULAR GENETICS

One of the main limitations of present strategies for *in vitro* selection is the lack of knowledge of causal physiological or biochemical reactions responsible for the desired phenotype. At present, *in vitro* screening cannot become a general procedure since the diagnostic systems are too weak. Improved morphological or biological tests (phenotypic level), better chromosome identification (chromosome level) or more general immunological procedures (gene product level) are needed. However, none of these techniques, although already on the gene product level (proteins and directly responsible enzymes), deal with the causal agent, the DNA and differences in the nucleotide sequence, at the DNA level.

21.5.1 Use of molecular markers

Direct DNA diagnosis would allow the immediate detection of specificities or alterations of the plant genome. Gene diagnosis or genome diagnosis, using DNA hybridization techniques, are becoming a general tool for selection. Their use normally requires a substantial quantity of tissue from which sufficient amounts of DNA can be extracted. Thus, *in vitro* systems with small calli or even single cells and microspores are not ideal material on which to apply molecular probes. In 1985 the polymerase chain reaction (PCR) was developed (Saiki *et al.*, 1985), which allows hybridization in very small portions of plant material, as little as 1 mg fresh weight (Dehmer *et al.*, 1991). Thus, the presence of a defined nucleotide sequence can be determined at a very early stage of development. In combination with PCR, the whole spectrum of techniques making use of molecular markers can now be applied to *in vitro* selection.

21.5.2 Selection during gene transfer

In addition to marker-based *in vitro* selection, molecular genetics offers techniques for gene transfer. Almost all transformation systems make use of *in vitro* culture and *in vitro* selection. Most gene constructs transferred either by *Agrobacterium* or by more direct systems contain selectable markers, enabling selection for transformed cells or tissues very early and in huge populations, e.g. by use of antibiotic resistances or reporter genes.

In vitro selection and its direct application with a breeding programme is well exemplified by Horsch *et al.*(1987) who combined *in vitro* selection techniques for a non-selective herbicide (glyphosate), the isolation of the responsible gene and the transformation to different plant species. Further details of the mechanisms underlying resistance are given in Chapter 10.

21.6 *IN VITRO* SELECTION FOR PRIMARY AND SECONDARY PRODUCTS

Crop plants are not only grown to feed man and animal, but are also used for the production of various non-food compounds. Of particular importance are primary products such as oil, fat and starch. While the chances of increasing the production of such compounds via *in vitro* selection are rather limited, it is expected that *in vitro* selection might contribute to the area of secondary product formation. Plant cell cultures might have economic potential, particularly for pharmaceutical products, where a chemical synthesis is still not possible. *In vitro* production coupled to selection for high producers would have the following advantages: (1) since some pharmaceutical plants are collected in the wild, industry would be relieved of problems of supply; (2) the products can be easily adapted to specific demands; (3) secondary compounds produced *in vitro* under controlled conditions would guarantee uniform quality. Secondary plant products are formed and stored in special organs of the plant. The question as to whether such production can be optimized *in vitro* can, in principle, be answered by 'yes'. However, there are still more negative than positive examples for a reproducible selection of secondary products *in vitro*. The most prominent application of this technique is the production of shikonin in *Lithospermum* cell cultures. Cell-culture-derived shikonin became famous when lipsticks containing it as a dye came on the market in Japan as Biosticks (Fujita and Tabata, 1987). From *Digitalis lanata* two main alkaloids are isolated, which are both used as a heart medicine. By using *Digitalis* cell suspension cultures it has become possible to select cultures which modify digitoxin into digoxin (Luckner and Diettrich, 1990). Industry, however, is still reluctant to change the present production process and to adopt this *in vitro* procedure. The situation is similar in other medicinal plants such as *Catharanthus roseus* (Blom *et al.*, 1990).

Callus cultures are not uniform cell populations but contain variant cell types which can be physically isolated. Some of these isolated somaclones yield high-producing cell lines, with productions of the desired compound even higher than in the intact plant (Heinstein and Emery, 1988). For isolation of such lines, a cell suspension culture is plated on solid medium and colonies are isolated, some of which may be genetically different (Zenk *et al.*, 1977). Among the subcultures, high-producing genotypes can be detected. Although this method appeared to be the answer to high levels of secondary product formation, it was found that the high production rate is not always stable. Depending on this instability, the selection of high-producing lines can be time-consuming and uneconomical. Furthermore, cell cultures of higher plants grow rather slowly compared to bacteria, and the culture media are rather expensive. Thus, there is a trend today towards isolating the enzymes responsible for the production of the compound of interest, to establish the coding sequences, and to transfer the cloned gene into bacteria. Another strategy is to use storage organs, e.g. tubers of potato for the production of compounds such as technical enzymes (Wenzel and Willmitzer, 1988). The importance of *in vitro* cultures is consequently decreasing.

21.7 PERSPECTIVES

The most important advantages of *in vitro* cell and tissue culture for selection processes in plant breeding are: (1) freedom from the effects of climate and the natural environment, which makes it easier to measure slight quantitative differences in polygenically inherited traits horizontal or general disease resistances; (2) the ability to handle large numbers of individuals in a very small space; (3) the ability to work with microspores and haploids, the simpler genome of which allows the uncovering of recessive traits and additive characters within a relatively small population. Furthermore, *in vitro* selection is an integral part of all gene transformation experiments.

As for all types of selection described in Part Six, it is also true for *in vitro* selection that chaos and order (variability and selection) is the principal sequence and the prerequisite for progress in plant breeding. The variability, visible in the plant kingdom, which is the consequence of millions of years of evolution, is a result of the interaction between natural variability and natural selection. While natural variability documented in the gene pool is a direct basis for modern varieties, a direct use of

natural selection is rarely possible. Selection for desired traits, ideally already being used *in vitro*, e.g. DNA probes, will probably be easier than the strategy of gene transfer, if a character is available in the crop in a heritable form. The availability of such probes would make it possible in breeding programmes classically to combine the interesting characters and subsequently to select in the segregating generation. Using PCR, this might become possible in regenerating heterozygous microspore populations. Restriction fragment length polymorphism (RFLP) markers also provide an appropriate technique to locate quantitative trait loci (QTL) on the chromosomes. Hence, if the linkage between RFLP markers and the loci in question is close enough, the inheritance of a gene can be traced *in vitro* without selection at the whole plant level. Many of the present failures of *in vitro* selection will be overcome, as soon as these molecular techniques allow a better programming of selection strategies, strategies that not only select for the desired phenotype but also help to eliminate the undesirable changes at an early phase in the selection process.

Particularly as the agriculturally important characters are not coded by only one gene but are of polygenic origin, modern strategies for selection will be more effective in plant breeding than transformation. But once the biochemical pathways are better understood, the more direct alterations achieved via gene transfer have good chances to create useful transgenic plants.

Finally, we would emphasize here the necessity of growing all *in vitro* selected plants in the field. Very often a powerful *in vitro* resistance does not show up at the whole plant level under field conditions, even when calli subsequently induced a second time from later generations express the resistance again, as was shown, for example for 2,4-D resistant tobacco lines (Ono, 1979). Furthermore as pointed out by Miller *et al.* (1991) it is important to link *in vitro* approaches and conventional breeding programmes. It is the field test which will finally decide the value of any *in vitro* selection approach.

REFERENCES

Amrhein, N., Holländer-Czytko, H., Johänning, D., Schulz, A., Smart, C.C. and Steinrücken, H.C. (1987), Overproduction of 5-endolpyruvylshikimate-3-phosphate synthase in glyphosate-tolerant plant cell cultures. In *Plant Tissue and Cell Culture*, Green, C.E., Somer, D.A., Hackett, W.P. and Biesboer, D.D. (eds), A.R. Liss, New York, pp. 119–133.

Arcioni, S., Pezzotti, M. and Damiani, F. (1987), In vitro selection of alfalfa plants resistant to *Fusarium oxysporum* f. sp. *medicaginis*. *Theor. Appl. Genet.*, **74**, 700–705.

Bates, G.W., Hasenkampf, C.A., Contolini, C.L. and Piastuch, W.C. (1987), Asymmetric hybridization in *Nicotiana* by fusion of irradiated protoplasts. *Theor. Appl. Genet.*, **74**, 718–726.

Behnke, M. (1979), Selection of potato callus for resistance to culture filtrates of *Phytophthora infestans* and regeneration of resistant plants. *Theor. Appl. Genet.*, **55**, 69–71.

Behnke, M. (1980), Selection of dihaploid potato callus for resistance to culture filtrate of *Fusarium oxysporum*. *Z. Pflanzenzenziicht.*, **85**, 354–358.

Binarova, P., Nedelnik, M., Fellner, M. and Nedbalkova, B. (1990), Selection for resistance to filtrates of *Fusarium* spp. in embryogenic cell suspension culture of *Medicago sativa* L. *Plant Cell Tissue and Organ Culture*, **22**, 191–196.

Blom, T.J.M., Sierra, M., Iren, F. van, Verpoorte, R. and Libbenga, K.R. (1990), Accumulation of a malicine and serpentine in vacuoles isolated from *Catharanthus roseus* cell suspension cultures. In *Progress in Plant Cellular and Molecular Biology*, Nijkamp, H.J.J., Plas, L.H.W. van der and Aartrijk, J. van (eds), Kluver, Dordrecht, pp. 577–581.

Bouharmont, J. and Dabin, P. (1986), Somaclonal variation in some cultivars of *Fuchsia*. In *Somaclonal variations and crop improvement*, Semal, J. (ed.), Martinus Nijhoff Publishers, Dordrecht, pp. 257–259.

Branchard, M. (1982), In vitro culture of barley: a method to study *Rhynchosporium* scald disease and select plants resistant to the toxin rhynchosporoside. In *Variability in Plants Regenerated from Tissue Culture*, Early, E.D. and Demarly, Y. (eds), Praeger, New York, pp. 343–350.

Buiatti, M., Tesi, R. and Molino, M. (1969), A developmental study of induced somatic in *Gladiolus*. *Radiat. Bot.*, **9**, 39–43.

Chaleff, R.S. and Parsons, M.F. (1978), Direct selection in vitro for herbicide-resistant mutants of *Nicotiana tabacum*. *Proc. Natl. Acad. Sci. USA*, **75**, 5104–5107.

Chandler, S.F., Peak, K.V., Pua, E.C., Ragolsky, E. and Mandal, B.B. (1988), The effectiveness of selection for salinity tolerance using in vitro shoot cultures. *Bot. Gaz.*, **149**, 166–172.

Chawla, H.S. and Wenzel, G. (1987), In vitro selection of barley and wheat for resistance against *Helminthosporium sativum*. *Theor. Appl. Genet.*, **74**, 841–845.

Collin, H.A., Putwain, P.D. and Giffard, S.C. (1986), Enhancement of Asulam resistance in barley. In *Genetic Manipulation in Plant Breeding*, Horn, W., Jensen, C.J., Odenbach, W. and Schieder, O. (eds), Walter de Gruyter, Berlin, pp. 589–591.

Comai, L. and Stalker, D. (1986), Mechanism of action of herbicides and their molecular manipulation. *Oxford Surv. Plant Mol. Cell Biol.*, **3**, 167–195.

Conner, A.J. and Meredith, C.P. (1985), Strategies for the selection and characterisation of aluminium-resistant variants from cell cultures of *Nicotiana plumbaginifolia*. *Planta*, **166**, 466–473.

Daub, M.E. (1984), A cell culture approach for the development of disease resistance: studies on the phytotoxin cercosporin. *HortScience*, **19**, 383–387.

Daub, M.E. (1986), Tissue culture and the selection of resistance to pathogens. *Ann. Rev. Phytopathol.*, **24**, 159–186.

Deaton, W.R., Keyes, G.J. and Collins, G.B. (1982), Expressed resistance to black shank among tobacco callus culture. *Theor. Appl. Genet.*, **63**, 65–70.

Dehmer, K.J., Graner, A. and Wenzel, G. (1991), Screening for defined DNA sequences in minimal amounts of barley tissue by PCR. *Plant Breed.*, **107**, 70–72.

Dix, P.J. (1980), Environmental stress resistance. In *Plant Cell Cultures: Results and Perspectives*, Sala., F. *et al.* (eds), Elsevier, Amsterdam, pp. 183–186.

Dudits, D., Maroy, E., Praznovszky, T., Olah, Z., Gyorgyey, J. and Cella, R. (1987), Transfer of resistant traits from carrot into tobacco by asymmetric somatic hybridization: regeneration of fertile plants. *Proc. Natl. Acad. Sci. USA*, **84**, 8434–8438.

Foroughi-Wehr, B., Friedt, W., Schuchmann, R., Köhler, F. and Wenzel, G. (1986), In vitro selection for resistance. In *Somaclonal Variations and Crop Improvement*, Semal, J. (ed.), Martinus Nijhoff Publishers, Dordrecht, pp. 35–44.

Freytag, A.H., Wrather, J.A. and Erichsen, A.W. (1990), Salt tolerant sugarbeet progeny from tissue cultures challenged with multiple salt. *Plant Cell Rep.*, **8**, 647–650.

Fujita, Y. and Tabata, M. (1987), Secondary metabolites from plant cells – pharmaceutical applications and processes in commercial production, In *Plant Tissue and Cell Culture*, Green, C.E., Somer, D.A., Hackett, W.P. and Biesboer, D.D. (eds), A.R. Liss, New York, pp. 169–185.

Gamborg, O.L., Miller, R.A., and Ojima, K. (1968), Nutrient requirement of suspension cultures of soybean root cells. *Exp. Cell Res.*, **50**, 151–158.

Gengenbach, B., Green, G. and Donovan, C.M. (1977), Inheritance of selected pathotoxin resistance in maize plants regenerated from cell cultures. *Proc. Natl. Acad. Sci. USA*, **74**, 5113–5117.

Hammerschlag, F.A. (1988), Selection of peach cells for insensitivity to culture filtrates of *Xanthomonas campestris* pv. *pruni* and regeneration of resistant plants. *Theor. Appl. Genet.*, **76**, 865–869.

Hassan, N.S. and Wilkins, D.A. (1988), *In vitro* selection for salt tolerant lines in *Lycopersicon peruvianum*. *Plant Cell Rep.*, **7**, 463–466.

Heath-Pagliuso, S. and Rappaport, L. (1990), Somaclonal variant UC-T3: the expression of *Fusarium* wilt resistance in progeny arrays of celery, *Apium graveolens* L. *Theor. Appl. Genet.*, **80**, 390–394.

Heinstein, P. and Emery, A. (1988), Processes with plant cell cultures. In *Biotechnology*, Vol. 6b VCH, Rehm, H.J. (ed.), Weinheim, pp. 213–248.

Heinz, D.J., Krishnamurthi, M., Nickell, L.G. and Maretzki, A. (1977), Cell tissue and organ culture in sugar cane improvement. In *Plant Cell Tissue and Organ Culture*, Reinert, J. and Bajaj, Y.P.S. (eds), Springer Verlag, New York, pp. 3–17.

Helgeson, J.P., Kemp, J.D., Haberlach, G.T. and Maxwell, B.P. (1972), A tissue culture system for studying disease resistance: The black shank disease in tobacco callus cultures. *Phytopathology*, **62**, 1439–1443.

Hoffmann, F., Thomas, E. and Wenzel, G. (1982), Anther culture as a breeding tool in rape. II. Progeny analysis of androgenetic lines and of induced mutants from haploid cultures. *Theor. Appl. Genet.*, **61**, 225–232.

Horsch, R., Fraley, R., Rogers, S., Fry, J., Klee, H., Shah, D., McCormick, S., Niedermeyer, J. and Hoffmann, N. (1987), *Agrobacterium* mediated transformation of plants. In *Plant Tissue and Cell Culture*, Green, C.E., Somer, D.A., Hackett, W.P. and Biesboer, D.D. (eds), A.R. Liss, New York, pp. 317–329.

Hunold, R., Krämer, R., Kunert, R. and Peterka, H. (1990), *In vitro*-Selektion bei der Gerste (*Hordeum vulgare* L.) auf Resistenz gegen *Drechslera teres*. *J. Phytopathol.*, **129**, 291–302.

Jain, R.K., Jain, S., Nainawatee, H.S. and Chowdhury, J.B. (1990), Salt-tolerance in *Brassica juncea* L. I. *In vitro* selection, agronomic evaluation and genetic stability. *Euphytica*, **48**, 141–152.

Jang, J.C. and Tainter, F.H. (1991), Optimum tissue culture conditions for selection of resistance to *Phytophthora cinnamomi* in pine callus tissue. *Plant Cell Rep.*, **9**, 488–491.

Jordan, M.C. and McHughen, A. (1987), Selection for chlorsulfuron resistance in flax (*Linum usitatissimum*) cell culture. *J. Plant Physiol.*, **131**, 333–338.

Kim, Y.-H., Chung, T.-Y. and Choi, W.-Y. (1988), Increased regeneration from NaCl-tolerant callus in rice. *Euphytica*, **39**, 207–212.

Köhler, F., Wenzel, G., Abentum, I. and Glaser, H. (1986), Regeneration and selection of isolated microspores of barley. In *Genetic Manipulation in Plant Breeding*, Horn, W., Jensen, C.J., Odenbach, W. and Schieder, O. (eds), Walter de Gruyter, Berlin, pp. 315–317.

Lai, A., Crino, P. and Saccardo, F. (1988), *In vitro* selection of tomato regenerated plants resistant to culture filtrate of *Phytophthora infestans* (Mont.) De Bary. *Genet. Agraria*, **42**, 78.

Larkin, P. and Scowcroft, W.R. (1981), Somaclonal variation – a novel source of variability from cell cultures for plant improvement. *Theor. Appl. Genet.*, **60**, 197–214.

Lazar, M.D., Chen, T.H.H., Gusta, L.V. and Kartha, K.K. (1988), Somaclonal variation for freezing tolerance in a population derived from Nostar winter wheat. *Theor. Appl. Genet.*, **75**, 480–484.

Lind, V. and Wenzel, G. (1984), Test auf Mehltauresistenz für die Roggenzüchtung. *Nachrichtenbl. deut. Pflanzenschutzd.*, **36**, 17–20.

Luckner, M. and Diettrich, B. (1990), Principles regulating formation and activity of secondary metabolic enzymes in plant tissue and cell cultures. In *Progress in Plant Cellular and Molecular Biology*, Nijkamp, H.J.J., Plas, L.H.W. van der and Aartrijk, J. van (eds), Kluwer, Dordrecht, pp. 744–753.

Lupotto, E., Lusardi, M.C. and Mongodi, M. (1989), *In vitro* selection of maize (*Zea mays* L.) salt tolerant somaclones and plant regeneration. *J. Genet. Breed.*, **43**, 215–222.

MacLean, N.L. and Grant, W.F. (1987), Evaluation of birdsfoot-trefoil (*Lotus corniculatus*) regenerated plants following *in vitro* selection for herbicide tolerance. *Can. J. Bot.*, **65**, 1275–1280.

Mandal, B.B., Chandler, S.F., Ragolsky, E. and Thorpe, T.A. (1989), Evidence for physiological adaptation of *Brassica napus* cv. 'Westar' callus to sodium sulfate. *Phytomorphology*, **39**, 115–119.

Matern, U., Strobel, G. and Shepard, J. (1978), Reaction to phytotoxins in a potato population derived from mesophyll protoplasts. *Proc. Natl. Acad. Sci.*, **75**, 4935–4939.

McCoy, T.J. (1987), Characterization of alfalfa (*Medicago sativa* L.) plants regenerated from selected NaCl tolerant cell lines. *Plant Cell Rep.*, **6**, 417–422.

McDonald, M.V., Ingram, D.S. and Hodgkin, T. (1986), Selection *in vitro* for resistance to *Alternaria brassicola* in *Brassica napus* ssp. *oleifera*, winter oilseed rape. *Proc. Crucifer Genetics Workshop III*, Universtiy of Guelph, 66.

McHughen, A. (1987), Salt tolerance through increased vigour in a flax line (STS-II) selected for salt tolerance *in vitro*. *Theor. Appl. Genet.*, **74**, 727–732.

Menczel, L., Nagy, G. and Maliga, P. (1983), Transfer of cytoplasmic male sterility by selection for streptomycin resistance after protoplast fusion in *Nicotiana*. *Mol. Gen. Genet.*, **189**, 365–369.

Meredith, C.P. (1978), Selection and characterization of aluminium-resistant variants from tomato cell cultures. *Plant Sci. Lett.*, **12**, 25–34.

Meulemans, M. and Fouarge, G. (1986), Regeneration of potato somaclones and *in vitro* selection for resistance to *Phytophthora infestans* (Mont.) de Bary. *Mededelingen van de Faculteit Landbouwwetenschappen Rijksuniversiteit Gent*, **51**, 533–545.

Miedaner, T., Grossmann, F., Walther, H. and Wenzel, G. (1987), Bedingungen für künstliche Infektionen von Weizenkeimlingen mit *Fusarium culmorum*. *Nachrichtenbl. deut. Pflanzenschutzd.*, **39**, 49–53.

Miller, D.R., Waskom, R.M., Brick, M.A. and Chapman, P.L. (1991), Transferring *in vitro* technology to the field. *Biotechnology*, **9**, 143–146.

Morel, G. (1946), Essais de laboratoire sur le mildiou de la vigne. *Rev. Vitic.*, **93**, 210–213.

Msikita, W., Wilkinson, H.T. and Skirvin, R.M. (1990), Resistance of *in vitro*-derived cucumber plants to *Pythium aphanidermatum*. *HortScience*, **25**, 967–969.

Murashige, T. and Skoog, F. (1962), A revised medium for rapid growth and bio assays with tobacco tissue cultures. *Physiol. Plant.*, **15**, 473–497.

Newhouse, K.E., Schaefer, T.J. and Singh, B.K. (1991), Maize mutants that confer resistance to imidazolinone herbicides. *Symposium on Plant Breeding in the 1990s*, NC State University, Raleigh, 66.

Newsholme, D.M., McDonald, M.V. and Ingram, D.S. (1989), Studies of selection *in vitro* for novel resistance to phytotoxic products of *Leptosphaeria masculans* (Desm.) Ces. and De Not. in secondary embryogenic lines of *Brassica napus* ssp. *oleifera* (Metzg.) Sinsk., winter oilseed rape. *New Phytol.*, **113**, 117–126.

Ojima, K. and Ohira, K. (1982), Characterization and regeneration of an aluminium-tolerant variant from carrot cell culture. In *Proc. 5th Intl. Cong. Plant Tissue and Cell Culture*, Fujiwara, A. (ed.), Jap. Ass. Plant Tissue Culture, pp. 575–476.

Ono, H. (1979), Genetical and physiological investigations of a 2,4-D resistant cell line isolated from tissue culture in tobacco. *Sci. Rep. Fac. Agric. Kobe University*, **13**, 273–277.

Pauly, M.H., Shane, W.W. and Gengenbach, B.G. (1987), Selection for bacterial blight phytotoxin resistance in wheat tissue culture. *Crop Sci.*, **27**, 340–344.

Pelletier, G., Primard, C., Ferault, M., Vedel, F., Chetrit, P., Renard, M. and Delourme, R. (1988), Use of protoplasts in plant breeding: cytoplasmic aspects. *Plant Cell, Tissue and Organ Culture*, **12**, 173–180.

Pua, E.C. and Thorpe, T.A. (1986), Differential response of nonselected and Na$_2$SO$_4$-selected callus cultures of *Beta vulgaris* L. to salt stress. *J. Plant Physiol.*, **123**, 241–248.

Raghava Ram, N.V. and Nabors, M.W. (1985), Salinity tolerance. In *Biotechnology: Application and Research*, Cheremisinoff, P.N. and Oulette, R.P. (eds), Technomic Pub., Lanc, pp. 623–642.

Rey, P., Eymery, F. and Peltier, G. (1990), Atrazine and diuron resistant plants from photoautotrophic protoplast-derived cultures of *Nicotiana plumbaginifolia*. *Plant Cell Rep.*, **9**, 241–244.

Sabbah, S. and Tal, M. (1990), Development of callus and suspensions cultures of potato resistant to NaCl and mannitol and their response to stress. *Plant Cell, Tissue and Organ Culture*, **21**, 119–128.

Saiki, R.K., Scharf, S., Faloona, F., Mullis, K.B., Horn, G.T., Erlich, H.A. and Arnheim, N. (1985), Enzymatic amplification of ß-globin genome sequences and restriction site analysis for diagnosis of sickle cell anemia. *Science*, **230**, 1350–1354.

Saxena, P.K., Williams, D. and King, J. (1990), The selection of chlorsulfuron-resistant cell lines of independent origin from an embryogenic cell suspension culture of *Brassica napus* L. *Plant Sci.*, **69**, 231–237.

Schuchmann, R. (1985), *In vitro* Selektion auf *Fusarium*-Resistenz bei der Kartoffel. Ph . Thesis, TU-Munchen.

Shepard, J.F. (1981), Protoplasts as sources of disease resistance in plants. *Ann. Rev. Phytopathol.*, **19**, 145–166.

Sjödin, C. and Glimelius, K. (1989a), Differences in response to the toxin sirodesmin PL produced by *Phoma lingam* (Tode ex Fr.) Desm. on protoplasts, cell aggregates and intact plants of resistant and susceptible *Brassica* accessions. *Theor. Appl. Genet.*, **77**, 76–80.

Sjödin, C. and Glimelius, K. (1989b), Transfer of resistance against *Phoma lingam* to *Brassica napus* by asymmetric somatic hybridization combined with toxin selection. *Theor. Appl. Genet.*, **78**, 513–520.

Subhashini, K. and Reddy, G.M. (1989), *In vitro* selection for salinity and regeneration of plants in rice. *Curr. Sci.*, **58**, 584–586.

Swaaij, A.C. van, Jacobsen, E., Kiel, J.A.K.W. and Feenstra, W.J. (1986), Selection, characterization and regeneration of hydroxyproline-resistant cell lines of *Solanum tuberosum*: tolerance to NaCl and freezing stress. *Physiol. Plant.*, **68**, 359–366.

Swaaij, A.C. van, Talsma, K., Krijgsheld, H., Jacobsen, E. and Feenstra, W.J. (1987), Frost tolerance in cell culture of potato. *Physiol. Plant.*, **69**, 602–608.

Swanson, E.B., Coumans, M.P., Brown, G.L., Patel, J.D. and Beversdorf, W.D. (1988), The characterization of herbicide tolerant plants in *Brassica napus* L. after *in vitro* selection of microspores and protoplasts. *Plant Cell Rep.*, **7**, 83–87.

Swanson, E.B., Herrgesell, M.J., Arnoldo, M., Sippell, D.W. and Wong, R.S.C. (1989), Microspore mutagenesis and selection: canola plants with field tolerance to the imidazolinones. *Theor. Appl. Genet.*, **78**, 525–530.

Tanno-Suenaga, L., Ichikawa, H. and Imamura, J. (1988), Transfer of the CMS trait in *Daucus carota* L. by donor-recipient protoplast fusion. *Theor. Appl. Genet.*, **76**, 855–860.

Taylor, R.J., Ruby, C.L. and Secor, G.A. (1988), Assessment of field performance and soft rot resistance in a population of protoplast-derived potato clones. *Phytopathol.*, **78**, 1595–1603.

Taylor, R.J. and Secor, G.A. (1990), Potato protoplast-derived callus tissue challenged with *Erwinia carotovora* subsp. *carotovora*: Survival, growth and identification of resistant callus lines. *J. Phytopathol*, **129**, 228–236.

Vajrabhaya, M., Thanapaisal, T. and Vajrabhaya, T. (1989), Development of salt tolerant lines of 'KDML' and 'LPT' rice cultivars through tissue culture. *Plant Cell Rep.*, **8**, 411–414.

Vasil, I.K. (1990), The realities and challenges of plant biotechnology. *Biotechnology*, **8**, 296–301.

Vasil, I.K. and Vasil, V. (1980), Clonal propagation. *Int. Rev. Cytol.*, Suppl. **11A**, 145–173.

Wenzel, G. (1980), Protoplast techniques incorporated into applied breeding programmes. In *Advances in Protoplast Research*, Ferenczy, L. and Farkas, L. (eds), Pergamon Press, London, pp. 327–340.

Wenzel, G. (1985), Strategies in unconventional breeding for disease resistance. *Ann. Rev. Phytopathol.*, **23**, 149–172.

Wenzel, G., Debnath, S.C., Schuchmann, R. and Foroughi-Wehr, B. (1987), Combined application of classical and unconventional techniques in breeding for disease resistant potatoes. In *The Production of New Potato Varieties*, Jellis, G.J. and Richardson, D.E. (eds), Cambridge University Press, Cambridge, pp. 277–288.

Wenzel, G. and Foroughi-Wehr, B. (1990), Progeny tests of barley, wheat, and potato regenerated from cell cultures after *in vitro* selection for disease resistance. *Theor. Appl. Genet.*, **80**, 359–365.

Wenzel, G. and Willmitzer, L. (1988), Nutzung der Biotechnologie bei der Kartoffelzüchtung. *Kartoffelbau*, **39**, 190–194.

Wersuhn, G., Kirsch, K. and Gienapp, R. (1987), Herbicide tolerant regenerants of potato. *Theor. Appl. Genet.*, **74**, 480–482.

Wersuhn, G., Nhi, H.H., Tellhelm, E. and Reinke, G. (1988), Aluminium-tolerant regenerants from potato cell cultures. *Potato Res.*, **31**, 305–310.

Ye, I.M., Kao, K.N., Harvey, B.L. and Rossnagel, B.C. (1987), Screening salt tolerant barley genotypes via F_1 anther culture in salt stress media. *Theor. Appl. Genet.*, **74**, 426–429.

Zenk, M.H., El-Shagi, H., Arens, H., Stöckigt, J., Weiler, E.W., and Deus, B. (1977), Formation of indole alkaloids serpentine and ajmalicine in cell suspension cultures of *Catharanthus roseus*. In *Plant Tissue Culture and its Bio-technological Application*, Barz, W., Reinhard, E. and Zenk, M.H. (eds), Springer Verlag, Berlin, pp. 27–43.

Part Seven

Adaptation

22

Genotype × environment interaction and adaptation

I. Romagosa and P.N. Fox

22.1 INTRODUCTION

This chapter has two objectives: first, to discuss implications of genotype by environment interaction (G×E) for breeding and second, to present major statistical models for assessing genotypic adaptation. We concentrate on widely used methodologies and on that which we believe will be extensively applied, but acknowledge the subjective element in such decisions.

G×E is differential genotypic expression across environments. It reduces association between phenotypic and genotypic values, and may cause selections from one environment to perform poorly in another, forcing plant breeders to examine genotypic adaptation. Measuring G×E is also important to determine an optimum breeding strategy for releasing genotypes with adequate adaptation to target environments. Some prefer the term 'adaptation' in the context of spatial variation and use the term 'stability' for performance at a given site across years or management practices. Here, we use stability and adaptation to refer to both spatial and temporal dimensions, as most studies implicitly assume stability measured spatially and temporally to be expressions of the same phenomenon. The term 'site' is used for consistency in this chapter, but is con-

sidered a synonym for 'location'. Both indicate spatial variation. 'Environment' is a general term that covers conditions under which plants grow and may involve sites, years, management practices or a combination of these factors. Commonly, every site/year is considered a separate environment. 'Trial' or 'experiment' indicates a set of genotypes grown in one or more environments.

22.2 THE BIOLOGICAL COMPLEXITY OF GENOTYPE × ENVIRONMENT INTERACTION

Allard (1960) described the biological complexity underlying G×E: 'virtually all phenotypic effects are not related to the gene in any simple way. Rather they result from a chain of physico-chemical reactions and interactions initiated by genes but leading through complex chains of events, controlled or modified by other genes and the external environment, to the final phenotype'. In the context of biological complexity, it is significant that different GxE patterns were elicited by three mutant genes in near-isogenic barley lines (Molina-Cano *et al.*, 1990; Romagosa *et al.*, 1993). This example involving three loci could be thought of as the 'tip of the G×E iceberg'. Strong G×E effects will influence genotypic adaptation through the much larger number of genes below the waterline. The biological basis of G×E may not be understood because of both the underlying environmental and the genetic complexity.

Plant Breeding: Principles and prospects. Edited by M.D. Hayward, N.O. Bosemark and I. Romagosa. Published in 1993 by Chapman & Hall, London. ISBN 0 412 43390 7.

Thus GxE often appears intractable. However, Baker (1971) explained G×E for grain yield in a series of wheat trials in terms of rust reactions and an example below shows how triticale adaptation is largely controlled by a single wheat chromosome.

22.3 IMPLICATIONS FOR BREEDING PROGRAMMES

For breeders, agroecological diversity of environments where most field crops are grown represents a 'double-edged sword'. The diversity complicates breeding and testing of improved genotypes with adequate adaptation, but it also permits identification of extreme environmental conditions that guarantee selection pressure from important stresses.

There are two conceptually different approaches for studying G×E and adaptation. The more common is empirical and statistical and involves relating observed genotypic responses, usually in terms of yield, to a sample of environmental conditions. The analytical approach defines environments and phenotypes in terms of biotic and abiotic factors. In practice, most breeding programmes incorporate elements of both approaches. We emphasize the concept of stability through genetic buffering by means of genotype mixtures, multilines and populations (Simmonds, 1979), but detailed consideration is beyond the scope of this chapter.

Two types of genotypic stability were distinguished by Becker (1981). He applied the word 'biological' to stability in the homeostatic sense in which a genotype maintains a constant yield across environments. Statistically, this concept of stability is assessed by genotypic variance across environments. However, homeostatic stability tends to be undesirable in modern agriculture, since genotypes should respond to improved conditions. The need for genotypic responsiveness to favourable environments leads to the concept of 'agronomic' stability, by which a genotype is considered stable if it yields well relative to the productive potential of test environments. If agronomic stability is demonstrated for a wide range of environments, a genotype is defined as having general or wide adaptation. On the contrary, if agronomic stability is manifest over a limited range, a genotype has specific or narrow adaptation.

G×E is considered quantitative if the ranking of genotypes does not change from one environment to another, that is, if the differential response of one genotype compared with another is a matter of scale (Fig. 22.1). Such quantitative (non-crossover) interactions are less important to breeders than so-called qualitative interactions, in which genotypes change rank (Baker, 1988). Qualitative, or crossover, interactions complicate selection and identification of superior genotypes.

For trials in which the same sites and genotypes are included over years, the G×E term from the Analysis of Variance (section 22.4.1) may be partitioned into components due to genotypes × sites (G×S), genotypes × years (G×Y) and genotypes × sites × years (G×S×Y). If G×S is the important portion of G×E, then specific adaptation is exploitable by subdividing target areas into homogeneous regions that minimize G×E within regions. Where G×Y and G×S×Y terms dominate, no simplification involving spatial subdivision of breeding regions is available. Generally, testing over a representative range of conditions is the common strategy. Such wide testing necessitates description of genotypic response across environments. Breeders aim to cover a representative sample of spatial and temporal variation. Sometimes a breeder's selection environments in one year may have little relation to those experienced in the next (Fox *et al.*, 1985). The sampling problem associated with yearly variation would suggest testing for many crop cycles. However, to save time, breeders opt to substitute temporal variation with spatial variation, assuming that testing over a wide geographic range can ensure a parallel degree of temporal buffering capacity in their germplasm. The implicit assumption that both spatial and temporal buffering rely on the same mechanism has been validated in experiments with sorghum (Barah *et al.*, 1981) and rice (Flinn and Garrity, 1989). Defining long-term similarities with areas in other countries and continents and systematic exchange of data and

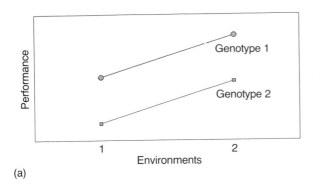

(a)

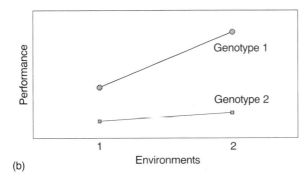

(b)

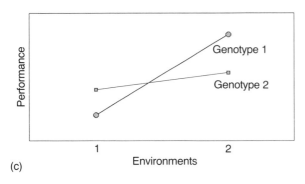

(c)

Fig. 22.1. The performance of two hypothetical genotypes in two environments, showing (a) no G×E and therefore homogeneous regressions, (b) 'quantitative' G×E, without reversal of rank of genotypes and (c) 'qualitative' G×E, with rank reversal.

germplasm with such areas may overcome the problem of sampling seasonal variation. Creative initiatives are required to establish such collaboration.

Particularly in subsistence agricultural systems, maximum yield *per se* may not be as important as the probability of achieving a certain minimum yield level. This involves a compromise between homeostatic and agronomic stability, that is, between responsiveness of yield and limited variation in yields across environments. Menz (1980) used stochastic dominance to examine risk-efficiency of wheats in international trials and concluded that recently bred materials were more risk-efficient, and thus more suitable for peasant farmers, than were previously developed genotypes at the time they entered international testing. There was a tendency for risk-efficient genotypes to have high mean yields, but exceptions existed. Menz maintained that a combination of high mean yield and risk-efficiency was a virtual guarantee of good performance across all environments.

For a single site, Binswanger and Barah (1980) considered a genotype to show maximum risk-efficiency over years if no other genotype could achieve the same average yield with lower standard deviation or the same standard deviation with higher average yield. They then suggested a methodology for accommodating trade-offs between average yield and variability in this situation, but did not extend the technique to risk-efficiency over sites. However, they provided a framework for comparing the mechanisms of spatial and temporal stability, concluding that both would be strongly related if stability is primarily a response to weather variables, but that spatial and temporal stability would differ if important stresses were specific to sites, especially if related to soil parameters. This is akin to a situation, detailed above, in which G×S variation is dominant. Desirable genotypes might combine insensitivity to weather-related factors with responsiveness to inputs. Binswanger and Barah did not consider possible interactions between weather and site-specific factors, usually manifest in large S×Y and G×S×Y terms.

A weighting approach, which uses a reference set of genotypes for handling unpredictable seasonal variation, has been suggested by Fox and Rosielle (1982b). A breeder would define a long-term target environment using relative yields or rankings from a reference set of genotypes grown for many years. Adjacent to every selection trial, the diverse reference genotypes would be grown to determine the proximity of the specific site-year combination to the long-term target, in terms of G×E. Then results from each site in a year would be weighted in accordance with the site's representativeness or proximity to the target. The concept of 'genotypic probes', or cultivars with known differential response to environmental stimuli, followed the reference genotype approach, but precluded genetic progress for unrecognized stresses. However, by analysing the response of each genotypic probe, it may be possible to determine the influence and timing of some environmental constraints.

There is controversy between wide testing of advanced lines *versus* exposure to a few key sites with defined stresses. We feel the two approaches complement each other. There is a chance of 'throwing out the baby with the bath water' in a strict key site approach for the reason given above, that is, genetic progress is minimized for unrecognized stresses. The case of nematode resistance in Australian wheats provides an example of the unwitting accumulation of resistance, in an aberrant site, for such an unrecognized factor (Rathjen and Pederson, 1986). 'Aberrant sites' may be more useful for crop improvement, in the long-term, than 'key sites', especially when underlying biological explanations for abnormal behaviour are revealed.

G×E is a major element in determining breeding strategies, but controversy exists over the existence of genotypes showing general adaptation. This argument has practical implications for breeding programmes. Should a breeding programme aim for wide or specific adaptation? Where should a site for selection be located? Should early generation selection be conducted in a stressed or stress-free environment? Should emphasis be towards multi-site testing of large numbers of advanced lines or towards early generation screening of fewer lines by means of trait-based selection?

Related to wide adaptation is the question of breeding site(s): can selection under optimum high-input environments identify genotypes adapted to more limiting environments? This is a vexed, at times even emotional, issue and generalizations for crop species, let alone plant breeding as a whole, are fraught with risk. We present some contrasting opinions and evidence and suggest that an understanding of the biology of the crop species involved may be the most potent weapon for resolving differences between proposed strategies. Ceccarelli (1989) argued that wide adaptation does not exist across different macro-agroecological environments and that selection for high yield potential has not increased yield under low inputs. Similarly, Lawn (1988) stated that high yield and agronomic stability were mutually exclusive over a wide range of environments. Patel *et al.* (1987) found that when barley populations were alternated between diverse sites, representing different zones of adaptation, natural selection lowered yields. Austin (see Chapter 23) discusses the issue of wide *vs* specific adaptation in an evolutionary context. However, many breeders would contest that accumulation of tolerances to a number of stresses is the key to wide adaptation and consequently selection in multiple environments is the best way to breed stable genotypes. In the First to the Fifteenth International Spring Wheat Yield Nurseries, grown at sites representative of all major wheat-growing areas, CIMMYT genotypes bred under high input environments were superior in yield and showed better adaptation than genotypes developed locally with or without CIMMYT germplasm (Pfeiffer and Braun, 1989). Poor adaptation of CIMMYT genotypes to specific environments generally reflected disease susceptibility, rather than biological inefficiency. On the contrary, Ceccarelli *et al.* (1987) concluded that barley genotypes bred for extreme conditions should be selected in these unfavourable conditions.

Success of CIMMYT wheats in combining high yield potential and wide adaptation involves: (1) large numbers of crosses; (2) the testing of advanced lines internationally; (3) continuous alternating selection cycles, now referred to as 'shuttle breeding', in environments which allow expression of high yield potential but differ in altitude, latitude, photoperiod, temperature, rainfall, soil-type and disease spectrum. General adaptation across large geographical and agroecological areas, perhaps with the exception of the very poorest environments, has been achieved. High yield potential of CIMMYT germplasm is expressed in Australian dryland conditions. More than any other major wheat exporting country, Australia, the driest continent, experiences large moisture deficiencies and improved drought tolerance has been an important objective since purposeful wheat breeding began (Marshall, 1987). The Australian average wheat yield is 1.5 t/ha. Brennan (1986) documented that, in 1983, half of the area sown to wheat in Australia was occupied by CIMMYT-derived varieties which showed a 7% yield advantage over other leading varieties and were estimated to have increased Australian yield by an average of 3.6%, ranging from 0.3% in the drier state of Western Australia to 7.2% in Queensland. Brennan (1986) states: 'On the basis of the pedigrees of Australian wheats, about two-thirds of this value can be attributed to the lines from CIMMYT'. Although Brennan showed that adoption of CIMMYT-derived materials was less dramatic in drier states, more recent data show rapidly changing trends in such areas as well. For example, R. Wilson (personal communication) estimated that the area sown to CIMMYT-derived germplasm in Western Australia increased from 9% in 1983 to 67% in 1990.

Genotypic performance is influenced by environments of early generation selection (Campbell and Lafever, 1977). Interplant competition in segregating populations of field crops is often not representative of commercial production, because breeders use low plant densities which are more manageable for selection. Choice of selection sites is particularly relevant in the case of production areas with variable levels of abiotic stress. Rathjen and Pederson (1986) reported the abandoning of one research station and the down-grading of another for selection trials, after examining genetic correlations between breeding sites and an extensive, and more commercially representative, network of recommendation trials.

Decisions on the need for separate breeding programmes for low-input environments depend on the extent to which yields in low- and high-input environments are under separate genetic control and on the accuracy of selection in the contrasting environments (Atlin and Frey, 1989). Before different programmes can be recommended, it must be demonstrated that alleles controlling yield in the two situations are different. However, such issues may be dynamic, in that breeders will strive to incorporate new alleles which are superior in both situations. Atlin and Frey (1989) remarked on the scarcity of estimates of genetic correlations between low- and high-input environments. Rosielle and Hamblin (1981) concluded on theoretical grounds that selection in stress environments, where genetic variance is small, will 'result in a reduced mean yield in non-stress environments and a decrease in mean yield. Selection for mean productivity will generally increase mean yield in both stress and non-stress environments unless the genetic correlation between yields in stress and non-stress environments is highly negative'. These conclusions have been supported experimentally. Richards (1982) found that selection in environments where water was not limiting was superior to selection under moisture stress in developing wheat germplasm that combines high genetic potential for yield with tolerance to moisture stress.

It has been suggested by Austin (Chapter 23) that breeding efficiency can be increased by early generation selection using trait-based selection, facilitated by integration of current knowledge of the physiology and genetics of crops. However, Whan *et al.* (1991) indicated that:

Many physiological studies are retrospective, involve only a few cultivars, and are designed to . . . provide a basic understanding of the plant processes that result in high yield. . . . We claim there has been a disproportionate emphasis in the past on physiological studies that have had little or no impact on the design and conduct of plant breeding programmes. To be useful to plant breeders, theories generated by physiologists must be validated, and their value demonstrated in populations of genetic material.

Marshall (1987) emphasized that, to be used in selection of segregating populations, any trait must be rapidly and cheaply measured. Efficiency of this selection depends upon understanding target environments in terms of biotic and abiotic constraints and their frequency and occurrence (Ludlow and Muchow, 1989). We feel that a distinction is warranted between traits for selection and traits for identification of parents. An example of the integration of physiology and breeding is the use of silking to anthesis interval under drought in tropical maize. The character influences yield potential and yield stability (Bolaños and Edmeades, 1988) and is a function of photosynthesis per plant and competition between the male and female inflorescences. The example highlights the species specific nature of stresses. Breeding gains require heritable variation for important characters. Artificially structured stress, commonly used for disease and insect reaction, may partially substitute multi-site, multi-year testing (Blum *et al.*, 1983; Eisemann *et al.*, 1990).

22.4 ANALYTICAL MODELS TO ASSESS STABILITY

Four statistical techniques for assessing the agronomic stability of a set of cultivars or advanced lines are reviewed: (1) partitioning of variance; (2) regression analysis; (3) non-parametric statistics; (4) multivariate techniques, with emphasis on pattern analysis and the additive main effects and multiplicative interaction (AMMI) model.

22.4.1 Analysis of variance

Significance tests from a combined analysis of variance are valid if error terms from different environments are homogeneous. If Bartlett's Test indicates heterogeneous variances, then data transformation or regrouping of environments into subsets with homogeneous variances is recommended (Steel and Torrie, 1980). However, plant breeders seldom apply Bartlett's Test. To avoid difficulties in interpretation of analyses, data should not be transformed unless experimental errors across environments are extremely diverse.

For any two-factor mixed model (fixed genotypes and random environments), the most commonly used combined analysis of variance is shown in Table 22.1.

Means adequately describe the potential of environments and the performance of genotypes in a trial when G×E is not significant. However, when the interaction is significant, main effects should be interpreted with caution and the nature of the interaction examined, as means often mask cases where genotypes perform well or poorly in subsets of sites. In analyses of variance, magnitudes of sums of squares of relevant terms as well as variance components are used to quantify sources of variation. Sums of squares attributable to a source of variation confound: (1) the nature of

Table 22.1. Mixed model (fixed genotype and random environment) analysis of variance for *g* genotypes at *e* locations with *r* replications at each site

Source of variation	Degrees of freedom	Mean squares	Expected mean squares	F-ratios
Total	$erg-1$			
Environ (E)	$e-1$	MS1	$\sigma_e^2 + g\sigma_{R(E)}^2 + rg\sigma_E^2$	MS1/MS2
Rep./E[a]	$e(r-1)$	MS2	$\sigma_e^2 + g\sigma_{R(E)}^2$	MS2/MS5
Genotypes (G)	$g-1$	MS3	$\sigma_e^2 + g\sigma_{GE}^2 + er\phi_G^2$	MS3/MS4
GxE	$(e-1)(g-1)$	MS4	$\sigma_e^2 + g\sigma_{GE}^2$	MS4/MS5
Error	$e(g-1)(r-1)$	MS5	σ_e^2	

[a] If replicates within environments are not separated from the error term, the environment main effect should then be tested against the error mean square.

the factor considered with respect to its ability to elicit variation; (2) the number of levels of the factor, e.g., the number of sites in a trial. However, variance components correct for the number of levels of factors and allow direct comparisons of estimates from sources with divergent numbers of sites and genotypes.

In a synthesis of regional and international variety trials for more than 20 species, DeLacy *et al.* (1990) surveyed the relative importance of sites, genotypes and G×E. They considered data from over 100 sites throughout the world and highlighted the ubiquitous nature of G×E for yield. In most yield trials, the proportion of sums of squares due to differences among sites ranged from 80 to 90% and variation due to G×E was usually larger than genotypic variation. The ratio of the G×E sum of squares to the sum of squares for genotypes was rarely less than 0.8 and reached 43. This ratio was preferred to variance components to avoid controversial assumptions as to whether genotypes and sites are random or fixed effects.

As a measurement of a genotype's stability, Plaisted and Peterson (1959) proposed averaging the variance components determined from each pairwise combination involving the genotype with every other in a trial. However, for a large number of genotypes, this method is laborious. The 'ecovalence' parameter of Wricke (1964) is the contribution of a genotype to the G×E term. Shukla (1972) modified the ecovalence term to give an unbiased estimate of the G×E variance for every genotype, allowing probability tests for separation of genotypes. The greater the magnitude of these estimates for a genotype, the poorer its general agronomic stability or adaptation. All methods ranked a series of genotypes in the same order (Lin *et al.*, 1986), but only Shukla's criterion is amenable to statistical testing. None of these methods has been extensively used in breeding programmes. Alternative techniques based on simple parameters from multi-environment analysis have been more recently proposed.

Often a breeder's assessment of the value of a genotype is based on comparison with one or more checks. Lin and Binns (1985) suggested using pairwise analyses of variance between known check cultivars and all tested genotypes to detect which genotypes show the same adaptation pattern as checks in regional variety trials. Genotypes that did not elicit significant G×E with a check were identified and their yields classified as significantly superior to, inferior to, or not different from the mean of a check. Cultivars with similar adaptation patterns to standard checks and higher yields across sites were identified for recommendation.

22.4.2 Regression

Simple linear regression provides a conceptual model for genotypic stability and is the most widely used and abused statistical technique in plant breeding. The method is often known as the Finlay and Wilkinson (1963) approach. The slopes of the regressions of individual genotype's yields against environmental mean yields, based on all genotypes in a trial, is determined and performance is explained in terms of: (1) main effects for genotypes and environments; (2) the product of environmental main effects multiplied by the regression coefficients of genotypes. The G×E term from analysis of variance is partitioned between heterogeneity of regressions and deviations from regressions. Statistical consideration of regression methodology and its extensions is given by Crossa (1990).

The idea of a line reflecting the manner in which a genotype responds to improving environmental conditions is intuitively appealing (Fig. 22.1), but a significant limitation is the juxtaposition of ecologically different environments with similar mean yields on the abscissa (Knight, 1970). This juxtaposition produces large deviations from fitted regression lines which are, nonetheless, often presented without indication of their fit to the experimental data. The proportion of G×E explained by regression, an overall measure of fit to the linear model, is more important than the statistical significance of the term for heterogeneity of regressions. This proportion is the sum of squares for the heterogeneity of regressions divided by the interaction sum of squares. Often, with many degrees of freedom for error, heterogeneity of regressions may be significant while accounting for a small fraction of G×E sums of squares. The likelihood of functions of environmental means or physical indices accounting for most of the G×E variation is highest when environmental variation is ecologically simple and can be adequately expressed on the one dimension of the abscissa. This might occur if, for example, available soil moisture were the sole major factor differentiating among environments. We suggest, as an approximate rule, that at least 50% of G×E should be explained before regression is useful. Seif and Pederson (1978) provided an example where variability in wheat yield was mainly attributable to spring rainfall, with regression against environmental means and regression against spring rainfall providing similar measures of adaptation. However, they did not report the percentage of interactions explained by regression.

If regression lines do not adequately represent data, points of intersection of such lines without consideration of their associated statistical errors, have limited biological meaning. However, spurious recommendations are often made to release different genotypes for zones with production levels alternatively above and below such points of intersection. Although successful application of the regression technique to barley data from Australia by Finlay and Wilkinson (1963) provided the major impetus for regression analysis, their success in accounting for about 80% of G×E may have resulted from an exceptional combination of factors. First, yield data were analysed for 277 random genotypes including well adapted, unadapted and genetic-marker material. Second, three relatively similar maritime sites were used. Third, rainfall may have been the single dominant environmental factor in Finlay and Wilkinson's study, as contrasting extremes of rainfall occurred in the three years sampled. Fourth, a logarithmic data transformation induced a high degree of linearity.

22.4.3 Non-parametric methods

Often, upon reading literature on analyses of ranking, one is surprised by the general lack of references to other studies. In a sense, it appears that breeders often 're-invent the wheel' in this context. We surmise that this phenomenon results from rejection of more conventional statistical analyses by breeders, who sense a growing loss of identity of their data the more the data are manipulated. By contrast, analysis of ranks represents a return to the foundation of the G×E problem. This is the basis for methods like stratified ranking (Fox *et al.*, 1990), which evaluates the proportion of sites where any genotype ranks in the top, middle or bottom third of the entries. A genotype usually found in the top third of entries across sites can be considered relatively well adapted.

Advantages of non-parametric techniques include: (1) freedom from assumptions concerning additivity of main effects, homogeneity of variances and linear response to increasing environmental yield potential; (2) insensitivity to errors of measurement; (3) measures of adaptation are not unduly influenced by genotypic performance in extreme environments. Baker (1988) and Huehn (1990) discussed tests for rank interactions.

A genotype is considered stable if its ranking is relatively consistent across environments. Ketata *et al.* (1989) proposed a ranking method, 'Consistency of Performance'. Plotting the average rank across sites against the standard deviation of ranks for every genotype allowed the allocation of genotypes among four classes: (1) consistently superior; (2) inconsistently superior; (3) consistently inferior; (4) inconsistently inferior. Cultivars showing general adaptation are found in the first category, whereas specific adaptation may be detected among genotypes that are inconsistently superior. Statistical tests for comparing such measures of genotypic stability are not readily available. Huehn (1990) discussed non-parametric methods for grouping environments and genotypes. Two environments, regardless of their yield level, may be considered alike for selection purposes if both rank genotypes similarly.

Therefore, the similarity of any two environments may be estimated by Spearman's Rank Correlation Coefficient. The average of all possible pairwise rank coefficients within a subset of sites, or the related Kendall's Coefficient of Concordance (W), is a measure of their cohesion. W can be used as a distance criterion for clustering environments or for grouping genotypes. Ranking does not, however, preserve the relative magnitude of the difference in yield between two consecutively ranked genotypes. Therefore, for clustering we prefer the environmental standardization of Fox and Rosielle (1982a) discussed below.

22.4.4 Multivariate exploration of relationships among sites and among genotypes

(a) Pattern Analysis of one attribute

Multivariate techniques may be applied to describe relationships among sites and among genotypes, typically using yield data from genotype × site matrices generated by breeding programmes in each crop cycle. Such multivariate methods originated in studies of numerical taxonomy and have been increasingly applied to data exploration in plant breeding since the landmark study of the Fourth International Spring Wheat Yield Nursery by Byth *et al.* (1976). The term 'Pattern Analysis' was coined to describe the parallel use of classification and ordination techniques to present the maximum variation from genotype × site matrices in a few dimensions. Classificatory techniques, such as 'Clustering', assume discontinuities within data, while 'Principal Components Analysis' (PCA) and other methods of ordination assume a continuous distribution. Thus classification and ordination emphasize slightly different features of data. No *a priori* assumptions, such as linearity, are made about the form of genotypic response. Instead, a genotype is described in multidimensional space, with each dimension representing a test environment, the coordinates for which are the yields produced. Conversely, sites are considered in multidimensional space with each dimension a genotype.

Formation of such relationships among sites can be likened to phytogeographical mapping, where the relative abundance of species integrates long-term environmental factors. In pattern analysis, the relative yields of a set of genotypes integrate the short-term interplay of biotic and abiotic influences during a crop cycle. The concept of separation of sites in multidimensional space can be illustrated by two-dimensional analogues (Fox and Rosielle, 1982a). Considering the performance of each of two varieties as the ordinate and the abscissa, a two-dimensional scattergram of relationships among sites can be obtained by plotting the sites (Fig. 22.2). The distance between two sites is the length of the straight line connecting them. In clustering, the squared Euclidean distance, equivalent to the square of this length but extended to many dimensions, is often used as a measure of dissimilarity.

The output from cluster analysis is usually displayed as a dendrogram or hierarchical tree. The complete dendrogram for genotypes (or sites) can be drawn with each genotype at the end of a branch of a tree to the left of the diagram. The horizontal axis, or the direction of growth of the tree, is proportional to squared Euclidean distance. Figure 22.3 is the dendrogram for 23 triticales from the Nineteenth International Triticale Yield Nursery (CIMMYT, 1990) based on yield at 64 sites. For 23 genotypes, the tips of the branches represent the 23-group, or lowest, level of the hierarchy. Each group is a single member. Moving right, the two most similar genotypes, the two uppermost ones in this case, fuse and their branches join. This first node is the 22-group level. Successive fusions of the branches decrease the number of groups until the 23 genotypes at the two-group level are divided between two trunks which join at the extreme right. Later fusions involve greater dissimilarity than earlier ones. At the two-group level, there is separation of complete and substituted genotypes. Complete triticales have seven pairs of rye chromosomes, whereas substituted types have one chromosomal pair from rye replaced by a pair of wheat chromosomes.

Such agglomerative procedures, which fuse successively upwards from the lowest level of

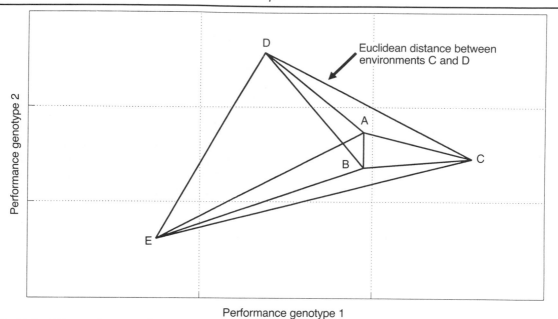

Fig. 22.2. Distance between five environments (A, B, C, D and E) in two dimensions representing genotypic performance. The dashed lines represent a hypothetical grouping structure.

the hierarchy, until one group is formed, are not influenced by the number of fusions presented in the dendrogram. The truncation question concerns presentation or subsequent analysis of the groups formed and not clustering itself. Vertical ordering of groups in dendrograms is partly arbitrary, as the dendrogram can be considered a freely rotating mobile. Although rotation changes the vertical order, it does not change the membership of a group.

As well as a similarity or dissimilarity measure, such as the squared Euclidean distance, a fusion strategy must be selected and Incremental Sums of Squares, or Ward's Method, is commonly used. At each fusion, the group formed is the one which minimizes the increment in the within-groups sums of squares. The method is space dilating. This means that, in Fig. 22.2, the first two environments to fuse, A and B, can be envisaged as being pushed together and then raised above the level of the plain containing the other environments, thus positioning them slightly further away from the rest. The practical result of space dilating algorithms is a tendency to form groups of uniform size and thus avoid

'chaining', or the formation of single-member groups. Chaining limits biological interpretation of relationships. To a degree, groups formed may be algorithm-dependent, but obtaining different patterns with different procedures may reflect the fact that there is no absolute, correct solution in a biological sense. Consider again Fig. 22.2, where a rectangular grid superimposed on the plain formed by the axes for the two genotypes could delineate groups of environments. A small displacement of the grid to the left would assign A, B and C to the same group. Group memberships are changed but relationships are not radically altered. In the same way, different biologically meaningful groupings are achievable, with slightly different features emphasized. We recommend however, that plant breeders use accepted procedures, such as those involving squared Euclidean distance and Incremental Sums of Squares, as biological interpretation will be more profitable than comparison of dissimilarity measures and algorithms. The alternatives to agglomerative strategies are divisive strategies, which assume all entities are clustered and investigate alternative options for

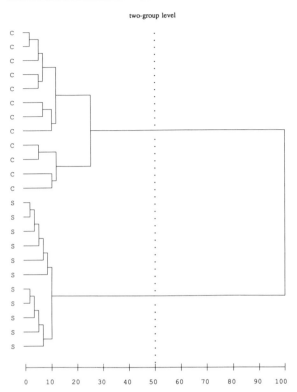

Fig. 22.3. Dendrogram of relationship among genotypes from the 19th International Triticale Yield Nursery for grain yield in 64 environments. The horizontal axis is proportional to squared Euclidean distance. Complete and substituted triticales are represented by C and S respectively.

separating them. Given the number of possibilities to investigate at each level of the hierarchy, computational costs of divisive procedures are prohibitive in plant breeding.

Turning to ordination, principal coordinates analysis computes the latent roots and latent vectors of a data matrix. Geometrically, the vector of the largest latent root defines a line, drawn through the sample mean, minimizing the sum of squares of the perpendicular distances from data points to the line. A second line, corresponding to the second largest root and vector, is drawn perpendicular to the first, such that the sum of squares of distances from the points to the plane so formed by the two lines is minimized. Similarly, further roots and vectors may be found. The largest root corres-

ponds to the first principal component, the second to the second principal component, and so on, explaining successively a decreasing proportion of the total variation. Algebraically, the latent roots generate the observable variances or correlations.

Multivariate methods usually present most of the total variation in a few dimensions, for instance, in a dendrogram or scattergram. The information is essentially that present in a correlation matrix, but the output is more readily interpreted. In practice, it is common to superimpose the groups obtained from Clustering on the Principal Components diagram. Differences between the two elements of Pattern Analysis are generally sufficiently small to accommodate this superimposition, again illustrating that alternative and compatible biological answers co-exist. This idea is integrated into another ordination technique, Factor Analysis, which assigns entities to main groups on the basis of primary factor loads, but accounts for affinities with other groups by secondary loads. Factor loads are akin to scores on Principal Components axes. Whereas cluster analysis assigns a site, for example, to one group only, factor analysis provides more flexibility by placing the site in a group while acknowledging that it could be considered a member of another group. Factor analysis may also leave entities unassigned to groups. It is somewhat a matter of personal choice whether this more plastic grouping process aids, or unduly complicates, interpretation.

Genotypic grouping does not indicate stability *per se*, but multivariate association of advanced lines with known varieties provides a reliable assessment of genotypic adaptation with one year's data from wide testing. Relationships show a high degree of repeatability from year to year. Traditionally, such trends required several years to become apparent. Menz (1980) indicated a tendency for groups formed by pattern analysis to contain genotypes with the same degree of risk-efficiency.

When considering patterns of relationships among sites for breeding purposes, site main effects are removed so that grouping is determined more by GxE effects than by production level (Brennan *et al.*, 1981; Fox and Rosielle,

1982). By contrast to relations among genotypes, repeatability of site groupings between
years is low. Intersite relationships based on
one year's data are dominated by unpredictable
weather patterns and are often not easily
interpreted. Pattern Analysis of data from
single years thus provides rapid assessment of
genotypic performance and a 'snap shot' of
oscillating intersite relationships. Lack of a
sufficient number of common varieties has
limited analyses across years. Peterson and
Pfeiffer (1989) overcame this problem by constructing a long-term correlation matrix of sites
using 17 years of yield data from the International Winter Wheat Production Nurseries.
They reasoned that the phenotypic correlation
between any pair of sites in a given year
measures biological similarity and, even though
entries change with time, pooling correlations
over years would provide a long-term average
measure of relatedness of sites. This could be
thought of, in the above terminology, as superimposition of a series of 'snap shots' to produce
the long-term picture. Factor analysis was used
to simplify this correlation matrix, and 56 sites
from 30 countries were allocated to seven mega-
environments for winter wheat production.

(b) Multi-attribute Pattern Analysis

Most techniques considered so far in this
section have examined two-way tables, generally genotypes by environments, for yield.
Kroonenberg and Basford (1989) extended
ordination to a further dimension, with genotypes by environments matrices considered for
more than one trait or attribute. Basford *et al.*
(1990) provided a parallel development for
classification. These methods can be considered
multivariate extensions of much earlier work on
selection indices which attempted to combine
multi-attribute information into a score. These
authors re-analysed regional soybean and
cotton data previously reported in conventional
Pattern Analysis studies of one attribute, but
had to ignore international data sets such as
those used by Byth *et al.* (1976) and Peterson
and Pfeiffer (1989). Multi-attribute tables from
international trials are dominated by missing
values. Yield is the most frequently scored

attribute, with other traits recorded less commonly and at the discretion of individual co-
operators. Resources are often unavailable to
score even those phenological notes of important interpretative value and many other traits,
particularly disease scores, are not universally
important. For both conventional and multi-
attribute pattern analysis, unbalanced analyses
are needed. However, the greater the proportion of missing cells, the greater are the risks in
procedures to allow for them. It remains to be
seen whether this method will gain acceptance
for exploration of unbalanced data from international trials. Perhaps the major application
will be the largely neglected analysis of secondary traits such as disease reactions. A subset of
sites reporting leaf rust in a wheat trial, for
example, could be analysed, using yield and
phenological data as secondary attributes.

(c) The Additive Main Effects and Multiplicative Interaction (AMMI) model

The Additive Main Effects and Multiplicative
Interaction (AMMI) model is a powerful analytical tool to interpret large genotype × environment × replicate tables without missing values
(Crossa *et al.*, 1991). It is an extension of
ordination, but is considered separately
because of biplot and data validation procedures discussed below. AMMI extracts genotype and environment main effects, then uses
PCA to explain pattern in the G×E, or residual,
matrix. Zobel *et al.* (1988) provided a scale for
PCA scores which allows estimation of specific
G×E interaction terms. The AMMI model for
the average yield, Y_{ij}, over replicates of the *i*th
genotype in the *j*th environment is:

$$Y_{ij} = \mu + G_i + E_j + \sum_{n=1}^{N} \lambda_n \gamma_{in} \delta_{jn} + \epsilon_{ij}$$

$$(22.1)$$

where μ is the overall mean, G_i and E_j are
genotypic and environmental main effects, N is
the number of PCA axes considered, λ_n is the
singular value of the *n*th PCA axis, γ_{in} and δ_{jn}
are scores for the *i*th genotype and *j*th environ-
ment on the *n*th PCA axis and ϵ_{ij} is the residual

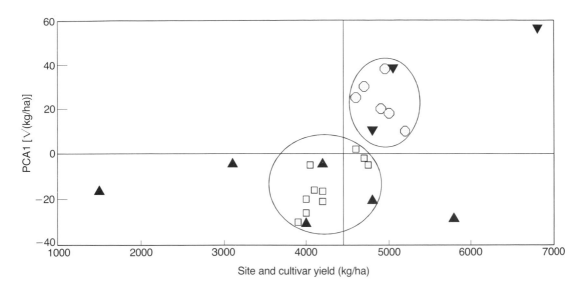

Fig. 22.4. Biplot of the unadjusted means (kg/ha) and the first PCA axis for interaction of six complete (○) and ten substituted (□) triticales grown at three sites with acid soils (▼) and six sites with basic soils (▲). The vertical line represents the grand mean of the experiment, the horizontal line is PCA1 = 0 (data from the Spanish system for varietal registration, 1988/89).

term which includes the experimental error. AMMI generates a family of models with different values of N. The simplest model, AMMI0 with N equal to zero, considers additive main effects, namely genotypic and environmental means, to explain the data matrix. AMMI0 thus ranks genotypes identically at each environment, ignoring G×E. The second model, AMMI1, considers main effects as well as one principal component axis to interpret the residual matrix. AMMI2 considers main effects plus two principal components axes for non-additive variation. Subsequent models consider cumulatively an additional principal component axis.

When one PCA axis accounts for most G×E, a feature of AMMI is the biplot procedure. Genotypes and environments are plotted on the same diagram, facilitating inference about specific interactions of individual genotypes and environments by using the sign and magnitude of PCA1 values. Any genotype with a PCA1 value close to zero shows general adaptation to the tested environments. A large genotypic PCA1 score reflects more specific

adaptation to environments with PCA1 scores of the same sign. Figure 22.4 shows the value of the biplot in assessing the nature of G×E using triticale data for varietal registration in Spain for the 1988/89 cycle. Six complete and 10 substituted triticales were grown in northwestern Spain, at three sites with acid soils and at six sites with basic soils. All complete triticales and sites with acid soils exhibited PCA1 values of the same sign, indicating that completes were better adapted to acid soils. Substituted triticales were better adapted to basic soils. AMMI1 explained 72% of the G×E sum of squares and thus provided a more adequate biological explanation of G×E than regression, which only accounted for 19% of interaction (Royo *et al.*, 1991). Extrapolating, we believe AMMI analysis will prove particularly useful when applied to across-year analyses with a higher element of unpredictability.

Two processes for determining the number of PCA axes to consider have been distinguished by Gauch and Zobel (1988). They used the terms 'postdictive' and 'predictive' accuracy. Postdiction is a descriptive procedure in which the best model is selected on the basis of variation explained by PCA axes. In prediction, individual replicates are allocated at random for each genotype x environment combination to either a data set for modelling or a set for validating the models. The sum of squared differences between validation data and predicted values, across the data matrix, is divided by the number of validation observations. The square root of this quantity is the root mean square predictive difference (RMS PD), for which smaller values indicate more accurate prediction. A final model, the so-called DATA model, similarly considers the ability of means of replicates of modelling data, for each genotype × environment combination, to predict the corresponding validation data, namely the other replicate(s). On the basis of the RMS PD, the best model is selected and re-applied to the data including all replicates.

Two major trends have emerged from AMMI analyses of yield trials. First, DATA is of low predictive accuracy compared with the other models. Second, additional PCA axes, after a certain number, decrease rather than increase predictive accuracy. PCA appears to extract 'pattern' in the first few axes, with subsequent axes being associated with 'noise'. For this reason, predictive modelling is preferred and considers fewer axes than postdiction (Gauch and Zobel, 1988).

The thrust for current research on AMMI analysis is empirical evidence from data validation that estimates from AMMI models may be more accurate than means across replicates (DATA) for estimating yields of genotype × environment combinations. Despite the evidence, use of AMMI adjustments in genotypic trials is controversial. The root of the controversy lies in using data from a number of different environments to adjust genotypic means at another. The theoretical basis for the increased predictive ability of AMMI analyses relates to the Stein Effect (Gauch, 1990) whereby accepting a small bias, from using AMMI estimates for genotype × site effects instead of the mean of the appropriate replicates, can produce a large gain in precision. This may be especially true for noisy data. In a trial of g genotypes in e environments with r replicates,

AMMI analysis uses all *ger* observations to estimate a genotype × environment combination. By contrast, the unbiased estimate is less precise because it is based on r data points. AMMI adjusted data with improved predictive precision can be analysed further by methods such as Pattern Analysis to study behaviour of groups of environments and genotypes. AMMI adjusted data showed clearer trends than analyses of raw data, but AMMI analysis should be compared with improved experimental designs as a means of improving precision (Crossa *et al.*, 1991).

In discussing the ability of multivariate models, such as AMMI, to selectively recover pattern as opposed to noise, Gauch and Zobel (1988) state: 'In essence, selectivity occurs because noise produces idiosyncratic variations in individual treatment means, whereas pattern originating from properties of the genotypes and sites produces coordinated variations in numerous treatment means'. However, Crossa *et al.* (1991) indicated the danger, from a biological standpoint, of over-riding a message that is asynchronous with the pattern. They suggest that a message from a single site with a specific stress may be lost among a majority of sites without this stress. The biologically meaningful GxE associated with the stress would be relegated to the residual term of the AMMI analysis. To guard against this possibility, Crossa *et al.* (1991) inspected the distribution of the AMMI adjustment to ascertain whether large adjustments were centred on asynchronous individual environments or genotypes. In their data, adjustments appeared to be evenly distributed, consistent with general noise reduction, but the inspection process lacked a rigorous statistical base. The biplot of Crossa *et al.* (1991) graphically indicated a common phenomenon in G×E studies: the factors causing differences in site mean yields have no association with factors causing G×S variation. Two major contrasting site groups, which were each relatively homogeneous for interactions with genotypes, covered the range of yields of sites in the trial.

Unresolved problems related to AMMI analysis remain. When the best AMMI model includes more than one PCA axis, assessment and presentation of genotypic stability are not as simple as the AMMI1 case. Crossa *et al.* (1991) analysed results of the Eighth Elite Selection Wheat Yield Trial (ESWYT). Eighteen genotypes at 25 sites from the Northern Hemisphere were studied and AMMI1 was the best predictive model. Analysis of Ninth ESWYT data, 29 genotypes by 55 sites (CIMMYT, 1991), identified AMMI4 as the best predictive model. The 55 sites were divided by Cluster Analysis between a relatively homogeneous group of 20 high yielding sites with minor stresses and 35 diverse sites, generally exhibiting more environmental stress. In an unpublished study, these two subgroups of sites were re-analysed separately using AMMI analysis. The group of 20 sites conformed to expectations, in that AMMI1, a simpler model than for the complete set, gave the best predictive success. However, for the other subgroup of sites, a more complex model, AMMI5, was selected. In the complete analysis, current wisdom would have deemed the variation explained by AMMI4 to be pattern and more complex variation considered by the fifth and subsequent principal components axes to be noise. However, a subset of data suggested the variation explained by AMMI5, because of its predictive value, was pattern.

This phenomenon of a more homogeneous subset, namely the 35 sites, indicating a more complex pattern than the complete set is paradoxical and suggests that the concept of noise extraction by AMMI analysis requires more investigation. We suggest that the designation of higher dimensions of PCA as associated with noise, because of lack of predictive ability, may in some cases be an artifact of applying models to situations beyond the limit of their capacity to explain interaction. Application to simpler subsets, as shown in the Ninth ESWYT, may allow better representation of data. In this context, it may be illuminating to re-examine the maize data reported by Crossa *et al.* (1990). For nine genotypes grown in 38 subtropical environments, G×E was highly significant, yet AMMI0 showed the best predictive value. This is alarming because it implies that the significant interaction is noise and that the best genotypic estimates a breeder has for an individual site are the genotypic main effects from

the combined analysis. We hypothesize that an application of AMMI analysis to less complex environmental subsets may reveal models in which interaction has predictive value.

The AMMI adjustment is selected on a per-matrix basis, i.e., based on data validation across all environments and genotypes. Per-environment selection of adjustment is worth investigation, as data precision varies from environment to environment. Data from environments for which DATA provides the greatest accuracy could be combined with AMMI adjusted data from other environments.

22.5 PERSPECTIVES

We expect gains from continued empirical breeding will be substantial as statistical methods, computerization and mechanization develop; while significant contributions will flow from a 'trait-oriented' approach as limiting stresses are identified. Maximum genetic gains will result from optimum allocation of resources between the two approaches.

The work by Finlay and Wilkinson (1963) was important in turning the attention of plant breeders towards adaptation and also has historical significance as the first results of the combination of mechanization and computerization in plant breeding. To conduct this analysis, punched cards had to be transferred between institutions. It is worth reflecting how subsequent advances in computerization have influenced plant breeding. Even though the numbers of plots have increased dramatically in many countries and computers are now widely available, analysis of results in mainstream breeding programmes has lagged behind.

Few breeders routinely analyse data for genotypic stability. Of those who do, the majority use regression analysis. Others consider G×E research a sideline and may not routinely apply their own published methodology. Nonetheless, many breeders develop a profound understanding of their environments and adaptation of their genetic materials. This appreciation may result from detailed field observations, often visually comparing newer lines to check varieties in paired plots at many sites. The assimilation of field observations frequently appears to be an intuitive process. However, we believe that statistical assessment of genotypic stability is needed, not to replace breeders' impressions, but to complement them. Easy computational and interpretable techniques, such as pairwise analyses of variance between known cultivars and advanced lines (Lin and Binns, 1985), consistency of performance (Ketata *et al.*, 1989) and stratified ranking (Fox *et al.*, 1990), could be applied in most breeding programmes. Speed is the essence and a breeder requires statistical outputs during field selection. More complex techniques, such as pattern analysis or AMMI models, will gain acceptance through better-documented, user-friendly software that handles missing values. Production of such mature software is a bottleneck. The very number of methods proposed mitigates against their widespread use. For many breeders, there are too many options to consider.

ACKNOWLEDGEMENTS

We thank colleagues too numerous to name for constructive criticism of this chapter.

REFERENCES

Allard, R.W. (1960), *Principles of Plant Breeding*, John Wiley, Chichester.

Atlin, G.N. and Frey, K.J. (1989), Breeding crop varieties for low-input agriculture. *Am. J. Alternative Agric.*, **4**, 53–58.

Baker, R.J. (1971), Effects of stem rust and leaf rust of wheat on genotype–environment interaction for yield. *Can. J. Plant Sci.*, **51**, 457–461.

Baker, R.J. (1988), Tests for crossover genotype–environmental interactions. *Can. J. Plant Sci.*, **68**, 405–410.

Barah, B.C., Binswanger, H.P., Rana, B.S. and Rao, N.G.P. (1981), The use of risk aversion in plant breeding; concept and application. *Euphytica*, **30**, 451–458.

Basford, K., Kroonenburg, P.M., DeLacy, I.H. and Lawrence, P.K. (1990), Multiattribute evaluation of regional cotton trials. *Theor. Appl. Genet.*, **79**, 225–234.

Becker, H.C. (1981), Correlations among some statistical measures of phenotypic stability. *Euphytica*, **30**, 835–840.

Binswanger, H.P. and Barah, B.C. (1980), *Yield Risk, Risk Aversion, and Genotype Selection: Conceptual Issues and Approaches*. Research Bulletin 3, ICRISAT, Patancheru.

Blum, A., Poyarkova, H., Golan, G. and Mayer, J. (1983), Chemical desiccation of wheat plants as a simulator of post-anthesis stress. I. Effects on translocation and kernel growth. *Field Crops Res.*, **6**, 51–58.

Bolaños, J. and Edmeades, G.O. (1988), CIMMYT's strategies in breeding for drought tolerance in tropical maize. In *Challenges in Dryland Agriculture – a Global Perspective*, Unger, P.W., Jordan, W.R., Sneed T.V. and Jensen, R.W. (eds), Texas Agricultural Experiment Station, College Station, pp. 752–754.

Brennan, J.P. (1986), *Impact of the Wheat Varieties from CIMMYT on Australian Wheat Production*. Agricultural Economics Bulletin 5, Department of Agriculture New South Wales.

Brennan, P.S., Byth, D.E., Drake, D.W., De Lacy, I.H. and Butler, D.G. (1981), Determination of the location and number of test environments for a wheat cultivar evaluation program. *Aust. J. Agric. Res.*, **32**, 189–201.

Byth, D.E., Eisemann, R.L. and De Lacy, I.H. (1976), Two-way pattern analysis of a large data set to evaluate genotypic adaptation. *Heredity*, **37**, 215–230.

Campbell, L.G. and Lafever, H.N. (1977), Cultivar × environment interaction in soft red winter wheat yield tests. *Crop Sci.*, **17**, 604–608.

Ceccarelli, S. (1989), Wide adaptation: How wide? *Euphytica*, **40**, 197–205.

Ceccarelli, S., Nachit, M.M., Ferrara, G.O., Mekni, M.S., Tahir, M., van Leur, J. and Srivastava, J.P. (1987), Breeding strategies for improving cereal yield and stability under drought. In *Drought Tolerance in Winter Cereals*, Srivastava, J.P., Porceddu, E., Acevedo, E. and Varma, S. (eds), John Wiley, Chichester, pp. 101–114.

CIMMYT (1990), *Results of 1987–1988 Triticale Nurseries*, CIMMYT, Mexico, D.F.

CIMMYT (1991), *Results of the Ninth Elite Selection Wheat Yield Trial 1987–88*, CIMMYT, Mexico, D.F.

Crossa, J. (1990), Statistical analyses of multilocation trials. *Adv. Agron.*, **44**, 55–85.

Crossa, J., Fox, P.N., Pfeiffer, W.H., Rajaram, S. and Gauch, H.G. (1991), AMMI adjustment for statistical analysis of an international wheat yield trial. *Theor. Appl. Genet.*, **81**, 27–37.

Crossa, J., Gauch, H.G. and Zobel, R.W. (1990), Additive main effects and multiplicative interaction analysis of two international maize cultivar trials. *Crop Sci.*, **30**, 493–500.

DeLacy, I.H., Eisemann, R.L. and Cooper, M. (1990), The importance of genotype-by-environment in regional variety trials. In *Genotype-by-Environment Interaction and Plant Breeding*, Kang, M.S. (ed.), Louisiana State University, Baton Rouge, pp. 287–300.

Eisemann, R.L., Cooper, M. and Woodruff, D.R. (1990), Beyond the analytical methodology – better interpretation and exploitation of genotype-by-environment interaction in breeding. In *Genotype-by-Environment Interaction and Plant Breeding*, Kang, M.S. (ed.), Louisiana State University, Baton Rouge, pp. 108–117.

Finlay, K.W. and Wilkinson, G.N. (1963), The analysis of adaptation in a plant-breeding programme. *Aust. J. Agric. Res.*, **14**, 742–754.

Flinn, J.C. and Garrity, D.P. (1989), Yield stability and modern rice technology. In *Variability in Grain Yields: Implications for Agricultural Research and Policy in Developing Countries*, Anderson, J.R. and Hazell, P.B.R. (eds), John Hopkins University Press, Baltimore, pp. 251–264.

Fox, P.N. and Rosielle, A.A. (1982a), Reducing the influence of environmental main-effects on pattern analysis of plant breeding environments. *Euphytica*, **31**, 645–656.

Fox, P.N. and Rosielle, A.A. (1982b) Reference sets of genotypes and selection for yield in unpredictable environments. *Crop Sci.*, **22**, 1171–1175.

Fox, P.N., Rosielle, A.A. and Boyd, W.J.R. (1985), The nature of genotype × environment interactions for wheat yield in Western Australia. *Field Crops Res.*, **11**, 387–398.

Fox, P.N., Skovmand, B., Thompson, B.K., Braun, H.-J. and Cormier, R. (1990), Yield and adaptation of hexaploid spring triticale. *Euphytica*, **47**, 57–64.

Gauch, H.G. (1990), Full and reduced models for yield trials. *Theor. Appl. Genet.*, **80**, 153–160.

Gauch, H.G. and Zobel, R.W. (1988), Predictive and postdictive success of statistical analyses of yield trials. *Theor. Appl. Genet.*, **76**, 1–10.

Huehn, M. (1990), Nonparametric estimation and testing of genotype × environment interactions by ranks. In *Genotype-by-Environment Interaction and Plant Breeding*, Kang, M.S. (ed.), Louisiana State University, Baton Rouge, pp. 69–93.

Ketata, H., Yau, S.K. and Nachit M. (1989), Relative consistency of performance across environments. Communicated to *Int. Symp. Physiol. Breed. Winter Cereals for Stressed Mediterranean Environ*, Montpellier, July 3–6, 1989.

Knight, R. (1970), The measure and interpretation of genotype-environment interactions. *Euphytica*, **19**, 225–235.

Kroonenburg, P.M. and Basford, K. (1989), An investigation of multi-attribute genotype response across environments using three-mode principal component analysis. *Euphytica*, **44**, 109–123.

Lawn, R.J. (1988), Breeding for improved plant performance in drought-prone environments. In *Drought Research Priorities for the Dry Land Tropics*, Bidinger, F.R., and Johansen, C. (eds), ICRISAT, Patancheru.

Lin, C.S. and Binns, M.R. (1985), Procedural approach for assessing cultivar × location data. Pairwise genotype-environment of test cultivars with check. *Can. J. Plant Sci.*, **65**, 1065–1071.

Lin, C.S., Binns, M.R. and Lefkovitch, L.P. (1986), Stability analysis: where do we stand. *Crop Sci.*, **26**, 894–900.

Ludlow, M.M. and Muchow, R.C. (1989), A critical evaluation of traits for improving crop yields in water-limited environments. *Adv. Agron.*, **43**, 107–153.

Marshall, D.R. (1987), Australian plant breeding strategies for rainfed areas. In *Drought Tolerance in Winter Cereals*, Srivastava, J.P., Porceddu, E., Acevedo, E. and Varma, S. (eds), John Wiley, Chichester, pp. 89–99.

Menz, K.M. (1980), A comparative analysis of wheat adaptation across international environments using stochastic dominance and pattern analysis. *Field Crops Res.*, **3**, 33–41

Molina-Cano, J.L., García del Moral, L.F., Ramos, J.M., García del Moral, M.B., Jimenez-Tejada, P., Romagosa, I. and Roca de Togores, F. (1990), Quantitative phenotypical expression of three mutant genes in barley and the basis for defining an ideotype for Mediterranean environments. *Theor. Appl. Genet.*, **80**, 762–768.

Patel, J.D., Reinbergs, E., Mather, D.E., Choo, T.M. and Sterling, J.D.E. (1987), Natural selection in a doubled-haploid mixture and a composite cross of barley. *Crop Sci.*, **27**, 474–479.

Peterson, C.J. and Pfeiffer, W.H. (1989), International winter wheat evaluation: relationships among test sites based on cultivar performance. *Crop Sci.*, **29**, 276–282.

Pfeiffer, W.H. and Braun, H.J. (1989) Yield stability in bread wheat. In *Variability in Grain Yields: Implications for Agricultural Research and Policy in Developing Countries*, Anderson, J.R. and Hazell, P.B.R. (eds), Johns Hopkins University Press, Baltimore, pp. 157–174.

Plaisted, R.L. and Peterson, L.C. (1959), A technique

for evaluation the ability of selection for yield consistency in different locations or seasons. *Am. Potato J.*, **36**, 381–385.

Rathjen, A.J. and Pederson, D.G. (1986), Selecting for improved grain yields in variable environments. *DSIR Plant Breeding Symposium. Special Publication 5*, N.Z. Agronomy Society.

Richards, R.A. (1982), Breeding and selection for drought resistance in wheat. In *Drought Resistance in Crops with Emphasis on Rice*, IRRI, Manila, pp. 303–316.

Romagosa, I., Fox, P.N., García del Moral, L.F., Ramos, J.M., García del Moral, B., Roca de Togores, F. and Molina-Cano, J.L. (1993) Integration of statistical and physiological analyses and adaptation of near-isogenic barley lines. *Theor. Appl. Genet.*, in press.

Roseille, A.A. and Hamblin, J. (1981), Theoretical aspects of selection for yield in stress and non-stress environments. *Crop Sci.*, **21**, 943–946.

Royo, C., Romagosa, I. and Rodriguez, A. (1991), Comparative adaptation of triticale and spring wheat in Spain. In *Proc. Second Int. Triticale Symp.*, CIMMYT, Mexico, D.F., pp. 593–597.

Seif, E. and Pederson, D.G. (1978), Effect of rainfall on the grain yield of spring wheat, with an application to the analysis of adaptation. *Aust. J. Agric. Res.*, **29**, 1107–1115.

Shukla, G.K. (1972), Some statistical aspects of partitioning genotype-environmental components of variability. *Heredity*, **29**, 237–245.

Simmonds, N.W. (1979), *Principles of Crop Improvement*, Longman Group Limited, London.

Steel, R.G.D. and Torrie, J.H. (1980), *Principles and Procedures of Statistics*, 2nd edn., McGraw-Hill, New York.

Whan, B.R., Anderson, W.K., Gilmour, R.F., Snelling, K.L., Regan, N.C. and Turner, N.C. (1991), A role for physiology in breeding for improved wheat yield under drought stress. In *Physiology-Breeding of Winter Cereals for Stressed Mediterranean Environments*, Acevedo, E., Conesa, A.P., Monneveux, P. and Srivastava, J.P. (eds), INRA-ICARDA, Versailles, pp. 179–194.

Wricke, G. (1964), Zur Berechnung der Ökovalenz bei Sommerweizen und Hafer. *Z. Pflanzenzüchtg.*, **52**, 127–138.

Zobel, R.W., Wright, M.J. and Gauch, H.G. (1988), Statistical analysis of a yield trial. *Agron. J.*, **80**, 388–393.

23

Augmenting yield-based selection

R.B. Austin

23.1 INTRODUCTION

Darwinian principles lead us to expect that in a stable agriculture cultivars of a crop species will eventually become optimally adapted to the environment in which they are grown. The corollary is that when agriculture is changing (inputs, farming systems or the climate) existing cultivars will no longer be optimally adapted. It is under such circumstances that plant breeding is beneficial (Fig. 23.1).

Selection for yield, pest and disease resistance and crop quality are the broad objectives of crop improvement programmes. For many decades, plant breeders have sought means of improving the efficiency of selection for these classes of characters. Although considerable improvements in the efficiency of selection have been made for characters other than yield, few, if any, alternatives to the direct selection for yield have become adopted, though many have been suggested. This may be limiting progress in improving yield in self-pollinating crops, as selection based on yield is only possible in the more advanced generations of a breeding programme, and high yielding individuals present in the early generations may have escaped recognition. This chapter considers some limitations of yield-based selection and how the effectiveness of selection may be improved by the exploitation of knowledge of the physiology of the crop and its responses to its environment.

Plant Breeding: Principles and prospects. Edited by M.D. Hayward, N.O. Bosemark and I. Romagosa. Published in 1993 by Chapman & Hall, London. ISBN 0 412 43390 7.

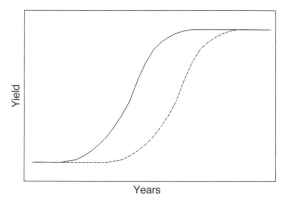

Fig. 23.1. Illustration of the relationships between the amelioration of the environment and the progress of crop improvement by breeding. Solid line represents the change with years in the yielding capacity of the environment resulting from progressive improvements in cultural practices, assuming that optimally adapted cultivars are being grown. Dashed line represents the change with years in the actual yield obtained with currently available cultivars. Note that the actual yield lags behind the yielding capacity of the environment because of the time taken to produce cultivars that are optimally adapted.

23.2 LIMITATIONS TO YIELD-BASED SELECTION

While breeding programmes in self-pollinating crops rely on selection for yield in advanced generations, most plant breeders also have a conceptual model of the plant type they consider to be ideal. Donald (1968) introduced the term 'ideotype' to describe this ideal plant type. As knowledge of the physiology of the crop improves, so does the accuracy with which the ideotype can be defined. Traits that are of little or negative value can be discarded and only those of value retained. Thus in practice, few, if any, breeding programmes are based purely on selection for yield, ignoring morphological features.

Knowledge of the environment and of the responses of the crop to elements of the environment (rainfall, temperature, etc.) is essential to define the limits of the area for which a breeder aims to produce improved cultivars. But, in the absence of such considerations, how may the target area be defined? Here, another Darwinian concept is valuable. Evolutionary theory and observation shows that, in general, increased fitness (yield, in the case of a crop) in a particular environment may be achieved at the expense of adaptability over a range of environments. However, there is no value in defining a target environment for a breeding programme in which the interannual variation in yield-determining elements of the environment is greater than the spatial variation in these elements. In principle it would be possible to delimit target environments using this criterion by analysis of the results from multilocation yield trials carried out over a period of years, and assessing genotype × environment interaction by statistical procedures (Finlay and Wilkinson, 1963). Target environments would then be defined as those over which the genotype × location interaction was not greater than the genotype × year interaction. In practice, target environments are defined by a combination of intuitive judgement, knowledge of the climate and soils and of the responses of the crop to edaphic and environmental factors as well as from the results of yield trials. In addition, political and economic considerations often impose limits to the target area.

Given a changing (usually improving) environment for agriculture and an appropriately defined target area, plant improvement programmes based largely on selection for high yield have been very successful. Examples are hybrid corn in the USA (Duvick, 1977; Russell, 1984), wheat in England (Austin *et al.*, 1989) and wheat and rice in irrigated agriculture in the subtropics and tropics, pioneered by plant breeders at CIMMYT in Mexico and IRRI in The Philippines, respectively. In other situations, such as wheat in Canada and Australia, upland rice in South East Asia and barley in the Middle East, progress has been much less, both in relative and absolute terms.

As further progress in producing high yielding cultivars by conventional, yield-based selection is requiring progressively larger breeding programmes, it is especially pertinent to examine whether the efficiency of selection can be increased by exploiting knowledge of the physiology and genetics of crops.

23.3 PHYSIOLOGICAL ANALYSIS OF TRAITS

In the present context, the terms trait, character and attribute are synonymous. They refer to any identifiable feature of a plant that may or may not affect its growth, development and yield. It is implied that the feature or property is heritable, though the heritability may be low.

23.3.1 Plants and crops as hierarchical systems

Most attempts to identify plant processes or biochemical steps in metabolism that limit growth, and to devise selection systems for increased yield based on such knowledge, have been unsuccessful. In general terms, this is because it is undesirable for the control of growth to be vested in any single step in metabolism, as plants need to respond to all factors in the environment, over time scales ranging from seconds to days, so as maximally to exploit any opportunities for increasing their growth and survival. This principle is well illustrated by the process of photosynthesis, which is influenced by numerous external and internal factors. A particular leaf, one component of a canopy, may at one moment need to maximize the efficiency with which it intercepts and absorbs light and uses it to reduce carbon dioxide. Within minutes, it may need to protect itself against the damaging effects of high irradiance and high rates of water loss. Clearly, the adaptive responses involved must be short term. They involve changes in stomatal conductance, thylakoid structure and the activity of the Calvin cycle enzymes. Longer-term adaptations, in response to consistently high (or low) irradiance, are also important, and usually involve changes in leaf anatomy, among other features. The capacity for adaptive change, both in the short and long term, is under genetic control. But because there needs to be a compromise between fitness and adaptability, some plants are best adapted to sunny conditions and others to shady conditions; relatively few to both sunny and shady conditions. The photosynthetic apparatus of sun and shade species differs in many quantitative respects, and correspondingly, its genetic control is complex.

The growth of plants proceeds in a hierarchical manner. In essence, this means that the growth of a given organ (leaf, fruit, etc.) depends not only on the current environmental conditions but on the size and growth of previously formed organs. In turn, this depends on previous environmental conditions. Figure 23.2 illustrates this for the

assimilation and recycling of carbon, but it applies equally for mineral nutrients and water. Taking wheat as an example, the size and number of the ears, and hence the potential yield, depends on the environment during the time when they are initiated early in the life cycle, and when the ears are growing, from the seedling stage until anthesis. For a given leaf, the conditions prevailing during its growth will determine its size and potential contribution to the photosynthesis of the plant as a whole but they will have their main influence on its rate of photosynthesis after it has fully expanded. Adverse conditions may partly be compensated for by good conditions later on, though only within limits and there will be certain critical stages during which irreversible and substantial damage will be done if the environmental conditions are adverse. Male and female sterility are obvious examples. Of course the extent to which the potential yield in grain crops is realized depends on conditions during grain filling, as well as on the size and photosynthetic activity of the canopy, which depends on the conditions which prevailed before grain filling commenced.

23.3.2 Identifying traits for analysis

Traits may be grouped into four classes:

1. Morphological and anatomical, for example plant height, leaf size and posture and stomatal frequency. Such characters are relatively simple to identify and quantify, do not change in the short term, and are often highly heritable. In practice, therefore, they are the most widely used by breeders. Their importance for growth and yield, however, is often imperfectly understood, and there are likely to be important interactions with other morphological and with physiological characters.

2. Compositional. Tests for quality characteristics are well established for most crops, especially those where the composition of the harvested product is relatively stable as

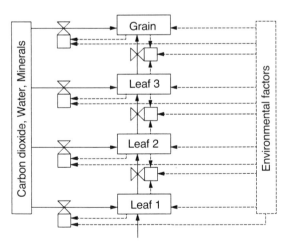

Fig. 23.2. Diagrammatic representation of the hierarchical interrelationships between the component organs of a simplified wheat plant. Boxes bounded by solid lines represent material (carbon dioxide, water, minerals or dry matter); those bounded by dashed lines represent information or states. Solid lines connecting boxes represent material flow; dashed lines represent the flow of information.

is the case with grain crops. Traits expressed in vegetative, growing tissue generally vary with age and environment and so standardization is necessary if the results are to be of value. Alternatively, corrections, based on correlations with age or environmental factors, may be used to adjust values to those equivalent to a standard state. Examples of compositional traits are proline, abscisic acid, sodium and protein concentration.

3. Process rates. Examples are respiration, photosynthesis, photoperiodic response, and winter hardiness. Process rates are very sensitive to current and previous environmental conditions, all of which must be carefully standardized if meaningful genetic comparisons are to be made. Winter hardiness tests in cereals are generally standardized in this way, and are well established and valuable procedures, although comparing genotypes which differ markedly in hardiness can be difficult because a common testing regime may not be appropriate. Screening for photosynthetic rate and photorespiration have been used on limited scales, especially for detecting large differences among species, but as discussed later, are of little value for examining variation within a species.

4. Process control. Examples are the activity of the Calvin cycle enzymes and stomatal aperture, which may control or limit photosynthesis, and nitrate reductase, which has been claimed to be a limiting step in nitrate assimilation. Most enzymes are very sensitive to environmental influences whether perceived directly or indirectly, and their amounts and activities can change dramatically even in the short term. It is also difficult to devise assays that will be meaningful in terms of whole plant performance, for activities are often determined at high substrate concentration and, for temperate crops, at unrealistically high temperatures. The enzymes present in grain, which may be important for quality, are exceptions, and tests for amylase activity form a routine part of the assessment of the suitability of wheat grain for bread making.

The choice of trait for genetical and physiological analysis will depend on the assessment the investigator makes as to the likely value of the trait in breeding. Sources of information on the value of traits are described in sections 23.3.3 and 23.3.4 of this chapter. The value of a trait will also be determined by the ease and speed with which it can be assessed, as discussed in section 23.5 of this chapter.

Yield component traits are obvious candidates for physiological and genetical analysis. While such traits have been the subject of numerous studies, the results are often equivocal and of limited value as guides to plant breeders. In part this is because of compensation. Compensation can be genetically determined and also be a consequence of environmental conditions. Taking wheat as an example, yield can be analysed in terms of three components, as follows: grain yield = number of ears × number of grains per ear × mean grain mass.

In a given environment, genotypes that produce many ears tend to have fewer grains per ear and sometimes lighter grains. Others, which may have similar yields, will produce fewer ears, but more grains per ear. Such associations are often strong, though not complete. Importantly, the associations may depend on environmental factors. Response to population density in cereals provides a good example of compensation caused by this cultural variable. In wheat, yield is usually independent of density over the range ca 100–1000 plants/m^2. For a given genotype, at low density, there will be fewer ears/m^2 but more grains per ear and somewhat heavier grains than at a higher density (Table 23.1). Thus any study

Table 23.1. Grain yield (g/m^2) and number of ears/ m^2 of three winter wheat cultivars grown at two population densities (100 and 400 plants/m^2) (unpublished data from a field experiment at Trumpington, UK)

Cultivar	Yield		Ears/m^2	
	100/m^2	*400/m^2*	*100/m^2*	*400/m^2*
Lilifen	622	439	275	373
Templar	781	683	305	443
Talent	660	585	367	590

designed to assess genetic variation in these yield components will be frustrated unless the comparisons are made at constant density. Also, the expression of genetic variation may depend on the density at which the assessments are made.

23.3.3 Evidence from comparisons of historical series of cultivars

Comparisons of old and contemporary cultivars can yield valuable information on the progress made by breeders in crop improvement, and on the trends in characters associated with yield improvement. Data from such trials cannot be used to predict with confidence whether past trends can be continued, and if so for how long, and this must remain a matter for judgement based on additional information.

As with cultivar trials generally, the consequences for yield of physiological and morphological characters may be masked or distorted if the cultivars are differentially affected by pests or diseases. The consequences may also depend on the environment and the cultural conditions under which the trials are carried out. This can be particularly important in situations where there has been a substantial change in husbandry associated with the introduction of the new cultivars (see Fig. 23.1 and Russell, 1984). Where there has been such change, the new cultivars may yield little more than the old ones under the original conditions, while the old ones may be greatly disadvantaged under the modern conditions. The availability of effective chemicals for the control of

many pests and diseases of crops makes it possible to avoid the complications of differential pest and disease incidence, but there are no means of avoiding other cultivar by husbandry interactions, except to prevent lodging in crops where this can occur. Thus it is desirable to carry out trials to assess such interactions, though this will at least double the work involved.

There are now several studies in which old and new cultivars have been compared. Austin *et al.* (1989) report a study on wheat in England and give references to similar studies in the USA, Mexico and Australia. Riggs *et al.* (1981) describe a study on barley and Russell (1984) and Duvick (1977) give the results of comparisons of a series of maize hybrids grown over several decades in the corn belt of the USA. Some results of the study of Austin *et al.* are given in Table 23.2. Apart from the considerably greater yield of the modern English winter wheat cultivars, the Table shows that they are shorter and flower earlier than their predecessors. Yet their total above ground dry matter production was little different from that of their predecessors, as has also been found in other studies with this crop. Little consistent difference in yield components is evident, though modern cultivars tended to have more ears and grains/m^2 than their predecessors. It must be emphasized that, in the strict sense, these associations between yield and other characters cannot be taken to be causal and neither is it possible to assess the relative importance for yield of changes in one character.

Nevertheless, the results from such trials

Table 23.2. Characteristics of groups of winter wheat cultivars assessed in field experiments at Trumpington, UK, in the harvest years 1984–1986

Cultivar group	No. in group	Grain yield, 0% moisture (g/m^2)	Biomass, 0% moisture (g/m^2)	Height to base of ear (cm)	Date of anthesis (June)	Duration of grain filling (days)
Very old	4	504	1500	145	21.7	48.4
Old	2	557	1541	134	20.2	47.5
Intermediate	2	669	1484	96	17.5	49.9
Modern	5	805	1588	78	16.0	50.7
Average s.e.		4	13	0.4	0.08	0.90

prompt the question: can past trends be continued and if so, for how long? It is interesting and informative to speculate on the answers. For wheat, Austin (1982) devised the scenario given in Table 23.3 and concluded that it was

Table 23.3. Distribution of dry matter (g/m^2) at harvest among different organs in four semi-dwarf cultivars of winter wheat grown at Trumpington, UK, in 1978, and a hypothetical distribution based on reducing stem and sheath dry weight to half its observed value (from Austin *et al.*, 1980; see text for further discussion)

	Observed	*Hypothetical*
Grain	707	896
Chaff (rachis, glume)	139	176
Leaf laminae	143	143
Stems and leaf sheaths	453	227
Total	1442	1442
Harvest index	49	62

unlikely that genetic gain in yield in this crop, associated with increased harvest index, could be as great as has occurred in the past and that the practical limit to harvest index would be about 55%. In turn, this focused attention on the ways in which the total dry matter production of the crop might be increased genetically.

23.3.4 Correlation analysis

This has been used and misused in numerous studies in attempts to identify traits correlated with high yield. It suffers from the disadvantage that the associations detected may not necessarily be causal. Nevertheless, when used correctly it can provide useful guides for further genetical and physiological analysis. At its simplest, trials are carried out with a number of genotypes and various characters scored or measured, and associations with yield sought. Ideally, a minimum of 20–30 genotypes needs to be compared, in suitably replicated trials, to reduce the likelihood of detecting fortuituous associations and to enable weak, though possibly biologically significant associations to be detected. Choice of genotype is important. The range in expected yield should be as large as possible, but the genotypes should also be at least moderately well adapted to the area in which the investigation is carried out. Of course these requirements tend to be mutually exclusive, so a compromise is necessary.

An example of a recent study is that of Acevedo *et al.* (1991) on barley grown in the Mediterranean environment of Syria. Some results from this study are given in Table 23.4.

Table 23.4. Simple correlations between grain yield and measured or scored traits for 35 two-rowed spring barley genotypes, computed on the mean results from trials harvested in Tel Hadya, Syria, in 1986 and 1987 (from Acevedo *et al.*, 1991)

Trait	*Correlation coefficient*	*Level of significance*[a]
Winter habit (score)	−0.13	n.s.
Seedling vigour (score)	0.53	***
Ground cover (score)	0.59	***
Days from seedling to ear emergence	−0.64	***
Days from seedling emergence to maturity	−0.37	*
Duration of grain filling (days)	0.45	**
Culm length (cm)	−0.15	n.s.
Peduncle length (cm)	−0.07	n.s.
Straw yield (g/m^2)	0.38	*
Number of ears/m^2	0.63	***
Number of grains per ear	0.29	n.s.
Mean grain mass (mg)	0.59	***

[a] n.s.: not significant; *, ** and ***: significance at $P = 0.05$, 0.01 and 0.001, respectively.

Because the study was carried out at sites differing in moisture availability, but which in other respects were similar, the study also provided information on characters especially important for yield at the driest sites. For two-rowed barley, good vigour at the seedling stage, early flowering and many ears/m^2 were all strongly associated with high yield. The results from this kind of study cannot be used to distinguish pleiotropy from linkage or simple association. Such assessments can only be made after crossing and selection, as described in the following section.

23.3.5 Preparation of experimental genotypes

The objective is to produce sets of genotypes contrasting in a character under study, no selection having been made for other characters, or for yield. The methods appropriate to a given crop and character depend on the breeding system of the crop and the mode of inheritance of the character. Here, procedures are outlined for a self-pollinated, inbreeding species.

For characters which are entirely or largely determined by a single gene with major effects, backcrossing can be used. An alternative is to maintain the locus in the heterozygous condition (discarding the homozygotes), whilst advancing the generations and increasing the homozygosity at other loci. After several generations, the homozygous segregants at the locus of interest are kept. The resulting lines will be near-isogenic except for the gene under study. This procedure has been used to produce near-isogenic lines of the dwarfing genes of wheat, Rht_1, Rht_2 and Rht_3.

Many characters are inherited quantitatively and are determined by genes at more than one locus. For self-pollinated crops, such characters can be handled as follows. Sets of genotypes contrasting in the expression of the characters, but otherwise well adapted to the locality are chosen. Pairs of contrasting genotypes are then crossed and the F_1 raised. Starting in the F_2 generation, selection for extreme genotypes, showing contrasting expression of the character, is carried out. It is essential that no selection is made for other characters, unless it is deliberate and the aim is to produce more than two (i.e. 4 or 8, etc.) phenotypic classes. By the F_4 or F_5 generation there will be little segregation for the character under study, or other characters. Taking into account replication of the crosses, the aim should be to produce ten or more 'low' lines and a similar number of 'high' lines. This means that many

more individuals will have to be selected in the F_2 and F_3 generations, as those which are intermediate in the following generations will need to be discarded. At this stage, the presumption is that the high and low lines differ only or mainly in respect of the character under study, and in characters, including yield, that are pleiotropically related to it.

For open pollinated crops, procedures that are appropriate to cultivar breeding have to be used. These will entail crossing, selfing for two or three generations prior to making composites from sets of inbred lines showing low or high expression of the chosen character (Crosbie *et al.*, 1980; Troyer and Larkins, 1985). Reselection from the composites and intermating to produce more advanced material is desirable. Again, the presumption is that the composites will be of comparable vigour to commercial ones or to F_1 hybrids, but will differ only in respect of the character under study, and others that are pleiotropically related to it. Appropriate variants of these methods have to be used when the species concerned is self-incompatible.

With some crops, for example barley and oilseed rape, it is possible to produce haploids and from these, homozygous diploid (DH) lines. Details of the procedures involved are given in Chapter 17. Selfed seed from the homozygous diploids is then grown. From the DH progeny produced, a subset is selected for contrasting expression of the character under study, and retained for evaluation.

23.3.6 Evaluation of experimental genotypes

From the previous section of this chapter, it will be assumed that there may be some 6–30 genotypes for evaluation. They will be reasonably well adapted to the area in which they are to be tested, and to the cultural practices used. Observations made during their production will have provided evidence of this and of any differential susceptibility to pests and diseases. Generally, it will be desirable to apply agrochemicals to control pests and diseases, and this is essential if susceptibility is linked to the character under study.

If an interaction between the character under study and either a cultural practice, a factor of the aerial environment or a soil factor is expected, it is highly desirable to design the evaluation procedure to enable the interaction to be detected. If an interaction with a cultural treatment such as spacing, planting date or fertilizer treatment is anticipated, this can be assessed by including it as a factor in the experiment. For this, a split-plot design will often be the most convenient, the main plots being the cultural treatment and the sub-plots, genotype. Such a design allows the most precise assessment of the overall effect of the character, and of the character × treatment interaction. A further level of splitting may also be desirable if the genotypes can be classified by attributes other than the one under study. This will be the case if the lines are derived from more than one cross, or, in the case of isogenic lines, if there are replicate lines and some basis for pairing them.

As with cultivar trials, there are often genotype × year (or season), or genotype × location interactions. In some cases these will have to be regarded as random. Where this is so, it will be sufficient to evaluate the genotypes over two or more years and locations, and average the results. Where an interaction is expected from prior experience, the trials should be dispersed in space so that they cover the range of variation in the relevant environmental factor or factors, which may be soil type, water availability, or less often, temperature or radiation receipt.

When carrying out the trials, it will be desirable to make appropriate observations of environmental variables, particularly if a character × environment interaction is expected. To detect pleiotropic associations with the character under study, it will be necessary to observe or measure other plant characters. Pleiotropic associations are very common: some examples are given in Table 23.5.

Finally, there must be sufficient replication to enable biologically or economically important effects to be detected as statistically significant. The plot size, shape and number of replicates chosen will inevitably be a compromise depending on the availability of seed, land,

Table 23.5. Some pleiotropic associations between traits

Crop species	Traits
Grasses, cereals, sugar beet	Small leaves, many leaves
Cereals	Many tillers, small leaves
Cereals	Large grains, few grains per ear
Cereals	Small flag leaves, high photosynthetic rate
Alfalfa, soybeans, rice	Thick leaves, high photosynthetic rate
Cereals	Early flowering, low biomass
Grasses, cereals	Large stomatal aperture, low stomatal density
Maize, barley	Poor seedling vigour, late maturity
Wheat, barley	Tall stature, late maturity
Wheat, barley	Tall stature, low harvest index
Wheat	Lax leaves, long leaves
Ryegrass	Erect leaves, long leaves
Cereals, grasses	Rapid growth in autumn, lack of winter hardiness

labour, the variability of the soil, and the number of genetic replicates of the character. The aim should be to detect as significant, differences in yield associated with the character under study of 4–5%.

Some examples of trait analysis using these procedures are given in Table 23.6.

23.3.7 Costs and benefits

Clearly, it is expensive to produce and evaluate experimental genotypes to obtain information on the value of a character for breeding. Is the value of the information worth the cost of acquiring it? The question is best addressed in the context of particular situations, as the following examples show.

Case 1. A crop species is being introduced into a new area and little is known about characters associated with high and stable yield
The first step in a plant improvement programme will be to introduce cultivars and

Table 23.6. Some examples of trait analysis

Crop	Character	Reference
Cotton	Drought tolerance	Quisenberry *et al.*(1985)
Maize	Cold tolerance	Crosbie *et al.* (1980)
Maize	Early flowering	Troyer and Larkins (1985)
Peanut	Yield components	Isleib and Wynne (1983)
Soybeans	Maturity, height, lodging, phenotypic score	Garland and Fehr (1981)
Sugar-beet	Tap-root:shoot ratio	Snyder (1985)
Tall fescue	Leaf area expansion rate	Reeder *et al.* (1984)
Wheat	Ears/m^2	Innes *et al.* (1981)
Wheat	Leaf posture	Innes and Blackwell (1983)
Wheat	Drought-induced abscisic acid accumulation	Innes *et al.* (1984)
Wheat	Height and date of ear emergence	Innes *et al.* (1985)

breeding lines from areas with similar climates, and compare them in observation plots or replicated trials. If records are taken of yield and of characters that are believed to be important for yield, the data can be used in a correlation analysis (section 23.3.4) and this will indicate which characters are associated with high yield, which are neutral, etc. Such information, refined in subsequent years by comparing a subset of the most promising material, and derivatives of it, will give a valuable guide for a breeding programme. A more elaborate approach will not be justified.

Case 2. An established breeding programme which is making good progress in improving yield

The breeders are likely to have identified the characters most important for high and stable yield and the most advanced breeding material will be approaching homozygosity for genes determining these characters. In some cases, the nature of the genetic control of the characters will be known (major genes, genes with small effects, etc.). There may be doubt as to the value of some characters. If the programme is well resourced, such characters could be studied by making and evaluating appropriate sets of genotypes. It is likely, however, that such characters will be of secondary importance and if so the effort involved will be large in relation to the probable benefit. Of course this may not be so if new variation is introduced, such as photoperiod insensitivity, novel dwarfing genes, etc. A rigorous assessment of such

new variation may save much wasted effort in conventional breeding, which might otherwise be based on incorrect assumptions on the value of the characters. Plant breeders may also require to know the appropriate degree of expression of quantitative characters where the extremes are likely to give poor performance. This may apply to characters such as maturity, height, and yield components.

Case 3. An established breeding programme which has been successful but where further improvement in yield has become increasingly difficult

As noted in section 23.1, a successful breeding programme in an environment (using the term in the broadest sense) which is stable will eventually reach the stage when no further improvement in yield is possible. As this stage is approached, breeding to improve yield will give progressively smaller returns. Some believe that this stage has been reached in wheat breeding in north-west Europe. The evidence for this is that, over the last four decades, there has been a rapid increase in the yield on a countrywide basis, and in the yielding capacity of new cultivars. It is perceived that weed, pest and disease control is close to complete and that the rates of application of fertilizer have stabilized, and that average yields have not increased in recent years (however, in trials, new cultivars continue to show a substantial advantage over their immediate predecessors). But for the sake of argument, let us assume that despite attempts to select for higher yield, new cultivars yield little, if at all, more than their predecessors. Is there no prospect of raising yield in the area? There are at least three possibilities. First, the yielding capacity of the environment may be increased for example by using more fertilizer, irrigation or other inputs. This would create new opportunities for breeding, as pointed out in section 23.1. Second, the target area could be subdivided so that the breeding was directed to producing cultivars for narrower ranges of environments, so that the yield penalties of wide adaptability could be exchanged for greater fitness. Third, the need for adaptability to the full range of year-to-year variation in climate could be reduced if it were possible to forecast with reasonable certainty the ensuing seasons' weather. In this case, cultivars could be produced that were especially suitable for dry seasons, or cool seasons, etc., and to complement them, others which were suitable for wet or hot seasons, etc. As with the previous option, this also involves trading adaptability for fitness. Of course, long-term weather forecasting is not nearly good enough to make this option practicable at present, and the other options would have to be assessed in the light of the financial costs of less centralized breeding programmes (in the case of the second option) and in the light of possible adverse environmental effects of increased inputs (in the case of the first option). Nevertheless, when the human population of the world creates a need for substantially higher crop yields than at present, these options will have to be considered seriously.

It is commonly found that the difference in yield between lines displaying high and low expression of a character is about 5–10%. Let us suppose that existing cultivars show intermediate expression of a set of characters. Let us also suppose that all these characters are independent in their effects on yield, and the effects additive. For illustration, if ten such characters are considered, yield would be increased by between 10×2.5 and 10×5, i.e. by between 25 and 50%, if a cultivar were bred which had high expression of all ten characters. Many breeders and physiologists would regard such increases as improbable (but see Table 23.2). However,

the illustration is intended to show that, even given careful assessments of the yield gains which could be obtained by selecting for individual traits, it would be unwise to make assumptions about additivity and independence, and that in practice, gains from simultaneous selection for several traits might be less than such assumptions would imply.

23.4 IDEOTYPES AND MODELS

The use of the term ideotype is attributed to Donald (1968). Some alternative expressions are: 'model plant', 'ideal model plant', and 'ideal plant type'. All refer to a conceptual plant that possesses characteristics which are considered ideal for a given environment. Quite distinct from ideotypes are simulation models, often referred to as 'models', or sometimes 'dynamic models'. Simulation models are sets of algebraic and differential equations which describe the rates of and interrelationships among the various component processes of plant growth and behaviour. The equations are solved numerically, rather than algebraically, to simulate the growth, development, and water and nutrient balance, etc., of a plant or crop. Among the best-known examples of simulation models are those used to explore the effects of varying nitrogen fertilizer and water availability on the growth of wheat (van Keulen and Seligman, 1987; Godwin et al., 1989). However, models can also be written to simulate the effects on growth and yield of varying plant characteristics (called genetic parameters in plant simulation models) and their interactions with environment.

It is important to remember that both ideotypes and simulation models are based on limited and imperfect information. As such, with improvements in knowledge of physiology and biochemistry in general, and of the species under study in particular, they can be continuously refined and improved.

23.4.1 Defining an ideotype

All plant breeders, whether or not they explicitly recognize it, have an ideotype in mind when they evaluate their material and choose plants for crosses. The ideotype will take account of the target environment that the breeder has defined, of agronomic practices, both current and prospective, of characterictics that determine crop quality, and of the need for pest and disease resistance. The ideotype is likely to be reliable only if it is based on a good understanding of the growth and development of the crop and its responses to environmental factors, rather than on intuition or prejudice.

Thus the first step in defining an ideotype is to define a target area and obtain information on its climate, including its variability; on soil types, their water-holding capacity and nutrient status. Given a particular crop species and some understanding of factors which affect and limit its growth and development, broad limits can be set to sowing and harvesting dates (in the case of annual crops), and to other aspects of its husbandry. Quality considerations may determine the limits to the size, shape, etc., of the harvested parts (e.g. roots, grain, tubers). Current agronomic practices may dictate limits to plant stature, branching and other characters. From this information, a list of desired characters can be drawn up. At this stage, complex characters such as heat tolerance and drought tolerance may have little meaning for the definition of the ideotype. The next step is to assess what changes in particular features might improve performance (yield and yield stability) in the target environment. Often, there will not be sufficient information available to make this assessment. If the features are perceived to be important, the procedures described in 23.3 may be used to obtain the information. Otherwise, it should be recognized for which characters there is no good information: selection for these characters may prove disadvantageous and should be avoided. Other considerations have a bearing on the choice of characters to include in the definition of an ideotype: these are discussed in 23.5.

23.4.2 *Simulation models as aids to trait analysis*

Provided that a simulation model includes a trait of interest and the model realistically

describes its effects and interactions with other characters and with the environment, the model can be used to explore, 'in the computer', the consequences for yield of varying the character, in a defined environment, and interactions with other plant characters and elements of the environment. This use of simulation models is sometimes known as sensitivity analysis. An example of the use of a simple model for the simulation of grain growth and yield of wheat is given by Austin (1982) and some predicted relationships are shown in Fig. 23.3. Leaf posture has been the subject of several simulation studies. The general conclusion is that in dense canopies, erect leaves are beneficial to canopy photosynthesis and yield, whereas in canopies likely to attain a leaf area index of less than 3–4, lax leaves are preferable.

At present, simulation models have some serious shortcomings when used as aids to the analysis of very dynamic features such as carbon allocation and photosynthesis. For example the method of allocating carbon to the

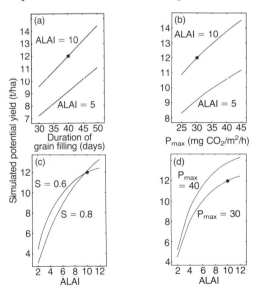

Fig. 23.3. Simulation of the effects on grain yield of wheat of varying (a) the duration of grain filling for two values of the leaf area index at anthesis (ALAI), (b) the light saturated rate of photosynthesis (P_{max}) for two values of ALAI, (c) ALAI for two values of the parameter *s* describing leaf posture ($s = 0.8$, erect leaves, $s = 0.6$, lax leaves) and (d) ALAI for two values of the light saturated rate of photosynthesis (adapted from Austin, 1982).

component organs is completely arbitary. This is because there is no adequate understanding of the basis of assimilate distribution, or of its dependence on environmental factors. Thus, models have to use empirically determined functions that are dependent on plant size and stage of development. Often, it is also assumed that assimilated carbon is distributed preferentially to the shoot, the roots receiving what is left over when the demands of the shoot have been satisfied. Clearly, this is an oversimplification. Another shortcoming is that it is only practicable to simulate growth, in terms of processes such as photosynthesis, respiration and nitrate uptake at the next lower level of organization. The relationships between these processes and environmental factors are described by simple equations. In reality, these processes are themselves complex, and subject to regulation and control by many internal and external factors. Taking photosynthesis as an example, much is known of the component processes and of their regulation by environmental factors in the very short (seconds), short (minutes), and longer (hours, days, weeks) term. However, present knowledge is not adequate to allow these relationships to be quantified and incorporated into simulation models. As a result, the photosynthetic properties of leaves are modelled as being essentially static, depending simply on the current light intensity. Yet it is known that the expression of genetic differences in the rates of photosynthesis of rice leaves, associated with leaf thickness and the amount of light harvesting chlorophyll *a* and *b*, depend on the previous light intensity (Tu *et al.*, 1988). Other features that it is difficult to model satisfactorily include water uptake and transpiration and the control of transpiration and leaf expansion by water status.

For all these reasons, simulation models as aids to character analysis have to be used with discretion. It is essential that the limitations of a chosen model are thoroughly understood before it is used to draw conclusions on the breeding value of a character.

In principle, the genetic parameters in a simulation model could be optimized by iterative procedures to predict the optimum expres-

sion of any combination of characters under study, as well as the characters varied one at a time. The procedure will be costly in computing time and the limitations of simulation models may make the results unreliable. Thus the procedure cannot be considered to be of practical value at present.

23.5 THE LOGISTICS OF TRAIT-BASED SELECTION

Numerous traits have been suggested to be of value in breeding programmes. Some can be easily and rapidly assessed and, if significant for yield, will be worthwhile selecting for routinely. Others, though purported to be significant for yield, prove to be of little value. This section considers the features of characters and the requirements of screening tests that have to be met if they are to form a useful part of a breeding programme. It has already been noted in section 23.3.2 that traits can be grouped into classes, differing in the ease and reliability with which they can be assessed.

To be of value in assessing a trait in a population of plants, a screening test must satisfy a number of criteria, as follows: (1) the character must be easier or more convenient to assess than yield itself; (2) there must be heritable variation in the character; (3) it should be possible to assess the character simply, rapidly and inexpensively, preferably at any time of the year; (4) there must be an appreciable genetic correlation between the attribute and yield under field conditions, and the correlation should preferably be causal; (5) other factors affecting the results must be easily controllable and/or their effects known. These criteria are discussed below.

As already noted, in the early stages of a breeding programme for a self-pollinated species, yield cannot be assessed with any degree of reliability and so a surrogate for yield, i.e. a set of screening tests, is necessary if the selection is to be better than taking a random subset of individuals in a given generation. Another feature of the early generations of a breeding programme is that many thousands of plants need to be examined. Thus rapid screen-

Table 23.7. Some traits for which screening has been proposed

Yield components	Moderately easy to assess; heritability varies from low to high; components usually negatively correlated
Height	Easy to assess; usually highly heritable; different dwarfing genes have different effects on yield and other characters
Leaf posture	Easy to assess though expression can be masked under poor cultural conditions. Except for extremes, consequences for yield tend to be minor
Stomatal frequency	Moderately easy to assess; usually strong negative correlation with stomatal pore size and leaf size. Consequences for stomatal conductance variable; consequences for water use efficiency negligible
Flowering date	Easy to assess; highly heritable; important consequences for yield depending on seasonal pattern of water availability
Winter hardiness	Difficult to assess reproducibly in the field; assessment in screening tests is expensive; moderately highly heritable
High photosynthetic rate	Very complex character; strong pleiotropic associations with other leaf characters; has not proved to be of value in screening for fast crop growth rate and high yield
Drought tolerance	Very complex character; needs to be analysed in terms of its components, but even these have not been sufficiently well established to be useful. Some components may be: leaf waxiness, diameter of xylem vessels in primary roots, capacity of roots and leaves to synthesize abscisic acid and proline when stressed. Carbon isotope discrimination may, however, prove a useful integrative measure of drought tolerance

ing is essential, particularly if the character can only be assessed over a limited period of the growth cycle. Clearly, morphological traits can be assessed rapidly and easily and should be screened for if there is good evidence of their value. Because expensive equipment is not required, the costs can be small. Assessment can often be done quite satisfactorily by ranking or scoring, rather than by actual measurement. In cereals this is the case for height, tillering, leaf size, leaf colour and for presence or absence of characters such as awns. The time needed for screening an individual is typically 1–5 s.

Modern portable porometers and gas analysers now make it possible to measure stomatal conductance and photosynthesis in 15–30 s and so the screening of up to a few thousand plants is possible in principle. However, as these 'characters' are influenced very markedly by many environmental factors in complex ways, they cannot be screened for meaningfully. Furthermore, there is no convincing evidence that high yield is correlated with high photosynthetic rate measured on single leaves

(Nelson, 1988), or that genetic variation in stomatal conductance within a species is correlated with water use efficiency. The same is true for screening tests based on enzyme activity, unless these tests are carried out under strictly defined conditions; even so, genetic associations with yield under field conditions are often weak or non existent. Some examples of traits for which screening has been proposed, with comments on their value in breeding programmes, are given in Table 23.7. Some of these traits are also discussed in Chapter 24.

23.6 EXPLOITING TRAITS IN ALIEN GERMPLASM

Alien germplasm usually refers to species that are different from the crop species. These may be wild relatives of the same genus, species of another genus in the same family, or totally unrelated organisms, for example, bacteria. In conventional breeding, traits in alien germplasm are only accessible if the alien species can be sexually hybridized with the cultivated species. In practice, there is a further constraint: it will only be feasible to transfer the trait if it is under fairly simple genetic control. This may be the case with resistance to some pests and diseases and with some traits that affect quality, but it will not usually apply for complex traits such as photosynthesis, transpiration, drought or salinity tolerance. Such complex traits usually comprise a syndrome of characters, each likely to be determined by several or many genes. The most that can be expected is to combine the genomes of the cultivated and the alien species by producing an amphiploid, and to select individuals with appropriate characteristics from progenies of amphiploid by amphiploid crosses. A prerequisite is that the amphiploids are at least partially fertile.

23.7 CONCLUSIONS AND PROSPECTS

With many annual crop species there have been enormous genetic gains in yielding capacity, as well as in their quality, pest and disease resistance. In part, this has come about because the breeding programmes have become progressively larger, both in terms of the number of crosses made and the number of plants examined. This increase in size had been made possible by improved mechanization and by the use of computers to store, classify and analyse records. For quality and for pest and disease resistance, more precise and rapid tests have been devised and have gained wide acceptance. By contrast, screening tests to improve the efficiency with which high yielding individuals can be identified in segregating progenies, have improved very little. For sound physiological reasons, tests that are 'snapshots' of plant function are extremely unlikely to give a satisfactory indication of plant performance over a long period, certainly not the entire growth cycle. In other words, yield is the best measure of integrated performance. The breeders eye, rapid and effective for assessing morphology, is still the mainstay of selection at this stage. It is greatly aided by undertaking the selection in appropriate conditions (locations, sowing dates, cultural conditions). Thus, a major contribution to increasing the efficiency of selection for yield in early generations will be to improve our knowledge of the physiology of the plant and its responses to the environment and to determine the most suitable conditions in which to observe morphology, growth and assess yield. This is of obvious importance in selecting for tolerance to adverse conditions, but it is also important in selecting for fitness in favourable environments with high inputs, and when wide adaptability is required.

REFERENCES

Acevedo, E., Craufurd, P.Q., Austin, R.B. and Perez-Marco, P. (1991), Traits associated with high yield in barley in low rainfall environments. *J. Agr. Sci.*, **116**, 23–36.

Austin, R.B. (1982), Crop characteristics and the potential yield of wheat. *J. Agr. Sci.*, **98**, 447–453.

Austin, R.B., Bingham, J., Blackwell, R.D., Evans, L.T., Ford, M.A., Morgan, C.L. and Taylor, M. (1980), Genetic improvements in winter wheat yields since 1900 and associated physiological changes. *J. Agr. Sci.*, **94**, 675–689.

Austin, R.B., Ford, M.A. and Morgan, C.L. (1989), Genetic improvement in the yield of winter wheat – a further evaluation. *J. Agr. Sci.*, **112**, 395–301.

Crosbie, T.M., Moss, J.J. and Smith, O.S. (1980), Comparison of gains predicted by several selection methods for cold tolerance traits in two maize populations. *Crop Sci.*, **20**, 649–655.

Donald, C.M. (1968), The breeding of crop ideotypes. *Euphytica*, **17**, 385–403.

Duvick, D.N. (1977), Genetic rates of gain in hybrid maize during the past 40 years. *Maydica*, **22**, 187–196.

Finlay, K.W. and Wilkinson, G.N. (1963), The analysis of adaptation in a plant breeding programme. *Aust. J. Agr. Res.*, **14**, 742–754.

Garland, M.L. and Fehr, W.R. (1981), Selection for agronomic traits in hill and row plots of soybeans. *Crop Sci.*, **21**, 591–595.

Godwin, D., Ritchie, J., Singh, U. and Hunt, L. (1989), *A Users Guide to CERES Wheat*, V2.10, International Fertilizer Development Center, Muscle Shoals, Alabama.

Innes, P. and Blackwell, R.D. (1983), Some effects of leaf posture on the yield and water economy of winter wheat. *J. Agr. Sci.*, **101**, 367–376.

Innes, P., Blackwell, R.D., Austin, R.B. and Ford, M.A. (1981), The effects of selection for number of ears on the yield and water economy of winter wheat. *J. Agr. Sci.*, **97**, 523–532.

Innes, P., Blackwell, R.D. and Quarrie, S.A. (1984), Some effects of genetic variation in drought induced abscisic acid accumulation on the yield and water use of spring wheat. *J. Agr. Sci.*, **102**, 341–351.

Innes, P., Hoogendoorn, J. and Blackwell, R.D. (1985), Effects of differences in date of ear emergence and height on yield of winter wheat. *J. Agr. Sci.*, **105**, 543–549.

Isleib, T.G. and Wynne, J.C. (1983), F₄ bulk testing in test crosses of peanut cultivars. *Crop Sci.*, **23**, 841–846.

Keulen, H. van and Seligman, N.G. (1987), *Simulation of Water Use, Nitrogen Nutrition and Growth of a Spring Wheat Crop*, Pudoc, Wageningen.

Nelson, C.J. (1988), Genetic associations between photosynthetic characteristics and yield : review of the evidence. *Plant Physiol. Biochem.*, **26**, 543–554.

Quisenberry, J.E., Wendt, C.W., Berlin, J.D. and McMichael, B.L. (1985), Potential for using leaf turgidity to select drought tolerance in cotton. *Crop Sci.*, **25**, 294–299.

Reeder, L.R. Jr, Sleper, D.A. and Nelson, C.J. (1984), Response to selection for leaf area expansion rate of tall fescue. *Crop Sci.*, **24**, 97–100.

Riggs, T.J., Hanson, P.R., Start, N.D., Miles, D.M., Morgan, C.L. and Ford, M.A. (1981), Comparison of spring barley varieties grown in England and Wales between 1880 and 1980. *J. Agr. Sci.*, **97**, 599–610.

Russell, W.A. (1984), Agronomic performance of maize cultivars representing different eras of breeding. *Maydica*, **29**, 375–390.

Snyder, F.W. (1985), Yield of a sugar beet hybrid from parents selected for a high tap root to leaf weight ratio. *Crop Sci.*, **25**, 87–89.

Troyer, A.F. and Larkins, J.R. (1985), Selection for early flowering in corn: 10 late synthetics. *Crop Sci.*, **25**, 695–697.

Tu, Z.P., Lin, X.Z., Huang, Q.M., Cai, W.J., Feng, H.Y. and Ye, L.Y. (1988), Photosynthetic characters of rice varieties in relation to growth irradiance. *Aust. J. Plant Physiol.*, **15**, 277–286.

24

Resistance to abiotic stresses

E. Acevedo and E. Fereres

24.1 INTRODUCTION

World food production is limited primarily by environmental stresses. It is very difficult to find 'stress free' areas where crops may approach their potential yields. Abiotic environmental factors are considered to be the main source (71%) of yield reductions (Boyer, 1982). In past decades, the primary approach for alleviation of environmental stress focused on modification of the environment through irrigation, soil amelioration, fertilizer use, etc. Economic and ecological constraints on environmental amelioration make the approach of genetic improvement of stress resistance a viable alternative for closing the gap between actual and potential crop yields in marginal areas. The world's cereal production requires a sustained increase of 3% in developing countries to maintain the current levels of nutrition to the year 2000 and beyond. Improving the stress resistance and yield stability of the major cereal crops and cropping systems, accounting for over 80% of the world's food production, is one approach to solving this crucial problem (Chapter 30 and McWilliam, 1989).

Resistance to abiotic stresses is part of the larger domain of environmental fitness of crops and cultivars. It is linked to the maximum or potential yield which is the yield that would be achieved in a particular environment in the absence of constraints to crop growth. The abiotic stresses are those which determine the

Plant Breeding: Principles and prospects. Edited by M.D. Hayward, N.O. Bosemark and I. Romagosa. Published in 1993 by Chapman & Hall, London. ISBN 0 412 43390 7.

extent to which the yield potential of a crop or a cultivar is realized if all other biotic limiting factors are controlled. Everywhere in the world there are environmental gradients, along which crops or cultivars are replaced by others which are better fitted.

Temperature and water are major factors affecting crop distribution. A typical example of water supply determining crop distribution is found in the cropping sequence, bread wheat, durum wheat and barley as rainfall decreases in a steep gradient from the Mediterranean sea inland in North Africa and South West Asia (Acevedo *et al.*, 1990). Good examples of crop resistance to limited water and to temperature extremes are usually found along those gradients. Soils may also pose additional major constraints to crop production. Indeed, soil pH, mineral deficiencies or toxicities, soil salinity or sodicity, and waterlogging all affect crop distribution and productivity. However, we will not discuss soil constraints in this chapter.

24.2 YIELD POTENTIAL

We define yield potential as the yield of a genotype growing in a particular environment to which it is adapted, in conditions under which soil nutrients and water are non-limiting and where other stresses are effectively controlled. By adaptation we essentially mean that the growth and development of the genotype is well-matched to the environment. As we use a given genotype for larger and larger areas, the possibility of encountering environmental variability and limiting factors for that genotype increases and hence its yield will decrease below potential levels. The range of environments to which genotypes are adapted varies. For example, photoperiod insensitive genotypes may be grown at various latitudes in contrast to genotypes which are sensitive to daylength.

A second reason to use the concept of yield potential in studies of stress resistance is that it provides the agronomist with an estimate of a yield ceiling attainable with present technology. Such an estimate is extremely useful to quantify the extent of yield reduction caused by the stresses in question. Also the perceived intensity of stress may influence the choice of cultivars relative to their yield potential. Third, it has been shown that, at least up up to a certain degree of stress, varieties with high yield potential may yield more than 'stress resistant' varieties (Chapter 22). Above a certain stress level the opposite may occur, a phenomenon known as 'crossover' (Ceccarelli, 1991). A fourth reason, which is not always realized, is that the major fraction of world food production for any crop comes from areas with good management and where environmental restrictions to yield are low. Table 24.1 illustrates this

Table 24.1. Percentage of developing world wheat production as function of major environments (CIMMYT source) and estimated difficulty for yield improvement

Environment	Wheat production[a]	Difficulty[b]
Irrigated	43	+
High rainfall	12	++
Mediterranean (drought)	10	+++
High temperatures	6	+++
Acid soils	2	++
High latitudes	5	++
Low temperatures	18	+++

[a] Approximate figures.
[b] Difficulty associated with yield improvement, either agronomical or genetical.

point for wheat and shows that the difficulty associated with genetic and agronomic yield improvement increases as the abiotic constraints become more severe.

The global increase in wheat yield potential today is of the order of 0.7% per annum

whereas the demand for wheat in the developing world is increasing at an annual rate of 3% (McWilliam, 1989). With continued population growth and wheat being incorporated as a basic staple in the diet of more and more people it is clear that a point will soon be reached when production does not satisfy demand. Wheat breeding, in combination with improved crop management techniques, has been extremely successful in the past in coping with an ever-increasing demand. Wheat scientists, however, have come to realize that the improvements in yield potential have occurred without increased crop growth or photosynthetic rates and without any important change in biomass. In fact, studies comparing historical series of cultivars under identical agronomy show a rather constant biomass (Austin *et al.*, 1980a). Only the biomass distribution to various plant organs has changed, notably to the grain. However, a reduction in the distribution of dry matter to roots, leaves, stems and reserves should decrease the ability of a genotype to compete with weeds, to recover from pests or diseases and to survive abiotic environmental stresses, and as a compensation requires better management (Evans, 1988). This may explain the crossover observed under highly stressed conditions, i.e. genotypes with very high yield potential performing poorly under severe stress conditions.

At this point, it should be emphasized that optimizing agronomic practices is the most powerful means of achieving higher and more stable yields where crops are subjected to environmental stress. However, given improved agronomy, two approaches may be conceptually followed to raise yields in the long run; one is based on increasing yield potential of broadly adapted varieties while the other relies on better exploitation of the adaptive features of genotypes by fitting cultivars to specific target environments.

24.2.1 Prospects of increasing yield potential

Various aspects of increasing yield potential are examined here such as growth and photosynthesis, changes in biomass distribution, regulatory processes and environmental responses.

(a) Photosynthesis

As already mentioned, genetic improvement has not increased growth rate. Relative growth rates of plants of comparable size are similar. In most cases, reported differences in biomass between genotypes are due to differences in growing period (Evans, 1984). Net photosynthesis may be expressed as:

$$NP = GP - PR - DR$$

where NP is the net or apparent photosynthesis, GP is the real or gross photosynthesis, PR is the photorespiration and DR is the dark respiration (Chapter 27). The importance to dry matter production of the relative rates of photorespiration, dark respiration and gross photosynthesis is therefore obvious. Keys and Whittingham (1981) show that in the field, photorespiration accounts for 25% of NP in C_3 crops. Respiratory losses, on the other hand, may be as large as 45% of the carbon fixed (Morgan and Austin, 1983). Using these figures it can be calculated that NP, PR and DR account for approximately 48, 12 and 40% of GP, respectively.

The photosynthetic efficiency or conversion of incident photosynthetically active radiation (PAR) into carbohydrates (CH_2O) in a C_3 crop at the field level is low. With an incident PAR of 100% there is a 15% loss due to reflection and transmission, a 30% loss due to light saturation of upper leaves, and a 40% loss due to respiration in the dark. The maximum measured efficiency of conversion of PAR to CH_2O is 17.7% taking into account losses of efficiency due to photorespiration and light respiration. This leaves a photosynthetic efficiency of 6.3% of incident PAR per mol of CO_2 incorporated. The approximate 6% operates only when the interception of PAR is complete and when environmental and endogenous conditions are optimum. In temperate climates for example, photosynthesis can be limited by low temperatures in the winter and by water deficits in the summer. The overall effect of such environmental stresses is to reduce the maximum photosynthetic efficiency to a yield of 2–4%, the reductions being greater for crops grown in marginal soils, with poor husbandry or with extreme stresses.

There is a linear relationship between final dry matter yield and cumulative intercepted total solar radiation. On average, dry matter is accumulated at a rate of 1.4 g/MJ of intercepted solar radiation (Monteith, 1977). When all other constraints are minimized, the capture of light becomes the overall limiting factor. There is little doubt that photosynthesis is correlated with growth. However, the relevant question is whether photosynthesis can be increased and if it will contribute to a correlated change in an economically important trait such as yield.

Several attempts have been made to select for increased photosynthetic rate (P_{max}) with the expectations of increasing biomass and yield. These efforts have generally been unsuccessful (Nelson, 1988). The reasons for failure include pleitropic associations, sink limitation, existence of linkage groups, environmental effects and others. In contrast to this, Mahon (1990) has shown that there is genotypic variability in carbon exchanges rates (CER) of peas, which is neither the result of different source–sink relations nor the result of different leaf area indices. He found CER to be highly heritable and largely additive. Austin (1990), however, emphasized that increases in grain yield in wheat breeding owes very little, if any, to increased photosynthetic capacity in spite of the fact that present cultivars in the United Kingdom yield some 50% more than do varieties grown at the beginning of this century. Only some recent wheat varieties produce a little more biomass than earlier varieties under some circumstances. Austin (1990) concluded that, in spite of the higher P_{max} of the diploid wheat (*Triticum urartu*, AA), and of a derived amphiploid obtained by crossing the diploid with durum wheat (6x, AAAABB) in comparison with *T. aestivum*, it seems difficult to genetically improve the photosynthetic characteristics to achieve greater biomass and yield potential (Chapter 27). An increase in photosynthetic rates would be dependent on an increase in the activity of carboxylating enzymes, a phenomenon which awaits further development. Whether increases in GP will result in yield increases remains an open question. Evans (1988) points out that selection for higher GP appears to be negatively correlated with yields, although other features of the photosynthetic system, namely the duration of photosynthetic activity of the leaves, have been improved along with crop yields.

(b) Partitioning of biomass

The increases of yield potential in the past can be mainly attributed to resistance to biotic stresses and to increases in partitioning towards harvestable yield (harvest index). A decreased vegetative growth in cereals has allowed a higher number of grains per inflorescence and per unit land area. The scope to continue this trend is limited (Chapter 23) but still important in many crops since the harvest index of most wheat varieties is below the 55–60% value believed to be the maximum. In fact, Evans (1988) argues that selecting for a higher number of grains per unit area may increase the photosynthetic rate due to greater assimilate demand by the sink. This would increase canopy photosynthesis and crop biomass in an indirect way. The 'stay green' feature or slow senescence of the crop canopy during the reproductive phase would also be of help during grain filling. The change in distribution of assimilates, which has led to greater harvest indices in modern cultivars, implies a substantial number of alterations in growth and developmental patterns, which we are far from understanding at present (Evans 1990).

Substantial efforts in plant breeding programmes are aimed at increasing yield potential of the major crops. For instance, wheat yield potential is being increased at a rate of 0.7% per year on a worldwide basis, which should have spillover effects on yields under stress. Such efforts normally emphasize wide adaptability of improved varieties, a requirement for a cultivar to be potentially useful in many parts of the world. Since wide adaptability is usually associated with relative insensitivity to environmental signals, it follows that there is additional room for improvement by better exploitation of specific environments. Local breeding programmes, which develop cultivars best fitted to

the relevant environments, offer a complementary strategy to sustain yield increases in areas subjected to substantial environmental stress.

24.3 IMPROVING ADAPTATION TO STRESSFUL ENVIRONMENTS

24.3.1 Quantitative assessment of the target environment

Environments which pose limitations to crop production are often characterized by the occurrence of more than one physical stress at the same time or throughout the growing cycle, even though one stress may dominate (e.g. drought, heat). In Mediterranean environments drought spells in cereal crops during winter may be associated with low temperatures and suboptimal radiation levels, whereas terminal drought is generally associated with above-optimal temperatures and excess radiation. Drought and salinity are stresses commonly occurring where irrigation is practised. The soil may impose additional constraints, such as high or low pH, aluminium toxicity, phosphorus and micronutrient deficiencies or toxicities as well as their interactions (e.g. phosphorus-induced zink deficiency in wheat).

In addition to the simultaneous occurrence of stresses, in rain-fed environments, some of them, notably drought, are unpredictable in occurrence, intensity and duration. Furthermore, the overall nature of particular stresses may be different. One hot environment with a high vapour pressure deficit is different from another having a low vapour pressure deficit. In the first case, transpiration is effective in keeping the plant organs below ambient temperature, whereas in the second case, the effect of transpirational cooling may be negligible. Water stress occurring in plants dependent on rainfall as the sole soil water source induces plant responses which are basically different from those arising from water stress that develops in plants which are grown on stored soil moisture or under other types of drought (Edmeades *et al.*, 1989).

It is within the above framework that the improvement of genotypes for adverse environments must be seen and understood. The genotypes to be grown in stressful environments need to be adapted to a mosaic of constraints. Because the constraints will vary in space and time, genotypes well adapted to a given pattern of constraints may not be equally adapted to a different pattern. It follows that the optimal combination of adaptation features may vary with stress environments. It is also evident that the quantitative characterization of the target environment is a prerequisite to defining such optimal combinations for cultivar improvement. This characterization beyond qualitative climatic and soils descriptions is a formidable task, often neglected by plant breeders and physiologists.

Environmental characterization may be done at different levels, from the climate of a region, to the year to year weather variation of a farm, to the soil profile features of a given field. Along these lines, the major contributions by agroclimatologists and soil scientists, which should be incorporated into integrated cultivar improvement programmes for stress resistance include: (1) techniques to compute rainfall probabilities; (2) soil water balance models (Angus, 1991) to quantify crop extractable water and, in conjuction with (1), drought patterns; (3) geographical information systems and remote sensing techniques that allow the characterization of the major environments and of their relative importance for ideotype design; (4) weather generators (Richardson, 1984) which can simulate daily weather parameters for many years, to quantify weather variability and climatic risk, even for areas with limited basic meteorological information.

The developments described above permit the unequivocal characterization, in terms of probabilities, of the environmental domain to which improved cultivars must be fitted. One prerequisite needed to take full advantage of such new developments is that of appropriate environmental data collection in breeders trials in order to increase the understanding of genotype × environment interactions beyond the traditional approaches based on statistical analyses (Chapter 22 and Shorter *et al.*, 1991).

24.3.2 Relations between stress resistance and yield

Plant growth and yield under stress have been the subject of much research. Morphological, physiological and anatomical characteristics have been identified which enable plants to grow and survive in stressed environments (Turner and Kramer, 1980; Taylor *et al.*, 1983; Srivastava *et al.*, 1987; Ludlow and Muchow, 1988; Schultze, 1988). Usually, a complex set of attributes is present in cereal species that grow under severe stress. This point is summarized in Table 24.2 for barley landraces which

Table 24.2. Attributes of 14 two-row pure lines isolated from Syrian landraces as compared with 23 two-row improved barleys (after Acevedo and Ceccarelli, 1989)

Higher grain yield under drought
Higher harvest index
Higher grain mass
Earlier heading
Shorter grain filling period
Higher drought resistance index
Prostrate growth habit in winter
Dark green leaves before stem extension
Light green leaves after stem extension
Shorter stature under drought
Higher tillering
Higher number of fertile spikes
Higher discrimination for ^{13}C
Lower transpiration efficiency
Medium vernalization requirement
Longer emergence to double ridge period
Shorter ear initiation and ear growth period

evolved in the Fertile Crescent. The two-row landraces have better adaptation to the low-rainfall (ca 250 mm) Mediterranean environments of Northern Syria than their two-row improved counterparts. It is a combination of the attributes (Table 24.2) that confers the adaptation; individually the traits appear to have relatively small effects in improving productivity, stress resistance and yield under stress (Fischer and Wood, 1979; Acevedo *et al.*, 1990). It seems difficult to find a single stress adaptive trait other than flowering time and yield potential that would be predictive of plant response to stress. Multiple selection criteria must be applied (Blum, 1979; Nass and Sterling, 1981; Acevedo and Ceccarelli, 1989).

Stress resistance is usually equated to maintaining yield under stress. However, higher yields and stability, may be the result of modified management practices and/or physiological mechanisms not necessarily associated whith the stress resistance of a genotype. Table 24.3 shows the effect of row spacing and

Table 24.3. Main effects of variety, row spacing and planting date on total above ground biomass (BIOY) (t/ha) and grain yield (GWT) (t/ha) in durum wheat grown in Northern Syria during the 1989/89 season[a]

Factor	Tel Hadya		Breda	
	BIOY	GWT	BIOY	GWT
Variety				
Om Rabi 14	4.8 b	1.6 a	1.6 a	0.4 a
Sebou	5.2 a	1.9 a	1.7 a	0.4 a
Korifla	5.1 ab	1.7 a	1.8 a	0.5 a
Row spacing				
10 cm	5.7 a	2.5 a	2.1 a	0.6 a
20 cm	5.0 b	2.3 a	1.6 b	0.4 b
40 cm	4.6 c	2.0 b	1.4 b	0.3 b
Planting date				
Early	6.7 a	2.4 a	3.9 a	0.7 a
Medium	4.3 b	1.5 b	1.9 b	0.5 b
Late	4.4 b	1.3 b	0.3 c	0.0 c

[a] Rainfall: Tel Hadya 239 mm + 50 mm irrigation, Breda 194 mm.
Means followed by a different letter within a column for a given factor differ significantly.

planting date on the average yield of three durum wheat varieties of almost identical phenology grown in stressed environments. Fertilizer use, particularly phosphorus, may also be extremely important in the yield of cereals in stressed Mediterranean environments (Cooper *et al.*, 1987). At the genotype level, a line with high yield potential will decrease its yield if subjected to stress but it may still yield more, up to a given stress level, than a genotype with a lower yield potential (Osmanzai *et al.*, 1987). Semidwarf wheat genotypes

produce more grain due to increased earliness (photoperiod insensitivity and low vernalization requirement) and a higher harvest index when compared to tall wheats (Laing and Fischer, 1977). This may not necessarily imply that semidwarfs tolerate severe (yields below 1 t/ha) stress better (Richards, 1982). Conversely, a genotype with a relatively long growing cycle may have stress resistance traits but will yield very poorly in a stressed, short-season environment. The quantification of stress resistance will continue to pose major problems as long as it is measured as grain yield under stress. Yield potential plus all the adaptive features of a genotype, whether they are related to a given stress resistance or not, influence yield under stress. Levitt (1980) distinguished between escape and resistance mechanisms through avoidance and tolerance when defining stress resistance. The problem is to find a way of measuring stress resistance which is independent of adaptational features unrelated to the stresses. Once this is done, specific traits could be unequivocally related to stress resistance.

The literature on plant water relations abounds with references to plant indicators of drought stress (Kozlowski, 1968; Turner and Kramer, 1980; Taylor *et al.*, 1983; Turner, 1986). The indicators consider either evapotranspiration, plant water status or accumulation of intermediary metabolites (Cooper *et al.* 1983; Pearson *et al.*, 1987) to either detect or quantify the water stress. It is unfortunate that none of them fulfill the basic requirement of being 'integrative' throughout the growing cycle. Many could provide an integrated value if continuous measurements are carried out but this is usually impractical and many times impossible (Acevedo, 1991a). The most integrative parameter indicative of plant stresses is biomass production. Since roots are difficult to measure, their assessment of biomass is generally limited to aboveground parts. However, a serious limitation is that biomass is difficult to measure on a large number of plots.

At present, in cereals, it appears unavoidable to define stress resistance by using grain yield. In a linear regression model, stress resistance would be equivalent to residual yield under stress after accounting for yield potential,

Table 24.4.   Attributes expected to lead to improved yield under stress in low rainfall Mediterranean environments

Purpose	Attribute	Screening tool
Maximize transpiration as a fraction of evapotranspiration	Fast ground cover	Visual
	Prostrate growth habit	Visual
	Good winter growth (early vigour)	Visual
Maximize photosynthesis under stress	High carboxylation efficiency	IRGA $\Delta^{13}C$
	Low non-stomatal effect on NP	IRGA $\Delta^{13}C$
	Stomatal adjustment to drought	Porometer Crop temperature
	Initial dark followed by light leaf colour	Visual
Maximize harvest index	High translocation of pre-anthesis assimilates to the ear	Chemical desiccation
Escape	Flowering date Short grain filling	Visual

phenology and experimental error (Bidinger *et al.*, 1987a,b). Other widely used yield-derived indices include the drought susceptibility index (S) of Fischer and Maurer, (1978) and the intercept of the joint regression analysis of genotype yield on site mean yield (Finlay and Wilkinson, 1963; Blum, 1983, 1988). It has been found, however, that yield derived stress indices have a relatively high genotype × environment interaction (Acevedo, 1991a) and that different indices give different rankings at the same site. The only index mentioned that is completely free of the influence of yield potential is the residual yield index of Bidinger *et al.* (1987b). It should be noted, that when the yield or yield-derived indices are used in the quantification of stress resistance, the results are relative to the nursery under study and should not be generalized.

24.4 SOURCES OF RESISTANCE TO ABIOTIC STRESSES

The identification of sources of resistance to abiotic stresses is a major task. They are expected to provide particular traits (genes) that would make the crop less vulnerable to stresses. The target environment, with its boundary conditions (major stresses, timing and intensity) affecting crop growth and development, needs to be defined. The sources of resistance require a high degree of physiological fitness to the target environment, such that all possible mechanisms of stress avoidance become available.

A conceptual model of yield determination, such as that developed by Fischer (1979, 1981) for wheat grown in Mediterranean environments, is useful for postulating stress-resistance traits that should be present in the source of stress resistance. As a first step, an ideotype based on the conceptual model can be formulated. This is presented in Table 24.4 for barley grown in the low rainfall areas of the Mediterranean environments of Northern Syria. Given the nature of these environments, growth should be maximized at times when the evaporative demand is at a minimum. This may require selecting for fast winter ground cover, rapid growth at low temperatures and frost resistance, such that intercepted radiation is increased at critical times when radiation is lowest and consequently photosynthesis per unit ground area is improved. By rapid establishment of ground cover, direct evaporation from the soil surface decreases, thus improving soil water infiltration or recharge, as well as evapotranspiration efficiency (Table 24.5). By

Table 24.5. Percentage difference between six barley (B) and four wheat (W) cultivars in ground cover and in the ratio of transpiration (T) to evapotranspiration (ET). The values are for the winter months of January and February, 40–80 days after emergence (after Acevedo and Ceccarelli, 1989)

Days after emergence	Ground cover (B-W)/B	T/ET (B-W)/B
40	16	63
50	17	52
60	12	46
70	10	20
80	9	14

having the attributes of fast ground cover, prostrate growth habit and good winter growth (early vigour), the genotypes should also increase the capture of limiting radiation and increase the efficiency of water use. Dark leaf colour will also favour light capturing by photosystem II in winter and light colour will avoid photoinhibition in spring (ICARDA, 1988b).

Low rainfall, rain-fed Mediterranean environments are characterized by crops growing on intermittent rainfall. Therefore, attributes that maximize photosynthesis under stress at the tissue level are needed (Table 24.4). Finally, the harvest index must be maximized. All stress escape mechanisms, such as appropriate flowering date and grain filling period, are also of major value to environmental fitness.

The value of the proposed traits needs assessment once the ideotype has been developed. A collection of genotypes should be assembled representing the variability in each of the traits and they should be grown in the target environments (Acevedo, 1991a) ideally under a stress intensity gradient. The correla-

tion of the traits with a measure of stress resistance should be observed. Those traits that are correlated with stress resistance are selected for further work. Table 24.6 shows some

Table 24.6. Broad sense heritablility of traits correlated with yield under stress (+,) = sign of correlation) in 20 genotypes of bread wheat at Breda and Bouider, Northern Syria

Trait	Sign of correlation	Heritability
Grain yield		0.28
No. fertile spikes	(+)	0.36
Grains/m^2	(+)	0.44
Grains/spike	(+)	0.24
Harvest index	(+)	0.48
Straw yield	(+)	0.41
Plant height	(+)	0.53
Peduncle length	(+)	0.58
Days to heading	(−)	0.93
Days to maturity	(−)	0.79
Pale leaf colour after anthesis	(+)	0.61
Growth vigour 5-leaf	(+)	0.43

identified, useful traits for bread wheat grown in stressed Mediterranean environments (Acevedo, 1991a).

The correlation analyses do not necessarily imply cause–effect relationships. It is thus imperative to verify the utility of the traits. This verification is best done by comparing near-isogenic lines differing in the trait under consideration. Their development, however, requires substantial breeding effort. A more pragmatic approach was proposed by Acevedo and Ceccarelli (1989) who suggested crossing genotypes diverging in four or five of the most important stress resistance traits. Individual F_2 plants are harvested and the F_3 seed grown and divergent selection carried out for the traits under study. Contrasts can be made for single traits in all possible combinations with the other traits, similarly for two or more traits. The traits are considered to be valuable, alone or in combination, if they increase the selection efficiency when compared to yield alone as the selection criteria. By regressing the values of F_4

on the F_3 families, estimates of realized herit-abilities are generated. Final yield testing is done in F_5 or F_6 using contrasting environ-ments. In the process just outlined, sources of stress resistance as well as stress resistant traits are identified.

24.5 SCREENING TESTS

Physiologists and breeders have been continu-ously confronted with the need to develop screening tests to identify stress-resistance traits as an aid to selection. Few of the many screening tests proposed have been adopted by breeders. This is probably to be expected since most screening tests discriminate a single attri-bute at a time. As discussed above, variation in stress resistance is likely to be due to the intregrative effects of many attributes and their interaction, most of them with small effects. The expression of the attributes is likely to vary with the degree and combination of stresses, age of the plant and tissue, and with the environment.

If screening tests are sought, the traits to be selected for must satisfy the following criteria: (1) there should be genetic variation within the germplasm pool for the trait under consider-ation; (2) the traits should have greater herit-abilities than yield itself; (3) they should be correlated with yield under stress or with a yield-based stress resistance index; (4) prefer-ably, they should be causally related to yield; (5) they should be economical, easy and rapid to assess.

Genetic variation has been observed in wheat for a number of adaptive traits related to resistance to environmental stress. They include maintenance of relatively higher leaf potential under soil water deficits, osmotic adjustment, tolerance to stress in plant or organ growth rate, plant recovery upon rehydration, tolerance in the photosynthetic system or its components, tolerance in enzyme activities, tolerance in translocation, stability of the cellu-lar membranes, proline accumulation, root growth and plant developmental or morpho-logical attributes such as leaf size, leaf area per plant, leaf orientation, tiller survival, epicutic-

ular wax content, and organ pubescence. With the probable exception of osmotic adjustment (Morgan, 1989) and stability of cellular mem-branes (Blum, 1988), most of this information will be of limited value to breeding work unless routine screening techniques for large numbers of genotypes become available and the relation-ships between crop performance under stress and the adaptive traits are clearly established.

24.6 USE OF SIMULATION MODELS IN BREEDING FOR STRESS RESISTANCE

Section 24.3.1 has emphasised the difficulties involved in the characterization of environmen-tal variability and the associated crop responses to environmental stress. Until now, the most powerful breeding approach to assess the value of certain genotypes or traits was the multi-year, multi-location fields trials followed by statistical analyses (Chapter 22). The large genotype × environment interaction of stressed environments (Rosielle and Hamblin, 1981) makes this empirical approach uncertain and relatively inefficient. Recent advances in the development of crop simulation models strongly suggest that they could be powerful tools in breeding programmes for resistance to environmental stress (Fereres, 1987).

At present, two applications of simulation models may be of value. On the one hand, the value of a trait may be assessed by simulating the presence or absence of such a trait (or of a certain level of expression) over a number of years in the relevant environment. Jordan *et al.* (1983) were among the first to evaluate the potential value of traits using a model. They assessed the role of osmotic adjustment against deep rooting in the design of specific breeding strategies for grain sorghum under drought stress. As simulation models become more sophisticated and mechanistic in nature by incorporating more basic physiology (Loomis *et al.*, 1990), it will be possible to test the value of more traits in a more realistic manner. Models will be particularly useful to evaluate several strategies based on combination of traits as affected by stress timing and severity. Alternative ideotypes could be evaluated to narrow down the number of potential options that need to be pursued in a breeding programme for stress resistance. On the other hand, simulation models may be used to broadly define the plant type that best fits a specific environment (Fereres, 1987) as well as to optimize agronomic practices such as planting date. Aspects such as optimum flowering date to avoid frost, earliness to escape drought, season length, etc., all in relation to a specific, marginal environment are now best evaluated in the crops where realistic models exist by simulating crop behaviour over a large number of years. Cumulative probability graphs of the various alternatives may be developed to quantify the performance of hypothetical cultivars in relation to environmental stress and climatic risk (Muchow *et al.*, 1991).

Although the above approaches can also be used in breeding for yield potential, simulation models are no doubt more important in the efforts to improve plant breeding for specific, suboptimal environments and their use should greatly improve the efficiency of such programmes in the future.

24.7 SELECTION FOR DROUGHT RESISTANCE

Traits such as dehydration avoidance and tolerance, have been found to be positively associated with yield under stress across genotypes of wheat (Keim and Kronstad, 1981; Sojka *et al.*, 1981) and barley (Acevedo and Ceccarelli, 1989; Acevedo *et al.*, 1991a). Dehydration avoidance is interpreted as the ability of genotypes to maintain a higher leaf water potential when grown under soil water deficits. Several traits contribute to dehydration avoidance. All those contributing to the improvement of the crop water status should be included, such as leaf rolling, best adapted root system, increased pubescence of the aerial organs, increased reflectance of incoming solar radiation, increased heat dissipation through decreased boundary layer resistance at the organ level (narrow leaves, presence of awns), etc. In nature, a better water balance is associated with

a higher proportion of energy dissipated as latent heat and hence a lower canopy temperature. This has prompted scientists to develop methods for fast crop temperature assessment (Blum *et al.*, 1978, 1982) and the method is being used in breeding programmes such as in wheat (Blum, 1988; ICARDA, 1988a), maize (Bolaños and Edmeades 1989) and others.

An important form of drought stress tolerance is the tolerance to post-anthesis stress. In wheat, as in other cereals, grain filling depends partly on actual photosynthesis during this period and partly on carbohydrates stored during pre-anthesis which are translocated from vegetative plant parts. Under conditions of terminal drought (and heat) stress, net photosynthesis decreases significantly in wheat (Acevedo, 1991b) and the proportion of translocation of stored soluble carbohydrates as a source for grain filling becomes larger (Austin *et al.*, 1977, 1980b; Richards and Townley-Smith, 1987). Values of 40–60% of final grain weight originating from stored pre-anthesis have been reported by these authors. Genetic variation exists within cereal crop species in the ability to sustain kernel growth by remobilization of reserves (Austin *et al.*, 1977; Acevedo and Ceccarelli, 1989). Blum *et al.* (1983) developed a technique to screen for this trait in wheat. They sprayed well watered plants with magnesium chlorate at the initial part of the linear phase of kernel growth (approximately 14 days after anthesis), when final cell number of the kernel has been determined. The desiccant kills all photosynthetic tissue including leaves, leaf sheaths, glumes and awns. The kernel weight at harvest of treated plots is then compared with that of non-treated, control plots. The technique is being applied experimentally in Australia using potassium iodide as a desiccant (Turner and Nicholas, 1987; N.C. Turner, personal communication).

Dehydration tolerance is related to cellular and subcellular processes. The stability of cell membranes under stress is central to cell performance. Measurements of membrane stability can be easily done with the electrolyte leakage test (Sullivan and Eastin, 1974) after subjecting plant tissue to stress. Chlorophyll fluorescence is also being used for this purpose.

It is difficult, however, to relate this type of test to plant production.

24.8 HEAT RESISTANCE

High temperature is an important production constraint in many tropical and subtropical environments. It may occur throughout the growing season, as in lowland tropics, or as terminal stress affecting the crop during grain filling in Mediterranean environments. Water and heat stresses are linked through the energy balance of plant organs. Excess net radiation may be predominantly dissipated via transpiration, such that the plant canopy may be several degrees below ambient temperature (the most important heat avoiding mechanism). If there is a water shortage and/or a low vapour pressure deficit, the plant organs will increase their temperature even above ambient.

Selection of heat resistant cereal genotypes is very much related to the ability of the genotypes to maintain the duration of growth periods; between germination to emergence (E), emergence to double ridges (GS1), double ridges to anthesis (GS2) and anthesis to maturity (GS3) (Shpiler and Blum, 1986; O'Toole and Stockle, 1991). Above normal temperatures speed up the growth and development phases, usually negatively affecting crop establishment, leaf area growth and development, ground cover by the crop, radiation interception and photosynthesis per unit ground area. The sensitivity to high temperatures varies with growth stages. Crop components directly related to yield, such as spike number, potential number of spikelets per spike, spike weight at anthesis, potential grain number and individual grain weight, are all decreased.

Factors and agronomic practices that would increase stand establishment and growth, or else decrease developmental rates would appear to favour grain yield at high temperatures. Escape mechanisms such as choosing a planting date that would produce anthesis during the coolest growing month (Fischer, 1985) helps in achieving higher yields as it does having genotypes that keep a lower temperature either by high transpiration rate and/or

reflective properties of the canopy (e.g. glaucousness of stems, leaves, and spikes) (Johnson et al., 1983).

A working hypothesis for identifying traits producing heat stress resistant genotypes was elaborated by Acevedo et al. (1991b). The proposed traits are presented in Table 24.7. The following conclusions were drawn by Acevedo et al. (1991b) for growing and improving wheat in hot environments and probably apply to most winter cereals:

1. Wheat should reach anthesis during a cool period.
2. Full ground cover and hence light interception should be attained early in the season, ideally one month before anthesis. Genotypic traits like fast leaf emergence (reduced phyllochron interval) and high leaf number are desirable. The terminal spikelet to anthesis stage should occur at a time when the leaf area index is high and the 'photothermal quotient', mean solar radiation/ (mean temperature – 4.5), is at a peak value for the season.
3. Some vernalization requirement should prolong the vegetative period and increase radiation interception in the early growth stages.
4. High transpirational cooling is desirable to avoid heat stress by lowering the canopy temperature.
5. Improved energy balance of heated organs via organ glaucousness appears desirable.
6. Longer grain filling period may help to maintain grain weight under heat stress.
7. Deep root systems may help in avoiding heat if there is moisture stored at depth in the soil profile.

As in the case of drought stress, sound agronomic practices are important to achieve high yields of winter crops grown under supraoptimal temperatures (Rawson, 1988).

Other than speeding up phenological processes, high temperatures have deleterious effects on photosynthesis, respiration and reproductive processes such as flower abscission, sterility and poor fruit set. At the cellular level, membrane stability at higher temperature

Table 24.7.   Physiological/morphological traits of crops of potential use in screening for resistance to high temperature stress[a]

Development stage	Trait	Screening tool	Screening environment
Emergence	Crop establishment	Plant (no./m²)	Growth chamber Early heat stress
Fifth leaf stage on the main shoot	Crop growth Radiation interception and photosynthesis/ ground area Long E-DR period Profuse root system	Visual Cover and vigour scores Days to DR (5 leaves) Low canopy temperature	Early heat stress Heat stress throughout
Anthesis	Low vernalization Timing of anthesis Radiation interception Potential grain number Stability of GS2 duration	Days to anthesis Ground cover Spike dry weight Duration DR-anthesis	Early heat stress Heat stress throughout Terminal heat stress
Yellow Spike	Photosynthesis/area Grain filling duration Grain filling rate Grain weight Good partitioning to the spike	Stay green DM-DA GW/ (DM-DA) 1000 kernel weight Harvest index	Terminal heat stress Heat stress throughout

[a] E = emergence; DR = double ridge stage; GS2 = growth stage 2; DA = days to anthesis, DM = days to maturity; GW = grain weight.

has been recognized as an important attribute which increases heat tolerance of genotypes. Conductometric measurements of solute leakage have been used and are popular to test for membrane thermostability as one factor affecting heat tolerance (Sullivan *et al.*, 1977; Oneoueme, 1979; Shanahan *et al.*, 1990). It would appear that once highly adapted material has been found for a given high temperature stress environment, the membrane stability test may be useful in detecting heat tolerant genotypes.

24.9 CONCLUSION AND PROSPECTS FOR BREEDING FOR ABIOTIC STRESSES

Abiotic stress resistance, as an important part of environmental fitness is bound to be location specific; thus, the appropriate traits should be assessed and verified in the target environment.

Consequently, environmental characterization using newly developed quantitative techniques is a prerequisite for any successful breeding programme. Except for rare exceptions, individual traits appear to be of limited value since resistance is governed by a complex of traits. Screening techniques which are field oriented are preferable except for a very limited number of cases such as membrane thermostability and osmotic adjustment. Divergent selection is suggested to establish the appropriate combination of traits for a given environment. The use of crop simulation models may aid ideotype definition, trait evaluation and assessment of alternative breeding strategies for stress resistance. Modellers will make important contributions to improving the selection efficiency of future breeding programmes.

Due to its complexity, we have not considered resistance to salinity stress. However, from a production standpoint, we support the argument of Richards (1983) that breeding for yield potential should have greater benefits in

salt-affected soils than breeding for salt resistance *per se*.

Breeding for abiotic stress resistance is
becoming more promising with the recognition
that selection should generally take place in the
stressed target environment, that stress resistance is confounded with environmental fitness
and that it is related to narrow adaptation.

REFERENCES

Acevedo, E. (1991a), Improvement of winter cereal
crops in Mediterranean environments. Use of
yield, morphological and physiological traits. In
*Physiology-Breeding of Winter Cereals for Stressed
Mediterranean Environments*, Acevedo, E.,
Conesa, A.P., Monneveux, P. and Srivastava,
J.P. (eds), INRA, Paris, pp. 273–305.

Acevedo, E. (1991b), Morphophysiological traits of
adaptation of cereals to Mediterranean environments. In *Improvement and Management of Winter
Cereals under Temperature, Drought and Salinity
Stresses*, Acevedo, E., Fereres, E., Gimenez, C.
and Srivastava, J.P. (eds), INIA, Madrid, pp. 85–
96.

Acevedo, E. and Ceccarelli, S. (1989), Role of
physiologist-breeder in a breeding program for
drought resistance conditions. In *Drought Resistance in Cereals*, Baker, F.W.G. (ed.), C.A.B.
International, Wallingford, pp. 117–139.

Acevedo, E., Craufurd, P.Q., Austin, R.B. and Perez-
Marco, P. (1991a), Traits associated with high
yield in barley in low-rainfall environments. *J.
Agr. Sci. Camb.*, **116**, 23–36.

Acevedo, E., Nachit, M. and Ortiz-Ferrara, G.
(1991b), Effects of heat stress on wheat and
possible selection tools for use in breeding
tolerance. In *Wheat for the Nontraditional, Warm
Areas*, CIMMYT, Mexico, D.F., pp. 401–421.

Acevedo, E., Perez-Marco, P. and van Oosterom, E.
(1990), Physiology of yield of wheat and barley
in stressed rainfed Mediterranean environments.
In *Proc. of the International Congress of Plant
Physiology*, Sinha, S.K., Sane, P.V., Bhangava,
S.C. and Agrawal, P.K. (eds), Society for
Plant Physiology and Biochemistry/Indian Agricultural Research Institute, New Delhi, pp. 117–
127.

Angus, J.F. (1991), The evolution of methods for
quantifying risk in water limited environments.
In *Climatic Risk in Crop Production: Models and
Management for the Semiarid Tropics and subtropics*,
Muchow, R.C. and Bellamy, J.A. (eds), C.A.B.
International, Wallingford, pp. 39–53.

Austin, R.B. (1990), Prospects for genetically increas-
ing the photosynthetic capacity of crops. In
*Prespectives in Biochemical and Genetic Regulation of
Photosynthesis*, Zelitch, I. (ed.), Alan R. Liss Inc.,
New York, pp. 395–409.

Austin, R.B., Bingham, J., Blackwell, R.D., Evans,
L.T., Ford, M.A., Morgan, C.L. and Taylor, M.
(1980a), Genetic improvements in winter wheat
yields since 1900 and associated physiological
changes. *J. Agr. Sci. Camb.*, **94**, 675–689.

Austin, R.B., Edrich, J.A., Ford, M.A. and Blackwell,
R.D. (1977), The fate of dry matter, carbohydrates and ^{14}C loss from leaves and stems of
wheat during grain filling. *Ann. Bot.*, **41**, 1309–
1321.

Austin, R.B., Morgan, R.B., Ford, M.A. and
Blackwell, R.D. (1980b), Contributions to grain
yield from pre-anthesis assimilation in tall and
dwarf barley genotypes in two contrasting seasons. *Ann. Bot.*, **45**, 309–319.

Bidinger, F.R., Mahalakshmi, V. and Rao, G.D.P.
(1987a), Assessment of drought resistance in
pearl millet (*Pennisetum americanum* (L.) Leeke). I.
Factors affecting yield under stress. *Aust. J. Agr.
Res.*, **38**, 37–48.

Bidinger, F.R., Mahalakshmi, Y. and Rao, G.D.P.
(1987b), Assessment of drought resistance in
pearl millet (*Pennisetum americanum* (L.) Leeke).
II. Estimation of genotype response to stress.
Aust. J. Agr. Res., **38**, 49–59.

Blum, A. (1979), Genetic improvement of drought
resistance in plants: a case of *Sorghum*. In *Stress
Physiology in Crop Plants*, Munsell, H. and
Staples, R.C. (eds), Wiley Interscience, New
York, pp. 242–445.

Blum, A. (1983), Genetic and physiological relationship in plant breeding for drought resistance.
Agr. Water Mgmt., **7**, 195–205.

Blum, A. (1988), *Plant Breeding for Stress Environments*,
CRC Press, Boca Ratón.

Blum, A., Mayer, J. and Golzan, G. (1982), Infrared
thermal sensing of plant canopies as a screening
technique for dehydration avoidance in wheat.
Field Crop. Res., **5**, 137–146.

Blum, A., Poiarkova, H., Golam, G. and Mayer, J.
(1983), Chemical dessication of wheat plants as a
simulator of post-anthesis stress. I. Effects on
translocation and kernel growth. *Field Crop. Res.*,
6, 51–58.

Blum, A., Schertz, K.F., Toler, R.W., Welch, R.I., Rosenow, D.T., Johnson, J.W. and Clark, L.E. (1978), Selection for drought avoidance in sorghum using aerial infrared photography. *Agron. J.*, **70**, 472–477.

Bolaños, J. and Edmeades, G.O. (1989), Cambios en la población tuxpeño después de ocho ciclos de mejoramiento, para resistencia a sequía. *Meeting at San Pedro Sula*, 2–9 April, 1989, PCMMA.

Boyer, J.S. (1982), Plant productivity and environment. *Science*, **218**, 443–448.

Ceccarelli, S. (1991), Selection for specific environments or wide adaptability. In *Improvement and Management of Winter Cereals under Temperature, Drought and Salinity Stress*, Acevedo, E., Fereres, E., Gimenez, C. and Srivastava, J.P. (eds), INIA, Madrid, pp. 227–237.

Cooper, P.J.M., Gregory, P.J., Tully, D. and Harris, H.C. (1987), Improving water use efficiency of annual crops in the rainfed farming systems of West Asia and North Africa. *Expl. Agr.*, **23**, 113–158.

Cooper, P.J.M., Keatinge, J.H.D. and Hughes, G. (1983), Crop evapotranspiration – a technique for calculation of its components by field measurements. *Field Crop. Res.*, **7**, 299–312.

Edmeades, G.O., Bolaños, J., Laffite, H.R., Rajaram, S., Pfeiffer, W. and Fischer, R.A. (1989), Traditional approaches to breeding for drought resistance in cereals. In *Drought Resistance in Cereals*, Baker, F.W.G. (ed.), C.A.B. International, Wallingford, pp. 27–52.

Evans, L.T. (1984), Physiological aspects of varietal improvement. In *Gene Manipulation in Plant Improvement*, Gustafson, J.P. (ed.), Plenum, New York, pp. 121–146.

Evans, L.T. (1988), Opportunities for increasing the yield potential of wheat. In *The Future Development of Maize and Wheat in the Third World*, CIMMYT, Mexico, D.F., pp. 79–93.

Evans, L.T. (1990), Assimilation, allocation, explanation, extrapolation. In *Theoretical Production Ecology: Reflections and Prospects*, Rabbinge, R., Goudriaan, J., van Keulen, H., Penning de Vries, F.W.T. and van Laar, H.H. (eds), Pudoc, Wageningen, pp. 77–87.

Fereres, E. (1987), Responses to Water Deficits in Relation to Breeding for Drought Resistance. In *Drought Tolerance in Winter Cereals*, Srivastava, J.P., Porceddu, E., Acevedo, E. and Varma, S. (eds), John Wiley, Chichester, pp. 263–273.

Finlay, K.W. and Wilkinson, G.N. (1963), The analysis of adaptation in a plant-breeding programme. *Aust. J. Agr. Res.*, **14**, 742–754.

Fischer, R.A. (1979), Growth and water limitations to dryland wheat yield in Australia: a physiological framework. *J. Aust. Inst. Agr. Sci.*, **45**, 83–94.

Fischer, R.A. (1981), Optimizing the use of water and nitrogen through breeding of crops. *Plant Soil*, **58**, 249–278.

Fischer, R.A. (1985), Physiological limitation to producing wheat in semitropical and tropical environments and possible selection criteria. In *Wheats for More Tropical Environments. A Proceeding of the International Symposium*, Villareal, L. (ed.), CIMMYT, Mexico, D.F., pp. 209–230.

Fischer, R.A. and Maurer, R. (1978), Drought resistance in spring wheat cultivars. I. Grain yield responses. *Aust. J. Agr. Res.*, **29**, 897–912.

Fischer, R.A. and Wood, J.R. (1979), Drought resistance in spring wheat cultivars. III. Yield association with morphophysiological traits. *Aust. J. Agr. Res.*, **30**, 1001–1020.

ICARDA (1988a), *Cereal Improvement Program, Annual Report*, ICARDA, Aleppo, pp. 39–53.

ICARDA (1988b), *Cereal Improvement Program, Annual Report*, ICARDA, Aleppo, pp. 86–113.

Johnson, D.A., Richards, R.A. and Turner, N.C. (1983), Yield, water relations, gas exchange and surface reflectance of near isogenic wheat lines differing in glaucousness. *Crop Sci.*, **23**, 318–325.

Jordan, W.R., Dugas, W.A. jr and Shouse, P.J. (1983), Strategies for crop improvement for drought-prone regions. In *Plant Production and Management under Drought Conditions*, Stone, J.F. and Willbis, W.O. (eds), Elsevier Science Publisher, Amsterdam.

Keim, D.L. and Kronstad, W.E. (1981), Drought response of winter wheat cultivars grown under field stress conditions. *Crop Sci.*, **21**, 11–15.

Keys, A.J. and Whittingham, C.P. (1981), Photorespiratory carbon dioxide loss. In *Physiological Processes Limiting Plant Productivity*, Johnson, C.B. (ed.), Butterworths, London.

Kozlowski, T.T. (1968), *Water Deficit and Plant Growth*, Vol. I, Academic Press, New York.

Laing, D.R. and Fischer, R.A. (1977), Adaptation of semidwarf cultivars to rainfed conditions. *Euphytica*, **26**, 129–139.

Levitt, J. (1980), Stress terminology. In *Adaptation of Plants to Water and High Temperature Stress*, Turner, N.C. and Kramer, P.J. (eds), John Wiley, New York, pp. 437–439.

Loomis, R.S., Luo, Y. and Kooman, P.L. (1990), Integration of activity in the higher plant. In *Theoretical Production, Ecology: Reflections and*

Prospects, Rabbinge, R., Groudriaan, J., van Keulen, H., Penning de Vries, F.W.T and van Laar, H.H. (eds), Pudoc, Wageningen, pp. 105–124.

Ludlow, M.M. and Muchow, R.C. (1988), Critical evaluation of the possibilities for modifying crops for high production per unit precipitation. In *Drought Research Priorities for the Dryland Tropics*, Bidinger, F.R. and Johanson, C. (eds), ICRISAT, Hyderabad, pp. 179–211.

Mahon, J.D. (1990), Photosynthesis and crop productivity. In *Perspectives in Biochemical and Genetic Regulation of Photosynthesis*, Zelitch, I. (ed.), Alan R. Liss Inc., New York, pp. 379–394.

McWilliam, J.R. (1989), The dimensions of drought. In *Drought Resistance in Cereals*, Baker, F.W.G. (ed.), C.A.B. International, Wallingford, pp. 1–111.

Monteith, J.L. (1977), Climate and the efficiency of crop production in Britain. *Philos. Trans. R. Soc. London B*, **281**, 277–294.

Morgan, C.L. and Austin, R.B. (1983), Respiratory loss of recently assimilated carbon in wheat. *Ann. Bot.*, **51**, 85–95.

Morgan, J.M. (1989), Physiological traits of drought resistance. In *Drought Resistance in Cereals*, Baker, F.W.G. (ed.), C.A.B. International, Wallingford, pp. 53–64.

Muchow, R.C., Hammer, G.L. and Carberry, P.S. (1991), Optimising Crop and Cultivar Selection in Response to Climatic Risk. In *Climatic Risk in Crop Production: Models and Management for the Semiarid Tropics and Subtropics*, Muchow, R.C. and Bellamy, J.A. (eds), C.A.B. International, Wallingford, pp. 235–262.

Nass, H.G. and Sterling, J.D.E. (1981), Comparison of tests characterizing varieties of barley and wheat for moisture stress resistance. *Can. J. Plant Sci.*, **61**, 283–289.

Nelson, C.J. (1988), Genetic associations between photosynthetic characteristics and yield: review of the evidence. *Plant Physiol Biochem.*, **26**, 243–254.

Oneoueme, I.C. (1979), Rapid, plant conserving estimation of heat tolerance in plants. *J. Agr. Sci.*, **92**, 527–536.

Osmanzai, M., Rajaram, S. and Knapp, E.B. (1987), Breeding for moisture-stresses areas. In *Drought Tolerance in Winter Cereals*, Srivastava, J.P., Porceddu, E., Acevedo, E. and Varma, S. (eds), John Wiley, Chichester, pp. 151–161.

O'Toole J.C. and Stockle, C.D. (1991), The role of conceptual and simulation modeling in plant breeding. In *Improvement and Management of*

Winter Cereals under Temperature, Drought and Salinity Stresses, Acevedo, E., Fereres, E., Gimenez, C. and Srivastava, J.P. (eds), INIA, Madrid, pp. 205–225.

Pearson, J., Smirnoff, N., Stewart, G.R. and Turner, L.B. (1987), Nitrogen metabolism in relation to water stress. In *Drought Tolerance in Winter Cereals*, Srivastava, J.P., Porceddu, E., Acevedo, E. and Varma, S. (eds), John Wiley and Sons, Chichester, pp. 241–262.

Rawson, H.M. (1988), Effects of high temperature on the development and yield of wheat and practices to reduce deleterious effects. In *Wheat Production Constraints in Tropical Environments*, Klatt, A.R. (ed.), CIMMYT, Mexico, D.F., pp. 44–62.

Richards, R.A. (1982), Breeding and selecting for drought resistance in wheat. In *Drought Resistance in Crops with Emphasis on Rice*, IRRI, Los Baños, pp. 303–316.

Richards, R.A. (1983), Should selection for yield in wheat be done on saline or non-saline soils? *Euphytica*, **32**, 431–438.

Richards, R.A. and Townley-Smith, T.F. (1987), Variation in leaf area development and its effect on water use, yield and harvest index of droughted wheat. *Aust. J. Agr. Res.*, **38**, 983–992.

Richardson, C.W. (1984), *WGEN: A Model for Generating Daily Weather Variables*, Bull. 8, USDA-ARS, Washington DC.

Rosielle, A.A. and Hamblin, J. (1981), Theorical aspects of selection for yield in a stress and non-stress environments. *Crop Sci.*, **21**, 943–946.

Schultze, E.D. (1988), Adaptation mechanisms of non cultivated arid-zone plants: useful lessons for agriculture? In *Drought Research Priorities for the Dryland Tropics*, Bidinger, F.R. and Johanson, C. (eds), ICRISAT, Hyderabad, pp. 159–177.

Shanahan, J.F., Edwards, I.B., Quick, J.S. and Fenwick, J.R. (1990), Membrane thermostability and heat tolerance of spring wheat. *Crop Sci.*, **30**, 247–251.

Shorter, R., Lawn, R.J. and Hammer, G.L. (1991), Improving genotypic adaptation in crops – a role for breeders, physiologists and modellers. *Exp. Agr.*, **27**, 155–175.

Shpiler, L. and Blum, A. (1986), Differential reaction of wheat cultivars to hot environments. *Euphytica*, **35**, 483–492.

Sojka, R.E., Stolzy, L.H. and Fischer, R.A. (1981), Seasonal drought response of selected wheat cultivars. *Agr. J.*, **73**, 838–844.

Srivastava, J.P., Porceddu, E., Acevedo, E. and Varma, S. (eds) (1987), *Drought Tolerance in Winter Cereals*, John Wiley, Chichester.

Sullivan, C.Y. and Eastin, J.D. (1974), Plant physiological responses to water stress. *Agr. Meteor.*, **14**, 113–127.

Sullivan, C.Y., Norcio, N.V. and Eastin, J.D. (1977), Plant responses to high temperature. In *Genetic Diversity in Plants*, Amir, M., Askel, R. and van Borstel, R.C. (eds), Plenum Press, New York, p. 301.

Taylor, H.M., Jordan, W.R. and Sinclair, T.R. (1983), *Limitation to Efficient Water Use in Crop Production*. ASA, Madison.

Turner, N.C. (1986), Crop water deficits: A decade of progress. *Adv. Agron.* **40**, 1–51.

Turner, N.C. and Kramer, P.J. (eds) (1980), *Adaptation of Plants to Water and High Temperature Stress*. John Wiley, New York.

Turner, N.C. and Nicholas, M.E. (1987), Drought resistance of wheat for light-textured soils in Mediterranean climate. In *Drought Tolerance in Winter Cereals*, Srivastava, J.P., Porceddu, E., Acevedo, E. and Varma, S. (eds), John Wiley, Chichester, pp. 203–216.

25

Resistance to parasites

R.E. Niks, P.R. Ellis and J.E. Parlevliet

25.1 INTRODUCTION

There is an appreciable deficit between realized and potential crop yields. The record yield of wheat in the USA as of 1975 was 14.5 t/ha, whereas the average yield achieved was only 1.9 t/ha. A similar deficit exists for all agricultural and horticultural crops (Boyer, 1982).

Factors responsible for the yield losses are called stress factors. They can be of an abiotic nature (drought, cold, poor soil structure or fertility) or of a biotic nature (pathogens, pests, weeds). This chapter deals with the biotic stress factors caused by parasites (pathogens and pests).

The losses due to parasites are considerable. Cramer (1967) estimated that economic losses for the most important agricultural crops averaged 11.8% for diseases and 12.2% for pests relative to potential economic value. Boyer (1982) estimated for the USA that yield reductions (kg/ha) were 4.1% due to diseases and 2.6% due to pests. His values were expressed relative to record yields. Epidemics or pest outbreaks were responsible for the disappearance of whole crops in certain regions (Russell, 1978; James, 1981; Littlefield, 1981; Robinson, 1987). Crops may be protected against parasites by cultural practices, biological control agents, application of chemicals (pesticides) and the introduction of genetic resistance into the crop. This chapter discusses the most important

Plant Breeding: Principles and prospects. Edited by M.D. Hayward, N.O. Bosemark and I. Romagosa. Published in 1993 by Chapman & Hall, London. ISBN 0 412 43390 7.

aspects of breeding for genetic resistance to pathogens and pests.

25.2 BENEFITS AND DRAWBACKS OF RESISTANCE *VERSUS* CHEMICAL PROTECTION

The use of resistance in crop cultivars has many advantages over the use of pesticides. First, there is the obvious economic saving of the costs of pesticide application. The seed of resistant cultivars usually costs the farmer no more than that of susceptible cultivars. Even if the resistance does not protect the crop completely, partial resistance can lead to a marked reduction in the dose of pesticide required to provide satisfactory control. This has been clearly demonstrated in carrot to the carrot fly (*Psila rosae*) (Thompson *et al.*, 1980), and in potato to *Phytophthora infestans* (Fry, 1978).

Schalk and Ratcliffe (1976) estimated that without resistance in cultivars a 37% increase in the use of insecticide would be required to maintain the existing level of insect control. In France, the savings from the use of rootstocks with resistance to the grape phylloxera, *Daktulosphaira vitifoliae*, must be considerable. The rootstocks are reputed to have saved the European wine industry from total collapse. Luginbill (1969) estimated the cost of the research and development needed to produce cultivars resistant to the Hessian fly (*Mayetiola destructor*), wheat stem sawfly (*Cephus cinctus*), spotted alfalfa aphid (*Therioaphis maculata*) and European corn borer (*Ostrinia nubilialis*) over ten years to be US$ 9.3 million. The savings were estimated to total US$ 308 million annually.

This means a return on the investment in research and development of 300:1. Second, there is the great advantage of safety for the grower and consumer, because of reduced contact risks and reduced residues, respectively. Third, the use of plant resistance is usually compatible with biological pest control measures, whereas chemical protection usually precludes such environmentally desirable control methods.

The impact that resistant cultivars have on the environment is minimal in terms of adverse effects, whereas it can reduce the pollution caused by pesticides considerably. Extensive planting of resistant cultivars may have a considerable impact on the populations of certain pests. In the USA the widespread use of wheat cultivars with resistance to the Hessian fly reduced the population of this insect to only 1% of its former levels, and in some areas the fly was difficult to find (Luginbill, 1969).

The drawbacks of resistance are few. One important problem is that it takes a long time and a lot of effort to introduce resistance into a crop. The breeder must produce a cultivar which is acceptable to both grower and consumer. The cultivar should not only be resistant, but also produce a high yielding good quality crop. Moreover, most genes for resistance are effective against only one parasite species. Pesticides are usually effective against a broader spectrum of pests. The efforts to introduce and to maintain genes for resistance need to be repeated for each cultivar to be produced and for each target parasite species. Resistance and chemical protection have one problem in common, namely in many cases 'limited durability' of the effectiveness due to genetic adaptation by the pest or pathogen.

25.3 DEFENCE MECHANISMS IN PLANTS

In the plant kingdom there is a great diversity of strategies and mechanisms of defence against potential consumers. The defence mechanisms may be classified into three groups, namely, avoidance, resistance and tolerance (Parlevliet, 1981a).

25.3.1 Avoidance

Avoidance, described by entomologists as antixenosis, reduces the probability of contact between potential consumers and plants. It acts 'before' establishment of intimate contact between host and parasite, including feeding and oviposition. Avoidance mechanisms are almost exclusively directed against pests, particularly insects, mites and vertebrates. Pests combine the use of sense organs and the capacity to move actively in search of food substrates. They express complex behavioural patterns, governed by visual and chemical stimuli. As a consequence, there is a premium for plants to appear and smell as unattractive as possible. Examples of avoidance mechanisms include: (1) deviating leaf colour (cabbage/ cabbage aphid, *Brevicoryne brassicae*, and cabbage root fly, *Delia radicum*, Radcliffe and Chapman (1965) and Prokopy *et al.* (1983)); (2) pubescence (cotton/cotton boll weevil, *Heliothis zea*); (3) phenology (radish cultivars/cabbage root fly, Ellis *et al.* (1979)); (4) repellent odour (terpenes in cotton/cotton boll weevil, Hedin *et al.*, (1974)).

Avoidance to pathogens is rare. An example is the erect leaf habit in cereals, which reduces the deposition of leaf rust inoculum in comparison to crops with horizontal leaf habit (Gasowski, 1990). In the literature the term non-preference is also used as a synonym of avoidance. We prefer, however to use non-preference as a term to indicate the behavioural characteristic of the animal which determines its choice of plant.

25.3.2 Resistance

Resistance is the ability of the plant to reduce the growth and/or development of the parasite 'after' contact has been initiated or established. Resistance to pests, which is expressed as higher mortality and reduced reproduction is often called antibiosis. In this chapter we use the term resistance.

Resistance may be complete, when the growth and development of the parasites are blocked completely. Resistance may also be a quantitative character. If, after a certain exposure, plant genotypes differ in level of infection, despite the same initial inoculum, the genotypes with the lowest infection are the most resistant (= the least susceptible).

Mechanisms of resistance are at least as diverse as those of avoidance. The resistance mechanism may be present in the host regardless of the presence or absence of the parasite (passive), or may be activated as soon as the parasite tries to establish a parasitic relationship (active). In either case the resistance may be of a biochemical, physiological or of an anatomical/morphological nature. An example of a passive resistance of a biochemical nature is the presence of catechol and protocatechuic acid in the outer scales of onion bulbs which confer protection against the smudge disease, *Coletotrichum circinans* (Dixon, 1981). Passive resistance of an anatomical nature is examplified by the solidity of wheat stems which affects the success of the wheat stem sawfly (O'Keefe *et al.*, 1960).

Phytoalexin production in cells surrounding the site of injury or infection is an example of active biochemical resistance (Keen, 1986; Bailey, 1987; Gross, 1987). Phytoalexins are low-molecular-weight antimicrobial compounds which are more or less specific for the plant in which they are formed, but aspecific in their antimicrobial activity. Most plant species seem to be able to produce such phytoalexins, but they have rarely been demonstrated in members of the Gramineae.

Hypersensitivity is a very common form of active resistance. It may be elicited by parasites as diverse as viruses, bacteria, powdery and downy mildew fungi, rust fungi, nematodes and some insect species. Cells surrounding the place of infection become necrotic early on in the development of the parasite. Hypersensitivity is usually associated with physiological changes in the tissue, such as the production of phytoalexin and lignin.

Resistance may also be active and of an anatomical nature. The formation of plant cell wall thickenings (wall appositions or papillae) is the most important example. Plant cells may form such papillae at the point of contact with fungal mycelium, at places where the pathogen attempts cell wall penetration. Papillae are especially common in non-host resistance reactions to fungi (Heath, 1980; Sherwood and Vance, 1980), but occur also in race-non-specific quantitative resistance of barley (Niks, 1986) and wheat (Jacobs, 1989a,b) to their respective leaf rusts (*Puccinia hordei* and *P. recondita*).

Knowledge of the physiology and molecular genetics of defence mechanisms provides a basis for strategies to obtain resistance through genetic manipulation (section 25.6.2(b)).

25.3.3 Tolerance

The term tolerance is applied to the mechanisms by which plants reduce the extent of damage per unit parasite present. Tolerance does not reduce levels of infection in contrast with resistance and avoidance. A plant may support high populations of pests or pathogens (it is very susceptible) but may suffer relatively little damage in terms of yield reduction (it is tolerant). Just as susceptibility is the opposing term to resistance, so sensitivity is the opposite to tolerance. From this definition it is clear that the use of the term tolerance (by some scientists) as a synonym of incomplete forms of resistance must be strongly discouraged.

Tolerance may be due to (unknown) physiological factors. It is not clear, for instance, why some cucumber cultivars suffer more from a certain infestation of mites than other cucumber cultivars with a similar infestation (de Ponti, 1978). Rapid recovery following pest attack (e.g., grazing) may also be considered as tolerance. Another strategy resulting in tolerance is the so-called compensation. Taylor and Bardner (1968) reported that the diamondback moth, *Plutella maculipennis*, ate leaves clearly leaving the veins intact. These leaves grew larger and lived longer than corresponding leaves on intact plants. This compensatory growth prevented loss in yield and, in some cases, increased yield of turnips.

Plant virologists use the term tolerance in a different sense. They assess symptoms caused by viruses. Plants with little or no disease symptoms (and in most cases an unknown concentration of virus particles) are called tolerant. One of the reasons for the different use of the term tolerance by virologists is the difficulty in assessing the amount of virus in a plant. As a consequence, the degree of damage per unit of virus infection is also difficult to determine. Symptom expression is not necessarily related to the level of virus infection. A plant may fail to show symptoms even when the concentration of virus is relatively high. Such plants are called symptomless carriers. Frequently, however, plants with mild symptoms also have a low virus concentration. In such cases tolerance may also include resistance.

It is very difficult to determine the difference in tolerance between plant genotypes. One has to determine the amount of yield reduction per unit of infection. This means that the yield under disease-free conditions has to be compared with the yield of the affected crop. Also the level of infection in or on the crop must be assessed. This is a complicated and difficult exercise with large experimental errors. It is not surprising, therefore, that there have been very few noteworthy studies of 'true' tolerance.

25.4 PRINCIPAL PROBLEMS IN BREEDING FOR RESISTANCE TO PARASITES

Breeding for resistance to parasites has much in common with breeding for other characters. There is one important complication, however. Biotic stress is caused by a large diversity of organisms. These organisms belong to species that vary genetically at least as much as the crop species. Whether a crop exhibits resistance to a parasite or not depends both on the genotype(s) of the plant population, and on the genotypes that constitute the parasite population. This section discusses the most important consequences of this aspect.

25.4.1 Specificity of defence mechanisms

Almost all plant species are non-hosts for almost all parasites. Non-host status may result from the inability of the potential parasite to infect the plant (the parasite has no parasitic ability towards the plant species) combined with defence mechanisms in the plant directed against all possible pest and pathogen species. Defence mechanisms which are directed against many different parasite species are called general or broad defence mechanisms (Parlevliet, 1981b). Examples include the high tannin content in oaks, milk-fluids in *Asclepias* (Dussourd and Eisner, 1987), the production of phytoalexins following wounding in many plant species, and the bitterness in lupin (*Lupinus*). According to Wink (1988), sweet lupin cultivars are more susceptible to a broad spectrum of consumers, ranging from rodents to some fungi, than the bitter wild type. This illustrates the broad action spectrum of the bitterness.

The next level of specificity of defence is the resistance that is effective against only one parasite species. This has been called the pathogen-specific resistance (Parlevliet, 1981b). If the resistance in this category is effective against all genotypes of the parasite species without 'cultivar × isolate interactions', it is 'race-non-specific', also known as 'horizontal resistance' (Vanderplank, 1968). An example of this type of resistance is the partial resistance of barley to barley leaf rust (*Puccinia hordei*) (Table 25.1). Parlevliet (1978) defined partial resistance

Table 25.1. Percentage of leaf area of three barley cultivars covered with pustules of five isolates of barley leaf rust (*Puccinia hordei*), just before maturity (after Parlevliet, 1978)

Cultivar	Isolate				
	11–1	18	1–2	22	24
Berac	8.1	6.7	3.1	5.0[a]	0.9
Julia	4.5	12.1[b]	1.8	1.1	0.6
Vada	0.8	0.5	0.6	0.2	0.1

[a] In the case of true race-non-specific resistance this value should have been 2.0.
[b] In the case of true race-non-specific resistance this value should have been 3.5.

as resistance that is characterized by a reduced rate of epidemic development in spite of a susceptible infection type. The three cultivars listed in Table 25.1 have a considerable level of partial resistance. These three cultivars differ somewhat from each other, but very strongly from the extremely susceptible barley cultivar 'Akka' (Parlevliet, 1989). Under the same conditions 'Akka' would have had a leaf coverage of over 50% with all five isolates. The data in Table 25.1 also illustrate that most of the partial resistance is expressed irrespective of the isolate of the pathogen, but not all. There are small cultivar × isolate interactions. 'Julia' is slightly too susceptible to isolate 18 and so is 'Berac' to isolate 22. So, in all, the resistance behaves approximately race-non-specifically. This pattern seems to occur frequently within pathogen specific resistances.

Partial resistance is pathogen-species-specific (Gavinlertvatana and Wilcoxson, 1978; Parlevliet, 1983a). Cultivars that have a high level of partial resistance to one rust species may be very susceptible to another rust species, powdery mildew or other pathogen. Pathogen-specific resistance with the highest level of specificity is known as 'race-specific' or 'vertical resistance' (Vanderplank, 1968). This type of resistance is only effective against certain genotypes of a particular pathogen species. A plant genotype with one or more genes for race-specific resistance will be resistant to some, but susceptible to other isolates of the pathogen. This depends on the genotype of the isolate. Genes for race-specific resistance are effective against only one pathogen species. Usually the gene symbols refer to the English or scientific name of the causal organism. The *Sr*-genes in wheat confer only resistance to the stem rust (*Puccinia graminis*), the *Pc*-genes in oat to crown rust (*Puccinia coronata*).

25.4.2 Specificity of parasitic ability

Parasitic ability is the ability of a parasite to exploit a plant as a source of nutrients. This includes the ability to locate the host. With pests this occurs to a large extent through their sensorial perceptions. With pathogens the reproduction and spread of propagules must ensure a sufficient chance of encountering suitable host tissue. Germ tubes of stoma-penetrating fungi must be capable of using morphological clues on the surface of the host species in order to find the stomata (Littlefield and Heath, 1979). Parasitic ability also includes the ability to circumvent or to eliminate effects of defence mechanisms of the host plant. Insects that are parasitic on milkweed for instance, puncture the milk ducts upstream of the site where they will feed (Dussourd and Eisner, 1987). Biotrophic pathogens appear to induce susceptibility (anaesthetize) in the cells of their host species to prevent papilla formation or a hypersensitive reaction (Ouchi *et al.*, 1979; Heath, 1985; Niks, 1989). Defence mechanisms that are based on the occurrence of toxins may be dealt with by the parasite through tolerating or breaking down these toxins. The versatility of parasite species to cope with the defence mechanisms of different plant species differs considerably. Some parasite species attack a large number of host plants, others only a few or only one species. Even the ability to infect certain genotypes within a host species may vary among isolates, as may be concluded from our discussion of race-specific resistance.

(a) Generalists

Generalists are parasite species that can attack many host plants. In the case of insects the term 'polyphagous' is used. All plant species that can act as hosts of a particular parasite species form the host range of that parasite. Examples of typical generalists are the soil-borne fungus *Sclerotinia sclerotiorum*, which is reported to have a host range of hundreds of plant species from at least 64 families belonging to the gymnosperms, monocotyledons and dicotyledons (Parlevliet, 1986); the green peach aphid (*Myzus persicae*), known to have summer host species belonging to over 50 plant families (Hodkinson and Hughes, 1982); and the mite *Tetranychus urticae* (van de Vrie *et al.*, 1972). It seems, however, that in some cases, individuals of generalist species may adjust physiologically to the plant on which they feed. When they are transferred to another plant species they have problems in surviving, but their progeny may thrive on the new host species. An example of this phenomenon has been reported by Eriksson (1974) in the stem nematode, *Ditylenchus dipsaci*.

(b) Specialists

Many pests and pathogens are parasitic on only a restricted range of plant species. Such parasites are called 'oligophagous'. Examples are the cabbage-white butterflies (*Pieris*) on a number of Cruciferae, and *Phytophthora infestans* on a number of species of the Solanaceae, including potato and tomato. Extreme specialists, 'monophagous', have a host range of only one plant species. Examples of such very narrowly specialized parasites are the cannabis leaf miner (*Liriomyza cannabis*) on hemp, and the rust of groundnut, *Puccinia arachidis*, which is confined to the groundnut and its putative wild ancestor (Subrahmanyam *et al.*, 1983).

One may wonder what advantage there is in being a specialist. Why is the polyphagous locust as successful as the oligophagous Colorado potato beetle? A current theory is that specialization for certain plant species leads to a more efficient exploitation of those plants. The specialists may, for instance, develop an appropriate protective colour, or will break down or tolerate certain toxic compounds in the plant more effectively. Generalists will use their substrates less efficiently, but are more flexible in the choice of their hosts. It is difficult, however, to find evidence to support this hypothesis (Hare and Kennedy, 1986).

(c) Formae speciales

It is important to realize, that the host range of a pest or pathogen species may be much wider than the host range of an individual isolate. Powdery mildew (*Erysiphe graminis*), for example, infects many species in the Poaceae. Individual isolates, however, may be extreme specialists. Isolates that are pathogenic on cultivated barley fail or hardly infect the related *Hordeum bulbosum* (Jones and Pickering, 1978), and also hardly infect other grasses and cereals (Hardison, 1944; Mühle and Frauenstein, 1970) as far as is known. Other isolates may be confined to oats, rye or wheat, respectively. Determination of host ranges is difficult. Conclusions may differ according to the origin of the isolates and plant material (Mühle and Frauenstein, 1970; Eshed and Wahl, 1975), the number of accessions tested, the criteria used by the researcher to define plant susceptibility, and the environmental conditions under which the infection experiments are carried out (Ride, 1985; Niks, 1987).

This infraspecific specialization within a parasite species is very common. Isolates that have the same host range constitute a *forma specialis* (plural: *formae speciales*). *Formae speciales* are usually named after their host species or host genus. *Erysiphe graminis* is subdivided into many *formae speciales*, amongst others *Erisyphe graminis* f.sp. *hordei*, f.sp. *avenae*, and f.sp. *secalis*. An equivalent term is 'pathovar' (in bacteriology). The *formae speciales* of a pathogen species may or may not be closely enough related to allow gene flow between them.

(d) Races

Within pathogen species or *formae speciales*, individuals can be classified according to the host cultivars they can infect. They are said to be 'virulent' to these cultivars. Pathogen genotypes sharing a collection of cultivars to which they are virulent constitute a (physiological) race. With viruses the term 'strain' is used, with nematodes also the term 'pathotype' and with insects the term 'biotype' is used, but this term is also used in other contexts. The term 'isolate', used earlier in this chapter, indicates a collected (and maintained) sample of the pathogen, without any connotation about virulence pattern.

25.4.3 Inheritance of resistance

As early as 1905 Biffen reported that resistance in wheat to stripe rust (*Puccinia striiformis*) was inherited in a Mendelian way. An ever-increasing flow of reports made it clear that many cases of resistance were inherited in a simple way. Dominance is very common for resistance to pests and pathogens, especially with hypersensitive resistance against biotrophic fungi. Monogenic recessive resistance occurs much less frequently. In a few cases non-allelic gene interaction (epistasis) has been reported.

The number of different major genes for

Table 25.2. Number of major genes for resistance (R) known to exist in various host–parasite relationships

Host species	Disease, Parasite species	R
Groundnut	Groundnut rust, *Puccinia arachidis*	0
Grapevine	Downy mildew, *Plasmopara viticola*	0
Sugar beet	Beet cyst nematode, *Heterodera schachtii*	0
Sorghum	Milo disease, *Periconia circinata*	1
Cabbage	Cabbage yellows, *Fusarium oxysporum*	1
Lettuce	Downy mildew, *Bremia lactucae*	>16
Tomato	Leaf mold, *Fulvia fulva*	>10
Wheat	Stem rust, *Puccinia graminis*	>46
Potato	Downy mildew, *Phytophthora infestans*	>13
Wheat	Hessian fly, *Mayetiola destructor*	20

resistance that have been reported varies per pathosystem. Table 25.2 shows that in some pathosystems no major genes for resistance have been found. In many cases only one gene has been found and incorporated into modern cultivars, for example the recessive *pc* gene in sorghum to *Periconia circinata* (Table 25.2). In many other host-parasite systems several major genes are known, each conferring full or nearly full protection. These genes may be on one locus (multiple allelic series), on closely linked (complex) loci, or on loci that are less closely linked or not linked at all (Islam and Mayo, 1990). It is remarkable that genes for resistance are frequently reported to be clustered in linkage groups (Parlevliet, 1986).

Resistance may also be based on several genes each with a relatively small effect. Examples of such polygenically inherited resistance are the partial resistance in potato to *Phytophthora infestans*, in maize to *Puccinia sorghi*, and in barley to *Puccinia hordei* (Parlevliet and Zadoks, 1977). In these, but also in several other pathosystems both partial polygenic resistance and major-gene hypersensitive resistance occur. Cytoplasmically inherited resistance is extremely rare.

25.4.4 Gene-for-gene interaction

As was mentioned above, the outcome of an attempt to infect a plant by a potential parasite may depend both on the genotype of the plant and on the genotype of the parasite. This is especially true for race-specific resistance. Flor (1956) discovered that in the flax-flax rust (*Melampsora lini*) relationship the major genes for resistance in the plant interact very specifically with major genes for avirulence in the fungus. He concluded that for each gene conditioning resistance in the host there is a specific gene conditioning virulence in the parasite. This conclusion became known as the 'gene-for-gene' concept. In Flor's wording the emphasis is still on virulence. Resistance genes are ineffective if the pathogen carries the appropriate (usually recessive) allele for virulence. This aspect is also emphasized in more recent literature (e.g., Littlefield, 1981; Robinson, 1987). Nearly all scientists now agree that the specific interaction is between the (dominant) resistance alleles and the (dominant) avirulence alleles (Day, 1974; Ellingboe, 1976; Heath, 1981; Parlevliet, 1981a; Browder and Eversmeyer, 1986; Sidhu, 1987). One could reword the gene-for-gene concept as follows: 'Any resistance gene can act only if a locus in the pathogen carries a matching allele for avirulence'. Table 25.3 presents a simple example of a gene-for-gene relationship.

It is hypothesized that the dominant alleles for resistance and the dominant alleles for avirulence (inability of the parasite to infect certain plant genotypes) produce gene products that are able to recognize each other in each corresponding gene-for-gene pair. These products may form a hybrid molecule (dimer) that elicits the defence reaction in the plant (e.g. Hadwiger, 1988; Heath, 1981; Hofferek, 1983). This reaction often consists of a hypersensitive reaction.

Table 25.3. Interactions between two loci in the host and two loci in the pathogen according to gene-for-gene concept. Each locus has two alleles, one in capitals (for resistance in the plant and for avirulence in the pathogen) and one in lower case (for susceptibility in the plant and for virulence in the pathogen). The host is represented as a diploid homozygous crop, the pathogen is haploid (as with Ascomycetes such as powdery mildew). A + indicates compatibility (there is infection), a − indicates incompatibility

Host	Pathogen			
	AB	aB	Ab	ab
rrss	+	+	+	+
RRss	−	+	−[a]	+
rrSS	−	−	+	+
RRSS	−	−	−	+

[a] Differential interaction.

In the example shown in Table 25.3 the incompatibility would be triggered by dimer *R-A* and/or *S-B*. The combination *RRss* with *aB* is compatible, because *R* does not find the corresponding gene product *A*. The presence of *B* is not relevant if the corresponding gene product *S* is not present. Direct evidence for this molecular hypothesis is lacking. Only in tomato/*Fulvia fulva* has a putative avirulence gene product been isolated and characterized (de Wit *et al.*, 1987).

Genetic evidence for a gene-for-gene interaction is usually based on genetic studies in both the plant and the parasite. If in the parasite no genetic studies are feasible, e.g. in bacteria or imperfect fungi, the occurrence of a differential interaction is taken as an indication for gene-for-gene interaction. Differential interaction is expressed as a reversed ranking of degree of compatibility, as indicated in Table 25.3. Evidence for a gene-for-gene interaction has been found for all major classes of parasites, including bacteria, fungi, insects and higher plants (*Orobanche*) (Day, 1976; Sidhu, 1987). Recently, it has been possible to isolate and clone genes for avirulence from some plant pathogenic bacteria (*Pseudomonas syringae* and *Xanthomonas campestris*, e.g. Staskawicz *et al.*, 1984; Shintaku *et al.*, 1989). This exciting type of research may

be expected to provide more insight into the molecular basis of the gene-for-gene interaction, and may have immediate applications for resistance breeders (de Wit, 1992).

25.4.5 Durability of resistance

Since the parasite species is usually genetically diverse, some genotypes will be more successful on the widely-grown cultivars of the crop than others. Such differences would result in natural selection of more and more harmful parasite genotypes. This is especially dramatic with the race-specific resistance based on gene-for-gene interactions. As mentioned above, the hypothesis is that avirulence is triggered by a gene product of the parasite. It is easy to imagine that mutant parasite genotypes may arise in which the gene product is not present, or present in an ineffective form. Mutations like deletions of (part of) the avirulence gene would result in virulence. Such mutants would be able to infect the formerly resistant plant genotype. The resistance is 'broken down'.

Such adaptations in the parasite to newly introduced resistant cultivars are very commonly observed in pathogens, but less common in pests. The resistance of such cultivars is called 'ephemeral' or 'not durable'. Table 25.4

Table 25.4. Durability (in years) of resistance in cultivars of certain crops to some pathogens and pests

Crop	Cultivar	Parasite or pest	Duration
Wheat	Clement	Stripe rust	1
	Tadorna	Stripe rust	1
	Flevina	Stripe rust	5
	Manella	Stripe rust	14
	Felix	Stripe rust	18
Flax	gene *L9*	Flax rust	5
	gene *M*	Flax rust	13
Lima bean	Thatcher	Downy mildew	0
	Dover	Downy mildew	0
Apple	Winter Majetin	Woolly aphid	>148
	Northern Spy	Woolly aphid	130
Lettuce	Avoncrisp	Lettuce root aphid	>20
	Avondefiance	Lettuce root aphid	>20

presents the durability of resistance of several cultivars in terms of years before the appearance of virulent races. The Table shows that even within pathosystems the durability varies considerably. Resistance against biotrophic pathogens in particular is frequently ephemeral. Break-down of resistance has been reported in remarkably few cases of pests. According to Gallun (1972) mechanisms of avoidance (antixenosis) are seldom, if ever, overcome by new biotypes. In the case of insects, aphids in particular are known to adapt to resistant cultivars (van Emden, 1987). This is explained by the fact that many aphids breed parthenogenetically. A resistance-breaking individual genotype can reproduce itself identically in the absence of competition and thus quickly build up a virulent population. This applies to all organisms in which clonal reproduction is the major multiplication strategy, like the highly adaptable powdery and downy mildews and rust fungi. Resistance against viruses, however, is usually more durable (Parlevliet, 1991).

It is difficult to prove that a particular resistance is durable. Resistance is considered durable if it remains effective during prolonged and widespread use in an environment favourable to the parasite (Johnson, 1984). There are several factors determining the durability of resistance.

1. Some forms of resistance require complex adaptations in the pest or pathogen. The wheat stem sawfly would need to alter its behaviour drastically to be able to feed on the solid stemmed wheat cultivar Rescue.
2. The product of the avirulence gene may play a role in vital functions of the parasite. A deletion or other mutation of the gene, which could result in virulence, would reduce the fitness (e.g. physiological functions) of the parasite accordingly. The newly created virulent parasite strain would not be able to survive, especially when it has to compete with other, more vigorous genotypes of the pathogen. Vanderplank (1968) thought that pathogen races with many alleles for virulence were weaker than races with few alleles for virulence when they were growing on a plant without effective genes for resistance. This would lead to stabilizing selection, i.e. (near) disappearance of those races that possess 'unnecessary' alleles for virulence. There are, however, no indications that stabilizing selection occurs as a general phenomenon (Parlevliet, 1981c).
3. Soil- and splash-borne parasites may spread slowly from one area to the next. Rare virulent mutants require a long time to become widespread in the area in which the resistant cultivar is grown.
4. A resistant cultivar may possess two effective race-specific resistance genes. The parasite would need to produce two independent mutations simultaneously to break down the resistance of the cultivar. It is surmised that the resistance of wheat cultivars Manella and Felix to wheat stripe rust (Table 25.4) is based on such a double resistance (R.W. Stubbs, pers. comm.).
5. In annual crops that are grown on only a small acreage in a rather scattered pattern, the chances of a pathogen propagule reaching a field containing the crop are small. If a resistant cultivar is grown, the chance of a mutant propagule reaching this crop is even smaller. If this occurs, the pathogen may not succeed in finding a suitable alternative host after the harvest, and it will become extinct. This situation applies to the flax-flax rust situation in The Netherlands. The resistance of the conventional major-gene race-specific type became 'durable' (but not according to the definition of Johnson, 1984!) since the crop covered a small acreage (Parlevliet, 1991).

25.5 MAJOR CONSIDERATIONS IN BREEDING FOR RESISTANCE TO PARASITES

Simmonds (1983) pointed out that if a new character is added to a breeding programme, either gains in other characters will suffer (for example yield potential), or the programme will have to be expanded by a factor which is dependent on the selection rate. The breeder should seriously consider whether breeding for disease resistance is economically worthwhile. The decision depends mainly on the frequency and the extent of epidemics in the area where the crop is to be grown and on the economic damage inflicted by the parasite.

Next, a breeder should identify which type of defence mechanism is most suitable for introduction into the crop. He may chose major-gene resistance with complete expression. Advantages of this type of resistance are: (1) the simple inheritance, which is of course very desirable in a breeding programme; (2) the normally complete protection of the crop from the parasite. The latter aspect is especially important in horticultural crops, where produce often needs to be blemish-free to be marketable. Resistance against root parasites like maggots in cabbage need not necessarily be complete, since this crop withstands quite large populations of maggots, and the roots are not the marketable produce. In carrot, however, where the marketed portion of the plant is attacked, resistance against maggots of the carrot fly should be complete. Complete resistance is also very useful in crops which are exported, for example chrysanthemums. Traces of a parasite in such crops may prompt countries to close their borders to imports for quarantine reasons.

A risk in choosing major-genic complete resistance is that this type of resistance may turn out to be ephemeral. There are, however, cases where major-gene resistance has been (quite) durable (Johnson, 1984; Table 25.4). If ephemeral types of resistance are applied according to carefully designed strategies, they may be very useful.

Polygenically inherited forms of resistance are more difficult to handle in breeding programmes. Backcross programmes in particular to introduce polygenes from wild relatives of the crop are hardly or not feasible. The polygenically inherited partial resistance (Table 25.1) is nearly always durable (Parlevliet, 1983a). In many agricultural crops a low level of infection is acceptable, and partial resistance may be combined with other control measures, such as the application of pesticides (Fry, 1978).

If the crop is grown for animal or human consumption, the breeder should make sure that the mechanism of resistance against the parasite has no harmful side-effects on the consumer. In potato, resistance against the Colorado potato beetle can be developed by breeding for a high alkaloid content. However, the alkaloids are part of the broad defence system in potato and breeding may result in the production of resistant potato genotypes which are too toxic for human consumption (Kuhn and Gauhe, 1947).

An interesting recent development is the attention to 'tritrophic effects' of plant characters. Certain plant genotypes may allow predators to control pest species more effectively than others. This may be due to plant architecture (Kareiva and Sahakian, 1990) or to the release by the infested plant of specific volatiles that attract natural enemies of the pest (Dicke *et al.*, 1990; Turlings *et al.*, 1990). Breeding for such plant characters is especially promising in crops where biological control is practised.

25.6 SOURCES OF RESISTANCE

25.6.1 Conventional sources of resistance

After having chosen the objectives, the next step in a breeding programme for resistance is the identification of an appropriate source of resistance. The genotypic variation within the botanical species of the crop and often also within related species should be studied. Sources for resistance may be found in taxonomic groups that are more or less distantly related to the crop (Harlan and de Wet, 1971).

1. The cultivar itself. Genetic variation for resistance in a cultivar may be sufficient in cross-pollinating crops. Self-pollinating crops are almost always genetically too uniform to make selection meaningful. An exception is the resistance in sorghum to Milo disease (caused by *Periconia circinata*) (Schertz and Tai, 1969). The spontaneous mutation frequency of the gene for resistance *Pc* (susceptible) to *pc* (resistant) appears to result in a ratio of 1:8000 mutant:wild-type gametes. This resistance occurs in all sorghum cultivars at a very low frequency.

2. Commercial cultivars. Resistance can be derived from other modern cultivars and from advanced breeding lines. The advantage is that this material already has a genetic background with genes for high yield and high quality and resistance to other pathogens. If modern cultivars from other parts of the world are used, the genetic background may contain undesirable features, such as inappropriate day-length responses.

3. Other forms. In a number of crops, different forms can be discerned. These forms are termed *varietas* (plural: *varietates*). Examples are corn and sweet corn, both belonging to *Zea mays* and the various forms of kale and cabbage (*Brassica oleracea*). Few crosses are attempted between such forms, because the hybrid possesses a combination of genes not directly useful. Therefore, the forms constitute distinct gene pools. It is possible that one form has genes for resistance to a certain parasite that are not present in the other form.

4. Landraces. Landraces are old, locally grown cultivars. They usually consist of a mixture of various genotypes. On a world-wide scale they are replaced by modern, higher yielding cultivars. A lot of potentially-useful genetic material has been lost with the disappearance of landraces (Frankel, 1977).

5. Wild progenitors. In the centre of origin of the crop, the parasite and its host have been living together for very long periods. They probably co-evolved (Watson, 1970), resulting in great variation for resistance in the plant and for pathogenicity (parasitic ability) in the parasite. The Middle East region, for example, is a particularly rich source of resistance to rust and powdery mildew fungi of cereals. Cultivated barley, wheat and oats originated from wild progenitors in that area.

6. Related species can be used in crosses when the required resistance is not found in the crop species. The wild relative *Coffea canephora* of coffee is, as a species, highly resistant to the coffee rust (*Hemileia vastatrix*). Crosses with the cultivated coffee, *C. arabica*, are possible, and some genes for resistance have been introduced into coffee.

These genes (S_H6 to S_H9) are supposed to confer a more durable resistance than the S_H genes found in *C. arabica* (Kushalappa and Eskes, 1989). Genes for resistance derived from highly resistant related species do not necessarily confer more durable resistance than those found in the crop species itself (see below).

7. Related genera. In some cases, resistance can be introduced from related genera. Resistance to powdery mildew, yellow, leaf and stem rust has been transferred from rye (*Secale*) and wheat grass (*Agropyron*) into wheat (*Triticum*). This is technically a very difficult procedure, and the resistance does not appear to be more durable than the resistances present in wheat itself (Roelfs, 1988). This is somewhat surprising, since rye is believed to possess durable, high level resistance to at least wheat powdery mildew and yellow rust.

The potential sources of resistance described above are presented in the order in which complications for the breeder increase. The main problems are failure to secure crosses between the crop and the donor species, sterility of the interspecific or intergeneric hybrid, and poor intrachromosomal recombination (Harlan and de Wet, 1971). Many generations of backcrossing are needed to remove undesirable traits introduced together with the resistance (Knott, 1989).

25.6.2 Alternative possibilities of obtaining resistance

(a) Mutation breeding

A mutagenic treatment may convert a susceptible genotype into a resistant one. If the mutation is a point mutation, the resistant mutant will be identical to the original cultivar, except for its resistance. Usually, however, there are undesirable side-effects of the mutagenic treatment. Several other genes may also have undergone changes, or the mutation for resistance has undesirable pleiotropic effects. As a consequence, the selection of a resistant mutant should be followed by further breeding efforts to produce a commercially acceptable cultivar.

In barley it has been relatively easy to induce mutations for resistance to the barley powdery mildew (Jørgensen, 1983). The resistance has almost always been located on the *Ml-0* locus on chromosome 4. The mutants always have a lower yield than the original cultivar, probably because of spontaneous necrosis (in the absence of the pathogen). The degree of this necrosis and yield reduction appears to depend partly on the genetic background of the barley line (Schwarzbach, 1976; Kjaer *et al.*, 1990). A famous example of successful mutation breeding for resistance is the case of mint/*Verticillium albo-atrum*. It took about 15 years of extensive work to develop a commercially acceptable resistant cultivar (Murray, 1971).

(b) Genetic manipulation

Cell and molecular genetic techniques make it feasible to transfer single isolated genes from any organism to plants (transformation). Monocotyledonous plant species are much more difficult to transform than dicotyledonous species. A very serious limitation of genetic manipulation is the difficulty of finding useful specific genes (e.g., genes for resistance) in the donor organism. Until now, the search for genes for pathogen-species-specific resistance has not been successful. Using several molecular genetic techniques it should be feasible to isolate, for instance, *Dm* genes for downy mildew (*Bremia lactucae*) in lettuce (Landry *et al.*, 1987; Young, 1990). Such genes could be transferred to high quality cultivars that are currently susceptible.

In recent years, some promising techniques of obtaining resistance through genetic manipulation have been developed.

1. Isolation of genes for 'broad defence mechanisms' (section 25.4.1) may be feasible. If substances that are responsible for the resistance are primary gene products and if they are synthesized in large quantities sometime during the development of the plant tissue,

the messenger RNA (mRNA) can be isolated, copied into complementary DNA, cloned, and transferred to the crop plant. The introduction of a cowpea inhibitor gene into tobacco (Hilder *et al.*, 1987; Gatehouse and Hilder, 1988), discussed in Chapter 10, demonstrates the potential value of this approach.

2. Molecular techniques make it feasible to isolate genes responsible for producing toxins in animals or microbes, and to transfer them to the crop. A potentially useful donor for toxin genes is the bacterium *Bacillus thuringiensis*. This bacterium species produces proteins that are specifically toxic to insects. Some strains of the bacterium are toxic to beetles, others to butterfies and moths, others to flies. Such a toxin-producing gene has been introduced into tobacco and protects the transformed plants from the tobacco horn worm (*Manduca sexta*) (Vaeck *et al.*, 1987). However, there is a danger of the pest developing races which can overcome this resistance by negating the toxin (for more information on this approach see Chapter 10).

3. Transfer of parts of a viral genome has resulted in an increased level of resistance of plants to viruses. Several strategies have been developed (Chapter 10). The most successful strategy is the incorporation into the plant genome of viral capsid protein genes. The resulting resistance is called 'coat protein-mediated resistance'. It is effective against the virus from which the gene was derived and against related viruses (Beachy *et al.*, 1990). Another strategy is the incorporation of complementary DNAs of small satellite RNAs into the genome of the plant. This method has been shown to increase the resistance of tobacco to cucumber mosaic virus (Harrison *et al.*, 1987). Antisense RNAs of several plant viruses may also provide protection against virus infection (e.g., Hemenway *et al.*, 1988). Antisense RNA consists of the complementary sequence of (part of) the mRNA (or viral RNA) that is the template for the translation into proteins.

25.7 ASSESSMENT OF RESISTANCE

Avoidance and resistance both lead to reduced infection or colonization of the crop compared with a susceptible genotype. When assessing the populations of a parasite or symptoms of attack, it is not apparent whether avoidance or resistance is the cause of the differences observed. This distinction is not made in this section.

25.7.1 Infestation levels in the crop

In many cases, the level of infection or infestation in a crop can be assessed directly. One can count the number of aphids per plant or per leaf, the number of nematodes per root system, or the percentage of leaf area covered with powdery mildew fungus. In other cases only a part of the parasite is visible. With rust fungi only the fruiting bodies (sori or pustules), but not the mycelium can be seen. The number of pustules may be counted per unit leaf area, or the percentage leaf area covered with pustules (e.g., Parlevliet and van Ommeren, 1984).

Other pathogens cannot be observed directly. When they cause (necrotic) spots (lesions) on leaves, fruits or stems, the percentage of plant tissue covered with lesions can be assessed (Parlevliet, 1989).

It is difficult to assess the percentage of tissue affected by a pathogen or the area covered with lesions. The human eye tends to assess the percentage of 'infected' tissue if less than 50% of the leaf surface is covered, but the percentage of 'healthy' tissue if the cover is more than 50% (Horsfall and Cowling, 1978). Several scientists have designed 'assessment keys' to circumvent these problems (Cobb, 1892; James, 1971; Saari and Prescott, 1975). There are also computer programs that have been designed to train the eye of the scientist to assess percentages of infected tissue. An example is 'Distrain', developed by R. Tomelin and Th. Howell (ARS, USDA, Beltsville, Maryland).

Many parasites, which cause physiological disturbances of the whole plant, cannot be observed directly. They cause wilting, stunting, leaf rolling, mottling or discoloration. Examples of such pathogens are vascular wilt fungi, nematodes, certain insect larvae, viruses and bacterial pathogens. The relationship between the severity of disease symptoms and the population of parasite present (which is the true measure of resistance) is often rather poor. True disease symptoms tend not to be reliable parameters for assessing the populations of parasites (Parlevliet, 1989). Assessment keys for this category are descriptive. An example is the scale for cabbage/*Fusarium oxysporum* f.sp. *conglutinans* developed by Ramirez-Villupadua *et al.* (1985). If the selection in the breeding programme is on the basis of severity of disease symptoms, one may select for tolerance as well as for true resistance.

25.7.2 Assessment of the infection type

The infection type (IT) describes the reaction of the plant to the infection in terms of the amount of necrosis or chlorosis at the infection sites, and the rate of sporulation of the individual colonies. The IT is usually expressed on a 0 to 4 (Stakman *et al.*, 1962) or on a 0 to 9 scale (McNeal *et al.*, 1971). The IT screening is most useful for biotrophic pathogens, notably rust and powdery mildew fungi. These fungi exploit the living tissue of plants. In fully compatible (susceptible) interactions (IT4 or IT9 respectively), the rust pustules or mycelium networks develop without causing cell death in the plants. In less compatible interactions, more chlorosis or necrosis and smaller pustules or mycelia develop. In fully incompatible interactions only necrotic or chlorotic spots are visible, or sometimes no symptoms appear at all at the macroscopic level (IT0).

The IT partly depends on the stage of development of the plant, age of the leaf and environmental factors. Determination of the IT is usually easier in seedlings than in mature plants. In cases of ambiguous expression of the IT, the scale of ITs may be reduced to two or three classes, for instance R (resistant), MR (moderately resistant) and S (susceptible). As mentioned before, the IT is an important criterion for recognizing partial resistance as defined by Parlevliet (1978) (Table 25.1).

25.8 OVER- AND UNDERESTIMATION OF RESISTANCE

Differences in amount of infection between plant genotypes or populations as assessed may not always reliably reflect differences in resistance. This is particularly true for quantitative types of resistance. Several factors may lead to under- or overestimation of differences in resistance.

25.8.1 Differences in earliness of the plants

Differences in development rate between plants may hamper a reliable assessment of the differences in resistance. Early-maturing plants will tend to carry more pathogen at the assessment time than late maturing ones. The upper leaves or inflorescences of early plants will have been exposed to the pathogen for a longer time than those of late maturing plants. As a consequence, the level of resistance of late maturing plants tends to be overestimated relative to those of early plants. Indeed, late-maturing potato cultivars tend to be rated as more resistant to *Phytophthora infestans* than early maturing cultivars. This effect is probably partly due to differences in foliage growth characteristics (van Oijen, 1991).

For this reason it is useful to record not only the amount and type of infection, but also the stage of development of the lines. For small grains, keys to describe the development stage of the crop have been developed (Zadoks *et al.*, 1974; Tottman and Makepeace, 1979) for this purpose. It is useful to sow the plant material to be tested in earliness-groups. Early lines are compared with other early lines (and early checks!), and late lines with late lines.

25.8.2 Interplot interference

In the experimental situation, particularly in the early phases of the breeding programme, the plots tend to be small and they border each other. A moderately resistant genotype may be adjacent to a highly susceptible one. With polycyclic, air-borne pathogens (rust and powdery mildew fungi, *Phytophthora infestans* and some *Helminthosporium* species) the resistant line can receive a large amount of inoculum from its susceptible neighbour, whereas it exports hardly any inoculum. The susceptible line has a much larger export than import of inoculum. The level of resistance in the resistant lines will be underestimated, the level of the susceptible lines overestimated. This phenomenon is called 'interplot interference' (Bowen *et al.*, 1984; Parlevliet and Van Ommeren, 1984; Randle *et al.*, 1986). In the farmers' field situation, the resistant cultivar covers a much larger plot in which the in-flow of inoculum from possible susceptible cultivars in the surroundings is of negligible influence on the total amount of infection.

Interplot interference can be demonstrated directly. Parlevliet and van Ommeren (1984) assessed the amount of infection by *Puccinia hordei* in barley in various experimental situations. They planted very susceptible and partially resistant cultivars in isolated large plots ('farmers' fields') and in small adjacent plots. They concluded that the smaller the plots, the smaller the differences in amount of infection between the extremely susceptible cv. Akka and the partially resistant cv. Vada. They recommended either estimating the level of resistance in isolated plots or including control cultivars with various levels of partial resistance as a reference. Their results illustrate that a breeder who screens many different genotypes in small adjacent plots should take small differences in level of infection very seriously.

Interplot interference, however, is not equally important for all pathogen species. With leaf blotch (caused by *Septoria nodorum*), the effect of the lay-out of the field is far less important than with leaf rust or powdery mildew. The *Septoria* spores are mainly dispersed vertically by splash dispersion, rather than horizontally to neighbouring plants. In four-row plots the central two rows give a reliable estimation of the level of resistance (Scott and Benedikz, 1977). Recent experiments indicated that for stripe rust interplot interference is practically absent (Danial and Parlevliet, unpublished).

25.8.3 Amount of inoculum

Too low a dosage of inoculum tends to lead to an overestimation of the resistance. A low dosage of inoculum may result in a high frequency of escapes. Plants that are susceptible escape from infection and are qualified as resistant. This effect is especially important with parasites that have one generation per season, like many soil-borne parasites. Too high a dosage of inoculum may obscure quantitative differences in resistance. Complete resistance, however, will then show up very well.

25.8.4 Moment of assessment

Parasites that have many generations each season build up according to a more or less predictable pattern (Zadoks, 1972; Zadoks and Schein, 1979). First, the parasite population increases exponentially, but soon the population increase slows down due to the fact that the amount of plant tissue still available becomes a limiting factor. Incomplete forms of resistance will affect the rate of increase of the parasite population.

If plant populations are tested for quantitative resistance against such polycyclic parasites, the amount of infection should not be assessed too early in the development of the epidemic development. The population level of the parasite is then too low to allow detection of quantitative differences between the plant populations. An assessment which is too late in the epidemic development, may also not allow the maximum distinction between the fairly resistant and the very susceptible populations. Both may be almost completely occupied by the

parasite if the environmental conditions are favourable to the parasite.

Sometime between the onset and the end of the epidemic, however, the different rates of epidemic development due to different levels of resistance will show up. At this stage, the most susceptible populations are completely occupied by the parasite, whereas in the more resistant material the amount of parasite has not yet reached its maximum.

Quantitative resistance to polycyclic parasites may also be assessed in a monocyclic test in the glasshouse. With partial resistance of barley to leaf rust, the latent period is a very useful trait to measure (Parlevliet, 1975, 1979). This is the time lapse between the inoculation and the onset of sporulation. Parlevliet and van Ommeren (1975) found a very high positive correlation between level of partial resistance in the field and the latent period in a monocyclic glasshouse test. In order to measure or to estimate the latent period, the inoculated plants should not be observed too early (none of the plants will carry mature uredia) nor too late (all the uredia on all the plants are mature). There is a short period in which the differences in latent period are obvious.

Differences in latent period have been reported to be important components of resistance in several other pathosystems (e.g., Parlevliet, 1979; van der Zaag, 1959; Jones, 1978; Johnson, 1990).

25.8.5 Environmental conditions

The expression of resistance to parasites may depend on environmental conditions. If the tests for resistance are carried out under conditions that differ from those that prevail in the farmers' fields, the assessments of the resistance levels may be irrelevant. Temperature is one of the factors that influence the expression of resistance. This applies to both major and minor gene resistance. There are examples of resistance declining or disappearing with increasing temperature as well as decreasing temperatures. Resistance to the greenbug, *Schizaphis graminum*, has been observed to decline in sorghum at reduced temperatures

(Wood and Starks, 1972). The degree of such a reduction of resistance depends on the biotype of the insect. Resistance to the Hessian fly in wheat has been found to decline with increasing temperature (Cartwright *et al.*, 1946). The *Sr6* and *Sr15* genes in wheat for resistance against stem rust are only expressed at relatively low temperatures, the *Sr9b* gene only at relatively high temperatures (Samborski *et al.*, 1977; Gousseau *et al.*, 1985).

Another environmental factor that affects the expression of resistance is light intensity. The resistance of sugar beet to the green peach aphid has been shown to be expressed poorly under low light intensity (Lowe, 1974). Qualitative and quantitative aspects of plant nutrition can also influence the expression of resistance. Increased levels of nitrogen in soils can increase the population growth of several leaf fungi but is beneficial in controlling the take-all disease of cereals (Russell, 1978). Decreases in potassium levels have been shown to increase multiplication of aphids on Brussels sprouts (van Emden, 1987).

25.9 METHODS OF TESTING FOR RESISTANCE

Development of pathogens and pests greatly depends on environmental conditions. Also, the expression of resistance in the plants depends on environmental conditions (section 25.8). Consequently, screening for resistance under field conditions has certain drawbacks, due to dependence on the weather and other environmental conditions. In some years the pest or epidemic of the target parasite fails to develop and as a consequence, no selection for resistance is possible. Where field screening is carried out against commonly occurring pathogens, results depend on the races that happen to occur naturally. Advantages of testing under field conditions are the good representation of the situation under which the crop is grown commercially, and the relatively low costs.

In glasshouses, climate rooms or even petri dishes (*in vitro* testing) the conditions are more highly controlled but the situation is increasingly less representative of commercial growing conditions. There is, however, good control of the amount and identity of parasite inoculum, on its distribution over the plants, the growth stage of the plants and on other environmental factors. It is also possible to screen the plants for resistance independent of the growing season. A great number of tests have been developed to assess levels of resistance under controlled conditions.

25.9.1 Application of inoculum

(a) Field conditions

In the field, the soil may be inoculated with a pathogen or a pest population may be built up by repeated cropping of a susceptible host. It is difficult to achieve a homogeneous distribution of the inoculum, and unaffected plants may be escapes rather than resistant. With leaf pathogens and pests in the field, the inoculum can be applied by means of spreader rows. A susceptible cultivar is planted at regular distances in the experimental field, e.g. at the weather-side of the populations to be tested. The spreader rows may or may not be inoculated artificially by dusting or spraying of inoculum or by infestation with the pest.

(b) Controlled conditions

Inoculum may be applied to plants or plant parts. Spore suspensions of soil-borne pathogens may be applied to seeds or to roots that are cut back. Inoculum of air- or splash-borne leaf pathogens may be applied by dusting or by spraying. Special equipment has been designed to achieve an even distribution of the inoculum over plants (e.g., Eyal *et al.*, 1968). Inoculation with many viruses is performed by rubbing leaves of the plants with carborundum powder to create tiny wounds in the epidermis, through which the virus may enter. Inoculum may also be injected into the plant, as with the screening for resistance to the Dutch elm disease (*Cerato-cystis ulmi*). A disadvantage is that certain possible mechanisms of resistance (avoidance and resistance to penetration) may be by-passed and so remain undetected.

Screening for resistance to mobile pest species requires special provisions. Insects that are released in the glasshouse will disperse unevenly over a genetically diverse plant population. Factors such as light and plant height contribute to the uneven insect distribution. In addition, the insects will display a preference for certain plant genotypes (section 25.3). Plants that are (relatively) free of insects might be considered 'resistant'. In genetically diverse populations, however, the insects may show non-preference to some of the plant genotypes. For monocultures only avoidance mechanisms leading to non-acceptance are relevant. It may be decided to eliminate effects of non-preference. This can be achieved by the use of plant cages and leaf-cages. The pest is released in the cage, and has the choice to feed on that plant or to starve. However, it is always advisable to confirm glasshouse and laboratory test findings in the field as the elimination of any stage in the complicated behavioural sequence leading to selection and colonization by a pest may result in a failure to identify mechanisms of avoidance.

25.9.2 Composition of inoculum

It is obvious that with race-specific resistance (section 25.4.1), the identity of the isolate of the parasite is crucial. The identity of the races is determined by means of a differential series of cultivars. This series consists of plant genotypes that provide maximum discrimination between the possible races of the parasite. The differentials in the series may contain one or more major genes for resistance. The ideal differential series consists of near-isogenic lines each with one gene for resistance, but sharing the same genetic background. In some pathogen species, races are coded according to rules agreed upon internationally (e.g. Johnson *et al.*, 1972), in others according to a nationally or locally used code (for instance in *Bremia lactucae*). Races may also be designated by the genes for resistance that are not effective against them, as in *Phytophthora infestans* (Black *et al.*, 1953) and the Hessian fly (Gallun, 1972).

Mixtures of races are often used to select for race-non-specific resistance. Parlevliet (1983b), however, explained that a race mixture is not suitable for identifying such race-non-specific resistance. A plant genotype that is (race-specifically) resistant to all but one component of the inoculum mixture will be infected at a much lower level than a genotype with susceptibility to all components. The former genotype may give the impression of partial, race-non-specific resistance. When a breeder looks for new genes for complete resistance, irrespective of race-specificity, testing with a mixture of many races is justified.

25.9.3 Plant part and plant development stage

Whenever possible it is preferable to test plants in the seedling stage or to test detached parts of plants. They require less space than mature intact plants, and can be kept in controlled environment chambers to facilitate testing under well-defined conditions. An additional advantage of testing detached plant parts indoors may be that the parasite is not being introduced into the environment where costly parental material is maintained. Tests in apple orchards for resistance to *Nectria galligena*, for example, would have the disadvantage that the breeder is spreading this harmful pathogen among very valuable trees that may suffer for many years afterwards. An indoor test on detached branches (van de Weg, 1989) does not have this drawback. Resistance to leaf pathogens, and also to mites, is often tested on detached leaves, or on leaf discs. Usually the leaf (parts) are put in petri dishes on an agar substrate containing benzimidazole. This compound interferes with protein degradation, thus delaying yellowing of the plant parts (Wang *et al.*, 1961).

Resistance to air- or splash-borne pathogens is often tested on plants in the seedling stage. In spinach, for example, resistance to downy mildew is tested on the cotyledons (Hubbeling and Ester, 1978). However, there are several instances in which the level of resistance of plants in the seedling stage does not correlate well with that of mature plants. Also diseases of flowers and fruits cannot be assessed in the seedling stage. Under these circumstances mature-plant tests need to be used.

25.9.4 *In vitro* selection

In principle, cell and tissue culture techniques make it possible to select single cells, undifferentiated plant tissue or somatic embryos for resistance to pathogens. The method is described by Wenzel and Foroughi-Wehr in Chapter 21. Tissue or cell cultures are exposed to a selective agent (e.g., a toxin of pathogen origin that plays a role in the pathogenesis). The surviving cells, calli or embryos are regenerated to produce plants. These regenerants should be resistant to the factor that was used as a selective agent.

Advantages of this procedure include the possibility of testing many individuals under very well-defined conditions. Each petri dish or flask may contain millions of individuals. It is feasible to aim for very rare mutants. In theory, the cells that have been derived from one plant should be genetically identical. It has been demonstrated, however, that the tissue culture procedure may induce considerable genetic diversity (somaclonal variation) and may also uncover pre-existing intraplant genetic diversity (Larkin and Scowcroft, 1983). *In vitro* selection procedures may be combined with transformation procedures, e.g. by means of *Agrobacterium tumefaciens*.

The procedure has serious restrictions. Surviving cells may be physiological (rather than genetic) variants. As soon as the selective agent is removed, the resistance disappears. This phenomenon is called 'habituation' (Chaleff, 1981). The method also depends on a good and relevant selective agent. With toxin producing pathogens (necrotrophs) the toxin is a suitable agent. It can be added to the tissue culture medium. It is hardly conceivable, however, that selection for resistance to biotrophic pathogens is possible at tissue culture level. These pathogens do not produce toxins that are essential in the pathogenesis. The tissue in the cultures is also too undifferentiated to allow infection by

obligate biotrophic pathogens, possibly with the exception of some viruses (Daub, 1986; Kluge, 1986). *In vitro* selection is not suitable for detecting defence mechanisms that are based on qualities of intact and differentiated tissue, such as glandular hairs or the ability of cork formation. Resistance at the cellular level, on the other hand, is not necessarily expressed at the whole plant level. It is essential to check whether the regenerated plants raised from the surviving cells as well as their progenies are indeed resistant. Some cases of successful application of *in vitro* selection for resistance to parasites have been reported (Chapter 21 and Daub, 1986). It is remarkable that an *in vitro* procedure without administration of a selective agent may yield rather high frequencies of resistant (somaclonal) variants (Sacristán, 1982).

25.10 SELECTION

If resistance in the crop is essential, the screening procedure relatively cheap and/or the frequency of genes for the resistance relatively low, selection will be done early in the cycle. In self-pollinating crops this is the F_2, in vegetatively propagated crops the seedling population derived from the cross. In potato, for instance, selection for resistance to *Phytophthora infestans* and to virus Y^N is carried out as early as in the seedling generation.

When resistance is introduced from agronomically low standard material (e.g. related wild species), several cycles of backcrossing are required to get rid of undesirable genes that have been contributed by the donor of resistance. Knott (1989) found that undesirable traits in wheat were present in some cases even after ten backcrosses.

With cross-pollinating crops, the purpose is to raise the frequency of alleles for resistance in the population. During maintenance and seed multiplication repeated tests for resistance are necessary, to prevent a possible increase of the allele frequencies for susceptibility. Recurrent selection is a very effective way of increasing the level of resistance in the population of a genetically heterogeneous cross-pollinating crop. This approach has led to increased levels of resistance in alfalfa to several parasite species (Hanson *et al.*, 1972).

In general it can be said, that the stronger the selection for resistance, the higher the chance for ending up with monogenic, possibly not durable, types of resistance. If lack of durability of resistance is known to be a problem in a pathosystem, selection for partial, polygenic types of resistance is recommended (Parlevliet, 1989). Since partial resistance of this type is inherited oligo- or polygenically, it cannot be expected that one cycle of crossing and selection will lead to a satisfactory level of resistance. A series of recurrent selection cycles is required. If in each cycle only about 30% of the most susceptible plants are discarded, a substantial increase in level of partial resistance can be obtained (Parlevliet and Van Ommeren, 1988 a,b).

25.11 STRATEGIES FOR IMPROVING DURABILITY

Several strategies have been developed to utilize major genes for (probably ephemeral) resistance in such a way, that development of virulent races is prevented.

25.11.1 Multilines and cultivar mixtures

One approach is to increase the genetic diversity of the crop that is grown. In nature, stands of plants belonging to one species usually consist of many different genotypes. This complexity of natural plant populations is considered to be one of the causes of the apparent stability of natural plant–parasite systems. The highest degree of genetic complexity is achieved by a mixture of many genotypes. In self-pollinating crops this is not a matter of course. Multilines and cultivar mixtures have been advocated as a way of increasing the diversity for resistance but retaining the agronomic value of the crop (Jensen, 1952). The best known example is the multiline-breeding in oat with resistance to crown rust (Browning and Frey, 1981). It has been demonstrated that

multilines may also be effective against insects. In a mixture of biotype-specifically resistant and susceptible rice genotypes a disproportionally low amount of brown plant hoppers (*Nilaparvata lugens*) developed (Weerapat *et al.*, 1977). The multiline procedure may pay if the epidemic region is very large. Drawbacks such as the series of backcrosses that are needed to create the near-isogenic lines and the ease by which complex races develop have prevented widespread application of this strategy.

An alternative to the multiline is the cultivar mixture. Each cultivar carries a different, still effective race-specific major-gene for resistance. Wolfe (1985) concluded that a cultivar mixture could protect barley against powdery mildew.

25.11.2 Cultivar diversification

A less drastic method of increasing diversity in the host population is by cultivar diversification, advocated in, for example, the British and Dutch lists of recommended cultivars of barley and wheat. The method involves the cultivation of two or three cultivars on each farm, each cultivar carrying different genes for resistance to the target pathogens. The parasite will have to cope with various genes for resistance within the epidemic region. Parasite genotypes that are able to overcome the resistance of one of the cultivars in the 'patchwork' of the crop may not be able to infect the other cultivars of the crop, and as a consequence, epidemic build-up of the virulent race will be reduced or prevented.

25.11.3 Duplicate genes

When a cultivar carries two or more major genes to which no virulence is known to occur in the parasite population, the cultivar has a double protection against the parasite. A single mutation in the parasite from avirulence to virulence for one of the genes for resistance is not sufficient to restore a compatible relationship with the plant. There are indications that the relative durability of the resistance of the wheat cultivars 'Manella' and 'Felix' to stripe rust (Table 25.4) was due to such a double barrier (R.W. Stubbs, personal communication). Maximum benefit is obtained from this system if the resistance genes are used only in specific combinations and not singly in other cultivars. Occurrence of single genes in some cultivars allows the parasite to develop virulence through single step changes (Parlevliet, 1991).

25.11.4 Integrated control

If through various measures the parasite population is kept at a continuously low level, the durability of a race-specific resistance may increase considerably. The chances of a new race arising and of establishing itself are then much reduced. An example in which a combination of measures resulted in increased durability of resistance is the pathosystem of potato/ *Synchytrium endobioticum*. The fungus causes the wart disease, and is soil-borne. The resistance of the potato is race-specific, and various races are known (Hampson and Proudfoot, 1974). The measures that were taken in The Netherlands were: (1) the requirement by law that any new potato cultivars released should be fully resistant; (2) the prohibition of potato growing in fields in which new races had been discovered; (3) application of several phytosanitary control measures, in fields used for seed potato production. This combination of measures has been so successful, that the law, requiring potato cultivars to be resistant to wart disease was withdrawn in 1974.

Integrated control may not only increase the durability of resistance, but also reduce the need for the application of ecologically harmful pesticides. In glasshouse crop production, particularly, the use of natural enemies of pest species may be an important element of integrated control. Breeding for plant characteristics that promote the effectiveness of such predators may offer interesting results (section 25.5).

25.11.5 Conditions for the successful application of the strategies

Strategies designed to increase the durability of resistance genes can only be successful if certain conditions are met. Breeders and breeding institutes should co-operate closely within the region where the parasite is a problem, and use the genes for resistance in the way the strategy prescribes. Competition between breeding companies and political discord between countries pose serious limitations on the potential benefits of the various strategies. In most cases the co-operation of the farmers is also indispensable. Many of the strategies require a thorough knowledge of the numbers and identity of the genes for resistance that are used and that are held in stock, and a knowledge of the races that are present in the parasite population. It is very difficult to meet all these requirements. As a consequence, in most breeding programmes for resistance there is a complete absence of a strategy. Many valuable sources of resistance are being squandered in this way.

25.12 PERSPECTIVES

Increasing public concern over the use and long-term effects of pesticides will strongly influence crop protection policies in the future. As governments legislate to reduce the growers' dependence on chemicals, it is timely that students and scientists should receive up-to-date information on the benefits, drawbacks and potential of alternatives to pesticides.

It has been demonstrated that molecular approaches may make available novel types of, possibly durable, resistance (section 25.6.2 and Chapter 10). However, there is public concern about the ecological safety of such genetically modified crops. At present, it is questionable whether the genetically modified crops are ecologically more safe than pesticides.

One thing is certain and that is the use of (at least conventionally bred) resistant cultivars has many advantages and few drawbacks as a means of crop protection. Therefore, the prospects are excellent for the development of resistant-cultivars as an environmentally-acceptable basis for integrated programmes of pest and disease control.

REFERENCES

Bailey, J.A. (1987), Phytoalexins: a genetic view of their significance. In *Genetics and Plant Pathogenesis*, Day, P.R. and Jellis, G.J. (eds), Blackwell Science. Publ., Oxford, pp. 233–244.

Beachy, R.N., Loesch-Fries, S. and Tumer, N.E. (1990), Coat protein-mediated resistance against virus infection. *Ann. Rev. Phytopathol.*, **28**, 451–474.

Biffen, R.H. (1905), Mendel's laws of inheritance and wheat breeding. *J. Agr. Sci.*, **1**, 4–48.

Black, W., Mastenbroek, C., Mills, W.R. and Peterson, L.C. (1953), A proposal for an international nomenclature of races of *Phytophthora infestans* and of genes controlling immunity in *Solanum demissum*. *Euphytica*, **2**, 173–179.

Bowen, K.L., Teng, P.S. and Roelfs, A.P. (1984), Negative inter-plot interference in field experiments with leaf rust of wheat. *Phytopathology*, **74**, 1157–1161.

Boyer, J.S. (1982), Plant productivity and environment. *Science*, **218**, 443–448.

Browder, L.E. and Eversmeyer, M.G. (1986), Parasite: host specificity and resistance/susceptibility, two concepts, two perspectives. *Phytopathology*, **76**, 379–381.

Browning, J.A. and Frey, K.J. (1981), The multiline concept in theory and practice. In *Strategies for the Control of Cereal Disease*, Jenkyn, J.F. and Plumb, R.T. (eds), Blackwell, Oxford, pp. 37–46.

Cartwright, W.B., Caldwell, R.M. and Compton, L.E. (1946), Relation of temperature to the expression of resistance in wheats to Hessian fly. *J. Am. Soc. Agron.*, **38**, 259–263.

Chaleff, R.S. (1981), *Genetics of Higher Plants. Applications of Cell Culture*, Cambridge University. Press, Cambridge.

Cobb, N.A. (1892), Contribution to an economic knowledge of the Australian rusts (Uredineae). *Agr. Gaz. New South Wales*, **3**, 60–68.

Cramer, H.H. (1967), Plant protection and world crop production. *Pflanzenschutz Nachrichten*, Bayer 20.

Daub, M. (1986), Tissue culture and the selection of resistance to pathogens. *Ann. Rev. Phytopathol.*, **24**, 159–186.

Day, P.R. (1974), *Genetics of Host–Parasite Interaction*, W.H. Freeman, San Francisco.

Day, P.R. (1976), Gene functions in host–parasite systems. In *Specificity in Plant Diseases*, Wood, R.K.S. and Graniti, A. (eds), Plenum Press, New York, London, pp. 65–73.

Dicke, M., Sabelis, M.W., Takabayashi, J., Bruin, J. and Posthumus, M.A. (1990). Plant strategies of manipulating predator–prey interactions through allelochemicals: prospects for application in pest control. *J. Chem. Ecol.*, **16**, 3091–3118.

Dixon, G.R. (1981), *Vegetable Crop Diseases*, McMillan, London.

Dussourd, D.E. and Eisner, T. (1987), Vein-cutting behaviour: insect counterploy to the latex defense of plants. *Science*, **237**, 898–901.

Ellingboe, A.H. (1976), Genetics of host–parasite interactions. In *Physiological Plant Pathology Encyclopedia of Plant Physiology*, New Series, Vol. 4, Heitefuss, R. and Williams, P.H. (eds), Springer-Verlag, Berlin, Heidelberg, New York, pp. 761–778.

Ellis, P.R., Hardman, J.A., Crisp, P. and Johnson, A.G. (1979), The influence of plant age on resistance of radish to cabbage root fly egg-laying. *Ann. Appl. Biol.*, **93**, 125–131.

van Emden, H.F. (1987), Cultural methods: the plant. In *Integrated Pest Management*, Burn, A.J., Coaker, T.H. and Jephson, P.C. (eds), Academic Press, New York, pp. 27–68.

Eriksson, K.B. (1974), Intraspecific variation in *Ditylenchus dipsaci*. 1. Compatibility tests with races. *Nematologica*, **20**, 147–162.

Eshed, N. and Wahl, I. (1975), Role of wild grasses in epidemics of powdery mildew on small grains in Israel. *Phytopathology*, **65**, 57–63.

Eyal, Z., Clifford, B.C. and Caldwell, R.M. (1968), A settling tower for quantitative inoculation of leaf blades of mature small grain plants with urediospores. *Phytopathology*, **58**, 530–531.

Flor, H.H. (1956), The complementary genic systems in flax and flax rust. *Adv. Genet.*, **8**, 29–54.

Frankel, O.H. (1977), Genetic resources as the backbone of plant protection. In *Induced Mutations Against Plant Diseases*, Proc. 1977, IAEA, Vienna, pp. 3–12.

Fry, W.E. (1978), Quantification of general resistance of potato cultivars and fungicide effects for integrated control of potato late blight. *Phytopathology*, **68**, 1650–1655.

Gallun, R.L. (1972), Genetic interrelationships between host plants and insects. *J. Environ. Qual.*, **2**, 333–334.

Gatehouse, A.M.R. and Hilder, V. (1988), Introduction of genes conferring insect resistance. *Brighton Crop Protection Conference – Pests and Diseases*, Vol. 3, British Crop Protection Council, Brighton, pp. 1245–1254.

Gavinlertvatana, S. and Wilcoxson, R.D. (1978), Inheritance of slow rusting of spring wheat by *Puccinia recondita* f.sp. *tritici* and host parasite relationships. *Trans. Br. Mycol. Soc.*, **71**, 413–418.

Gousseau, H.D.M., Deverall, B.J. and McIntosh, R.A. (1985), Temperature sensitivity of the expression of resistance to *Puccinia graminis* conferred by the Sr15, Sr9b and Sr14 genes in wheat. *Physiol. Plant Pathol.*, **27**, 335–343.

Gross, D. (1987), Chemische Abwehrstoffe der Pflanze. *Biol. Rundsch.*, **25**, 225–237.

Hadwiger, L.A. (1988), Possible role of nuclear structure in disease resistance of plants. *Phytopathology*, **78**, 1009–1014.

Hampson, M.C. and Proudfoot, K.G. (1974), Potato wart disease, its introduction to North America, distribution and control problems in Newfoundland. *FAO Plant Prot. Bull.*, **22**, 53–64.

Hanson, C.H., Busbice, T.H., Hill, R.R. Jr, Hunt, O.J. and Oakes, A.J. (1972), Directed mass selection for developing multiple pest resistance and conserving germplasm in alfalfa. *J. Environ. Qual.*, **1**, 106–111.

Hardison, J.R. (1944), Specialization of pathogenicity in *Erysiphe graminis* on wild and cultivated grasses. *Phytopathology*, **34**, 1–20.

Hare, J.D. and Kennedy, G.G. (1986), Genetic variation in plant–insect associations: survival of *Leptinotarsa decemlineata* populations on *Solanum carolinense*. *Evolution*, **40**, 1031–1043.

Harlan, J.R. and de Wet, J.M.M. (1971), Toward a rational classification of cultivated plants. *Taxon*, **20**, 509–517.

Harrison, B.D., Mayo, M.A. and Baulcombe, D.C. (1987), Virus resistance in transgenic plants that express cucumber mosaic virus satellite RNA. *Nature*, **328**, 799–802.

Heath, M.C. (1980), Reactions on nonsuscepts to fungal pathogens. *Ann. Rev. Phytopathol.*, **18**, 211–236.

Heath, M.C. (1981), A generalized concept of host–parasite specificity. *Phytopathology*, **71**, 1121–1123.

Heath, M.C. (1985), Implications of nonhost resistance for understanding host–parasite interactions. In *Genetic Basis of Biochemical Mechanisms of Plant Disease*, Groth, J.V. and Bushnell, W.R. (eds), APS Press, St. Paul, pp. 25–42.

Hedin, P.A., Maxwell, F.G. and Jenkins, J.N. (1974), Insect plant attractants, feeding stimulants, repellents, deterrents and other related factors affecting insect behaviour. In *Proc. Summer Inst. Biol. Control Plant Insects and Diseases*, Maxwell, F.G. and Harris, F.A. (eds), Univ. Press Mississippi, Jackson, pp. 494–527.

Hemenway, C., Fang, R.-X., Kaniewski, J.J., Chua, N.-H. and Tuner, N.E. (1988), Analysis of the mechanism of protection in transgenic plants expressing the potato virus X coat protein or its antisense RNA. *EMBO J.*, **7**, 1273–1280.

Hilder, V.A., Gatehouse, A.M.R., Sheerman, S.E., Barker, R.F. and Boulter, D. (1987), A novel mechanism of insect resistance engineered into tobacco. *Nature*, **330**, 160–163.

Hodkinson, I.D. and Hughes, M.K. (1982), *Insect Herbivory*, Chapman and Hall, London.

Hofferek, H. (1983), Aspekte der Spezifität bei obligaten Wirt-Pathogen Interaktionen. In *Probleme der Resistenz von Pflanzen gegen Viren, bakterielle und pilzliche Krankheitserreger sowie tierische Schaderreger*, Tagungsbericht Nr. 216, Teil I, 115–128.

Horsfall, J.G. and Cowling, E.B. (1978), Pathometry: the measurement of plant diseases. In *Plant Disease: an Advanced Treatise*, Vol. II, Horsfall, J.G. and Cowling, E.B. (eds), Academic Press, New York, pp. 119–136.

Hubbeling, N. and Ester, A. (1978), Het toetsen van spinazie op resistentie tegen wolf. *Zaadbelangen*, **32**, 36–40.

Islam, M.R. and Mayo, G.M.E. (1990), A compendium on host genes in flax conferring resistance to flax rust. *Plant Breed.*, **104**, 89–100.

Jacobs, Th. (1989a), The occurrence of cell wall appositions in flag leaves of spring wheats, susceptible and partially resistant to wheat leaf rust. *J. Phytopathol.*, **127**, 239–249.

Jacobs, Th. (1989b), Haustorium formation and cell wall appositions in susceptible and partially resistant wheat and barley seedlings infected with wheat leaf rust. *J. Phytopathol.*, **127**, 250–261.

James, C. (1981), The costs of disease to world agriculture. *Seed Sci. Technol.*, **9**, 679–685.

James, W.C. (1971), An illustrated series of assessment keys for plant diseases: their preparation and usage. *Can. Plant Dis. Surv.*, **51**, 39–65.

Jensen, N.F. (1952), Intra-varietal diversification in oat breeding. *Agron. J.*, **44**, 30–34.

Johnson, D.A. (1990), Development of rust on asparagus cultivars after inoculation with basidiospores, aeciospores and urediniospores of *Puccinia asparagi*. *Phytopathology*, **80**, 321–325.

Johnson, R. (1984), A critical analysis of durable resistance. *Ann. Rev. Phytopathol.*, **22**, 309–330.

Johnson, R., Stubbs, R.W., Fuchs, E. and Chamberlain, N.H. (1972), Nomenclature for physiologic races of *Puccinia striiformis* infecting wheat. *Trans. Br. Mycol. Soc.*, **58**, 475–480.

Jones, I.T. (1978), Components of adult plant resistance to powdery mildew (*Erysiphe graminis* f.sp. *avenae*) in oats. *Ann. Appl. Biol.*, **90**, 233–239.

Jones, I.T. and Pickering, R.A. (1978), The mildew resistance of *Hordeum bulbosum* and its transference into *H. vulgare* genotypes. *Ann. Appl. Biol.*, **88**, 295–298.

Jørgensen, J.H. (1983), Experience and conclusions from the work at Risø on induced mutations for powdery mildew resistance in barley. In *Induced mutations for disease resistance in crop plants II*, IAEA, Vienna, pp. 73–87.

Kareiva, P. and Sahakian, R. (1990), Tritrophic effects of a simple architectural mutation in pea plants. *Nature*, **345**, 433–434.

Keen, N.T. (1986), Phytoalexins and their involvement in plant disease resistance. *IOWA State J. Res.*, **60**, 477–499.

Kjaer, B., Jensen, H.P., Jensen, J. and Jørgensen, J.H. (1990), Associations between ml-o powdery mildew resistance genes and agronomic traits in barley. *Euphytica*, **46**, 185–193.

Kluge, S. (1986), Pflanzliche Protoplasten als Wirtszellen für Viren und Viroide. *Biol. Rundsch.*, **24**, 85–99.

Knott, D.R. (1989), The effect of transfers of alien genes for leaf rust resistance on the agronomic and quality characteristics of wheat. *Euphytica*, **44**, 65–72.

Kuhn, R. and Gauhe, A. (1947), Über die Bedeutung des Demissins für die Resistenz von *Solanum demissum* gegen die Larven des Kartoffelkäfers. *Zeitschr. f. Naturforsch.*, **2b**, 407–408.

Kushalappa, A.C. and Eskes, A.B. (1989), Advances in coffee rust research. *Ann. Rev. Phytopathol.*, **27**, 503–531.

Landry, B.S., Kesseli, R.V., Farrara, B. and Michelmore, R.W. (1987), A genetic map of lettuce (*Lactuca sativa* L.) with restriction fragment length polymorphism, isozyme, disease resistance and morphological markers. *Genetics*, **116**, 331–337.

Larkin, P.J. and Scowcroft, W.R. (1983), Somaclonal variation and crop improvement. In *Genetic Engineering of Plants: an Agricultural Perspective*, Basic Life Science, Vol. 26, Kosuge, T., Meredith, C.P. and Hollaender, A. (eds), Plenum Press, New York, London, pp. 289–314.

Littlefield, L.J. (1981), *Biology of the Plant Rusts*, Iowa State University Press, Ames.

Littlefield, L.J. and Heath, M.C. (1979), *Ultrastructure of Rust Fungi*, Academic Press, New York/San Francisco, London.

Lowe, H.J.B. (1974), Testing sugar beet for aphid resistance in the glasshouse: a method and some limiting factors. *Z. Angew. Entomol.*, **76**, 311–321.

Luginbill, P. (1969), Developing resistant plants – the ideal method of controlling insects. *USDA Prod. Res. Rep.*, No. 111.

McNeal, F.H., Konzak, C.F., Smith, E.P., Tate, W.S. and Russell, T.S. (1971), *A Uniform System for Recording and Processing Cereal Research Data*, USDA, Agric. Res. Serv., Washington, DC., ARS 34–121.

Mühle, E. and Frauenstein, K. (1970), Untersuchungen zur physiologischen Spezialisierung von *Erysiphe graminis* DC. *Theor. Appl. Genet.*, **40**, 56–58.

Murray, M.J. (1971), Additional observations on mutation breeding to obtain *Verticillium* resistant strains of peppermint. In *Mutation Breeding Disease Resistance*, IAEA, Vienna, pp. 171–195.

Niks, R.E. (1986), Failure of haustorial development as a factor in slow growth and development of *Puccinia hordei* in partially resistant barley seedlings. *Physiol. Mol. Plant Pathol.*, **28**, 309–322.

Niks, R.E. (1987), Nonhost plant species as donors for resistance to pathogens with narrow host range. I. Determination of nonhost status. *Euphytica*, **36**, 841–852.

Niks, R.E. (1989), Induced accessibility and inaccessibility of barley cells in seedling leaves inoculated with two leaf rust species. *J. Phytopathol.*, **124**, 296–308.

van Oijen, M. (1991), Light use efficiencies of potato cultivars with late bight (*Phytophthora infestans*). *Potato Res.*, **34**, 123–132.

O'Keefe, L.E., Callenbach, J.A. and Lebsock, K.L. (1960), Effect of culm solidness on the survival of the wheat stem sawfly. *J. Econ. Entomol.*, 53, 244–246.

Ouchi, S., Hibino, C., Oku, H., Fujiwara, M. and Nakabayashi, H. (1979), The induction of resistance or susceptibility. In *Recognition and Specificity in Plant Host–Parasite Interactions*, Daly, J.M. and Uritani, I. (eds), Tokyo University, Tokyo, pp. 49–65.

Parlevliet, J.E. (1975), Partial resistance of barley to leaf rust, *Puccinia hordei*. I. Effect of cultivar and development stage on latent period. *Euphytica*, **24**, 21–27.

Parlevliet, J.E. (1978), Race-specific aspects of polygenic resistance of barley to leaf rust, *Puccinia hordei. Neth. J. Plant Pathol.*, **84**, 121–126.

Parlevliet, J.E. (1979), Components of resistance that reduce the rate of epidemic development. *Ann. Rev. Phytopathol.*, **17**, 203–222.

Parlevliet, J.E. (1981a), Disease resistance in plants and its consequences for breeding. *Proc. Plant Breeding Symp. II*, Frey, K.J. (ed.), Iowa State University, Ames, pp. 309–364.

Parlevliet, J.E. (1981b), Race-non-specific disease resistance. In *Strategies for the Control of Cereal Disease*, Jenkyn, J.F. and Plumb, R.T. (eds), Blackwell, Oxford, pp. 47–54.

Parlevliet, J.E. (1981c), Stabilizing selection in crop pathosystems: an empty concept or a reality? *Euphytica*, **30**, 259–269.

Parlevliet, J.E. (1983a), Models explaining the specificity and durability of host resistance derived from the observations on the barley – *Puccinia hordei* system. In *Durable Resistance in Crops*, Lamberti, F., Wallen, J.M. and Graaff, N.A. van der (eds), Plenum Press, New York, pp. 57–80.

Parlevliet, J.E. (1983b), Can horizontal resistance be recognized in the presence of vertical resistance in plants exposed to a mixture of pathogen races? *Phytopathology*, **73**, 379.

Parlevliet, J.E. (1986), Coevolution of host resistance and pathogen virulence: possible implications for taxonomy. In *Coevolution and Systematics*, Stone, A.R. and Hawksworth, D.L. (eds), Systematics association, Clarendon Press, Oxford, pp. 19–34.

Parlevliet, J.E. (1989), Identification and evaluation of quantitative resistance. In *Plant Disease Epidemiology*, Vol. 2, *Genetics, Resistance, and Management*, Leonard, K.J. and Fry, W.E. (eds), McGraw-Hill, New York, pp. 215–248.

Parlevliet, J.E. (1991), Durable resistance to pathogens and how to breed for it. In *Breeding for Stress Tolerance in Cool Season Food Legumes*, ICARDA, University of Napels, in press.

Parlevliet, J.E. and van Ommeren, A. (1975), Partial resistance of barley to leaf rust, *Puccinia hordei*. II. Relationship between field trials, micro plot tests and latent period. *Euphytica*, **24**, 293–303.

Parlevliet, J.E. and van Ommeren, A. (1984), Interplot interference and the assessment of barley cultivars for partial resistance to leaf rust, *Puccinia hordei. Euphytica*, **33**, 685–697.

Parlevliet, J.E. and van Ommeren, A. (1988a), Accumulation of partial resistance in barley to barley leaf rust and powdery mildew through recurrent selection against susceptibility. *Euphytica*, **37**, 261–274.

Parlevliet, J.E. and van Ommeren, A. (1988b), Recurrent selection for grain yield in early generations of two barley populations. *Euphytica*, **38**, 175–184.

Parlevliet, J.E. and Zadoks, J.C. (1977), The integrated concept of disease resistance: a new view including horizontal and vertical resistance in plants. *Euphytica*, **26**, 5–21.

de Ponti, O.M.B. (1978), Resistance in *Cucumis sativus* L. to *Tetranychus urticae* Koch. 3. Search for sources of resistance. *Euphytica*, **27**, 167–176.

Prokopy, R.J., Collier, R.H. and Finch, S. (1983), Leaf color used by cabbage root flies to distinguish among host plants. *Science*, **221**, 190–192.

Radcliffe, E.B. and Chapman, R.K. (1965), Seasonal shifts in the relative resistance to insect attack of eight commercial cabbage varieties. *Ann. Entomol. Soc. Am.*, **58**, 892–897.

Ramirez–Villupadua, J., Endo, R.M., Bosland, P. and Williams, P.H. (1985), A new race of *Fusarium oxysporum* f.sp. *conglutinans* that attacks cabbage with type A resistance. *Plant Dis.*, **69**, 612–613.

Randle, W.M., Davis, D.W. and Groth, J.V. (1986), Interplot interference in field plots with leaf rust of maize. *J. Am. Soc. Hort. Sci.*, **111**, 297–300.

Ride, J.P. (1985), Non-host resistance to fungi. In *Mechanisms of Resistance to Plant Diseases*, Fraser, R.S.S. (ed.), Martinus Nijhoff/Dr W. Junk Publ., The Hague, pp. 29–61.

Robinson, R.A. (1987), *Host Management in Crop Pathosystems*, MacMillan, New York, London.

Roelfs, A.P. (1988), Resistance to leaf and stem rusts in wheat. In *Breeding Strategies for Resistance to the Rusts of Wheat*, Simmonds, N.W. and Rajaram, S. (eds), CIMMYT, Mexico, D.F., pp. 10–22.

Russell, G.E. (1978), *Plant Breeding for Pest and Disease Resistance*, Butterworth, London.

Saari, E.E. and Prescott, J.M. (1975), A scale for appraising the foliar intensity of wheat diseases. *Plant Dis. Reporter*, **59**, 377–380.

Sacristán, M.D. (1982), Resistance responses to *Phoma lingam* of plants regenerated from selected cell and embryogenic cultures of haploid *Brassica napus. Theor. Appl. Genet.*, **61**, 193–200.

Samborski, D.J., Kim, W.K., Rohringer, R., Howes, N.K. and Baker, R.J. (1977), Histological studies on host cell necrosis conditioned by the Sr6 gene for resistance in wheat to stem rust. *Can. J. Bot.*, **55**, 1445–1452.

Schalk, J.M. and Ratcliffe, R.H. (1976), Evaluation of ARS program on alternative methods of insect control: host plant resistance to insects. *Bull. Entomol. Soc. Am.*, **22**, 7–10.

Schertz, F.K. and Tai, Y.P. (1969), Inheritance of reaction of *Sorghum bicolor* (L.) Moench to toxin produced by *Periconia circinata* (Mang.) Sacc. *Crop Sci.*, **9**, 621–624.

Schwarzbach, E. (1976), The pleiotropic effects of the *ml-o* gene and their implications in breeding. In *Barley Genetics III*, Proc. 3rd Int. Barley Genetics Symp., Gaul, H. (ed.), Verlag Karl Thiemig, München, pp. 440–445.

Scott, P.R. and Benedikz, P.W. (1977), Field techniques for assessing the reaction of winter wheat cultivars to *Septoria nodorum*. *Ann. Appl. Biol.*, **85**, 345–358.

Sherwood, R.T. and Vance, C.P. (1980), Resistance to fungal penetration in Gramineae. *Phytopathology*, **70**, 273–279.

Shintaku, M.H., Kluepfel, D.A., Yacoub, A. and Patil, S.S. (1989), Cloning and partial characterization of an avirulence determinant from race 1 of *Pseudomonas syringae* pv. *phaseolicola*. *Physiol. Mol. Plant Pathol.*, **35**, 313–322.

Sidhu, G.S. (1987), Host–parasite genetics. In *Plant Breeding Reviews*, Vol. 5, Janick, J. (ed.), Van Nostrand Reinhold Comp., New York, pp. 393–433.

Simmonds, N.W. (1983), Strategy of disease resistance breeding. *FAO Plant Prot. Bull.*, **31**, 2–10.

Stakman, E.C., Stewart, D.M. and Loegering, W.Q. (1962), Identification of physiologic races of *Puccinia graminis* var. *tritici*. *USDA-ARS Tech. Bull.*, No. E617.

Staskawicz, B.J., Dahlbeck, D. and Keen, N.T. (1984), Cloned avirulence gene of *Pseudomonas syringae* pv *glycinea* determines race-specific incompatibility on *Glycine max* (L.) Merr. *Proc. Natl. Acad. Sci. USA*, **81**, 6024–6028.

Subrahmanyam, P., Moss, J.P. and Rao, V.R. (1983), Resistance to peanut rust in wild *Arachis* species. *Plant Dis.*, **67**, 209–212.

Taylor, W.E. and Bardner, R. (1968), Effects of feeding by larvae of *Phaedon cochleariae* (F.) and *Plutella maculipennis* (Curt.) on the yield of radish and turnip plants. *Ann. Appl. Biol.*, **62**, 249–254.

Thompson, A.R., Ellis, P.R., Percivall, A.L. and Hardman, J.A. (1980), Carrot fly. Integrated control using insecticides with carrot cultivars of differing susceptibilities. *1979 Rep. Natl. Vegetable Res. Stn.*, p. 30.

Tottman, D.R. and Makepeace, R.J. (1979), An explanation of the decimal code for the growth stages of cereals, with illustrations. *Ann. Appl. Biol.*, **93**, 221–234.

Turlings, T.C.J., Tumlinson, J.H. and Lewis, W.J. (1990), Exploitation of herbivore-induced plant odors by host-seeking parasitic wasps. *Science*, **250**, 1251–1253.

Vaeck, M., Reynaerts, A., Hofte, H., Jansens, S., de Beuckleer, M., Dean, C., Zabeau, M., van Montagu, M. and Leemans, J. (1987), Transgenic plants protected from insect attack. *Nature*, **328**, 33–37.

Vanderplank, J.E. (1968), *Disease Resistance in Plants*, Academic Press, New York, London.

van der Vrie, E., McMurtry, J.A. and Huffaker, C.B. (1972), Ecology of Tetranychid mites and their natural enemies: a review. III. Biology, ecology and pest status and host-plant relations of Tetranychids. *Hilgardia*, **41**, 343–432.

Wang, D., Hao, M.S.H. and Waygood, E.R. (1961), Effect of benzimidazole analogues on stem rust and chlorophyll metabolism. *Can. J. Bot.*, **39**, 1029–1036.

Watson, I.A. (1970), The utilization of wild species in the breeding of cultivated crops resistant to plant pathogens. In *Genetic Resources in Plants – their Exploration and Conservation*, Frankel, O.H. and Bennett, E. (eds), Blackwell, Oxford, pp. 441–457.

Weerapat, P., Purivirojkul, W. and Chaturonrangsri, T. (1977), Mixing rice varieties to combat brown planthopper. *Int. Rice Res. Newsletter*, **2**(1), 3.

van de Weg, W.E. (1989), Screening for resistance to *Nectria galligena* Bres. in cut shoots of apple. *Euphytica*, **42**, 233–240.

Wink, R. (1988), Plant breeding: importance of plant secondary metabolites for protection against pathogens and herbivores. *Theor. Appl. Genet.*, **75**, 225–233.

de Wit, P.J.G.M. (1992) New tendencies in phytopathology around the year 2000. *Med. Fac. Landbouw Univ. Gent.*, **57/2a**, 97–107.

de Wit, P.J.G.M., Hofman, A.E., Velthuis, G.C.M. and Toma, I.M.J. (1987), Specificity of active defense responses in plant–fungus interactions: tomato–*Cladosporium fulvum*, a case study. *Plant Physiol. Biochem.*, **25**, 345–351.

Wolfe, M.S. (1985), The current status and prospects of multiline cultivars and variety mixtures for disease resistance. *Ann. Rev. Phytopathol.*, **23**, 251–273.

Wood, E.A. and Starks, K.J. (1972), Effect of temperature and host plant interaction on the biology of three biotypes of the greenbug. *Environ. Entomol.*, **1**, 230–234.

Young, N.D. (1990), Potential applications of map-based cloning to plant pathology. *Physiol. Mol. Plant Pathol.*, **37**, 81–94.

Zaag, D.E. van der (1959), Some observations on breeding for resistance to *Phytophthora infestans*. *Eur. Potato. J.*, **2**, 292–299.

Zadoks, J.C. (1972), Reflections on disease resistance in annual crops. In *Biology of Rust Resistance in*

Forest Trees, Bingham, R.T., Hoff, R.J. and McDonald, G.I. (eds), Proc. NATO IUFRO Advanced Study Inst., U.S. Dep. Agr., Forest Service, Misc. Publ. 1221, pp. 43–63.

Zadoks, J.C., Chang, T.T. and Konzak, C.F. (1974), A decimal code for the growth stages of cereals. *Weed Res.*, **14**, 415–421.

Zadoks, J.C. and Schein, R.D. (1979), *Epidemiology and Plant Disease Management*, Oxford University Press, New York.

Part Eight
Specific Trait Breeding

26

Breeding for improved symbiotic nitrogen fixation

L.R. Mytton and L. Skøt

26.1 INTRODUCTION

All crops are limited in the range of agricultural environments in which they have economic value. Crop improvement therefore rests on developments which enhance our ability to match genotypes to environments. Progress in agriculture can be made either by modifying genotypes to improve economic performance in particular environments or by altering the environment to make it better suited to the genotype. Both approaches are important and interrelated since each needs to be constantly cross-referenced to the other for the whole to succeed.

The need to cross-reference is brought about by unpredictable reactions between plants and environmental variables. These effects are generally identified as genotype × environment interactions (G×E). If plants reacted additively to environmental change, broad referencing would be unnecessary. Alterations in productivity would be proportional to changes in limiting factors and measurements made over a restricted range of environment would have predictive value for all others. Non-additive reactions are the basis of poor heritabilities. They are important in all breeding programmes but are of special significance in developing better symbiotic associations.

Plant Breeding: Principles and prospects. Edited by M.D. Hayward, N.O. Bosemark and I. Romagosa. Published in 1993 by Chapman & Hall, London. ISBN 0 412 43390 7.

Symbioses create unique problems because the phenotype is the product of an intimate association between different organisms. As a result, the development of the symbiotic phenotype can be influenced by genes which affect each organism's independent existence as well as those which determine their joint association. Such factors have the potential to produce levels of complexity between genotypes and environments that are reflected in the phenotype which are not easily unravelled and thus not easily stabilized. It is, perhaps, important to emphasize at this point that the agronomic objective is the phenotype rather than the genotype. Inferences about genotype are made from phenotype and assessment of the strength and stability of this relationship is of central importance to the speed and precision with which genetic patterns can be altered in populations. Thus, the first task in any breeding programme is to identify and quantify phenotypic variations in relevant traits. Methods must then be devised which will reveal and quantify the genetic basis of these differences. It is also reasonable to assume that the more we understand about the genetic and environmental mechanisms which control and regulate the phenotype the better will be our ability to manipulate them to a particular end.

This chapter will therefore consider methods of measuring phenotypic variation in symbiotic traits, demonstrate ways of revealing and quantifying the genetic basis of these differences and discuss how the information revealed can be used to develop strategies for improving the symbiotic phenotype (i.e. the legume plant in association with the *Rhizobium* bacterium). The examples used will rely on the legume–*Rhizobium* symbiosis, but the principles involved will be applicable to any symbiotic association. We also aim to identify topics where more information is needed and discuss the benefits which this could produce.

26.2 MEASUREMENT OF PHENOTYPIC VARIATION IN SYMBIOTIC TRAITS

It is self-evident that such measurements require the presence of both partners. Despite the apparent obviousness of this requirement it is surprising how often it is overlooked, especially in the early stages of exotic germplasm evaluation.

Plant introduction has been an important and productive method of extending the range of variation available to breeders, but proper field evaluation of new legume germplasm cannot be made if the soil does not contain appropriate rhizobia. This problem is not always easily solved. The various legume genera and species have evolved different degrees of specialization in their compatibility with rhizobia. Early experiments on the transfer of nodule isolates between species revealed limitations to exchange and led to the cross-inoculation taxonomic grouping of rhizobia (Fred *et al.*, 1932). Although the taxonomy of *Rhizobium* has developed considerably in recent years (Table 26.1; Jordan, 1984; see also summary by Sprent and Sprent, 1990), the old cross-inoculation groupings are relevant to the agronomic assessment of legume germplasm because they provide a practical guide to the *Rhizobium* requirements of particular legume genera and species. Generally speaking, it is a simple matter to introduce appropriate rhizobia into the soil once the need for them has been identified and suitable strains have been located. Commercial sources are sometimes available but, even when they are not, *Rhizobium* is easily isolated from root nodules and can readily be cultured (for methods see Vincent, 1970). When collecting novel germplasm it would be sensible to collect

Table 26.1. *Rhizobium* species and their host legumes

Microsymbiont	Host plant
Rhizobium leguminosarum	
bv. *viceae*	Pea, vetch, broad bean
bv. *trifolii*	Clover
bv. *phaseoli*	French bean
R. meliloti	Alfalfa, *Melilotus*
R. loti	Birdsfoot trefoil
R. fredii	Soybean
Bradyrhizobium japonicum	Soybean
B. spp.	Mungbean, lupin

nodules or soil to ensure that appropriate *Rhizobium* germplasm is available if required. If there is any doubt about the ability of the host to nodulate and fix nitrogen during the early stages of plant evaluation, growing the plants with and without added nitrate would be valuable. This should quantify the genetic potential of the plant and reveal any problems with fixation. Such difficulties can then be dealt with at a later stage. However, undomesticated legumes sometimes grow more poorly with fertilizer than with nodule nitrogen. Fay *et al.* (1991) recently evaluated 64 *Trifolium* species using the above approach. Results were generally satisfactory, but a small number of species responded adversely to added nitrate and the method did not therefore indicate their true potential.

26.2.1 Measuring nitrogen fixation

Accurate field measurement of nitrogen fixation is surprisingly difficult. The three methods most commonly used are as follows.

1. Acetylene reduction. The 'standard' method of acetylene reduction involves incubating fixing systems in a closed vessel of known volume containing 10% acetylene and measuring ethylene accumulation by gas chromatography (Hardy *et al.*, 1973). However, in recent years a number of problems have been identified in using this method with legumes (Witty and Minchin, 1988). It has a high potential for error when used to

make comparisons between different genotypes or different treatments, or even the same genotypes in the same treatments, if subjected to differential disturbance.

2. Nitrogen difference. This method relies on comparing total N accumulation in the fixing plant with that in a non-fixing reference plant. Its accuracy rests on the assumptions that the soil N pool under both plants is equal and that their quantitative recovery of soil N is also comparable. The validity of these assumptions needs careful checking when using the method.

3. ^{15}N fertilizer addition. The naturally occurring heavy isotope ^{15}N provides an alternative method of quantifying fixation. The approach used is essentially the same as for the nitrogen difference method, but in addition, fertilizer enriched with ^{15}N is used to 'label' the nitrogen in the rooting medium. This method shares a number of problems with the nitrogen difference approach (for a review see Witty and Giller, 1991), but it has the advantage that its accuracy is less affected by differences in quantitative recovery so long as both treatments recover ^{15}N and ^{14}N in the proportions available in the soil. However, it is more costly and laborious, it requires access to a mass spectrometer for analysis, and the assumptions on which it is based are not easily validated in the field.

In practical terms there may be much to recommend dispensing with elaborate methods of assessing fixation and relying instead on simple indirect methods. When legumes are grown in soils of low nitrogen status, there is invariably a good correlation between fixation and dry matter production. Available soil nitrogen can be reduced by previous cropping with plants which have a high nitrogen demand, or by incorporating low-N organic matter such as chopped straw. In the early stages of assessment, measurement of simple parameters such as dry weight, grain yield and total nitrogen are not only straightforward but may also be of direct relevance to the objectives of the breeding programme. However, it should be recognized that the role of the plant in fixing nitrogen is confounded with its capacity to use it. This

suggests that steps should be taken to screen out symbioses which are limited by carbon fixation, because it will be difficult to improve if the plant has no capacity to use more nitrogen.

26.3 CONVENTIONAL BREEDING APPROACHES TO IMPROVEMENT OF SYMBIOTIC NITROGEN FIXATION

26.3.1 Partitioning the effects of plant genotype, *Rhizobium* genotype and the environment

When dealing with the interrelationship of several factors, the effects of each on the others can only be measured accurately where all combinations are measured simultaneously. Such factorial experiments provide a wide inductive basis to the main effect comparisons and they reveal and quantify interactions between the various factors. Thus, the ideal approach to partitioning the major factors influencing symbiotic variability comprises a factorial arrangement of diverse macro- and microsymbionts in a range of appropriate environments.

Table 26.2 provides a summary of factorial experiments which reveal and quantify plant–*Rhizobium* genotypic effects in nitrogen fixation.

Table 26.2. Estimated values of components of variation as percentage of phenotypic variance of sample populations of three legume species grown with various *Rhizobium* strains

	V. faba	*T. repens*	*M. sativa*	
	Fixed[a]	Fixed[a]	Fixed[a]	Mixed[a]
Source of variation				
Host varieties	11.8	33.5	10.2	4.8
Rhizobium strains	8.9	5.0	19.8	21.0
Strain × variety	73.8	23.0	5.7	6.0
Residual	5.3	33.5	63.4	68.2

[a] Analysis model

The results are synopses of previously published data taken from experiments on field bean (*Vicia faba*), white clover (*Trifolium repens*) and lucerne (*Medicago sativa*) (Mytton, 1975; Mytton and De Felice, 1977; Mytton *et al.*, 1984). Two points emerge clearly from these experiments. First, there is a wide range of phenotypic variation in these symbioses, much of which is genotypically based. Second, in all the species examined, a substantial proportion of this variation springs from host × strain interactions. Although much is now known about the genetics of nitrogen fixation in microorganisms, including rhizobia (for reviews see Djordjevic *et al.*, 1987; Rolfe and Gresshoff, 1988; Long, 1989a,b; Martinez *et al.*, 1990), these factorial experiments suggest that more attention should be directed at understanding plant mechanisms which control and regulate the activity of the established nodule (see also Mytton *et al.*, 1992). Precise expression of *Rhizobium* genes concerned with fixation is clearly a quantitative rather than a qualitative trait. Thus, decisions on selection and breeding strategies to improve fixation need to be made in the light of information on heritabilities. The data presented here allow comparisons of phenotypic and genotypic variances and can therefore be used to determine broad sense heritabilities (h_b^2), i.e. the ratio of genotypic to phenotypic variance. In these experiments it can be inferred that both cultivar and strain differences are genetic in origin, but the factorial arrangement of the symbionts and the statistical treatment of the data reveals the extent to which the phenotype is dependent upon joint rather than independent effects. Taking the *Medicago* data as an example, if only the genetic variation of plant origin is considered to be broadly heritable, then heritability (h_b^2) is 5% which is a low value. If plant cultivar × *Rhizobium* strain interactions are also heritable, then there is an increase in h_b^2 (up to 11%). Finally, if *Rhizobium* strain differences are also considered as heritable, h_b^2 increases to 32%. Unfortunately, under many agricultural conditions, *Rhizobium* strain differences cannot be regarded as heritable. This is because we lack precise knowledge of the traits which determine success and we have little information on their underlying genetic control. As a consequence, it has proved difficult to control nodulation by particular strains in mixed populations. It follows from this, that a fixed effects model analysis of variance is inappropriate for estimating stability of such genotypic effects. The rhizobia tested must be regarded as a random sample of populations at large and this alters the expectation of mean squares in the plant cultivar main effects comparison. Thus, the mixed effects model analysis of variance is a more appropriate estimate of stability of cultivar effects and by this analysis cultivar differences account for only 4.8% of total variance. Variation from uncontrolled sources is also high (68%) and host strain interaction effects are slightly larger (6%) than cultivar main effects. If we consider that these data derive from an experiment carried out under controlled conditions, the results emphasize the difficulty of identifying superior genotypes with any degree of certainty. The low heritabilities inevitably indicate that, under field conditions, with their more variable environments and uncontrolled mixed populations of rhizobia, the rate of genetic advance through simple selection will, at best, be slow. Nevertheless, the data and their interpretation point to two possible strategies for improving the symbiosis. These are better general symbiotic competence and improved specific symbiotic competence which are analogous to the concept of specific and general combining ability in plant hybridization programmes.

(a) Improving general symbiotic competence

Both the plant and the *Rhizobium* genotype can be appropriate targets for this approach. At present the plant is probably the most logical target, because methods for establishing and maintaining new plant populations as stable components of agricultural systems are well defined. This advantage cannot reliably be conferred on *Rhizobium* populations in soil, but this latter approach should not be neglected.

An essential component of screening and testing for general competence would be to examine the reaction of plants to a range of rhizobia. However, this creates problems in defining a sample which will be generally representative of rhizobia in agriculture. One approach would be to examine the reaction of plants to *Rhizobium* populations from various locations. Mytton and Livsey (1983) examined the reaction of five white clover cultivars (*T. repens*) to rhizobia from five locations. In general, no cultivar was outstanding with all rhizobia and no population was excellent with all cultivars. An extreme example of the problem is illustrated in the sample of data shown in Table 26.3. The outstanding perform-

Table 26.3. Average dry matter yield (mg) of two clover cultivars H grown with *Rhizobium* isolates from two different locations

Site	Cultivar		Site mean
	S184	Menna	
Reading	7.5	12.6	10.05
Swindon	21.8	5.4	13.60
Cultivar mean	14.65	9.60	

ance of cv. 'S184' with rhizobia from Swindon compared with the poor performance of cv. 'Menna' with the same population should be noted. Nevertheless, some cultivars did reasonably well with all populations, indicating a potential in the species for development of general symbiotic competence. This finding has also been confirmed by more recent studies on a range of African *Trifolium* species (Mytton *et al.*, 1988).

So far, all of the examples cited suggest that plant genotype × *Rhizobium* genotype interactions are a major reason for poor heritabilities. However, by analogy with analysis of genotype × environment (G×E) interaction, it is possible to demonstrate that some of this interactive variation can be regarded as usefully heritable. Breese (1969) used the joint regression analysis proposed by Finlay and Wilkinson (1963) to demonstrate heritabilities of G×E effects in cocksfoot (*Dactylis glomerata*). This approach has been adapted for analysis of symbiotic interactions (Mytton *et al.*, 1984). In a factorial arrangement of plants and rhizobia, each *Rhizobium* strain can be regarded as an unspecified genetic component (designated 'R-

genetic') whose effect is measured by the average performance with all plant cultivars under test. This provides a quantitatively scaled R-genetic index which can be arranged from poorest to best. The differential performance of cultivars can then be computed against this index as linear regressions of individual cultivar performance with each *Rhizobium* strain. Interactions can then be represented graphically and may be revealed as being of two types. First, the linear regression coefficient for each cultivar represents its symbiotic competence with *Rhizobium* strains of differing relative growth capacity. Second, departures from linearity indicate associations which can be regarded as genetically anomalous and, thus, represent specific interactions rather than general responses. Figure 26.1 shows an example of such an analysis (Mytton *et al.*, 1984) taken from an experiment on lucerne (*M. sativa*). It illustrates how the variable symbiotic responses result in interactions. Substantial differences in regression values are associated with lines which intersect. Thus, the ranked order of cultivars alters at various points along the R-genetic index. This behaviour is now revealed to be less random than would appear from the raw data. In this example, about half the variation attributable to host × strain interactions was explained by differences between the slopes of linear regressions so that a significant portion of what previously appeared to be random and uninterpretable effects, is now revealed as a series of undefined, but generalizable, and thus predictable, differences between the responses of cultivars to changes in the genetic make-up of rhizobia. The value of this approach is perhaps best illustrated by the performance of Siriver. It has the highest mean productivity of all cultivars tested and it also has a regression line of low slope with a relatively low standard error and is thus revealed to have the desired trait of high N-fixing ability stably expressed over a range of genetically variable *Rhizobium* strains. This could doubtless be developed further by use of established selection and hybridization procedures. It is appropriate here to give some consideration to the mechanism which gives Siriver its symbiotic stability. It could have arisen by the development of a

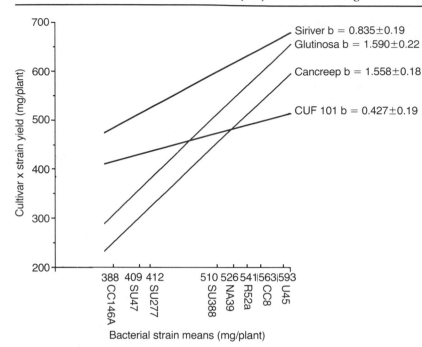

Fig. 26.1. Regression of individual cultivar yields on mean yield of all cultivars with each of eight strains of *R. meliloti.*

mixture of differentially adapted individuals permitting contrasting genotypes within the cultivar to form a series of outstandingly effective symbiotic associations with a range of *Rhizobium* genotypes. This would widen average effectiveness and raise mean productivity but variability would also be increased. Interestingly, this cultivar had low variability, so its improved performance appears to result from better individual buffering rather than from improved population buffering. However, better population buffering is an approach to improving symbiotic performance which would be worth exploring. This would simply require the development of mixtures of plant genotypes which were differentially adapted to a wide range of *Rhizobium* genotypes. This would result in a wide range of phenotypes which would not be appropriate in a grain legume crop, but would be less of a problem in a forage crop where poor phenotypes would be eliminated by competition.

(b) Improving specific symbiotic competence

The data presented above show that phenotypic differences in nitrogen fixation are not vested wholly in the plant genotype, or the *Rhizobium* genotype but arise to a greater or lesser extent from joint action between specific individuals. It is therefore self-evident that the most rapid and efficient method of exploiting this variation is by coincidental selection of both genotypes, with the aim of producing a highly co-adapted inoculation package. However, this approach can only succeed if some method can be found of ensuring that the selected rhizobia can preferentially nodulate the selected host in the field. In *Rhizobium*-free soil this should be reasonably straightforward, but even here difficulties can arise. Interactions can occur between soils, or soil factors, which make some *Rhizobium* strains unsuccessful (Mytton and Hughes, 1984; Sprent and Sprent, 1990). The factors involved are not well understood and more information in this area would be valuable. These problems are exacerbated in soils which contain indigenous rhizobia because in these situations there is also competition for nodule sites. Some genetic mechanism is therefore required to overcome these problems.

The original concept of *Rhizobium* species was based on the ability of host plants successfully to exchange rhizobia. Extensive testing revealed legume–*Rhizobium* groupings which were generally so strongly interactive as to preclude frequent and fully compatible exchange of partners. This indicates a genetically determined recognition mechanism operating between symbionts. Finding and enhancing this trait within species should help to ensure that the 'right' rhizobia re-nodulate the 'right' host. Mytton and De Felice (1977) showed the existence of this useful trait in white clover. The experiment involved screening samples of a single white clover population against four individual *Rhizobium* genotypes known to fix different amounts of nitrogen and which therefore produce differences in dry matter production. The results were compared with those produced by plants grown with 50:50 mixtures of the strains in all possible pair combinations. It was found that some components of *Rhizobium* strain mixtures influenced plant yield more than others indicating differential nodulating ability between strains. There was also considerable plant-to-plant variation within treatments indicating possible modification of this trait by the plant genotype. This was further investigated by cloning selected plants and reinfecting them with pure and mixed cultures of *Rhizobium*. An example of the effects revealed is shown in Table 26.4. The first two

Table 26.4. Dry matter yield (mg) (mean of three vegetative propagules) of selected plant genotypes grown with pure and mixed cultures of two *Rhizobium* strains

Plant genotype	Selected as characteristic of strain	Grown with Rhizobium strains			Genotype mean
		r8	r6	r8+r6	
A1	r8	44	20	44	36
A2	r8	32	19	30	27
B1	r6	40	26	25	30
B2	r6	32	11	15	19
B3	r6	33	21	18	24
B4	r6	45	19	25	30
Strain mean		37	19	28.5	

columns of data show that the four plant genotypes behave similarly within each strain

treatment but there is a distinct difference between the two *Rhizobium* strain with 'r8' being about twice as effective as 'r6'. When the plants are grown in a mixture of the *Rhizobium* strains differences emerged between plant genotypes with the 'A' plants being about twice as productive as the 'B1' plants. The analysis in Table 26.5 confirms that a significant genetic differences exist between plant genotypes when exposed to the mixture of rhizobia. This

Table 26.5. Estimates of plant genotypic and environmental components of variation for six selected S.100 genotypes grown with *Rhizobium* strains r6, r8 and a mixture of the two

Strain	MS for genotypes[a] (Vp)	MS for varieties within genotypes (Ve)	Genotypic variance (Vg)	Heritability
r6	70.5 ns	26.6	–	
r8	107.8 ns	65.2	–	
r6+r8	319.0***	27.4	16.2	0.37

[a] ns, non significant; ***, $P \leq 0.001$.

effect occurs in the strain mixture because the 'B' plants are nodulated almost exclusively by strain 'r6' whereas the 'A' plants are nodulated mostly by strain 'r8'. The plant genotype is clearly having a strong influence over which strain it is nodulated by. The heritability of this effect has been confirmed in subsequent breeding experiments by Hardarson and Jones (1979). The mechanisms involved in competition will be covered in more detail in a later section.

The information considered above has been used to develop an experimental breeding strategy (Mytton *et al.*, 1988). The approach is as follows; clover plants are grown in soil which is relatively low in soluble nitrogen, but is well supplied with other nutrients so that fixation will correlate with productivity. The soil has a large natural population of rhizobia which has been further supplemented by introducing a wide range of strains from other locations. Phenotypic selection is done on the basis of dry matter production although a check is also kept on protein content. The best nodules are also selected from the best plants. Rhizobia are isolated, multiplied and reintroduced into the

Table 26.6.   Response to selection for increased yield and nitrogen fixation in white clover (g DM/plant)

Generation	Growing conditions			
	Fixed N	% increase	Combined N	% increase
Lines selected for increased yield with *Rhizobium*				
Base pop.	5.59		7.76	
F_2	7.44	33.1	8.50	9.5
Base pop.	5.76		7.90	
F_3	8.75	51.9	8.64	9.4
Lines selected for increased yield with combined nitrogen				
Base pop.	5.59		7.76	
F_2	6.80	21.7	8.73	12.5
Base pop.	5.76		7.90	
F_3	7.36	27.7	8.98	13.7

soil as each screening cycle proceeds. An identical complementary programme is carried out by screening in soil supplemented with non-limiting applications of soluble nitrogen. Progeny from both programmes are tested with and without soluble nitrogen. A sample of results is shown in Table 26.6. In some selection lines yield in the F_2 and F_3 generations was respectively 33% and 51% better than the base population. The same material grown with non-limiting nitrate was only 9% better than the base population in both generations. This indicates some general improvement in yield capacity regardless of nitrogen source, but it is evident that the greatest advance under selection has been in the capacity to fix nitrogen. F_3 Progeny in the complementary high N programme showed a 14% improvement in yield when grown under high nitrogen and a 28% improvement with *Rhizobium*. This suggests that improving plant vigour, and hence the demand for nitrogen, can bring about worthwhile improvements in fixation, but these effects were not as large as those achieved in the fixation programme.

These experiments clearly show that the base population contains two types of potentially high yielding plant; those which are significantly limited by fixation and those which have little capacity to exploit additional nitrogen without further improvements in carbon fixation capacity. Carbon limited plants impede the rate of genetic advance under selection and must be eliminated by screening under high

nitrate. The breeding programme is carried out by growing plants in soil with a natural population of rhizobia. Both plant and *Rhizobium* genotypes are selected and carried forward for further cycles of hybridization and reselection.

The potential of this approach has been demonstrated in hill land inoculation trials. Young and Mytton (1983) have shown that in situations where populations of native rhizobia are poor, genetic matching of host plants with inoculant strains can be used to substantially manipulate yield of specific cultivars. Typical results are shown in Figure 26.2. More recent studies confirm that the approach has general relevance, even in situations where native *Rhizobium* populations are large and reasonably effective (Mytton, 1987, 1989). However, trials of various plant–*Rhizobium* combinations at different locations also show that environmental factors can alter responses. This is a common problem in breeding programmes and due consideration must be given to it in selection procedures.

26.3.2 Environmental effects

Precise definitions of the effects of environmental variables on the survival, nodulating ability, and effectiveness of host–*Rhizobium* associations under field conditions are not easily determined, but some quantification of these effects must accompany attempts at genetic

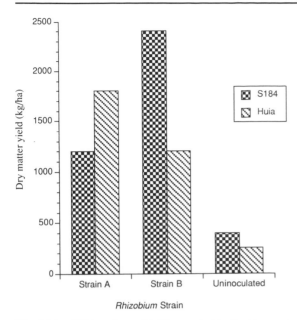

Fig. 26.2. White clover DM harvested in first harvest year (three cuts).

improvement if the results are to be of general relevance to agriculture.

The data of Mytton *et al.* (1988) provide an example of problems that environmental variables can cause (Table 26.7). The experiment examined various plant–*Rhizobium* associations on two soil types. Clover swards were established using either full cultivation or surface seeding techniques inoculated with rhizobia specifically adapted to the cultivar and the environment (Mytton and Hughes, 1984). As can be seen from Table 26.7 the clover responded to inoculation on both soils but the effect was strongly modified by cultivation. On soil A, inoculation was most beneficial in the cultivated treatment, whereas on soil B, it was best on the uncultivated treatment. The effect is

undoubtedly mediated by soil nitrogen, which in turn depends upon the interrelationship between soil type and cultivation. This serves as an illustration of the potential complexities which can arise out of genotype × environment interactions. In the absence of a full understanding of the basis of such effects, the traditional approach to dealing with them has been to evaluate material in a range of appropriate environments. At present, there is little option but to adopt this procedure for symbiotic studies. Sylvester-Bradley *et al.* (1986) have shown that undisturbed soil cores taken from appropriate field sites provide a very useful method of evaluating legume–*Rhizobium* associations under realistic conditions. Figure 26.3 shows the value of this system. The difference between the two control treatments indicates the limitations to growth imposed by the native rhizobia and the response to inoculant strains is judged against the controls. Strain 21 is the only one which has real impact on phenotypic performance. Each of the inoculant strains carries a genetic marker (antibiotic resistance). This facilitates identification of nodule occupancy; strain 21 occupied over 75% of nodules indicating strong competitive ability against a large native population. Like many of the methods outlined so far, this approach is essentially empirical and relies on drawing general inferences about genetic control of traits. Such methods are, however, both convenient and powerful because they allow an assessment of the stability of specific traits in situations which mimic the natural environment. Until we fully understand the factors influencing gene expression, linking phenotype and genotype will remain a relatively uncertain exercise. As such there will be a continuing need for rigorous field assessment of novel genotypes. However, the rapidly developing

Table 26.7. Effect of cultivation, applied nitrogen and inoculation on clover yield (kg/ha) on two different upland soils

	Soil A		Soil B	
	Uninoculated	*Inoculated*	*Uninoculated*	*Inoculated*
Uncultivated	9	68	8	220
Cultivated	17	144	26	82

Fig. 26.3. Effect of *Rhizobium* strain on clover growth. Left to right: strain 9, native rhizobia control, strain 20, strain 21, native and 250 kg N/ha.

field of molecular biology presents ever better methods for identifying and manipulating genes and for revealing the biological processes and mechanisms which they control and regulate. Exploiting this knowledge and integrating it with established methods of altering genes and genetic patterns in populations provides us with new powerful tools for rapid advances in many areas of plant breeding.

26.4 MOLECULAR APPROACHES TO IMPROVEMENT OF SYMBIOSIS

26.4.1 Genetic engineering in *Rhizobium*

The distinguishing character of *Rhizobium*, *Bradyrhizobium* and *Azorhizobium* is that they are able to enter into a nitrogen-fixing symbiotic relationship with plants belonging to the legume family (Fabaceae). The only non-legume known to form a symbiotic relationship with *Bradyrhizobium* spp. is *Parasponia* (Ulmaceae) (Trinick, 1979). *Rhizobium* differs from *Bradyrhizobium* by not only having different host species (Table 26.1) and faster growth rate, but also by having one or more large plasmids varying in size between 100 kilobase pairs (kb) and 500 kb, although *R. meliloti* is known to harbour megaplasmids of up to 1500 kb (Fig. 26.4).

The genetic analysis of *Rhizobium* began in earnest when Beringer (1974) found that the broad host range R plasmids of incompatibility group P could be transferred between rhizobia. Subsequently, Beringer *et al.* (1978b) showed that some R factors had chromosome mobiliz-

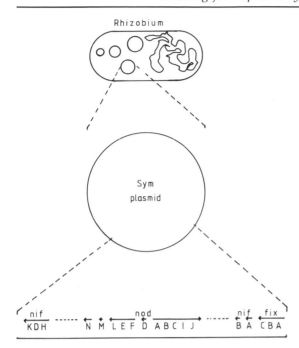

Fig. 26.4. Schematic representation of a typical *Rhizobium leguminosarum* strain with a 4000 kb chromosome and between two and seven large plasmids. One of these, the symbiotic plasmid, contains genes necessary for nodulation including host specificity, and various genes essential for nitrogen fixation and nodule integrity (see text).

ing ability. Such R factors as well as R' factors carrying chromosomal segments were used to construct genetic linkage maps of *R. leguminosarum* bv. *viciae*, *R. leguminosarum* bv. *trifolii* and *R. meliloti* (Kondorosi *et al.*, 1980). The next development occurred when it was discovered that *R. leguminosarum* carries large plasmids, some of which were self-transmissible (Hirsch, 1979). Plasmid transfer between rhizobia demonstrated that nodulation host range and nitrogen fixation ability is plasmid borne (Johnston *et al.*, 1978). Since then, numerous studies have shown that genes essential for nitrogen fixation, as well as for nodulation, are located on large plasmids in *R. leguminosarum*, *R. meliloti* and *R. fredii*, whereas *B. japonicum* carries them on the chromosome. The *R. loti* strains investigated so far harbour a single large plasmid which does not carry nitrogen fixation or nodulation DNA sequences (Pankhurst *et al.*,

1983; Skøt, unpublished). Figure 26.4 illustrates the organization of the symbiotic genes in *Rhizobium*. The nitrogen fixation genes are divided into two groups; the *nif* genes all have an homologous counterpart from the *nif* gene cluster of the free-living diazotroph *Klebsiella pneumoniae*. Examples are *nif H, D, K*, and *A*. The first three encode the three subunits of the nitrogenase enzyme complex and the last one encodes a key transcriptional regulator responsible for the activation of *nif H, D, K* and other *nif* genes. Expression of *nif A* is in turn responding to changes in the intracellular O_2 concentration via a complex regulatory circuitry. The *fix* genes are also involved in building or maintaining the nitrogen fixation apparatus in the root nodules but have no homologues in *K. pneumoniae*. Examples are *fix A, B, C*, the gene products of which may be involved in diverting electrons to nitrogenase. The nodulation (*nod*) genes have been extensively studied. They will be considered in more detail in section 26.4.4, but the regulatory *nod D* gene provides an appropriate example for illustration. It encodes a polypeptide which, when combined with certain flavonoids excreted by the legume root system, activates the other *nod* gene operons. Many of the other *nod* genes are responsible for the synthesis of an oligosaccharide signal molecule which affects root hair curling and cortical cell division, these being the first visible signs of nodule formation. Host specificity is built into this signal molecule through modifications to the oligosaccharide structure. Several excellent, recent reviews have dealt with *Rhizobium* genetics in general and symbiotic genes in particular (Djordjevic *et al.*, 1987; Rolfe and Gresshoff, 1988; Long, 1989b; Martinez *et al.*, 1990), so here we will concentrate on the methods and tools available for the manipulation of the *Rhizobium* genome, and on the cases which appear to be promising with respect to improvement in the symbiotic performance of *Rhizobium*.

26.4.2 Methods and tools

Transfer of genes between bacteria is obtained by using plasmids. However, the most commonly used plasmid cloning vectors such as pBR322, pUC (Chapter 8) and their derivatives have a restricted host range and are for use mainly in *Escherichia coli*. To overcome this problem a number of promiscuous plasmid cloning vectors have been developed for use in non-enteric Gram-negative bacteria such as *Rhizobium* (Franklin and Spooner, 1989). However, until recently it was not possible to transform *Rhizobium* directly in the same manner as *E. coli*. Consequently, the cloning and physical analysis is carried out with *E. coli* as the host. The plasmids are then transferred by conjugation to *Rhizobium*.

Another tool which has been extremely useful for insertional mutagenesis of the rhizobial genome is transposons. The most widely used is *Tn5*, and several suicide vectors have been developed for its delivery to *Rhizobium* (Beringer *et al.*, 1978a; Simon *et al.*, 1983).

A new generation of broad host range vectors have been constructed for use in Gram-negative bacteria including *Rhizobium*. Some of these carry various promoters to facilitate expression of foreign genes (Labes *et al.*, 1990). Others in the same family of plasmids carry promoterless reporter genes such as *lacZ*, *phoA* or *luxAB* for analysis of promoter sequences (Ditta *et al.*, 1985; Zaat *et al.*, 1987; Labes *et al.*, 1990). Thus, an ever-increasing range of vectors and transposons are becoming available for the analysis and manipulation of the rhizobial genome. Furthermore, reports are emerging, which describe the efficient direct DNA transformation of *Bradyrhizobium* (Hattermann and Stacey, 1990) using electroporation. Hitherto, the introduction of DNA to rhizobia has relied on conjugative transfer of plasmids. When electroporation becomes routine for rhizobial transformation, *Rhizobium* will be almost as easy to manipulate as *E. coli*.

26.4.3 Case histories

(a) Hydrogenase

The key enzyme in the *Rhizobium*–legume symbiosis is nitrogenase. It catalyses the reduction of atmospheric nitrogen (N_2) to ammonia (NH_3). One characteristic of the nitrogenase complex is its extreme sensitivity to oxygen (O_2). Another is its high requirement for energy (ATP), as well as reduction equivalents (NADH), to drive the reaction. Concomitantly, with the reduction of N_2 is the reduction of protons (H^+) to hydrogen (H_2) with at least 1 mol H_2 formed for each mol of N_2 reduced. This hydrogen evolution is responsible for the loss of at least 25% of energy in the form of ATP and NADH invested by the nodule in N_2 reduction and constitutes a substantial drain on the supply of photosynthate to the nodules. Certain strains of *B. japonicum* and *R. leguminosarum* bv. *viciae* do, however, possess a hydrogen uptake enzyme which catalyses the oxidation of H_2 to H^+ regaining the reduction equivalents, but not the ATP, thus saving some of the otherwise lost energy (Schubert and Evans, 1976). The hydrogen uptake gene (*hup*) has been isolated from *B. japonicum* (Cantrell *et al.*, 1983) and from *R. leguminosarum* bv. *viciae* (Leyva *et al.*, 1987). The wild-type isolates of *B. japonicum* which carry the *hup* gene are able to recycle all the H_2 developed by nitrogenase, thus recapturing almost 43% of the energy lost by H_2 evolution, whereas the *R. leguminosarum* bv. *viciae hup*+ rarely recycles all evolved H_2 (Schubert and Evans, 1976; DeJong *et al.*, 1982; Leyva *et al.*, 1987). The two genes show extensive homology. If the photosynthate supply to root nodules imposes limitations on host plant growth and development (discussed by Vance and Heichel, 1991), it follows that the *Hup*+ phenotype should be advantageous to the symbiosis. In comparisons between two otherwise identical *B. japonicum* strains, it was found that the *Hup*+ strains gave rise to a higher host plant dry weight yield. Theoretical considerations by Evans and his group suggested that dry matter and yield improvements of about 10% could be expected from recycling evolved H_2. The experimental evidence so far indicate that this is indeed achievable (Evans *et al.*, 1987).

In contrast to *B. japonicum* the *hup* gene from *R. leguminosarum* bv. *viciae* is located on the symbiotic plasmid (Brewin *et al.*, 1980). Transfer of this plasmid to other strains and subsequent comparative experiments with pea as the host plant initially suggested that the $Hup^{\pm}$ phenotype also improved the symbiotic performance (DeJong *et al.*, 1982). However, it was later shown that the improved performance was due to other, as yet unidentified, factors on this particular symbiotic plasmid (Cunningham *et al.*, 1985). As mentioned above, this *hup* gene is rarely capable of total H_2 recycling. Only recently has this *hup* gene been isolated and cloned from *R. leguminosarum* bv. *viciae*. Perhaps the reintroduction of this gene on a high copy number plasmid to *Rhizobium* is a way of obtaining full H_2 recycling in the pea-*Rhizobium* symbiosis.

(b) Dicarboxylic acid transport and oxygen in root nodules

The energy supplied to *Rhizobium* bacteroids for maintaining root nodules and for nitrogen fixation comes from the host plant photosynthates. It is generally accepted that bacteroids in the root-nodules derive their energy from sucrose which is first metabolized to dicarboxylates (Vance and Heichel, 1991). Mutations in the dicarboxylic acid transport genes (*dct*) result not only in inability to use such compounds as the sole carbon source in the non-symbiotic state, but also in non-fixing root nodules (Ronson *et al.*, 1981). Introduction of cloned *dct* genes into *R. meliloti* and *B. japonicum* have resulted in both increased dicarboxylic acid transport activity, and nitrogenase activity in the free living state (O'Gara *et al.*, 1988). Whether this result can be extrapolated to symbiotic N_2 fixation remains to be seen. The case for such work still rests on the argument that energy supply to root nodules is a limiting factor for N_2 fixation. This brings us to a paradoxical problem for the root nodules. Oxygen is needed to convert the carbohydrates into energy and reductants. On the other hand, the nitrogenase enzyme complex is extremely sensitive to O_2, and is inactivated if exposed to it (Robson and Postgate, 1980). The legume root nodule has solved the problem using two strategies. First, the root nodules contain leghaemoglobin, a haemoprotein very similar to

the oxygen carrying haemoglobin in animals (Jensen *et al.*, 1981). This protein facilitates oxygen transport to bacteroids without damaging nitrogenase (Wittenberg *et al.*, 1974). Second, over the last 7–8 years a lot of evidence has accumulated in support of the hypothesis that a variable oxygen diffusion barrier exists in root nodules (for a review see Witty *et al.*, 1986). This barrier can open and close in response to variations in the nodules' supply of photosynthate, and thus exert an influence on nitrogenase activity. The location of the barrier is thought to be in the innermost layers of the cortex surrounding the infected zone of the root nodule. This is supported by measurements of oxygen concentrations in root nodules with microelectrodes (Tjepkema and Yocum, 1974; Witty *et al.*, 1987). Microscopic studies have also suggested this location because the air spaces in this region appear to be much less evident than in the outer cortex and the infected zone (Witty *et al.*, 1987). The changes in the barrier are thought to be obtained by changes in the degree to which water occupies the air spaces in the inner cortex (Witty *et al.*, 1986). Exactly how these changes happen is not yet clear, but recent evidence suggest that a glycoprotein, first identified as a major component of the infection thread matrix of root nodules (Vandenbosch *et al.*, 1989), plays a role. This glycoprotein is also present in the intercellular spaces of the inner cortex of root nodules. Furthermore, soybeans grown in the presence of varying oxygen concentrations displayed correlating differences in the degree to which intercellular spaces were occupied by the glycoprotein (James *et al.*, 1991). Cloning of the host plant gene encoding the protein part of the glycoprotein would enable us to manipulate its expression by using specific promoters and/or antisense RNA strategies, thus providing further information about its role in oxygen diffusion into root nodules. A better understanding of exactly how the barrier operates and how this relates to carbohydrate supply and utilization should provide new insights for constructing symbioses with improved nitrogen fixing capacities.

(c) Introducing foreign genes to
Rhizobium

Instead of manipulating existing genes involved in symbiotic N_2 fixation, foreign genes can be introduced to the root nodule via *Rhizobium*. One such example concerns the root nodule weevil *Sitona* spp., which as adults eat the foliage of various legume plants including pea, broad bean, alfalfa and white clover. The larvae specifically invade and consume the root nodules. This can result in severely reduced nitrogen fixation activity and yield (Nielsen, 1990). The Gram-positive bacterium *Bacillus thuringiensis* produces crystal protein toxins when it sporulates. These proteins are toxic to various insect orders depending on the specific *Bacillus* strain they come from. The *B. thuringiensis* ssp. *tenebrionis* (*B.t.t.*) toxin is active against beetle (Coleoptera) larvae. The gene encoding this toxin was cloned and transferred to *Rhizobium*, and the transgenic strains had insecticidal activity (Skøt *et al.*, 1990). Similar results have been obtained with another toxin gene and other rhizobia (Nambiar *et al.*, 1990). Expression of such genes can be maximized specifically in the root nodule, in the rhizosphere or constitutively by fusing the appropriate promoter to the coding sequence. Furthermore, the methodology is also available for integrating such constructs stably into the genome of *Rhizobium*. Such work is indeed in progress as far as the *B.t.t.* toxin gene (*cryIIIA*) is concerned (Skøt, 1990).

Several strategies can be used to ensure integration of a gene construct into the genome. First, the construct can be cloned into a transposon such as *Tn5*, which is then transferred to *Rhizobium* inserting itself as described earlier. A number of suicide plasmid vectors designed for the delivery of *Tn5* to *Rhizobium* and other Gram-negative bacteria are available. Very recently, a derivative of *Tn5* has been made in which a multiple cloning site as well as the ß-galactosidase complementation system for direct selection of recombinants is present, thus facilitating the cloning of DNA fragments (Penfold and Pemberton, 1990).

Second, the DNA of interest can be flanked at both ends by DNA from the target organism.

The whole cassette can then be transferred to *Rhizobium* on a suicide vector such as pSUP202. If a suitable selectable marker (such as kanamycin resistance) is also present in the cassette, homologous recombination can be selected for. This should ensure a more stable integration of the DNA than transposons, but this method has the drawback that different flanking sequences may be necessary to ensure recombination into different target strains. As with transposon insertion one must also ensure that the insertion takes place in a symbiotically silent region.

The advantage of integrating DNA into the genome of the target strain is that it is not as easily lost as some of the broad host range vectors currently in use particularly during the symbiotic phase where frequency of loss of such plasmids can approach 80–100% from nodule isolates (O'Gara, personal communication; Skøt, unpublished). Furthermore, it has been reported that some broad host range vectors by themselves have a depressing effect on symbiotic performance (Cannon *et al.*, 1988). On the other hand, high copy number vectors may help to increase expression of foreign genes, whereas chromosomal integration normally only results in one copy.

26.4.4 Competition and release

The benefits and problems of selecting or developing competitive rhizobia have been considered in section 26.3.1(b). A major difficulty in dealing with this problem has been the lack of information on the mechanisms which determine success in colonizing soils and competing for nodule sites. In recent years there has been substantial progress in this field of research (for reviews see Dowling and Broughton, 1986 and Triplett, 1990a). Here we will concentrate on a few of the possible target mechanisms for manipulating nodulation competitiveness in rhizobia.

First, bacteriocin production has been implicated in competitiveness, particularly the anti-rhizobial trifolitoxin. This is produced by a *Fix⁻R.l.trifolii* strain (Schwinghamer and Belkengren, 1968) which is very competitive when co-inoculated with bacteriocin sensitive strains. The gene encoding the trifolitoxin polypeptide (*tfx*) was cloned and transferred to other *R. leguminosarum* bv. *trifolii* strains (Triplett and Barta, 1987; Triplett, 1990b) which then became competitive. If these results can be repeated in the field this could become an important contribution to the solution of the competitiveness problem.

Another possible strategy is to manipulate the rhizobial nodulation genes. Flavonoids exuded by the legume host plant roots together with the regulatory *nodD* gene of *Rhizobium* are responsible for the induction of *nod* genes (Firmin *et al.*, 1986; Peters *et al.*, 1986; Redmond *et al.*, 1986), which in turn are responsible for the production of a sulphated and acylated oligosaccharide signal molecule which induces root hair deformation and cortical cell divisions (Lerouge *et al.*, 1990). Several recent reviews have dealt with this subject exhaustively (Downie and Johnston, 1988; Long, 1989a,b; Rolfe and Gresshoff, 1988). However, some flavonoids act as inhibitors in certain symbioses, whereas they are inducers in others (Firmin *et al.*, 1986). Furthermore, experimental evidence suggests that the amounts of flavonoids in the root exudate can be a limiting factor for nodulation and subsequent nitrogen fixation in alfalfa (Kapulnik *et al.*, 1987).

The regulatory *nodD* gene interacts with flavonoids and binds to the promoter regions (*nod* box) of the other *nod* genes to induce their transcription. The gene product has different domains which are responsible for either binding of flavonoids or binding to DNA. Mutants affecting certain domains of the protein have been made. Some of those can induce *nod* genes independently of flavonoids (Downie and Johnston, 1988). A *Rhizobium* strain containing such a mutation could be used as inoculant together with flavonoid inhibitors to reduce competition from indigenous strains. Indeed, Cunningham *et al.* (1991) demonstrated the possibility of using flavonoid applications to manipulate interstrain competition. However, flavonoids are very expensive and also very unstable, so application of such compounds does not appear to be a practical option.

The host specific nodulation gene *nodX* and the corresponding host plant gene *sym-2* was used by Fobert *et al.* (1991) to create matching pea–*R. leguminosarum* strain combinations. Strains possessing the *nodX* gene will only nodulate pea lines homozygous for the *sym-2* mutation. Greenhouse tests demonstrated that such strains were able to out-compete the control strain lacking the *nodX* gene.

A possible alternative to the nodulation genes are the so-called rhizopines. These are *O*-methyl-inosamines which can be synthesized and catabolized exclusively by certain *Rhizobium* strains and not by others. Ononitol (4-*O*-methyl-*myo*inositol) and *O*-methyl-*scyllo*-inositol were found to occur in pea root nodules formed by certain *R. leguminosarum* bv. *viciae* strains and not in the roots (Skøt and Egsgaard, 1984). Subsequently, it was discovered that a few *R. meliloti* strains were able to synthesize and catabolize the closely related compound *O*-methyl-*scyllo*-inosamine (Murphy *et al.*, 1987). The *moc* and *mos* genes responsible for this phenotype are located on the sym-plasmid of *R. meliloti* and they are symbiotically regulated in a similar manner to the *nif* genes (Murphy *et al.*, 1988). This suggests that these compounds play a role in the symbiosis. One possibility is that they serve as a specific carbon and nitrogen source for the bacteroids inside the root nodules, but only a few rhizobia possess the *moc* and *mos* genes, and their inactivation by mutagenesis did not affect their symbiotic phenotype (Murphy *et al.*, 1987). It is conceivable that strains with *moc* and *mos* genes can have a competitive advantage by giving them an exclusive carbon and nitrogen source in the rhizosphere. However, since the *mos* (synthesis) genes are under *nifA* control, rhizopines can only be synthesized in root nodules already formed by such strains, thus limiting their usefulness in competition terms.

The successful use of improved *Rhizobium* strains as inocula in the field depends on their ability to compete. Another potential barrier to the use of genetically manipulated strains is the question of their impact on the environment. The survival and persistence of genetically engineered micro-organisms and the possible spread of foreign DNA to the indigenous soil microflora are among the questions currently being addressed (Tiedje *et al.*, 1989) as well as the effects on the size and distribution of the existing microflora. With the development of new and improved methods for detection of DNA and micro-organisms from soil this area of research looks promising.

In conclusion there appears to be considerable scope for improving the legume–*Rhizobium* symbiosis and it is evident that combining conventional selection and breeding methods with genetic engineering approaches is likely to provide the most productive way forward (Hodgson and Stacey, 1986; Paau, 1991). We have discussed here several promising developments, and others are under way which have the potential to overcome the competition problem. As our knowledge of the signalling events taking place during nodulation improves, no doubt other ways of manipulating this important aspect of the symbiosis will become apparent.

26.5 ENGINEERING HOST TRAITS

With N_2 fixation in plants there are two possible broad objectives: either to improve the existing symbioses of which the most important ones are the legume–*Rhizobium* associations, or to try to introduce the ability to fix nitrogen into other plant species, notably the cereals. This idea has been around for a long time, but is far from being realized. The first objective, however, can be achieved by traditional selection and breeding methods or by genetic engineering methods, whereas the latter objective is only possible to achieve by genetic engineering. In the legume 30–40 proteins have been identified as being synthesized only in developing and/or functioning root nodules (Stougaard *et al.*, 1986). These proteins have been named nodulins. They have been divided in two groups, the early nodulins which are synthesized in the early stages of nodule formation, and the late nodulins which are synthesized at the onset of nitrogen fixation activity. The majority of the nodulins belongs to the late category. The most abundant of them is leghaemoglobin whose role is to facilitate the diffusion of O_2, in the

infected region of the nodule, from the intercellular spaces into the cells (Wittenberg *et al.*, 1974). The function of some other nodulins is also known. These are glutamine synthetase (Cullimore *et al.*, 1984) and sucrose synthase (Thummler and Verma, 1987) and in soybean and other ureide exporting root nodules, uricase (Bergman *et al.*, 1983). They are involved in the assimilation of fixed nitrogen and the supply of energy and carbon skeletons. These are some of the most abundant late nodulins, but there are many more whose function remains to be determined.

The early nodulins appear to be involved in either the infection process or in nodule morphogenesis. The best studied early nodulin is *Enod2*, found in pea, alfalfa and soybean root nodules. cDNA and genomic clones have been isolated and characterized (Franssen *et al.*, 1987; Gloudemans and Bisseling, 1989). The presence of repeat units of proline-rich amino acid sequences suggests that it is a hydroxyproline rich glycoprotein (HRGP) associated with cell walls (Gloudemans and Bisseling, 1989). *In situ* hybridization with antisense RNA probes have shown that *Enod2* is expressed specifically in the inner cortex of soybean and pea root nodules (Van de Wiel *et al.*, 1990). This location makes it another possible candidate for involvement in the creation of the oxygen diffusion barrier mentioned earlier (Witty *et al.*, 1986). Other early nodulin genes have been isolated and a fuller account of nodulin genes and their regulation can be found elsewhere (Gloudemans and Bisseling, 1989; Sanchez *et al.*, 1991). As more and more of these genes are isolated and characterized, their function and regulation will be elucidated through manipulation. By constructing chimeric genes consisting of nodulin promoter regions fused to a reporter gene, such as β-glucuronidase (*gus*), chloroamphenicol acetyl transferase (*cat*) or luciferase (*luc*), their expression can be easily monitored in transgenic legumes.

Transgenic legumes are obtained by using *Agrobacterium tumefaciens* or *A. rhizogenes*. In the process of tumour or hairy root formation induced by these bacteria part of their Ti or Ri plasmid (T-DNA) is transferred to the host plant genome. This feature has been exploited in the creation of disarmed binary vectors carrying cloning sites and dominant selectable markers such as kanamycin or hygromycin resistance (Weising *et al.*, 1988). After cloning of the chimeric reporter gene into such a vector and transfer into *Agrobacterium* transgenic plants are regenerated. A limitation to this approach is the few legume species amenable to *Agrobacterium* mediated transformation. A relatively recent account of the status of legume transformation can be found in Nisbett and Webb (1990). Progress has been made since then, most significantly with pea (*Pisum sativum*) (Puonti-Kaerlas *et al.*, 1989; De Kathen and Jacobsen, 1990; Nauerby *et al.*, 1991). However, the only legume routinely used for this purpose is *Lotus corniculatus* (birdsfoot trefoil) regenerated from hairy root cultures induced by *A. rhizogenes*. It was first used to study the regulation of the soybean leghaemoglobin *lbc*$_3$ gene in transgenic *Lotus* root nodules (Stougaard *et al.*, 1986). Since then, similar studies have been performed with the nodule enhanced glutamine synthetase gene *gln* (Forde *et al.*, 1989), the early nodulin gene *Enod2* (deBruijn *et al.*, 1990) and nodulin *N23* (Jørgensen *et al.*, 1988).

An alternative approach to identification of plant genes involved in nodule formation and nitrogen fixation is gene tagging with transposable elements or T-DNA (reviewed by Walden and Schell, 1990) (see Chapter 8). The principle of transposon mutagenesis in plants is the same as in bacteria such as *Rhizobium* with *Tn5* mutagenesis. By inserting a segment of foreign DNA into a gene, its structure or regulation is disrupted. One difference is that plants are diploid so that recessive mutants (which are likely to be the most frequent) can only be scored in their homozygous progeny. Another difference is the size of the plant genome which obviously requires a large number of individual insertion events in order to isolate a specific mutant. In the case of plant encoded nodulation and nitrogen fixation genes a transformable legume with a small diploid genome and a short life cycle is desirable. A candidate which fulfils these requirements is *Lotus japonicus*. It is a diploid relative of *L. corniculatus* and has a haploid genome size of about 0.5 pg, is self-fertile and has a generation time of about 12

weeks (Stougaard, personal communication). A programme combining transposon mutagenesis with conventional chemical mutagenesis as well as RFLP and/or RAPD mapping would facilitate localization and eventual cloning and characterization of host plant genes involved in nodule formation and nitrogen fixation. Knowledge and understanding of how such genes are regulated and of their biochemical function will eventually lead to ways of manipulating them in order to modify and improve nodulation or fixation properties.

26.6 NEW NITROGEN-FIXING PLANTS

Knowledge of the 'symbiotic' plant genes is also imperative if the possibility of transferring the nitrogen fixation ability to other crop species is to become a reality. There are two ways of approaching this: (1) introduction and manipulation of host plant genes involved in nodulation and fixation in order to make a new host plant susceptible to *Rhizobium* nodulation; or (2) introduction of the *nif* genes, e.g. from *Klebsiella pneumonia*, to a new host plant. It is becoming increasingly clear that some of the legume host nodulin genes have homologous counterparts in many non-legume plants. A prominent example is the leghaemoglobin gene family (Landsmann *et al.*, 1986). Furthermore, expression of certain nodulin genes in plant parts other than the nodule have been demonstrated (deBruijn *et al.*, 1990). Perhaps many non legumes possess most 'nodulin' genes so that 'only' regulation of their expression needs to be modified. Indeed, nodule-like structures containing rhizobia have been obtained in rice and wheat (Bender *et al.*, 1990), and treatment of root hairs from rice and oil seed rape plants with cell wall degrading enzymes followed by inoculation with rhizobia in polyethylene glycol (PEG) has also resulted in the development of nodule-like structures (Al-Mallah *et al.*, 1989; Cocking *et al.*, 1990). Such results have been taken as an indication that perhaps only a few genes involved in the infection process needs to be manipulated, whereas most of the genes involved in nodule morphogenesis are functional once the correct trigger signal(s) are able to act (Cocking *et al.*, 1990). However, none of the reported nodule-like structures have shown any measurable nitrogenase activity. This suggests that other host genes necessary for N_2 fixation are not properly regulated or present. Clearly, a lot of work remains to be done on host plant nodulation and nitrogen fixation before any rational manipulations are possible. There is, however, no fundamental biological reason apparent to suggest that *Rhizobium*-induced nitrogen fixing nodules can not be formed on non-legumes (Sprent and Sprent, 1990).

Introduction of the *nif* genes from *Klebsiella pneumonia* into a suitable non-legume host plant is certainly also achievable with present plant transformation technology. The problem is expression. Prokaryotic genes often have different codon usages and sometimes also contain RNA destabilizing sequences, all factors which could contribute to prevent expression (Beringer and Hirsch, 1984). It has been suggested that by inserting the *nif* cluster, plus regulating genes, into the chloroplast genome problems with expression might be overcome (Postgate, 1987). The argument is that the chloroplast genome is similar to that of prokaryotes in many respects, and that the energy supply to the fixation process is much more favourable than in the roots, because of photosynthetic activity in the chloroplasts. However, the O_2 evolution occurring in chloroplasts during daytime as a result of photosynthesis would pose a serious problem to the O_2 sensitive nitrogenase enzyme, and even in the night the problem would be there because of the lack of a nodule structure with its oxygen diffusion barrier. Maybe these problems can be overcome, but it seems unlikely that such nitrogen fixing non-legumes are an immediate prospect.

With expanding knowledge of the signalling events taking place during *Rhizobium*-induced nodule formation, and in view of the likelihood that nodulin-like genes are present in non-legumes, the creation of symbiotic associations in non-legumes becomes more feasible. However, although the prospects of achieving this objective are improving, there are still important areas of information which remain obscure

and which provide testing intellectual challenges to achieving the goal of engineering new nitrogen-fixing plants.

REFERENCES

Al-Mallah, M.K., Davey, M.R. and Cocking, E.C. (1989), Formation of nodular structures on rice seedlings by rhizobia. *J. Exp. Bot.*, **40**, 473–478.

Bender, G.L., Preston, L., Barnard, D. and Rolfe, B.G. (1990), Formation of nodule-like structures on the roots of the non-legumes rice and wheat. In *Nitrogen Fixation: Achievements and Objectives*, Gresshoff, P.M., Roth, L.E., Stacey, G. and Newton, W.E. (eds), Chapman and Hall, New York, London, p. 825.

Bergman, K., Preddie, E. and Verma, D.P.S. (1983), Nodulin-35: a subunit of specific uricase (uricase II) induced and localized in the uninfected cells of soybean nodules. *EMBO J.*, **2**, 2333–2339.

Beringer, J.E. (1974), R factor transfer in *Rhizobium leguminosarum*. *J.Gen Microbiol.*, **84**, 188–198.

Beringer, J.E., Beynon, J.L., Buchanan-Wollaston, A.V. and Johnston, A.W.B. (1978a), Transfer of the drug-resistance transposon Tn5 to *Rhizobium*. *Nature*, **276**, 633–634.

Beringer, J.E. and Hirsch, P.R. (1984), Genetic engineering and nitrogen fixation. In *Biotechnology and Genetic Engineering Reviews*, Vol. 1, Intercept Ltd, Newcastle, pp. 65–88.

Beringer, J.E., Hoggan, S.A. and Johnston, A.W.B. (1978b), Linkage mapping in *Rhizobium leguminosarum* by means of R plasmid-mediated recombination. *J. Gen. Microbiol.*, **104**, 201–207.

Breese, E.L. (1969), The measurement and significance of genotype-environment interactions in grasses. *Heredity*, **31**, 181–200.

Brewin, N.J., DeJong, T.M., Phillips, D.A. and Johnston, A.W.B. (1980), Co-transfer of determinants for hydrogenase activity and nodulation ability in *Rhizobium leguminosarum*. *Nature*, **288**, 77–79.

Cannon, F.C., Beynon, J., Hankinson, T., Kwiatkowski, R., Legocki, R.P., Ratcliffe, H., Ronson, C., Szeto, W. and Williams, M. (1988), Increasing biological nitrogen fixation by genetic manipulation. In *Nitrogen Fixation: Hundred Years After*, Bothe, H., deBruijn, F.J. and Newton, W.E. (eds), Gustav Fischer, Stuttgart, New York, pp. 735–740.

Cantrell, M.A., Haugland, R.A. and Evans, H.J. (1983), Construction of a *Rhizobium japonicum* gene bank and use in isolation of a hydrogen uptake gene. *Proc. Natl. Acad. Sci. USA*, **80**, 181–185.

Cocking, E.C., Al-Mallah, M.K., Benson, E. and Davey, M.R. (1990), Nodulation of non-legumes by rhizobia. In *Nitrogen Fixation: Achievements and Objectives*, Gresshoff, P.M., Roth, L.E., Stacey, G. and Newton, W.E. (eds), Chapman and Hall, New York, London, pp. 813–823.

Cullimore, J.V., Gebhardt, C., Seardamin, R., Miflin, B.J., Laidler, K.B. and Barker, R.F. (1984), Glutamine synthetase of *Phaseolus vulgaris* L.: organ-specific expression of a multigene family. *J. Mol. Appl. Genet.*, **2**, 589–599.

Cunningham, S., Kollmeyer, W.D. and Stacey, G. (1991), Chemical control of interstrain competition for soybean nodulation by *Bradyrhizobium japonicum. Appl. Environ. Microbiol.*, **57**, 1886–1892.

Cunningham, S.D., Kapulnik, Y., Kagan, S.A., Brewin, N.J. and Johnston, A.W.B. (1985), Hup activity determined by pRL6JI does not increase N_2 fixation. In *Nitrogen Fixation Research Progress*, Evans, H.J., Bottomly, P.J. and Newton, W.E. (eds), Martinus Nijhoff Publishers, Dordrecht, p. 227.

De Kathen, A. and Jacobsen, H-J. (1990), *Agrobacterium tumefaciens*-mediated transformation of *Pisum sativum* L. using binary and cointegrate vectors. *Plant Cell Rep.*, **9**, 276–279.

deBruijn, F.J., Szabados, L. and Schell, J. (1990), Chimeric genes and transgenic plants are used to study the regulation of genes involved in symbiotic plant–microbe interactions (nodulin genes). *Dev. Genet.*, **11**, 182–196.

DeJong, T.M., Brewin, N.J., Johnston, A.W.B. and Phillips, D.A. (1982), Improvement of symbiotic properties in *Rhizobium leguminosarum* by plasmid transfer. *J. Gen. Microbiol.*, **128**, 1829–1838.

Ditta, G., Schmidhauser, T., Yakobsen, E., Lu, P., Liang, X., Finlay, D.R., Guiney, D. and Helinski, D.R. (1985), Plasmids related to the broad host range vector, pRK290, useful for gene cloning and for monitoring gene expression. *Plasmid*, **13**, 149–153.

Djordjevic, M.A., Gabriel, D.W. and Rolfe, B.G. (1987), *Rhizobium* – the refined parasite of legumes. *Ann. Rev. Phytopathol.*, **25**, 145–168.

Dowling, D.N. and Broughton, W.J. (1986), Competition for nodulation of legumes. *Ann. Rev. Microbiol.*, **40**, 131–157.

Downie, J.A. and Johnston, A.W.B. (1988), Nodulation of legumes by *Rhizobium*. *Plant, Cell and Environment*, **11**, 403–412.

Evans, H.J., Harker, R.P., Papen, H., Russell, S.A.,

Hanus, F.J. and Zuber, M. (1987), Physiology, biochemistry and genetics of the uptake hydrogenase in rhizobia. *Ann. Rev. Microbiol.*, **41**, 335–361.

Fay, M.J., Mytton, L.R. and Dale, P.J. (1991), Germplasm assessment in *Trifolium* species. *Plant Breeding*, **106**, 226–234.

Finlay, K.W. and Wilkinson, G.N. (1963), The analysis of adaptation in a plant breeding programme. *Aust J. Agr. Res.*, **14**, 742–754.

Firmin, J.L., Wilson, K.E., Rossen, L. and Johnston, A.W.B. (1986), Flavonoid activation of nodulation genes in *Rhizobium* reversed by other compounds present in plants. *Nature*, **324**, 90–92.

Fobert, P.R., Roy, N., Nash, J.H.E. and Iyer, V.N. (1991), Procedure for obtaining efficient root nodulation of a pea cultivar by a desired *Rhizobium* strain and preempting nodulation by other strains. *Appl. Environ. Microbiol.*, **57**, 1590–1594.

Forde, B.G., Day, H.M., Turton, J.F., Wen-Jun, S., Cullimore, J.V. and Oliver, J.E. (1989), Two glutamine synthetase genes from *Phaseolus vulgaris* L. display contrasting developmental and spatial patterns of expression in transgenic *Lotus corniculatus* plants. *Plant Cell*, **1**, 391–401.

Franklin, F.C.H. and Spooner, R. (1989), Broad host range cloning vectors. In *Promiscuous Plasmids of Gram-Negative Bacteria*, Thomas, C.M. (ed.), Academic Press, London, San Diego, pp. 247–267.

Franssen, H.J., Nap, J.P., Gloudemans, T., Stikema, W., Van Dam, H., Govers, F., Louwerse, J., Van Kammen, A. and Bisseling, T. (1987), Characterization of cDNA for nodulin-75 of soybean: A gene product involved in early stages of root nodule development. *Proc. Natl. Acad. Sci. USA*, **84**, 4495–4499.

Fred, E.B., Baldwin, I.L. and McCoy, E. (1932), Root nodule bacteria and leguminous plants. *Studies in Science*, No. 5, University of Wisconsin, Madison.

Gloudemans, T. and Bisseling, T. (1989), Plant gene expression in early stages of *Rhizobium*-legume symbiosis. *Plant Sci.*, **65**, 1–14.

Hardarson, G. and Jones, D.G. (1979), The inheritance of preference for strains of *Rhizobium trifolii* by white clover (*T. repens*). *Ann. Appl. Biol.*, **92**, 329–333.

Hardy, R.W.F., Burnes, R.C. and Holsten, R.D. (1973), Applications of acetylene–ethylene assay for measurement of nitrogen fixation. *Soil Biol. Biochem.*, **5**, 47–81.

Hattermann, D.R. and Stacey, G. (1990), Efficient DNA transformation of *Bradyrhizobium japonicum* by electroporation. *Appl. Environ. Microbiol.*, **56**, 833–836.

Hirsch, P.R. (1979), Plasmid-determined bacteriocin production by *Rhizobium leguminosarum*. *J. Gen. Microbiol.*, **113**, 219–228.

Hodgson, A.L.M. and Stacey, G. (1986), Potential for *Rhizobium* improvement. *CRC Crit. Rev. Biotechnol.*, **4**, 1–74.

James, E.K., Sprent, J.I., Minchin, F.R. and Brewin, N.J. (1991), Intercellular location of glycoprotein in soybean nodules: effect of altered rhizosphere oxygen concentration. *Plant Cell Environ*, **14**, 467–476.

Jensen, E.Ø., Paludan, K., Hyldig-Nielsen, J.J., Jørgensen, P. and Marcker, K.A. (1981), The structure of a chromosomal leghaemoglobin gene from soybean. *Nature*, **291**, 677–679.

Johnston, A.W.B., Beynon, J.L., Buchanan-Wollaston, A.V., Hirsch, P.R. and Beringer, J.E. (1978), High frequency transfer of nodulating ability between strains and species of *Rhizobium*. *Nature*, **276**, 634–636.

Jordan, D.C. (1984), Family III. Rhizobiaceae. In *Bergey's Manual of Systematic Bacteriology*, Vol. 1, Krieg, N.R. and Holt, J.G. (eds), Williams and Wilkins, Baltimore/London, pp. 234–244.

Jørgensen, J.E., Stougaard, J., Marcker, A. and Marcker, K.A. (1988), Root nodule-specific gene regulation: Analysis of the soybean nodulin N23 gene promoter in heterologous symbiotic systems. *Nucleic Acids Res.*, **16**, 39–50.

Kapulnik, Y., Joseph, C.M. and Phillips, D.A. (1987), Flavone limitations to root nodulation and symbiotic nitrogen fixation in alfalfa. *Plant Physiol.*, **84**, 1193–1196.

Kondorosi, A., Vincze, E., Johnston, A.W.B. and Beringer, J.E. (1980), A comparison of three *Rhizobium* linkage maps. *Mol. Gen. Genet.*, **178**, 403–408.

Labes, M., Puhler, A. and Simon, R. (1990), A new family of RSF1010-derived expression and lac-fusion broad-host-range vectors for Gram-negative bacteria. *Gene*, **89**, 37–46.

Landsmann, J., Dennis, E.S., Higgins, T.J.V., Appleby, C.A., Kortt, A.A. and Peacock, W.J. (1986), Common evolutionary origin of legume and non-legume plant haemoglobins. *Nature*, **324**, 166–168.

Lerouge, P., Roche, P., Faucher, C., Maillet, F., Truchet, G., Prome, J.C. and Denarie, J. (1990), Symbiotic host-specificity of *Rhizobium meliloti* is determined by a sulphated and acylated glucosamine oligosaccharide signal. *Nature*, **344**, 781–784.

Leyva, A., Palacios, J.M., Mozo, T. and Ruiz-Argüeso, T. (1987), Cloning and characterization of hydrogen uptake genes from *Rhizobium leguminosarum*. *J. Bacteriol.*, **169**, 4929–4934.

Long, S.R. (1989a), *Rhizobium*–legume nodulation: life together in the underground. *Cell*, **56**, 203–214.

Long, S.R. (1989b), *Rhizobium* genetics. *Ann. Rev. Genet.*, **23**, 483–506.

Martinez, E., Romero, D. and Palacios, R. (1990), The *Rhizobium* genome. *Crit. Rev. Plant Sci.*, **9**, 59–93.

Murphy, P.J., Heycke, N., Banfalvi, Z., Tate, M.E., deBruijn, F., Kondorosi, A., Temp, J. and Schell, J. (1987), Genes for the catabolism and synthesis of an opine-like compound in *Rhizobium meliloti* are closely linked and on the sym plasmid. *Proc. Natl. Acad. Sci. USA*, **84**, 493–497.

Murphy, P.J., Heycke, N., Trenz, S.P., Ratet, P., deBruijn, F. and Schell, J. (1988), Synthesis of an opine-like compound, a rhizopine, in alfalfa nodules is symbiotically regulated. *Proc. Natl. Acad. Sci. USA*, **85**, 9133–9137.

Mytton, L.R. (1975), Plant genotype × *Rhizobium* strain interactions in white clover. *Ann. Appl. Biol.*, **80**, 103–107.

Mytton, L.R. (1987), *Annual Report*, IGAP, Welsh Plant Breeding Station, p. 18.

Mytton, L.R. (1989), *Annual Report*, IGAP, Welsh Plant Breeding Station, p. 20.

Mytton, L.R., Brockwell, J. and Gibson, A.H. (1984), The potential for breeding an improved lucerne-*Rhizobium* symbiosis. 1. Assessment of genetic variation. *Euphytica*, **33**, 401–410.

Mytton, L.R., Cresswell, A. and Colbourn, P. (1992), Improvement in soil water percolation associated with white clover. *Grass and Forage Science*, in press.

Mytton, L.R. and de Felice, J. (1977), The effect of mixtures of *Rhizobium* strains on the dry matter production of white clover grown agar. *Ann. App. Biol.*, **87**, 83–93.

Mytton, L.R. and Hughes, D.M. (1984), Inoculation of white clover with different strains of *Rhizobium trifolii* on a mineral hill soil. *J. Agr. Sci. Camb.*, **102**, 455–459.

Mytton, L.R., Hughes, D.M. and Kahurananga, J. (1988), Host-*Rhizobium* relationships and their implications for legume breeding. In *Nitrogen Fixation by Legumes in Mediterraneum Agriculture*, Beck, D.P. and Materon, L.A. (eds), Nijhoff, Dordrecht, pp. 131–143.

Mytton, L.R. and Livsey, C.J. (1983), Specific and general effectiveness of *Rhizobium trifolii* populations from different agricultural locations. *Plant and Soil*, **73**, 299–305.

Nambiar, P.T.C., Ma, S.W. and Iyer, V.N. (1990), Limiting and insect infestation of nitrogen-fixing root nodules of the pigeon pea (*Cajanus cajan*) by engineering the expression of an entomocidal gene in its root nodules. *Appl. Environ. Microbiol.*, **56**, 2866–2869.

Nauerby, B., Madsen, M., Christiansen, J. and Wyndaele, R. (1991), A rapid and efficient regeneration system for pea (*Pisum sativum*), suitable for transformation. *Plant Cell Rep.*, **9**, 676–679.

Nielsen, B.S. (1990), Yield responses of *Vicia faba* to infestation levels of *Sitona lineatus* L. (Col. Curculionidae). *J. Appl. Entomol.*, **110**, 398–407.

Nisbett, G.S. and Webb, K.J. (1990), Transformation in legumes. In *Biotechnology in Agriculture and Forestry*, Vol. 10, *Legumes and Oilseed Crops I*, Bajaj, Y.P.S. (ed.), Springer-Verlag, Berlin, Heidelberg, pp. 38–48.

O'Gara, F., Birkenhead, K., Wang, Y.P., Condon, C. and Manian, S.S. (1988), Dicarboxylic acid utilisation and nitrogen fixation efficiency in *Rhizobium*–legume symbiosis. In *Physiological Limitations and the Genetic Improvement of Symbiotic Nitrogen Fixation*, O'Gara, F., Manian, S. and Drevon, J.J. (eds), Academic Publishers, Kluwer, pp. 149–157.

Paau, A.S. (1991), Improvement of *Rhizobium* inoculants by mutation, genetic engineering and formulation. *Biotech. Adv.*, **9**, 173–184.

Pankhurst, C.E., Broughton, W.J. and Wieneke, U. (1983), Transfer of an indigenous plasmid of *Rhizobium loti* to other rhizobia and *Agrobacterium tumefaciens*. *J. Gen. Microbiol.*, **129**, 2535–2543.

Penfold, R.J. and Pemberton, J.M. (1990), A transposon vector for broad host range cloning. *Nucleic Acids Res.*, **18**, 5913.

Peters, N.K., Frost, J.W. and Long, S.R. (1986), A plant flavone, luteolin induces expression of *Rhizobium meliloti* nodulation genes. *Science*, **233**, 977–980.

Postgate, J. (1987), Prospects for the improvement of biological nitrogen fixation. *J. Appl. Bacteriol.*, Symposium Supplement 85S–91S.

Puonti-Kaerlas, J., Stabel, P. and Eriksson, T. (1989), Transformation of pea (*Pisum sativum* L.) by *Agrobacterium tumefaciens*. *Plant Cell Rep.*, **8**, 321–324.

Redmond, J.W., Batley, M., Djordjevic, M.A., Innes, R.W., Kuempel, P.L. and Rolfe, B.G. (1986), Flavones induce expression of nodulation genes in *Rhizobium*. *Nature*, **323**, 632–635.

Robson, R.L. and Postgate, J.R. (1980), Oxygen and hydrogen in biological nitrogen fixation. *Annu. Rev. Microbiol.*, **34**, 183–207.

Rolfe, B.G. and Gresshoff, P.M. (1988), Genetic analysis of legume nodule initiation. *Annu. Rev. Plant Physiol. Plant Mol. Biol.*, **39**, 297–319.

Ronson, C.W., Lyttleton, P. and Robertson, J.G. (1981), C_4-dicarboxylate transport mutants of *Rhizobium trifolii* form ineffective nodules on *Trifolium repens*. *Proc. Natl. Acad. Sci. USA*, **78**, 4284–4288.

Sanchez, F., Padilla, J.E., Perez, H. and Lara, M. (1991), Control of nodulin genes in root-nodule development and metabolism. *Annu. Rev. Plant Physiol. Plant Mol. Biol.*, **42**, 507–528.

Schubert, K.R. and Evans, H.J. (1976), Hydrogen evolution: a major factor affecting the efficiency of nitrogen fixation in nodulated symbionts. *Proc. Natl. Acad. Sci. USA*, **773**, 1207–1211.

Schwinghamer, E.A. and Belkengren, R.P. (1968), Inhibition of rhizobia by a strain of *Rhizobium trifolii*: some properties of the antibiotic and of the strain. *Arch. Mikrobiol.*, **64**, 130–145.

Simon, R., Priefer, U. and Puhler, A. (1983), A broad host range mobilisation system for *in vivo* genetic engineering: transposon mutagenesis in Gram-negative bacteria. *Biotechnology*, **1**, 784–791.

Skøt, L. (1990), Insecticidal activity of *Rhizobium* strains containing the δ-endotoxin gene from *Bacillus thuringiensis*. *Aspects Appl. Biol.*, **24**, 101–108.

Skøt, L. and Egsgaard, H. (1984), Identification of ononitol and O-methyl-scyllo-inositol in pea root nodules. *Planta*, **161**, 32–36.

Skøt, L., Harrison, S.P., Nath, A., Mytton, L.R. and Clifford, B.C. (1990), Expression of insecticidal activity in *Rhizobium* containing the δ-endotoxin gene cloned from *Bacillus thuringiensis* subsp. *tenebrionis*. *Plant Soil*, **127**, 285–295.

Sprent, J.I. and Sprent, P. (1990), *Nitrogen Fixing Organisms*, Chapman and Hall, London, New York.

Stougaard Jensen, J., Marcker, K.A., Otten, L. and Schell, J. (1986), Nodule-specific expression of a chimaeric soybean leghaemoglobin gene in transgenic *Lotus corniculatus*. *Nature*, **321**, 669–674.

Sylvester-Bradley, R., Mosquera, D. and Mendez, J.E. (1986), Selection of rhizobia for inoculation of forage legumes in savanna and rain-forest soils of tropical America. In *The Nitrogen Fixing Potential of Legumes in Semi-Arid Environments*, ICARDA-UNDP Workshop, Aleppo.

Thummler, F. and Verma, D.P.S. (1987), Nodulin-100

of soybean is the subunit of sucrose synthase regulated by the availability of free heme in nodules. *J. Biol. Chem.*, **262**, 14730–14736.

Tiedje, J.M., Colwell, R.K., Grossman, Y.L., Hodson, R.E., Lenski, R.E., Mack, R.N. and Regal, P.J. (1989), The planned introduction of genetically engineered organisms: ecological considerations and recommendations. *Ecology*, **70**, 298–315.

Tjepkema, J.D. and Yocum, C.S. (1974), Measurement of oxygen partial pressure within soybean root nodules by oxygen microelectrodes. *Planta*, **119**, 351–360.

Trinick, M.J. (1979) Structure of nitrogen-fixing nodules formed by *Rhizobium* on roots of *Parasponia andersonii* Planch. *Can. J. Microbiol.*, **25**, 565–578.

Triplett, E.W. (1990a), The molecular genetics of nodulation competitiveness in *Rhizobium* and *Bradyrhizobium*. *Mol. Plant–Microbe Interact.*, **3**, 199–206.

Triplett, E.W. (1990b), Construction of a symbiotically effective strain of *Rhizobium leguminosarum* bv. *trifolii* with increased nodulation competitiveness. *Appl. Environ. Microbiol.*, **56**, 98–103.

Triplett, E.W. and Barta, T.M. (1987), Trifolitoxin production and nodulation are necessary for the expression of superior nodulation competitiveness by *Rhizobium leguminosarum* bv. *trifolii* strain T24 on clover. *Plant Physiol.*, **85**, 335–342.

Van de Wiel, C., Scheres, B., Franssen, H., van Lerop, M.J., van Lammeren, A., van Kammen, A. and Bisseling, T. (1990), The early nodulin transcript Enod2 is located in the nodule-specific parenchyma (inner cortex) of pea and soybean nodules. *EMBO J.*, **9**, 1–7.

Vance, C.P. and Heichel, G.H. (1991), Carbon in N_2 fixation: limitation or exquisite adaptation. *Annu. Rev. Plant Physiol. Plant Mol. Biol.*, **42**, 373–392.

VandenBosch, K.A., Bradley, D.J., Knox, J.P., Perotto, S., Butcher, G.W. and Brewin, N.J. (1989), Common components of the infection thread matrix and the intercellular space identified by immunocytochemical analysis of pea nodules and uninfected roots. *EMBO J.*, **8**, 335–342.

Vincent J.M. (1970), A manual for the practical study of root-nodule bacteria. *IBP Handbook*, No. 15, Scientific Publications, Oxford, Edinburgh, pp. 164.

Walden, R. and Schell, J. (1990), Techniques in plant molecular biology – progress and problems. *Eur. J. Biochem.*, **192**, 563–576.

Weising, K., Schell, J. and Kahl, G. (1988), Foreign genes in plants: Transfer, structure, expression, and applications. *Ann. Rev. Genet.*, **22**, 421–477.

Wittenberg, J.B., Bergersen, F.J., Appleby, C.A. and Turner, G.L. (1974), Facilitated oxygen diffusion. *J. Biol. Chem.*, **249**, 4057–4066.

Witty, J. F. and Giller, K.E. (1991), Evaluation of errors in the measurement of biological nitrogen fixation using ^{15}N fertilizer. In *Proc. Symposium on Use of Stable Isotopes in Soil Fertility and Plant Nutrition*, IAEA/FAO, Vienna, pp. 59–72.

Witty, J.F. and Minchin, F.R. (1988), Measurement of nitrogen fixation by the acetylene reduction assay: Myths and mysteries. In *Nitrogen Fixation by Legumes in Mediterraneum Agriculture*, Beck, D.P. and Materon, L.A. (eds), Nijhoff, Dordrecht, pp. 331–344.

Witty, J.F., Minchin, F.R., Skøt, L. and Sheehy, J.E. (1986), Nitrogen fixation and oxygen in legume root nodules. In *Oxford Surveys of Plant Molecular and Cell Biology*, Vol. 3, Miflin, B.J. (ed.), Oxford University Press, Oxford, pp. 275–314.

Witty, J.F., Skøt, L. and Revsbech, N.P. (1987), Direct evidence for changes in the resistance of legume root nodules to O_2 diffusion. *J. Exp. Bot.*, **38**, 1129–1140.

Young, N.R. and Mytton, L.R. (1983), The response of white clover to different strains of *Rhizobium trifolii* in hill land reseeding. *Grass and Forage Science*, **38**, 13–19.

Zaat, S.A.J., Wijffelman, C.A., Spaink, H.P., Van Brussel, A.A.N., Okker, R.J.H. and Lugtenberg, B.J.J. (1987), Induction of the nodA promoter of *Rhizobium leguminosarum* sym plasmid pRL1JI by plant flavanones and flavones. *J. Bacteriol.*, **169**, 198–204.

27

Photosynthetic and respiratory efficiency

J. Azcón-Bieto and A. Caballero

27.1 INTRODUCTION

The growth rate of a plant depends on many environmental and endogenous factors, but can ultimately be defined in terms of carbon by the balance between photosynthetic CO_2 assimilation and the respiratory costs, and also by a carbon partitioning factor, as modelled by Masle *et al.* (1990). Theoretically, it is possible to improve plant growth (and thus productivity and final biological or economical yield) either by increasing the amount of photosynthesis, by reducing 'unnecessary' respiratory costs, or by allocating more carbon into appropriate sinks (optimization of 'assimilate processing').

One way of increasing photosynthesis is by improving the efficiency of this process either at the photochemical level (light capture, electron transport, ATP synthesis) or at the biochemical level (CO_2 fixation or assimilation reactions, photorespiration, utilization of assimilates) (Neyra, 1985). The rate of net photosynthesis or CO_2 assimilation per unit leaf area (A) can usefully integrate the aforementioned photochemical and biochemical components of photosynthesis, together with the leaf gas (CO_2 and water vapour) diffusion processes (see Sharkey, 1985). This parameter A is relatively easy to measure (e.g. using gas exchange techniques; see Pearcy *et al.*, 1989) and under-

stand when the external conditions are well controlled, and thus A seemed an obvious candidate for detecting plants with higher photosynthetic efficiency. For example, the normally higher leaf photosynthesis rates of C_4 compared to C_3 plants (interspecific variation) have been associated with higher growth rates under their optimal respective environmental conditions (see Hay and Walker, 1989; Dennis and Turpin, 1990).

Paradoxically, selection for higher or maximal leaf net photosynthesis rates within a given species (intraspecific variation) was often not associated with higher productivity or yield (see reviews by Elmore, 1980; Nelson, 1988; Austin, 1990). However, positive relationships between leaf net photosynthesis rate per unit area and growth (i.e. relative growth ratio, biological yield) have also been reported in several genera, species and cultivars (e.g. *Populus*: Isebrands *et al.*, 1988; *Pisum sativum*: Mahon, 1990; *Sorghum bicolor*: Peng *et al.*, 1991; *Zea mays*: Kelly, 1988; Rocher *et al.*, 1989). Thus, it can be concluded that no single general trend exists between the photosynthesis rate per unit leaf area and growth, and this correlation needs to be established in every species. However, as expected, it is now generally accepted that the total amount of photosynthesis by the plant canopy during the growth period and the biological yield are well correlated (Zelitch, 1982; Ashley and Boerma, 1989). A possible solution to this paradox can be related to the fact that instantaneous adult leaf photosynthesis rates and total photosynthesis of a plant

Plant Breeding: Principles and prospects. Edited by M.D. Hayward, N.O. Bosemark and I. Romagosa. Published in 1993 by Chapman & Hall, London. ISBN 0 412 43390 7.

or a crop are not necessarily correlated, given that photosynthesis is also a function of total leaf area, leaf duration, and of the total intercepted radiation by the canopy (Monteith, 1981; Bugbee and Salisbury, 1988). Thus, genetic variation in A may be compensated for by morphological changes in total leaf area and light interception properties in some cases (Elmore, 1980; Nelson, 1988). On the other hand, short-term measurements of the rate of net photosynthesis during a given developmental stage of the plant do not generally reflect the amount of photosynthesis during the complete growing season (Zelitch, 1982).

Changes in 'dark' respiration could also significantly contribute to the plant growth efficiency. It is well known that about 40–50% of the carbon fixed by photosynthesis is normally released by respiration on a daily basis (Lambers, 1985; Amthor, 1989). Reduction of the respiration rate seems to offer possibilities of small, but significant increases in yield in some species (Wilson, 1975, 1982; Lambers, 1985; Nelson, 1988; Amthor, 1989; Hay and Walker, 1989). Bugbee and Salisbury (1988) considered that carbon use efficiency, which is directly related to the efficiency of respiration, can be significantly increased in plants. However, it is still a matter of discussion among physiologists and biochemists as to whether a true 'wasteful' component of respiration really exists (see ap Rees, 1988; Azcón-Bieto, 1992). This question should be clarified in order to use respiration as a valuable process for plant yield improvement.

Yield is a complex character which depends on many genes and thus is determined by the interaction of many physiological and biochemical processes which could be manipulated through plant breeding and genetics. However, the physiological (e.g. process rate) and biochemical (e.g. enzyme activity) bases of yield need to be better understood in order to give useful selection tools to plant breeders (see Harper *et al.*, 1985). Ultimately, yield can be defined in a simple way as a function of the net amount of carbon fixed by the system, that is as a function of the balance between photosynthesis and respiration. A central question in this chapter is to know whether plant productivity

and yield (outputs) can be significantly increased by manipulating one or more reactions contributing to photosynthetic and respiratory efficiency, and, subsequently, how successfully to manipulate these reactions. The improvement of photosynthetic and respiratory efficiency might also help to reduce the elevated inputs of modern agriculture (e.g. fertilizers, water, pesticides, etc.) and thus to optimize the input/output ratio with consequent benefits for health and environmental aspects.

27.2 THE LIMITS OF PLANT PRODUCTIVITY

A relevant question before considering any possible biochemical or genetic manipulation of photosynthesis and respiration is whether plant productivity is photosynthetically (or carbon) limited. The answer is definitely in the affirmative because plant or crop yield can be greatly increased at high irradiance and/or high CO_2 in the absence of any stress (Bugbee and Salisbury, 1988; Allen, 1990; Campbell *et al.*, 1990). These factors are known to increase the rate of net CO_2 assimilation of single leaves (Sharkey, 1985). Wheat plants have been grown hydroponically at very high integrated irradiances and high CO_2 under constant, non-stressing conditions, to find the limits of potential productivity in this crop (Bugbee and Salisbury, 1988). They concluded that wheat is physiologically and genetically capable of much higher productivity and photosynthetic efficiency than the highest values recorded in the field. One of the reasons for the lower efficiency under normal conditions is because the available light limits potential productivity (see Monteith, 1981). High CO_2 concentrations improve photosynthetic efficiency by increasing the quantum yield of photosynthesis and by decreasing photorespiration (Ogren, 1984). Bugbee and Salisbury (1988) also estimated that carbon use efficiency in growth and maintenance requirements is another important factor in maximum productivity, although it has been less well studied.

In addition to light, CO_2 and respiration, potential crop yield is often limited by carbon partitioning into photosynthetic products (e.g. starch, sucrose) and their distribution and utilization at the plant level (Gifford and Evans, 1981; Wardlaw, 1990). It is now well accepted that leaf photosynthesis can be inhibited by excess carbohydrates in the leaves (Azcón-Bieto, 1983; Sharkey, 1985; Foyer, 1988; Stitt and Quick, 1989; Stitt *et al.*, 1990). A mechanism of this 'feedback' inhibition may involve intracellular inorganic phosphate limitation (Foyer, 1988; Stitt and Quick, 1989) and decrease of photosynthetic enzymes (Stitt *et al.*, 1990). By studying the O_2 sensitivity of photosynthesis, Sage and Sharkey (1987) estimated that feedback limitations can play a role in determining carbon gain in the field, especially at low temperature, and these limitations will become more important as the atmospheric CO_2 concentration increases. Obviously, the combination of higher photosynthetic efficiency with increased capacity to utilize additional assimilate produced, would positively affect potential yield.

It has been suggested by Stitt and Quick (1989) that manipulation of carbon partitioning is possible, by designing mutants with changed enzyme activities to alter the carbon fluxes through the different pathways. Crucial or key enzymes affecting the control of carbon partitioning pathways need to be identified in order to affect significantly agricultural traits. One such enzyme could be sucrose-phosphate synthase, a key enzyme for sucrose biosynthesis (Stitt and Quick, 1989), which has been found to correlate well with the growth rate in maize (Rocher *et al.*, 1989). However, the manner in which carbon partitioning can control growth at the whole plant level is still not sufficiently well understood (Wardlaw, 1990).

Several authors suggested the use of molecular biology techniques for the improvement of photosynthetic and respiratory efficiency, on the basis already discussed, and this, together with the development of transformation techniques for plants, opens up new perspectives for improvement (Austin, 1988; Dyer, 1989; Somerville, 1990). For instance, the use of antisense technology (Chapters 8 and 10) can be a powerful tool to evaluate the modifications of enzymes (e.g. concentration) of metabolic pathways (Rodermel *et al.*, 1988; Dyer, 1989; Stitt and Quick, 1989; Mol *et al.*, 1990; Stitt *et al.*, 1991). The utilization of the numerous photosynthetic mutants already existing (Somerville, 1986a) is useful for molecular biologists. An advantage of this novel approach for plant improvement, in comparison with that of conventional breeding, is the shorter time-scale needed for obtaining new genotypes. However, conventional breeding techniques should still not be overlooked (although it has to be admitted that they have not so far been very successful in relation to photosynthesis) because it is possible to breed directly for increased growth rate in some species (e.g. oats, Frey, 1988) following well defined methodologies, and also for higher maximal leaf photosynthesis rate (Austin, 1990). In summary, the scope for optimizing photosynthesis and respiration in relation to yield apparently exists, but the specific lines of action (e.g. physiological rates, photochemical and/or biochemical reactions) need to be very well defined in order to obtain effective improvements.

27.3 IMPROVEMENT OF PHOTOSYNTHETIC EFFICIENCY

The experimental evidence suggests that there is a large inter- and intraspecific genetic variability in the rate of leaf net photosynthesis, but attempts to use this variability in breeding programmes have not generally been satisfactory (Austin, 1988; Nelson, 1988). For example, the production of hybrids between C_3 and C_4 species with the objective of obtaining higher rates of photosynthesis than the parental C_3 plants have not given the expected results (Osmond *et al.*, 1980). Screening procedures based on survival under CO_2 compensation point conditions for finding lines or hybrids of C_3 plants with higher photosynthesis and/or lower photorespiration have similarly not been successful (Menz *et al.*, 1969; Nelson *et al.*, 1975; Medrano and Primo-Millo, 1985; Delgado *et al.*, 1992). These screening methods were popular in the past because of their simplicity and possibility of large-scale application, and fuelled hopes of overcoming the main technical limitations of selection methods based on direct mesurements of photosynthesis rate (Nasyrov, 1978).

It is generally accepted that the main achievements of breeders in increasing yield have been through changes in plant development and morphology and not through direct changes in photosynthesis rate or properties (Chapters 23 and 24 and Austin, 1988). A classical example can be found in the evolution of wheat species: the high-yielding modern hexaploid wheat cultivars have a clearly different morphology (related to dwarfing genes), higher harvest index (that is, greater proportion of total dry mass in the grain) and paradoxically, lower light-saturated rates of photosynthesis per unit area than primitive diploid wheat species (Austin, 1988, 1990). Another interesting example of morphological changes is the selection for cell or even leaf size, which can affect the concentration of photosynthetic machinery and the gaseous diffusion pathways in leaves (Nelson, 1988; Kaminski *et al.*, 1990).

In our view, the identification and manipulation of key or limiting photosynthetic reactions should be the main objective for improving the efficiency of photosynthesis at the general level, although selection for higher or maximal leaf photosynthesis rates can be useful in some well-defined cases.

27.3.1 Photochemical reactions

Independently of the frequent limitation to photosynthesis by the availability of light under natural conditions, the manipulation of individual photochemical reactions seems to offer less possibilities for significantly increasing the efficiency of photosynthesis than the manipulation of the so-called dark reactions (section 27.3.2). The 'quantum yield' (i.e. the number of molecules of CO_2 fixed or O_2 evolved per photon used in photosynthesis) is a parameter that reflects very well the efficiency of the photochemical apparatus. The maximal values measured in leaves can be very close to the theoretical maximum of the Z-scheme (D.A. Walker, personal communication), and are not

likely to be further increased (Horton *et al.*, 1988).

However, the quantum yield is very sensitive to stress factors (e.g. water stress, low temperature, photoinhibition, herbicides), reflecting the lower efficiency of the regulation of the excitation energy utilization and dissipation by the photosynthetic membrane, and it has been suggested that minimizing the reduction of quantum yield values under stress should be advantageous for yield (Horton *et al.*, 1988; Ort and Baker, 1988). Thus, rather than increasing the efficiency of the photochemical reactions *per se*, it can be more useful and feasible to avoid a suboptimal use, or even the destruction of the photochemical apparatus when stress conditions are present. On the other hand, the optimization of light availability to the plant through morphological and physiological changes (e.g. increasing leaf area and duration, improving the angle and orientation of leaves, etc) would also positively affect the efficiency of light utilization at the plant or crop level in the absence of changes in the properties of the photochemical apparatus. Gutschick (1988) has suggested on the basis of a simulation model, to select plants with decreased leaf chlorophyll content and thus leaf absorbance, to improve light distribution in the lower canopy, which would result in some increase in total canopy photosynthesis. In conclusion, the efficiency of the photochemical apparatus seems to be quite good, and efforts should thus be devoted to optimize the use and maintenance of this apparatus under adverse environmental situations.

27.3.2 Dark reactions

The biochemical reactions leading to the fixation of CO_2 into carbon products have shown higher genetic variability than the photochemical reactions, offering more possibilities for directly increasing the efficiency of photosynthesis (Nelson, 1988). The enzyme ribulose-1,5-bisphosphate-carboxylase-oxygenase (rubisco) catalyses perhaps the most important limiting reaction of photosynthesis, and may determinate ultimate agronomic yield. Rubisco has received a lot of attention, and now there is a considerable knowledge about its structure, catalysis mechanism, regulation, genetics, biosynthesis and assembly (Bogorad and Rodermel, 1990; Gutteridge, 1990).

The oxygenase reaction of rubisco competes with the carboxylase reaction depending on the relative concentrations of O_2 and CO_2 around the active enzyme sites, which is the starting point of photorespiration, a process resulting in a significant net CO_2 loss for the plant (Ogren, 1984). Studies with mutants deficient in photorespiratory enzymes clearly suggested that the only essential function of photorespiration is the recovery by the chloroplast of a large part of the carbon initially lost from this organelle as a result of the oxygenase activity (Ogren, 1984; Somerville, 1986a). Thus, the manipulation of the carboxylase/oxygenase ratio should be a promising way to greatly increase the rate of net photosynthesis on a leaf basis. Additional advantages of a higher carboxylase/ oxygenase ratio can be, a better carboxylating activity at lower intercellular CO_2 concentrations, an improved water use efficiency and the optimization of nitrogen investment in rubisco protein (Gutteridge, 1990). The carboxylase/oxygenase ratio shows considerable variability between several photosynthetic organisms (Jordan and Ogren, 1981), but attempts to find important changes within the same species have not yet been succesful. For instance, manipulation of the enzyme's structure by site-specific mutagenesis has not resulted in improved carboxylase activity without negatively affecting other important enzyme characteristics, such as the turnover rate of catalysis (Somerville, 1986b, 1990; Gutteridge, 1990). Selection of mutant plants for long-term survival or resistance to atmospheres with either low CO_2 or very high O_2 have not resulted in changes in the carboxylase/oxygenase ratio or in the CO_2 compensation point (for example, the case of selected haploid tobacco plants: Medrano and Primo-Millo, 1985; Zelitch, 1989; Delgado *et al.*, 1992; D.A. Lawlor, A.J. Keys, E. Delgado and H. Medrano, unpublished). Interestingly, Zelitch (1989) has found that tobacco plants selected for higher O_2-resistant photosynthesis presented higher levels of catalase, which may reduce the release of photorespiratory CO_2 at higher temperatures by decarboxylation reactions other than those mediated by glycine decarboxylase. As a result, the net photosynthesis rate can be slightly improved at higher temperatures.

Selection of plants with increased leaf levels of rubisco has been suggested as a breeding objective (Nelson, 1988). However, the available evidence suggests that rubisco is not a good general biochemical marker for yield (Rocher *et al.*, 1989; Loza-Tavera *et al.*, 1990). One possible reason for this is that rubisco appears to be in excess in some leaf tissues, as suggested by the data of Stitt *et al.* (1991) obtained by using a series of transgenic tobacco plants transformed with an 'antisense' rubisco gene. Given the great abundance of rubisco in the leaf cell, a significant increase in the concentration of this enzyme (in the absence of changes in kinetic characteristics as previously mentioned) would result in a large investment of energy and nitrogen in protein synthesis, and certainly agronomic yield would not be increased in the same proportion (Nelson, 1988). On the other hand, the respiratory costs of maintaining plant structures with high nitrogen content are large and can negatively affect yield if rubisco protein is not used efficiently in photosynthesis.

27.4 IMPROVEMENT OF RESPIRATORY EFFICIENCY

Historically, respiration was mainly regarded as 'a path of energy degradation', but now there is better knowledge of the biochemistry of this process and its integration with the rest of plant cell metabolism (e.g. photosynthesis, nitrogen metabolism; see Dennis and Turpin, 1990). Plant growth would not be possible without respiration because carbohydrates and other substrates could not be processed into biosynthetic pathways, due to the lack of ATP and intermediate metabolites. As a result of the necessary respiratory activity, carbon is significantly lost from the whole plant (including illuminated leaves) and returned to the atmosphere as CO_2, and the question is whether or not there is a significant 'wasteful' loss of carbon not coupled to any benefit for the plant.

Respiration is considered to be a rather efficient process from the energetic point of view, and improvement of its efficiency seems difficult and has not sufficiently attracted the attention of breeders and molecular biologists. However, there are several respiratory reactions which are not so energetically efficient, the most important involving the cyanide-resistant alternative oxidase, which is the terminal end of a non-phosphorylating mitochondrial electron transport pathway. The function of the alternative pathway is not well established at present (except thermogenesis in some plants; for a review see Lambers, 1985). An interesting hypothesis suggests that the alternative oxidase could wastefully burn carbohydrate present in excess to the growth, maintenance, ion transport and other requirements, working as an 'energy overflow' (Lambers, 1985). However, this function could well be included in a more general one considering that the alternative oxidase works as a valve within the respiratory chain for increasing 'flexibility' of electron transport (Palet *et al.*, 1991).

An empirical equation was developed by McCree (1970) in which a substantial portion of respiration of white clover plants was proportional to the daily net carbon gain by photosynthesis, whereas the remainder was proportional to dry weight of living material of the plant. This division of respiration led to the functional concept of a portion of respiration associated with 'growth' (and directly linked to photosynthesis) and another portion linked to the 'maintenance' of the life functions of the plant. Considerable experimental evidence agrees with the essential basis of this respiration model (Amthor, 1989), but from the strictly biochemical point of view, the distinction between growth and maintenance respiration is not clear. The relationship between these two components of respiration and the existence of non-phosphorylating pathways in plant mitochondria, which consume carbohydrate wastefully from an energetic point of view, is also not clear at the moment. It would be too simple to say that the alternative oxidase is truly wasteful respiration, in addition to

growth and maintenance respiration, but more experimental evidence is required.

It was considered by ap Rees (1988) that oxidation of a certain amount of substrate, not coupled to any growth, maintenance or other requirement, can be within the limits of biochemical regulation. This means that control is not 100% effective, but it is not evidence of wasteful respiration. However, there can be situations in which agronomically wasteful coupled respiration can occur, and not necessarily linked to the operation of the alternative oxidase. For instance, when the proportion of photosynthates invested for maintenance requirements of the existing structure (e.g. protein turnover) proportionally increases in relation to the photosynthates invested in new growth, then respiration can be wasteful for productivity but is still energetically well coupled and within the limits imposed by metabolic regulation. Imbalances between the amount of photosynthates used in maintenance and growth occur with increasing plant or crop age (Biscoe and Gallagher, 1977; Amthor, 1989), or when the growth of sink tissues (e.g. wheat grain, Pearman *et al.*, 1981) is not adequate and under several stresses (Amthor, 1989; Taylor, 1989; Masle *et al.*, 1990), because plant growth rate is reduced and frequently the absolute maintenance costs increase to cope with the stress.

On the basis of a predictive mathematical model, Masle *et al.* (1990) concluded that 'any environmental conditions which reduce relative growth rate should cause greater respiratory losses as a proportion of carbon fixation' (and *vice versa*). Reduction of maintenance cost of the structure is necessary for adaptation of plants from sun to shade conditions, e.g. shade plants normally have lower amounts of many photosynthetic enzymes, including rubisco (Etherington, 1982). Plant species differ greatly in the investment in structural tissue, and also in the maintenance respiration coefficient (Etherington, 1982; Amthor, 1989), and furthermore, there is some evidence indicating that reduction of adult tissue maintenance requirements by genetic selection in some species (e.g. *Lolium perenne*) may have a significant positive impact on productivity (see reviews by Amthor, 1989;

Hay and Walker, 1989). A better understanding of how photosynthates can be efficiently used for the growth, maintenance, and other energy requirements of plants could be useful significantly to optimize plant growth and productivity in different environments, as predicted by carbon balance models (Masle *et al.*, 1990).

Considerable intraspecific variability of the rate of adult leaf respiration per unit dry weight (and also per unit area) has been found in several species: perennial ryegrass (Wilson, 1975, 1982; Day *et al.*, 1985), maize (Heichel, 1971), wheat (Fig. 27.1), tall fescue (Volenec *et al.*, 1984), tobacco (Fig. 27.2) and peanuts (M. Ribas-Carbó and J.A. Berry, unpublished). Pearman *et al.* (1981) reported large varietal

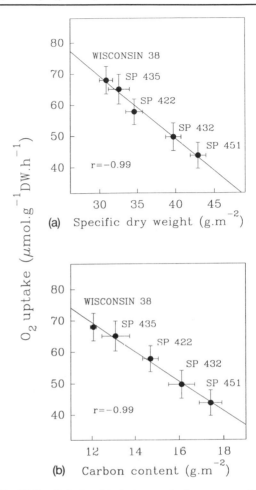

Fig. 27.2. Relationship between dark oxygen uptake per unit dry weight and either specific dry weight (a) or carbon content per unit area (b) in adult leaves of several *Nicotiana tabacum* lines. Four self-pollinated (SP), doubled haploid tobacco lines (SP422, SP432, SP435 and SP451), selected as haploids by survival in a low CO_2 atmosphere (Medrano and Primo-Millo, 1985), and the parental cultivar 'Wisconsin-38' were grown from seed in a growth room kept at high CO_2 levels (600–700 ppm) (for further details see Delgado *et al.*, 1992).

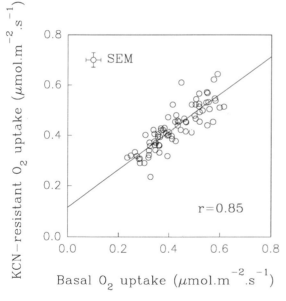

Fig. 27.1. Relationship between the respiration rate and the cyanide-resistant respiration rate in 64 hexaploid and 14 tetraploid *Triticum* cultivars. Oxygen uptake was measured with an oxygen electrode at 25°C. Cyanide concentration was 0.5 mm. Every point corresponds to the mean values (*n* = 4–6) for a single cultivar. The results suggested that there is a considerable variability in the rates of respiration and cyanide-resistant respiration (the latter is an estimate of the capacity of the alternative pathway), and both rates were very significantly correlated (X. Aranda and J. Azcón-Bieto, unpublished).

differences in wheat ear respiration during grain growth. In some cases, the genetic variability of tissue respiration has been negatively correlated with plant yield or size (Heichel, 1971; Pearman *et al.*, 1981; Wilson, 1982; Delgado *et al.*, 1992). X. Aranda and J. Azcón-Bieto (unpublished) have found a strong negative relationship between dark respiration of whole young seedlings (including the root) and the total seedling weight in tobacco lines differing in leaf respiration (Delgado *et al.*, 1992).

The dark respiration rate of adult tissues normally reflects a large maintenance component (Robson and Parsons, 1981; Wilson, 1982; Volenec *et al.*, 1984), it is correlated with nitrogen and protein content (Wilson, 1982; Volenec *et al.*, 1984; Delgado *et al.*, 1992), and is dominated by the cytochrome pathway, especially after several hours in darkness (Azcón-Bieto *et al.*, 1983; Day *et al.*, 1985; Delgado *et al.*, 1992). Robson and Parsons (1981) suggested the use of leaf respiration rate after several hours in darkness as a selection criterion for lower maintenance respiration. In summary, the rate of mature tissue respiration seems to be a parameter with sufficient intraspecific variability to be useful for breeding programmes. The biochemical basis of the varietal differences in leaf respiration is not well known but in ryegrass could be related to changes in the level of some glycolytic enzymes (Rainey *et al.*, 1990).

In addition to maintenance respiration (which mainly includes the cytochrome path), the alternative oxidase has been postulated to be a possible target for manipulation of respiration for increasing yield, given its non-energy conserving nature. It has been reported that some cytoplasmic male sterile (CMS) lines may lack or have little alternative oxidase activity, and had higher yields than fertile lines of the same species having the alternative oxidase (Musgrave *et al.*, 1986a; Connett and Hanson, 1990), especially when grown under conditions favouring large carbohydrate accumulation (e.g. high atmospheric CO_2, Musgrave *et al.*, 1986b). However, other authors were unable to reproduce many of these results with the same or other lines because the alternative oxidase was always present in large amounts, independent of the male fertility of the line (Obenland *et al.*, 1988; Van Dijk and Kuiper, 1989; Farineau *et al.*, 1990; Hakansson *et al.*, 1990). So far, it has been very difficult to find genotypes without alternative oxidase.

Recently, the protein nature of the alternative oxidase has been demonstrated (Elthon *et al.*,

1989), opening the possibility for genetic manipulation. In this way, the likely use in the near future of antisense gene technology for changing the enzymic levels of alternative oxidase in tissues (L. McIntosh, personal communication) could be very useful for determining the importance of this oxidase for agronomic productivity and perhaps for understanding its function.

The interesting observation that high atmospheric CO_2 significantly reduces tissue respiration rate in some plants (Gifford *et al.*, 1985; Reuveni and Gale, 1985; see also Amthor, 1991), and the fact that this reduction can be associated with increased yield (Reuveni and Gale, 1985; B.G. Drake, personal communication) confirms the idea that some portion of respiration may be manipulated for improving the carbon balance (Azcón-Bieto, 1992). Long-term (3–5 years) experiments of CO_2 enrichment in field grown plants also showed important reductions in tissue respiration in C_3 plants but not in C_4 plants (Table 27.1). The respiration

Table 27.1. Respiratory activities of adult green tissues and extracted cytochrome *c* oxidase from *Scirpus olneyi* (C_3) and *Spartina patens* (C_4) plants grown during five years at either ambient or elevated CO_2 levels, inside open top chambers in the field. Temperature during measurements was 25°C. The values shown are means ± SE of 5–11 replicates (J. Azcón-Bieto, W. Doherty and B.G. Drake, unpublished)

CO_2 concentration during growth	Oxygen uptake ($\mu mol/g$ dry weight h)	
	Tissue	*Cytochrome* c *oxidase*
Stems of *Scirpus olneyi* (C_3)		
350 ppm	9.4 ± 0.8	41.2 ± 4.5
700 ppm	5.3 ± 0.5	18.0 ± 2.4
Leaves of *Spartina patens* (C_4)		
350 ppm	12.6 ± 1.3	250 ± 18
700 ppm	13.4 ± 0.8	235 ± 17

changes in plants grown in high CO_2 levels were correlated with decreases in the activities of the cytochrome pathway and of the enzyme cytochrome *c* oxidase (J. Azcón-Bieto, W. Doherty and B.G. Drake, unpublished; Table 27.1). The alternative oxidase capacity was unchanged by the CO_2 treatment (data not shown). These results suggest that cytochrome *c* oxidase levels may contribute to the control of respiration rate in plants grown under natural conditions. The possibility that cytochrome *c* oxidase is an important physiological limiting factor for respiration should be further investigated.

27.5 PROSPECTS

On the basis of the evidence reviewed, we suggest that manipulation of the efficiency of photosynthesis and respiration can be a useful strategy for increasing plant yield, the main targets being the manipulation of rubisco kinetics, the improvement of the quantum yield under stress, and the decrease in the rate of maintenance respiration in adult tissues. The possibility that manipulation of the enzyme levels of the alternative oxidase and cytochrome *c* oxidase could be useful for optimizing yield should be further explored.

ACKNOWLEDGEMENTS

We are grateful to Dr B.G. Drake and Dr H. Medrano for useful discussions.

REFERENCES

Allen, L.H. (1990), Plant responses to rising carbon dioxide and potential interactions with air pollutants. *J. Environ. Qual.*, **19**, 15–34.

Amthor, J.S. (1989), *Respiration and Crop Productivity*, Springer-Verlag, New York.

Amthor, J.S. (1991), Respiration in a future, higher-CO_2 world: opinion. *Plant Cell Environ.*, **14**, 13–20.

ap Rees, T. (1988), Hexose phosphate metabolism in non-photosynthetic tissues of higher plants. In *The Biochemistry of Plants*, Vol. 14, Preiss, J. (ed.), Academic Press, New York, pp. 1–33.

Ashley, D.A. and Boerma, H.R. (1989), Canopy photosynthesis and its association with seed yield in advanced generations of a soybean cross. *Crop Sci.*, **29**, 1042–1045.

Austin, R.B. (1988), New opportunities in breeding. *HortScience*, **23**, 41–45.

Austin, R.B. (1990), Prospects for genetically increasing the photosynthetic capacity of crops. In *Perspectives in Biochemical and Genetic Regulation of Photosynthesis*, Zelitch, I. (ed.), Alan R. Liss Inc., New York, pp. 395–409.

Azcón-Bieto, J. (1983), Inhibition of photosynthesis by carbohydrates in wheat leaves. *Plant Physiol.*, **73**, 681–686.

Azcón-Bieto, J. (1992), Relationships between photosynthesis and respiration in the dark in plants. In *Trends in Photosynthesis Research*, Barber, J., Guerrero, M.G. and Medrano, H. (eds), Intercept Ltd, UK, pp. 241–255.

Azcón-Bieto. J., Lambers, H. and Day, D.A. (1983), Effect of photosynthesis and carbohydrate status on respiratory rates and the involvement of the alternative pathway in leaf respiration. *Plant Physiol.*, **72**, 598–603.

Biscoe, P.V. and Gallagher, J.N. (1977), Weather, dry matter production and yield. In *Environmental Effects on Crop Physiology*, Landsberg, J.J. and Cutting, C.V. (eds), Academic Press, London, pp. 75–100.

Bogorad, L. and Rodermel, S.R. (1990), Molecular developmental biology of ribulose bisphosphate carboxylase. In *Perspectives in Biochemical and Genetic Regulation of Photosynthesis*, Zelitch, I. (ed.), Alan R. Liss Inc., New York, pp. 119–132.

Bugbee, B.G. and Salisbury, F.B. (1988), Exploring the limits of crop productivity. I. Photosynthetic efficiency of wheat in high irradiance environments. *Plant Physiol.*, **88**, 869–878.

Campbell, W.J., Allen, L.H. and Bowes, G. (1990), Response of soybean canopy photosynthesis to CO_2 concentration, light, and temperature. *J. Exp. Bot.*, **41**, 427–433.

Connett, M.B. and Hanson, M.R. (1990), Differential mitochondrial electron transport through the cyanide-sensitive and cyanide-insensitive pathways in isonuclear lines of cytoplasmic male sterile, male fertile, and restored *Petunia*. *Plant Physiol.*, **93**, 1634–1640.

Day, D.A., De Vos, O.C., Wilson, D. and Lambers, H. (1985), Regulation of respiration in the leaves and roots of two *Lolium perenne* populations with contrasting mature tissue respiration rates and crop yields. *Plant Physiol.*, **78**, 678–683.

Delgado, E., Azcón-Bieto, J., Aranda, X., Palazón, J. and Medrano, H. (1992), Leaf photosynthesis and respiration of high CO_2 grown tobacco plants selected for survival under CO_2 compensation point conditions. *Plant Physiol.*, **98**, 949–954.

Dennis, D.T. and Turpin, D.H. (1990), *Plant Physiology, Biochemistry and Molecular Biology*, Longman Scientific & Technical, Singapore.

Dyer, T.A. (1989), The use of transgenic plants to manipulate photosynthetic processes. In *Techniques and New Developments in Photosynthesis Research*, Barber, J. and Malkin, R. (eds), Plenum Press, New York, pp. 437–447.

Elmore, C.D. (1980), The paradox of no correlation between leaf photosynthetic rates and crop yields. In *Predicting Photosynthesis for Ecosystem Models*, Vol. 2, Hesketh, J.D. and Jones, J.W. (eds), CRC Press, Boca Ratón, pp. 155–167.

Elthon, T.E., Nickels, R.L. and McIntosh, L. (1989), Monoclonal antibodies to the alternative oxidase of higher plant mitochondria. *Plant Physiol.*, **89**, 1311–1317.

Etherington, J.R. (1982), *Environment and Plant Ecology*, 2nd edn, John Wiley, Chichester.

Farineau, J., Pascal, L. and Pelletier, G. (1990), Study of respiratory and photosynthetic activities in several cytoplasmic hybrids of rapeseed with cytoplasmic male sterility. *Plant Physiol. Biochem.*, **28**, 333–342.

Foyer, C.H. (1988), Feedback inhibition of photosynthesis through source-sink regulation in leaves. *Plant Physiol. Biochem.*, **26**, 483–492.

Frey, K.J. (1988), Increasing grain yield of oats. *Plant Physiol. Biochem.*, **26**, 539–542.

Gifford, R.M. and Evans, L.T. (1981), Photosynthesis, carbon partitioning and yield. *Annu. Rev. Plant Physiol.*, **32**, 485–509.

Gifford, R.M., Lambers, H. and Morison, J.I.L. (1985), Respiration of crop species under CO_2 enrichment. *Physiol. Plant.*, **63**, 351–356.

Gutschick, V.P. (1988), Optimization of specific leaf mass, internal CO_2 concentration, and chlorophyll content in crop canopies. *Plant Physiol. Biochem.*, **26**, 525–537.

Gutteridge, S. (1990), Limitations of the primary events of CO_2 fixation in photosynthetic organisms – The structure and mechanism of rubisco. *Biochim. Biophys. Acta*, **1015**, 1–14.

Hakansson, G., Glimelius, K. and Bonnett, H.T. (1990), Respiration in cells and mitochondria of male-fertile and male-sterile *Nicotiana* spp. *Plant Physiol.*, **93**, 367–373.

Harper, J.E., Schrader, L.E. and Howell, R.W. (1985), *Exploitation of Physiological and Genetic Variability*

to *Enhance Crop Productivity*, American Society of Plant Physiologists, Waverly Press, Baltimore.

Hay, R.K.M. and Walker, A.J. (1989), *An Introduction to the Physiology of Crop Yield*, Longman Scientific & Technical, Hong Kong.

Heichel, G.H. (1971), Confirming measurements of respiration and photosynthesis with dry matter accumulation. *Photosynthetica*, **5**, 93–98.

Horton, P., Oxborough, K., Rees, D. and Scholes, J.D. (1988), Regulation of the photochemical efficiency of photosystem II; consequences for the light response of field photosynthesis. *Plant Physiol. Biochem.*, **26**, 453–469.

Isebrands, J.G., Ceulemans, R. and Wiard, B. (1988), Genetic variation in photosynthetic traits among *Populus* clones in relation to yield. *Plant Physiol. Biochem.*, **26**, 427–437.

Jordan, D.B. and Ogren, W.L. (1981), Species variation in the specificity of ribulose bisphosphate carboxylase/oxygenase. *Nature*, **219**, 513–515.

Kaminski, A., Austin, R.B., Ford, M.A. and Morgan, C.L. (1990), Flag leaf anatomy of *Triticum* and *Aegilops* species in relation to photosynthesis rate. *Ann. Bot.*, **66**, 359–365.

Kelly, H. (1988), Corn bred successfully for higher photosynthesis rate. *Agr. Res.*, **36**, 13–14.

Lambers, H. (1985), Respiration in intact plants and tissues: its regulation and dependence on environmental factors, metabolism and invaded organisms. In *Encyclopedia of Plant Physiology, New Series*, Vol. 18, Douce, R. and Day, D.A. (eds), Springer-Verlag, Berlin, pp. 418–473.

Loza-Tavera, H., Martínez-Barajas, E. and Sánchez-de-Jiménez, E. (1990), Regulation of ribulose-1,5-bisphosphate carboxylase expression in second leaves of maize seedlings from low and high yield populations. *Plant Physiol.*, **93**, 541–548.

Mahon, J.D. (1990), Photosynthetic carbon dioxide exchange, leaf area, and growth of field-grown pea genotypes. *Crop Sci.*, **30**, 1093–1098.

Masle, J., Farquhar, G.D. and Gifford, R.M. (1990), Growth and carbon economy of wheat seedlings as affected by soil resistance to penetration and ambient partial pressure of CO_2. *Aust. J. Plant Physiol.*, **17**, 465–487.

McCree, K.J. (1970), An equation for the rate of respiration of white clover plants grown under controlled conditions. In *Prediction and Measurement of Photosynthetic Productivity*, Setlik, I. (ed.), Centre for Agricultural Publishing and Documentation (PUDOC), Wageningen, pp. 221–229.

Medrano, H. and Primo-Millo, E. (1985), Selection of *Nicotiana tabacum* haploids of high photosynthetic efficiency. *Plant Physiol.*, **79**, 505–508.

Menz, K.M., Moss, D.N., Cannell, R.Q. and Brun, W.A. (1969), Screening for photosynthetic efficiency. *Crop Sci.*, **9**, 692–694.

Mol, J.N.M., van der Krol, A.R., van Tunen, A.J., van Blokland, R., de Lange, P. and Stuitje, A.R. (1990), Regulation of plant gene expression by antisense RNA. *FEBS Lett.*, **268**, 427–430.

Monteith, J.L. (1981), Does light limit crop production? In *Physiological Processes Limiting Plant Productivity*, Johnson, C.B. (ed.), Butterworths, London, pp. 23–38.

Musgrave, M.E., Antonovics, J. and Siedow, J.N. (1986a), Is male-sterility in plants related to lack of cyanide-resistant respiration in tissues? *Plant Sci.*, **44**, 7–11.

Musgrave, M.E., Strain, B.R. and Siedow, J.N. (1986b), Response of two pea hybrids to CO_2 enrichment: a test of the energy overflow hypothesis for alternative respiration. *Proc. Natl. Acad. Sci. USA*, **83**, 8157–8161.

Nasyrov, Y.S. (1978), Genetic control of photosynthesis and improving of crop productivity. *Annu. Rev. Plant Physiol.*, **29**, 215–237.

Nelson, C.J. (1988), Genetic association between photosynthetic characteristics and yield: review of the evidence. *Plant Physiol. Biochem.*, **26**, 543–554.

Nelson, C.J., Asay, K.H. and Patton, L.D. (1975), Photosynthetic responses of tall fescue to selection for longevity below the CO_2 compensation point. *Crop Sci*, **15**, 629–633.

Neyra, C.A. (1985), *Biochemical Basis of Plant Breeding*, Vol. I, *Carbon Metabolism*, CRC Press, Boca Ratón.

Obenland, D., Hiser, C., McIntosh, L., Shibles, R. and Stewart, C.R. (1988), Occurrence of alternative respiratory capacity in soybean and pea. *Plant Physiol.*, **88**, 528–531.

Ogren, W.L. (1984), Photorespiration: pathways, regulation and modification. *Annu. Rev. Plant Physiol.*, **35**, 415–442.

Ort, D.R. and Baker, N.R. (1988), Consideration of photosynthetic efficiency at low light as a major determinant of crop photosynthetic performance. *Plant Physiol. Biochem.*, **26**, 555–565.

Osmond, C.B., Bjorkman, O. and Anderson, D.J. (1980), *Physiological Processes in Plant Ecology: Towards a Synthesis with Atriplex*, Springer-Verlag, Berlin.

Palet, A., Ribas-Carbó, M., Argilés, J.M. and Azcón-Bieto, J. (1991), Short-term effects of carbon dioxide on carnation callus cell respiration. *Plant Physiol.*, **96**, 467–472.

Pearcy, R.W., Ehleringer, J., Mooney, H.A. and Rundel, P.W. (1989), *Plant Physiological Ecology. Field Methods and Instrumentation*, Chapman and Hall, London.

Pearman, I., Thomas, S.M. and Thorne, G.N. (1981), Dark respiration of several varieties of winter wheat given different amounts of nitrogen fertilizer. *Ann. Bot.*, **47**, 535–546.

Peng, S., Krieg, D.R. and Girma, F.S. (1991), Leaf photosynthetic rate is correlated with biomass and grain production in grain sorghum lines. *Photosynthesis Res.*, **28**, 1–7.

Rainey, D.Y., Mitton, J.B., Monson, R.K. and Wilson, D. (1990), Effects of selection for dark respiration rate on enzyme genotypes in *Lolium perenne*. *Ann. Bot.*, **66**, 649–654.

Reuveni, J. and Gale, J. (1985), The effect of high levels of carbon dioxide on dark respiration and growth of plants. *Plant Cell Environ.*, **8**, 623–628.

Robson, M.J. and Parsons, A.J. (1981), Respiratory efflux of carbon dioxide from mature and meristematic tissue of uniculm barley during eighty hours of continuous darkness. *Ann. Bot.*, **48**, 727–731.

Rocher, J.P., Prioul, J.L., Lecharny, A., Reyss, A. and Joussaume, M. (1989), Genetic variability in carbon fixation, sucrose-P-synthase and ADP glucose pyrophosphorylase in maize plants of differing growth rate. *Plant Physiol.*, **89**, 416–420.

Rodermel, S.R., Abbott, M.S. and Bogorad, L. (1988), Nuclear- organelle interactions: nuclear antisense gene inhibits ribulose bisphosphate carboxylase enzyme levels in transformed tobacco plants. *Cell*, **55**, 673–681.

Sage, R.F. and Sharkey, T.D. (1987), The effect of temperature on the occurrence of O_2 and CO_2 insensitive photosynthesis in field grown plants. *Plant Physiol.*, **84**, 658–664.

Sharkey, T.D. (1985), Photosynthesis in intact leaves of C_3 plants: physics, physiology and rate limitations. *Bot. Rev.*, **51**, 53–106.

Somerville, C.R. (1986a), Analysis of photosynthesis with mutants of higher plants and algae. *Annu. Rev. Plant Physiol.*, **37**, 247–274.

Somerville, C.R. (1986b), Future prospects for genetic manipulation of rubisco. *Phil. Trans. R. Soc. London Ser. B*, **313**, 305–324.

Somerville, C.R. (1990), The biochemical basis for plant improvement. In *Plant Physiology, Biochemistry and Molecular Biology*, Dennis, D.T. and Turpin, D.H. (eds), Longman Scientific & Technical, Singapore, pp. 495–501.

Stitt, M. and Quick, W.P. (1989), Photosynthetic carbon partitioning – its regulation and possibilities for manipulation. *Physiol. Plant.*, **77**, 633–641.

Stitt, M., Quick, W.P., Schurr, U., Schulze, E.D., Rodermel, S.R. and Bogorad, L. (1991), Decreased ribulose-1,5-bisphosphate carboxylase-oxygenase in transgenic tobacco transformed with 'antisense' rbcS. II. Flux-control coefficients for photosynthesis in varying light, CO_2, and air humidity. *Planta*, 183, 555–566.

Stitt, M., Von Schaewen, A. and Willmitzer, L. (1990), 'Sink' regulation of photosynthetic metabolism in transgenic tobacco plants expressing yeast invertase in their cell wall involves a decrease of the Calvin-cycle enzymes and an increase of glycolytic enzymes. *Planta*, **183**, 40–50.

Taylor, G.J. (1989), Maximum potential growth rate and allocation of respiratory energy as related to stress tolerance in plants. *Plant Physiol. Biochem.*, 27, 605–611.

Van Dijk, H. and Kuiper, P.J.C. (1989), No evidence for a pleiotropic relationship between male sterility and cyanide-resistant respiration in *Plantago lanceolata*. *Physiol. Plant.*, **77**, 579–586.

Volenec, J.J., Nelson, C.J. and Sleper, D.A. (1984), Influence of temperature on leaf dark respiration of diverse tall fescue genotypes. *Crop Sci.*, **24**, 907–912.

Wardlaw, I.F. (1990), The control of carbon partitioning in plants. *New Phytol.*, **116**, 341–381.

Wilson, D. (1975), Variation in leaf respiration in relation to growth and photosynthesis of *Lolium*. *Ann. Appl. Biol.*, **80**, 323–338.

Wilson, D. (1982), Response to selection for dark respiration rate of mature leaves in *Lolium perenne* and its effect on growth of young plants and simulated swards. *Ann. Bot.*, **49**, 303–312.

Zelitch, I. (1982), The close relationship between net photosynthesis and crop yield. *BioScience*, **32**, 796–802.

Zelitch, I. (1989), Selection and characterization of tobacco plants with novel O_2-resistant photosynthesis. *Plant Physiol.*, **90**, 1457–1464.

28

Breeding for efficient root systems

J.M. Clarke and T.N. McCaig

28.1 INTRODUCTION

There is no reason to doubt that root system traits, as with traits associated with the aerial portion of the plant, can be genetically manipulated if one is willing to make the required effort. A review by O'Toole and Bland (1987) documented evidence of variation within crop species in factors such as root system size and weight, depth of penetration in the soil, distribution within the soil profile, and rate at which individual root xylem axes transport water. Some of these traits have been shown to be amenable to selection. However, it must be determined what manipulations are beneficial, and whether these manipulations are practical in crop breeding.

Various workers have suggested that breeding for particular root system traits would improve adaptation of crops to specific environmental conditions such as drought. However, it seems that few breeders have accepted these recommendations, principally because there is insufficient evidence to warrant such a course of action. It may seem to physiologists that breeders have a cynical view of the possibilities of using physiological selection criteria, but it must be understood that addition of further selection criteria will only be contemplated if they improve selection efficiency. Allard and Bradshaw (1964) went so far as to question whether understanding the biochemical, morphological, and physiological basis of genotype response to environment is of any value in breeding programmes with the primary purpose of developing superior cultivars. Similarly, Marshall (1987) stated that 'empirical' breeding is likely to continue to be very important. Only solid evidence of the worth of selection for physiological or morphological traits will encourage their incorporation into practical breeding programmes.

The evidence required includes heritabilities of the traits, phenotypic and genotypic correlations between the traits and agronomic traits such as yield, and some indication of the complexity of genetic control. At the same time, an understanding must be developed of the effect of manipulation of a root trait on the expression of other important traits. For example, the effect of increased size of the root system on aboveground biomass must be determined. Review of the literature on root traits of crop species indicates very few instances of complete evaluation of either physiological aspects or of breeding strategies to incorporate such traits into commercial cultivars. This may be partly due to the difficulty and expense of studying roots, but also implies a lack of communication among physiologists, breeder/physiologists, and practical breeders. It is important that all scientists involved in such investigations plan and conduct their research in a co-ordinated fashion that will provide the required information in the shortest possible time.

In light of the current lack of detailed understanding of which root system characteristics would be desirable for specific environments, our discussion will have to be broader than

Plant Breeding: Principles and prospects. Edited by M.D. Hayward, N.O. Bosemark and I. Romagosa. Published in 1993 by Chapman & Hall, London. ISBN 0 412 43390 7.

genetic manipulation *per se*. It will be useful to mould the discussion around a logical approach to the problem, from development of ideotypes, proof of the relationship of the proposed traits to crop yield, through to development of breeding strategies.

28.2 A SYSTEMATIC APPROACH

28.2.1 Possible approaches

Several workers have suggested a systematic approach to the determination of crop morphological, physiological and phenological characteristics associated with crop yield in water-limiting environments (Zobel, 1983; Clarke, 1987). Similar procedures would apply for other environmental constraints. In the case of roots, O'Toole and Bland (1987) proposed the required evaluation steps of definition of limitations, determination of genetic variation, hybridization and selection, and field evaluation of resultant genotypes. Development and assessment of breeding strategies should be added to this list.

The basic assumption when one undertakes to manipulate root systems is that there is, in fact, a specific root system for each environment which optimizes our demands for a final economic product, based upon acceptable management practices. Therefore, the first step must be to define the optimum as clearly as possible in terms of agronomic, environmental and economic considerations. For example, is stable yield more important than high yield potential; is end-use quality important, or is there a requirement to limit costly inputs or hazardous chemicals; is the goal to maximize profits over some time frame? When the goal has been defined, a specific plant or root ideotype which best utilizes available resources to achieve the goal can be proposed. In the initial design of the ideotype, one's thinking should not be too limited by current methodologies, because these will change as research progresses.

With an ideotype defined, the next logical step is to identify genotypes showing genetic variation for the traits considered important. When there is no variation in locally adapted genotypes, both evaluation of the traits and their manipulation will be complicated, as discussed below (section 28.2.4). If genetic manipulation of particular traits looks worthwhile, research can then progress through hybridization, determination of inheritance, and development of breeding strategies. This latter phase is of greatest interest to breeders, and yet remains one of the least-emphasized aspects of the evaluation of physiological traits.

It is not necessary to execute each of these steps consecutively. In fact, it may be desirable after definition of the ideotype and preliminary evaluations to conduct several of the phases concurrently. For example, hybridization at early stages in the evaluation would permit study of the root trait in an adapted genetic background, assessment of its interaction with other important traits, and preliminary determination of its genetic control.

28.2.2 Methods for study of roots

It is beyond the scope of this chapter to discuss in detail methodologies for studying root systems. However, it is important to review briefly several techniques that show some promise for evaluating large enough numbers of plants for at least limited genetic and selection studies. Evaluation of methods for assessing root system traits must include proof that there is reasonable agreement between measurements made in controlled environments and those made in the field, or between seedling root characteristics and those in more mature plants.

Potential methods for the study of roots are no fewer than those that have been applied to the aboveground portion of the plant. There is an unfortunate tendency to associate 'root methods' with only those that determine gross physical features such as the presence or absence of a trait, root length, and so on. The degree of difficulty or cost of the method should determine the breeding strategy rather than *vice versa*. Because of the inherent problems of dealing with roots, there is also a tendency to conclude that relatively easily measured traits are desirable components of the ideotype. For example, it is probably not coincidence that root size, length and geometry are prime features of many ideotypes, and are also the easiest traits to measure.

Methodologies for study of root systems have ranged from tedious manual excavations in the field (Pavlychenko, 1937), through pot studies in sand, hydroponic culture (MacKey, 1973), and video monitoring of root growth *in situ* (Upchurch and Richie, 1983). Current methods of studying roots tend to be labour-intensive and have, therefore, seldom been exploited in the large studies required to determine the potential for breeding for particular traits.

Hydroponic systems (MacKey, 1973; Sullivan and Ross, 1979; Ekanayake *et al.*, 1985a) in which plants are grown individually in vertical or angled tubes appear to have the potential for screening moderate numbers of lines with relatively low labour requirement. Ekanayake *et al.* (1985a) used a system capable of growing 110 individual plants, whereas those of MacKey (1973) and Sullivan and Ross (1979) were about five times larger. If such systems are utilized with only liquid media, roots can be readily observed and sampled without the rigorous washing required when solid rooting media are used. However, MacKey (1973) suggested that root growth is less and shoot growth is greater in liquid than in solid media. This would obviously affect measurements of root/shoot ratio, and it must be determined whether factors such as root depth and distribution in liquid media correlate with the situation in the field. Sullivan and Ross (1979) reported that measurements of rooting characters and root/shoot ratio measured in liquid hydroponic media ranked genotypes similarly to response to drought in the field. Recently, Mian *et al.* (1990) reported high correlations between root fresh weight of hydroponically grown wheat (*Triticum aestivum* L.) seedlings (4 weeks) and root length density of relatively mature (Zadok stage 85) field grown material.

The resistance required to pull plants from the soil is related to root system size in maize (*Zea mays* L.) and rice (*Oryza sativa* L.), and constitutes another simple, rapid assessment method. In rice, the force required to pull seedlings from paddy soil was correlated with root weight and branching (O'Toole and Soemartono, 1981). Pulling force in maize has been related to improved lodging resistance (Arihara and Crosbie, 1982) and root system size in relation to corn rootworm resistance (Peters *et al.*, 1982). Root-pulling is relatively easy and can be partially mechanized.

Brar *et al.* (1990) evaluated seedling growth of legume species in growth pouches and slanted tubes filled with sand. These techniques allowed handling of large numbers of plants. Root growth at 10 days in the pouches was correlated with rooting depth in the field at 32 days after planting ($r = 0.87**$), whereas correlations for root growth in the tubes for various intervals with field rooting depth ranged from $0.62**$ to $0.71**$. In sorghum (*Sorghum bicolor* L.) there was a poor relationship, however, between seminal root length and total plant root length or volume (Blum *et al.*, 1977). Installation of minirhizotron tubes to observe roots *in situ* in the field (Bragg *et al.*, 1983; Upchurch and Richie, 1983; ASA, 1987) is probably too cumbersome and expensive for routine screening, but could be useful for developing correlations between controlled-environment and field results.

28.2.3 Formulation of an ideotype

The root system ideotype may be one that absorbs water, nitrogen or other nutrients more efficiently, resists pathogens, penetrates deeper into the soil, has lower respiratory losses, has high tolerance to certain elements, has a specific geometry, or has other useful biochemical and physiological characters. Extensive research may be required before an ideotype can be adequately defined. Formulation of hypotheses concerning optimal root systems to overcome particular environmental problems such as drought can be approached by studying weather, edaphic, hydrologic, and crop yield data. From these data it may be possible to determine the crop developmental stage at which water deficit stress is greatest, and to propose a plant ideotype with greater tolerance or better avoidance of the stress. The ideotype is likely to vary, of course, from one environment to another.

The timing of precipitation deficits and the use of stored soil water in relation to crop phenology have led research workers to differing plant ideotypes. These can be divided into two basic types: increased rooting depth and/or rate of root growth, *versus* restriction of the water transport rates of roots to ration water through the growth period.

Based on studies of the root systems of limited numbers of inbred lines of wheat (Hurd, 1964, 1968) and the observation that stored soil water remained at depths of 0.6–1.2 m at maturity, Hurd (1974) hypothesized that in dry environments, genotypes with more extensive root systems would have a yield advantage over those with smaller root systems. Similar views frequently appear in the literature dealing with drought resistance of crops. Conversely, Passioura (1972), suggested that reduction in the rate of water uptake by roots would conserve water for later stages of development and improve yield potential under dry growing conditions. These approaches may both be valid in different environmental situations: utilization of stored soil water at depths of up to 1 m or more in the former case, and shallow soil water profiles in the latter.

Opinion is still divided as to which approach is preferable. With regard to increasing rooting depth or efficiency of water absorption at depth, Passioura (1983) raised the valid argument that it must first be shown that this water is actually available on a continuing basis. Hydrologic studies must be undertaken to determine if the water used by deeply rooted genotypes is recharged, at least in a reasonable proportion of seasons, prior to the start of the next crop cycle. For example, it may take five or more years to recharge soil water profiles following deeply rooted alfalfa (*Medicago sativa* L.) crops grown in dry environments (Brown, 1964). It may be that much of the water below average rooting depth represents the accumulation of small, unused residual water over several years (Passioura, 1983). Derera *et al.* (1969) concluded that breeding for deeper or more prolific rooting of wheat would be of little value because they found little genetic variability in rate of penetration of wheat roots, and low heritability of root system weight. They also noted that increasing root system size would divert photosynthate from shoot growth.

Similarly, doubt has been expressed about the desirability of the Passioura approach of reducing root system effectiveness by increasing the hydraulic resistance to water flow in the root (Taylor and Klepper, 1978; O'Brien, 1979). O'Brien (1979) suggested that not enough is known about the optimal root system to justify a particular breeding approach. More recently, Siddique *et al.* (1990) expressed a similar sentiment, although their results were somewhat supportive of Passioura's approach.

Implicit in the development of ideotypes is the assumption that the environment for which we are breeding is well known and perhaps homogeneous. But what is really meant, for example, by 'dry' or 'drought-prone' environments? Temporal and spatial variability in precipitation tends to be large in many environments. Therefore, the mean precipitation as well as its standard deviation are important considerations in defining the 'target population of environments' (Comstock, 1977) for which improved cultivars are being developed.

Genotype–environment interactions are a further problem in definition of the ideotype. Particular traits, such as an extensive root system and high root/shoot ratio, may be desirable in the driest seasons at a particular site, but a disadvantage in seasons with average to above average precipitation. The Passioura ideotype addresses this problem by relying on increased hydraulic conductivity of the seminal root system. This system is of paramount importance in dry conditions that limit tillering and development of nodal roots, but water and nutrient uptake are not restricted in favourable growing conditions because development of a prolific nodal root system by-passes the seminal system.

Other ideotypes have been suggested to address problems such as biotic stresses. In maize, for example, increased root thickness appears to confer tolerance to the corn rootworm (Peters *et al.*, 1982). Similarly, resistance to root rot may be a desirable feature in wheat ideotypes because of the negative relationship between grain yield potential and root rot incidence (Duczek, 1989). Root system type of cultivated alfalfas was changed from a tap root to a creeping form by transferring the trait from *Medicago falcata* L. (Heinrichs, 1963), thereby improving stand longevity under grazing.

As will be seen below, very little is known concerning the relationship between most root traits and the desired economic components of crops, and so we may be forced to rely on intuition when formulating ideotypes. The success of breeding for root traits depends on their correlation with the desired agronomic traits or economic products. If one chooses root traits for use as potential selection criteria only on the basis of ease of measurement, the correlation may be so low as to ensure failure. Of course, the cost of screening for particular traits is an important consideration. Therefore, the traits which are chosen must ultimately be a compromise between what the ideotype should be and what can be measured. However, it must be remembered that a compromise has been made, and that the ideotype should be modified as improved methods become available.

28.2.4 Assessment of proposed traits

This topic encompasses much of the literature on root systems. A few examples of such studies are warranted to illustrate the complexity of the task and some of the potential pitfalls. A first requirement is to have plant material differing for the desired traits. Typically, researchers begin by evaluating inbred lines or cultivars that appear to show variable responses to the environmental constraint in question. In a few instances, isogenic lines or populations known to differ in root system traits or performance in adverse environments might be available. To test his hypothesis regarding the slowing of water use during vegetative growth, Passioura (1972) used a single cultivar that had been physically modified to limit the number of seminal roots, whereas Cornish (1981) demonstrated that sow-

ing depth could be used to modify length of the subcrown internode and rate of water use.

The discrete evaluation steps of proving the relationship, looking for genetic variability, hybridizing and field-testing proposed for evaluation of root traits (O'Toole and Bland, 1987) often seem to become stalled on attempts to show a relationship between the trait of interest and yield potential in small numbers of inbred lines. After preliminary investigations show some evidence that a particular trait contributes to yield potential, studies must be expanded to either large numbers of diverse inbred lines, a series of isogenic lines or populations, or segregating lines (Richards, 1987) to test the relationship more fully. If using inbred lines, it is imperative to use large enough numbers to minimize interactions between the trait of interest and other factors such as phenology. Divergent selections and isogenic lines are not without drawbacks either, since factors such as linkage and epistasis must be considered. Use of several genetic backgrounds in the development of isogenic lines is helpful.

An important factor to consider when proposing to modify root systems is the consequence of such actions on the aerial portion of the plant. Increasing root weight, for example, will divert assimilates from tops to roots and so change the root/shoot ratio. The energy cost to the plant for producing roots may be at least double that for producing shoot dry matter (Passioura, 1983). There is presumably an optimal root/shoot ratio for each set of environmental conditions (Passioura, 1983). In dry environments this balance must be such that shoot growth, the primary factor in water loss, is matched to the size, distribution, and efficiency of the root system that supplies water.

Root/shoot ratio has been studied extensively in wheat in relation to dwarfing genes. Some of the most studied genes in wheat include the height-reducing Rht_1 and Rht_2 genes from 'Norin 10'. The shoots of genotypes carrying these genes do not elongate in response to endogenous or applied gibberellic acid (GA). Since GA is involved with many growth and development functions, it is natural to ask what effects these genes have on root systems. Unfortunately there is no clear answer to this question and it emphasizes the problems involved in the study of interactions among traits. The conflict results from a combination of the difficulty of studying roots, the high inherent variation normally associated with roots, environmental interactions, and interactions between the *Rht* genes and the genetic background into which they are incorporated. MacKey (1973) concluded, based on hydroponics experiments, that the 'Norin 10' genes decreased both shoot and root biomass equally.

We have carried out several controlled-environment experiments (hydroponically in sand) involving the four gene combinations $Rht_1 Rht_2$ (dwarf); $Rht_1 rht_2$ /$rht_1 Rht_2$ (semi-dwarf); $rht_1 rht_2$ (tall)) in several diverse genetic backgrounds (T.N. McCaig, unpublished). These results contradict those of MacKey (1973) and suggest that the Rht_1 and Rht_2 genes decrease shoot biomass slightly with no effect on root biomass; the result is higher root/shoot ratios, especially in two-gene dwarf genotypes (Fig. 28.1). These results underline the importance of differentiating between genetic effects on the root system and those that influence root/shoot ratios through an effect on shoot growth.

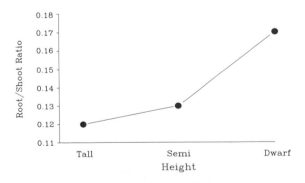

Fig. 28.1. Effect of the Rht_1 and Rht_2 height-reducing genes on the root/shoot ratios of tall ($rht_1 rht_2$), semidwarf ($Rht_1 rht_2/rht_1 Rht_2$) and dwarf ($Rht_1 Rht_2$) wheat at anthesis (means of controlled-environment experiments involving genotypes from four genetic backgrounds).

28.2.5 Examples of genetic manipulation

There are relatively few examples in the literature of detailed, long-term efforts to genetically manipulate root systems to evaluate specific traits. An exception is the work by Richards and Passioura (1981b, 1989) to reduce seminal root xylem vessel diameter in wheat to test the hypothesis that a reduced rate of water use would be beneficial under drought (Passioura, 1972). This research demonstrates the high degree of effort that is required to prove association between a morphological and an agronomic trait.

Preliminary investigations demonstrated the need to maintain consistent environmental conditions and to use seed of uniform size to maximize the expression of heritable differences in seminal root morphology (Richards and Passioura, 1981a). Richards and Passioura (1981b) then surveyed over 1000 accessions of wheat, including diploid, tetraploid, and hexaploid forms, for maximum xylem vessel diameter, number of seminal root axes, and proportion of seminal axes with multiple metaxylem vessels. They concluded that maximum xylem vessel diameter was the most amenable to selection, because this trait had greater genetic variation than either number of axes or proportion of multiple metaxylem vessels. Narrow-sense heritability for the vessel diameter trait ranged from 0.38 to 0.78, with a mean of 0.52 (Richards and Passioura, 1981b). As will be discussed below (see section 28.3.2 of this chapter), these heritabilities, in practical terms, are probably lower. A single cycle of divergent selection for maximum xylem vessel diameter produced highly significant (P < 0.01) differences in diameter in several populations.

Richards and Passioura (1981b) were unable to find maximum xylem vessel diameters of less than 60μm in modern wheat cultivars, so embarked on a breeding programme to transfer narrow diameter vessels (50μm) from a landrace wheat from Turkey into two well-adapted Australian cultivars with vessel diameters of 60 and 65μm (Richards and Passioura, 1989). In BC$_3$-derived lines, narrow vessel selections outyielded unselected lines in one genetic background, but not in the other (Table 28.1). For the most part, however, the recurrent parents had higher grain yields than the narrow vessel selections. BC_5 lines had recovered the yield potential of the recurrent parents, but in only two of the tests did narrow-vessel lines yield more than unselected lines (Table 28.1). In another environment with low levels of stored soil water, the narrow vessel selections did not show higher yields than the unselected lines (Richards and Passioura, 1989).

Staying with the theme of roots in association with water relations, there are indications that characteristics of the root system of rice are related to genotype performance under water-deficit situations. Ekanayake *et al.* (1985a) found that maximum root length, thickness, and root system volume, as measured in a hydroponic culture system, were correlated (-0.46^* to -0.52^{**}) with a field drought recovery score. Heritabilities of the three traits, measured as 2/3 of the regression coefficient (following the suggestion of Smith and Kinman, 1965), of F_3 on F_2 parental values were 0.53, 0.61, and 0.18, respectively. This work (Ekanayake *et al.*, 1985a) was not carried far enough to determine the improvement in yield under dry conditions that could be achieved by selection for a trait such as root length. The authors did observe that despite a general positive association between plant height and root system depth, it was possible to obtain short-statured segregants with deep root systems.

In other research with rice Ekanayake *et al.* (1985b), found that root-pull resistance of rice plants was correlated with a drought severity score (-0.39^* to -0.41^{**}). Heritability measured and expressed as in the previous study (Ekanayake *et al.*, 1985a) was 0.39–0.47.

Maize roots have been investigated extensively, primarily from the point of view of resistance to the corn rootworm and to lodging, as noted above. Both of these factors may be related to crop biomass or grain yield when the stresses are present, whereas selection for yield in the absence of the stresses may not produce a correlated response in either.

Kevern and Hallauer (1983) investigated root-pull resistance of maize in relation to root lodging. Heritability estimates for pre- and

Table 28.1. Grain yields of BC_3 and BC_5 wheat lines selected for narrow maximum seminal root xylem vessel diameter in comparison to unselected controls and the two recurrent parents 'Kite' and 'Cook' (from Richards and Passioura, 1989)

Year	Grain yield relative to recurrent parent (%)			
	Kite		Cook	
	Narrow	Unsel.	Narrow	Unsel.
BC_3 lines				
1981	90	81*	98	93
1982	79	73*	80	77
1983 site 1	106	95*	98	98
site 2	94	89*	88	85
BC_5 lines				
1984 site 1	99	103	103	101
site 2	108	111	101	96*
1985	104	97*	99	102

* Unselected significantly different from narrow selections ($P < 0.05$.

post-anthesis pull resistance and root lodging were 0.76, 0.81, and 0.74, respectively. The genetic correlation between pre-anthesis (-0.20) and post-anthesis (-0.28) pull-resistance and yield was quite low. Consequently, the correlated response of root lodging through selection for pre-anthesis and post-anthesis pull-resistance was 26% and 35%, respectively, of the response achieved by direct selection for root lodging. The authors concluded that although the correlated response to selection for root-pull resistance was lower than the direct response, selection for root-pull resistance would contribute to reduced root lodging, and would permit selection progress in the absence of environmental conditions conducive to the expression of root lodging.

28.3 EVALUATION OF BREEDING STRATEGIES

Evaluation and development of a breeding strategy is a crucial final step in the evaluation of a proposed trait, without which the trait is unlikely to be considered by breeders. Evaluation of strategies is time-consuming and costly, but time and money could be saved if preliminary aspects of the evaluation were conducted concomitantly with trait assessment. The complexity of the inheritance of a trait has an impact on efforts to evaluate its relationship to yield or other agronomic traits, and on the ease of its manipulation through breeding.

28.3.1 Genetic control of root traits

Determination of the genetic control of traits related to root biomass is difficult because of plant factors such as the competition between root and shoot for photosynthate, and the interaction of genotype with environment. Numerous short-term studies using generation means analysis have been conducted, and conclusions drawn regarding dominance *vs* additive effects in the genetic control of root traits. Relatively few workers, however, have determined the number of genes involved in controlling particular roots traits, or their chromosomal locations.

Subcrown internode length of several cultivars of common wheat seemed to be controlled by either one or two genes (McKenzie, 1971). Control of coleoptile length is somewhat more complicated, being influenced by genes on 3 to 7 chromosomes in crosses to the 'Chinese Spring' monosomic series (Allan and Vogel, 1964). In durum wheat (*T. turgidum* L. var. *durum*), Scarascia-Mugnozza and Porceddu (1973) concluded that the same trait seemed to be under polygenic control. Richards and Passioura (1981b) observed that the control of maximum xylem vessel diameter, number of seminal root axes, and number of metaxylem vessels of common wheat was multigenic. There is evidence that root/shoot ratio of wheat is influenced by single genes. In winter wheat, for example, Monyo and Whittington (1970) showed that single genes controlling growth period affected both root and shoot, although polygenic systems were also involved. The complexity of the inheritance of a trait affects evaluation of its relationship to yield and other

agronomic traits, and on the ease of its manipulation through breeding.

28.3.2 Heritabilities and correlations

As will be seen below, phenotypic and genotypic correlations and heritabilities are required to assess potential selection traits. Numerous heritability estimates for root traits have been published; a few examples are summarized in Table 28.2. Some of the estimates were not corrected for inbreeding as suggested by Smith and Kinman (1965); this correction has been made, where appropriate, to facilitate comparison of the published values. Some of the heritabilities are high enough to make them of potential interest for manipulation.

Although many heritability estimates have been published for root traits, few correlations have been measured, particularly in relation to root traits and yield under drought. Hurd (1968) pointed out that he did not consider it possible to show a correlation between root length or weight and grain yield of wheat. It is difficult to establish meaningful correlations between physiological traits and yield because an agronomic trait such as yield is the result of the action of many genes, each with relatively small individual effect and each interacting with environment.

Derera *et al.* (1969) reported a correlation of 0.73** ($n = 15$) between number of nodal roots and grain yield of wheat in an artificially drought–stressed environment. Kevern and Hallauer (1983) measured correlations of 0.03–0.40** between pre-anthesis root-pull resistance and yield, and −0.57**–0.09 between post-anthesis root-pull and yield in maize. In another maize experiment, correlations of post-anthesis root-pull resistance with yield ranged from 0.22–0.31 (Peters *et al.*, 1982).

Environment will no doubt have a large effect on the magnitude and sign of correlations of yield and root traits. Negative correlations between yield and root mass could be envisaged in well-watered environments, whereas the reverse could be observed in environments with a water deficit. This would complicate both

Table 28.2. Examples of heritabilities of root traits in several crop species

Trait	Heritability		Reference
	Reported	*Corrected[a]*	
Wheat			
Xylem vessel diameter	0.52	0.35	Richards and Passioura (1981b)
Number of seminal roots	0.34	0.23	Richards and Passioura (1981b)
Multiple metaxylem ves.	0.49	0.33	Richards and Passioura (1981b)
Coleoptile internode length	0.74	0.39	Gul and Allan (1978)
Coleoptile internode length	0.17–0.53		Porceddu *et al.* (1978)
Root length	0.14–0.51		Porceddu *et al.* (1978)
Root:shoot ratio	0.32		Kazemi *et al.* (1978)
Oat			
Seedling root length	0.83	0.44	Barbour and Murphy (1984)
Rice			
Maximum root length	0.35		Ekanayake *et al.* (1985a)
Root thickness	0.61		Ekanayake *et al.* (1985a)
Root volume	0.18		Ekanayake *et al.* (1985a)
Root dry weight	0.56		Ekanayake *et al.* (1985a)
Root-pull resistance	0.39–0.47		Ekanayake *et al.* (1985b)
Maize			
Root-pull – seedling	0.56		Arihara and Crosbie (1982)
– preanthesis	0.76		Kevern and Hallauer (1983)
– postanthesis	0.81		Kevern and Hallauer (1983)
White clover			
Tap root diameter	0.54		Woodfield and Caradus (1990)
Tap root dry weight	0.56		Woodfield and Caradus (1990)
Tap root number	0.08		Woodfield and Caradus (1990)

[a] Corrected for degree of inbreeding (Smith and Kinman, 1965).

site to site and year to year comparisons within sites.

28.3.3 Selection strategies

The possible manipulation of a root system trait to improve yield or another agronomic trait can be evaluated in terms of three basic approaches. The first is to use correlated responses, since selection for yield in a particular environment will perhaps change the root trait in the desired direction. Choice of parental genotypes for hybridization is an example of this approach. Indirect selection, that is, selection for the root trait rather than yield, is a second possibility. This concept is actually closely related to the former, but the selection is considered in the opposite manner, that is for the secondary trait rather than yield. A third approach is index selection for both the root trait and yield. These approaches can be evaluated using quantitative genetic principles, but in practical terms, economic factors such as relative costs must be considered as well.

Screening and selection of potential parents carrying desired traits is frequently suggested for incorporation of physiological or morphological traits into new cultivars. Following crossing with adapted local parents, normal selection procedures are followed; it is assumed that the traits of interest will be selected through their correlated relationship to the character, such as yield, for which selection is being practised. Hurd (1971), for example, suggested such an approach in breeding for drought resistance in wheat, and demonstrated it in the development of the cultivar 'Wascana' (Hurd and Townley-Smith, 1972). Zobel (1983) called this process 'associative breeding'.

Non-adapted parents, although they may carry putative drought tolerance or other desirable traits, may be difficult to use in a breeding programme. An example of the effect of a non-adapted parent on yield performance of a cross can be seen in the work of Richards and Passioura (1989) described above; only after five backcrosses was the yield of the adapted local parents recovered. In addition to backcrossing, double or three-way crosses may be useful for transfer of physiological traits from introduced to adapted genotypes (Clarke and Townley-Smith, 1984).

How effectively the parental selection and normal agronomic evaluation approach incorporates the desired root traits depends upon the heritability of the trait and its genetic correlation with yield. For economic reasons, the breeder may be willing to accept modest improvement of a trait such as rooting depth through selection for yield. The correlated response to selection for an indirect trait, X, through selection for a direct trait, Y, can be calculated as follows (Falconer, 1988):

$$CR_X = ih_y\, h_x\, r_g\, \sigma_p x \qquad (28.1)$$

where i is the standardized selection differential, h_y and h_x are the square roots of the heritabilities of the two traits, r_g is the genotypic correlation between them, and $\sigma_p x$ is the phenotypic standard deviation of X.

Kevern and Hallauer (1983) calculated the expected improvement in root lodging through selection for pre-anthesis root-pulling resistance of maize. The average heritabilities were 0.76 and 0.74 for root-pull resistance and lodging, respectively, and the average genetic correlation between them was −0.27. With a 10% selection intensity, increased pulling resistance reduced root lodging by 5.6%. Direct selection for root lodging at the same intensity gave a predicted change of 21.6%.

Reliance on the correlated response of the non-selected trait will clearly be less efficient than direct selection unless the heritabilities and genetic correlations are high. However, where the physiological or morphological trait is difficult to select for, the correlated response approach has merit, permitting the manipulation of complex traits that would be prohibitively expensive to breed for directly (Zobel, 1983). Further, selection for yield will presumably optimize the combinations of traits, such as root/shoot ratio, in the environments in which selection is carried out. Blum (1979) proposed development of cultivars that are buffered against adverse environmental conditions; careful screening and selection of parents might be the easiest way to achieve this goal. Detailed knowledge of desirable traits and their interactions with environment is required, however, to permit an informed choice of prospective parents. Hurd (1974) stressed the need to evaluate carefully a small number of targeted crosses among well-known parents rather than superficially evaluating a large number of crosses among unknown parents.

It is worth noting that the predicted correlated responses will not always be obtained in the course of actual selection. First, measured genetic correlations are often imprecise, and they change during the course of selection (Falconer, 1988). Second, the consistency of the environments in which selection is practised will affect the relationship between the traits, since high yield may be associated with deep rooting in the dry but not the wetter seasons encountered during the selection cycle. One would have to be quite sure of the worth of a root trait to follow the suggestion of O'Toole and Bland (1987) to select parents, select for agronomic traits in early generations, and eliminate advanced lines on the basis of the desired root trait.

Now consider indirect selection for improvement in a primary agronomic trait by way of a physiological or morphological trait. Searle (1965) suggested that the relative efficiencies of direct and indirect selection must be compared before indirect selection can be advocated. This can be done if the heritabilities of the two traits and the genetic correlation between them are known. For example, Searle (1965) described a relative selection efficiency parameter (RSE) for direct selection for a basic trait Y *versus* indirect selection for Y through an alternative trait X:

$$RSE(X,Y,y) = r_g \sqrt{(h_x^2/h_y^2)} \qquad (28.2)$$

where y is the genotype corresponding to Y, r_g is the genetic correlation between Y and X, and h_x^2 and h_y^2 are the heritabilities of X and Y, respectively. RSE > 1 suggests that indirect selection for Y by way of X would be advantageous. It is useful to explore the relationships among the components of RSE to illustrate the basic conditions that must be met before indirect selection can be advocated.

If the ratio of h_x^2/h_y^2 is large, RSE will be greater than 1 at lower r_g than is the case when the ratio is small. In many of the examples above, heritability of root traits was relatively high, whereas most estimates of the heritability of yield tend to be relatively low. Where the heritability of yield is say 0.2 and that of the root trait is 0.6, RSE will exceed 1 at $r_g > 0.6$. Contrast this with $h_y^2 = 0.4$ and $h_x^2 = 0.6$, where RSE does not exceed 1 until $r_g > 0.8$. Clearly, the heritability of the indirect trait must be very much greater than that of the direct trait, or the genetic correlation between yield and the root trait has to be strong, to consider indirect selection for yield via the root trait.

Searle (1965) observed that the error of RSE can be large, so even where RSE substantially exceeds 1 there may be some question as to the desirability of using the alternate trait. The error tends to be high because there are large changes in RSE with small changes in $\sqrt{(h_x^2/h_y^2)}$; precise measurements of h_y^2 and h_x^2 are required to improve the precision of RSE. Consider an example from the literature. Ekanayake *et al.* (1985a) measured several root traits in rice; using the trait root dry weight (F_3/F_2 heritability 0.61) as an indirect selection criterion for root volume (F_3/F_2 heritability 0.18) gives an RSE of 1.57 ± 0.33. If dry weight is easier to measure than volume, it would be advantageous to select for weight to improve volume.

Desirability of selection for yield *versus* a root trait will have an economic consideration as well. Even if RSE were somewhat greater than 1, the cost of selecting for the root trait *vs* the

cost of determining yield would have to be taken into account. Consider a situation where the cost of testing for the indirect root trait is 1.5 times greater than that of testing for yield. The breeder must consider the best allocation of resources, that is, to commit them to selection for the root trait or to increased yield testing. Response to selection (R) depends on the selection intensity (*i*), the square root of the heritability of the trait (*h*), and the measured genotypic standard deviation (σ_g) as follows:

$$R = ih\sigma_g \qquad (28.3)$$

Baker (1984) indicated that h can be increased by increasing the number of replications of each line because of the following relationship:

$$h_m^2 = rh_s^2/[1+h_s^2(r-1)] \qquad (28.4)$$

where h_m^2 = heritability of means of lines, h_s^2 = heritability on a single plot basis, and r = the number of replications. With constant *i* and σ_g, increasing r by 1.5 (say from 4 to 6 replications) will improve R by a factor of approximately 1.1. Therefore, the breeder must actually compare response to indirect selection for the root trait with response to yield selection with a 1.5 increase in resource allocation to yield testing.

Under some circumstances a breeder might consider use of the indirect selection trait even if RSE is somewhat less than one. For example, deep rooting might be desirable in a dry environment, but consistently dry environments might not be available for yield selection. Indirect selection for yield by selection for deep rooting could produce a correlated improvement in yield potential in the dry environment. However, there is a strong possibility that traits other than the root trait under selection would not be optimized as they would be through direct selection for yield in the target environment. Index selection for both yield and the root trait would perhaps be more useful in this case than indirect selection alone.

An index combining the direct and indirect traits is always better than using the direct trait alone (Searle, 1965), with the magnitude of the improvement depending on the phenotypic correlation between the direct and indirect traits. When the basic trait Y and the indirect trait X are combined into an index, I, the relative selection efficiency becomes:

$$\mathrm{RSE}(I,Y,y) = \sqrt{[1 + (r_p)^2/(1 - r_p^2)]} \quad (28.5)$$

where p is $\mathrm{RSE}(X,Y,y)$ (Equation 28.2), and r_p is the phenotypic correlation between X and Y (Searle, 1965). This function is never less than 1, and may be quite large if the phenotypic correlation is high. The value $\mathrm{RSE}(I,Y,y)$ is equivalent to the ratio of selection responses of index selection to simple selection given by Falconer (1988).

Baker (1986) noted that r_g is greater in magnitude than r_p; in practice, r_g may be considerably greater than r_p, especially when r_p is small (e.g. Pascal *et al.*, 1988). Where the ratio of the heritabilities is low, i.e. when the heritability of the direct trait approaches that of the indirect trait, index selection will be very little better than direct selection for Y. Even when the heritability of X is three times greater than that of Y, index selection is not substantially better than direct selection unless r_p is greater than about 0.4. Therefore, given the apparently weak correlations between root traits and yield, the ratio of index/direct selection or of $\mathrm{RSE}(I,Y,y)$ will often be very little greater than 1.

For example, Kevern and Hallauer (1983) found high and similar heritabilities for yield (0.79) and pre-anthesis root-pull resistance (0.76) in maize, and a relatively low (0.20) phenotypic correlation between the traits. Selection for yield via an index combining yield and root-pull resistance would be no better than selection for yield alone because $\mathrm{RSE}(I,Y,y)$ equals 1 under these conditions. Even where preliminary studies predict that selection with an index would be beneficial, it is still necessary to prove this in actual selection studies. Falconer (1988) cautioned that the genetic correlation between two traits will eventually become negative as selection progresses, and further, the response to selection in the desired direction may not be as good as anticipated due to factors causing asymmetry in the selection response. Also, unless heritabilities and inter-trait correlations are measured in numerous

crosses and environments, erroneous conclusions could be drawn.

As with indirect selection, there will be an economic consideration. In the case of an index, the total cost of testing will include the cost for yield testing plus that of the additional traits. Adding traits that can be visually observed on existing yield plots to an index has little impact on the cost of testing. However, if the cost of screening a root trait is high, as it would be to test for rooting depth, the efficiency of index selection would have to be substantially better than could be achieved by expanded yield testing alone.

The parental selection approach suggested by Hurd (1971) will probably be the most effective means of incorporating root and other difficult to screen traits into new cultivars. Where the trait is transferred from a non-adapted background, pre–breeding, backcrossing, or complex crosses will be required.

28.4 FUTURE DIRECTIONS

Root systems have been a consideration in the breeding of crops. Examples include the alteration of root system morphology in alfalfa through selection (Heinrichs, 1963), and transfer of a deep, extensive rooting pattern into an improved wheat cultivar through parental selection (Hurd and Townley-Smith, 1972). Ample evidence of genetic variation in root traits has been published. However, in most crops, there is insufficient information to recommend particular breeding strategies for manipulating root systems for the improvement of factors such as yield under drought stress. This is largely due to the lack of information on ideotypes, interactions among traits, and response to selection.

The situation is unlikely to improve if physiologists and breeders maintain their traditional courses of action, the former suggesting breeding for roots on the basis of studies of limited numbers of genotypes, and the latter being too busy with conventional breeding to develop a breeding strategy for root traits. We re-emphasize Hurd's (1971) suggestion of the need for a co-ordinated approach, and concomitant study of the physiology, genetics, and breeding potential of putative traits.

Concurrent physiological and genetic studies could speed up trait evaluation. It is worth considering whether this process could be enhanced by using new genetic and breeding technologies. One of the major problems in evaluation of complexly inherited physiological or morphological traits, particularly when they are difficult to measure, is the development of homozygous lines for study. Screening the large populations required can be very tedious and expensive.

Homozygosity can be attained rapidly in species in which haploidy can be readily induced. Doubling the chromosome number of induced haploids produces homozygous lines which can then be used to assess the effects of the different alleles on agronomic performance, or to backcross the traits of interest into adapted genotypes.

Some of the genome mapping projects underway in various crop plants may also provide techniques that will simplify screening for relatively simply-inherited, but difficult to measure traits. The restriction fragment length polymorphism (RFLP) maps being generated could be used to identify the genotype of segregating lines, thus facilitating the development of homozygous lines carrying the trait of interest. Tanksley and Hewitt (1988), however, pointed out that the effects of traits transferred in this way still must be evaluated in a range of genetic backgrounds. In fact, the restrictions of the classical methods of producing homozygous lines apply to the new biotechnologies, including the need to determine the interaction of the trait of interest with other important traits. Although *in vitro* techniques for haploid production, mutant selection, and vegetative propagation of crops such as cereals are progressing, the transfer of modified genes is still in the future (Chapter 9 and Lörz *et al.*, 1988).

REFERENCES

Allan, R.E. and Vogel, O.A. (1964), F$_2$ monosomic analysis of coleoptile and first-leaf development in two series of wheat crosses. *Crop Sci.*, **4**, 338–339.

Allard, R.W. and Bradshaw, A.D. (1964), Implications of genotype–environmental interactions in applied plant breeding. *Crop Sci.*, **4**, 503–508.

Arihara, J. and Crosbie, T.M. (1982), Relationships among seedling and mature root system traits of maize. *Crop Sci.*, **22**, 1197–1202.

ASA Special Publ. #50 (1987), *Minirhizotron Observation Tubes: Methods and Applications for Measuring Rhizosphere Dynamics*, Taylor, H.M. (ed.), American Society for Agronomy, Madison.

Baker, R.J. (1984), Quantitative genetic principles in plant breeding. In *Gene Manipulation in Plant Improvement*, 16th Proc. Stadler Genetics Symposium, Gustafson, J.P. (ed.), Plenum Press, New York, pp. 147–176.

Baker, R.J. (1986), *Selection Indices in Plant Breeding*, CRC Press, Boca Raton.

Barbour, N.W. and Murphy, C.F. (1984), Field evaluation of seedling root length selection in oats. *Crop Sci.*, **24**, 165–169.

Blum, A. (1979), Genetic improvement of drought resistance in crop plants: a case for sorghum. In *Stress Physiology in Crop Plants*, Mussell, H. and Staples, R.C. (eds), Wiley, New York.

Blum, A., Arkin, G.F. and Jordan, W.R. (1977), Sorghum root morphogenesis and growth. I. Effect of maturity genes. *Crop Sci.*, **17**, 149–153.

Bragg, P.L., Govi, G. and Cannell, R.Q. (1983), A comparison of methods, including angled and vertical minirhizontrons, for studying root growth and distribution in a spring oat crop. *Plant and Soil*, **73**, 435–440.

Brar, G.S., Matches, A.G., Taylor, H.M., McMichael, B.L. and Gomez, J.F. (1990), Two methods for characterizing rooting depth of forage-legume seedlings in the field. *Crop Sci.*, **30**, 414–417.

Brown, P.L. (1964), *Legumes and grasses in dryland cropping systems in the Northern and central Great Plains*, USDA Misc. Publ. No. 952, Washington, DC.

Clarke, J.M. (1987), Use of physiological and morphological traits in breeding programmes to improve drought resistance of cereals. In *Drought Tolerance in Winter Cereals*, Srivastava, J.P., Porceddu, E., Acevedo, E. and Varma, S. (eds), Wiley, Chichester, pp. 171–189.

Clarke, J.M. and Townley-Smith, T.F. (1984), Screening and selection techniques for improving drought resistance. In *Crop Breeding: A Contemporary Basis*, Vose, P.B. and Blixt, S.G. (eds), Pergamon Press, Oxford, pp. 137–162.

Comstock, R.E. (1977), Quantitative genetics and the design of breeding programs. In *Proc. Int. Conference on Quantitative Genet.*, Pollack, E., Kempthorne, O. and Bailey, T.B. (eds), Iowa State University Press, Ames, pp. 705–718.

Cornish, P.S. (1981), Resistance to water flow in the intracoleoptile internode of wheat. *Plant and Soil*, **59**, 119–125.

Derera, N.F., Marshall, D.R. and Balaam, L.N. (1969), Genetic variability in root development in relation to drought tolerance in spring wheats. *Expl. Agric.*, **5**, 327–337.

Duczek, L.J. (1989), Relationship between common root rot (*Cochliobolus sativus*) and tillering in spring wheat, *Can. J. Plant Pathol.*, **11**, 39–44.

Ekanayake, I.J., Garrity, D.P., Masajo, T.M. and O'Toole, J.C. (1985b), Root pulling resistance in rice: Inheritance and association with drought tolerance. *Euphytica*, **34**, 905–913.

Ekanayake, I.J., O'Toole, J.C., Garrity, J.P. and Masajo, T.M. (1985a), Inheritance of root characters and their relations to drought resistance in rice. *Crop Sci.*, **25**, 927–933.

Falconer, D.S. (1988), *Introduction to Quantitative Genetics*, 3rd edn., Longman, Burnt Mill.

Gul, A. and Allan, R.E. (1978), Inheritance of subcrown internode length, crown depth, and crown-tissue regrowth in a winter wheat cross. *Crop Sci.*, **18**, 438–443.

Heinrichs, D.H. (1963), Creeping alfalfas. *Adv. Agron.*, **15**, 317–337.

Hurd, E.A. (1964), Root study of three wheat varieties and their resistance to drought and damage by soil cracking. *Can. J. Plant Sci.*, **44**, 240–248.

Hurd, E.A. (1968), Growth of roots of seven varieties of spring wheat at high and low moisture levels. *Agron. J.*, **60**, 201–205.

Hurd, E.A. (1971), Can we breed for drought resistance? In *Drought Injury and Resistance in Crops*, Larson, K.L. and Eastin, J.D. (eds), CSSA Spec. Publ. No. 2, Crop Sci. Soc. Amer., Madison, pp. 77–78.

Hurd, E.A. (1974), Phenotype and drought tolerance in wheat. *Agric. Meteor.*, **14**, 39–55.

Hurd, E.A. and Townley-Smith, T.F. (1972), Techniques used in producing Wascana wheat. *Can. J. Plant Sci.*, **52**, 689–691.

Kazemi, H., Chapman, S.R. and McNeal, F.H. (1978), Components of genetic variance for root/shoot ratio in spring wheat. In *Proc. 5th Int. Wheat Genet. Symp.*, Indian Society of Genetics and Plant Breeding, New Delhi, pp. 597–605.

Kevern, T.C. and Hallauer, A.R. (1983), Relation of vertical root-pull resistance and flowering in maize. *Crop Sci.*, **23**, 357–363.

Lörz, H., Gobel, E. and Brown, P. (1988), Advances in tissue culture and progress towards genetic transformations of cereals. *Plant Breed.*, **100**, 1–25.

MacKey, J. (1973), The wheat root. In *Proc. 4th Int. Wheat Genetics Symp.*, Sears, E.R. and Sears, L.M.S. (eds), University of Missouri, Columbia, pp. 827–842.

Marshall, D.R. (1987), Australian plant breeding strategies for rainfed areas. In *Drought Tolerance in Winter Cereals*, Srivastava, J.P., Porceddu, E., Acevedo, E. and Varma, S. (eds), Wiley, Chichester, pp. 89–99.

McKenzie, H. (1971), Inheritance of subcrown internode length in four spring wheat cultivars. *Can. J. Plant Sci.*, **51**, 133–141.

Mian, M.A.R., Nafziger, E.D., Teyker, R.H. and Kolb, F.L. (1990), Root growth of wheat genotypes in different media. *Agron. Abstr.*, p. 101.

Monyo, J.H. and Whittington, W.J. (1970), Genetic analysis of root growth in wheat. *J. Agric. Sci. Camb.*, **74**, 329–338.

O'Brien, L. (1979), Genetic variability of root growth in wheat (*Triticum aestivum* L.). *Aust. J. Agric. Res.*, **30**, 587–595.

O'Toole, J.C. and Bland, W.L. (1987), Genotypic variation in crop plant root systems. *Advan. Agron.*, **41**, 91–145.

O'Toole, J.C. and Soemartono (1981), Evaluation of a simple technique for characterizing rice root systems in relation to drought resistance. *Euphytica*, **30**, 283–290.

Pascal, M.F., Ecochard, R. and Planchon, C. (1988), Morphological and physiological selection criteria in barley (*Hordeum vulgare*) using estimates of heritability, correlations and selection response. *Field Crops Res.*, **19**, 135–149.

Passioura, J.B. (1972), The effect of root geometry on the yield of wheat growing on stored water. *Aust. J. Agric. Res.*, **23**, 745–752.

Passioura, J.B. (1983), Roots and drought resistance. *Agric. Water Man.*, **7**, 265–280.

Pavlychenko, T.K. (1937), Quantitative study of the entire root system of weeds and crop plants under field conditions. *Ecology*, **18**, 62–69.

Peters, D.W., Shank, D.B. and Nyquist, W.E. (1982), Root pulling resistance and its relationship to grain yield in F_1 hybrids of maize. *Crop Sci.*, **22**, 1112–1114.

Porceddu, E., Martignano, F. and Scarascia Mugnozza, G.T. (1978), Intrapopulation variation of coleoptile and primary root length in durum wheat. In *Proc. 5th Int. Wheat Genet. Symp.*, Indian Society of Genetics and Plant Breeding, New Delhi, pp. 852–860.

Richards, R.A. (1987), Physiology and the breeding of winter-grown cereals for dry areas. In *Drought Tolerance in Winter Cereals*, Srivastava, J.P., Porceddu, E., Acevedo, E. and Varma, S. (eds), Wiley, Chichester, pp. 133–150.

Richards, R.A. and Passioura, J.B. (1981a), Seminal root morphology and water use of wheat. I. Environmental effects. *Crop Sci.*, **21**, 249–252.

Richards, R.A. and Passioura, J.B. (1981b), Seminal root morphology and water use of wheat. II. Genetic variation. *Crop Sci.*, **21**, 253–255.

Richards, R.A. and Passioura, J.B. (1989), A breeding program to reduce the diameter of the major xylem vessel in the seminal roots of wheat and its effect on grain yield in rain-fed environments. *Aust. J. Agric. Res.*, **40**, 943–950.

Scarascia-Mugnozza, G.T. and Porceddu, E. (1973), Genetic studies of coleoptile length in *T. durum*. In *Proc. 4th Int. Wheat Genetics Symposium*, Sears, E.R. and Sears, L.M.S. (eds), University of Missouri, Columbia, pp. 301–308.

Searle, S.R. (1965), The value of indirect selection. I. Mass selection. *Biometrics*, **21**, 682–707.

Siddique, K.H.M., Belfort, R.K. and Tennant, D. (1990), Root:shoot ratios of old and modern, tall and semi-dwarf wheats in a mediterranean environment. *Plant Soil*, **121**, 89–98.

Smith, J.D. and Kinman, M.L. (1965), The use of parent-offspring regression as an estimator of heritability. *Crop Sci.*, **5**, 595–596.

Sullivan, C.Y. and Ross, W.M. (1979), Selecting for drought and heat resistance in grain sorghum. In *Stress Physiology in Crop Plants*, Mussell, H. and Staples, R.C. (eds), Wiley, New York, pp. 263–281.

Tanksley, S.D. and Hewitt, J. (1988), Use of molecular markers in breeding for soluble solids content in tomato – a re-examination. *Theor. Appl. Genet.*, **75**, 811–823.

Taylor, H.M. and Klepper, B. (1978), The role of rooting characteristics in the supply of water to plants. *Adv. Agron.*, **30**, 99–128.

Upchurch, D.R. and Richie, J.T. (1983), Root observations using a video recording system in mini-rhizotrons. *Agron. J.*, **75**, 1009–1015.

Woodfield, D.R. and Caradus, J.R. (1990), Estimates of heritability for, and relationships between, root and shoot characters of white clover. II. Regression of progeny on mid-parent. *Euphytica*, **46**, 211–215.

Zobel, R.W. (1983), Crop manipulation for efficient use of water: Constraints and potential techniques in breeding for efficient water use. In *Limitations to Efficient Water Use*, Taylor, H.M., Jordan, W.R. and Sinclair, T.R. (eds), ASA-CSSA-SSSA, Madison, pp. 381–392.

29

On the utilization of renewable plant resources

L. Munck

29.1 INTRODUCTION

> ... wholeness is what is real, and ... fragmentation is the response of this whole to man's action ...
>
> *David Bohm, 1980*

The formidable success of classical science and technology through the last 300 years, in exploiting fossil resources in industrializing agriculture, has brought a wide range of inventions and insights, in the form of marketable products supporting, solving and explaining many detailed functions in our daily life which are most attractive and impressive for the individuals, stimulating consumption beyond basic needs. At present, we are concerned about problems with regard to overpopulation and the environment related to the consumption of all kinds of resources which we are unable to tackle systematically because we have not yet developed a realistic primary hypothesis on 'the metabolism' of our world.

In a recent essay (Munck, 1991), a conceptual issue of how to address another closely related global problem, that of the environment, is discussed. Using the same framework, the present chapter aims at surveying and ordering inputs for the exploitation of agricultural plant resources. The headings of the present chapter mirror the stages of the selec-

Plant Breeding: Principles and prospects. Edited by M.D. Hayward, N.O. Bosemark and I. Romagosa. Published in 1993 by Chapman & Hall, London. ISBN 0 412 43390 7.

tion cycle (Fig. 29.1), which experimental man (e.g. a plant breeder) according to the model uses to achieve and judge information, starting with a 'survey' stage generating a 'primary provisional hypothesis' based on a holistic concept (e.g. utilization of plants as a renewable resource of material and energy). With this information in mind, the selector focuses on 'specific analyses of partite elements' in a local area (e.g. production of maize for starch manufacture) and develops screening methods (e.g. for starch) to select candidate objects (e.g. starch-rich maize plants) which he can further test to form a 'secondary hypothesis' in the local area to be executed (e.g. how to realize that genes for high starch content could be combined with high yield). In a 'primary evaluation' in the global conceptual area, results from several different secondary selection hypotheses (e.g. concerning markets, suitable maize varieties, plant husbandry regimes, transportation, starch processing conditions and quality as well as environmental gains and drawbacks) are integrated with the earlier provisional primary hypothesis to obtain a new improved one. Thus, the perspective of man as selector along the production chains in his double capacity of producer and consumer amplifies the market concept: the invisible hand of Adam Smith. Obviously, money sums up in each step in the production chain a highly visible secondary selection criterion which, however, has to be supplemented by a quality estimate to be operative. The resulting specialization makes an overview of the production chains, including

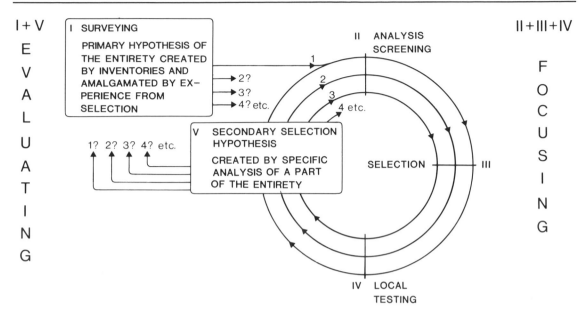

I + V

E
V
A
L
U
A
T
I
N
G

II + III + IV

F
O
C
U
S
I
N
G

Fig. 29.1. The selection cycle (Munck, 1991a).

the environmental consequences difficult for the actors, however; what is hidden for one selector is often visible for another and *vice versa*. This chapter aims to exemplify within the plant utilization area the options and limitations of the primary principle which drives the market: 'man as selector'. Such an approach is necessary to be able to discuss those criteria for the exploitation of plant resources which should be included in the new provisional primary hypothesis now raised by the market forces, emphasizing sustainability and recycling (The Brundtland Report, 1987), aiming at alleviating the problems of our relation with the wholeness.

29.2 SURVEYING

In spite (or because?) of the present information explosion, data concerning the collective experience of man on how to utilize a renewable plant resource are widely spread and are often not collated and acknowledged by those who are conducting research in this field. Systematic inventories should be made, the scope of which may only be hinted at here. As a general rule in

the surveying stage, all kinds of information together with notes about its precision and feasibility should be registered in a database. They should all be tried to be understood rather than judged and even if unused, they should be allowed to be incorporated in the database of the primary evaluation area to be readily available.

29.2.1 Monitoring the environment

Chlorophyll has a characteristic fluorescence spectrum which reflects the state of energy conversion in the plant and is affected by environmental factors such as temperature, pests, and pollution. With today's satellites, these changes can be precisely monitored by remote sensing techniques. Similarly, world meteorology can be monitored by weather satellites including the state of the ozone layer so important as a filter to protect life from aggressive short UV-radiation. A lot of such data is already recorded and evaluated at several computer centres throughout the world. Still, we do not seem to have a firm prognosis of what these changes may imply for the direction of the environment of the world other than the fact that the human influence is considerable. The issue is now how are we going to handle this wealth of data to draw realistic conclusions of our course and how could the plant option be utilized to improve our prospects?

29.2.2 Surveying plant and plant-based technologies based on ethnographic and archeological/historical evidence

Resources must be allocated to register and systematize the dwindling locally available experience of the utilization of plants, including the cultural, medical, nutritional and technological aspects. Thus, the first civilizations were dependent on stores of hard grains (Steensberg, 1989) which were instrumental, through the need for housekeeping records, in creating a written language and a mathematical concept, including money (Steensberg, 1986).

As noted by Clarke (1990), even the past inhabitants of the Polynesian Islands were not always able to live in balance with nature but were occasionally overexploiting their environment by burning and overpopulation. However, by learning from the local environmental catastrophes which they had induced themselves, they were able to temper their way of living in reaching a new balance point with nature, finding measures for population control as well as methods for obtaining an endurable exploitation of nature. This realization that man in earlier times in his exploitation of nature was able to incorporate a moderation in his living as a part of his culture to secure a reasonable future for his offspring is worthwhile to focus on in our present efforts.

29.2.3 Surveying plant resources and their potential use

As mentioned by Wilson (1988), the destruction of the forests, especially the rich ecosystems of the tropics containing several hundred thousand plant species, deprives us of valuable genetic material of ecological, medical, nutri-tional, technological and aesthetic value. Thus, man has through history utilized about 7000 kinds of food plants, only 20 of which now supply 90% of the world's food resources and just three (wheat, maize and rice) supplying more than half. It has been estimated (Bell, 1988) that more than 24,000 species of the world's flora have been used in medicine and a great many more have found uses in the control of insects, nematodes, fungi and fish and in other technical ways. The remaining potential is thus great if we can prevent its eradication by moderating our living.

29.2.4 Surveying the historical role of plants and agriculture in the industrialization process in developing and industrialized countries

In particular, interdisciplinary projects should with new eyes reassess the industrialization process to learn the lessons of which, we in the western world already seem to have forgotten. The agricultural food production and the military armament areas have been the two major developmental stimuli leading to a wealth of spin off inventions which are now generally used by man and society.

The global introduction of a monetary economy has stimulated specialization both locally and between countries leading to an extremely elaborate production network and a similarly complicated bureaucracy where quality has to be controlled each time products or work are exchanged for money. Thus on average, the Danish farmer, getting 27% of the market sales price of his produce, only retains about 2% of the same price for his private consumption. The other 98% is used to employ people and to drive the complicated food production chains, not only producing products directly used and consumed in agriculture but also its supply industry, the food-processing industry as well as in the organizations for distribution and sale, including governmental regulating bodies and research. The energy to drive and develop this impressive network comes almost exclusively from fossil and nuclear sources, whereas the management of this developmental potential has been implemented by 'man as selector', heavily involving the monetary principle.

In Europe, in contrast to Africa, we have been able to improve our consumer-oriented industry parallel and in close contact with agriculture and the agriculture-oriented industry, boosting crop yield to levels exceeding our greatest expectations. Thus, we have established strong economic bonds between the rural and the urban areas.

In Tanzania, sacks of charcoal can often be seen by the roadside, sold by local people for 30 shillings to the truck drivers to be transported to Dar-el-Salam. Surprisingly, Tanzania is an exporter of charcoal for cooking to Saudi Arabia, receiving in return petrol products e.g. to drive trucks transporting charcoal to the harbour. This is indeed a lucid example on how the monetary system and the fossil fuels haul out the subsistence farmer from his ecological niche and effectively destroy the production capability of the renewable resources, including the last remains of the forest. African cities today belong commercially more to the US and Europe than to Africa. Attractive, subventioned imported grains such as wheat and maize, which are difficult to grow in the arid parts of Africa, are imported and sold cheaply, thus destroying the incentive for the farmers to produce and sell a surplus of local hardy grains, such as sorghum, for processing and consumption in the urban areas. Thus the subsistence farmers continue to grow sorghum and millets for their own needs and to make charcoal from the forest, often illegally, since this is the only way to obtain cash other than growing maize and wheat which are unreliable.

It should be possible to create some of the necessary options for a change by introducing technology for producing convenient sorghum and millet foods in urban areas in order to open a local cash market for the farmer for the African grains (Eggum *et al.*, 1982; Munck *et al.*, 1982). However, a great number of developing countries have in spite (or because?) of inputs from industrialized countries by and large failed to copy the industrial revolution of the west to utilize the fossil energy option, to establish and boost the turnover in trade and services between the local, rural and central, urban areas as well as between agriculture and industry.

29.2.5 Surveying the energy issue as a marker of technical and economic options and constraints

The earth, is to be looked on as an open system which receives the major part of its energy from a star, the sun, a nuclear reactor, placed at a safe distance. Photosynthetic organisms on earth transferring the energy from the sunlight into chemical bonds are by and large responsible for our present ecological platform with its plant material (about 150 billion ton dry matter), including the oxygen which we breathe and the vast fossil energy resources generated through geological time.

Through the last 100 years, fossil hydrocarbon materials have substituted many products and services formerly supplied by plant resources for heating, cooking, transportation, buildings, clothing, etc. The imbalance stands clear. While agriculture in the western world produces 1.1 billion tons of starch in grains, we are extracting 2.9 billion tons of crude oil corresponding to a more than six times higher energy value (Roels, 1986).

The costs of oil in relation to the plant commodity maize has increased about 8 times during the period from 1973 (Fig. 29.2) which has initiated a new interest in considering the option for non-food utilization of plant material.

The waste of energy in the modern industrial society is monumental because of cheap oil prices. However, because of the lack of energy management there are great possibilities to save energy and reduce pollution by better logistics, planning and conservation techniques, again closing the biological production circuits which have been broken up by technology based on fossil fuels resulting in the piling up of waste.

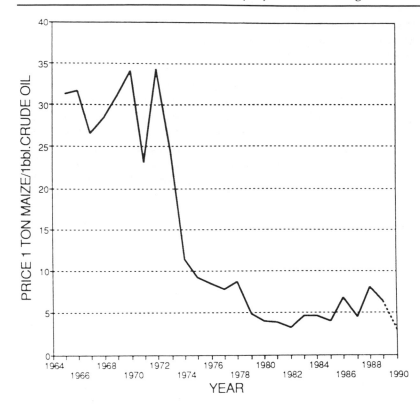

Fig. 29.2. Price relation between world market price of maize and crude oil 1964–1990 (Bjørn Petersen and Munck, 1993).

29.3 GENERATING A PROVISIONAL PRIMARY SELECTION HYPOTHESIS ON THE BASIS OF THE SURVEYS

We can thus tentatively conclude:

1. life on earth and our civilization are based on plants and will be continuously dependent on a diversified wild flora and agriculture;
2. securing optimal living conditions for plants is thus a necessary prerequisite for our existence;
3. there is no way at the present level of consumption of hydrocarbons that plants could substitute for fossil resources;
4. the present consumption of fossil fuels is wasteful and too high; our present resources of fossil energy and minerals should be used to change our production structure to conserve energy and support a sustainable plant production and not as of now to degrade the capability of the environment to produce our

renewable plant resources, which in fact is our basic capital;

5. it should be possible by fundamentally changing our concepts and habits to reduce the energy consumption of the industrialized world to one fifth of the present level without essentially influencing the quality of life;
6. a major part of the agricultural plant material which by and large is not directly used for foods could be exploited in the non-food industry with a positive impact on employment and environment, stimulating and crosslinking the economies of the local rural and central urban areas.

In the following, we shall focus on how science and technology could contribute by selecting specific means by which plants could potentially be utilized in the non-food industry to fulfil some of the above claims. After envisaging some packages of technology which could be worthwhile to consider in the local secondary area, we will come back to the global primary area for a final tentative evaluation of a given example.

29.4 FOCUSING ON ANALYSIS OF PARTITE ELEMENTS AND SELECTION AND TESTING OF TECHNOLOGICAL MEANS

29.4.1 Fractionation of whole plants into botanical, cellular and chemical components and the potential use of the components

From an industrial point of view, a plant is a fascinating raw material containing the building blocks for a wide variety of food and non-food commodities of major importance to our daily life. However, the crop as a whole is a heterogeneous raw material. It consists of various botanical organ constituents such as root, leaf, stem and seed (hull, pericarp, aleurone, endosperm and germ) possessing different cellular (e.g. starch granules and cell walls) and chemical (e.g. carbohydrates, proteins and fats) components applicable to different purposes.

Therefore, an optimal total utilization will imply a number of processes which are used to divide the crop into its main components (Rexen and Munck, 1984; Munck and Rexen, 1990).

These intermediates can be potentially utilized by industry in a wealth of possibilities as is indicated by the example of maize utilization which displays the strategy for the further presentation (Fig. 29.3).

In the following, statistics, mainly from the EEC, will be cited as examples elucidating the present relative importance of different productions which are also by and large relevant in other parts of the industrialized world such as in North America.

29.4.2 Utilization of carbohydrates

The main carbohydrates suitable for industrial use are the polymers starch, cellulose and β-glucans based on glucose units and hemicellulose also involving pentose sugars. The former is contained in a cellular organelle, the starch granule, while the others are integrated together with lignin in a firm complex, constituting the plant cell wall. Whereas starch is

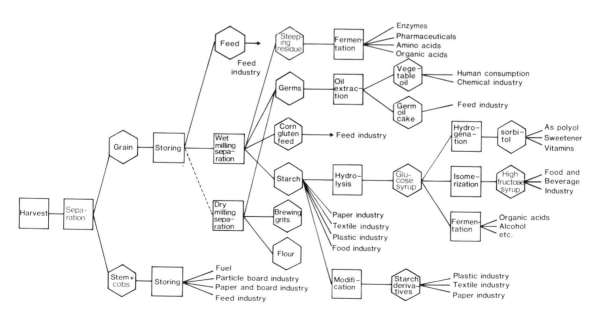

Fig. 29.3. Potential applications for maize as an example of whole crop utilization.

relatively easy to extract and purify, cellulose needs hard chemical processes to be extracted from the lignin complex. Another class of carbohydrates important for industry are sugars produced for example by sugar beets.

(a) Starch

Starch is a remarkable natural polymer of α-glucose residues which can be manufactured quite pure (99%) and with a potentially very wide use in industry.

The main raw materials for starch production are maize (EEC production 3.5 million ton in 1988), potatoes (1.0 million ton) and wheat (1.1 million ton). Wheat starch manufacturing in the EEC has recently increased because of raw material intervention price and a high market price of the protein co-product, gluten. However, maize has an advantage over wheat because its starch grains are uniform, whereas wheat starch manufacture is more complex because of a class of small starch granules, which are difficult to purify from protein, resulting in a B-starch fraction of inferior quality. The maize germ co-product is a valuable raw material for oil extraction and the concentrated maize steeping liquor is a preferred raw material for fermentation of, for example penicillin. Potatoes give a high yield of starch per hectare but are difficult to store in the campaign-driven potato starch industry which also has high costs for effluent treatment. There is now a tendency to combine potato and wheat starch manufacture, using the same machinery, to be able to produce starch throughout the whole year.

Starch consists of a linear glucose polymer amylose and a branched component called amylopectin, the balance of which determines the swelling and gelatinization properties important for the utilization of starch (Light, 1990). Thus, the paper and board industry, consuming about half of the approximately 1.5 million ton of starch used for non-food purposes in the EEC, prefers a starch with a high amount of the more hydrophobic and crystalline amylose starch as an additive. There is a great genetic variation in amylose/amylopectin ratio in maize and potato which

can be utilized by breeders, while wheat in this respect is less flexible.

The natural polymer starch may be modified thermally or combined in many different ways by means of organic chemistry with chemicals made from fossil resources or from plant material, resulting in a great number of potential products such as films, adhesives, absorbants, surfactants, etc. (Fig. 29.4) (Kempf, 1990), a few of which will be dealt with as examples in the following.

Starch is generally cheaper than paper pulp by one-third to one-half and thus about half the starch produced in the EC is used in the paper industry. Intensive research for new starch products designed for the paper and board industry will make it advantageous to add starch to paper and board from the current 2–5% to a potential of about 30%. Experiments with new laboratory-made starch derivatives have also revealed that such products can improve the strength of short-fibred pulp such as straw pulp. Hence, it may be possible that the future paper industry will have access to starch-derived products which makes it economical to increase the content of short-fibred pulp, e.g. from straw, in writing and printing paper without any deterioration of the paper quality.

About one-third of the starch production in

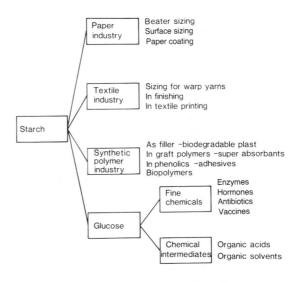

Fig. 29.4. Non-food application of starch.

the EC is used in the chemical industry. The natural polymer starch is much cheaper on the world market than the synthetic polymers polypropylene and polyethylene by a quarter to a third. A wide range of combination products can be produced with starch and chemicals from fossil sources.

Graft polymers built on starch reacting with vinyl or acrylic monomers represent a category of promising combination products between a natural product and a chemically derived one from the petrol industry. Such products have been prepared from both granular starch and from gelatinized starch (Otey and Doane, 1980). The graft polymers may be used for production of plastic films, moulded products, etc. Such a solid polymer of starch with acrylonitrile can absorb many hundred times its weight of water without being dissolved. It has been named 'super slurper'. There are several potential applications (Olsen, 1983) for super slurpers. The most promising areas seem to be agricultural applications such as for seed and root coating and as an additive to fast draining soils, conserving water, for cultivation of plants under desert conditions. Large-scale field trials with maize, soybeans and cotton seeds coated with super slurper have shown increased germination and yield. Up to a 30% increase of the yield has been reported for maize (1.3 ton of maize/ha) in areas with an extremely dry climate. The areas of personal and medical care are also of interest to the producers of super slurper. It is already being used in a number of personal care products such as baby powder, baby napkins, sanitary towels, etc. which are marketed in the US. In hospitals, super slurper is used as an additive to absorbent soft goods such as disposable diapers, bandages and hospital bed pads.

Phenol formaldehyde polymers belong to the oldest synthetic polymers still produced (world consumption about 2 million ton). Starch can substitute not only formaldehyde (-CHO group) but also to a large extent phenol (-OH group) in phenolformaldehyde polymers (Kock et al., 1983; Woelk, 1981). The reduction in formaldehyde has the benefit of reducing the emission of formaldehyde from for example phenolic-glued wood or straw boards which is a significant environmental problem in indoor locations.

Otey and Doane (1980) have shown that crosslinked starch xanthate is useful as a replacement for carbon black in rubber now originating from petroleum (US consumption 1.5 million tons). Another promising application for starch xanthates is for encapsulation of e.g. pesticides. With a similar technique it should be possible to produce slow-release chemical fertilizers to increase efficiency in plant husbandry, thus avoiding pollution.

(b) Lignocelluloses

Historically, cellulose fibres (consisting of glucose with β-glucosidic bonds) from annual crops have been of major industrial importance. The use of wood for the production of paper is a relatively recent invention. Papyrus was used consistently for thousands of years and, as late as a century ago, the most common materials for paper were cotton, flax, hemp and cereal straw. The main problem in paper manufacture of cereal straw is that it is lower than wood in cellulose content and the silicon in the straw hampers the recycling of chemicals and utilization of energy in the process liquor by adhering to the boiler. In Europe during World War II, there was a short revival of interest in straw as a cellulose source, but the inadequate, unreliable supply of poor quality straw to the small obsolete polluting straw pulping plants meant that they could not compete with the huge wood pulp plants, e.g. from North America and Scandinavia, which could afford continuously to improve their processes at considerable costs to meet the increasingly rigorous regulations on pollution.

However, in Denmark since World War II straw pulping technology has improved considerably, partly by solving the silicon problem. A straw pulp of excellent quality, competing with the wood-based products, is now produced in Denmark in a closed recycling process which saves energy and chemicals (Kaas Hansen, 1990). This industry has a capacity of 100000 tons of straw per year and an efficiently organized feedstock supply system. The straw is transported to the factory from distances of up to 150 km, the average transportation distance being 80–90 km. The total number of suppliers are 2000 farmers. With new resource-saving recycling processes, the present straw-based paper industry now located mainly in the developing countries, could be upgraded and enlarged.

Figure 29.5 outlines the separation of cereal (barley) straw and the chemical composition of the botanical fractions. Internodes (the stems) have a higher content of lignocellulose (approaching that of wood) and a lower content of ash and silicon which is of importance for the paper pulp-processing industry compared to

the leaf fraction which has an increased content of protein and digestible matter, favouring its use as a feed for ruminants, but detracting from its use as a raw material for paper pulp. The leaf fraction could also be used for compost or as a fuel, fired as a dust, or as pellets.

A simple, mechanical disc mill process has been developed for separation of leaves and internodes (Rexen and Munck, 1984; Bjørn Petersen and Munck, 1991) where the leaf fraction can be sifted from the internode (stem) fraction which comes out as coarse chips after milling. Thus, since straw is a surplus product, it seems logical to us to reinvestigate the possibilities of substantially supplementing domestic fibre production by using cereal straw and other agricultural fibre crops. Hardwood fibres from birch, etc. and straw fibres are similar in fibre length but the latter are much more slender.

The smaller diameter of the straw fibre seems to have no adverse effect on the technological properties of paper derived from it. Paper from good quality straw pulp has the same quality as paper from good quality hardwood pulp. This is in accordance with the general assumption that a good fibre quality is characterized by a high ratio between fibre length and fibre width. Research on the quality of straw (internodes) for paper is thus a new discipline for future plant breeding. We have found variation between barley lines in a small study with regard to differences in α-cellulose (39.5–42.7) and average length-weighted fibre length (0.83–1.02) (Bjørn Petersen and Munck, 1991). Other sources of cellulose fibres are specialized fibre crops like cotton, hemp, flax, etc. which are used predominantly by the textile industry but which can also be used for paper and fibre board when special qualities are needed.

It is technically possible to produce particle boards of excellent qualities from straw inter-nodes alone (Rexen and Munck, 1984). The leaf fraction of the straw impairs the strength characteristics of the fibre boards. Both pilot plant and industrial-scale experiments have been performed, and the general conclusion is that the quality of straw particle boards is at least as high as that of good standard quality wood-based particle boards. We have shown

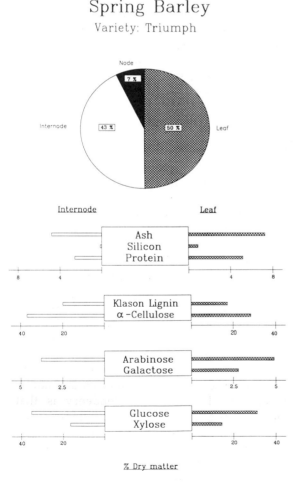

Fig. 29.5. The botanical and chemical composition of barley straw.

that it is possible to obtain a glue suitable for straw fibre board production based on wheat flour modified with phenol and formaldehyde as described earlier. As with paper, it seems potentially feasible given future research, to make fibre boards from almost 100% agriculturally produced raw materials. As with starch there is an impressive range of useful chemical derivatives from cellulose, such as carboxymethylcellulose (used for example as an additive in paints) and viscose (a textile fibre).

(c) Sugars

In 1988, 97.5 million ton of raw sugar (saccharose) were produced on a world basis. Monomeric (e.g. glucose and pentose) and dimeric sugars (e.g. saccharose) are produced in agriculture through three main sources: from starch after hydrolysis, from the hydrolysis of lignocellulosic material and from sugar crops such as sugar beet. Starch is easily hydrolysed into glucose by mild heat treatment under acid conditions or by industrially produced enzymes.

Cereal straw contains approximately 70% carbohydrates, primarily cellulose and pentosans, which under quite harsh chemical conditions can be hydrolysed to low-molecular weight sugars, mostly glucose and xylose. Enzyme processes yield pure products and do not damage the lignin as does the chemical processing involving acids, but their glucose conversion rate is only about 50–60%, and the feedstock usually requires extensive chemical pretreatment and long reaction times. Organosolvent processes, e.g. including phenol, are attractive to dissolve the lignin, enabling a total fractionation of the material with a minimum of side reactions.

There are today in western industry, market applications where the whole plant crop as a raw material is competitive with fossil raw materials, as was the case for some of the processes described under the starch section. If the price of fossil hydrocarbons was kept at a high level (US$ 50–100 per barrel of crude oil) for a long time (5–10 years), the renewable plant sources would have a fair chance of making a significant contribution. In developing countries lacking fossil fuels and hard currency, the range of economic applications is today much wider but implementation is difficult due to lack of infrastructure and because of unawareness of the potential.

In principle, the glucose molecule could be transformed to almost any kind of commodity chemical as well as many fine chemicals and pharmaceuticals, but, mainly for economic reasons, most of the transformations involving simple base chemicals like ethylene and ethanol on a large scale are unrealistic even at high oil prices (Rexen and Munck, 1984). However, more complex derivatives of starch and cellulose, which retain some of the unique properties of these natural polymers, have a much better chance to compete (Rexen et al., 1988).

New trade policies such as that introduced by the EEC in 1986 (Gray, 1990) where the non-food industry utilizing grain can obtain its raw materials at world market prices is further stimulating the interest of the chemical industry to use agricultural raw materials in production processes.

29.4.3 Pharmaceuticals

Most pharmaceuticals are expensive products produced in small volumes. Those derived from fermentation of substrates made from plants include vitamins, antibiotics, hormones, etc. The application of agricultural plant sources in the pharmaceutical industry, although representing a considerable end-product value, implies with regard to the tonnage of raw material consumption much less than the other applications. Outside the regular agricultural sector, about 25% of medical prescriptions is dependent on medicinal plants (Turner and MacLean, 1988), but their added value and potential are created by the industry and not by agriculture. The prevalent tendency is that chemists are getting their inspiration for finding new medicines by screening and selecting a large range of plants for active substances. In this respect the tropical vegetation is a most rich but a precious source. When the feasibility of a new product has been asserted, the pharmaceutical industry tries to synthesize it by organic chemical or molecular biological means instead of producing it by growing the plant. However, this is not always successful as is exemplified by the case of steroids (world consumption 1979 approx. 2000 ton) which are purified from wild yams (*Dioscorea*) and from the sisal plant (*Agave*) for further chemical modification.

29.4.4 Oils and fats

In total, about 6 million tons of edible plant oil is consumed in the EC (1982) annually, and about 3 million tons of vegetable and animal oil is used annually for technical purposes, which is a higher volume than the non-food starch consumption (1.5 million ton). The industrial uses are concentrated in the following areas: (1) 40% in soaps and detergents (cocos and tallow); (2) 22% in plastic derivatives (plasticizers); (3) 13% in paint and varnish (drying oils); (4) 11% in food (emulsifiers); (5) 11% in lubricants; (6) 3% for other uses. Different oil plant species are used for each purpose due to the composition of the oils. Out of a total global vegetable fat and oil production of 42.5 million ton, soybeans represent 33%, sunflower 13% and rape seed 9.5%, i.e. more than half the global production, leaving 21% for palm oil and coconut (with short-chained fatty acids) and the rest to small commodities (Eerkens, 1983).

For many reasons (crop rotation and diversification into short-chained fatty acids as well as present total deficit in the EC in vegetable oil and protein), it is desirable to increase vegetable oil production (Röbbelen *et al.*, 1989). Thus any substantial improvement in the situation calls for more alternatives. The following untraditional oil crops are examples of potential 'industrial' oil crops. *Crambe* oils have high contents of long-chain fatty acids and could replace high erucic rape seed oils for lubricant production. *Cuphea* oils are interesting because of their high content of short-chain fatty acids (capylic, capric, lauric and myristic acids) which may be used in synthetic ester lubricants. The oils might serve as substitutes for coconut oil as a source of 12 carbon compounds in surfactant applications. Epoxy oils in *Stokesia* are possible replacements for epoxidized soybean oil for

applications in paints. The *Lesquerella* oils (lesquerolic acids) have hydroxy groups as well as double bonds and they could partly replace castor oil.

The introduction of new culture plants is due to adaptation and yield problems a long-term breeding project with a time horizon of at least 20 years. Alternatively it seems more feasible in a shorter perspective to improve further existing well-adapted species by breeding, using both conventional and molecular techniques, for a tailored fatty acid composition for industrial needs. There are thus available high-yielding rape seed lines with a very high content of the long-chained erucic acid which is ideal for lubrication applications where full biological degradability is desired.

29.4.5 Proteins

Proteins mainly from oil crops are to a large extent used industrially for feed purposes although some soy and gluten protein is processed for direct human consumption. The potential increase in domestic production of vegetable oil and starch from cereals would yield significant additional amounts of protein concentrates. A representative part (30–85% depending on the kind of animal production) of the protein consumed in feeds is supplied as a part of cereals, grass, roots and other feed products. It has been possible to improve the protein feeding quality of maize, sorghum and barley (Bang-Olsen *et al.*, 1991) by use of the so-called high-lysine varieties reviewed by Munck (1992). This further emphasizes the conceptual and marketing difficulties encountered when launching such inventions in the present market.

29.5 GENERATING A RANGE OF SECONDARY SELECTION HYPOTHESES: A SYNTHESIS EFFORT IN THE LOCAL AREA

We must acknowledge the extreme complexity of derived characteristics which we try to select from the wholeness by manipulating parts of it intentionally, leading, however, to unintended consequences of environmental implication (e.g. Munck, 1991). In this chapter, it has been hinted how the survey stage generated a primary provisional hypothesis on utilizing the renewable plant resources and how science and technology by focusing on analysis have accumulated a specific knowledge. We will now try to use our experimentally generated knowledge in the local areas of the wholeness, may it be in a certain production chain starting with a certain kind of raw material or conceptually e.g. how to utilize data in production control and plant breeding. On the basis of this synthesis in the local area, we will try to generate a range of secondary selection hypotheses of how partite elements and concepts are related in the local conceptual area.

Example I. Compensating the human imperfections of evaluating a multifaceted reality by applying logistics coded in data programmes controlled by man

Although the human brain can with high precision deduce differences between two complex structures by comparison, it has large difficulties in tackling a multivariate situation of nonlinear characters. There are data techniques available which allow us to overcome these weaknesses but the conceptual advantages have still not reached society by introducing new conclusions and strategies.

In classic 'hard statistics', a great number of prerequisites are set up for accepting data regarding distribution and orthogonality. It is also very difficult in classical statistics to handle many sets of interrelated factors. With the help of the new 'soft statistics' such as the user friendly 'Unscrambler' data program developed by Martens and Næs (1989), including PLS (Partial Least Square) and factor analysis (Principal Component Analysis, PCA), we can now predict and define which factors are interrelated and which are independent, and with the help of the computer, including a graphical display, surpass the capacity limit of the human mind where man can only visualize a few criteria at the same time. These methods cannot, however, substitute for low quality data but they

can point out effectively where bad data reside so that the reliability of data measurement and collection can be improved. In plant breeding and industry, a large number of control analyses have been accumulated giving information of a range of different aspects of, for example, a production chain, which are difficult to overview. The increasing costs of quality control, where demands for new analysis, also in plant breeding, steadily appear, makes it necessary to rationalize the old ones.

Let us as an example investigate 186 different commercial lots of malts from four different barley varieties analysed in the laboratory with the international EBC (European Brewery Convention) Congress mashing method with a total of 11 different parameters to simulate the brewhouse process (Table 29.1). The purpose of this step in the brewing process is to produce a sugar-rich extract from barley called wort by exploiting the enzymes produced by germinated barley (malt). The wort is later fermented into beer by yeast. This enzymatic degradation of starch is done by extracting the crushed malt with hot water. With the PCA technique, the different kinds of analytical parameters (Table 29.1) are assorted in groups or families (Fig. 29.6) as principal components or latent factors.

The evaluation of the malts can be explained as follows (Table 29.1):

1. Latent factor 1 explains 30% of the total varition and consists of wort factors associated with enzyme activity, namely extract (fine grind character in Table 29.1), soluble nitrogen in wort, Kolbach Index (nitrogen retention in wort from malt) and wort colour.
2. Latent factor 2 explains 26% of the information and consists of two subgroups of factors with opposite signs related to the physical structure of barley (malt). The first subgroup, where the values increase with improved quality, is malt friability (a hardness milling test), malt modification, including its homogeneity (detecting cell wall breakdown visually in UV-light by staining with a fluorochrome). The second subgroup, where the values decrease with improved quality, is composed of extract difference

Table 29.1. Analyses of commercial malts from 1988 (data compiled by J. Brandt, Carlsberg Research Laboratory) (see also Figs 29.6 and 29.7)

Character	Barley variety			
	a 55	*g* 20	*t* 43	*m* 68
1 Malt extract fine	80.8±1.0	81.6±0.5	81.5±0.6	78.6±0.6
2 Difference fine/coarse	2.5	1.7±0.5	1.7±0.4	2.4±0.64
3 Wort colour	4.4±0.7	4.1±0.4	3.6±0.36	2.9±0.26
4 Soluble nitrogen in wort	0.71±0.05	0.77±0.05	0.74±0.05	0.55±0.024
5 Viscosity in wort	1.55	1.50±0.02	1.49±0.026	1.50±0.03
6 Homogeneity	48±13	55±11	59±11	65±10
7 β-Glucan in wort	354±190	251±66	262±81	194±83
8 Malt modification (%)	84	87±4	90±4.5	92±4.5
9 Friability	81	86±3	84±3	85±3
10 Kolbach Index	40.6±3	41.9±1.8	40.3±2.6	31.2±1.7
11 Nitrogen in malt	1.76±0.09	1.84	1.85±0.07	1.70±0.04

(between fine and coarse malt grind) together with β-glucan content in wort (originally a constituent of the barley endosperm cell wall) and viscosity in wort caused by the viscous β-glucan.

3. Latent factor 3, nitrogen in malt/barley, is independent of the other component groups and gives the third principal component, explaining for 12% of the variation.

It is seen that the factors related to malt structure belong to the same component and

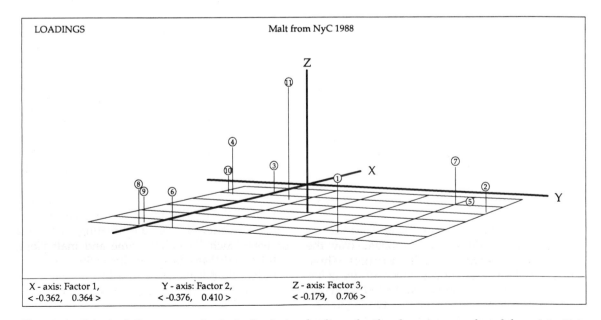

LOADINGS   Malt from NyC 1988

X - axis: Factor 1,	Y - axis: Factor 2,	Z - axis: Factor 3,
< -0.362, 0.364 >	< -0.376, 0.410 >	< -0.179, 0.706 >

Fig. 29.6. Principal Component Analysis displaying loadings for the characters numbered from 1 to 11 in Table 29.1, X-axis, factor 1 (<−0.362, 0.364>) characters (1, 3, 4, 10); Y-axis, factor 2 (<−0.376, 0.4107>) characters (2, 5, 6, 7, 8, 9); Z-axis, factor 3 (<−0.179, 0.706>) character (11).

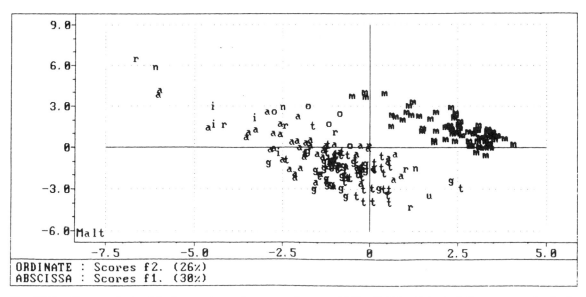

Fig. 29.7. Data analysis of barley varieties (a, g, t and m, see Table 29.1) by Principal Component Analysis (PCA) on malt parameters. Principal component 1 (enzyme activity) = abscissa; principal component 2 (physical structure) = ordinate.

are thus expressing the same principle originally related to the cellular component barley endosperm cell wall thickness and cell size. It is logical that the malt enzyme dependent factors are belonging to the same component (phenomenon) whereas nitrogen content in this material of malt surprisingly expresses itself as a separate factor. Behind the 11 anonymous laboratory analyses thus lies an implicate order (Bohm, 1980) of three major hidden (latent) factors or principles, the concept of which now can be communicated and further explored, e.g. by scientists, plant breeders and engineers, to find the three best analyses which represent the three major principles of malt quality, substituting for the elaborate set of 11 analyses now used.

The latent factors could also be used as fingerprints to identify in a holistic way the malting performance of barley varieties. Thus, in Fig. 29.7, displaying a plot of the scores for factor 1 *versus* the scores for factor 2 for all malts, the barley variety 'm' (Table 29.1) is completely separate from the other three varieties with regard to component 1, the enzyme activity complex. Such a clear-cut differenti-

ation is not easy to envisage with classical statistics as seen from the means and standard deviations of the 11 analyses in Table 29.1 or with regression analysis in pairs of different combinations of parameters. The PCA analysis visualizes in an instructive way, the uniqueness of the malting performance of barley variety 'm', enabling the maltster without any elaborate knowledge of statistics and computer programs to contemplate in detail the malting and brewing performance otherwise hidden in the complex of laboratory data of Table 29.1. The outcome of this evaluation can thus be communicated to the barley producer and to the plant breeder.

Obviously the evaluation would have been still more representative if performance 'data from the process' of the maltings and the brewery such as malting time and malt yield, extract yield, viscosity, β-glucan in wort, colour, lautering time, filter pressure, etc. had been included. The major problem here is to make the brewing batch process sufficiently translucent, i.e. being able to take into account in the computer the blending of the different pure malts and the mixing of brews (worts) and beers so that the 'fingerprints' of each pure malt batch and its barley raw material in the process could be expressed and communicated to the farmers and the plant breeders. This is the essence of the 'unscrambler' symbol where the eggs are reconstituted (unscrambled) from the 'scrambled' omelette. In modern computerized breweries such a reconstitution is today partly possible, but it is almost impossible for a traditional brewery.

To make any process, production chain and society structure translucent, technologically and economically is a conceptual, administrative and technological problem, placing demands on the consistency of the concept, and thus human individuals, on-line sensors, equipment and computer power. The above-described 'unscrambler' program can, with modification, be used for analysis and central process evaluation and prediction. Such programs (PLS) are also today used decentrally in 'black box systems' built on fluorescence (Munck, 1989) and near infrared (NIR) instruments in non-destructive more or less remote sensing systems where for example, complex spectra from wheat flours could be used to deduce their botanical composition to control a wheat mill. Similarly, complex data from satellites and ground stations could be processed to envisage the global state of plant ecology and meteorology and the hidden latent factors. However, in order to use these extensions of our senses beneficially, we must realize and respect the nature of the built-in limitations and uncertainties in our monitoring technology and conceptually in our brains, languages and data programmes as well as in the world itself.

Example 2. Establishing an improved utilization of straw from annual crops as paper, fibre board, cattle feed and energy

A technological process is economically based on the principle that a given raw material can be separated (and modified) into fractions which have built-in specific functions of added value for their further use. Thus, leaves from wheat straw have a higher feed value for ruminants than unseparated straw due to increased levels of digestible nutrients, while the complementary fraction, the wheat stem, has an improved

value compared with whole straw for the cellulose industry because of a higher content of cellulose. However, for the present straw cellulose industry to be able to compete seriously, requires, as has been done by the wood-based cellulose industry, an integration between raw material production, handling, transportation, processing and marketing, the logistics of which are by and large unheard of in the agricultural sector with the exception of the sugar beet industry (Munck and Rexen, 1990).

Example 3. Establishing a simplified method to extract oil from oil crops and starch from cereals to be used decentrally

Vegetable oil extraction factories are at present huge centralized industries which extract and process the oil with explosive petroleum chemicals which need certain precautions. A technique has recently been developed by Novo Industries (Sejr Olsen, 1990) of treating crushed oil seeds with water and cell wall dissolving enzymes and afterwards fractionating the crude oil from water with a centrifuge, like separating cream from milk. If feasible, this process would allow a decentralization to smaller units of the first step of the vegetable oil process, the extraction, and leave the co-products, including the protein cake, for local utilization by animal husbandry farmers. Likewise as outlined by Munck *et al.* (1988), there are developed disc mill dry-milling techniques which together with new wet-milling concepts would allow a decentralization of starch manufacture from wheat and barley.

Example 4. The concept of the agricultural refinery to bridge decentralized agriculture and centralized industry

At a higher degree of complexity whilst still considering the local area concept, we can envisage how to apply the straw, oil and starch separation techniques together with other agro-industrial options, as is done for example in Denmark. The success of this agriculture-based industry will depend on a steady supply of cheap raw materials in quantities and with properties that are specially suited to individual biotechnological processes. The need for co-ordination between agricultural production and industrial demand for raw materials must, therefore, be strongly stressed. Thus, there will in the future be a need for integrated planning and development of total agricultural plant material production systems for food, feed and industrial non-food commodities like chemicals and fibre products. This integration could be stimulated by the establishment of the necessary intermediates, 'agricultural refineries', between agriculture and industry (Fig. 29.8) (Rexen and Munck, 1984).

When discussing total utilization of crops, one is tempted to make comparison with crude oil processing. It is characteristic that the oil refinery possesses a great deal of flexibility both technically and economically. The proportion between the hundreds of different product streams can easily be changed according to actual demands and price relationships on the market, including a range of different raw materials. A similar flexibility could be achieved with regard to crops if agricultural refineries were established as intermediates between agriculture and industry.

The present agricultural system in northern industrial countries such as Denmark is plagued by the serious structural disadvantages listed in Table 29.2. The concept of the agricul-

tural refinery is launched as a tentative attempt to alleviate as many of these drawbacks as possible.

The traditional combine harvest system is not suited to total crop harvest. The combiner driven by cheap fossil fuels leaves the straw in the field for later collection by baling machines or degrades straw to a waste product which has to be eliminated at a minimum of cost, e.g. by burning. The system is dependent on the weather because of late harvest and the quality of grain and straw is reduced and plant material is lost in the field (Rexen and Munck, 1984). An alternative to the combine system was introduced around 1970 in Denmark and Sweden (Rexen, 1971; Munck, 1972). The idea was to adopt the same system as used in the green crop drying industry, employing pneumatic drum dryers. The system is based on a self-propelled chopper which in one step cuts the whole crop, chops it coarsely and blows it into a container attached to the harvester. After drying, straw and grain are separated in a stationary separator (situated at the large drum dryer plant). This system has some obvious advantages compared with traditional combiner harvest: (1) it is independent of weather; (2) the field is cleared in one step; (3) a higher yield of both grain and straw is achieved; (4) the straw quality is much better due to less microbiological degradation; (5) it can also be used (after some development) to harvest, dry and process a wide range of non-cereal crops, including grass, legumes and oilseeds. By using the leaf meal fraction for heating, the drying process will be selfsupplying in energy. The whole-crop harvesting machine concept can further be rationalized by realizing the principle of the machine tool carrier (Persson, 1990), a tractor which can carry not only the chopper but all kinds of ploughs, spreaders, etc., so that the major machinery is used for 2000 hours per year instead of 400 hours.

From 1985 to 1987, a feasibility study was sponsored by the Danish Government, including a few experiments, on a provisional agricultural refinery at the island of Bornholm, Denmark (Rexen, 1990). The refinery consists of an old green meal processing factory with large buildings where three large drum dryers are

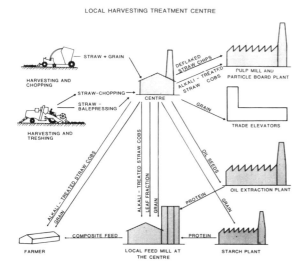

Fig. 29.8. The agricultural refinery concept.

Table 29.2. Disadvantages of the present agricultural system in northern Europe

It is unbalanced from the point of view of a product (surpluses and deficits) and from a local production distribution, e.g. imbalance between plant and husbandry agriculture

Specialization leads to monoculture and disease problems

Mismanagement of plant nutrition is expensive and polluting: excessive and uncontrolled distribution of untreated local animal manure and of over-soluble chemical fertilizers

Disease problems are counteracted at a high price with pesticides

Erosion and soil packing problems affect yield

Too inefficient use of expensive farming equipment and energy

Combine harvesting seeds alone leave potentially valuable straw in the fields

Strong seasonal concentration of plant husbandry work prevents efficient utilization of rural manpower

Lack of drying and other conservation facilities makes in many places plant production too dependent on weather, resulting in loss of quality. This is a particular problem for industrial utilization of crops where constant, defined quality is important

Agricultural land is becoming too expensive because of competition with alternative uses

Mechanization makes agriculture less labour intensive and the lack of alternative local work leads to a depopulated countryside with negative side-effects on rural society and the environment

Centralized agro-industries such as feed, sugar and starch factories result in excessive transportation costs

The industrial structure and distorted price relationships due to subventions and taxes prevent the efficient utilization of whole plants for non-food purposes

The cited inefficiencies result in unnecessarily high energy costs per unit of produce, retard restructuring to deficit crops, increase pollution and brings a higher price of agricultural products than necessary, blocking for alternative uses of land and products, which could reduce the surplus

installed. Within 100M of this centre are a local feed factory and a particle board factory which now uses wood chips but which for reasons of economy and supply would like to switch to straw chips. The site provides cereal processing facilities which could also include a starch and an oil extraction unit for further diversification of the agricultural refinery. An economic study supported the feasibility of the agricultural refinery principle in the long perspective (Andreasen *et al.*, 1988). Its economy rests on: (1) increased efficiency in agricultural crop harvesting and handling; (2) optimal utilization of the capital invested in machinery and equip-

ment; (3) optimal utilization of all botanical components of the plant material, including the purpose for energy; (4) extended growth season through a diversified plant cropping system enabling a continuous operation (1–3 shifts for 5–8 months); (5) optimal division of labour between decentralized agriculture and centralized industry, improving efficiency by decreasing transportation costs and pollution. Thus to be useful, several agricultural refineries, each based on many farmers, must be integrated in a network together with a wide range of industries, which can utilize whole plants for a broad variety of crops for many different purposes.

Different geographic conditions and cultures may require entirely different implementation of the agricultural refinery concept as discussed by Munck and Rexen (1985) for a rice-producing area in a developing country.

29.6 ATTEMPTING AN EVALUATION IN THE PRIMARY AREA BY COMBINING INFORMATION FROM SURVEYS WITH RESULTS FROM THE SECONDARY EVALUATION AREA

We are now cautiously entering the primary evaluation area of the wholeness carrying with us experience from surveys expressed in a provisional primary hypothesis as well as results selected in several cycles from local experiments and analyses expressed in provisional secondary hypotheses. The strategy (Munck, 1991) of ordering and selecting in the primary area is completely different from that of the secondary one. While it is highly recommendable in thinking and experimentation to go to the extremes in the local secondary area to obtain information, this is not advisable in the global primary, first because it is here often impossible to carry out controlled experiments and second if done, the economical risks may be devastating. Instead the main function, by evaluating in the primary area, is checking the 'consistency' of the concepts generated in the different secondary selection cycles. By further reasoning, this should create a high level evaluation apparatus which, in a realistic way, can absorb the knowledge.

Let us take the straw cellulose case as an example of reasoning in the primary area. In the temperate regions of the northern hemisphere, forestry is to a great extent industrialized using large amounts of machinery and fossil fuels. There are only small areas of natural forests left. There is an increasing concern among the public about the positive contribution of the forests to the environment, including its usefulness for recreation and production of food (game, berries, etc.). The present tropical forests are vanishing at a steadily accelerating rate in countries such as Indonesia and Tanzania, leading to a deterioration in the climate,

irreversibly damaging the great biological diversity of the forests, and the living conditions and productivity of the adjacent farm land. This negative development is partly due to an increase in the population pressure turning forest into farm land but also due to charcoal burning, as in Tanzania and because of regular mining of timber and wood chips for export to industrialized countries such as is the case in Indonesia.

In order to alleviate these negative trends which will only stop when all natural or ecologically managed forests are gone and when large parts of the landscape, including agriculture, lie eroded and unproductive, one has both to recognize the positive side of the monetary system, to put up a highly visible and attractive secondary selection criterion for the individual producer and consumer to obtain cash to survive and to avert the negative side, the resulting latent environmental implications, by changing the incentive of the consumer to another object to be selected (e.g. from earning cash by making charcoal from wood to selling locally produced agricultural products to industry and towns). The lignocelluloses produced in agriculture come into focus here. They are, however, at present conceptually downgraded. 150 years ago, all paper was made from annual crops like cereal straw which was used whole. When developing the wood pulping technology, a simple innovation was made by separating the needles and leaves from the stem of the trees which made wood superior to whole straw in cellulose content. An elaborate centralized industrial system driven by fossil fuels for extracting and transporting wood from forests was developed (e.g. in Scandinavia) as opposite to decentralized agriculture where nobody designed systems which in this scale could absorb fossil energy to harvest and handle straw, compress and transport it to the industry. Consequently, straw was considered an inferior raw material for paper. As discussed here, it is also quite easy to separate straw into a leaf and a stem fraction, the latter fraction having a comparable cellulose content to that of wood. The straw-based paper pulp factory in Denmark (Kaas Hansen, 1990) producing in competition with the world

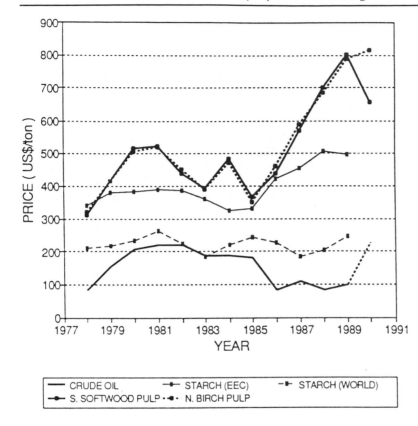

Fig. 29.9. Price of commodities 1978–1990 (Eurostat) (Bjørn Petersen and Munck, 1993).

market a high quality paper pulp shows that the straw option is already on the border of being competitive. However, this comparatively small factory (one tenth of a wood pulp plant) cannot alone afford to implement the concept of agricultural refineries discussed above and to use the leaf/stem fractionation process which in combination could greatly improve economy, including the reduction of costs of meeting the antipollution standard imposed by government.

World market prices for wood pulp have for the last 10 years fluctuated between 400 and 800 US$/ton compared to around 200 US$/ton for starch reflecting the high costs of extracting the wood from the forests (including the process costs) compared to the agro-industrial product starch (Fig. 29.9) (Bjørn Petersen and Munck, 1993).

Today in the world, 141.4 million ton (1988) of paper pulp is produced from wood compared to 11 million ton from annual plants, including

straw. Thus only 7% of the total paper pulp production comes from agriculture. A more diversified use of lignocellulose raw materials could probably help in rescuing some of the precious tropical forests and allow globally for a more generous attitude towards a sustainable forestry in general. Here the industrial aspect has to live in balance with other means for its utilization, such as food production, material for handicrafts, for recreation and for environmental reasons, thus creating the highest possible production value from and integrated standpoint. Also if straw use could be further realized, it would add to the diversification of agriculture and create more employment and cash for farmers, which would be especially valuable in developing countries.

The paper and board sector in the EC is a pronounced deficit sector (about 40% self-sufficiency) constituting a potential expansion of EC agriculture. In the EC today (1990), only about 0.2 million ton of paper pulp is produced from straw. Although the potential of an upgraded straw-based production chain seems obvious, of course only as a supplement to the wood based, its implementation is hindered first because of the division of labour in the world trade as codified in the GATT negotiations where EC is considered a major importer of lignocellulose products, vegetable oil seeds and feed proteins to balance its export of cars, machinery, etc.; second because the major wood pulp producing companies have bought up the major part of the paper industry and thus control them to get an outlet for wood pulp. It is thus less likely that somebody from the market in Europe should take the chance and develop a modern production chain for straw chips and paper pulp from straw, including agricultural refineries, and similarily the incentive is rather slow in other industrial countries or in the developing world. It can be concluded that in order for a technology to take off and compete, there must be a basic infrastructure, including a concept on which a certain critical number of people depend for their welfare and who can communicate its advantages to the public and to the legislators. The economic laws tell us that the global agricultural lignocellulosic potential which exceeds 1.5 billion ton/year will be utilized in large scale for paper pulp, particle board and energy first when almost all the natural and ecologically managed wood sources and habitats are gone and the environment is further irreversibly deteriorated.

Obviously the conceptual side (Munck, 1991) is as fundamental as the technical and economical sides in starting up a new production chain. In fact launching a new product, the penetration of the market both conceptually and technologically is the prerequisite for economy. Thus the concept 'a forest' could in a short perspective be looked upon as a mine or a source of money, in a medium perspective as a monoculture plantation for exploitation and in a longer perspective as a renewable resource with a diversified broad production of different products with many 'customers'. Similarly the concept 'straw' has to different people different implications such as debris, fertilizer, cattle feed and bedding material, fuel, roofing, material for

handicrafts and industrial raw material. By placing the conceptual components: 'a forest: a renewable, diversified production source', together with the component 'straw: an industrial raw material for industry and fuel', in the same area of the wholeness, we have obtained a new improved primary hypothesis, indicating that the two renewable plant resources could be combined together, stabilizing the ecological platform and the total output of benefits from plant production, while still satisfying the needs for paper made from both wood and straw. However, there is still at least one third component missing, namely how to visualize this combination as an advantage, both ethically and monetarily, for the producers and consumers involved in the production chains. As discussed, we have in fact, all, great difficulties in making such judgements and in overviewing the implications, which are the latent consequences of our selection of resources visible to us, just as the manager of the maltings and the plant breeder had difficulties in judging from the 11 malt analyses the three latent factors hidden in the previously described barley malt quality case which, however, could be easily revealed by the 'unscrambler' conceptualization. Tentatively we could in the paper pulp production case, envisage three major latent areas: (1) the forestry industry complex; (2) the agriculture (now also including some pulp and paper plants); (3) the consumer market; three different conceptual worlds to be harmonized as one to improve our relation to the real world. This conclusion has to be communicated with a minimum of contradictions by a dialogue with the selectors (consumers) who are those who in the end are going to implement it. By getting the concept accepted as their own provisional primary selection hypothesis for the advancement and survival of society, we are creating confidence in our future comparable with that of monetary stability. Only on this basis can we identify and implement the necessary structural changes of our society as, for example, identified in the Brundtland Report (1987).

The issue of the exploitation of the renewable plant resources is a global question addressing the metabolism of the world with many facets which we have to admit that we never will be able to overview nor control in all its aspects. Impedingly however, the public is used to the fact that all questions addressing problems of general interest 'in principle' are considered officially possible to answer and to solve by technical and economic means. It is understandable that this partly unrealistic optimism is founded on impressive technological development since the industrial revolution backed by classical science. This science sometimes appears as a sort of planned economy of the mind, mistaking the human concept for the real world. Realizing the advantages and drawbacks of our present course, we have to develop further from the static, deterministic world of classical science, adopting a new paradigm of science, as discussed by Bohm (1980) and Prigogine and Stengers (1984). Here we have to start to acknowledge the uncertainties and the inherent limitations in conceptualization by shifting the paradigm of science, now concentrating on apparently static objects, to include the dynamic world of latent factors, some of which we call environmental effects. With our new concepts, the computer and with the mass communication media, we will be able to advance much further in a concerted primary evaluation in our culture but we will never find the bottom. Our new insights defining the degree and character of the uncertainties involved by the mass action of the more than 5 billion human selectors of resources will give us a new impetus to form a strategy: a, for the time, viable, provisional primary hypothesis of how to act and communicate. By knowing ourselves, we should be able to correct our course as did the old Polynesians, who could find methods to control overpopulation and overexploitation of their habitat (Clarke, 1990). Recycling and utilization of the renewable plant resources leading to habits restricting our excess needs while improving employment are key concepts to achieve this purpose. However, this experience has to be communicated as a paradigm of a structural development within a limited framework, in a dialogue on ethical grounds as reflected in monetary tools, emphasizing population control, recycling, sustainability and quality definition and control, to receive a maximum of confidence from the functional units by men and women as selectors to be included as their own primary hypothesis for the advancement and survival of society. Thus, biological and electronic circuits could be connected by the selectors, co-ordinating in a self-regulating system within defined borders production and information in a new, more harmonious, post-industrial culture.

ACKNOWLEDGEMENTS

This chapter is based on research from 1980 to 1988, kindly supported by the Commission of the European Communities (DG VI and DG XII) and the Carlsberg Research Laboratory with the enthusiastic participation of Finn Rexen (technology), Pernille Bjørn Petersen (lignocellulose chemistry) and Bent Petersen (economy). Jan Brandt has kindly supported with the 'unscrambler' computation example and Kirsten Kirkegaard assisted in developing the manuscript.

REFERENCES

Andreasen, L., Munck, L., Petersen, B. and Rexen, F. (1988), *Projekt Biomasseraffinaderi Bornholm*, Rapport fra forsøgsarbejdet 1986/87, Carlsberg Research Laboratory, Copenhagen.

Bang-Olsen, K., Stilling, B. and Munck, L. (1991), The feasibility of high-lysine barley breeding – a summary. In *Barley Genetics VI*, Vol. I, Munck, L. (ed.), Munksgaard International Publishers Ltd, Copenhagen, pp. 139–142.

Bell, E.A. (1988), Extensions of traditional technology. In *Towards an Agro-Industrial Future*, Monograph Series No. 8, Royal Agricultural Society of England, Kenilworth, pp. 75–83.

Bjørn Petersen, P. and Munck, L. (1993), Whole crop utilization of barley, including new potential uses. In *Barley: Chemistry and Technology*, AACC Monograph on Barley, MacGregor, A.W. and Bhatty, R.S. (eds), American Association of Cereal Chemists, St. Paul.

Bohm, D. (1980), *Wholeness and the Implicate Order*, Ark Paperbacks, London.

Brundtland Report, The (1987), United Nations, World Commission on Environment and Development, Oxford University Press, Oxford.

Clarke, W.C. (1990), Learning from the past: traditional knowledge and systemable development, *The Contemporary Pacific*, University of Hawaii Press, 2(2), pp. 233–253.

Eerkens, I.C. (1983), Unilever research lab. In *Old and New Industrial Crops*, Proc. EEC Workshop, Wageningen.

Eggum, B.O., Knudsen, K.E.B., Munck, L., Axtell, J. and Mukuru, S.Z. (1982), Milling and nutritional value of sorghum in Tanzania. In *Proc. International Symposium on Sorghum Grain Quality*, 1981, (Rooney W. and Murty D.S. (eds), ICRRISAT, Patancheru, pp. 211–225.

Gray, P.S. (1990) Agricultural raw materials for industrial use. In *Agricultural Refineries – A Bridge from Farm to Industry*, EUR 11583 EN, Munck, L. and Rexen, F. (eds), The Commission of the European Communities, Brussels, pp. 43–52.

Kaas Hansen, S. (1990), Pulp and paper from straw: the Fredericia mill. In *Agricultural Refineries – A Bridge from Farm to Industry*, EUR 11583 EN, Munck, L. and Rexen, F. (eds), The Commission of the European Communities, Brussels, pp. 105–114.

Kempf, W. (1990), Present and future possibilities of utilizing starch and starch products in non-food areas. In *Agricultural Refineries – A Bridge from Farm to Industry*, EUR 11583 EN, Munck, L. and Rexen, F. (eds), The Commission of the European Communities, Brussels, pp. 71–83.

Kock, H., Kranse, F., Steffan, R. and Woelk, H.U. (1983), Herstellung von Phenolharzen unter Verwendung von Stärkeprodukten. *Die Stärke*, 35, 304–313.

Light, J.M. (1990), Modified food starches: why, what, where, and how. *Cereal Foods World*, 35, 1081–1092.

Martens, H. and Næs, T. (1989), *Multivariate Calibration*, John Wiley, Chichester.

Munck, L. (1972), Improvement of nutritional value in cereals. *Hereditas*, 72, 1–128.

Munck, L. (ed.) (1989), *Fluorescence Analysis in Foods*, Longman Scientific and Technical, Harlow.

Munck, L. (1991), Man as selector – A Darwinian boomerang striking through natural selection. In *Environmental Concerns – An Interdisciplinary Exercise*, Proc. Symposium held by The Danish Academy of Technical Sciences, 17–21 September, Elsinore, 1990, Hansen, J.Aa. (ed.), Elsevier Science Publishers, Amsterdam, pp. 211–227.

Munck, L. (1991) The case of high-lysine barley. In *Barley: Genetics, Molecular Biology and Biotechnology*, Shewry, P.R. (ed.), CAB International, Wallingford, pp. 573–601.

Munck, L., Knudsen, K.E.B. and Axtell, J.D. (1982), Industrial milling of sorghum in the 1980s. In *Sorghum in the Eighties*, Proc. International Symposium on Sorghum, ICRISAT, Patancheru, 1981, pp. 565–570.

Munck, L. and Rexen, F. (1985), Increasing income and employment in rice farming areas – Role of whole plant utilization and mini rice refineries. In *Impact of Science on Rice*, IRRI, Manila, pp. 271–280.

Munck, L., Rexen, F. and Haastrup Pedersen, L. (1988), Cereal starches within the European community – Agricultural production, dry and wet milling and potential use in industry. *Die Stärke*, 40, 81–87.

Munck, L. and Rexen, F. (eds) (1990) *Agricultural Refineries – A Bridge from Farm to Industry*, EUR 11583 EN, The Commission of the European Communities, Brussels.

Olsen, J. (1983), *Væskeabsorberende Stivelse: Super Slurper*, Teknologirådetsprojekt nr. 1331001–82274, Copenhagen.

Otey, F.H. and Doane, W.M. (1980), *Chemicals from Starch*, Report from Northern Regional Research Center, Peoria.

Persson, L.-I. (1990), Whole crop harvesting – Possible today, necessary tomorrow. In *Agricultural Refineries – A Bridge from Farm to Industry*, EUR 11583 EN, Munck, L. and Rexen, F. (eds), The Commission of the European Communities, Brussels, pp. 191–205.

Prigogine, I. and Stengers, I. (1984), *Order out of Chaos – Man's New Dialogue with Nature*, Fontana Paperpacks, London.

Rexen, F. (1971), Fremtidens tekniske system til høst, bjergning og konservering og oplagring af kornafgrøder. *Tidsskrift for Landøkonomi*, 5, 183–233.

Rexen, F. (1990), A new project in Denmark – the biorefinery at Bornholm. In *Agricultural Refineries – A Bridge from Farm to Industry*, EUR 11583 EN, Munck, L. and Rexen, F. (eds), The Commission of the European Communities, Brussels, pp. 115–122.

Rexen, F. and Munck, L. (1984), *Cereal Crops for Industrial Use in Europe*, EUR 9617 EN, Report prepared for The Commission of the European Communities, Brussels.

Rexen, F., Pedersen, P.B. and Munck, L. (1988), Exploitation of cellulose and starch polymers from annual crops. *TIBTECH*, 6, 204–205.

Roels, J.A. (1986), 'Grain' as basic resource for the

fermentation industry. In *Proc. Int. Grain Forum*, 10–12 November, 1986, Int. Grain Forum, Amsterdam, pp. 149–162.

Röbbelen, G., Downey, R.K. and Ashri, A. (eds) (1989), *Oil Crops of the World: Their Breeding and Utilization*, McGraw-Hill, New York.

Sejr Olsen, H. (1990), Aqueous enzymatic extraction of rapeseed oil. In *Agricultural Refineries – A Bridge from Farm to Industry*, EUR 11583 EN, Munck, L. and Rexen, F. (eds), The Commission of the European Communities, Brussels, pp. 137–146.

Steensberg, A. (1986), *Man – The Manipulator*, The Royal Danish Society of Science and Letters, Copenhagen.

Steensberg, A. (1989), *Hard Grains, Irrigation,* *Numerals and Script in the Rise of Civilization*, The Royal Danish Society of Science and Letters, Copenhagen.

Turner, M.K. and MacLean, C.W. (1988), Pharmaceuticals from agricultural sources – An industry view. In *Towards an AgroIndustrial Future*, Monograph Series No. 8, Royal Agricultural Society of England, Kenilworth, pp. 93–107.

Wilson, E.O. (1988), The biodiversity crisis: a challenge to science. In *Systematic Botany – A Key Science for Tropical Research and Documentation*, Hedberg, I. (ed.), Almqvist & Wiksell International, Stockholm, pp. 5–12.

Woelk, H.U. (1981), Stärke als Chemierohstoff. Möglichkeiten und Grenzen. *Die Stärke*, **33**(12), 397–408.

Part Nine

Epilogue

30

The need for a comprehensive plant breeding strategy

N.O. Bosemark

30.1 POPULATION GROWTH, WORLD FOOD SUPPLY AND A SUSTAINABLE AGRICULTURE

Most of the increase in agricultural production in the developing countries in the years following the Second World War resulted from clearing new land and extending traditional farming to increasingly marginal soils. As a result, by the early 1960s, virtually all land that could be readily brought into production was being farmed. With only marginal increase in agricultural production and continued rapid population growth, famine began to occur with increasing frequency in developing countries in Asia, Latin America and Africa. That the threat of large-scale starvation has been averted must be largely credited to the plant research institutes CIMMYT (Centro Internacional de Mejoramiento de Maiz y Trigo) in Mexico and IRRI (International Rice Research Institute) in the Philippines. The plant breeders at these centres developed high-yielding, semi-dwarf varieties of wheat and rice and introduced them, together with fertilizers, pesticides and modern agricultural techniques, to developing countries in the tropics. Through these efforts, known as the 'Green Revolution', world agricultural output rose by 25% between 1972 and 1982. In the

Plant Breeding: Principles and prospects. Edited by M.D. Hayward, N.O. Bosemark and I. Romagosa. Published in 1993 by Chapman & Hall, London. ISBN 0 412 43390 7.

same period of time production in the developing countries increased by 33% (Avery, 1985).

In the industrialized countries the increase was more modest and on an average amounted to 18%. Still, in many countries this increased production soon exceeded, not only domestic demands but also export opportunities, resulting in a continuing accumulation of surpluses. This development was the result of greater and more efficient use of capital and purchased inputs, better adapted and higher yielding varieties, more use of irrigation, but also, and not the least important, a price support system, the cost of which has become a major concern in many countries. Thus, in spite of an increase in world population of 1.8 billion people, the feared food shortage over much of the developing world did not materialize, and available food even rose from an average of 2320 calories per capita in 1961 to 2660 in 1983 (FAO, 1988).

However, since the early 1980s global food production is again stagnating with an average annual increase in the period 1984 to 1990 amounting to less than 1%. In the same period of time world population growth was nearly 2%, making the long-term food situation even more critical than in the 1960s. In the developing countries, yield increases, which can be credited to the Green Revolution, have been restricted to areas with relatively fertile soils, irrigation water and farmers who can afford to buy the necessary fertilizers and other inputs. However, areas with nutrient-poor, drought-prone soils are much more extensive than are fertile zones. Today, more than 1 billion people depend on these marginal lands for their livelihood, and most of the 2 billion children, expected to be born by the year 2000, will grow up in countries where nutrition is already inadequate and likely to deteriorate still further (Plucknett and Smith, 1982).

At the same time, there is a growing awareness in the industrialized countries of the limitations and environmental consequences of the present energy intensive farming systems. While in the 1970s attention was focused on industrial pollution, we are now equally concerned about erosion and land degradation, the effects of surplus fertilizers and pesticides that contaminate rivers, lakes and the ground water and the loss of biodiversity. Instead of the previous encouragement of input-intensive farming practices, aiming at steadily increasing yields, farmers are now being told to use less fertilizers and pesticides, to manage the land in a more sustainable way and to increase quality rather than yield. To bring down the surpluses the price supports are reduced, quotas are imposed for certain crops and farmers are paid for taking land out of production. At the same time, we have recently been reminded of how vulnerable food production is even in the developed countries. Thus, at the beginning of 1990, after the severe droughts in the United States and China, global food reserves were down from 461 million tons, estimated to be enough to feed the world for 102 days, to 290 million tons, enough for only 62 days, and very close to the 60 days considered by the UN and FAO to be an absolute minimum (Brown, 1991). Of particular concern is that even the very good harvest in 1990 did not increase the food safety margin by more than a few days. If, and when, the much debated greenhouse effect sets in, droughts and heat-waves may occur with increasing frequency, making it still more difficult to rebuild depleted food stocks. Since global grain production is no longer increasing but rather shows signs of diminishing, there is no guarantee that food production will manage to keep pace with population growth in the years to come. With a population of 5.2 billion today, which will have grown to over 6 billion by the year 2000 and is likely to be nearly 11 billion by the year 2050, the world may instead soon be facing a gigantic food-supply problem (Freeman, 1990).

It is against this complex, changeable and frequently contradictory background that one has to judge current and future agricultural policies and developments in both the industrialized and the developing countries. Thus, although the world food situation is deteriorating and likely to become critical in a couple of decades, in the industrialized and affluent countries in Europe and North America, food production is being curbed because surpluses cannot be sold at acceptable prices. While there is general agreement that the developing countries will have to become capable of feeding their own growing populations, and that prolonged, unconditional food support may in this respect be counter-productive, the affluent countries have an obligation to maintain a high production to be able to provide the food and technological assistance required to stimulate and support the development of a productive and sustainable agriculture in the developing world. To achieve such a development will require crop varieties and cropping systems adapted to marginal lands and capable of gradually reconstituting overutilized and nutrient-deficient soils. In the development of such systems it is important to take advantage of traditional methods that contribute to recirculation and conservation of natural resources, and to combine them with new technology.

Also in the industrialized countries agriculture will have to change. Both overproduction and environmental concerns are making it increasingly difficult for indebted farmers to improve their competitiveness and compensate for increased costs by maximizing production through increased inputs or by transferring to non-quota crops. Instead, the efficiency of production has to increase. Apart from the possibility of deliberately producing less per hectare with better total economy, which has its obvious limitations, there are only two alternatives that will achieve the goal to reduce production and at the same time maintain and increase production efficiency. One alternative, and the only one seriously addressed so far, is to reduce the acreage devoted to agriculture and to restrict food production to the best soils. The other alternative requires a reorientation of farm businesses in such a way that less land is used for traditional food and feed production and the surplus is instead used to grow various raw materials for industrial uses, as discussed in Chapter 29. Although it is only possible to introduce new crops for industrial purposes gradually over a long period of time, as an alternative to restrictive agricultural policies, it would avoid social problems caused by laying down farms, reduce the dependence on fossil fuels, allow for more varied crop rotations and, not the least important, retain in production crop land which may well be required for food production in the not too distant future. However, for agriculture to provide not only adequate amounts of food and feed, but also industrial raw materials at prices making them competitive with oil and other fossil feed stocks, yield and productivity will have to increase substantially. At the same time, ecological stability must be maintained and natural resources preserved. Considering that in the industrialized world yields of all major crops have been steadily increasing over the past 50 years, this appears not to be an unrealistic expectation. Most of these yield increases have been due to the identification and adjustment of controllable stress factors limiting yield, mainly available water and plant nutrients, weed competition, pests and diseases. The yield increases that can be credited to plant breeding have been estimated to range from 35 to 50%. Although yields are likely to continue to increase in the years to come, to predict the magnitude of these increases, and how they will relate to better plant husbandry and progress in crop variety improvement, respectively, is more difficult. However, it appears likely that, in the developed countries, yield increases due to better tillage and soil preparation, more irrigation, more fertilizers and more effective herbicides, fungicides and insecticides, will in general be smaller than in the past. To maintain the present rate of yield increase we shall thus have to rely more on the contribution of plant breeding, especially breeding for resistance to pests and diseases and yield limiting abiotic stress conditions. That the potential for such

yield increases is huge has been demonstrated by Boyer (1982), who studied eight major crops in the United States and found that on an average only 22% of the estimated yield potential was realized in the farmers' fields, and that as much as 70% was lost due to unidentified environmental stress factors. The problem is that the mechanisms underlying tolerance or resistance to most abiotic stress factors are poorly understood, and resistance thus difficult to select for. However, as has been emphasized in Chapters 24 and 25, recent scientific advances are making exploration and manipulation of these mechanisms more feasible, and are likely, eventually, to make possible large gains in productivity. With limited possibilities to increase the various inputs mentioned earlier, better adapted crop varieties are the cheapest, most reliable and environmentally safest way to increase productivity and secure the world's food supply.

30.2 THE ROLE OF PLANT BREEDING

For plant breeding to be able to contribute both to a more secure world food supply and to sustainable agricultural systems, in the future plant breeders will have to: (1) increase crop yields and product quality still further; (2) develop varieties with wide adaptation as well as varieties for local, specific environments; (3) produce varieties which have improved resistance to various abiotic stress conditions and which make better use of crop inputs; (4) develop varieties with better resistance to pests and diseases and thereby less dependence on agrochemicals; (5) provide crops and varieties with a wider range of end uses. As in the past, the main breeding objective has to be increased yield and quality of products; or to be more precise, to produce maximum yield of saleable products at an economical level of input and with a minimum of negative environmental effects. Although the other breeding objectives mentioned are often major or even dominant features of a breeding programme, they also, with the exception of the last one, indirectly affect yield and frequently also quality.

30.2.1 Breeding strategy and resource allocation

With the steadily increasing commercial competition in plant breeding, a breeder has to produce and market a continuous flow of varieties with improved performance and incorporating new desirable traits such as resistance to diseases and pests or improved quality.

For varietal development procedures to result in rapid genetic gain it is necessary to apply a high selection pressure on individual genotypes. As a consequence they lead to a rapid restriction of genetic variability and thereby also to restricted possibilities for long-term genetic advance. On the contrary, a low or moderate selection pressure imposed on broad base populations is unlikely quickly to produce varieties but favours recombination and retention of genetic variability and thereby long-term genetic progress. This fundamental difference between short-term and intermediate to long-term genetic improvement, makes it necessary to separate these breeding operations. Figure 30.1 illustrates such a separation of the operations related to short-term varietal production, intermediate-term population improvement and long-term genetic resources development. The nature and volume of the genetic materials employed, the breeding efforts spent, the financial resources allocated and the classical selection methods most commonly employed in the different operations, are all indicated in Fig. 30.1. Although such a resource allocation may vary considerably from crop to crop, the principles and operational structure should be applicable to both self- and cross-pollinated species and be relevant to all breeding objectives mentioned earlier.

In the following this structure will be used as a backbone of classical plant breeding, to discuss how some recent developments, both in

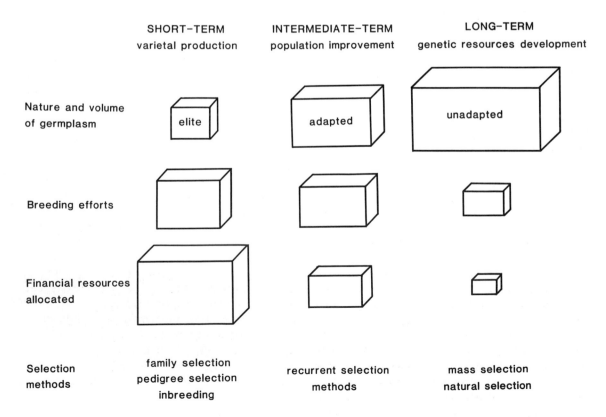

Fig. 30.1. Resource allocation in plant breeding.

the classical field and in cell biology, molecular biology and genetic engineering, when integrated with the classical methods, are likely to impact on practical plant breeding in the years to come.

30.2.2 Marker assisted selection in short-term variety development

As indicated in Fig. 30.1, short-term breeding operations are designed to develop quickly competitive varieties and hybrids and frequently involve pedigree and backcross selection to produce new varieties or lines or to incorporate a new trait into an existing and otherwise desirable genotype. Of particular interest in such contexts is the access to RFLPs and other molecular markers, described in Chapter 19. Where found to be genetically correlated with a qualitative or quantitative trait, such markers may be used as indirect selection criteria and will greatly increase the speed and precision with which desired traits can be transferred both between lines and cultivars and from various genetic resources to cultivars. The quick and efficient transfer of genes governing quantitative as well as qualitative traits to elite inbreds, will be an extremely useful supplement to the traditional recycling of inbreds, used in hybrid breeding in maize and elsewhere. Detailed linkage maps, heavily saturated with RFLP markers, now exist for several crops, including tomato, potato, maize, barley, oilseed rape and sugar beet, and have already been used to identify and facilitate the transfer of genes associated with insect resistance (Nienhuis *et al.*, 1987), virus resistance (Young and Tanksley, 1989), water use efficiency (Martin *et al.*, 1989), resistance to fungal pathogens (Bentolila *et al.*, 1991) and tolerance to low-nutrient stress (Reiter *et al.*, 1991). The fact that all major groups of biotic and abiotic stress factors, to which improved resistance will have to be obtained in the years to come, are already represented among the examples is most encouraging. Other breeding operations in which RFLP markers are helpful include estimations of genetic diversity of germplasm and fingerprinting of varieties and breed-

ing lines. In the not too distant future they may also help in the dissection of quantitatively inherited traits into single gene components.

Of all developments in molecular biology so far, that of RFLPs and other DNA-markers is likely to be the one most universally useful to the plant breeder. Most major plant breeding organizations have equipped themselves with the laboratory and computer facilities required for a cost-effective, large-scale implementation of marker-based methodologies in practical breeding. As pointed out by Beckmann and Soller (1989), this development will permit varieties to be produced more quickly than before and result in a much more rapid turn over of improved varieties and a shorter life-span for any given variety, both factors likely to favour the large breeding organizations at the expense of the smaller.

30.2.3 Intermediate- to long-term population improvement

The efficient, stepwise upgrading of selected recombinant line varieties and elite inbreds, made possible through RFLP-assisted backcrossing, is likely to lead to an increased use of this breeding method. However, too strong an emphasis on single-cross source populations, may result in a narrowing of the genetic base of the breeding programme. It is therefore important that parallel with refinements in pedigree and backcross breeding systems, go the development and continuous improvement of source populations designed to capitalize on a broad, adapted gene pool as well as the genetic variability available in primitive cultivars and wild relatives of crop species.

Having been originally developed to improve natural and synthetic populations of maize as sources of inbred lines, recurrent selection methods are now used, not only in cross-pollinated species, but also, increasingly, in self-pollinated crops. As discussed in Chapter 18, which gives a detailed account of a number of intra- and interpopulation recurrent selection methods, the choice of selection procedure should be based on the kind of gene action governing the trait to be selected for, and the type of variety aimed at. Thus, in a cross-pollinated species where dominance is important but hybrids cannot readily be produced, intrapopulation selection methods are recommended, whereas, when the aim is to develop hybrid varieties and heterosis is important, the best choice should be reciprocal recurrent selection. However, in spite of this, commercial breeders only rarely use reciprocal recurrent selection, but instead rely on phenotypic half-sib, full-sib or S_1-progeny selection. Although this choice is probably more often governed by practical considerations and limited resources than by experimental or theoretical evidence, it is supported by a series of studies (Jinks, 1983), which suggest that overdominance is not a major factor in heterosis, which is instead due to the accumulation of favourable dispersed dominant genes. As described in Chapter 2, with this genetic background to heterosis, it should in principle be possible to derive lines as good as the best hybrids, in cross-pollinated crops, although, with a high number of loci and linkage effects, the likelihood of obtaining such lines may be exceedingly small.

Although the principles of recurrent selection are equally valid for autogamous species as for allogamous, the difficulties in producing the necessary intercrossings in each selection cycle have discouraged its use in self-pollinated crops. However, as described in Chapter 14, in self-pollinated species nuclear male sterility may be introduced into a population to facilitate the crossings in the recombination phase, so-called male sterile facilitated recurrent selection (MSFRS). Correspondingly, in cross-pollinated species, where self-incompatibility prevents production of selfed seed, the introduction of an S^F gene and a gene for nuclear male sterility will facilitate both selfing and S_1-selection and recombination.

As indicated in Fig. 30.1, recurrent selection methods are not intended to be used alone but should be integrated with other selection methods. Since all recurrent selection involves testing of progenies of selected individuals, superior progenies may enter a pedigree breeding programme aimed at developing recombinant line varieties or inbred lines for use in hybrid production and thus provide the inflow of new variability into the short-term varietal production programme previously mentioned.

In crops where large-scale haplo-diploidization is possible recurrent selection on the phenotypes of doubled haploid (DH) lines is theoretically the most efficient intrapopulation selection method (Gallais, 1988, 1989). In a self-pollinated crop like barley, superior DH-lines may directly qualify as commercial varieties (see also Chapter 17).

While of relevance in virtually all crop improvement, recurrent selection methods have not been as widely used by commercial breeders as might have been expected, due to their complexity and cost and the many and frequently contradictory reports of their effectiveness in different contexts. However, with indications that simple and straightforward intrapopulation methods are likely to be the best in most genetic situations, and irrespective of the type of variety aimed at, the use of such methods for population improvement is likely to increase considerably in the years to come.

30.2.4 Long-term genetic resources development

Long-term genetic resources development aims at making use of sexually compatible germplasm belonging to the primary gene pool of a crop. Although the volume of material initially sampled or available in gene banks may be large, in the hands of the breeder it usually boils down to a limited number of populations, based on crosses between adapted and primitive germplasm, and each with one or more potentially useful characteristics. Since such material is usually a long way from commercial use, it is likely to receive only a small share of the breeding effort and an even smaller share of the financial resources. Consequently the selection methods need to be cheap and the selection cycles short. In most cases recurrent mass selection and/or half-sib selection should be appropriate. Until improved by several cycles of selection genetic resources populations need to be kept separate. However, once the performance approaches that of adapted breeding material, germplasm from these populations may be fed into the elite populations. If nuclear male sterility has been introduced in both kinds of populations, crosses between MS-plants from the elite population and the genetic resources population can be evaluated as half-sib or S_1-families in strict comparison with crosses within the elite population (Bosemark, 1989). This makes it possible to assess the value of the introduced germplasm before it is integrated into the elite population.

The time and effort required to develop the kind of genetic resources populations discussed obviously depends both on the crop and the material that originally went into populations. However, experiences with sugar beet suggest that even the wild *Beta* species are not as inaccessible as might be thought, and that relatively few cycles of recurrent mass and half-sib selection in sugar beet × *B. maritima* crosses may result in populations sufficiently adapted and domesticated to be used as germplasm feeder populations as outlined above (Bosemark, 1989). Even so, due to the steadily increasing pressure on commercial breeders to produce new varieties, most of them will be reluctant to engage in this kind of work or will do so in a very limited way. In this situation it is important that both commercial breeders and researchers and breeders in governmental institutes support and participate in the Crop Genetic Resources Networks which are currently being organized for several crops, with the assistance of IBPGR, and which aim at organizing co-operative, pre-competitive research and pre-breeding utilizing primitive or wild germplasm resources.

30.2.5 Novel, genetically engineered male sterility systems

In principle any crop species showing pronounced inbreeding depression should be a potential candidate for hybrid breeding. So far, in many agricultural and horticultural crops that meet this requirement, hybrid development is hampered or prevented by the lack of reliable and cost effective methods of pollination control. However, in the near future, the possibility of introducing male sterility, with

the help of recombinant-DNA techniques, should make possible both hybrid seed production and the development of efficient population breeding systems in a wide variety of cross-pollinated crops (Chapters 10 and 14). This is of importance since hybrid varieties offer many advantages over synthetics and other open-pollinated varieties. Thus, besides offering higher yield potential and improved crop uniformity, hybrid breeding makes it possible to combine in a single hybrid genes for disease resistance present in the parental inbreds governing resistance to different diseases as well as offering protection against different races of the same pathogen. Similarly genes governing different traits and introduced into parental lines through genetic transformation, can be readily combined in a hybrid. Finally, as discussed earlier, elite inbred lines for hybrid production can be quickly and efficiently upgraded via marker-assisted backcrossing, resulting in similarly improved hybrid cultivars.

Although the importance of developments in pollination control is unquestionable, it should be kept in mind that access to a reliable male sterility system for a crop does not in itself guarantee that hybrid varieties will be a success. As emphasized by Simmonds (1989), plant breeding is an economic activity and plant breeding operations should be judged both on their biological and economic values. To exemplify this Simmonds uses hybrid wheat and shows that even with an efficient chemical hybridizing agent or male sterility system, the relatively moderate levels of heterosis and the high seed production costs, make it highly unlikely that hybrid wheat will ever be a viable proposition. However, even if the seed production could be made cheaper and the level of heterosis increased, the long-term prospects of hybrid wheat may still be questioned. This is due to the fact that a superior hybrid wheat variety is most likely also a superior cross for subsequent development of recombinant line varieties, which traditional wheat breeders would take immediate advantage of. As a result, the yield of traditional pure line varieties might well increase at a faster rate than would be the case for the hybrids, resulting in a shrinking yield advantage for the hybrids. Although wheat may be considered a special case, being a strictly self-pollinated crop, the same considerations are also likely to be valid for other predominantly self-pollinated species such as oilseed rape (Chapter 2).

30.2.6 The likely impact of genetic engineering

Although genetic engineering is unlikely to play a major role in crop improvement for at least another 10 years, it is important for an assessment of its likely long-term impact to observe that the emphasis of current research and development work, as well as the products of such work already obtained, very largely concern resistance to pests and diseases, environmental stress factors and various quality traits (Chapter 16). These traits are precisely those, discussed earlier, which will reduce dependence on agrochemicals, result in increased yields and improved product quality and thus contribute to a more sustainable agriculture. Integrated with classical breeding methods, molecular techniques will speed up breeding programmes and make possible also the development of crop varieties with a wider range of end uses or even totally new crops for specific industrial uses.

Over the past few years, field tests with genetically engineered plants have been conducted in increasing numbers in several countries. Thus, from 1986 until the end of 1991 around 390 field tests have been carried out world wide (J. Duesing, personal communication). For example, in the period June–August 1991 the US Department of Agriculture issued 23 permits for field trials with genetically modified plants including seven with resistance to different viruses, three with insect resistance, four with herbicide resistance and one each with bacterial resistance, fungal resistance and frost resistance. The remaining six cases concerned improved nutrition, metabolism and shelf-life, flower colour and marker genes (Anonymous, 1991a).

Although, the advancement in molecular technologies almost entirely emanate from research in the industrialized countries, they may in a longer perspective be of even greater significance to the people in the developing world. Recognizing this, in the last five years, the Rockefeller Foundation has spent more than US$ 33 million in support of research to improve rice through biotechnology techniques, including genetic engineering. Besides IRRI, support has been given to almost 100 laboratories and researchers throughout the world, two-thirds of which are located in developing countries (Anonymous, 1991b).

At the same time as the opportunities and benefits of molecular biology and genetic engineering to plant breeding are becoming more visible, so also are the costs and effort required to bring the products of the technology to the market.

Contrary to molecular markers, which are solely an assisting technology and to which no objections can be raised, genetic engineering is often perceived as unnatural and risky and is also frequently associated with chemical and pharmaceutical industries, believed to be more concerned with safeguarding their own interests than those of the farmers and consumers. To gain public acceptance of genetically engineered plants, and food based on such plants, requires risk assessment and regulations which can be widely understood and legally enforced. However, rules and regulations should be based on known scientific facts and cover the product rather than the technology used to bring it about. Even so, to overcome public distrust is likely to take time and require a considerable educational effort from both the scientific community and the industry.

30.3 THE IMPORTANCE OF A BALANCED APPROACH

In spite of the importance of biotechnology and genetic engineering in changing future production methods in agriculture and in breeding new crop varieties, biotechnology and genetic engineering are not going to take the place of traditional plant breeding research and breeding methodology, nor diminish their value. On the contrary, further refinement of these methods and better knowledge of classical genetics are requisites for a rational use of new tools such as molecular markers. It is therefore important that the less glamorous classical areas of plant breeding research continue to receive support and that such studies, as far as possible, are closely co-ordinated with cell and molecular biology research.

To arrive at sustainable agricultural systems, which are both highly efficient and productive, will require not only better crop varieties but also a more flexible and innovative plant husbandry than we have to-day. Instead of relying on a limited number of cash crops, we need crop rotations with a diversity of crops, including new industrial crops, designed to economize with fertilizers, reduce problems from weeds, diseases and pests and to minimize soil erosion and leakage of nutrients. To this should be added a truly integrated pest control which combines the effects of resistant varieties with those of precise timing of plant husbandry measures, including judicious use of chemical pesticides.

REFERENCES

Anonymous (1991a), US Department of Agriculture permits for field trials of genetically engineered plants issued in June–August 1991. *AgBiotech. News and Information*, **3**, 972.

Anonymous (1991b), IRRI major actor in RF biotech. network. *AgBiotech. News and Information*, **3**, 971–972.

Avery, D. (1985), U.S. farm dilemma: the global bad news is wrong. *Science*, **230**, 408–412.

Beckmann, J.S. and Soller, M. (1989), Genomic genetics in plant breeding. *Vortr. Planzenzüchtg.*, **16**, 91–106.

Bentolila, S., Guitton, C., Bouvet, N., Sailland, A., Nykaza, S. and Freyssinet, G. (1991), Identification of an RFLP marker tightly linked to the Ht1 gene in maize. *Theor. Appl. Genet.*, **82**, 393–398.

Bosemark, N.O. (1989), Prospects of beet breeding and use of genetic resources. *Report of an International Beta Genetic Resources Workshop in Wageningen*, IBPGR, Rome, pp. 89–98.

Boyer, J.S. (1982), Plant productivity and environment. *Science*, **218**, 443–448.

Brown, L. (1991), The new world order. In *State of the World, A Worldwatch Institute Report on Progress Toward a Sustainable Society*, W.W. Norton & Company, New York, London, pp. 3–20.

FAO (1988), *World Agriculture: Toward 2000. An FAO Study*, Alexandratos, N. (ed.), New York University Press, New York.

Freeman, O.L. (1990), Meeting the food needs of the coming decade. *The Futurist*, Nov.–Dec. 1990, 15–20.

Gallais, A. (1988), A method of line development using doubled haploids: the single double haploid descent recurrent selection. *Theor. Appl. Genet.*, **75**, 330–332.

Gallais, A. (1989), Lines *versus* hybrids. – The choice of the optimum type of variety. Contribution of quantitative genetics and selection theory. *Vortr. Pflanzenzüchtg.*, **16**, 69–80.

Jinks, J.L. (1983), Biometrical genetics of heterosis. In *Heterosis – Reappraisal of theory and practice*, Frankel, R. (ed.), Springer-Verlag, Berlin, Heidelberg, New York, Tokyo, pp. 1–46.

Martin, B., Nienhuis, J., King, G. and Schaefer, A. (1989), Restriction fragment length polymorphisms associated with water use efficiency in tomato. *Science*, **243**, 1725–1728.

Nienhuis, J., Helentjaris, T., Slocum, M., Ruggero, B. and Schaefer, A. (1987), Restriction fragment length polymorphism analysis of loci associated with insect resistance in tomato. *Crop Sci.*, **27**, 797–803.

Plucknett, D.L. and Smith, N.J.H. (1982), Agricultural research and Third World food production. *Science*, **217**, 215–220.

Reiter, R.S., Coors, J.G., Sussman, M.R. and Gabelman, W.H. (1991), Genetic analysis of tolerance to low-phosphorus stress in maize using restriction fragment length polymorphisms. *Theor. Appl. Genet.*, **82**, 561–568.

Simmonds, N.W. (1989), Economic aspects of plant breeding with especial reference to economic index selection. *Res. Dev. Ag.*, **6**, 55–62.

Young, N.D. and Tanksley, S.D. (1989), RFLP analysis of the size of chromosomal segments retained around the *Tm-2* locus of tomato during backcross breeding. *Theor. Appl. Genet.*, **77**, 353–359.

Species index

Subject index